MW00332589

TODAY'S TECHNICIAN ™

Shop Manual for

Automotive Engine Performance

Fourth Edition

TODAY'S TECHNICIAN ™

Shop Manual for
Automotive Engine Performance
Fourth Edition

■

Ken Pickerill
Ivy Tech Community College of Indiana
Sellersburg, IN

THOMSON

DELMAR LEARNING ™

Australia Canada Mexico Singapore Spain United Kingdom United States

THOMSON

DELMAR LEARNING

Today's Technician: Automotive Engine Performance Fourth Edition Shop Manual

Ken Pickerill

Vice President, Technology & Trades SBU:

Alar Elken

Editorial Director:

Sandy Clark

Senior Acquisitions Editor:

David Boelio

Developmental Editor:

Matthew Thouin

Marketing Director:

David Garza

Channel Manager:

Bill Lawrensen

Marketing Coordinator:

Mark Pierro

Production Director:

Mary Ellen Black

Project Editor:

Barbara L. Diaz

Art & Design Specialist:

Cheri Plasse

COPYRIGHT © 2006 Thomson Delmar Learning. Thomson, the Star Logo, and Delmar Learning are trademarks used herein under license.

Printed in the United States of America
1 2 3 4 5 XX 07 06 05

For more information contact
Thomson Delmar Learning
Executive Woods
5 Maxwell Drive, PO Box 8007,
Clifton Park, NY 12065-8007
Or find us on the World Wide Web at
www.delmarlearning.com

ALL RIGHTS RESERVED. No part of this work covered by the copyright hereon may be reproduced in any form or by any means—graphic, electronic, or mechanical, including photocopying, recording, taping, Web distribution, or information storage and retrieval systems—without the written permission of the publisher.

For permission to use material from the text or product, contact us by
Tel.   (800) 730-2214
Fax   (800) 730-2215
www.thomsonrights.com

Library of Congress Cataloging-in-Publication Data:

Pickerill, Ken.
    Shop manual for automotive engine performance / Ken Pickerill. — 4th ed.
        p. cm.
Rev. ed. of: Shop manual for automotive engine performance / Douglas Vidler. 3rd ed. 2003.
    ISBN 0-4180-0062-0 (sm&cm : alk. paper) — ISBN 0-4180-0064-7 (iml)
        1. Automobiles—Motors—Modification. 2. Automobiles—Performance. I. Title: Automotive engine performance. II. Vidler, Douglas. Shop manual for automotive engine performance. III. Title.

TL210.V5 2006
629.25'04'0288—dc22   2005052885

NOTICE TO THE READER

Publisher does not warrant or guarantee any of the products described herein or perform any independent analysis in connection with any of the product information contained herein. Publisher does not assume, and expressly disclaims, any obligation to obtain and include information other than that provided to it by the manufacturer.

The reader is expressly warned to consider and adopt all safety precautions that might be indicated by the activities herein and to avoid all potential hazards. By following the instructions contained herein, the reader willingly assumes all risks in connection with such instructions.

The publisher makes no representation or warranties of any kind, including but not limited to, the warranties of fitness for particular purpose or merchantability, nor are any such representations implied with respect to the material set forth herein, and the publisher takes no responsibility with respect to such material. The publisher shall not be liable for any special, consequential, or exemplary damages resulting, in whole or part, from the readers' use of, or reliance upon, this material.

CONTENTS

Photo Sequence Contents viii

Job Sheets Contents ix

Dedication x

Preface xi

CHAPTER 1

Tools and Safety 1

Introduction 1 ● Personal Safety 2 ● Lifting and Carrying 3 ● Occupational Safety and Health Act 4 ● Shop Hazards 5 ● Hand Tool Safety 6 ● Power Tool Safety 7 ● Compressed Air Equipment Safety 12 ● Lift Safety 13 ● Jack and Jack Stand Safety 15 ● Cleaning Equipment Safety 16 ● Vehicle Operation 18 ● Work Area Safety 20 ● Hazardous Waste Disposal 23 ● Hand Tools and Diagnostic Equipment 26 ● Employer and Employee Obligations 28 ● National Institute for Automotive Service Excellence (ASE) Certification 30 ● Terms to Know 31 ● ASE-Style Review Questions 32 ● Job Sheets 33

CHAPTER 2

Typical Shop Procedures and Equipment 37

Introduction 37 ● Units of Measure 42 ● Engine Diagnostic Tools 43 ● Electrical Diagnostic Tools 50 ● Ignition System Tools 55 ● Fuel System Tools 59 ● Scan Tools 62 ● Oscilloscopes 65 ● Exhaust Analyzers 67 ● Engine Analyzer 68 ● Miscellaneous Engine Performance Tools 70 ● Service Information 71 ● Engine Tune-Up 76 ● Case Study 77 ● Terms to Know 78 ● ASE-Style Review Questions 78 ● Job Sheets 81

CHAPTER 3

General Engine Condition Diagnosis 85

General Diagnostic Procedure 85 ● Engine Leak Diagnosis 85 ● Engine Noise Diagnosis 89 ● Engine Exhaust Diagnosis 92 ● Diagnosis of Oil Consumption 93 ● Engine Oil Pressure Tests 93 ● Engine Temperature Tests 95 ● Cooling System Inspection and Diagnosis 96 ● Vacuum Tests 103 ● Exhaust Gas Analyzer 104 ● Engine Power Balance Test 109 ● Compression Tests 110 ● Cylinder Leakage Test 114 ● Valve Timing Checks 117 ● Valve Adjustment 117 ● Case Study 120 ● Terms to Know 121 ● ASE-Style Review Questions 121 ● ASE Challenge Questions 122 ● Job Sheets 123

CHAPTER 4

Basic Electrical Tests and Service 129

Basic Electrical Diagnosis 129 ● Electrical Test Equipment 134 ● Digital Multimeter (DMM) Usage 139 ● Lab Scope Usage 142 ● Logic Probes 148 ● Basic Electrical Troubleshooting 149 ● Basic Electrical Repairs 150 ● Testing Basic Electrical Components 155 ● Battery Diagnosis and Service 161 ● Charging a Battery 166 ● Starter Diagnosis and Service 170 ● AC Generator Diagnosis and Service 173 ● Case Study 179 ● Terms to Know 179 ● ASE-Style Review Questions 179 ● ASE Challenge Questions 181 ● Job Sheets 183

CHAPTER 5 | **Intake and Exhaust System Diagnosis and Service** | **191**

Air Filters 192 • Engine Vacuum 195 • Exhaust System Service 201 • Catalytic Converters 203 • Replacing Exhaust System Components 208 • Turbocharger Diagnosis 211 • Supercharger Diagnosis and Service 220 • Case Study 224 • Terms to Know 225 • ASE-Style Review Questions 225 • ASE Challenge Questions 226 • Job Sheets 227

CHAPTER 6 | **Engine Control System Diagnosis and Service** | **235**

Electronic Service Precautions 236 • Basic Diagnosis of Electronic Engine Control Systems 237 • Service Bulletin Information 240 • Self-Diagnostic Systems 243 • Using a Scanner 246 • Retrieving Trouble Codes 250 • Diagnosis of Computer Voltage Supply and Ground Wires 267 • Testing Input Sensors 273 • Switches 271 • Variable Resistor-Type Sensors 275 • Generating Sensors 282 • Mass Air Flow Sensors 295 • Case Study 301 • Terms to Know 302 • ASE-Style Review Questions 302 • ASE Challenge Questions 303 • Job Sheets 305

CHAPTER 7 | **Servicing Computer Outputs and Networks** | **315**

Introduction 315 • Diagnosing Computer Outputs 315 • Computer Networks 320 • Terms to Know 325 • ASE-Style Review Questions 326 • ASE Challenge Questions 327 • Job Sheets 329

CHAPTER 8 | **On-Board Diagnostic System Diagnosis and Service** | **333**

OBD II PCM 335 • OBD II Diagnostics 337 • OBD II Terminology 353 • Malfunction Indicator Lamp 355 • Freeze Frame 359 • Diagnostic Trouble Codes 359 • Data Link Connector 362 • Diagnostic Software 363 • Adaptive Fuel Control Strategy 364 • OBD II Monitor Test Results 365 • Intermittent Faults 366 • Repairing the System 369 • Verifying Vehicle Repair 375 • Case Study 375 • Terms to Know 376 • ASE-Style Review Questions 376 • ASE Challenge Questions 377 • Job Sheets 379

CHAPTER 9 | **Diagnosing Related Systems** | **383**

Introduction 383 • Clutches 384 • Manual Transmissions 386 • Automatic Transmissions 386 • Driveline 394 • Heating and Air Conditioning 395 • Speed Control Systems 396 • Brake Systems 396 • Suspension and Steering Systems 399 • Wheels and Tires 402 • Case Study 403 • Terms to Know 404 • ASE-Style Review Questions 404 • ASE Challenge Questions 405 • Job Sheets 407

CHAPTER 10 | **Fuel System Diagnosis and Service** | **411**

Introduction 411 • Alcohol in Fuel Test 412 • Fuel System Pressure Relief 412 • Fuel Tanks 413 • Fuel Lines 418 • Fuel Filters 422 • Fuel Pumps 425 • Fuel Injection Systems 432 • Evaporative Emissions 443 • Terms to Know 446 • ASE-Style Review Questions 447 • ASE Challenge Questions 448 • Job Sheets 449

CHAPTER 11

Electronic Fuel Injection Diagnosis and Service 453

Introduction 453 ● Preliminary Checks 454 ● Basic EFI System Checks 457 ● Air System Checks 459 ● Fuel System Checks 464 ● Injector Checks 467 ● Injector Service 477 ● Fuel Rail, Injector, and Regulator Service 479 ● Idle Speed Checks 482 ● Idle Speed Controls 485 ● Case Study 492 ● Terms to Know 493 ● ASE-Style Review Questions 493 ● ASE Challenge Questions 495 ● Job Sheets 497

CHAPTER 12

Distributor Ignition System Diagnosis and Service 503

Introduction 503 ● General Ignition System Diagnosis 504 ● Visual Inspection 507 ● Diagnosing with an Engine Analyzer 512 ● Oscilloscope Testing 518 ● Individual Component Testing 536 ● No-Start Diagnosis 543 ● Distributor Service 555 ● Ignition Timing 561 ● Case Study 567 ● Terms to Know 567 ● ASE-Style Review Questions 568 ● ASE Challenge Questions 569 ● Job Sheet 571

CHAPTER 13

Electronic Ignition Systems (EI) Diagnosis and Service 583

Introduction 583 ● Effects of Improper Ignition Timing 583 ● Visual Inspection 585 ● General Diagnostics 591 ● Chrysler EI Systems 600 ● Ford EI Systems 604 ● General Motors EI Systems 608 ● Mitsubishi EI Systems 616 ● Nissan EI Systems 617 ● Toyota EI Systems 618 ● Case Study 620 ● Terms to Know 621 ● ASE-Style Review Questions 621 ● ASE Challenge Questions 622 ● Job Sheet 623

CHAPTER 14

Emission Control System Diagnosis and Service 625

Introduction 625 ● Emissions Testing 626 ● Evaporative Emission Control Diagnosis and Service 632 ● PCV System Diagnosis and Service 637 ● Spark Control Systems 640 ● EGR System Diagnosis and Service 643 ● Electronic EGR Controls 650 ● Catalytic Converter Diagnosis 654 ● Air Injection System Diagnosis and Service 656 ● Terms to Know 661 ● ASE-Style Review Questions 661 ● ASE Challenge Questions 663 ● Job Sheets 665

Appendix A ASE Practice Examination 677

Appendix B Metric Conversion Chart 682

Appendix C Engine Performance Special Tool Suppliers 683

Glossary 685

Index 701

Photo Sequences

1 Typical Procedure for Diagnosing Engine, Ignition, Electrical, and Fuel Systems with an Engine Analyzer **69**

2 Typical Procedure for Testing Engine Oil Pressure **94**

3 Typical Procedure for Testing a Cooling System **97**

4 Conducting a Cylinder Compression Test **111**

5 Performing a Voltage Drop Test **135**

6 Typical Procedure for Testing Battery Capacity **167**

7 Using a Volt and Amp Tester to Test the Charging System **175**

8 Typical Procedure for Catalytic Converter Diagnosis **206**

9 Typical Procedure for Inspecting Turbochargers and Testing Boost Pressure **214**

10 Replacing a PROM **242**

11 Diagnosis with a Scan Tester **249**

12 Performing Ford Flash Code Diagnosis, Key On, Engine Off (KOEO), and Key On, Engine Running (KOER) Tests **258**

13 Testing a Ford MAP Sensor **293**

14 Testing Computer Outputs Using a Scan Tool **319**

15 Reprogramming an OBD II PCM **336**

16 Comparing O_2 Signals **358**

17 Diagnosing an Electronic Automatic Transmission **392**

18 Typical Procedure for Relieving Fuel Pressure and Removing Fuel Filter **424**

19 Checking the Fuel Pressure on a PFI System **429**

20 Removing and Replacing a Fuel Injector on a PFI System **441**

21 Typical Procedure for Testing Injector Balance **470**

22 Performing a Scan Tester Diagnosis of an Idle Air Control Motor **487**

23 Determining the Cause of a High Firing Line **532**

24 Timing the Distributor to the Engine **560**

25 Coil Current Ramping on a Waste Spark Ignition System **589**

26 Typical Procedure for Diagnosing Knock Sensors and Knock Sensor Modules **642**

27 Diagnosing an EGR Vacuum Regulator Solenoid **652**

Job Sheets

1 Shop Safety Survey **33**

2 Working Safely Around Air Bags **35**

3 Use of a Voltmeter **81**

4 Use of an Ohmmeter **83**

5 Cooling System Inspection and Diagnosis **123**

6 Conduct a Cylinder Power Balance Test **125**

7 Conduct a Cylinder Compression and Leakage Test **127**

8 Using a DSO on Sensors and Switches **183**

9 Inspecting a Battery **185**

10 Testing the Battery's Capacity **187**

11 Testing Charging System Output **189**

12 Check Engine Manifold Vacuum **227**

13 Test a Catalytic Converter for Efficiency **231**

14 Retrieve Codes from the Computer of an OBD I Engine Control System **305**

15 Test an ECT Sensor **307**

16 Check the Operation of a TP Sensor **309**

17 Test an O_2 Sensor **311**

18 Testing a MAP Sensor **313**

19 EGR Actuation **329**

20 Computer Control of the Fuel Injector **331**

21 Conduct a Diagnostic Check on an Engine Equipped with OBD II **379**

22 Monitor the Adaptive Fuel Strategy on an OBD II-Equipped Engine **381**

23 Road Test a Vehicle to Check the Operation of the Automatic Transmission **407**

24 Inspecting Drive Belts **409**

25 Relieving Pressure in an EFI System **449**

26 Testing Fuel Pressure on an EFI System **451**

27 Visually Inspect an EFI System **497**

28 Check the Operation of the Fuel Injectors on an Engine **499**

29 Conduct an Injector Balance Test **501**

30 Scope Testing an Ignition System **571**

31 Testing an Ignition Coil **573**

32 Individual Component Testing **575**

33 Setting Ignition Timing **579**

34 Visually Inspect a Distributor Ignition System **581**

35 Visually Inspect and Test an Electronic Ignition (EI) System **625**

36 Check the Emission Levels on an Engine **665**

37 Check the Operation of a PCV System **671**

38 Check the Operation of an EGR Valve **673**

DEDICATION

I want to dedicate this book to the memory of my father, Miles K. Pickerill.
Dad taught me love, honesty, and craftsmanship.

To Peggy, my wife of 29 years, and my two sons, Adam and Eric.
My family has shown nothing but support for me through the process of revising this book.

Thanks go out to:

The late John Riddle, who took me in as a young apprentice many years ago.

The management and staff of Clapp Auto Group in Clarksville, in particular
Dennis Price, who always believed in my ability.

Ken Pickerill

PREFACE

Thanks to the support the Today's Technician™ series has received from those who teach automotive technology, Thomson Delmar Learning, the leader in automotive related textbooks, is able to live up to its promise to provide new editions of the series regularly. We have listened and responded to our critics and our fans and present this new revised fourth edition. By revising our series regularly, we can and will respond to changes in the industry, changes in technology, changes in the certification process, and to the ever-changing needs of those who teach automotive technology.

The Today's Technician™ series, by Thomson Delmar Learning, features textbooks that cover all mechanical and electrical systems of automobiles and light trucks (whereas the Heavy-duty Trucks portion of the series does the same for Heavy-duty vehicles). Principally, the individual titles correspond to the eight main areas of ASE (National Institute for Automotive Service Excellence) certification. Additional titles include remedial skills and theories common to all of the certification areas and advanced or specific subject areas that reflect the latest technological trends. Each text is divided into two volumes: a Classroom Manual and a Shop Manual.

Unlike yesterday's mechanic, the technician of today and for the future must know the underlying theory of all automotive systems and be able to service and maintain those systems. Dividing the material into two manuals provides the reader with the information needed to begin a successful career as an automotive technician without interrupting the learning process by mixing cognitive and performance learning objectives into one volume.

The design of Thomson Delmar Learning's Today's Technician™ series was based on features that are known to promote improved student learning. The design was further enhanced by a careful study of survey results, in which the respondents were asked to value particular features. Some of these features can be found in other textbooks, whereas others are unique to this series.

Each Classroom Manual contains the principles of operation for each system and subsystem. The Classroom Manual also contains discussions on design variations of key components used by the different vehicle manufacturers. This volume is organized to build on basic facts and theories. The primary objective of this volume is to allow the reader to gain an understanding of how each system and subsystem operates. This understanding is necessary to diagnose the complex automobiles of today and tomorrow. Although the basics contained in the Classroom Manual provide the knowledge needed for diagnostics, diagnostic procedures appear only in the Shop Manual. An understanding of the basics is also a requirement for competence in the skill areas covered in the Shop Manual.

A spiral bound Shop Manual covers the "how-tos." This volume includes step-by-step instructions for diagnostic and repair procedures. Photo Sequences are used to illustrate some of the common service procedures. Other common procedures are listed and are accompanied with line drawings and photos that allow the reader to visualize and conceptualize the finest details of the procedure. This volume also contains the reasons for performing the procedures as well as when that particular service is appropriate.

The two volumes are designed to be used together and are arranged in corresponding chapters. Not only are the chapters in the volumes linked together, but also the contents of the chapters are linked. This linking of content is evidenced by marginal callouts that refer the reader to the chapter and page that the same topic is addressed in the other volume. This feature is valuable to instructors. Without this feature, users of other two-volume textbooks must search the index or table of contents to locate supporting information in the other volume. This is not only cumbersome but also creates additional work for an instructor when planning the presentation of material and when making reading assignments. It is also valuable to the students; with page references, they also know exactly where to look for supportive information.

Both volumes contain clear and thoughtfully selected illustrations, many of which are original drawings or photos specially prepared for inclusion in this series. This means that the art is a vital part of each textbook and not merely inserted to increase the numbers of illustrations.

The page layout used in this series is designed to include information that would otherwise break up the flow of information presented to the reader. The main body of the text includes all of the "need-to-know" information and illustrations. In the wide side margins of each page are many of the special features of the series. Items that are truly "nice-to-know" information include simple examples of concepts just introduced in the text, explanations or definitions of terms that will not be defined in the glossary, examples of common trade jargon used to describe a part or operation, and exceptions to the norm explained in the text. Many textbooks attempt to include this type of information and insert it in the main body of the text; this tends to interrupt the thought process and cannot be pedagogically justified. By placing this information off to the side of the main text, the reader can select when to refer to it.

Jack Erjavec
Series Editor

Classroom Manual

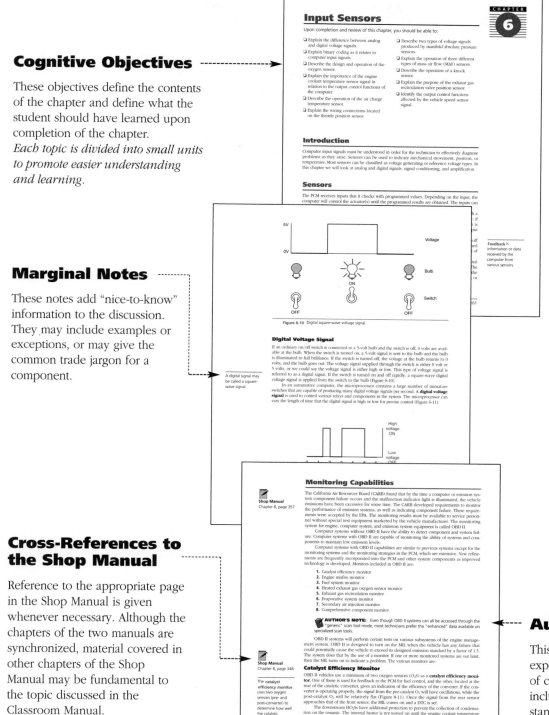

Cognitive Objectives

These objectives define the contents of the chapter and define what the student should have learned upon completion of the chapter.
Each topic is divided into small units to promote easier understanding and learning.

Marginal Notes

These notes add "nice-to-know" information to the discussion. They may include examples or exceptions, or may give the common trade jargon for a component.

Cross-References to the Shop Manual

Reference to the appropriate page in the Shop Manual is given whenever necessary. Although the chapters of the two manuals are synchronized, material covered in other chapters of the Shop Manual may be fundamental to the topic discussed in the Classroom Manual.

Author's Notes

This feature includes simple explanations, stories, or examples of complex topics. These are included to help students understand difficult concepts.

A Bit of History

This feature gives the student a sense of the evolution of the automobile. This feature not only contains nice-to-know information, but also should spark some interest in the subject matter.

Terms to Know List

A list of new terms appears next to the Summary.

Review Questions

Short answer essay, fill-in-the-blank, and multiple-choice questions are found at the end of each chapter. These questions are designed to accurately assess the student's competence in the stated objectives at the beginning of the chapter.

Terms to Know Definitions

New terms are pulled out into the margin and defined.

Summaries

Each chapter concludes with a summary of key points from the chapter. These are designed to help the reader review the chapter contents.

Ignition Coil Action

All ignition systems share a number of common components. Some, such as the battery and ignition switch, perform simple functions. The battery supplies battery voltage to the ignition primary circuit. Current flows when the ignition switch is in the "start" or the "run" position.

To generate a spark to begin combustion, the ignition system must deliver high voltage to the spark plugs. Because the amount of voltage required to bridge the gap of the spark plug varies with the operating conditions, most late-model vehicles can easily supply 30,000 to 60,000 volts to force a spark across the air gap. Since the battery delivers 12 volts, a method of stepping up the voltage must be used. Multiplying battery voltage is the job of a coil.

Shop Manual
Chapter 12, page 522

A BIT OF HISTORY

A turning point in the automotive industry came in 1902 when the Humber, an automobile from England, used a magneto electric ignition system.

The ignition coil is a **pulse transformer**. It transforms battery voltage into short bursts of high voltage. As explained previously, when a wire is moved through a magnetic field, voltage is induced in the wire. The inverse of this principle is also true—when a magnetic field moves across a wire, voltage is induced in the wire.

If a wire is bent into loops forming a coil and a magnetic field is passed through the coil, an equal amount of voltage is generated in each loop of wire. The more loops of wire in the coil, the greater the total voltage induced.

Also, the faster the magnetic field moves through the coil, even higher voltage is induced in the coil. If the speed of the magnetic field is doubled, the voltage output doubles.

An ignition coil uses these principles and has two coils of wire wrapped around an iron core. An iron or steel core is used because it has low **inductive reluctance**. The primary coil winding is normally composed of 100 to 200 turns of 20-gauge wire. This coil of wire conducts battery current. When a current is passing through the primary coil, it magnetizes the iron core. The strength of the magnet depends on the number of wire loops and the amount of current flowing through those loops. The secondary coil of wires may consist of 15,000 to 25,000 or more turns of very fine copper wire.

Because of the effects of counter EMF (electromotive force) on the current flowing through the primary winding, it takes some time for the coil to become fully magnetized or saturated. Therefore, current flows in the primary winding for some time between firings of the

---

triggering is located above the pickup plate in the distributor and is referred to as the reference pickup. The second pickup unit is positioned below the plate. A ring with two notches is attached to the distributor shaft and rotates through the lower pickup unit. This lower pickup is called the synchronizer (SYNC) pickup.

In other designs, the two pickup units are mounted below the pickup plate and one set of blades rotates through both Hall effect units. The shutter blade representing number one cylinder has a large opening in the center of the blade. When this blade rotates through the SYNC pickup, a different signal is produced compared to the other blades. This number one blade signal informs the PCM when to activate the injectors.

Summary

Terms to Know

After top dead center (ATDC)
Basic timing
Before top dead center (BTDC)
Bypass
Computer-controlled dwell (CCD)
Crankshaft position sensor
Distributor ignition (DI)
Dwell
Dwell time
Electronic control unit (ECU)
Electronic spark timing (EST)
Gap
Heat range
High Energy Ignition (HEI)
Ignition timing
Inductive reluctance
Initial timing
Lookup tables
Photoelectric sensor
Points
Primary circuit
Profile ignition pickup
Pulse transformer

❑ The ignition system supplies high voltage to the spark plugs to ignite the air-fuel mixture in the combustion chambers.

❑ The arrival of the spark is timed to coincide with the compression stroke of the piston. This basic timing can be advanced or retarded under certain conditions such as high engine rpm or extremely light or heavy engine loads.

❑ The ignition system has two interconnected electrical circuits: a primary circuit and a secondary circuit.

❑ The primary circuit supplies low voltage to the primary winding of the ignition coil. This creates a magnetic field in the coil.

❑ The electronic control unit (ECU), or module, opens and closes the primary circuit.

❑ A switching device interrupts primary current flow, collapsing the magnetic field and creating a high-voltage surge in the ignition coil secondary winding.

❑ The switching device used in electronic systems is a module.

❑ The secondary circuit carries high-voltage surges to the spark plugs. On some systems, the circuit runs from the ignition coil through a distributor to the spark plugs.

❑ Spark plugs provide an electrode gap in each combustion chamber, and the secondary current flows across this gap to ignite the air-fuel mixture in the combustion chamber.

❑ EI systems provide longer spark duration at the spark plug electrodes than conventional electronic ignition systems, and this helps to fire leaner air-fuel ratios in today's engines.

❑ The ignition coil primary winding contains a few hundred turns of heavy wire.

❑ The secondary coil winding contains thousands of turns of very fine wire.

❑ Dwell is the length of time that the primary circuit is turned on prior to each cylinder firing.

❑ Ignition timing is directly related to the position of the crankshaft. Magnetic pulse generators and Hall effect sensors are the most widely used engine position sensors. They generate an electrical signal at certain times during crankshaft rotation. This signal triggers the electronic switching device to control ignition timing.

❑ The distributor may house the device plus centrifugal or vacuum timing advance mechanisms. Some systems locate the switching device outside the distributor housing.

❑ The distributor may house the triggering device (position sensor) and module.

❑ The module can be located outside of the distributor.

❑ The computer receives input from numerous sensors. Based on this data, the computer determines the optimum firing time and signals an ignition module to activate the secondary circuit at the precise time

---

Review Questions

Short Answer Essays

1. Describe the design of the primary and secondary coil windings.
2. Explain how the voltage is induced in the distributor pickup coil as the reluctor high point approaches alignment with the pickup coil.
3. Define spark plug reach.
4. Describe the type of driving conditions when a colder range spark plug may be required.
5. Explain how the high voltage is induced in the secondary coil winding in an electronic ignition system.
6. Explain why dwell time is important to ignition system operation.
7. Name the three major functions of an ignition system.
8. What is the basic difference between the primary and secondary ignition circuits?
9. Name the engine operating conditions that affect ignition timing requirements.
10. What primary role does a rotor have in the ignition system?

Fill-in-the-Blanks

1. Modern ignition cables contain fiber cores that act as a _____ in the secondary circuit to cut down on radio and television interference and reduce spark plug wear.
2. The ends of the secondary coil winding are usually connected to the terminal in the coil tower and one of the _____ terminals.
3. The arrival of the spark is timed to coincide with the _____ stroke of the piston.
4. Basic ignition timing is typically _____ with increases in engine speed and _____ with heavy engine load.
5. The ignition system has two interconnected electrical circuits: a _____ circuit and a _____ circuit.
6. The _____ interrupts primary current flow, collapsing the magnetic field and creating a high-voltage surge in the ignition coil secondary winding.
7. The switching device used in electronic systems is a(n) _____.
8. _____ _____ and _____ sensors are the most widely used engine position sensors.
9. A cold-range spark plug has a _____ heat path compared to a hot-range spark plug.
10. Computer-controlled ignition systems rely on the inputs from various _____ to control ignition timing.

Terms to Know continued

Push start
Reference
Reference low
Saturation
Schmitt trigger
Secondary circuit
Spark out (SPOUT)
Thick film integrated (TFI) IV
Top dead center (TDC)
Triggering device

Shop Manual

To stress the importance of safe work habits, the Shop Manual dedicates one full chapter to safety. Other important features of this manual include:

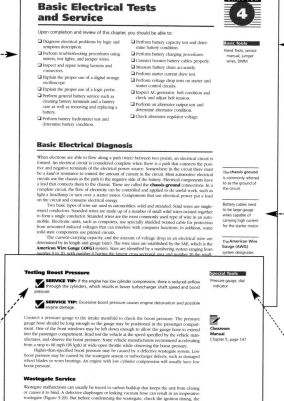

Performance-Based Objectives

These objectives define the contents of the chapter and define what the student should have learned upon completion of the chapter. These objectives also correspond with the list of required tasks for ASE certification. *Each ASE task is addressed.*

Although this textbook is not designed to simply prepare someone for the certification exams, it is organized around the ASE task list. These tasks are defined generically when the procedure is commonly followed and specifically when the procedure is unique for specific vehicle models. Imported and domestic model automobiles and light trucks are included in the procedures.

Service Tips

Whenever a short-cut or special procedure is appropriate, it is described in the text. These tips are generally those things commonly done by experienced technicians.

Basic Tools Lists

Each chapter begins with a list of the basic tools needed to perform the tasks included in the chapter.

Marginal Notes

These notes add "nice-to-know" information to the discussion. They may include examples or exceptions, or may give the common trade jargon for a component.

Special Tools Lists

Whenever a special tool is required to complete a task, it is listed in the margin next to the procedure.

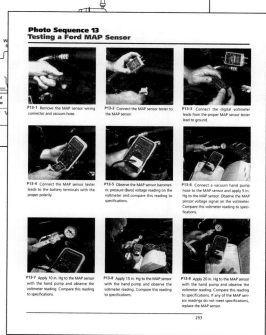

Photo Sequences

Many procedures are illustrated in detailed Photo Sequences. These detailed photographs show the students what to expect when they perform particular procedures. They also can provide the student a familiarity with a system or type of equipment which the school may not have.

Cautions and Warnings

Throughout the text, warnings are given to alert the reader to potentially hazardous materials or unsafe conditions. Cautions are given to advise the student of things that can go wrong if instructions are not followed or if a nonacceptable part or tool is used.

Cross-References to the Classroom Manual

Reference to the appropriate page in the Classroom Manual is given whenever necessary. Although the chapters of the two manuals are synchronized, material covered in other chapters of the Classroom Manual may be fundamental to the topic discussed in the Shop Manual.

Customer Care

This feature highlights those little things a technician can do or say to enhance customer relations.

Job Sheets

Located at the end of each chapter, the Job Sheets provide a format for students to perform procedures covered in the chapter. A reference to the ASE Task addressed by the procedure is referenced on the Job Sheet.

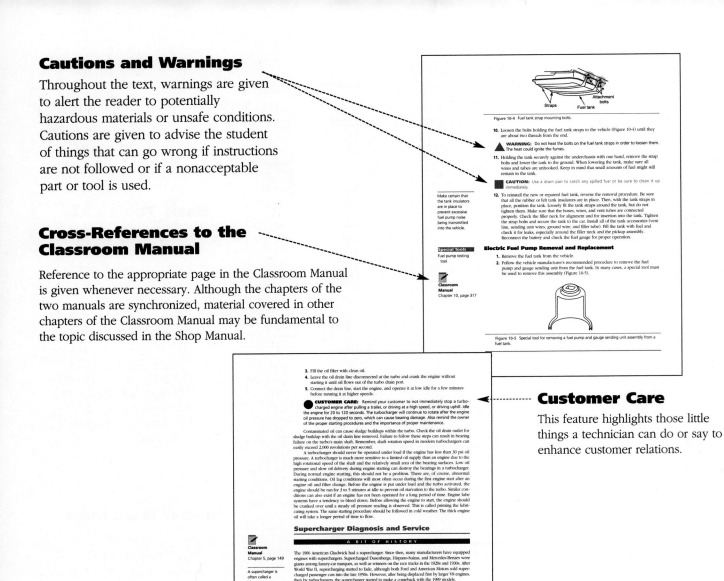

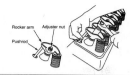

Rocker arm   Adjuster nut

Pushrod

Figure 3-24 Rocker arm adjustment with hydraulic lifters and individual rocker arms.

while rotating the pushrod (Figure 3-24). Continue rotating the rocker arm nut until the end of the rocker arm contacts the end of the valve stem and the pushrod becomes more difficult to turn. Continue turning the rocker arm nut clockwise the number of turns specified in the service manual. In some engines, this specification is one turn plus or minus one-quarter turn.

CASE STUDY

A police car was brought to the shop with the explanation that a local transmission shop had rebuilt the transmission three times and still has a shudder in fourth gear. The technician drove the car and discovered that it did indeed have a pronounced shutter from about 45-55 mph in high gear. If you placed your foot on the brake pedal just hard enough to release the torque converter clutch but not hard enough to apply the brakes, the shudder was unnoticeable. Several technicians have been mistaken in deciding whether the problem was engine or transmission related, especially with torque converter clutch. A torque converter clutch locks the engine and transmission together mechanically just like a manual transmission. Any vibration or irregularity in the operation of the transmission or engine can definitely be felt, especially in a low torque overdrive gear range. To make the problem worse, there were several cases of bad converters that caused the same problem. Another common cause of the shudder was fuel injector tip

Terms to Know

Drive lugs   Electronic distributorless ignition   Variable rate sensor signal (VRS)
                               system (EDIS)

ASE-Style Review Questions

1. While discussing the primary and secondary ignition circuits:
   Technician A says the primary controls the secondary.
   Technician B says the primary circuit also controls ignition timing in the secondary circuit.
   Who is correct?
   A. A only   C. Both A and B
   B. B only   D. Neither A nor B

2. While discussing EI ignition systems:
   Technician A says base timing is not adjustable on EI systems.
   Technician B says that the air gap is adjustable on some crank sensors but timing is not affected.
   Who is correct?
   A. A only   C. Both A and B
   B. B only   D. Neither A nor B

3. *Technician A* says a logic probe can be used to check a Hall effect sensor.
   Technician B says the red LED should light on both end terminals with the key on.
   Who is correct?
   A. A only   C. Both A and B
   B. B only   D. Neither A nor B

4. Two technicians are discussing a Nissan EI system:
   Technician A says that the coil used for this system should have a primary resistance of 2 ohms.
   Technician B says the secondary circuit should have between 7,000 and 8,000 ohms resistance.
   Who is correct?
   A. A only   C. Both A and B
   B. B only   D. Neither A nor B

5. While discussing waste spark EI systems:
   Technician A says a high resistance spark plug can affect the firing of its companion spark plug.
   Technician B says improper spark plug torque can cause an engine misfire.
   Who is correct?
   A. A only   C. Both A and B
   B. B only   D. Neither A nor B

6. While discussing how to test a crankshaft position sensor:
   Technician A says a logic probe can be used.
   Technician B says a DMM can be used.
   Who is correct?
   A. A only   C. Both A and B
   B. B only   D. Neither A nor B

7. While discussing the possible causes for a no-start condition on an EI-equipped engine:
   Technician A says a shorted crankshaft sensor may prevent the engine from starting.
   Technician B says a shorted spark plug may stop the engine from starting.
   Who is correct?
   A. A only   C. Both A and B
   B. B only   D. Neither A nor B

8. While discussing the diagnosis of an EI system in which the crankshaft and camshaft sensor tests are satisfactory but a spark tester connected from the spark plug wires to ground does not fire:
   Technician A says the coil assembly may be defective.
   Technician B says the voltage supply wire to the coil assembly may be open.
   Wh
   A.
   B.

ASE Challenge Questions

1. A V-6 engine has oil leaking from the vicinity of the lower front engine covers. The most probable cause of the leak is:
   A. Worn front main seal
   B. Timing belt cover
   C. Valve cover(s)
   D. Oil pressure sending unit

2. The customer states the coolant level must be topped off every few days. There are no visible leaks.
   Technician A says this may be caused by a bad radiator cap.
   Technician B says a stuck open thermostat will cause this problem.

4. During an engine vacuum test, the gauge reading fluctuates between 11 and 16 in. Hg. This indicates service should be performed on:
   A. Improper idle mixture
   B. The valve and valve springs
   C. Fuel injectors
   D. Either A or C

5. A vacuum leak is suspected for a rough idle concern. Using a four-gas infrared exhaust analyzer, *Technician A* says O₂ will be higher than normal.
   Technician B says CO will be higher than normal.
   Who is correct?
   A. A only   C. Both A and B
   B. B only   D. Neither A nor B

APPENDIX A

ASE PRACTICE EXAMINATION

1. *Technician A* says a hydrometer reading of 1.200 at 80°F means the battery must be recharged before performing a capacity test.
   Technician B says a capacity test can be correctly performed with the battery cables connected.
   Who is correct?
   A. A only   C. Both A and B
   B. B only   D. Neither A nor B

2. *Technician A* says engine detonation may be caused by low octane fuel.
   Technician B says that detonation can be caused by engine overheating.
   Who is correct?
   A. A only   C. Both A and B
   B. B only   D. Neither A nor B

3. The PCV system is being discussed.
   Technician A says oil in the crankcase breather filter will confirm that the PCV valve is clogged.
   Technician B says the PCV valve must be disconnected to be checked with the engine operating.
   Who is correct?
   A. A only   C. Both A and B
   B. B only   D. Neither A nor B

4. An EI-equipped vehicle will not start.
   Technician A says a good first step is to make certain that the MIL is illuminated with the key on.
   Technician B says to check for spark at more than one plug wire during diagnosis.
   Who is correct?
   A. A only   C. Both A and B
   B. B only   D. Neither A nor B

5. Electronic fuel injection systems are being discussed.
   Technician A says a hard to start engine may have an open electrical fuel pump relay bypass circuit.
   Technician B says a hesitation when accelerating from idle may be caused by dirty injectors.
   Who is correct?
   A. A only   C. Both A and B
   B. B only   D. Neither A nor B

6. A vehicle backfires during almost every deceleration.
   Technician A says a secondary air injection diverter valve may be causing the backfire.
   Technician B says this condition may be caused by a low performing secondary air injection pump.
   Who is correct?
   A. A only   C. Both A and B
   B. B only   D. Neither A nor B

7. A vehicle has a consistent slow- or no-crank condition. This may be caused by:
   A. A slipping AC generator drive belt
   B. A defective AC generator
   C. A low regulator voltage
   D. Any of the above

8. A fuel-injected vehicle comes in with a miss on acceleration.
   Technician A says that there could be an ignition problem.
   Technician B says that the intake to MAF duct could be opening up at a tear on acceleration.
   Who is correct?
   A. A only   C. Both A and B
   B. B only   D. Neither A nor B

9. *Technician A* says static electricity generated by clothing in contact with vehicle upholstery must be discharged before beginning work on electronic devices.
   Technician B says the best control for static electricity is the wearing of a waist or wrist grounding strap.
   Who is correct?
   A. A only   C. Both A and B
   B. B only   D. Neither A nor B

10. *Technician A* says a camshaft installed one tooth retarded can be compensated by adjusting the ignition's base timing.
    Technician B says to use the starter to crank the engine with the timing belt removed to determine if the engine is causing interference or free-wheeling.
    Who is correct?
    A. A only   C. Both A and B
    B. B only   D. Neither A nor B

11. A fuel-injected engine runs rough and sluggish on acceleration. When checking the fuel pressure, the technician notices that the fuel pressure is at specifications during idle, but falls off on acceleration.
    Technician A says the fuel pump could be faulty.
    Technician B says the fuel filter could be clogged.
    Who is correct?
    A. A only   C. Both A and B
    B. B only   D. Neither A nor B

677

Case Studies

Case Studies concentrate on the ability to properly diagnose the systems. Beginning with Chapter 3, each chapter ends with a case study in which a vehicle has a problem, and the logic used by a technician to solve the problem is explained.

ASE-Style Review Questions

Each chapter contains ASE-style review questions that reflect the performance-based objectives listed at the beginning of the chapter. These questions can be used to review the chapter as well as to prepare for the ASE certification exam.

ASE Practice Examination

A 50 question ASE practice exam, located in the appendix, is included to test students on the contents of the Shop Manual.

Terms to Know List

Terms in this list can be found in the Glossary at the end of the manual.

ASE Challenge Questions

Each technical chapter ends with five ASE challenge challenge questions. These are not more review questions, rather they test the students' ability to apply general knowledge to the contents of the chapter.

Reviewers

The author and publisher would like to extend a special thanks to the following reviewers for their contributions to this text:

C. Neel Flannagan
Aiken Technical College
Graniteville, SC

Ronald D. Finney
Ivy Tech Community College
Indianapolis, IN

Robert Gibbens
North Central Kansas Technical College
Beloit, KS

Mike Huneke
Texas State Technical College
Waco, TX

Andy O'Neal
University of Northwestern Ohio
Lima, OH

Shane Sampson
Western Iowa Tech Community College
Sioux City, IA

Don Sykora
Morton College
Cicero, IL

Gary Tucker
Oklahoma City Community College
Oklahoma City, OK

Dan Wilson
Westwood College
Denver, CO

Paul Witek
Western Technical Institute
El Paso, TX

Alexander Wong
Sierra College
Rocklin, CA

Tools and Safety

Upon completion and review of this chapter, you should be able to:

❑ Understand the importance of safety and accident prevention in an automotive shop.

❑ Explain the basic principles of personal safety, including protective eye wear, clothing, gloves, shoes, and hearing protection.

❑ Demonstrate proper lifting procedures and precautions.

❑ Explain the procedures and precautions for safely using tools and equipment.

❑ Follow safety precautions regarding the use of power tools.

❑ Demonstrate proper safety precautions during the use of compressed air equipment.

❑ Demonstrate proper vehicle lift operating and safety procedures.

❑ Observe all safety precautions when hydraulic tools are used in the automotive shop.

❑ Follow the recommended procedure while operating hydraulic tools such as presses, floor jacks, and vehicle lifts to perform automotive service tasks.

❑ Follow safety precautions while using cleaning equipment in the automotive shop.

❑ Observe all shop rules when working in the shop.

❑ Operate vehicles in the shop according to shop driving rules.

❑ Explain what should be done to maintain a safe work area, including handling vehicles in the shop and venting carbon monoxide gases.

❑ Fulfill employee obligations when working in the shop.

❑ Describe the Automotive Service Excellence (ASE) technician testing and certification process, including the eight areas of certification.

❑ Follow safety precautions while handling hazardous waste materials.

❑ Dispose of hazardous waste materials in accordance with state and federal regulations.

Introduction

Safety and accident prevention must be a top priority in all automotive shops. There is great potential for serious accidents simply because of the nature of the business and the equipment used. In fact, the automotive repair industry is rated as one of the most dangerous occupations in the country.

Vehicles, equipment, and many parts are very heavy, and parts often fit tightly together. Many components become hot during operation and high fluid pressures can build up inside the cooling system, fuel system, or battery. Batteries contain highly corrosive and potentially explosive acids. Fuels and cleaning solvents are flammable. Exhaust fumes are poisonous. During some repairs, technicians can be exposed to harmful dust particles and vapors.

Good safety practices eliminate these potential dangers. A careless attitude and poor work habits invite disaster. Shop accidents can cause serious injury, temporary or permanent disability, and death. Safety is a very serious matter. Both the employer and employees must work together to protect the health and welfare of all who work in the shop.

This chapter contains many safety guidelines concerning personal safety, tools and equipment, and work area, as well as some of the responsibilities you have as an employee and responsibilities your employer has to you. In addition to these rules, special warnings have been used throughout this book to alert you to situations where carelessness could result in personal injury. Finally, when working on cars, always follow safety guidelines given in service manuals and other technical literature. They are there for your protection.

Personal Safety

Personal safety simply involves those precautions you take to protect yourself from injury. These include wearing protective gear, dressing for safety, and correctly handling tools and equipment.

Eye Protection

Your eyes can become infected or permanently damaged by many things in a shop. Some repair procedures, such as grinding, result in tiny particles of metal and dust that are thrown off at very high speeds. These metal and dirt particles can easily get into your eyes, causing scratches or cuts on your eyeball. Pressurized gases and liquids escaping a ruptured hose or fuel line fitting can spray a great distance. If these chemicals get into your eye, they can cause blindness. Dirt and sharp bits of corroded metal can easily fall down into your eyes while you are working under a vehicle.

Eye protection should be worn whenever you are exposed to these risks. To be safe, you should wear **safety glasses** whenever you are working in the shop. There are many types of eye protection available (Figure 1-1). To provide adequate eye protection, safety glasses have lenses made of safety glass. They also offer some sort of side protection. Regular prescription glasses do not offer sufficient protection and therefore should not be worn as a substitute for safety glasses.

Wearing safety glasses at all times is a good habit to have. To help develop this habit, wear safety glasses that fit well and feel comfortable.

If chemicals such as battery acid, fuel, or solvents get into your eyes, flush them continuously with clean water. Have someone call a doctor and get medical help immediately.

> If you wear prescription glasses, wear safety goggles that will securely fit over your glasses.

Clothing

Your clothing should be well fitted and comfortable but made with strong material. Loose, baggy clothing can easily become caught by moving parts and machinery. Neckties should not be worn. Some technicians prefer to wear coveralls or shop coats to protect their personal clothing. Cutoffs and short pants are not appropriate for shop work.

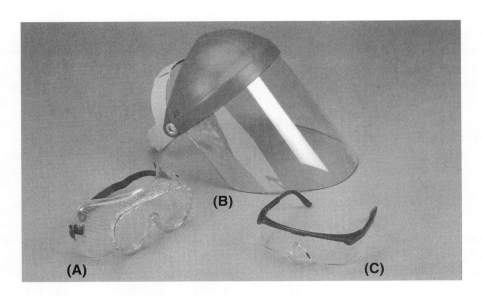

Figure 1-1 (A) Safety goggles, (B) Face shield, (C) Safety glasses.

Automotive work involves the handling of many heavy objects that can be accidentally dropped on your feet or toes. Always wear leather or similar material shoes or boots with non-slip soles. Steel-tipped safety shoes can give added protection for your feet. Jogging or basketball shoes, street shoes, and sandals are not appropriate in the shop.

Good hand protection is often overlooked. A scrape, cut, or burn can limit your effectiveness at work for many days. A well-fitted pair of heavy work gloves should be worn during operations such as grinding and welding or when handling high-temperature components. Always wear approved rubber gloves when handling caustic chemicals. Caustic chemicals are strong and dangerous chemicals. They can easily burn your skin. Be very careful when handling this type of chemical.

Caustic chemicals can burn through cloth and skin.

Ear Protection

Exposure to very loud noise levels for extended periods of time can lead to a loss of hearing. Air wrenches, engines running under a load, and vehicles running in enclosed areas can all generate annoying and harmful levels of noise. Simple ear plugs or earphone-type protectors should be worn in constantly noisy environments.

Hair and Jewelry

Long hair and loose, hanging jewelry can create the same type of hazard as loose-fitting clothing. They can become caught in moving engine parts and machinery. If you have long hair, tie it back or cover it with a cap.

Never wear rings, watches, bracelets, or neck chains. These can easily get caught in moving parts and cause serious injury, especially when working on or near batteries and electrical systems.

Other Personal Safety Warnings

❏ Never smoke while working on a vehicle or while working with any machine in the shop.
❏ Playing around or horseplay is not fun when it sends someone to the hospital. Such things as air nozzle fights, creeper races, or practical jokes have no place in the shop.
❏ To prevent serious burns, keep your skin away from hot metal parts such as the radiator, exhaust manifold, tailpipe, catalytic converter, and muffler.
❏ When working with a hydraulic press, make sure that the pressure is applied in a safe manner. It is generally wise to stand to the side when operating the press. Always wear safety glasses.
❏ Properly store all parts and tools by putting them away in a place where people will not trip over them. This practice not only cuts down on injuries, it also reduces time wasted looking for a misplaced part or tool.

Lifting and Carrying

Knowing the proper way to lift heavy objects is important. You should also use back protection devices when you are lifting a heavy object. Always lift and work within your ability and ask others to help when you are not sure that you can handle the size or weight of an object. Even small, compact parts can be surprisingly heavy or unbalanced. Think about how you are going to lift something before beginning. When lifting any object, follow these steps.

1. If the object is to be carried, be sure your path is free from loose parts or tools.
2. Place your feet close to the object. Position your feet so that you will be able to maintain a good balance.

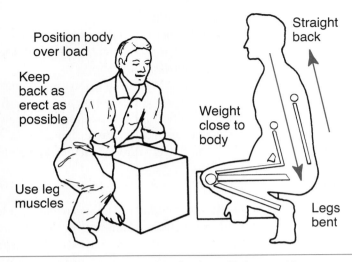

Figure 1-2 Use your leg muscles, never your back, to lift heavy objects.

3. Keep your back and elbows as straight as possible. Bend your knees until your hands reach the best place to get a strong grip on the object (Figure 1-2).
4. If the part is in a cardboard box, make sure the box is in good condition. Old, damp, or poorly sealed boxes will tear and the part will fall out.
5. Firmly grasp the object or container. Never try to change your grip as you move the load.
6. Keep the object close to your body and lift it up by straightening your legs. Use your leg muscles, not your back muscles.
7. If you must change your direction of travel, never twist your body. Turn your whole body, including your feet.
8. When placing the object on a shelf or counter, do not bend forward. Place the edge of the load on the shelf and slide it forward. Be careful not to pinch your fingers.
9. When setting a load down, bend your knees and keep your back straight. Never bend forward—this strains the back muscles.
10. Set the object onto blocks of wood to protect your fingers when lowering something heavy onto the floor.

Occupational Safety and Health Act

The **Occupational Safety and Health Act (OSHA)** was passed by the United States government in 1970. The purposes of this legislation are these:

The **Occupational Safety and Health Act (OSHA)** regulates working conditions in the United States.

1. To assist and encourage the citizens of the United States in their efforts to assure safe and healthful working conditions by providing research, information, education, and training in the field of occupational safety and health.
2. To assure safe and healthful working conditions for working men and women by authorizing enforcement of the standards developed under the Act.

Since approximately 25 percent of workers are exposed to health and safety hazards on the job, the OSHA is necessary to monitor, control, and educate workers regarding health and safety in the workplace. Employers and employees should be familiar with **Workplace Hazardous Materials Information Systems (WHMIS)**.

Shop Hazards

Service technicians and students encounter many hazards in an automotive shop. When these hazards are known, basic shop safety rules and procedures must be followed to avoid personal injury. Some of the hazards in an automotive shop are listed below:

Shop hazards must be recognized and avoided to prevent personal injury.

1. Flammable liquids such as gasoline and paint must be handled and stored properly in special metal cabinets.

2. Flammable materials such as oily rags must be stored properly in special covered receptacles (Figure 1-3) to avoid a fire hazard.

3. Batteries contain a corrosive sulfuric acid solution and produce explosive hydrogen gas while charging.

4. Loose sewer and drain covers may cause foot or toe injuries.

5. Caustic liquids such as those in hot cleaning tanks are harmful to skin and eyes.

6. High-pressure air in the shop compressed air system can be very dangerous if it penetrates the skin and enters the bloodstream.

7. Frayed cords on electric equipment and lights may result in severe electrical shock.

8. Hazardous waste material such as batteries and the caustic cleaning solution from a hot or cold cleaning tank must be handled properly to avoid harmful effects.

9. Carbon monoxide from vehicle exhaust is poisonous.

10. Loose clothing, jewelry, or long hair may become entangled in rotating parts on equipment or vehicles, resulting in serious injury.

11. Dust and vapors generated during some repair jobs are harmful. Asbestos dust generated during brake lining service and clutch service is a contributor to lung cancer.

12. High noise levels from shop equipment such as an air chisel may be harmful to the ears.

13. Oil, grease, water, or parts cleaning solutions on shop floors may cause someone to slip and fall, resulting in serious injury.

AUTHOR'S NOTE: The old saying, "Had I known I was going to live this long, I would have taken better care of myself" should serve as a friendly reminder to all technicians. We do preventive maintenance on our customers' cars, right? So why not take the same measures to prevent temporary or permanent injury? Taking care of yourself in the shop in even the simplest way can reward you greatly later in life. This means wearing gloves, eye protection, ear protection, and dust masks when appropriate. Try to anticipate the potential risks for an injury while performing a particular service on a vehicle, then act on methods of protection that will mitigate the risks.

Figure 1-3 Keep combustibles in safety containers.

Hand Tool Safety

Many shop accidents are caused by improper use and care of hand tools. These hand tool safety steps must be followed:

1. Keep tools clean and in good condition. Worn tools may slip and result in hand injury. If a hammer is used with a loose head, the head may fly off and cause personal injury or vehicle damage. Keep all hand tools free of grease and in good condition. Tools that slip can cause cuts and bruises. If a tool slips and falls into a moving part, it can fly out and cause serious injury.

2. Use the proper tool for the job. Make sure the tool is of professional quality. Using poorly made tools or the wrong tools can damage parts, the tool itself, or cause injury. Do not use broken or damaged tools.

3. Be careful when using sharp or pointed tools. Do not place sharp tools or other sharp objects in your pockets. They can stab or cut your skin, ruin automotive upholstery, or scratch a painted surface.

4. Tool tips that are intended to be sharp should be kept in a sharp condition. Sharp tools such as chisels will do the job faster with less effort. Dull tools can be more dangerous than sharp tools.

5. Never use a tool for which it is not intended, such as using a screwdriver as a prybar. The blade or blade tip can break and cause personal injury.

6. Use only impact sockets that are designed for use with an impact wrench. Ordinary sockets can break or shatter if used improperly (Figure 1-4).

Figure 1-4 A worn-out impact socket must be replaced.

Power Tool Safety

Power tools are operated by an outside source of power, such as electricity, compressed air, or hydraulic pressure. Safety around power tools is very important. Serious injury can result from carelessness. Always wear safety glasses when using power tools.

If the tool is electrically powered, make sure it is properly grounded. Check the wiring for cracks in the insulation and bare wires before using it. Also, when using electrical power tools, never stand on a wet or damp floor. Disconnect the power source before performing any service on the machine or tool. Before plugging in any electric tool, make sure the switch is off to prevent serious injury. When you are through using the tool, turn it off and unplug it. Never leave a running power tool unattended. When you leave, turn it off.

When using power equipment on a small part, never hold the part in your hand. Always mount the part in a bench vise or use vise grip pliers. Never try to use a machine or tool beyond its stated capacity or for operations requiring more than the rated power of the tool.

When working with larger power tools, like bench or floor equipment, check the machines for signs of damage before using them. Place all safety guards into position (Figure 1-5). A safety guard is a protective cover over a moving part. It is designed to prevent injury. Wear safety glasses or a face shield. Make sure there are no people or parts around the machine before starting it up. Keep your hands and clothing away from the moving parts. Maintain a balanced stance while using the machine.

Hydraulic Press

WARNING: When operating a hydraulic press, always be sure that the components being pressed are properly supported on the press bed with steel supports.

CAUTION: When using a hydraulic press, never operate the pump handle when the pressure gauge exceeds the maximum pressure rating of the press. If this pressure is exceeded, some part of the press may suddenly break and cause severe personal injury.

When two components have a tight precision fit between them, a **hydraulic press** is used to separate them or press them together. The hydraulic press rests on the shop floor. An adjustable steel beam bed is retained to the lower press frame with heavy steel pins. A hydraulic cylinder

Figure 1-5 Safety guards on a bench grinder.

Figure 1-6 A hydraulic press.

and ram is mounted on the top part of the press with the ram facing downward toward the press bed (Figure 1-6). The component being pressed is placed on the press bed with the appropriate steel supports. A hand-operated hydraulic pump is mounted on the side of the press. When the handle is pumped, hydraulic fluid is forced into the cylinder, and the ram is extended against the component on the press bed to complete the pressing operation. A pressure gauge on the press indicates the pressure applied from the hand pump to the cylinder. The press frame is designed for a certain maximum pressure. This pressure must not be exceeded during operation.

Electrical Safety

1. Frayed cords on electrical equipment must be replaced or repaired immediately.
2. All electric cords from lights and electric equipment must have a ground connection. The ground connector is the round terminal in a three-prong electrical plug. Do not use a two-prong adapter to plug in a three-prong electrical cord. Three-prong electrical outlets should be mandatory in all shops.
3. Do not leave electrical equipment running and unattended.

Battery Safety

Batteries give off hydrogen and oxygen gases, which are explosive! Use extreme caution when working around or handling batteries.

1. Always wear eye protection when working near a battery.
2. Do not smoke or allow any open flame around a battery.
3. Do not connect or disconnect a battery charger while the charger is in the "on" position. In addition, it is a good practice to unplug the battery charger from the 110-volt AC outlet before connecting or disconnecting the charger from the battery.

4. Observe proper polarity when installing a battery, connecting a battery charger, or jump-starting the vehicle.

5. Do not allow the battery to exceed 125°F while charging the battery.

6. When using load-test equipment, be certain the load controller is off prior to connecting or disconnecting the unit from the battery.

7. If battery acid contacts your skin, wash immediately for at least 10 minutes.

8. Battery acid can harm skin, eyes, clothing, and some paint finishes.

9. Neutralize any battery acid spills with baking soda.

Gasoline Safety

Gasoline is a very flammable liquid! One exploding gallon of gasoline has energy equal to fourteen sticks of dynamite. It is the expanding vapors from gasoline that are extremely dangerous. These vapors are present even in cold temperatures. Vapors formed in gasoline tanks on cars are controlled, but vapors from a gasoline storage container may escape from the can, resulting in a hazardous situation. Therefore, gasoline storage containers must be placed in a well-ventilated space.

Approved gasoline storage cans have a flash-arresting screen at the outlet (Figure 1-7). These screens prevent external ignition sources from igniting the gasoline within the can while the gasoline is being poured. Follow these safety precautions regarding gasoline containers:

1. Always use approved gasoline containers that are designed specifically for gasoline and properly labeled as such.

2. Do not fill gasoline containers completely full. Always leave the level of gasoline at least one inch from the top of the container. This action allows expansion of the gasoline at higher temperatures. If gasoline containers are completely full, gasoline vapors will expand when the temperature increases. This expansion forces gasoline from the can and creates a dangerous spill.

3. If gasoline containers must be stored, place them in a well-ventilated area such as a storage shed. Do not store gasoline containers in your home or in the trunk of a vehicle.

4. When a gasoline container must be transported, be sure it is secured against upsets.

5. Do not store a partially filled gasoline container for long periods of time because it may give off vapors and produce a potential danger.

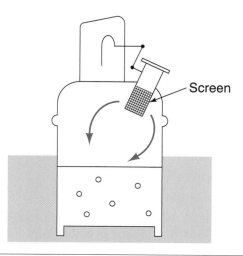

Screen

Figure 1-7 Approved gasoline container.

6. Never leave gasoline containers open except while filling or pouring gasoline from the container.

7. Do not prime an engine with gasoline while cranking the engine.

8. Never use gasoline as a cleaning agent.

9. Avoid contact with unprotected skin.

Refrigerant Safety

It is illegal to knowingly discharge or vent air-conditioning refrigerant into the atmosphere. When servicing an air-conditioning system, recover and/or recycle refrigerant using approved, certified equipment. These services must be performed by an EPA-recognized, certified technician following specific procedures. When working around refrigerant or air-conditioning service equipment, it is important to wear full eye protection. Refrigerant can cause immediate blindness.

General Shop Safety

1. All sewer covers must fit properly and be kept securely in place.

2. Always wear a face shield, protective gloves, and protective clothing when necessary. Gloves should be worn when working with solvents and caustic solutions, handling hot metal, or grinding metal. Various types of protective gloves are available. Shop coats and coveralls are the most common types of protective clothing.

3. Never direct high-pressure air from an air gun against human flesh or *any* part of the body. If this action is allowed, air may penetrate the skin and enter the bloodstream, causing serious health problems or death. Always keep air hoses in good condition. If an end blows off an air hose, the hose may whip around and result in personal injury. Use only Occupational Safety and Health Act (OSHA) approved air gun nozzles.

4. Handle all hazardous waste materials according to state and federal regulations. (These regulations are explained later in this chapter.)

5. Always place a shop exhaust hose on the tailpipe of a vehicle if the engine is running in the shop and be sure the shop exhaust fan is turned on.

6. Keep hands, long hair, jewelry, and tools away from rotating parts such as fan blades and belts on running engines. Remember that an electric-drive fan may start turning without warning. Avoid wearing rings, watches, or bracelets, especially when working on or around batteries or other electrical systems.

7. When servicing brakes or clutches from manual transmissions, always clean asbestos dust from these components with an approved asbestos dust parts washer, and wear a dust mask.

8. Always use the correct tool for the job. For example, never strike a hardened steel component, such as a piston pin, with a steel hammer. This type of component may shatter and fragments may penetrate eyes or skin.

9. Follow the car manufacturer's recommended service procedures.

10. Be sure that the shop has adequate ventilation.

11. Make sure the work area has adequate lighting.

12. Use trouble lights with steel or plastic cages around the bulb. If an unprotected bulb breaks, it may ignite flammable materials in the area.

13. When servicing a vehicle, always apply the parking brake and place the transmission in park with an automatic transmission or neutral with a manual transmission if the engine is running. When the engine is stopped, place the transmission in park with an automatic transmission or either reverse or first gear with a manual transmission.

14. Avoid working on a vehicle parked on an incline.

15. *Never work under a vehicle unless the vehicle chassis is supported securely on jack stands.*

16. When one end of a vehicle is raised, place wheel chocks on both sides of the wheels remaining on the floor.

17. Be sure that you know the location of shop first-aid kits, eyewash fountains, and fire extinguishers.

18. Collect oil, fuel, brake fluid, and other liquids in the proper safety containers.

19. Use only approved cleaning fluids and equipment. Do not use gasoline to clean parts.

20. Obey all state and federal safety, fire, and hazardous material regulations.

21. Always operate equipment according to the equipment manufacturer's recommended procedure.

22. Do not operate equipment unless you are familiar with the correct operating procedure.

23. Do not leave running equipment unattended.

24. Be sure the safety shields are in place on rotating equipment.

25. All shop equipment must have regularly scheduled maintenance and adjustment.

26. Some shops have safety lines around equipment. Always work within these lines when operating equipment.

27. Be sure that shop heating equipment is well ventilated.

28. Do not run in the shop or engage in horseplay.

29. Post emergency phone numbers near the phone. These numbers should include a doctor, ambulance, fire department, hospital, and police.

30. Do not place hydraulic jack handles where someone can trip over them.

31. Keep aisles clear of debris.

32. Use caution when working near a raised lift. Be aware of the available clearance.

Fire Safety

1. Familiarize yourself with the location and operation of all shop fire extinguishers.

2. If a fire extinguisher is used, report it to management so the extinguisher can be recharged.

3. Do not use any type of open flame heater to heat the work area.

4. Do not turn on the ignition switch or crank the engine with a gasoline line disconnected.

5. Store all combustible materials such as gasoline, paint, and oily rags in approved safety containers.

6. Clean up gasoline, oil, or grease spills immediately.

7. Always wear clean shop clothes. Do not wear oil-soaked clothes.

8. Do not allow sparks and flames near batteries.

9. Welding tanks must be securely fastened in an upright position.

10. Do not block doors, stairways, or exits.

11. Do not smoke when working on vehicles.

12. Do not smoke or create sparks near flammable materials or liquids.

13. Store combustible shop supplies such as paint in a closed steel cabinet.

14. Gasoline must be kept in approved safety containers.

15. If a gasoline tank is removed from a vehicle, do not drag the tank on the shop floor.

16. Know the approved fire escape route from your classroom or shop to the outside of the building.

17. If a fire occurs, do not open doors or windows. This action creates extra draft which makes the fire worse.

18. Do not put water on a gasoline fire because the water will make the fire worse.

19. Call the fire department as soon as a fire begins and then attempt to extinguish the fire.

20. If possible, stand 6 to 10 feet from the fire and aim the fire extinguisher nozzle at the base of the fire with a sweeping action.

21. If a fire produces a lot of smoke in the room, remain close to the floor to obtain oxygen and avoid breathing smoke.

22. If the fire is too hot or the smoke makes breathing difficult, get out of the building.

23. Do not re-enter a burning building.

24. Keep solvent containers covered except when pouring from one container to another. When flammable liquids are transferred from bulk storage, the bulk container should be grounded to a permanent shop fixture such as a metal pipe. During this transfer process, the bulk container should be grounded to the portable container. These ground wires prevent the buildup of a static electric charge, which could result in a spark and disastrous explosion. Always discard or clean empty solvent containers, because fumes in these containers are a fire hazard.

25. Familiarize yourself with different types of fires and fire extinguishers and know the type of extinguisher to use on each fire.

Compressed Air Equipment Safety

Tools that use compressed air are called **pneumatic tools**. Compressed air is used to inflate tires, apply paint, and drive tools. Compressed air can be dangerous when it is not used properly. The shop air supply contains high-pressure air in the shop compressor and air lines. Serious injury or property damage may result from careless operation of compressed air equipment. Follow these guidelines when working with compressed air:

1. Safety glasses or a face shield should be worn for all shop tasks, including those tasks involving the use of compressed air equipment.

2. Wear ear protection when using compressed air equipment.

3. Always maintain air hoses and fittings in good condition. If an end suddenly blows off an air hose, the hose will whip around, and this may cause personal injury.

4. Do not direct compressed air against the skin. This air may penetrate the skin, especially through small cuts or scratches. If compressed air penetrates the skin and enters the bloodstream, it can be fatal or cause serious health complications. Use only air gun nozzles approved by OSHA.

5. Do not use an air gun to blow off clothing or hair.

6. Do not clean the workbench or floor with compressed air. This action may blow very small parts against your skin or into your eye. Small parts blown by compressed air may cause vehicle damage. For example, if the car in the next stall has the air cleaner removed, a small part may go into the throttle body. When the engine is started, this

part will likely be pulled into the cylinder by engine vacuum and will penetrate through the top of a piston.

7. Never spin bearings with compressed air because the bearing will rotate at extremely high speed. Under this condition, the bearing may be damaged or it may disintegrate causing personal injury.

8. All pneumatic tools must be operated according to the manufacturer's recommended operating procedure.

9. Follow the equipment manufacturer's recommended maintenance schedule for all compressed air equipment.

Lift Safety

A **lift** is used to raise a vehicle so a technician can work under the vehicle. The lift arms must be placed under the car at the manufacturer's recommended lift points. There are several types of vehicle lifts in use today. Categories of lifts include single, double, and four-post varieties. Typically, single-post lifts are considered an in-ground type with hydraulic post and lines located directly under the lift. Above-ground types include twin and four posts with adjustable arms and contact pads or drive-on runways. Others consist of a scissors-type design (Figure 1-8). Some lifts have an electric motor that drives a hydraulic pump to create fluid pressure and force the lift upward. Other lifts use air pressure from the shop air supply to force the lift upward. If shop air pressure is used for this purpose, the air pressure is applied to fluid in the lift cylinder. A control lever or switch is placed near the lift. The control lever supplies shop air pressure to the lift cylinder, and the switch turns on the lift pump motor. Always be sure that the safety lock is engaged after the lift is raised. When the safety lock is released, a release lever is operated slowly to lower the vehicle.

Always be careful when raising a vehicle on a lift or a hoist. Adapters and hoist plates must be positioned correctly on twin-post and rail-type lifts to prevent damage to the underbody of the vehicle. There are specific lift points. These points allow the weight of the vehicle to be

Figure 1-8 A scissors-type, drive-on lift.

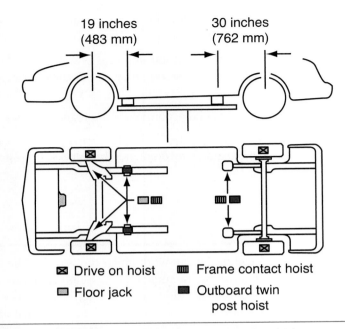

19 inches (483 mm)   30 inches (762 mm)

☒ Drive on hoist   ⠿ Frame contact hoist

▢ Floor jack   ▣ Outboard twin post hoist

Figure 1-9 Hoisting and lifting points for a typical unibody vehicle.

evenly supported by the adapters or hoist plates. The correct lift points can be found in the vehicle's service manual. Figure 1-9 shows typical locations for frame and unibody cars. These diagrams are for illustration only. Always follow the manufacturer's instructions. Before operating any lift or hoist, carefully read the operating manual and follow the operating instructions.

▲ **WARNING:** Never use a lift or jack to move something heavier than it is designed for. Always check the rating before using a lift or jack. If a jack is rated for 2 tons, do not attempt to use it for a job requiring 5 tons. It is dangerous for you and the vehicle.

When guiding someone driving a vehicle onto a lift, be sure to stand off to the driver's side of the vehicle rather than in front of it. Use clear, deliberate hand signals and/or verbal instructions to the driver to indicate vehicle direction. In the event of an unexpected action by the vehicle, always leave yourself a clear path. It is a good practice to check the clearance underneath the vehicle prior to driving over the lift. Low-hanging accessories or exhaust systems may be damaged when coming in contact with the lift.

Before driving a vehicle over a lift, position the arms and supports to provide an unobstructed clearance. Do not hit or run over lift arms, adapters, or axle supports. This could damage the lift, vehicle, or tires.

Position the lift supports to contact the vehicle at its lifting points. Raise the lift until the supports contact the vehicle. Then, check the supports to make sure they are in full contact with the vehicle. Raise the lift to the desired working height.

Make sure the vehicle's doors, hood, and trunk are closed before raising the vehicle. Never raise a car with someone inside.

▲ **WARNING:** Before working under a car, make sure the lift's locking device is engaged.

After lifting a vehicle to the desired height, always lower it onto its mechanical safeties. On some vehicles, the removal (or installation) of components can cause a critical shift of the vehicle's weight, which may cause the vehicle to be unstable on the lift. Refer to the vehicle's service manual for the recommended procedures to prevent this from happening.

Make sure tool trays, stands, and other equipment are removed from under the vehicle. Release the lift's locking devices according to the instructions before attempting to lower the lift.

Jack and Jack Stand Safety

An automobile can be raised off the ground by a hydraulic jack (Figure 1-10). **Jack stands** (Figure 1-11) are supports of different heights that sit on the floor. They are placed under a sturdy chassis member, such as the frame or axle housing, to support the vehicle. Like jacks, jack stands also have a capacity rating. Always use the correct rating of jack stand.

A **floor jack** is a portable unit mounted on wheels. The lifting pad on the jack is placed under the chassis of the vehicle, and the jack handle is operated with a pumping action. This jack handle operation forces fluid into a hydraulic cylinder in the jack, and the cylinder extends to force the jack lift pad upward and lift the vehicle. Always be sure that the lift pad is positioned securely under one of the car manufacturer's recommended lift points. To release the hydraulic pressure and lower the vehicle, the handle or release lever must be turned slowly.

The jack should be removed after the jack stands are set in place. This eliminates a hazard, such as a jack handle sticking out into a walkway. A jack handle that is bumped or kicked can cause a tripping accident or cause the vehicle to fall. Never use a jack by itself to support an automobile. Always use a jack stand with the jack as a safety precaution. Make sure the jack stands are properly placed under the vehicle.

Accidents involving the use of floor jacks and jack stands may be avoided if these safety precautions are followed:

1. Never work under a vehicle unless jack stands are placed securely under the vehicle chassis and the vehicle is resting on these stands.

Jack stands are also called safety stands.

As the vehicle is lifted, the floor jack will roll under the load. Make sure that the wheels on the floor jack are free to roll and are not blocked by gravel or other obstructions on the shop floor. If the jack cannot roll, the lifting plate may be pulled off the vehicle lifting point.

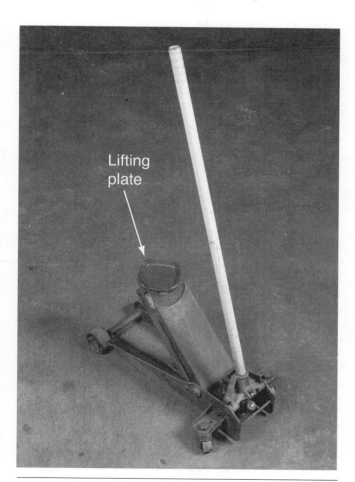

Lifting plate

Figure 1-10 Typical hydraulic floor jack.

Figure 1-11 Support stands.

2. Prior to lifting a vehicle with a floor jack, be sure that the jack lift pad is positioned securely under a recommended lift point on the vehicle. Lifting the front end of a vehicle with the jack placed under a radiator support may cause severe damage to the radiator and support.

3. Position the jack stands under a strong chassis member, such as the frame or axle housing. The jack stands must contact the vehicle manufacturer's recommended lift points.

4. Since the floor jack is on wheels, the vehicle and jack tend to move as the vehicle is lowered from a floor jack onto jack stands. Always be sure the jack stands remain under the chassis member during this operation, and be sure the jack stands do not tip. All the jack stand legs must remain in contact with the shop floor.

Cleaning Equipment Safety

All technicians are required to clean parts during their normal work routines. Face shields and protective gloves must be worn while operating cleaning equipment. The solution in hot and cold cleaning tanks may be caustic, and contact between this solution and skin or eyes must be avoided. Parts cleaning often creates a slippery floor, and care must be taken when walking in the parts cleaning area. The floor in this area should be cleaned frequently. When the caustic cleaning solution in hot or cold cleaning tanks is replaced, environmental regulations require that the old solution be handled as hazardous waste. Use caution when placing aluminum or aluminum alloy parts in a cleaning solution. Caustic solutions will damage these components by reacting with the aluminum. Always follow the cleaning equipment manufacturer's recommendations. Parts cleaning is a necessary step in most repair procedures. Cleaning automotive parts can be divided into four basic categories.

Chemical cleaning relies primarily on some type of chemical action to remove dirt, grease, scale, paint, or rust. A combination of heat, agitation, mechanical scrubbing, or washing may also be used to help remove dirt. Chemical cleaning equipment includes small parts washers, hot/cold tanks, pressure washers, spray washers, and salt baths.

Some parts washers provide electro-mechanical agitation of the parts to provide improved cleaning action (Figure 1-12). These parts washers may be heated with gas or electricity, and various water-based hot tank cleaning solutions are available depending on the type of metals being cleaned. For example, Kleer-Flo Greasoff™ Number 1 powdered detergent is available for cleaning iron and steel. Nonheated electromechanical parts washers are also available, and these washers use cold cleaning solutions such as Kleer-Flo Degreasol™ formulas.

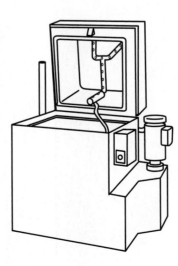

Figure 1-12 A parts washer with an electro-mechanical agitator.

Many cleaning solutions, such as Kleer-Flo Degreasol™ 99R, contain no ingredients listed as hazardous by the Environmental Protection Agency's Resource Conservation and Recovery Act (RCRA). This cleaning solution is a blend of sulphur-free hydrocarbons, wetting agents, and detergents. Degreasol™ 99R does not contain aromatic or chlorinated solvents, and it conforms to California's Rule 66 for clean air. Always use the cleaning solution recommended by the equipment manufacturer.

Some parts washers have an agitated immersion chamber under the shelves, which provides thorough parts cleaning. Folding work shelves provide a large upper cleaning area with a constant flow of solution from the dispensing hose. This cold parts washer operates on Degreasol™ 99R cleaning solution.

An aqueous parts cleaning tank uses a water-based, environmentally friendly cleaning solution such as Greasoff™ 2, rather than traditional solvents. The immersion tank is heated and agitated for effective parts cleaning (Figure 1-13). A sparger bar pumps a constant flow of cleaning solution across the surface to push floating oils away, and an integral skimmer removes these oils. This action prevents floating surface oils from redepositing on cleaned parts.

Thermal cleaning relies on heat, which bakes off or oxidizes the dirt. Thermal cleaning leaves an ash residue on the surface that must be removed by an additional cleaning process such as airless shot blasting or spray washing.

Abrasive cleaning relies on physical abrasion to clean the surface. This includes everything from a wire brush to glass bead blasting, airless steel shot blasting, abrasive tumbling, and vibratory cleaning. Chemical in-tank solution sonic cleaning might also be included here because it relies on the scrubbing action of ultrasonic sound waves to loosen surface contaminants.

Steam cleaning uses hot water vapor mixed with chemical cleaning agents to clean dirt from an object. After steam cleaning, the object should be thoroughly hosed down with clean water then air dried.

There are several reasons why steam cleaning has rapidly declined in recent years. Concerns about the environment have led to mandates that a closed-loop system be used for steam cleaning. This means that the runoff from the cleaning process must be collected and contained within the steam-cleaning system. The runoff cannot flow into a public sewage system.

Steam cleaning is normally done in a noncongested portion of the shop or in a separate building. Care must be taken to protect all painted surfaces and exposed skin. These could come into contact with the steam's heat and chemicals, and injury or damage will result. In addition, care must also be taken when working on the slippery floor that this process creates.

Before using a steam cleaning machine, check the electrical cords. Pay special attention to the grounding connector at the plug. If the machine is not properly grounded, there is a great possibility of getting an electrical shock. Finally, steam cleaning takes a great deal of time and work. Most shops cannot justify the labor cost for using an open steam cleaning system.

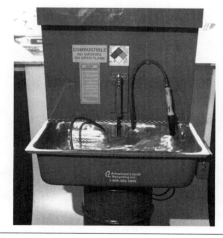

Figure 1-13 An aqueous parts cleaning tank.

Vehicle Operation

When the customer brings a vehicle in for service, certain shop safety guidelines should be followed. For example, when moving a car into the shop, check the brakes before beginning. Then buckle the safety belt. Drive carefully in and around the shop. Make sure no one is near, that the way is clear, and that there are no tools or parts under the car before you start the engine.

When road testing the car, obey all traffic laws. Drive only as far as is necessary to check the automobile. Never make excessively quick starts, turn corners too quickly, or drive faster than conditions or speed limits allow.

If the engine must be running while working on the car, block the wheels to prevent the car from moving. Place the transmission into park for automatic transmissions or in neutral for manual transmissions. Set the emergency brake. Never stand directly in front of or behind a running vehicle.

Run the engine only in a well-ventilated area to avoid the danger of poisonous **carbon monoxide (CO)** in the engine exhaust. If the shop is equipped with an exhaust ventilation system (Figure 1-14), use it. If not, use a hose and direct the exhaust out of the building.

> ⚠️ **WARNING:** Never run the engine in a vehicle inside the shop without an exhaust hose connected to the tailpipe.

Vehicle exhaust contains small amounts of carbon monoxide, which is a poisonous gas. Strong concentrations of carbon monoxide may be fatal for human beings. All shop personnel are responsible for air quality in the shop. Shop management is responsible for providing an adequate exhaust system to remove exhaust fumes from the maximum number of vehicles that may be running in the shop at the same time. Technicians should never run a vehicle in the shop unless a shop exhaust hose is installed on the tailpipe of the vehicle. The exhaust fan must be switched on to remove exhaust fumes.

If shop heaters or furnaces have restricted chimneys, they release carbon monoxide emissions into the shop air. Therefore, chimneys should be checked periodically for restriction and proper ventilation.

Monitors are available to measure the level of carbon monoxide in the shop. Some of these monitors read the amount of carbon monoxide present in the shop air while other monitors provide an audible alarm if the concentration of carbon monoxide exceeds the danger level.

Diesel exhaust contains some carbon monoxide, but particulates are also present in the exhaust from these engines. **Particulates** are small carbon particles, that can be harmful to the lungs.

The **sulfuric acid** solution in car batteries is a very corrosive, poisonous liquid. If a battery is charged with a fast charger at a high rate for a period of time, the battery becomes hot, and the sulfuric acid solution begins to boil. Under this condition, the battery may emit a strong sulfuric acid smell, and these fumes may be harmful to the lungs. If this condition occurs in the shop, the battery charger should be turned off or the charger rate should be reduced considerably.

Carbon monoxide is an odorless, poisonous gas. When it is breathed in, it can cause headaches, nausea, ringing in your ears, tiredness, and heart flutter. A heavy amount of CO can kill you.

Figure 1-14 When running an engine in a shop, make sure the vehicle's exhaust is connected to the shop's exhaust ventilation system.

When an automotive battery is charged, hydrogen gas and oxygen gas escape from the battery. If these gases are combined, they form water, but hydrogen gas by itself is very explosive. While a battery is charged, sparks, flames, and other sources of ignition must not be allowed near the battery.

Air Bag Safety

Before working on the steering column or dash areas, make certain that the air bag system has been disarmed according to the manufacturer's recommended procedure. Some general points to keep in mind are:

1. The air bag can deploy for several minutes even after the battery has been disconnected.
2. NEVER test an air batg circuit using a test lamp or self-powered test lamp.
3. Never splite into the air bag harness. The harness can be identified by its bright yellow color and identification markings.
4. Always carry the air bag with the trim cover facing away from your body.
5. Do not place the air bag on a bench with the trim cover down.
6. If the steering column is removed from the vehicle, do not stand the column with the air bag facing down.
7. Do not deploy an air bag by any procedure other than the manufacturer's recommendations.
8. If it is necessary to handle a deployed air bag, make certain to take precautions against the residue left behind; it can cause skin irritation.

Personal Protection

1. Always wear safety glasses or a face shield in the shop (Figure 1-15).
2. Wear ear plugs or covers if high noise levels are encountered.

Personal injury, vehicle damage, and property damage must be avoided by following safety rules regarding personal protection, substance abuse, electrical safety, gasoline safety, housekeeping safety, fire safety, and general shop safety.

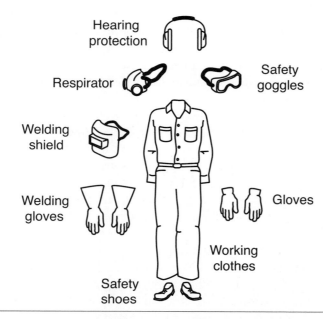

Figure 1-15 Shop safety equipment, including safety goggles, respirator, welding shied, proper work clothes, ear protection, gloves, and safety shoes.

3. Always wear boots or shoes that provide adequate foot protection. Safety work boots or shoes with steel toe caps are best for working in the automotive shop. Most safety shoes also have slip-resistant soles. Footwear must protect against heavy falling objects, flying sparks, and corrosive liquids. Soles on footwear must protect against punctures by sharp objects. Sneakers and street shoes are not recommended in the shop.

4. Do not wear watches, jewelry, or rings when working on a vehicle. Severe burns occur when jewelry makes contact between an electric terminal and ground. Jewelry may catch on an object, resulting in painful injury.

5. Do not wear loose clothing, and keep long hair tied behind your head. Loose clothing or long hair is easily entangled in rotating parts.

6. Wear a respirator to protect your lungs when working in dusty conditions.

Smoking, Alcohol, and Drugs in the Shop

The improper or excessive use of alcoholic beverages and/or drugs may be referred to as substance abuse.

Do not smoke when working in the shop. If the shop has designated smoking areas, smoke only in these areas. Do not smoke in customers' cars. Nonsmokers may not appreciate cigarette odor in their cars. A spark from a cigarette or lighter may ignite flammable materials in the workplace. The use of drugs or alcohol must be avoided while working in the shop. Even a small amount of drugs or alcohol affects reaction time. In an emergency situation, slow reaction time may cause personal injury. If a heavy object falls off the workbench and your reaction time is slowed by drugs or alcohol, you may not get your foot out of the way in time, resulting in foot injury. When a fire starts in the workplace and you are a few seconds slower getting a fire extinguisher into operation because of alcohol or drug use, it could make the difference between extinguishing a fire and having expensive fire damage.

Work Area Safety

CAUTION: Always know the location of all safety equipment in the shop and be familiar with the operation of this equipment.

Shop rules, vehicle operation in the shop, and shop housekeeping are serious business. Each year a significant number of technicians are injured and vehicles are damaged by disregarding shop rules, careless vehicle operation, and sloppy housekeeping.

Your work area should be kept clean and safe. The floor and bench tops should be kept clean, dry, and orderly. Any oil, coolant, or grease on the floor can make it slippery. Slips can result in serious injuries. To clean up oil, use a commercial oil absorbent. Keep all water off the floor. Water is slippery on smooth floors, and electricity flows well through water. Aisles and walkways should be kept clean and wide enough to easily move through. Make sure the work areas around machines are large enough to safely operate the machine.

Proper ventilation of space heaters, used in some shops, is necessary to reduce the CO levels in the shop. Also, proper ventilation is very important in areas where volatile solvents and chemicals are used. A volatile liquid is one that vaporizes very quickly.

Keep an up-to-date list of emergency telephone numbers clearly posted next to the telephone. These numbers should include a doctor, hospital, and fire and police departments. Also, the work area should have a first-aid kit for treating minor injuries. There should also be eye flushing kits readily available.

Gasoline is a highly flammable volatile liquid. Always keep gasoline or diesel fuel in an approved safety can and never use it to clean your hands or tools. Oily rags should also be stored in an approved metal container. When these oily, greasy, or paint-soaked rags are left lying about or are not stored properly, they can cause spontaneous combustion. Spontaneous combustion results in a fire that starts by itself without a match.

Make sure that all drain covers are snugly in place. Open drains or covers that are not flush to the floor can cause toe, ankle, and leg injuries.

Figure 1-16 Flammable liquids should be stored in safety-approved containers.

Handle all solvents (or any liquids) with care to avoid spillage. Keep all solvent containers closed, except when pouring. Be extra careful when transferring flammable materials from bulk storage (Figure 1-16). Static electricity can build up enough to create a spark that could cause an explosion. Discard or clean all empty solvent containers. Solvent fumes in the bottom of these containers are very flammable. Never light matches or smoke near flammable solvents and chemicals, including battery acids. Solvent and other combustible materials must be stored in approved and designated storage cabinets or rooms (Figure 1-17). Storage rooms should have adequate ventilation.

Oil absorbent must be treated as hazardous waste.

Figure 1-17 Store combustible materials in approved safety cabinets.

	Class of Fire	Typical Fuel Involved	Type of Extinguisher
Class **A** Fires (green)	**For Ordinary Combustibles** Put out a Class A fire by lowering its temperature or by coating the burning combustibles.	Wood Paper Cloth Rubber Plastics Rubbish Upholstery	Water[1] Foam* Multipurpose dry chemical[4]
Class **B** Fires (red)	**For Flammable Liquids** Put out a Class B fire by smothering it. Use an extinguisher that gives a blanketing flame-interrupting effect; cover whole flaming liquid surface.	Gasoline Oil Grease Paint Lighter fluid	Foam* Carbon dioxide[5] Halogenated agent[6] Standard dry chemical[2] Purple K dry chemical[3] Multipurpose dry chemical[4]
Class **C** Fires (blue)	**For Electrical Equipment** Put out a Class C fire by shutting off power as quickly as possible and by always using a nonconducting extinguishing agent to prevent electric shock.	Motors Appliances Wiring Fuse boxes Switchboards	Carbon dioxide[5] Halogenated agent[6] Standard dry chemical[2] Purple K dry chemical[3] Multipurpose dry chemical[4]
Class **D** Fires (yellow)	**For Combustible Metals** Put out a Class D fire of metal chips, turnings, or shaving by smothering or coating with a specially designed extinguishing agent.	Aluminum Magnesium Potassium Sodium Titanium Zirconium	Dry powder extinguishers and agents only

*Cartridge-operated water, foam, and soda-acid types of extinguishers are no longer manufactured. These extinguishers should be removed from service when they become due for their next hydrostatic pressure test.

Notes:
(1) Freeze in low temperatures unless treated with antifreeze solution, usually weighs over 20 pounds, and is heavier than any other extinguisher mentioned.
(2) Also called ordinary or regular dry chemical. (solution bicarbonate)
(3) Has the greatest initial fire-stopping power of the extinguishers mentioned for class B fires. Be sure to clean residue immediately after using the extinguisher so sprayed surfaces will not be damaged. (potassium bicarbonate)
(4) The only extinguishers that fight A, B, and C class fires. However, they should not be used on fires in liquified fat or oil of appreciable depth. Be sure to clean residue immediately after using the extinguisher so sprayed surfaces will not be damaged. (ammonium phosphates)
(5) Use with caution in unventilated, confined spaces.
(6) May cause injury to the operator if the extinguishing agent (a gas) or the gases produced when the agent is applied to a fire is inhaled.

Figure 1-18 Guide to fire extinguisher selection.

Know where the fire extinguishers are and what types of fires they put out (Figure 1-18). A multipurpose dry chemical fire extinguisher will put out ordinary combustibles, flammable liquids, and electrical fires. Never put water on a gasoline fire. The water will just spread the fire. Use a fire extinguisher to smother the flames. Remember, during a fire, never open doors or windows unless it is absolutely necessary; the extra draft will only make the fire worse. A good rule is to call the fire department first and then attempt to extinguish the fire.

To extinguish a fire, stand 6 to 10 feet from the fire. Hold the extinguisher firmly in an upright position. Aim the nozzle at the base and use a side-to-side motion, sweeping the entire width of the fire. Stay low to avoid inhaling the smoke. If it becomes too hot or too smoky, get out. Remember, never go back into a burning building for anything.

First-Aid Kits

First-aid kits should be clearly identified and conveniently located (Figure 1-19). These kits contain such items as bandages and ointment required for minor cuts. All shop personnel must be familiar with the location of first-aid kits. At least one of the shop personnel should have basic first-aid training, and this person should be in charge of administering first aid and keeping first-aid kits filled.

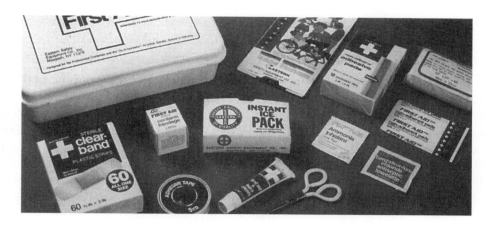

Figure 1-19 First-aid kit.

Hazardous Waste Disposal

Hazardous waste materials in automotive shops are chemicals or components that the shop no longer needs and that pose a danger to the environment and people if they are disposed of in ordinary garbage cans or sewers. However, it should be noted that no material is considered hazardous waste until the shop has finished using it and is ready to dispose of it. When handling any hazardous waste, always wear proper protective clothing that may include a respirator and gloves (Figure 1-20), and use equipment detailed in the right-to-know laws. The **Environmental Protection Agency (EPA)** publishes a list of hazardous materials that is included in the Code of Federal Regulations. Waste is considered hazardous if it is included on the EPA list of hazardous materials or has one or more of these characteristics:

1. *Reactive.* Any material that reacts violently with water or other chemicals is considered hazardous. When exposed to low-pH acid solutions, if a material releases cyanide gas, hydrogen sulphide gas, or similar gases, it is hazardous.

2. *Corrosive.* If a material burns the skin or dissolves metals and other materials, it is considered hazardous.

Figure 1-20 Wear recommended safety clothing and equipment when handling hazarous materials.

The **Environmental Protection Agency (EPA)** is a government agency that oversees activities and laws designed to protect water, air, land, and the public from harmful pollution or contamination.

A material that reacts violently with water or other chemicals is referred to as **reactive**.

Corrosive materials burn the skin or dissolve metals or other materials.

23

Materials are **toxic** if they exceed levels greater than 100 times that in drinking water.

3. *Toxic.* Materials are hazardous if they leach one or more of eight heavy metals in concentrations greater than 100 times primary drinking water standard.

4. *Ignitable.* A liquid is hazardous if it has a flash point below 140°F (60°C), and a solid is hazardous if it ignites spontaneously.

The automotive service industry is considered a generator of hazardous wastes. However, the vehicles it services are the real generators. Once you drain oil from an engine, you have generated the waste and now become responsible for the proper disposal of this hazardous waste. There are many other wastes that need to be handled properly after you have removed them, such as batteries, brake fluid, and transmission fluid.

Engine coolant should not be allowed to go down sewage drains. This is also true for all liquids drained from a car. Coolant should be captured and recycled or disposed of properly.

Filters for fluids (transmission, fuel, and oil filters) also need to be handled in designated ways. Used filters need to be drained and then crushed or disposed of in a special shipping barrel. Most regulations demand that oil filters be drained for at least 24 hours before they are disposed of or crushed.

Federal and state laws control the disposal of hazardous waste materials. Every shop employee must be familiar with these laws. Hazardous waste disposal laws include the **Resource Conservation and Recovery Act (RCRA)**. This law basically states that hazardous material users are responsible for hazardous materials from the time they become a waste until the proper waste disposal is completed. Many automotive shops hire an independent hazardous waste hauler to dispose of hazardous waste material (Figure 1-21). The shop owner or manager should have a written contract with the hazardous waste hauler. Rather than have hazardous waste material hauled to an approved hazardous waste disposal site, a shop may choose to recycle the material in the shop. Therefore, the user must store hazardous waste material properly and safely and be responsible for the transportation of this material until it arrives at an approved hazardous waste disposal site and is processed according to the law.

The RCRA controls these types of automotive waste:

1. Paint and body repair products waste
2. Solvents for parts and equipment cleaning
3. Batteries and battery acid
4. Mild acids used for metal cleaning and preparation
5. Waste oil, engine coolants, or antifreeze
6. Air-conditioning refrigerants
7. Engine oil filters

Never, under any circumstances, use these methods to dispose of hazardous waste material:

1. Pour hazardous wastes on weeds to kill them.
2. Pour hazardous wastes on gravel streets to prevent dust.
3. Throw hazardous wastes in a dumpster.

Figure 1-21 Hazardous waste hauler.

4. Dispose of hazardous wastes anywhere but an approved disposal site.

5. Pour hazardous wastes down sewers, toilets, sinks, or floor drains.

6. Bury hazardous wastes in the ground.

The **right-to-know laws** state that employees have a right to know when the materials they use at work are hazardous. The right-to-know laws started with the **Hazard Communication Standard** published by OSHA in 1983. This document was originally intended for chemical companies and manufacturers that required employees to handle hazardous materials in their work situation. At the present time, most states have established their own right-to-know laws. Meanwhile, the federal courts have decided to apply these laws to all companies, including automotive service shops. Under the right-to-know laws, the employer has three responsibilities regarding the handling of hazardous materials by its employees.

First, all employees must be trained about the types of hazardous materials they will encounter in the workplace. The employees must be informed about their rights under legislation regarding the handling of hazardous materials. All hazardous materials must be properly labeled, and information about each hazardous material must be posted on **material safety data sheets (MSDS)** available from the manufacturer (Figure 1-22). In Canada, MSDS sheets are called workplace hazardous materials information systems (WHMIS).

```
HEXANE
==========================================================
MSDS Safety Information
==========================================================
Ingredients
==========================================================
Name: HEXANE (N_HEXANE)
% Wt: >97
OSHA PEL: 500 PPM
ACGIH TLV: 50 PPM
EPA Rpt Qty: 1 LB
DOT Rpt Qty: 1 LB
==========================================================
Health Hazards Data
==========================================================
LD50 LC50 Mixture: LD50:(ORAL,RAT) 28.7 KG/MG
Route Of Entry Inds _ Inhalation: YES
Skin: YES
Ingestion: YES
Carcinogenicity Inds _ NTP: NO
IARC: NO
OSHA: NO
Effects of Exposure: ACUTE:INHALATION AND INGESTION ARE HARMFUL AND MAY BE FATAL.
INHALATION AND INGESTION MAY CAUSE HEADACHE, NAUSEA, VOMITING, DIZZINESS, IRRITATION
OF RESPIRATORY TRACT, GASTROINTESTINAL IRRITATION AND UNCONSCIOUSNESS. CONTACT
W/SKIN AND EYES  MAY CAUSE IRRITATION. PROLONGED SKIN MAY RESULT IN DERMATITIS (EFTS
OF OVEREXP)
Signs And Symptons Of Overexposure: HLTH HAZ:CHRONIC:MAY INCLUDE CENTRAL
NERVOUS SYSTEM DEPRESSION.
Medical Cond Aggravated By Exposure: NONE IDENTIFIED.
First Aid: CALL A PHYSICIAN. INGEST:DO NOT INDUCE VOMITING. INHAL:REMOVE TO FRESH AIR. IF
NOT BREATHING, GIVE ARTIFICIAL RESPIRATION. IF BREATHING IS DIFFICULT, GIVE OXYGEN.
EYES:IMMED FLUSH W/PLENTY OF WATER FOR AT LEAST 15 MINS. SKIN:IMMED FLUSH W/P LENTY
OF WATER FOR AT LEAST 15 MINS WHILE REMOVING CONTAMD CLTHG & SHOES. WASH CLOTHING
BEFORE REUSE.
==========================================================
Handling and Disposal
==========================================================
Spill Release Procedures: WEAR NIOSH/MSHA SCBA & FULL PROT CLTHG. SHUT OFF
IGNIT SOURCES:NO FLAMES, SMKNG/FLAMES IN AREA. STOP LEAK IF YOU CAN DO SO W/OUT
HARM. USE WATER SPRAY TO REDUCE VAPS. TAKE UP W/SAND OR OTHER NON_COMBUST MATL &
PLACE INTO CNTNR FOR LATER (SU PDAT)
Neutralizing Agent: NONE SPECIFIED BY MANUFACTURER.
Waste Disposal Methods: DISPOSE IN ACCORDANCE WITH ALL APPLICABLE FEDERAL, STATE AND
LOCAL ENVIRONMENTAL REGULATIONS. EPA HAZARDOUS WASTE NUMBER:D001 (IGNITABLE
WASTE).
Handling And Storage Precautions: BOND AND GROUND CONTAINERS WHEN TRANSFERRING LIQUID.
KEEP CONTAINER TIGHTLY CLOSED.
Other Precautions: USE GENERAL OR LOCAL EXHAUST VENTILATION TO MEET
TLVREQUIREMENTS. STORAGE COLOR CODE RED (FLAMMABLE).
==========================================================
Fire and Explosion Hazard Information
==========================================================
Flash Point Method: CC
Flash Point Text:  9F,_23C
Lower Limits: 1.2%
Upper Limits: 77.7%
Extinguishing Media: USE ALCOHOL FOAM, DRY CHEMICAL OR CARBON DIOXIDE. (WATER MAY BE
INEFFECTIVE).
Fire Fighting Procedures: USE NIOSH/MSHA APPROVED SCBA & FULL PROTECTIVE
  EQUIPMENT (FP N).
Unusual Fire/Explosion Hazard: VAP MAY FORM ALONG SURFS TO DIST IGNIT SOURCES & FLASH
BACK. CONT W/STRONG OXIDIZERS MAY CAUSE FIRE. TOX GASES PRDCED MAY INCL:CARBON
MONOXIDE, CARBON DIOXIDE.
==========================================================
```

Figure 1-22 Material safety data sheets (MSDS) to inform employees about hazardous materials. N-Hexane is commonly used in solvents and many brake cleaning products.

The employer has a responsibility to place MSDS where they are easily accessible by all employees. The MSDS provide extensive information about the hazardous material such as:

1. Chemical name
2. Physical characteristics
3. Protective equipment required for handling
4. Explosion and fire hazards
5. Other incompatible materials
6. Health hazards such as signs and symptoms of exposure, medical conditions aggravated by exposure, and emergency and first-aid procedures
7. Safe handling precautions
8. Spill and leak procedures

Second, the employer has a responsibility to make sure that all hazardous materials are properly labeled. The label information must include health, fire, and reactivity hazards posed by the material, as well as the protective equipment necessary to handle the material. The manufacturer must supply all warning and precautionary information about hazardous materials, and this information must be read and understood by the employee before handling the material.

Third, employers are responsible for maintaining permanent files regarding hazardous materials. These files must include information on hazardous materials in the shop, proof of employee training programs, and information about accidents such as spills or leaks of hazardous materials. The employer's files must also include proof that employees' requests for hazardous material information such as MSDS have been met. A general right-to-know compliance procedure manual must be maintained by the employer.

Hand Tools and Diagnostic Equipment

Automotive engine performance is among the most challenging areas of automotive service and encompasses many different theories and service aspects. A thorough understanding of theory and application of engine mechanics, chemistry, mathematics, electricity, physics, computer operation, safety, and more must be practiced regularly. Part of your responsibility as a professional is to remain current in advancing service technology, tool and equipment usage, as well as any regulatory practices that you need to adhere to including any local, state, or federal licensing or certification mandates. Performing safe, accurate, effective, and efficient service on an automobile—specifically in the area of engine performance—will require a wide array of tools and equipment. Tools are designed to be work savers. When used as designed, they will reduce the amount of time or effort used in performing a service operation. In addition, many tool manufacturers are producing hand tools that are ergonomically designed and comfortable to use to help reduce stress and fatigue on muscles and joints. You also need various types of specialty tools in order to perform different diagnostic tests. Tool and equipment manufacturers are continually upgrading and introducing new products to keep pace with automobile changes. You may invest in some of these tools and diagnostic equipment, but some will belong to the shop due to their cost. The following brief list provides you, the technician, with both common and specialty tools that are typically used in the area of automotive engine performance.

1. Basic hand tools. The basic hand tool set generally includes common tool favorites such as:
   ❏ a variety of metric and SAE-size wrenches and sockets.
   ❏ screwdrivers, Torx® drivers, hex-key wrenches.
   ❏ hammers, chisels, and punches.

❑ ratchets and extensions.

❑ an assortment of various pliers.

❑ a gear puller set, files, and saws.

❑ a stethoscope.

❑ power tools and torque wrenches.

2. Safety and protection equipment:

❑ Gloves, hand protection

❑ Safety glasses, earplugs

❑ Fender and seat covers

❑ First-aid kit

3. Measuring devices:

❑ Feeler gauges, spark plug gapping tool

❑ Fuel pressure gauges, vacuum gauge

❑ Thermometers, dwell-tachometer

❑ Various micrometers, dial indicators

❑ An engine compression tester

4. Electrical tools:

❑ A test light, jumper leads

❑ A circuit tester, logic probe, and test probes

❑ A soldering gun, wire repair kit

❑ Crimping and stripping tools

❑ A battery tester; battery service tools

5. Diagnostic test equipment:

❑ A graphing and/or digital multi-meter (DMM); lab scope

❑ A scan tool, diagnostic engine analyzer

❑ An infrared exhaust gas analyzer, timing light

❑ A cylinder leakage tester, vacuum hand pump

❑ A fuel injector tester, injector test lights, spark tester

❑ Test probes; charging system tester

6. Service information:

❑ Service manuals; time, parts, and labor guides

❑ Computer-based service information systems, CD-ROM

❑ Reference manuals, hotline support assistance

Power tools are operated by outside power sources.

Employer and Employee Obligations

When you begin employment, you enter into a business agreement with your employer. A business agreement involves an exchange of goods or services that have value. Although the automotive technician may not have a written agreement with his or her employer, the technician exchanges time, skills, and effort for money paid by the employer. Both the employee and the employer have obligations. The automotive technician's obligations include the following:

1. *Quality.* Each repair job should be a quality job! Work should never be done in a careless manner. Nothing improves customer relations like quality workmanship.

2. *Productivity.* As an automotive technician, you have a responsibility to your employer to make the best possible use of time on the job. Each job should be done in a reasonable length of time. Employees are paid for their skills, effort, and time.

3. *Teamwork.* The shop staff is a team, and everyone, including technicians and management personnel, are team members. You should cooperate with, and care about, other team members. Each member of the team should strive for harmonious relations with fellow workers. Cooperative teamwork helps to improve shop efficiency, productivity, and customer relations. Customers may be turned off by bickering between shop personnel.

4. *Honesty.* Employers and customers expect and deserve honesty from automotive technicians. Honesty creates a feeling of trust between technicians, employers, and customers.

5. *Loyalty.* As an employee, you are obliged to act in the best interests of your employer, both on and off the job.

6. *Attitude.* Employees should maintain a positive attitude at all times. As in other professions, automotive technicians have days when it may be difficult to maintain a positive attitude. For example, there will be days when the technical problems on a certain vehicle are difficult to solve. However, developing a negative attitude certainly will not help the situation! A positive attitude has a positive effect on the job situation as well as on the customer and employer.

7. *Responsibility.* You are responsible for your conduct on the job and your work-related obligations. These obligations include always maintaining good workmanship and customer relations. Attention to details, such as always placing fender and seat covers on customer vehicles prior to driving or working on the vehicle, greatly improve customer relations.

8. *Following directions.* All of us like to do things our way. Such action may not be in the best interests of the shop and, as an employee, you have an obligation to follow the supervisor's directions.

9. *Punctuality and regular attendance.* Employees have an obligation to be on time for work and to be regular in attendance on the job. It is very difficult for a business to operate successfully if they cannot count on their employees to be on the job at the appointed time.

10. *Regulations.* Automotive technicians should be familiar with all state and federal regulations pertaining to their job situation such as the OSHA and hazardous waste disposal laws. In Canada, employees should be familiar with WHMIS.

Employer-to-employee obligations include:

1. *Wages.* The employer has a responsibility to inform the employee regarding the exact amount of financial remuneration he or she will receive and when that wage will be paid.

2. *Fringe benefits.* A detailed description of all fringe benefits should be provided by the employer. These benefits may include holiday pay, sickness and accident insurance, and pension plans.

3. *Working conditions.* A clean, safe workplace must be provided by the employer. The shop must have adequate safety equipment and first-aid supplies. Employers must be certain that all shop personnel maintain the shop area and equipment to provide adequate safety and a healthy workplace atmosphere.

4. *Employee instruction.* Employers must provide employees with clear job descriptions and be sure that each worker is aware of his or her obligations.

5. *Employee supervision.* Employers should inform their workers regarding the responsibilities of their immediate supervisors and other management personnel.

6. *Employee training.* Employers must make sure that employees are familiar with the safe operation of all the equipment that they are required to use in their job situation. Since automotive technology is changing rapidly, employers should provide regular update training for their technicians. Under the right-to-know laws, employers are required to inform all employees about hazardous materials in the shop. Employees should be familiar with WHMIS, which detail the labeling and handling of hazardous waste and the health problems if exposed to hazardous waste.

An automotive technician has specific responsibilities regarding each job performed on a customer's vehicle:

1. Do every job to the best of your ability. There is no place in the automotive service industry for careless workmanship! Automotive technicians and students must realize they have a very responsible job. During many repair jobs, you, as a student or technician working on a customer's vehicle, actually have the customer's life and the safety of his or her vehicle in your hands. For example, if you are doing a brake job and leave the wheel nuts loose on one wheel, that wheel may fall off the vehicle at high speed. This could result in serious personal injury for the customer and others, plus extensive vehicle damage. If this type of disaster occurs, the individual who worked on the vehicle and the shop may be involved in a very expensive legal action. As a student or technician working on customer vehicles, you are responsible for the safety of every vehicle that you work on! Even when careless work does not create a safety hazard, it leads to dissatisfied customers who often take their business to another shop, and nobody benefits when that happens.

2. Treat customers fairly and honestly on every repair job. Do not install parts that are unnecessary to complete the repair job.

3. Use published specifications; do not guess at adjustments.

4. Follow the service procedures in the service manual provided by the vehicle manufacturer or an independent manual publisher.

5. When the repair job is completed, always be sure the customer's complaint has been corrected.

6. Do not be too concerned with work speed when you begin working as an automotive technician. Speed comes with experience.

National Institute for Automotive Service Excellence (ASE) Certification

The **National Institute for Automotive Service Excellence (ASE)** has provided voluntary testing and certification of automotive technicians on a national basis for many years. The image of the automotive service industry has been enhanced by the ASE certification program. More than 265,000 ASE-certified automotive technicians now work in a wide variety of automotive service shops. The ASE provides certification in these eight basic areas of automotive repair:

1. Engine repair
2. Automatic transmissions/transaxles
3. Manual drivetrain and axles
4. Suspension and steering systems
5. Brake systems
6. Electrical systems
7. Heating and air-conditioning systems
8. Engine performance

The Advanced Engine Performance Test is best known as the L1 test.

In addition to these eight basic tests, a technician can take advanced tests (e.g., Advanced Engine Performance) and become certified in that area. The advanced tests require the technician to have particular basic certifications before attempting to be certified at an advanced level.

A technician may take the ASE test and become certified in any or all of the eight basic areas. When a technician passes an ASE test in one of the eight areas, an Automotive Technician's shoulder patch is issued by the ASE. If a technician passes all eight tests, he or she receives a Master Technician's shoulder patch (Figure 1-23). Retesting at 5-year intervals is required to remain certified.

The certification test in each of the eight areas contains forty to eighty multiple choice questions. The test questions are written by a panel of automotive service experts from various areas of automotive service such as automotive instructors, service managers, automotive manufacturers' representatives, test equipment representatives, and certified technicians. The test questions are pretested and checked for quality by a national sample of technicians. Most questions have the *Technician A* and *Technician B* format similar to the questions at the end of each chapter in this book. ASE regulations demand that each technician must have 2 years of working experience in the automotive service industry prior to taking a certification test or tests. However, relevant formal

Figure 1-23 ASE certification patches.

ENGINE PERFORMANCE TEST SPECIFICATIONS		
Content area	Questions in test	Percentage of test
A. General engine diagnosis	19	24.0%
B. Ignition system diagnosis and repair	16	20.0%
C. Fuel, air induction, and exhaust system diagnosis and repair	21	26.0%
D. Emissions control system diagnosis and repair	19	24.0%
1. Positive crankcase ventilation (1)		
2. Spark timing controls (3)		
3. Idle speed controls (3)		
4. Exhaust gas recirculation (4)		
5. Exhaust gas treatment (2)		
6. Inlet air temperature controls (2)		
7. Intake manifold temperature controls (2)		
8. Fuel vapor controls (2)		
E. Engine related service	2	2.5%
F. Engine electrical system diagnosis and repair	3	3.5%
1. Battery (1)		
2. Starting system (1)		
3. Charging system (1)		
Total	**80**	**100%**

Figure 1-24 Engine performance test specifications.

training may be substituted for one of the years of working experience. Contact the ASE for details regarding this substitution. The contents of the Engine Performance test are listed in Figure 1-24.

Shops that employ ASE-certified technicians display an official ASE blue seal of excellence. This blue seal increases the customer's awareness of the shop's commitment to quality service and the competency of certified technicians.

Terms to Know

Abrasive cleaning
Carbon monoxide (CO)
Chemical cleaning
Environmental Protection Agency (EPA)
Floor jack
Hazard Communication Standard
Hydraulic press
Jack stands
Lift

Material safety data sheets (MSDS)
National Institute for Automotive Service Excellence (ASE)
Occupational Safety and Health Act (OSHA)
Particulates
Pneumatic tools
Power tools

Resource Conservation and Recovery (RCRA)
Right-to-know laws
Safety glasses
Sulfuric acid
Thermal cleaning
Workplace Hazardous Materials Information Systems (WHMIS)

ASE-Style Review Questions

1. *Technician A* says it is recommended that you wear shoes with non-slip soles in the shop.
Technician B says steel-toed shoes offer the best foot protection.
Who is correct?
   A. A only
   B. B only
   C. Both A and B
   D. Neither A nor B

2. *Technician A* says that some machines can be routinely used beyond their stated capacity.
Technician B says that a power tool can be left running unattended if the technician puts up a power "on" sign.
Who is correct?
   A. A only
   B. B only
   C. Both A and B
   D. Neither A nor B

3. *Technician A* ties his long hair behind his head while working in the shop.
Technician B covers her long hair with a brimless cap.
Who is correct?
   A. A only
   B. B only
   C. Both A and B
   D. Neither A nor B

4. *Technician A* uses compressed air to blow dirt from his clothes and hair.
Technician B says this should only be done outside.
Who is correct?
   A. A only
   B. B only
   C. Both A and B
   D. Neither A nor B

5. While discussing the proper way to lift heavy objects:
Technician A says you should bend your back to pick up a heavy object.
Technician B says you should bend your knees to pick up a heavy object.
Who is correct?
   A. A only
   B. B only
   C. Both A and B
   D. Neither A nor B

6. While discussing shop cleaning equipment safety:
Technician A says some hot tanks contain caustic solutions.
Technician B says some metals such as aluminum may dissolve in hot tanks.
Who is correct?
   A. A only
   B. B only
   C. Both A and B
   D. Neither A nor B

7. While discussing shop rules:
Technician A says breathing carbon monoxide may cause arthritis.
Technician B says breathing carbon monoxide may cause headaches.
Who is correct?
   A. A only
   B. B only
   C. Both A and B
   D. Neither A nor B

8. While discussing air quality:
Technician A says a restricted chimney on a shop furnace may cause carbon monoxide gas in the shop.
Technician B says monitors are available to measure the level of carbon monoxide in the shop air.
Who is correct?
   A. A only
   B. B only
   C. Both A and B
   D. Neither A nor B

9. While discussing air quality:
Technician A says a battery gives off hydrogen gas during the charging process.
Technician B says a battery gives off oxygen gas during the charging process.
Who is correct?
   A. A only
   B. B only
   C. Both A and B
   D. Neither A nor B

10. While discussing parts cleaning:
Technician A says thermal cleaning relies on a chemical reaction to break down dirt and deposits.
Technician B says using abrasive cleaning to clean the surface relies on physical abrasion from a brush, glass bead blasting, or similar methods.
Who is correct?
   A. A only
   B. B only
   C. Both A and B
   D. Neither A nor B

Job Sheet 1

1

Name _____ Date _____

Shop Safety Survey

As a professional technician, safety should be one of your first concerns. This job sheet will increase your awareness of shop safety rules and safety equipment. As you survey your shop area and answer the following questions, you will learn how to evaluate the safeness of your workplace.

Procedures

Your instructor will review your progress throughout this worksheet and should sign off the sheet when you complete it.

Task Completed

1. Before you begin to evaluate your work area, evaluate yourself. Are you dressed to work safely? ☐ Yes ☐ No

   If no, what is wrong? _____

2. Are your safety glasses OSHA approved? ☐ Yes ☐ No

   Do they have side protection shields? ☐ Yes ☐ No

3. Watch around the shop and note any area that poses a potential safety hazard or is an area that you should be aware of.

   _____

   _____

   _____

   Any true hazards should be brought to the attention of the instructor immediately.   ☐

4. Are there safety areas marked around grinders and other machinery? ☐ Yes ☐ No

5. What is the line air pressure in the shop? ___ psi
   What should it be? ____ psi

6. Where are the tools stored in the shop? _____

   _____

7. If you could, how would you improve the tool storage area?

   _____

   _____

8. What types of hoists are used in the shop? _____

   _____

9. Ask your instructor to demonstrate the proper use of the hoist.   ☐

10. Where is the first-aid kit(s) kept in the work area?

    _____

    _____

11. What is the shop's procedure for dealing with an accident?

_____

_____

_____

☐   **12.** Have your instructor supply you with a vehicle make, model, and year. Using the appropriate service manual, find the location of the correct lifting points for that vehicle. On the rear of this sheet, draw a simple figure showing where these lift points are.

Instructor's Response _____

_____

_____

Job Sheet 2

Name _____ Date _____

Working Safely Around Air Bags

Upon completion of this job sheet, you should be able to work safely around and with air bag systems.

Tools and Materials

A vehicle with air bags   Safety glasses
Service manual for the above vehicle   A DMM
Component locator for the above vehicle

Describe the vehicle being worked on.

Year _____ Make _____ Model _____

VIN _____ Engine type and size _____

Procedures

1. Locate the information about the air bag system in the service manual. How are the critical parts of the system identified in the vehicle?

   _____

   _____

2. List the main components of the air bag system and describe their location.

   _____

   _____

3. There are some very important guidelines to follow when working with and around air bag systems. These are listed below with some key words left out. Read through these and fill in the blanks with the correct words.

   a. Wear _____ _____ when servicing an air bag system and when handling an air bag module.

   b. Wait at least _____ minutes after disconnecting the battery before beginning any service. The reserve _____ module is capable of storing enough energy to deploy the air bag for up to _____ minutes after battery voltage is lost.

   c. Always handle all _____ and other components with extreme care. Never strike or jar a sensor, especially when the battery is connected. This can cause deployment of the air bag.

   d. Never carry an air bag module by its _____ or _____, and, when carrying it, always face the trim and air bag _____ from your body. When placing a module on a bench, always face the trim and air bag _____.

e. Deployed air bags may have a powdery residue on them. _____ is produced by the deployment reaction and is converted to _____ when it comes in contact with the moisture in the atmosphere. Although it is unlikely that harmful chemicals will still be on the bag, it is wise to wear _____ _____ and _____ when handling a deployed air bag. Immediately wash your hands after handling a deployed air bag.

f. A live air bag must be _____ before it is disposed. A deployed air bag should be disposed of in a manner consistent with the _____ and manufacturer's procedures.

g. Never use a battery- or AC-powered _____, _____, or any other type of test equipment in the system unless the manufacturer specifically says to. Never probe with a _____ _____ for voltage.

Instructor's Response

_____

_____

_____

Typical Shop Procedures and Equipment

Upon completion and review of this chapter, you should be able to:

❑ Understand basic diagnostic procedures.

❑ Describe customer service expectations.

❑ List the basic units of measure for length, volume, and mass in the two measuring systems.

❑ Name the diagnostic tools and equipment commonly used in vehicle repair work.

❑ Describe the basic applications and operation of these tools.

❑ Explain basic infrared analyzer operation.

❑ Describe what a waveform on an oscilloscope displays.

❑ Describe the major test capabilities of an engine analyzer.

❑ Explain the use of a logic probe.

❑ Describe how you would find specific information in a manufacturer's service information.

❑ Explain how you would locate specifications and special tool lists in a manufacturer's service information.

❑ Describe the information contained in an electronic data system and explain how this information is displayed.

❑ Describe the purposes of a repair order and explain how this order is processed by various personnel in an automotive shop.

❑ Explain the purpose of an engine tune-up.

Introduction

During the last decade, advances in technology have made today's vehicles quite advanced machines. In the past, vehicles were diagnosed using a fairly straightforward cause-and-effect procedure. Each system was treated as a separate, independent system isolated from the others. An engine-related problem, for example, was usually confined to just the engine, while transmission-related problems were confined to the transmission, and so on. In the case of a vehicle with black smoke spewing from the tailpipe, the focus was on the fuel delivery system. A misfire was usually associated with the ignition system with its respective components. Although still somewhat applicable to today's systems, there is a major difference in how the technician must approach the problem. This chapter will cover typical diagnostic procedures, shop procedures, and technician responsibilities and behaviors. Additionally, it will cover general tools and equipment application procedures specific to engine performance.

Most vehicles are now managed by one or more computers that integrate the operations of several systems together. Generically speaking, the powertrain control module (PCM) in many modern vehicles controls the ignition, fuel, and emission systems; engine and transmission; brakes and traction control; and entertainment and climate controls. While these systems still operate independently to a point, they are also interrelated to other systems via the computer. The technician must determine early in the diagnostic process what system could be affecting another seemingly unrelated system. Today's electronically controlled vehicle appears to be very complex, but it can be looked at in relatively simple, easy-to-understand segments. It is critical the technician understand the basic theory and operating principles of the systems in the vehicle—mechanical or electrical. In doing so, the insights for accurate diagnosis are formed. Regardless of the symptom, concern, or system, a specific guideline for a uniform diagnostic procedure must be established. Although finding the problem in a reasonable time-frame is important, the accuracy of the diagnosis is *more* important. As the skill level and experience of the technician improves over time, so will the speed at which problems are diagnosed. Technicians should practice the "fix-it-right-the-first-time" concept, thus eliminating a frustrated customer returning repeatedly for the same problem.

A Systematic Approach for Effective Diagnosis

The diagnostic procedure has evolved and is now a result of a critical thought process instead of a mechanical process. In other words, rather than assuming a specific component is faulty and then replacing it, the diagnostic thought process proves *why* you are replacing it based on factual test results. Using a symptom to establish a diagnosis is important; however, acting on the symptom alone without performing qualifying tests can lead you down a costly and incorrect path. Assumptions must be proved out. "Magic quick fixes" do not always work. Certain steps need to be followed when addressing any problem with the vehicle. The following systematic approach for an effective diagnosis gives the technician a strategy and process to follow that puts things in order. It can apply to just about anything related to the vehicle, from fluid leaks to driveability problems. Throughout this book, you may find yourself revisiting this section for reference.

1. *Questions and answers.* This is the time that the customer will communicate the symptom that has led to the vehicle being brought to you. Usually, a specific event occurred that appeared out of the ordinary to the customer. How the customer tells you of the problem may require you to be a bit of an interrogator and a detective. During your questioning, you may reveal an important piece of information that may not have occurred to the customer to tell you. Even minor information can be valuable. Also, refrain from letting the customer diagnose the problem for you. They are not always correct, or they could be making an assumption based on partial experience or facts. Use a worksheet to remind you of the specific questions you need to ask (Figure 2-1).

2. *Clarify the complaint.* Repeat to the customer what you understand of the complaint. Use specific explanations in the dialog and avoid using technical jargon that may confuse the customer. Although many terms are commonly used throughout our industry, few customers actually understand their true meaning. For example, a customer may request an alignment because they know it has something to do with the wheels and steering, but what if the customer asked for an alignment because the car vibrated at speeds over 40 mph? Obviously, an alignment would not help this situation. Imagine the multitude of descriptors that could be used to explain engine or driveability-related problems, or anything else concerning the vehicle.

3. *Climate, speed, time, distance, and action.* These five categories can further aid your diagnosis by understanding exactly what was happening at the time the problem occurred. If necessary, ask specific questions about the climate. Was it a hot or cold day? Was it raining? Ask about the vehicle speed. Was the vehicle at idle, city, or cruise speed? Next, find out when it happened. Did the problem occur at the first start of the day, or after a hot-soak period? How often did it happen? How far did the customer drive the vehicle? Did the problem begin just out of the driveway or after a long distance? What action was the vehicle performing during the problem? Was the vehicle stopped, accelerating, turning, going uphill, pulling a load, etc.? Also, if applicable, find out if any aftermarket equipment or accessories are installed or if other service has been done recently.

4. *Verify the complaint.* This step is nothing more than confirming what is fact. This verifies that the problem exists and matches the description. It may require a test drive, perhaps

Customer worksheet for driveability problems

Please fill out the following worksheet to the best of your ability and attach to the repair order. This form will be used by the technicians to better understand your problem.

Please circle all that apply

1. Is the check engine light:
   • always on
   • on sometimes
   • never on

2. If the check engine light does come on:
   • the engine runs noticeably worse
   • the engine runs fine even when the light is on

3. What is the nature of the problem you are having?
   • Hard starting
   • Odors   Please describe _____
   • Fuel economy
   • Spark knock or pinging on acceleration
   • Rough idle
   • Hesitation (slow from stops)
   • Surging
   • Engine noises
   • No power
   • Surging
   • Engine miss

When does the problem occur?

Always	Hot	Deceleration/coasting
Usually	Cold	Wet weather
Seldom	During acceleration	Under load

Figure 2-1 A customer worksheet can help a technician duplicate a customer concern.

with the customer, to duplicate the symptom. It may be necessary to attempt to duplicate the conditions as experienced by the customer when the problem occurred. Of course, depending on the situation, the vehicle may have to be left for a period of time in order to recreate the condition and problem, perhaps even overnight.

5. *Utilize the resources that are available to you.* Whenever possible, you should use all available information and other resources to effectively diagnose the problem, the least of which is your own experience. Consult the service manual or a service information system to know how to apply specific tests to the vehicle. In some cases, it will be beneficial to research any technical service bulletins (TSBs) that may apply to the vehicle. Also, as simple as it may sound, do not overlook the owner's manual that came with the vehicle. Other resources will include proper tools and specific test equipment needed to perform tests and repairs. As you work into a more complex test procedure, it is best to follow the recommended factory test procedure for that particular vehicle, which on occasion may require specialty tools of some type. Do not hesitate to ask for assistance.

6. *Begin with an area test.* The area test gives you the broadest and quickest overview and assessment of the systems in question. You may choose a test that was based on a symptom or a suspected fault. This gives you the most general of starting points. The purpose is to isolate one system from another. For example, in the case of a no-start complaint, you need to determine if the problem is mechanical, electrical, ignition, or fuel related. By performing an area test, you will determine the systems that are working and those that are not. Therefore, through an elimination process, you can begin to focus in on the specific system that is most likely causing the complaint or symptom. The objective is to exclude all other components that have nothing to do with the problem and determine the root cause.

7. *Next, perform the system test.* Now that you know what is and is not working, your attention will focus on the system that is initially determined to be at fault. This is because you tested the other systems with no problems found during the area test. The system you will be testing is probably made up of several components or sub-systems. In the case of an ignition system problem, you will need to perform a few tests to determine which part of the ignition system is not functioning. For example, is the problem in the primary or secondary ignition? As another example, suppose our discussion was related to the mechanical condition of the engine. If an area test revealed low engine vacuum, the system test may include a compression test to determine if an internal or external problem exists.

8. *Pinpoint or component test.* After determining the system fault, this diagnostic step will lead you to the end result, or very close to it. With the problem isolated to a given system, the pinpoint test determines the component that is most likely the reason for the failure. It is nothing more than continuing the elimination process. Depending on the situation, some tests are done with the component installed in the vehicle, while others need to be tested independently, with the component removed from the vehicle. These tests may be as simple as a visual inspection of a component after disassembly, a precise measurement of a part, or a dynamic test with a test instrument. This testing process helps to validate the performance potential of the unit or part in question. As you can see, the situations will vary, but the concept and process is the same, regardless of the problem.

9. *Approval and repair.* After determining the problem, advise the customer of the cost and get approval for the completion of the repair. Explain what is needed and why. As complex as the systems are, it is possible to uncover other items that need to be addressed before, during, or after the repair of the original problem. Be sure you communicate this to the customer ahead of time.

10. *Confirm and retest.* Make it a mandatory practice to confirm your repairs. Prove the repair you made solved the problem. The repairs you make should restore the vehicle to designed performance levels. If other maintenance is required to prevent a repeat of the failure, it is best to do it before returning the vehicle to the customer.

The above procedure helps establish consistency in the way you begin the diagnosis and how you communicate with the customer. Even if you are not in direct contact with the customer, you still need to be involved with the service writer to impart clear information to the customer.

Customer Service and Vehicle Care

Customers are the purpose of your business. They ultimately pay the bills and your wages. Because of basic human nature, you will encounter a variety of people and behaviors. They are human beings, just like you. The type of customer service you deliver can make the difference between a good shop and a great shop. Obviously, customers want to be treated with courtesy and respect. They expect proper and thorough service to be performed on their vehicle at a fair price, in a reasonable time. No one wants to pay for something that is not needed. When discussing a problem with customers, be understanding of their needs. Taking a few extra moments or a few extra steps will win customers over. Be accurate in your diagnosis, the estimated charges for the repair, and when the vehicle will be ready. If you run into an unforeseen problem, be upfront and honest with customers immediately. They would rather know about it early than be unpleasantly surprised later.

You should follow some common-sense procedures when handling a customer's vehicle. The following are some simple yet important things to remember when servicing a customer's vehicle:

❑ Perform a quick safety check before test driving the vehicle (brakes, steering, etc.)

❑ Use seat covers, steering wheel covers, and floor mats while test driving

❑ Observe all speed limits

❑ Do not operate the vehicle beyond its designed capacity or ability

❑ Upon entering the shop, roll the driver's side window down to avoid accidentally locking the keys in the car

❑ Use clean fender covers

❑ Remove any tools that may be in your pockets before getting into the vehicle

❑ Perform a basic, courtesy safety check—lights, tires, inflation, wiper blades, fluid leaks, etc.

❑ Do not smoke in or around the vehicle

❑ Do not eat or drink in the vehicle

❑ Do not change any radio settings (unless of course you are working on it)

❑ Return the seat to its original position if you moved it

❑ Remove any evidence of work being performed (remove tools, papers, parts boxes, etc.)

❑ Reset radio, clock, and other memory-related devices if the battery was disconnected, or use auxiliary power to keep memories alive

❑ If it is a practice in your service facility, clean the windshield when repairs are completed

❑ Clean any grease or dirt smudges and/or finger prints from the interior or exterior

❑ Secure and lock the vehicle when work is completed—bring the keys inside unless otherwise instructed by the customer

❑ Promptly call the customer when repairs are completed

Being a technician does not begin with opening the hood nor does it end with closing the hood. Your demonstration of quality customer service and vehicle care throughout the entire process, in addition to your technical aptitude, appearance, and work ethic, defines true professionalism.

Repairing the modern automobile requires the use of various tools. Many of these tools are common hand and power tools used every day by a technician. Other tools are very specialized and are only for specific repairs on specific systems and/or vehicles. Since units of measurement play such an important part in tool selection and in diagnosing automotive problems, this chapter begins with a presentation of measuring systems. A description of the measuring systems and some important measuring tools follow.

Units of Measure

The **United States Customary system (USC)** measuring system is commonly referred to as the English system.

Two systems of weights and measures are commonly used in the United States. One system of weights and measures is the **United States Customary system** (**USC**). Well-known measurements for length in the USC system are the inch, foot, yard, and mile. In this system, the quart and gallon are common measurements for volume, and ounce, pound, and ton are measurements for weight. A second system of weights and measures is the **International System (SI)**, and is also referred to as the metric system.

In the USC system, the basic linear measurement is the yard, whereas the corresponding linear measurement in the metric system is the meter (Figure 2-2). Each unit of measurement in the metric system is related to the other metric units by a factor of 10. Thus, every metric unit can be multiplied or divided by 10 to obtain larger units (multiples) or smaller units (submultiples). For example, the meter may be divided by 10 to obtain centimeters (1/100 meter) or millimeters (1/1,000 meter).

One meter is a little more than three inches longer than a yard.

Service technicians must be able to work with both the USC and the metric system. One meter (m) in the metric system is equal to 39.37 inches (in.) in the USC system. Some common equivalents between the metric and USC systems are these:

1 meter (m) = 39.37 inches
1 centimeter (cm) = 0.3937 inch
1 millimeter (mm) = 0.03937 inch
1 inch = 2.54 cm
1 inch = 25.4 mm

 SERVICE TIP: A metric conversion calculator provides fast, accurate metric-to-USC conversions or vice versa.

In the USC system, phrases such as $^1/_8$ of an inch are used for measurements. The metric system uses a set of prefixes. For example, in the word kilometer, the prefix kilo indicates 1,000, and this prefix indicates there are 1,000 meters in a kilometer. Common prefixes in the metric system follow:

NAME	SYMBOL	MEANING
mega	M	one million
kilo	k	one thousand
hecto	h	one hundred
deca	da	ten
deci	d	one tenth of
centi	c	one hundredth of
milli	m	one thousandth of
micro	μ	one millionth of

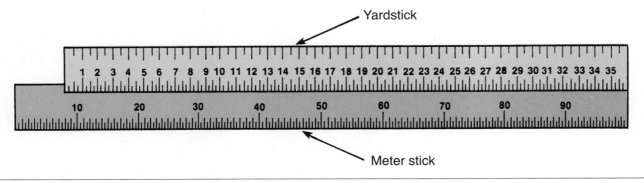

Figure 2-2 A meter is slightly longer than a yard.

Measurement of Mass

In the metric system, mass is measured in grams, kilograms, or tonnes. One thousand grams (g) = 1 kilogram (kg). In the USC system, mass is measured in ounces, pounds, or tons. When converting pounds to kilograms, 1 pound = 0.453 kilogram.

Measurement of Length

In the metric system, length is measured in millimeters, centimeters, meters, or kilometers. Ten millimeters (mm) = 1 centimeter (cm). In the USC system, length is measured in inches, feet, yards, or miles. When distance conversions are made between the two systems, some of the conversion factors are these:

    1 inch = 25.4 millimeters
    1 foot = 30.48 centimeter
    1 yard = 0.91 meter
    1 mile = 1.60 kilometers

Measurement of Volume

In the metric system, volume is measured in milliliters, cubic centimeters, and liters. One cubic centimeter = 1 milliliter. If a cube has a length, depth, and height of 10 centimeters (cm), the volume of the cube is 10 cm $\times$ 10 cm $\times$ 10 cm = 1,000 cm^3 = 1 liter. When volume conversions are made between the two systems, 1 cubic inch = 16.38 cubic centimeters. If an engine has a displacement of 350 cubic inches, 350 $\times$ 16.38 = 5,733 cubic centimeters, and 5,733 / 1,000 = 5.7 liters.

Cubic inch displacement is normally listed as cid, whereas cubic centimeters are shown as cc or L for liters.

Engine Diagnostic Tools

As the trend toward the integration of ignition, fuel, and emission systems progresses, diagnostic test equipment must also keep up with these changes. New tools and techniques are constantly being developed to diagnose electronic engine control systems.

Today's technician must not only keep up with changes in automotive technology, but must also keep up with the new testing procedures and specialized diagnostic equipment. To be successful, a shop must continuously invest in this equipment and the training necessary to troubleshoot today's electronic engine systems.

Regardless of how automated new test equipment can be or how new tests are developed for new systems, one thing will probably never change—the need to understand and perform fundamental tests. Because of all the advances in the automobile, some technicians sometimes make problems more difficult than they have to be. Basic engine, electrical, fuel, and ignition tests still need to be performed. New test equipment and analyzers can make the job go faster and easier and even help the technician's diagnostic accuracy, but when beginning any diagnostic procedure, *do not forget the basics!*

Compression Testers

Internal combustion engines depend on compression of the air-fuel mixture to maximize the power produced by the engine. The upward movement of the piston on the compression stroke compresses the air and fuel mixture within the combustion chamber. The air-fuel mixture becomes hotter as it is compressed. The hot mixture is easier to ignite, and when ignited it will generate much more power than the same mixture at a lower temperature.

If the combustion chamber leaks, some of the air-fuel mixture will escape when it is compressed, resulting in a loss of power and a waste of fuel. Cylinder heat is also lost. The leaks can be caused by burned valves, a blown head gasket, worn rings, slipped timing belt or chain, worn valve seats, a cracked head, and more.

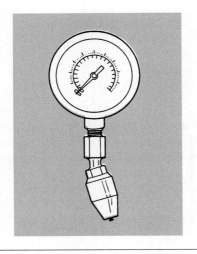

Figure 2-3 Push-in compression gauge.

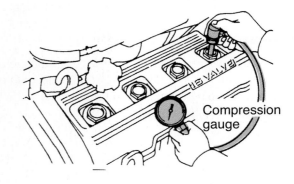

Compression gauge

Figure 2-4 A compression tester in place.

A compression test checks the cylinder's sealing ability.

An engine with poor compression (lower compression pressure due to leaks in the cylinder) will not run correctly and cannot be tuned to factory specifications. To see if a driveability problem is caused by poor compression, a compression test is performed.

A **compression gauge** is used to check cylinder compression. The dial face on the typical compression gauge indicates pressure in both **pounds per square inch (psi)** and **kilopascals (kPa)**. The range is usually 0 to 300 psi and 0 to 2,100 kPa.

There are two basic types of compression gauges: the push-in gauge (Figure 2-3) and a screw-in gauge.

The push-in type has a short stem that is either straight or bent at a 45-degree angle. The stem ends in a tapered rubber tip that fits any size spark plug hole. The rubber tip is placed in the spark plug hole after the spark plugs have been removed and is held there while the engine is cranked through several compression cycles. Although simple to use, the push-in gauge may give inaccurate readings if it is not held tightly in the hole.

The screw-in gauge has a long, flexible hose that ends in a threaded adapter (Figure 2-4). This type compression tester is often used because its flexible hose can reach into areas that are difficult to reach with a push-in type tester. The threaded adapters are changeable and come in several thread sizes to fit 10 mm, 12 mm, 14 mm, and 18 mm diameter holes. The adapters screw into the spark plug holes in place of the spark plugs.

Most compression gauges have a vent valve that holds the highest pressure reading on its meter. Opening the valve releases the pressure when the test is complete.

Cylinder Leakage Tester

Cylinder leakage is the amount of air or volume lost from a sealed cylinder measured in percent.

A cylinder leakage test determines the point of cylinder leakage.

If a compression test shows that any of the cylinders are leaking, a **cylinder leakage** test can be performed to measure the percentage of compression lost and help locate the source of leakage.

A cylinder leakage tester (Figure 2-5) applies compressed air to a cylinder through the spark plug hole. Before the air is applied to the cylinder, the piston of that cylinder must be at top dead center (TDC) on its compression stroke. A threaded adapter on the end of the air pressure hose screws into the spark plug hole. The source of the compressed air is normally the shop's compressed air system. A pressure regulator in the tester controls the pressure applied to the cylinder. An analog gauge registers the percentage of air pressure lost from the cylinder when the compressed air is applied. The scale on the dial face reads 0 to 100 percent.

Figure 2-5 A cylinder leakage tester is connected to the cylinder through the spark plug hole.

A zero reading means that there is no leakage from the cylinder. Readings of 100 percent would indicate that the cylinder will not hold any pressure. The location of the compression leak can be found by listening and feeling around various parts of the engine.

Most vehicles, even new cars, experience some leakage around the rings. Up to 20 percent is considered acceptable during the leakage test. When the engine is actually running, the rings will seal much better and the actual percent of leakage will be lower. However, there should be no leakage around the valves or the head gasket.

Vacuum Gauge

Measuring intake manifold vacuum is another way to diagnose the condition of an engine. Manifold vacuum is tested with a vacuum gauge (Figure 2-6). **Vacuum** is formed on a piston's intake stroke. As the piston moves down, it lowers the pressure of the air in the cylinder if the cylinder is sealed. This lower cylinder pressure is called engine vacuum. If there is a leak, atmospheric pressure will force air into the cylinder and the resultant pressure will not be as low. The reason atmospheric pressure will enter is simply that whenever there is a low and high pressure, the high pressure will always move toward the low pressure.

Vacuum is best defined as any pressure lower than atmospheric pressure. Atmospheric pressure is the pressure of the air around us.

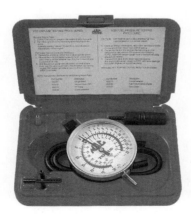

Figure 2-6 Vacuum gauge with line adapters.

The vacuum gauge measures the difference in pressure between intake manifold vacuum and atmospheric pressure. If the manifold pressure is lower than atmospheric pressure, a vacuum exists. Vacuum is measured in inches of mercury (in. Hg), kilopascals (kPa), or millimeters of mercury (mm Hg).

To measure vacuum, a flexible hose on the vacuum gauge is connected to a source of manifold vacuum, either on the manifold or a point below the throttle plates. Sometimes this requires removing a plug from the manifold and installing a special fitting.

The test is made with the engine cranking and/or running. A good vacuum reading is typically at least 16 in. Hg. However, a reading of 15 to 20 in. Hg (50 to 65 kPa) is normally acceptable. Since the intake stroke of each cylinder occurs at a different time, the production of vacuum occurs in pulses. If the amount of vacuum produced by each cylinder is the same, the vacuum gauge will show a steady reading. If one or more cylinders are producing different amounts of vacuum, the gauge will show a fluctuating reading.

Low or fluctuating readings can indicate many different problems. For example, a low, steady reading might be caused by retarded ignition timing or incorrect valve timing. A sharp vacuum drop at regular intervals might be caused by a burned intake valve.

Vacuum Pumps

There are many vacuum-operated devices and vacuum switches in cars. These devices use engine vacuum to cause a mechanical action or to switch something on or off. The tool used to test vacuum-actuated components is the vacuum pump (Figure 2-7). There are two types of vacuum pumps: an electrical operated pump and a hand-held pump. The hand-held pump is most often used for diagnostics. A hand-held vacuum pump consists of a hand pump, a vacuum gauge, and a length of rubber hose used to attach the pump to the component being tested. Tests with the vacuum pump can usually be performed without removing the component from the car.

When the handles of the pump are squeezed together, a piston inside the pump body draws air out of the component being tested. The partial vacuum created by the pump is registered on the pump's vacuum gauge. While forming a vacuum in a component, watch the action of the component. The vacuum level needed to actuate a given component should be compared to the specifications given in the factory service manual.

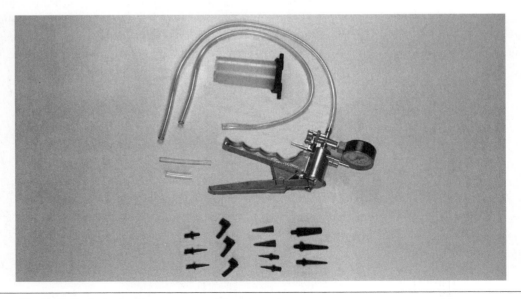

Figure 2-7 Typical vacuum pump with accessories.

The vacuum pump is also commonly used to locate vacuum leaks. This is done by connecting the vacuum pump to a suspect vacuum hose or component and applying vacuum. If the needle on the vacuum gauge begins to drop after the vacuum is applied, a leak exists somewhere in the system.

Vacuum Leak Detector

Low compression pressures might be revealed by a compression check, a cylinder leak down test, or a manifold vacuum test. However, finding the location of a vacuum leak can often be very difficult.

A simple but time-consuming way to find leaks in a vacuum system is to check each component and vacuum hose with a vacuum pump. Simply apply vacuum to the suspected area and watch the gauge for any loss of vacuum. A good vacuum component will hold the vacuum that is applied to it.

Another method of leak detection is done by using an ultrasonic leak detector (Figure 2-8). Air rushing through a vacuum leak creates a high-frequency sound, higher than the range of human hearing. An ultrasonic leak detector is designed to hear the frequencies of the leak. When the tool is passed over a leak, the detector responds to the high-frequency sound by emitting a warning beep. Some detectors also have a series of light-emitting diodes (LEDs) that light up as the frequencies are received. The closer the detector is moved to the leak, the more LEDs light up or the faster the beeping occurs. This allows the technician to zero in on the leak. An ultrasonic leak detector can sense leaks as small as $1/500$ inch and accurately locate the leak to within $1/16$ inch.

Smoke Tester

Another way of detecting vacuum leaks that is gaining popularity is the smoke leak tester (Figure 2-9). A hose is connected from the machine to a hose on the intake manifold. The machine manufactures a non-toxic chemical smoke that can be introduced into the manifold. A high-intensity lamp helps to spot small leaks. The smoke machines are also popular for leak detection tasks in the evaporative emission system which can set a check engine lamp with a leak as small as 0.020 inch in diameter. These testers can also find leaks in several other systems such as exhaust.

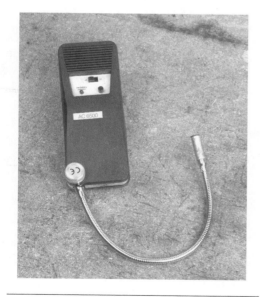

Figure 2-8 An ultrasonic vacuum leak detector.

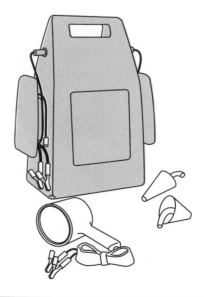

Figure 2-9 Finding small leaks can be made easier with a smoke machine.

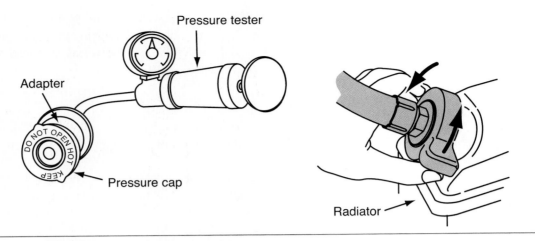

Figure 2-10   Cooling system pressure tester.

Cooling System Pressure Tester

A cooling system pressure tester contains a hand pump and a pressure gauge. A hose is connected from the hand pump to a special adapter that fits on the radiator filler neck (Figure 2-10). This tester is used to pressurize the cooling system and check for coolant leaks. Additional adapters are available to connect the tester to the radiator cap. With the tester connected to the radiator cap, the pressure relief action of the cap may be checked.

Coolant Hydrometer

A coolant hydrometer is used to check the effectiveness of antifreeze in the coolant. This tester contains a pick-up hose, coolant reservoir, and squeeze bulb. The pick-up hose is placed in the radiator coolant. When the squeeze bulb is squeezed and released, coolant is drawn into the reservoir. As coolant enters the reservoir, a pivoted float moves upward with the coolant level. A pointer on the float indicates the freezing point of the coolant on a scale located on the reservoir housing (Figure 2-11).

Oil Pressure Gauge

The oil pressure gauge may be connected to the engine to check the oil pressure (Figure 2-12). Various fittings are usually supplied with the oil pressure gauge to fit different openings in the lubrication system.

Figure 2-11   Coolant hydrometer.

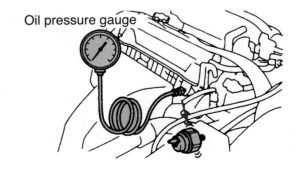

Figure 2-12   Oil pressure gauge threaded into the oil pressure sending unit's bore.

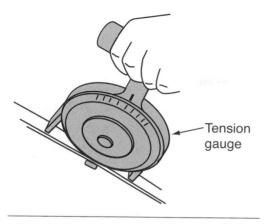

Figure 2-13 Belt tension gauge.

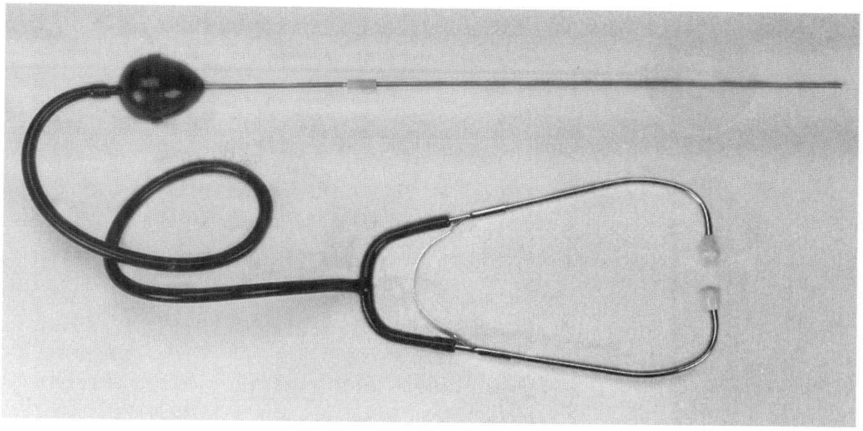

Figure 2-14 Stethoscope.

Belt Tension Gauge

A belt tension gauge is used to measure drive belt tension. The belt tension gauge is installed over the belt and indicates the amount of belt tension (Figure 2-13).

Stethoscope

A stethoscope is used to locate the source of engine and other noises. The stethoscope pickup is placed on the suspected component, and the stethoscope receptacles are placed in the technician's ears (Figure 2-14). Amplified stethoscopes are also available.

Feeler Gauge

A **feeler gauge** is a thin strip of metal with a precision thickness. Feeler gauges with different thicknesses are usually sold in sets. These gauges are pivoted on one end and are mounted inside a metal holder. When a specific feeler gauge is required, that feeler gauge is pivoted out of the set (Figure 2-15).

Feeler gauges are used to make precision measurements of small gaps. Individual feeler gauges are used for certain service operations such as measuring piston clearance in a cylinder. Feeler gauge thickness in the English system is measured in thousandths of an inch, whereas metric feeler gauge thickness is measured in millimeters (mm). Round wire feeler gauges are recommended for measuring spark plug gaps (Figure 2-16). Depending on the application, you may need to use *non-metallic* gauges.

A **feeler gauge** is a device of a specific thickness used to measure the distance between two components.

Figure 2-15 Flat feeler gauge set.

Figure 2-16 Round wire-type feeler gauge set for spark plugs.

Electrical Diagnostic Tools

> ■ **CAUTION:** Do not use any type of test light or circuit tester to diagnose automotive air bag systems. Use only the vehicle manufacturer's recommended equipment on these systems. Many different tools are used to diagnose electrical problems. Those discussed here are the most common. Always use the correct tool and procedure for testing an electrical or electronic circuit.

Circuit Testers

Circuit testers (Figure 2-17) are used to identify shorted and open circuits in any electrical circuit. Low-voltage testers are used to troubleshoot 6- to 12-volt circuits. High-voltage circuit testers diagnose primary and secondary ignition circuits.

A circuit tester looks like a stubby ice pick. Its handle is transparent and contains a light bulb. A probe extends from one end of the handle and a ground clip and wire from the other end. When the ground clip is attached to a good ground and the probe is touched to a live connector, the bulb in the handle will light up. If the bulb does not light, voltage is not available at the connector.

A self-powered circuit tester (Figure 2-18) is used on open circuits. This tester also has a clip and probe end, and it has a small internal battery. When the ground clip is attached to the negative side of a component and the probe is touched to the positive side, the lamp will light if there is continuity in the circuit. If an open circuit exists, the light will not be illuminated.

> ■ **CAUTION:** Do not use a conventional 12-V test light to diagnose components and wires in computer systems. The current draw of these test lights may damage computers and computer system components. High-impedance test lights are available for diagnosing computer systems. Always be sure the test light you are using is recommended by the tester manufacturer for testing computer systems.

Electrical Test Meters

Several meters are used to test and diagnose electrical systems. These are the voltmeter, ohmmeter, and ammeter.

A circuit tester is commonly called a test light.

A self-powered test light is called a **continuity tester**.

Typically the negative side of a component is called the ground side, while the positive side is called the power or feed side.

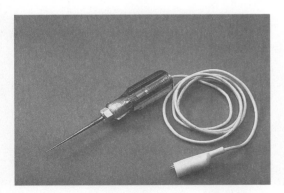

Figure 2-17 Typical circuit tester, commonly called a test light.

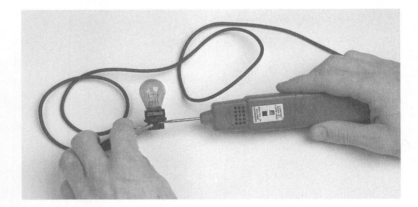

Figure 2-18 Typical self-powered circuit tester, commonly called a continuity tester or self-powered test lamp.

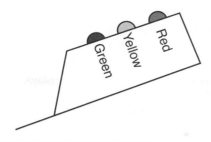

Figure 2-19 The logic probe will turn the red LED on with a positive voltage over 10; the green LED with voltage under 4.

Logic Probe

The logic probe (Figure 2-19) can be used on sensitive electronic circuits with very low amperage. The logic probe requires a connection to both the positive and negative cables of the battery. If a voltage detected is over 10 volts, the red LED comes on, if under 4 volts, the green LED comes on. If the voltage changes or pulses, the yellow LED flashes along with the red or green LED.

Voltmeter

The **voltmeter** measures the voltage available at any point in an electrical system. For example, it can be used to measure the voltage available at the battery. It can also be used to test the voltage available at the terminals of any component or connector. A voltmeter can also be used to test voltage drop across an electrical circuit, component, switch, or connector.

A voltmeter has two leads: a red positive lead and a black negative lead. The red lead should be connected to the positive side of the circuit or component. The black should be connected to ground or to the negative side of the component. Voltmeters should always be connected across the circuit being tested.

Ohmmeter

An **ohmmeter** measures resistance to current flow in a circuit. In contrast to the voltmeter which uses the voltage available in the circuit, the ohmmeter is battery powered. The circuit being tested must be open. If the power is on in the circuit, the ohmmeter will be damaged.

The two leads of the ohmmeter are placed across or in parallel with the circuit or component being tested. The red lead is placed on the positive side of the circuit and the black lead is placed on the negative side of the circuit. The meter sends current through the component and determines the amount of resistance based on the voltage dropped across the load. The scale of an ohmmeter reads from 0 to infinity. A 0 reading means there is no resistance in the circuit and may indicate a short in a component that should show a specific resistance. An infinity reading indicates a number higher than the meter can measure. This is usually an indication of an open circuit or faulty component.

Ammeter

An **ammeter** measures current flow in a circuit. Current is measured in amperes. Unlike the voltmeter and ohmmeter, the ammeter must be placed into the circuit or in series with the circuit being tested. This normally requires disconnecting a wire or connector from a component and connecting the ammeter between the wire or connector and the component. The red lead of the ammeter should always be connected to the side of the connector closest to the positive side of the battery and the black lead should be connected to the other side. Do not connect an ammeter into a circuit that may have more amps than the meter is rated. Most meters are protected with a fuse; however, it is better to avoid the risk of meter damage if you are not sure about the circuit you are testing.

It is much easier to test current using an ammeter with an inductive pickup. The pick-up clamps around the wire or cable being tested. The ammeter determines amperage based on the magnetic field created by the current flowing through the wire. This type of pickup eliminates the need to separate the circuit to insert the meter.

Volt/Ampere Tester

A **volt/ampere tester** (**VAT**) is used to test batteries, starting systems, and charging systems. The tester contains a voltmeter, ammeter, and carbon pile. The carbon pile is a variable resistor. A knob on the tester allows the technician to vary the resistance of the pile. When the tester is attached to the battery, the carbon pile will draw current out of the battery. The ammeter will read the amount of current draw. When testing a battery, the resistance of the carbon pile must be adjusted to match the ratings of the battery. Newer VATs have more automated capabilities for quicker test results (Figure 2-20). These menu-driven units feature reduced testing time on batteries in various stages of charge, and automatic loading of charging systems. They work on all 12-volt to 30-volt systems. Results can be printed for comparison testing or customer review.

Figure 2-21 shows a small battery tester. This battery tester does not have a carbon pile. However, it can be used to test circuit voltage and to perform battery load tests on 6- and 12-volt batteries. It can also be used to check the condition of battery cables and connectors and to check running voltage.

Figure 2-20 Battery, starting and charging system tester.

Figure 2-21 Typical battery tester.

Multimeter

It is not necessary for a technician to own separate voltmeters, ohmmeters, and ammeters. These meters are combined in a single tool called a **multimeter**. A multimeter is one of the most versatile tools used in diagnosing engine performance and electrical systems. The most commonly used multimeter is the **digital multimeter (DMM)**, or digital volt ohmmeter. This meter does not use a sweeping needle and scales to display the measurement. Rather, it displays the measurement digitally on the meter (Figure 2-22).

■ **CAUTION:** A high-impedance digital multimeter must be used to test the voltage of some components and systems such as an oxygen (O_2) sensor circuit. If a low-impedance analog meter is used in this type of circuit, the current flow through the meter is high enough to damage the sensor. Always use the type of meter specified by the vehicle manufacturer.

Multimeters provide these readings on several different scales:

1. DC volts
2. AC volts
3. Ohms
4. Amperes
5. Milliamperes
6. Diodes

■ **CAUTION:** Always be sure the proper scale is selected on the multimeter and that the correct lead connections are completed for the component or system being tested. Improper multimeter lead connections or scale selections may blow the internal fuse in the meter or cause meter damage.

Since multimeters do not have heavy leads, the highest ammeter scale on this type of meter is normally 10 amperes. A control knob on the front of the multimeter must be rotated to the desired reading and scale (Figure 2-23). Some multimeters are autoranging, which means the

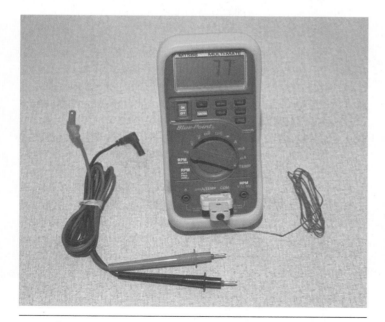

Figure 2-22 The digital volt/ohmmeter can be used to test voltage and resistance.

Figure 2-23 This multimeter has 18 test ranges that have to be selected by the technician since it is not auto ranging.

meter automatically switches to a higher scale if the reading goes above the value of the scale being used. For example, if the meter is reading on the 10-volt scale and the leads are connected to a 12-volt battery, the meter automatically changes to the next highest scale. If the multimeter is not autoranging, the technician must select the proper scale for the component or circuit being tested. When using a meter that is not auto-ranging, being on the wrong scale can create some confusion. An easy way to get to the correct scale is a method called "scaling down." Simply make your DVOM connection and begin with the highest scale. Begin to scale down by turning the scale selector knob one range at a time. When the meter displays an "over-limit" indicator, scale up to the next highest range.

Top-of-the-line multimeters are multifunctional. Most test DC and AC volts, ohms, and amperes. There are usually several test ranges provided for each of these functions. Many meters have a diode testing function. This typically uses a similar setup as an ohmmeter test. During the diode test, the meter applies a small voltage to the diode to act on the boundary layer in the diode. That is why there is usually a separate meter selection for this test. In addition to these basic electrical tests, multimeters also test engine revolutions per minute (rpm), ignition dwell, diode condition, distributor conditions, frequency, and even temperature. Figure 2-24 shows some of the many tests that can be performed with a multimeter.

Multimeters are available with either analog or digital displays. Analog meters enjoyed wide popularity prior to electronic control systems. They are still favored by technicians who have used them for years and prefer the visual reference provided by a moving needle.

However, there are several drawbacks to using most analog meters for testing electronic control systems. Many electronic components require very precise test results. Digital meters can measure volts, ohms, or amperes in tenths and hundredths. Some have multiple test ranges that must be manually selected. Others are auto ranging.

Another problem with analog meters is their low internal resistance (input impedance). The low input impedance allows too much current to flow through circuits and should not be used on delicate electronic devices.

For our purposes, impedance is best defined as operating resistance.

System/ Component	Measurement Type			
	Voltage Presence and Level	Voltage Drop	Current	Resistance
Charging System				
Alternators	•		•	
Regulators	•			
Diodes		•		•
Connectors	•	•		
Starting System				
Batteries	•	•	•	
Starters		•	•	
Solenoids	•	•		•
Connectors	•	•		•
Cables	•	•	•	•
Ignition System				
Coils	•			•
Connectors	•	•		•
Distributor Caps				•
Plug Wires				•
Rotors				•
Magnetic pickup	•			•

Figure 2-24 Electrical testing with a multimeter.

Figure 2-25 Graphing digital multimeter.

Digital meters have a **high input impedance**, usually at least 10 megohms. Metered voltage for resistance tests is well below 5 volts, reducing the risk of damage to sensitive components and delicate computer circuits.

Graphing Digital Multimeter

Other DMMs offer enhanced diagnostic features that help the technician capture and read data more than one way. The graphing multimeter can create a waveform or display a signal's history over a given time span. Figure 2-25 is a unit that displays measurements graphically, digitally, or through a history graph. It has a diagnostic database including various pattern samples for reference. Its testing capabilities include AC and DC volts, amps and ohms, vacuum and pressure, primary and secondary ignition, frequency, and pulse width.

Ignition System Tools

Some of the tools used to troubleshoot ignition systems have been in use for years. Others are relatively new, having been developed in response to electronic ignition and computer-controlled ignition systems.

Tach-Dwell Meter

A tool commonly used for tune up and engine diagnosis on older engines is the tach-dwell meter. This meter is a combination tachometer and dwellmeter. A **tachometer** measures engine speed. Engine speed is measured in rpm (revolutions per minute). A **dwell meter** measures the time a circuit is on. On ignition systems, the dwell time is the degree of crankshaft rotation during which the primary circuit is on. A small internal dry-cell battery powers most tach-dwell meters. The red tach-dwell meter lead is connected to the negative primary coil terminal, and the black meter lead is connected to ground. A switch on the meter must be set in the rpm or dwell position. On distributorless ignition systems, a special tachometer lead may be provided in the ignition system for the tach-dwell meter connection.

Distributorless ignitions and many distributor ignitions do not have adjustable timing.

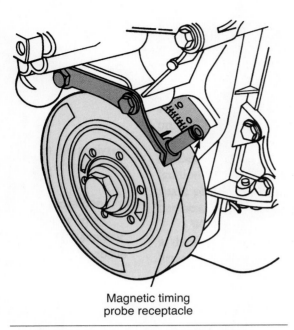

Magnetic timing
probe receptacle

Figure 2-26 Location of the magnetic timing probe receptacle.

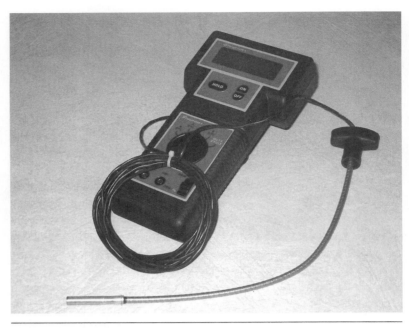

Figure 2-27 A digital tachometer with a magnetic probe.

Tachometers

Digital tachometers are available with an inductive pickup that is clamped over the number one spark plug wire. These meters provide an rpm reading from the speed of spark plug firings. This type of tachometer is suitable for distributorless ignition systems.

Many engines have a magnetic probe receptacle mounted above the crankshaft pulley (Figure 2-26). Digital tachometers are available with a magnetic pickup that fits in this magnetic probe receptacle. Each timing mark rotation past the pickup sends a pulse signal to the meter (Figure 2-27). The digital tachometer provides an rpm reading from these pulses. This type of digital tachometer may be used on diesel engines. The magnetic probe receptacle may also be used for engine timing purposes.

Timing Light

A timing light is essential for checking the ignition timing in relation to crankshaft position. Two leads on the timing light must be connected to the battery terminals with the correct polarity. Most timing lights have an inductive clamp that fits over the number one spark plug wire (Figure 2-28).

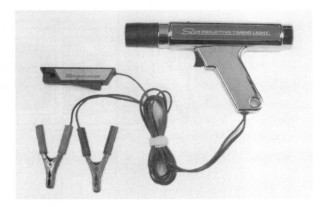

Figure 2-28 A timing light with an inductive pickup.

Older timing lights have a lead that goes in series between the number one spark plug wire and the spark plug. A trigger on the timing light acts as an off/on switch. When the trigger is pulled with the engine running, the timing light emits a beam of light each time the spark plug fires.

The timing marks are usually located on the crankshaft pulley or on the flywheel. A stationary pointer, line, or notch is positioned above the rotating timing marks. The timing marks are lines on the crankshaft pulley or flywheel that represent various degrees of crankshaft rotation when the number one piston is **before top dead center (BTDC)** on the compression stroke. The TDC crankshaft position and the degrees are usually identified in the group of timing marks. Some timing marks include degree lines representing the **after top dead center (ATDC)** crankshaft position (Figure 2-29).

Before checking the ignition timing, complete all the vehicle manufacturer's recommended procedures. On fuel injected engines, special timing procedures are required, such as disconnecting a timing connector to be sure the computer does not provide any spark advance while checking basic ignition timing. These timing instructions are usually provided on the underhood emission label.

Many timing lights have a timing advance knob that may be used to check spark advance. This knob has an index line and a degree scale surrounding the knob. Prior to checking the spark advance, the basic timing should be checked. Accelerate the engine to 2,500 rpm or the speed recommended by the vehicle manufacturer. While maintaining this rpm, slowly rotate the advance knob toward the advanced position until the timing marks come back to the basic timing position. Under this condition, the index mark on the advance knob is pointing to the number of degrees advance provided by the computer or distributor advances. The reading on the degree scale may be compared to the vehicle manufacturer's specifications to determine if the spark advance is correct.

Timing lights can also be used for several quick tests. For example, connect the timing light inductive pick up around each spark plug wire and observe the light. If the timing light flashes, there is current flowing through the spark plug wire.

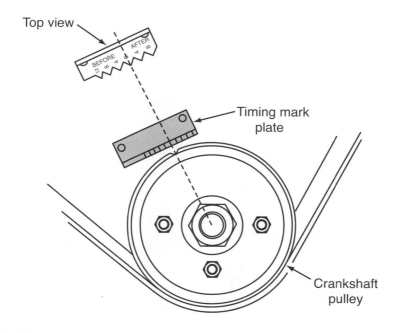

Figure 2-29 Timing reference marks.

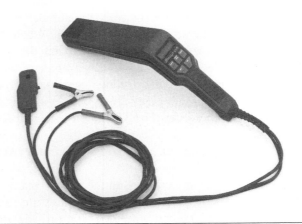

Figure 2-30 Digital advance-type timing light.

Some timing lights have a digital reading in the back of the light that displays the number of degrees advance as the engine is accelerated (Figure 2-30). When the trigger on the light is squeezed, the light flashes and the digital display reads the degrees of spark advance. If the trigger is released, the digital reading indicates engine rpm.

Magnetic Timing Probe

A timing light is not necessary for tuning many cars. A magnetic probe receptacle on the crankshaft position sensor allows the ignition timing to be electronically monitored with a magnetic timing probe. Changes in the magnetic field create electrical pulses in the tip of the timing probe. These pulses are monitored by the timing meter to determine crankshaft position and ignition timing. The timing indicated by this type of equipment is extremely accurate.

The only trick to using a magnetic timing probe is correcting the offset of the probe receptacle. The receptacle is usually situated on either side of the vehicle's coil pickup. The degree of offset must be factored into the timing probe's reading or the timing readout will be inaccurate by that many degrees. The timing meter must be programmed for the degree of offset specified by the car manufacturer.

Spark Testers

The spark tester shown in Figure 2-31 is used to stress the ignition system for testing in the shop. A well-performing ignition system should be able to jump the gap of the tester. This

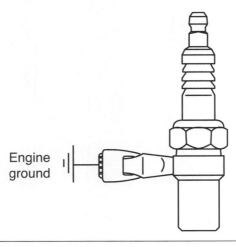

Engine
ground

Figure 2-31 A spark tester is used to check coil output.

eliminates having to use a grounded screwdriver or other makeshift device to check for sufficient spark. Remember, there are several types of spark testers from small engine to electronic ignition, so make sure you have the right tester for the vehicle.

Fuel System Tools

The proper amount of fuel is essential to the proper operation of an engine. Therefore, misadjusted or faulty fuel system components will adversely affect engine performance. The following tools are used to test the parts of the fuel system.

Pressure Gauge

A **pressure gauge** (Figure 2-32) is used to measure the pressure in the fuel system. This tester is very important for diagnosing fuel injection systems. These systems rely on very high fuel pressures, from 35 to 70 psi. A drop in fuel pressure will reduce the amount of fuel delivered to the injectors and result in a lean air-fuel mixture.

A fuel pressure gauge is used to check the discharge pressure of fuel pumps, the regulated pressure of fuel injection systems, and injector pressure drop. This test can identify faulty pumps, regulators, or injectors and can identify restrictions present in the fuel delivery system. Restrictions are typically caused by dirty fuel filters, collapsed hoses, or damaged fuel lines.

Some fuel pressure gauges also have a valve and outlet hose for testing fuel pump discharge volume. The manufacturer's specification for discharge volume will be given as a number of pints or liters of fuel that should be delivered in a certain number of seconds.

⚠️ **WARNING:** While testing fuel pressure, be careful not to spill gasoline. Gasoline spills may cause explosions and fires, resulting in serious personal injury and property damage.

Injector Balance Tester

The injector balance tester is used to test the injectors in a port fuel injected engine for proper operation. A fuel pressure gauge is also used during the injector balance test. The injector balance

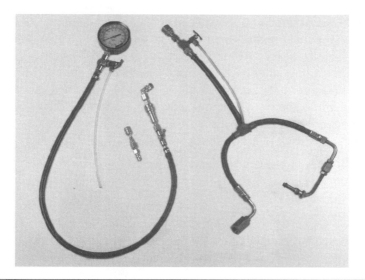

Figure 2-32 Fuel pressure gauge and various adapters.

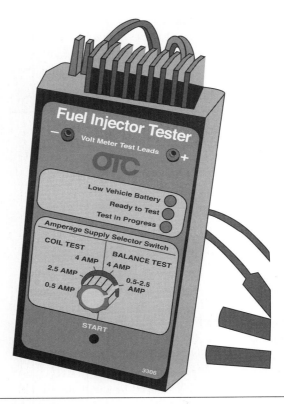

Figure 2-33 An injector tester.

tester contains a timing circuit, and some injector balance testers have an off/on switch. A pair of leads on the tester must be connected to the battery with the correct polarity (Figure 2-33). The injector terminals are disconnected, and a second double lead on the tester is attached to the injector terminals.

Before the injector balance test, the fuel pressure gauge is connected to the Schrader valve on the fuel rail, and the ignition switch should be cycled two or three times until the specified fuel pressure is indicated on the pressure gauge. When the tester push button is depressed, the tester energizes the injector winding for a specific length of time and the technician records the pressure decrease on the fuel pressure gauge. This procedure is repeated on each injector.

Some vehicle manufacturers provide a specification of 3 psi (20 kPa) maximum difference between the pressure readings after each injector is energized. If the injector orifice is restricted, there is not much pressure decrease when the injector is energized. Acceleration stumbles, engine stalling, and erratic idle operation are caused by restricted injector orifices. The injector plunger is sticking open if excessive pressure drop occurs when the injector is energized. Sticking injector plungers may result in a rich air-fuel mixture.

> **CAUTION:** Electronic fuel injection systems are pressurized, and these systems require depressurizing prior to fuel pressure testing and other service procedures.

Injector Tester

If the injector tester has the capability of testing the voltage drop on each injector's coil, move the selector on the tool to "coil test" and select the correct amperage level from the manufacturer's specification. Connect the multimeter leads to the appropriate sockets on the tester. Hit the button on the tester and the tester will turn the injector on and leave it on instead of pulsing as it does for the balance test. If the coil is shorting or opening intermittently, this test should find the fault. Record the voltage drop on the injector coil and compare it to specifications.

The balance test energizes the injectors 100 times each to check the pintle operation.

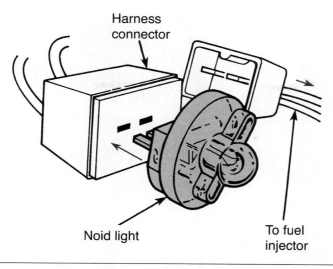

Harness
connector

Noid light

To fuel
injector

Figure 2-34 A noid light is used to check for voltage pulses at the fuel rail harness.

Injector Circuit Test Light

A special test light called a **noid light** can be used to determine if a fuel injector is receiving its proper voltage pulse from the computer. The wiring harness connector is disconnected from the injector and the noid light is plugged into the connector (Figure 2-34). After disabling the ignition to prevent starting, the engine is turned over by the starter motor. The noid light will flash rapidly if the voltage signal is present. No flash usually indicates an open in the power feed or ground circuit to the injector. It should be noted that there are several types of noid lights for specific applications. The connectors are different from one injector to the next. The noid lights may also have different resistance values to match the fuel injector's resistance. Check the manufacturer's application instructions.

A **noid light** is used to determine if a fuel injector is receiving the proper voltage pulse.

Fuel Injector Cleaner

Fuel injectors spray a certain amount of fuel into the intake system. If the fuel pressure is low, not enough fuel will be sprayed. This is also true if the fuel injector is dirty. Normally, clogged injectors are the result of inconsistencies in gasoline detergent levels and the high sulfur content of gasoline. When these sensitive fuel injectors become partially clogged, fuel flow is restricted. Spray patterns are altered, causing poor performance and reduced fuel economy.

The solution to a sulfated and/or plugged fuel injector is to clean it, not replace it. There are several kinds of fuel injector cleaners. One is a pressure tank. A mixture of solvent and unleaded gasoline is placed in the tank, following the manufacturer's instructions for mixing, quantity, and safe handling. The vehicle's fuel pump must be disabled, and on some vehicles the fuel line must be blocked between the pressure regulator and the return line. Then, the hose on the pressure tank is connected to the service port in the fuel system. The in-line valve is then partially opened and the engine is started. It should run at approximately 2,000 rpm for about 10 minutes to thoroughly clean the injectors.

Figure 2-35 Fuel injector cleaner using the shop's compressed air.

Figure 2-36 A fuel injector and carbon cleaning system.

Another type is a pressurized canister (Figure 2-35) in which the solvent solution is pre-mixed. Use of the canister-type cleaner is similar to this procedure but does not require mixing or pumping.

The canister is connected to the injection system's servicing fitting, and the valve on the canister is opened. The engine is started and allowed to run until it dies. Then, the canister is discarded.

Several other fuel injector cleaning methods are available. Check with the vehicle manufacturer for their recommended process. Fuel injection cleaning methods have been improved over the past few years due to new cleaning agents and service equipment. Carbon buildup can be as much of a problem as dirty injectors. Carbon can absorb fuel and affect fuel mixture management by the computer. The unit in Figure 2-36 cleans the fuel rail, injectors, intake manifold, plenum, intake runners, and throttle body. In addition, it cleans the carbon on the back of the intake valve, inside the combustion chamber, and on the oxygen sensor. This method is an alternative to the physical disassembly of an engine to remove carbon deposits.

Scan Tools

A scan tool may be called a scanner or a scan tester.

The introduction of computer-controlled ignition and fuel systems brought with it the need for tools capable of troubleshooting electronic engine control systems. There are a variety of computer scan tools available today that do just that. A **scan tool** (Figure 2-37) is a microprocessor designed to communicate with the vehicle's computer. Connected to the computer through diagnostic connectors, a scan tool can access trouble codes, run tests to check system operations, and monitor the activity of the system. Trouble codes and test results are displayed on an LED screen or can be printed out on the scanner printer.

Figure 2-37 A typical scan tool and accessories.

Scan tools will retrieve fault codes from a computer's memory and digitally display these codes on the tool. A scan tool may also perform many other diagnostic functions depending on the year and make of the vehicle. Most scan tools have removable modules that are updated each year. These modules are designed to test the computer systems on various makes of vehicles. For example, some scan testers have a 3-in-1 module that tests the computer systems on Chrysler, Ford, and General Motors vehicles. A 10-in-1 module is also available to diagnose computer systems on vehicles from ten different manufacturers. These modules plug into the scan tool.

Scanners are capable of testing many onboard computer systems such as engine computers, antilock brake computers, air bag computers, and suspension computers, depending on the year and make of the vehicle and the type of scan tester. In many cases, the technician must select the computer system to be tested with the scanner after it has been connected to the vehicle.

The scan tool is connected to specific diagnostic connectors on various vehicles. Some manufacturers have one diagnostic connector. This connects the data wire from each onboard computer to a specific terminal in this connector. Other vehicle manufacturers have several different diagnostic connectors on each vehicle, and each of these connectors may be connected to one or more onboard computers. A set of connectors is supplied with the scanner to allow tester connection to various diagnostic connectors on different vehicles (Figure 2-38).

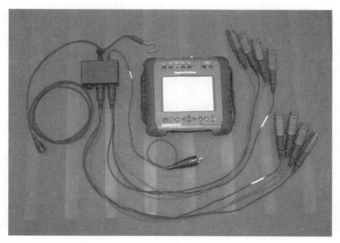

Figure 2-38 A PDA can be equipped to work as a scan tool. Special adapters and software are required.

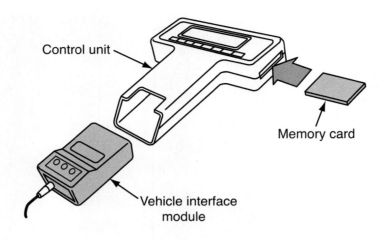

Control unit

Memory card

Vehicle interface
module

Figure 2-39 Scanner with memory card.

The scanner must be programmed for the model year, make of vehicle, and type of engine. With some scan testers, this selection is made by pressing the appropriate buttons on the tester as directed by the digital tester display. On other scan testers, the appropriate memory card must be installed in the tester for the vehicle being tested (Figure 2-39). Some scan testers have a built-in printer to print test results, while other scan testers may be connected to an external printer.

As automotive computer systems become more complex, the diagnostic capabilities of scan testers continue to expand. Many scan testers now have the capability to store, or "freeze," data into the tester during a road test and play back this data when the vehicle is returned to the shop.

Some scan testers now display diagnostic information based on the fault code in the computer memory. Service bulletins published by the scan tester manufacturer may be indexed by the tester after the vehicle information is entered in the tester. Other scan testers will display sensor specifications for the vehicle being tested.

In addition to the computer systems listed above, scanners also interface directly with the body control computer. Depending on the scanner and the vehicle's programmed capabilities, the technician could have access to additional testing capabilities for climate controls, entertainment systems, lights, power accessories, the instrument panel, and other systems. Testing the second-generation on-board diagnostic systems (OBD II) is also done with the scanner. OBD II systems can offer enhanced testing capabilities in addition to the more common tests. The scanner processes information at a speed that is controlled primarily by the vehicle's computer. The latest scanners feature not only the conventional display of vehicle data, but can also graphically display one or more parameters. This can be done on the scanner or through software installed on a desktop or laptop computer. This method makes it easier to see many data signals at the same time, either digitally or graphically. Remember that the trouble codes are the result of a problem.

A **breakout box** is connected in series between the vehicle's computer and wiring harness to perform computer circuit tests.

Trouble codes are only set by the vehicle's computer when a voltage signal is entirely out of its normal range or when there is a disagreement between two or more inputs. The codes help technicians identify the cause of the problem when this is the case. If a signal is within its normal range but is sending false information, the vehicle's computer may not display a trouble code. However, a driveability problem will still exist. As an aid to identify this type of problem, most manufacturers recommend that the signals to and from the computer be carefully looked at. Ford, for example, incorporates the use of a **breakout box** (Figure 2–40). A breakout box allows the technician to check voltage and resistance readings between specific points within the computer's wiring harness.

Figure 2-40 The wiring harness breakout box allows for detailed testing of individual circuits.

Oscilloscopes

The **oscilloscope** (Figure 2-41) is still one of the most important diagnostic tools and performs some very valuable tests on today's electronic systems. An oscilloscope converts electrical signals into a visual image representing voltage changes over a specific period of time. This information is displayed on a cathode ray tube (CRT) in the form of a continuous voltage line called a **wave-form** pattern or **trace** (Figure 2-42).

Although there are many designs of scopes, all automotive scopes can be classified into one of two categories: low-voltage or high-voltage. High-voltage scopes are used to monitor the activity of the ignition system and are quite valuable diagnostic tools. Low-voltage scopes are also extremely valuable diagnostic tools. They are used to monitor the activity of the inputs and outputs of a computerized system.

An **oscilloscope** is a type of voltmeter that displays voltage over time graphically with a trace.

The term oscilloscope is usually replaced with the word scope.

A low-voltage scope is normally called a lab scope.

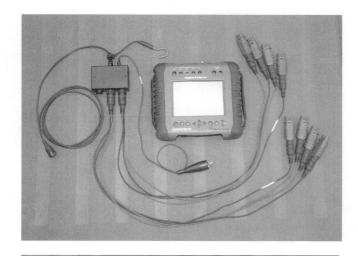

Figure 2-41 A hand-held ignition/lab scope.

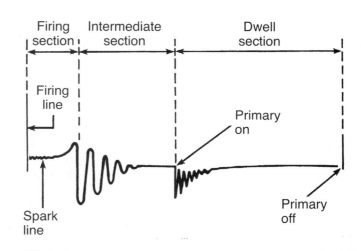

Figure 2-42 Typical oscilloscope waveform pattern.

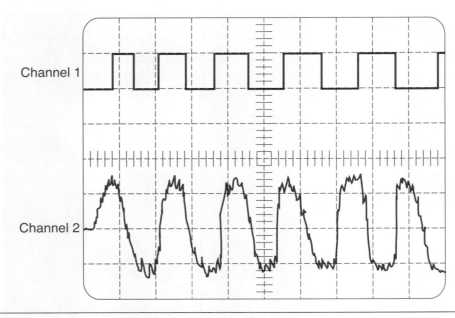

Figure 2-43 Two wave forms from a dual-trace lab scope.

Dual trace oscilloscopes (Figure 2-43) can display two different waveform patterns at the same time. For example, dual-trace oscilloscopes can show how fuel-injector pulse width affects the oxygen sensor voltage signal and other highly useful cause-and-effect and relational electrical tests.

An oscilloscope may be considered as a very fast reacting voltmeter that reads and displays voltages in a system. These voltage readings appear as a voltage trace on the oscilloscope screen.

An upward movement of the voltage trace on an oscilloscope screen indicates an increase in voltage, and a downward movement of this trace represents a decrease in voltage. As the voltage trace moves across an oscilloscope screen, it represents a specific length of time (Figure 2-44).

The size and clarity of the displayed waveform is dependent on the voltage scale and the time reference selected by the operator. Most scopes are equipped with controls that allow voltage and time interval selection. It is important when choosing the scales to remember that a scope displays voltage over time.

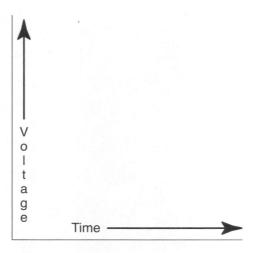

Figure 2-44 Voltage and time axis on a scope's screen.

Exhaust Analyzers

Exhaust analyzers are very valuable diagnostic tools. By looking at the quality of an engine's exhaust, a technician is able to look at the effects of the combustion process. Any defect can cause a change in exhaust quality. The amount and type of change serves as the basis of diagnostic work.

Early emission analyzers measured the amount of hydrocarbons (HC) and carbon monoxide (CO) in the exhaust. Hydrocarbons in the exhaust are raw, unburned fuel. HC emissions indicate that complete combustion is not occurring in the engine. Emissions analyzers measure HC in parts per million (ppm) or grams per mile (g/mi). Carbon monoxide is an odorless, toxic gas that is the product of combustion and is typically caused by a lack of air or excessive fuel. CO is measured as a percent of the total exhaust.

The levels of HC, O_2, and CO_2 in the exhaust are a direct indication of engine performance. A high level of hydrocarbons could indicate a fouled spark plug, a defective spark plug wire, or a burned valve. All of these problems would decrease the effectiveness of combustion and allow fuel to leave the cylinder unburned. A high CO level indicates an excessively rich air-fuel mixture caused by a restriction in the air intake system or too much fuel being delivered to the cylinders.

Many of the emission control devices that have been added to vehicles over the past 30 years have decreased the amount of HC and CO in the exhaust. This is especially true of catalytic converters. These devices alter the contents of the exhaust. Therefore, checking the HC and CO contents in the exhaust may not be a true indication of the operation of an engine.

The manufacturers of exhaust analyzers have altered their machines so that they can look at the efficiency of an engine despite the effectiveness of the emission controls. These machines are four-gas exhaust analyzers. In addition to measuring HC and CO levels, a four-gas exhaust analyzer also monitors carbon dioxide (CO_2) and oxygen (O_2) levels in the exhaust. The latter two gases are changed only slightly by the emission controls and therefore can be used to check engine efficiency. Many exhaust analyzers are now available that measure a fifth gas, oxides of nitrogen (NO_x). In addition to being used in certain emission testing programs, five-gas analyzers (Figure 2-45) can be used to diagnose the following conditions:

❏ Rich or lean mixtures
❏ Catalytic converter malfunction
❏ Faulty injectors
❏ Leaking exhaust gas recirculation (EGR) valves
❏ Leaks or restrictions in the exhaust system
❏ Air pump malfunctions
❏ Blown head gaskets
❏ Intake manifold leaks
❏ Excessive misfire
❏ Excessive spark advance

Exhaust analyzers are often called infrared testers. This is because many analyzers use infrared light to analyze the exhaust gases.

Figure 2-45 Exhaust emission analyzer (5 gas).

Engine Analyzer

When performing a complete engine performance analysis, an engine analyzer is used. An engine analyzer houses all of the necessary test equipment. Although the term "engine analyzer" is often loosely applied to any multipurpose test meter, a complete engine analyzer will incorporate most, if not all, of the test instruments mentioned in this chapter. Most engine analyzers are based on a computer that guides a technician through the tests. Most will also do the work of the following tools:

- ❏ Compression gauge
- ❏ Pressure gauge
- ❏ Vacuum gauge
- ❏ Vacuum pump
- ❏ Tachometer
- ❏ Timing light/probe
- ❏ Voltmeter
- ❏ Ohmmeter
- ❏ Ammeter
- ❏ Oscilloscope
- ❏ Computer scan tool
- ❏ Emissions analyzer

With an engine analyzer, you can perform tests on the battery, starting system, charging system, primary and secondary ignition circuits, electronic control systems, fuel system, emissions system, and the engine assembly. Photo Sequence 1 takes you through the different steps for using an engine analyzer to check out an engine and its systems. The analyzer is connected to these systems by a variety of leads, inductive clamps, probes, and connectors. The data received from these connections is processed by several computers within the analyzer.

Some computerized engine analyzers are programmed with specifications for specific model vehicles. Diagnostic trouble codes have also been loaded into the analyzer's memory. Based on the input from the leads and connectors, the microprocessors can identify worn, misadjusted, or faulty components in all major engine systems. The analyzer will also list the probable causes of specific performance problems and will prompt, or guide, the technician step by step through a troubleshooting procedure designed to verify and correct the problem.

Vehicle information can be entered into the analyzer on a computer-like keyboard. Specifications, commands, and test results are displayed on the CRT screen. Some analyzers will graphically display test results on their CRT screen. The analyzer's printer can also print out copies of the information that appears on the screen (Figure 2-46).

Many engine analyzers will perform a complete series of tests and record the results automatically. The analyzer compares all the test results to the vehicle manufacturer's specifications. When the test series is completed, the analyzer prints a report indicating those readings that were

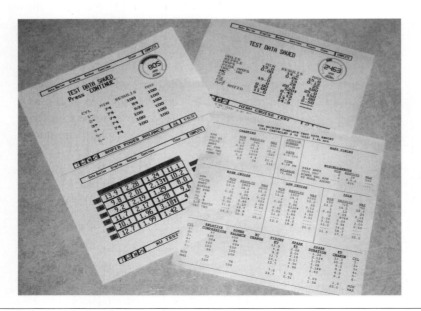

Figure 2-46 Typical engine analyzer's printouts.

Photo Sequence 1
Typical Procedure for Diagnosing Engine, Ignition, Electrical, and Fuel Systems with an Engine Analyzer

P1-1 Connect the analyzer leads and hoses to the engine according to the directions given by the tester's manufacturer.

P1-2 With the engine at normal operating temperature, enter the necessary information in the analyzer regarding the vehicle being tested.

P1-3 Perform a visual inspection of the vehicle according to the menu on the tester. Then enter the results into the analyzer.

P1-4 Perform battery and cranking tests, and observe the results on the screen.

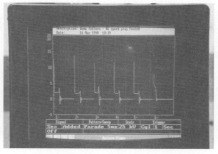

P1-5 Perform primary ignition circuit tests and check secondary kV, and observe the results on the screen.

P1-6 Perform a cylinder performance test and observe the results on the analyzer screen. Look for imbalance problems.

P1-7 Perform a complete exhaust gas analysis and observe the results. Determine if previously noted faults would cause the readings.

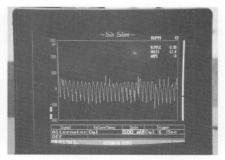

P1-8 Display the charging system's waveform and check the generator's condition.

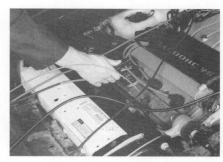

P1-9 Turn the engine off and disconnect the analyzer leads and hoses. Summarize the results, including all that is good and all that needs further testing.

not within specifications. Many analyzers also provide diagnostic assistance for the problems indicated by the readings that were not within specifications.

However, the technician may select any test function, or functions, separately. Engine analyzer test capabilities vary depending on the equipment manufacturer, and technicians must familiarize themselves with the engine analyzer in their shop. For example, some analyzers have oscilloscope patterns and digital ignition readings, and others use digital ignition readings exclusively.

Some engine analyzers have vehicle specifications on diskette and the technician enters the necessary information, such as model year and engine size, for the vehicle being tested. Specifications may be updated simply be obtaining a new diskette from the equipment manufacturer. (See Photo Sequence 1.)

A phone modem or high-speed Internet connection is possible with some engine analyzers to provide networking and communication capabilities. This connectivity allows the technician to transmit all technical reports and pattern reports of a specific problem vehicle to off-location technical support teams or access on-line service information.

Miscellaneous Engine Performance Tools

Oxygen Sensor Service Tool

Most oxygen sensors are difficult to access with an open-end wrench, and the ring prevents the use of a normal deep socket. There are special sockets that have been developed especially for the oxygen sensor as shown in Figure 2-47.

Spark Plug Thread Taps

Spark plug thread taps are meant to repair lightly damaged threads in a cylinder head. They can also help repair damaged O_2 sensor threads. No one wants to use these tools, but sometimes spark plugs seize in the cylinder head if the technician before you was not too careful about how the spark plugs were installed. A set of spark plug taps is shown in Figure 2-48.

Figure 2-47 Oxygen sensor socket.

Figure 2-48 Spark plug thread taps.

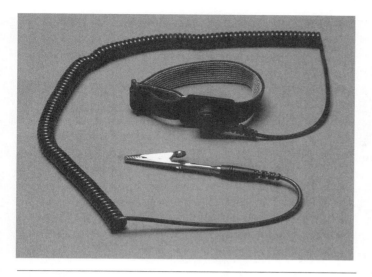

Figure 2-49 A grounding bracelet is attached to a vehicle ground and the technician's wrist to bleed off possible static charges while working on sensitive electronics.

Figure 2-50 Pinch-off pliers.

Static Protection Straps

Sensitive electronic components can be destroyed by static electricity. Many technicians have unknowingly damaged PCM's digital dashes, radios, and control modules. Sometimes the damage is not immediately apparent. In order to avoid static electricity damage, it is wise to wear a static strap. The static strap drains any buildup of electrical charge on your body. A picture of the wrist strap style is shown in Figure 2-49. The grounding clip is connected to a good ground on the vehicle.

Pinch-off Pliers

Pinch-off pliers are meant to be used on the older style air injection reaction system plumbing to close off air to the catalytic converter for testing. They could also be used to close off vacuum lines. Never use these pliers on rubber break lines nor nylon fuel lines since they will be damaged. Figure 2-50 shows a picture of pinch-off pliers.

Service Information

The most important tool of the trade is information. There are many different sources of information that should be available to automotive technicians.

Vehicle Identification

One of the first steps you need to take when servicing any vehicle is to find out exactly what you are working on. The year, make, and model are a good start, but due to the various engines, transmissions, accessories, and mid-year production changes, you need to find out about the *exact* vehicle. The vehicle identification number (VIN) contains all the data about that specific

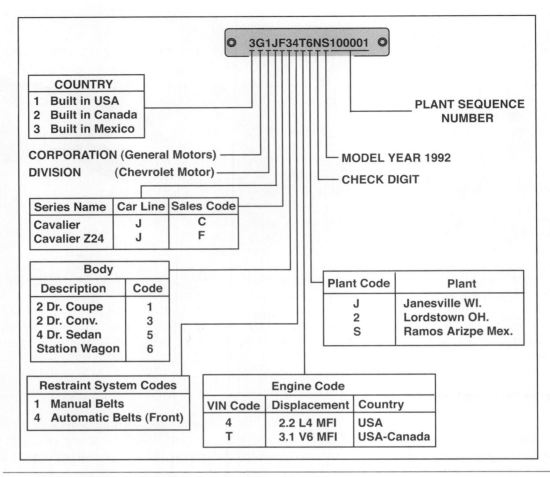

Figure 2-51 Vehicle identification number.

vehicle (Figure 2-51). It indicates the manufacturer, engine size and type, vehicle model information, and the serial number. Most any service information system, be it paper-based or an electronic version, will usually reference at least a portion of the VIN. Other vehicle information can be found in several other places. The owner's manual has a great deal of useful information regarding accessory operation and how to reset memory devices, recommended service schedules, and even part numbers for service parts. The underhood decal (Figure 2-52) also has infor-

Vehicle Emission Control Information - XXXXXXX
Engine Family XXXXXXXXX
Evaporative Family XXXX
Displacement XXXXcm^3
                        CATALYST
Tune-up specifications
 Tune-up conditions:
   Engine at normal operating temperature, all
   accessories turned off, cooling fan off,
   transmission in neutral
Adjustments to be made in accordance with
indications given in shop manual

Idle speed	5 speed transmission	800 50 rpm
	Automatic transmission	800 50 rpm
Ignition timing at idle		15 2 BTDC
Valve lash	Setting points between camshaft and rocker arm	In. 0.17 0.02mm cold
		Ex. 0.19 0.02mm cold
No other adjustments needed		

Figure 2-52 Underhood decal.

mation about the emission control system and perhaps some minor specifications. The driver's door placard contains information about the build date, vehicle weight, tire size, and inflation information. Elsewhere in the vehicle may be an accessory and equipment build code placard. These are not always out in the open where they are readily visible. It contains a summary, by code number, of all the standard equipment and options the vehicle was built with. For example, it will include the actual type of entertainment system, the exact type of springs used, and so on.

Automotive Service Information

Most service information that the manufacturers release is located on the Internet or on DVD ROM and accessed by the technician at the dealership. The format is generally similar to that which the manufacturer previously used in their service manuals. By releasing new information on the Internet, there is no lag between finding new service techniques and implementing them in the service information. DVDs can be released monthly to keep information fresh.

The service information includes specifications and service procedures for all service requirements on the vehicle. Torque specifications, measurements, service techniques, are included in service information, but theory is generally excluded from the information because it is commonly assumed that technicians using the information are experienced.

There are as many types of service information as there are manufacturers, so we will not attempt to describe them in detail. Usually the electronic service information has the technician choose a vehicle from a drop-down menu then proceed to the specific vehicle system in question. The pages of the service manual are usually **hyperlinked**, meaning that relevant information located elsewhere can be quickly brought onto the screen instead of turning through pages in a book (Figure 2-53).

PCM Quick Test

1. Complete preliminary checks

Visual inspection for obvious problems
Electrical connections, vacuum lines, air intake system for leaks and restrictions, quality of fuel and cooling system for proper operation

2. Search the TSB database for any related concerns

3. If the scan tool cannot access the PCM, go to ET1

Were any DTCs present?

Yes	No
If engine runs rough at idle and DTCs are present: GO to Section 4 (Powertrain DTC charts) for direction to service DTCs after noting the following: Any DTCs Retrieve any freeze frame information	GO TO SYMPTOM PRESENT NO DTC SET CHARTS

Figure 2-53 Typical electronic information page showing hyperlinks in blue.

⚠️ **WARNING:** Always follow the service procedures in the vehicle manufacturer's service manual for the vehicle on which you are working. The use of improper service procedures may result in personal injury and/or vehicle damage.

Automotive manufacturers also publish a series of technician reference books. The publications provide general instructions on the service and repair of their vehicles with their recommended techniques.

Aftermarket Suppliers' Guides and Catalogs

Many of the larger parts manufacturers have excellent guides on the various parts they manufacture or supply. They also provide updated service bulletins on their products.

General and Specialty Repair Manuals

These are published by independent or aftermarket companies rather than the manufacturers. However, they pay for and receive most of their information from the car makers. They may contain component information, diagnostic steps, repair procedures, and specifications for several car makes in one book. Information is usually condensed and is more general in nature than the manufacturer's manuals. The condensed format allows for more coverage in less space and therefore is not always specific. They also may contain several years of models as well as several car makes in one book (Figure 2-54).

■ **CAUTION:** Always use specifications in the vehicle manufacturer's service manual or a general service manual. Guessing at specifications may result in damaged automotive components.

Flat-Rate Manuals

Flat-rate manuals contain figures relating to the length of time a specific repair is supposed to require. Normally they also contain a parts list with approximate or exact parts prices. They are excellent for making cost estimates and are published by the manufacturers and independents.

General repair manuals may be referred to as specialty manuals or by the publisher of the manual: Mitchell, Motor, Chilton, and so on.

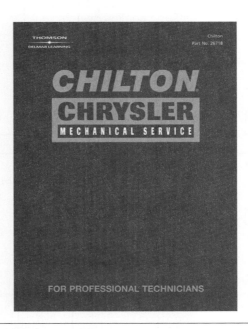

Figure 2-54 Aftermarket service manual. This particular manual is dedicated to Ford vehicles.

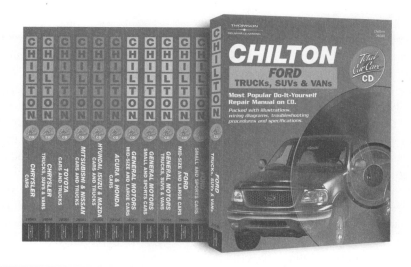

Figure 2-55 Examples of CDs for an automotive electronic data system.

Electronic Data Systems

Electronic data systems replace publications such as service manuals, flat-rate manuals, and service bulletins. In an electronic data system, the information is contained on **compact discs (CD-ROMs)**, or digital video discs (DVDs). A typical electronic data system contains flat rate information, parts information, service bulletins, and service information. Can you imagine the number of pages required to publish this information for ten model years on domestic vehicles and most imported vehicles? Some electronic data systems contain 120,000 pages of information on four CDs (Figure 2-55) and up to six times that on a DVD!

The service bulletins in an electronic data system cover all automotive systems for ten model years, including the current model year. Domestic and imported vehicle information is included in these systems. Much of the service information is in the form of diagnostic charts. Most electronic information systems are available on a subscription basis to ensure timely and regular updates.

The shop must have a CD drive and a compatible computer to display the electronic data system information. Information from the CD may be printed out by the printer connected to the computer. The technician simply enters the make of vehicle, model year, type of engine, and any other requested information to obtain data on a particular vehicle. Some electronic data systems allow the shop to order parts electronically from a supplier with a compatible interconnected computer.

One great advantage of this system is that finding information is easier and quicker. The disks are normally updated monthly and not only contain the most recent service bulletins but also engineering and field service fixes. Other sources for up-to-date technical information are trade magazines and trade associations.

Hotline Services

There are many companies that provide on-line help to technicians. As the complexity of the automobile grows, so do the popularity of these services. Two of the most commonly used hotlines are the Delphi TechSource Assistance Center and Identifix Telediagnosis, which have ASE-certified specialists available for inquiries. These experts are technicians familiar with the different systems of certain manufacturers. Armed with various factory service manuals, general service manuals, service bulletins, and electronic data sources, these brand experts give information to technicians to help them through diagnostic and repair procedures.

Engine Tune-Up

The term engine tune up is often misunderstood. Even to automotive technicians, the words "engine tune up" do not always mean the same thing. The tune-up procedure may vary depending on the make of vehicle, type of test equipment in the shop, and the shop policy. Many customers may not fully understand what is done to their vehicle during an engine tune up.

In general terms, the purpose of an engine tune up is to restore or maintain the original performance and economy of the engine. When an engine tune up is performed, the customer and the technician are basically concerned about obtaining satisfactory performance and economy from the engine. The technician and the customer should understand the limitations of an engine tune up. After an engine tune up, they can only expect to obtain the original performance and economy designed into the vehicle by the manufacturer. Even the original performance and economy may be difficult to obtain because of normal wear on engine components. In some cases, the customer has a performance or an economy problem, and the technician performs an engine tune up to restore the original performance and economy of the engine. Some customers may have an engine tune up performed as preventive maintenance. In this situation, their vehicle operation is satisfactory, but they have a tune up performed to maintain the engine in good running condition.

During the 1980s and 1990s, an electronics revolution had occurred in the automotive industry, and the pace of this revolution continues to accelerate. Automotive manufacturers can now design and build a new model in less than 3 years. A few years ago, the manufacturers required about 7 years to complete the same job. This accelerated production speed is possible because of computer-assisted engineering and the team concept in which all the engineers, including those from suppliers, work as a project team. Vehicle manufacturers are now able to introduce more new and revised models in a shorter time period, and each of these new models is equipped with some innovative electronic components and systems.

This revolution in automotive engineering and manufacturing has brought about a simultaneous revolution in the automotive service industry, especially in the tune-up area. In the 1960s, the average recommended spark plug replacement interval was 10,000 miles (16,000 km). The use of unleaded fuel and the more accurate air-fuel mixture supplied by electronic fuel injection systems have greatly reduced spark plug carbon deposits. Now the average spark plug replacement interval is 30,000 miles (48,000 km). More recently, new ignition system designs and the use of platinum-tipped spark plugs have again increased the service intervals anywhere from 60,000 miles to 100,000 miles. Of course, problems with components still can occur, and when they do, service is needed immediately.

There is no standard, universal tune-up procedure. As mentioned previously, tune-up procedures vary for a number of reasons. Many shops advertise a special price for a specific tune-up procedure, and the technicians in the shop follow the advertised procedure. If the technicians perform more tests or services than advertised, they take longer to complete the job, and profit margins are reduced. Therefore, the technicians are restricted to the advertised procedure, and any additional service must be approved by the customer and considered a separate operation. The average tune up includes all or some of these tests and services:

1. Road test the vehicle if necessary to verify certain problems.
2. Inspect hoses and belts and replace as necessary.
3. Check all fluid levels.
4. Check cooling fan operation.
5. Inspect the general condition of vacuum hoses and electrical wires.
6. Clean battery terminals, and check cable condition.
7. Perform battery load test and perform battery hydrometer test if possible; determine battery condition.

8. Check alternator maximum output and normal voltage setting; determine charging system condition.

9. Check starting motor current draw; determine starting motor condition.

10. Perform compression test or cylinder power balance test.

11. Replace spark plugs.

12. Test spark plug wires and replace as required.

13. Test and inspect distributor cap and rotor and replace as necessary (distributor-type ignition system).

14. Test maximum coil voltage; replace coil if necessary.

15. Check ignition timing; adjust if necessary.

16. Test fuel pump pressure and replace fuel filter.

17. Check and replace air filter.

18. Check heated air inlet system operation.

19. Check injector balance.

20. Inspect fuel system for leaks.

21. Clean injectors and intake system; adjust throttle plate angle.

22. Check emission devices such as PCV, EGR, and canister purge systems for proper operation; check exhaust emissions; check catalytic converter.

23. Check engine computer system for fault codes.

24. Road test the vehicle to be sure there are no performance problems.

CASE STUDY

The word "diagnosis" is commonly used in automotive textbooks and service manuals. However, it is a term that is seldom explained. It is more than following a series of interrelated steps in order to find the solution to a specific condition. Diagnosis is a way of looking at systems that are not functioning the way they should and finding out why. It is knowing how the system should work and deciding whether or not it is working correctly. Through an understanding of the purpose and operation of the car's system, a technician can accurately diagnose problems.

Most good diagnosticians use the same basic diagnostic procedure. Because of its logical approach, this procedure can quickly lead a technician to the cause of a problem. Accurate diagnostics include the following steps:

1. Gather information about the condition or problem.

2. Verify that the condition exists.

3. Thoroughly define what the problem is and when it occurs.

4. Determine the possible causes of the problem.

5. Isolate the problem by general testing.

6. Test to pinpoint the cause of the problem.

7. Repair the problem and verify the repair.

Most service manuals contain diagnostic charts that help technicians pinpoint the cause of the problem. The steps of diagnostic charts or trees are designed to be followed in order. These charts contain the most probable causes of a particular problem and simple tests that lead to the exact cause.

Terms to Know

After top dead center (ATDC)	Feeler gauge	Pounds per square inch (psi)
Ammeter	Flat-rate manuals	Pressure gauge
Before top dead center (BTDC)	High input impedance	Scan tool
Breakout box	Hyperlinked	Tachometer
Compact discs (CD-ROMs)	International System (SI)	Trace
Compression gauge	Kilopascals (kPa)	United States Customary system (USC)
Continuity tester	Multimeter	Vacuum
Cylinder leakage	Noid light	Volt/ampere test (VAT)
Digital multimeter (DMM)	Ohmmeter	Voltmeter
Dual trace oscilloscope	Oscilloscope	Waveform
Dwell meter		

ASE-Style Review Questions

1. While discussing infrared analyzers:
   Technician A says hydrocarbon (HC) emissions are affected by the catalytic converter.
   Technician B says carbon monoxide (CO) emissions are affected by the catalytic converter.
   Who is correct?
   A. A only **C.** Both A and B
   B. B only **D.** Neither A nor B

2. While discussing engine analyzers:
   Technician A says some engine analyzers use vehicle specifications contained on diskette.
   Technician B says some engine analyzers have a phone modem that provides networking with off-location technical support groups.
   Who is correct?
   A. A only **C.** Both A and B
   B. B only **D.** Neither A nor B

3. While discussing injector balance testers:
   Technician A says the injector balance tester contains a polarity sensor.
   Technician B says the injector balance tester contains a timing circuit.
   Who is correct?
   A. A only **C.** Both A and B
   B. B only **D.** Neither A nor B

4. While discussing injector cleaning:
   Technician A says the fuel return line must be blocked during the injector cleaning process on some vehicles.
   Technician B says the electric fuel pump must be disabled during the injector cleaning process.
   Who is correct?
   A. A only **C.** Both A and B
   B. B only **D.** Neither A nor B

5. While discussing circuit testers:
   Technician A says a conventional 12-volt test light may be used to diagnose automotive computer circuits.
   Technician B says a self-powered test light may be used to diagnose an air bag circuit.
   Who is correct?
   A. A only **C.** Both A and B
   B. B only **D.** Neither A nor B

6. While discussing oscilloscopes:
   Technician A says the upward voltage traces on an oscilloscope screen indicate a specific length of time.
   Technician B says the cathode ray tube (CRT) in an oscilloscope is like a very fast reacting voltmeter.
   Who is correct?
   A. A only **C.** Both A and B
   B. B only **D.** Neither A nor B

7. While discussing engine tune-up purpose:
   Technician A says the purpose of an engine tune up is to improve the vehicle manufacturer's original performance and economy.
   Technician B says in some cases the purpose of an engine tune-up is to maintain the engine in satisfactory running condition and prevent engine performance and economy problems in future driving miles.
   Who is correct?
   A. A only **C.** Both A and B
   B. B only **D.** Neither A nor B

8. While discussing general service manuals:
Technician A says general service manuals usually include information on several model years.
Technician B says general service manuals are usually more detailed and extensive compared to vehicle manufacturers' service manuals.
Who is correct?

A. A only   **C.** Both A and B
B. B only   **D.** Neither A nor B

9. While discussing electronic data systems:
Technician A says an electronic data system contains information only on vehicles in the current model year.
Technician B says an electronic data system may contain flat rate information, parts information, service bulletins, and service information on compact discs (CDs).
Who is correct?

A. A only   **C.** Both A and B
B. B only   **D.** Neither A nor B

10. *Technician A* says that a push-in compression gauge must be held tightly in the spark plug hole in order to achieve an accurate reading.
Technician B says the gauge will read a vacuum when the cylinder is on the compression stroke.
Who is correct?

A. A only   **C.** Both A and B
B. B only   **D.** Neither A nor B

Job Sheet 3

3

Name _____ Date _____

Use of a Voltmeter

Upon completion of this job sheet, you should be able to measure available voltage and voltage drop.

ASE Correlation

This job sheet is related to the ASE Engine Performance Test's content area: computerized engine controls diagnosis and repair; task: obtain and interpret digital multimeter readings; and the ASE Electrical/Electronic System Test's content area: general electrical system diagnosis; task: check voltage and voltage drop in electrical/electronic circuits using a digital multimeter; determine needed repairs.

Tools and Materials

A vehicle
A DMM
Wiring diagram for vehicle
Basic hand tools

Procedures

Task Completed

1. Set the DMM to the appropriate scale to read 12 volts DC. ☐

2. Connect the meter across the battery (positive to positive and negative to negative).
   What is your reading on the meter? _____ volts

3. With the meter still connected across the battery, turn the vehicle's headlights on.
   What is your reading on the meter? _____ volts

4. Keep the headlights on.
   Connect the positive lead of the meter to the point on the vehicle where the battery's ground cable attaches to the frame. Keep the negative lead where it is.
   What is your reading on the meter? _____ volts
   What is being measured? _____

5. Disconnect the meter from the battery and turn off the headlights. ☐

6. Refer to the correct wiring diagram and determine what wire at the right headlight delivers current to the lamp when the headlights are on and low beams selected.
   Color of the wire _____

7. From the wiring diagram, identify where the headlight is grounded.
   Place of ground _____

8. Connect the negative lead of the meter to the point where the headlight is grounded. ☐

9. Connect the positive lead of the meter to the power input of the headlight. ☐

10. Turn the headlights on.

What is your reading on the meter? _____ volts

What is being measured? _____

11. What is the difference between the reading here and the battery's voltage?

_____ volts

12. Explain why there is a difference.

_____

_____

_____

_____

Instructor's Response _____

_____

_____

Job Sheet 4

4

Name _____ Date _____

Use of an Ohmmeter

Upon completion of this job sheet, you should be able to check continuity of a circuit and measure resistance on a variety of components.

ASE Correlation

This job sheet is related to the ASE Engine Performance Test's content area: computerized engine controls diagnosis and repair, task: obtain and interpret digital multimeter readings; and the ASE Electrical/Electronic System Test's content area: general electrical system diagnosis; task: check voltage and voltage drop in electrical/electronic circuits using a digital multimeter; determine needed repairs.

Tools and Materials

A vehicle
A DMM
Wiring diagram for the vehicle

NOTE: An ohmmeter works by sending a small amount of current through the path to be measured. Because of this, all circuits and components being tested must be disconnected from power. An ohmmeter must never be connected to an energized circuit; doing so may damage the meter. The safest way to measure ohms is to disconnect the negative battery cable before taking resistance readings.

Procedures

Task Completed

1. Locate the fuse panel or power distribution box.

2. With no power to the fuses, check the resistance of each fuse. Summarize your findings: ☐

   _____

   _____

   _____

   _____

   _____

3. Connect the leads of the digital meter across the negative battery cable.

   Your reading is: _____ ohms

4. Disconnect the wires leading to the ignition coil.

   Connect the leads of the digital meter across the terminals of the coil.

   Your reading is: _____ ohms

5. Reconnect the wires to the coil.

   Carefully remove one spark plug wire from the spark plug and the ignition coil or distributor cap.

   Connect the leads of the digital meter across the wire.

   Your reading is: _____ ohms

6. Carefully reinstall the spark plug wire. Locate the cigar lighter inside the vehicle.

Connect the leads of the digital meter from the heating coil to its case.

Your reading is: _____ ohms

7. Refer to the service manual and find out how to remove the bulb in the dome light. Remove it.

Connect the leads of the digital meter across the bulb.

Your reading is: _____ ohms

8. Reinstall the bulb. Remove the rear brake light bulb.

Connect the leads of the digital meter across the bulb.

Your reading is: _____ ohms

9. Reinstall the bulb. Disconnect the wire connector to one of the headlights. From the wiring diagram, identify which terminals are for low beam operation.

Connect the leads of the digital meter across the low beam terminals.

Your reading is: _____ ohms

10. From the wiring diagram, identify which terminals are for high beam operation.

Connect the leads of the digital meter across the high beam terminals.

Your reading is: _____ ohms

11. You measured the resistance across several different light bulbs. On each you should have read a different amount of resistance. Based on your findings, which light bulb would be the brightest and which would be the dimmest? Explain why.

_____

_____

_____

_____

Instructor's Response

_____

_____

General Engine Condition Diagnosis

Upon completion and review of this chapter, you should be able to:

❏ List the steps in a general diagnostic procedure that may be used in any diagnostic situation.

❏ Diagnose fuel leaks and determine needed repairs.

❏ Diagnose engine oil leaks and determine needed repairs.

❏ Diagnose engine coolant leaks and determine necessary repairs.

❏ Diagnose engine exhaust odor, color, and noise and determine needed repairs.

❏ Diagnose engine noises and vibration problems and determine the necessary repairs.

❏ Test engine oil pressure and determine the cause of low oil pressure.

❏ Diagnose engine overheating problems.

❏ Pressure test the cooling system.

❏ Diagnose engine defects from intake manifold vacuum readings.

❏ Perform an engine power balance test and determine the needed repairs.

❏ Diagnose engine problems with an exhaust gas analyzer.

❏ Perform an engine compression test and determine the necessary repairs.

❏ Perform a cylinder leakage test and determine the needed repairs.

❏ Perform valve adjustments on mechanical and hydraulic lifters.

❏ Check valve timing.

Basic Tools

Basic tool set, appropriate service manuals, fender covers, safety glasses

General Diagnostic Procedure

As you will probably continue to witness, automotive technology will become more complex with every model year. An automotive service technician's skill and knowledge must grow with these changes. The fundamentals of effective diagnostic skills become increasingly important.

   The purpose of this chapter is to give you the skills to test the engine. If we begin diagnosing a driveability problem with the assumption that the engine is fine when it is not fine, we will waste a great deal of time and will probably become frustrated. The key to diagnostics is to know what tests to conduct, and when to conduct them. To know this, you must understand the system and test.

Engine Leak Diagnosis

Engine leaks not only cause unsightly messes on the engine, engine compartment, and driveway, they can also be an indication of a bigger problem. This big problem can cause safety, driveability, and durability concerns. It is extremely important that you are able to identify the type of fluid that is leaving the mess and the possible causes for the leakage. On computer-controlled engines, leaks can and do cause driveability problems.

   ⚠ **WARNING:** Never attempt to repair a metal fuel tank until it is drained, removed, and steamed out for at least 30 minutes. Gasoline fumes left in gasoline tanks are extremely dangerous. When ignited, these fumes will cause a very serious explosion and fire, resulting in personal injury and property damage.

WARNING: Always store gasoline drained from a fuel tank in approved gasoline safety cans.

WARNING: While removing and replacing a gasoline tank and handling gasoline, be sure there are no sources of ignition in the area. Do not smoke while performing these operations, and be sure nobody else smokes in the area. Do not drag the gasoline tank on the floor.

WARNING: If gasoline is spilled on the shop floor, place approved absorbent material on it immediately. Dispose of the absorbent material in approved waste containers.

Fuel Leaks

An exhaust gas analyzer can be used to find the area of a fuel leak. Slowly move the probe along the fuel lines. A high HC reading indicates the area of the leak.

Engine fuel leaks are expensive and dangerous, and they should be corrected immediately when they are detected. Fuel leaks are expensive because they waste fuel. Since gasoline is very explosive; fuel leaks are extremely dangerous because any spark near a fuel leak may start a fire or cause an explosion. If gasoline odor occurs inside or near a vehicle, it should be inspected for fuel leaks immediately. To locate a fuel leak, inspect these fuel system components:

1. Fuel tank. If a fuel tank is leaking, it must be removed and repaired. Many fuel tanks do not have a drain plug, so the gasoline must be pumped from the tank with a hand-operated pump.
2. Fuel tank filler cap.
3. Fuel lines and filter.
4. Vapor recovery system lines.
5. Carburetor.
6. Pressure regulator, fuel rail, and injectors (fuel injected engines).

Engine Oil Leak Diagnosis

Classroom Manual
Chapter 3, page 62

Engine oil leaks may cause an engine to run out of oil, resulting in serious engine damage. Therefore, oil leaks should not be ignored. If an engine has an oil leak, the oil level should be checked often and oil should be added as required until the necessary repairs are completed. When it is difficult to locate the exact cause of an oil leak, the engine should be cleaned and then operated while the technician watches for the source of the oil leak. Oil leaks may be difficult to locate because the oil runs down the engine surfaces. For example, the oil from a leak at a rocker arm cover may run down the side of the engine and appear as an oil pan leak. However, closer examination of the engine will indicate that the oil leak is coming from the rocker arm cover. Another method used to detect oil leaks is to add a special dye to the crankcase. The dye may be easier to see because it is colored. Another type of dye that is available for oil leak detection can be easily seen with a black light. It may be necessary to run the vehicle for a period of time, perhaps having the customer return in a few days, to allow the leak to show up adequately. Whenever any additive is used, be sure to follow the manufacturer's instructions and verify that its use will not interfere with any vehicle warranties. The possible causes of oil leaks are:

1. Rear main crankshaft bearing seal
2. Expansion plug in rear camshaft bearing
3. Rear oil gallery plug
4. Oil pan
5. Oil filter
6. Rocker arm covers
7. Intake manifold front and rear gaskets (V-type engines)

8. Mechanical fuel pump gasket or worn fuel pump pivot pin
9. Timing gear cover or seal
10. Front main bearing
11. Oil pressure sending unit
12. Distributor O-ring or gasket
13. Engine casting porosity
14. Oil cooler or lines (where applicable)

Engine Coolant Leak Diagnosis

When a coolant leak causes low coolant level, the engine quickly overheats and severe engine damage may occur. Therefore, coolant leaks must not be ignored. Coolant leaks may occur at these locations:

1. Upper radiator hose
2. Lower radiator hose
3. Heater hoses
4. Bypass hose
5. Water pump
6. Engine expansion plugs or block heater
7. Radiator
8. Thermostat housing
9. Heater core
10. Internal engine leaks caused by leaking intake manifold gasket, head gasket, or cracked engine components

Classroom Manual
Chapter 3, page 52

▲ **WARNING:** Never loosen a radiator cap on a warm or hot engine. When the cap is loosened, the pressure on the coolant is released and the coolant suddenly boils. This action may cause a technician or anyone standing nearby to be seriously burned.

A pressure tester may be used to determine if a coolant leak is present and to locate the source of the leak. Remove the radiator cap and install the pressure tester on the radiator filler neck (Figure 3-1). Operate the pump on the tester until the rated pressure on the radiator cap appears on the tester gauge. Leave this pressure applied to the cooling system for several minutes, then check the gauge pressure. If there is no drop in gauge pressure during this time, the cooling system does not have a leak. If the pressure drops, the cooling system has a leak. Normally the source of the leak is evident by coolant spraying out of the leak. Inspect the entire

Special Tools

Cooling system pressure tester

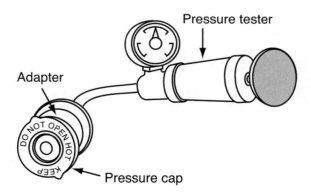

Pressure tester
Adapter
Pressure cap

Figure 3-1 Cooling system pressure tester.

cooling system, and repair any leaks as required. Check the floor of the vehicle for any indication of coolant leaking out of the heater core and dripping onto the floor. Heater core leaks usually cause an odor inside the passenger compartment and fogging of the windshield.

The radiator pressure cap may be connected to the cooling system pressure tester with a special adapter. When the tester pump is operated, the gauge on the tester indicates the pressure required to open the cap pressure relief valve. This pressure should be the same as the pressure rating stamped on the cap.

If the pressure drops but no external leaks are evident, the spark plugs should be removed to check for coolant in the cylinders. After the spark plugs are removed, disable the injection and ignition systems. Crank the engine while checking for any sign of coolant discharging from the spark plug openings. If coolant is discharged from a spark plug opening, the head gasket, cylinder head, or cylinder wall is leaking or cracked.

When there are coolant leaks in the combustion chamber, bubbles may appear in the coolant at the radiator filler neck with the engine running. A two-gas or four-gas infrared exhaust gas analyzer can be used to check for coolant leaks in the combustion chamber. With the radiator cap removed, place the sample probe of the exhaust analyzer at the top of the radiator filler neck (Figure 3-2) or at the top of the coolant recovery tank opening if the radiator has no cap. Use extreme caution when opening the radiator cap, especially if the engine is already hot. If there is pressure in the system, allow it to cool down before doing this procedure. Be certain the sample probe does not come in contact with any liquid by inadvertently immersing it in the coolant. The analyzer can be damaged if coolant is sucked into the unit. Once the probe is in place, run the engine at 1,500 to 2,000 rpm and observe the readings. Allow a few seconds for the sample to be read and displayed on the unit. If you are using a two-gas infrared, look for hydrocarbons (HC) to show a reading. You will notice an increase in HC if there is a combustion leak. If you have a four-gas unit, look for an increase in carbon dioxide (CO_2). CO_2 is produced as a result of combustion, and if it is present in the cooling system, it is there because of an internal combustion chamber leak.

If the coolant level continues to go down with no indication of external leaks, the coolant may be leaking through a cracked block into the oil pan. If this type of leak is suspected, check the dipstick first. Drain the engine oil and check for signs of coolant in the oil if further inspection is needed.

Special Tools

Exhaust gas analyzer

The hydrocarbon (HC) meter on an exhaust gas analyzer may be placed above the radiator filler neck with the engine running to check for a combustion chamber leak.

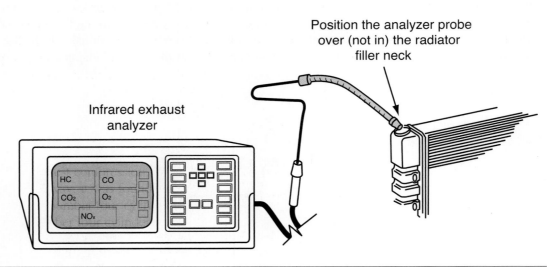

Figure 3-2 Testing for combustion chamber leaks with an exhaust gas analyzer.

Engine Noise Diagnosis

There are usually warning noises from the engine long before a serious engine failure occurs. When engine problems are detected early, the engine repairs may be much less expensive. Engine defects such as damaged pistons, worn rings, loose piston pins, worn crankshaft bearings, worn camshaft lobes, and loose and worn valve train components usually produce their own peculiar noises. Certain engine defects also cause a noise under specific engine operating conditions. A technician must be able to diagnose engine noises accurately to avoid unnecessary engine repairs.

Since it is sometimes difficult to determine the exact location of an engine noise, a stethoscope may be useful in diagnosing noise locations. The stethoscope probe is placed on or near the suspected component, and the ends of the stethoscope are installed in your ears. The stethoscope amplifies sound to assist in noise location. When the stethoscope probe is moved closer to the source of the noise, the sound is louder in your ears. If a stethoscope is not available, a length of hose placed from the suspected component to your ear amplifies sound and helps to locate the cause of the noise.

Special Tools

Stethoscope

> ⚠️ **WARNING:** When placing a stethoscope probe in various locations on a running engine, be careful not to catch the probe or your hands in moving components such as cooling fan blades and belts.

Since lack of lubrication is a common cause of engine noise, always check the engine oil level and condition prior to noise diagnosis. Carefully observe the oil for contamination by coolant or gasoline. During the diagnosis of engine noises, always operate the engine under the conditions when the noise occurs. Remember that aluminum engine components such as pistons expand more when heated than cast iron alloy components. Therefore, a noise caused by a piston defect may occur when the engine is cold but disappear when the engine reaches normal operating temperature.

Main Bearing Noise

Loose crankshaft main bearings cause a heavy thumping knock for a brief time when the engine is first started after it has been shut off for several hours. This noise may also be noticeable on hard acceleration depending on the amount of bearing wear. Worn main bearings or crankshaft journals or a lack of lubrication cause this problem. When the thrust surfaces of a crankshaft bearing are worn, the crankshaft has excessive endplay. Under this condition, a heavy thumping knock may occur at irregular intervals on acceleration. The main bearings must be replaced to correct this problem, and the crankshaft main bearing journals may require machining. If the journals are severely scored or burned, the crankshaft must be replaced.

Connecting Rod Bearing Noise

Loose connecting rod bearings cause a lighter, rapping noise at speeds above 35 mph (21 kph). The noise from a loose connecting rod bearing may vary from a light to a heavier rapping sound depending on the amount of bearing looseness. Connecting rod bearing noise may occur at idle if the bearing is very loose. When the spark plug is shorted out in the cylinder with the loose connecting rod bearing, the noise usually decreases. Worn connecting rod bearings or crankshaft journals or a lack of lubrication may cause this problem. To correct this noise, the connecting rod bearings require replacement, and the crankshaft journals should be machined. Severely worn or burned crankshaft journals require crankshaft replacement. The connecting rod alignment should also be checked.

Piston Slap

A piston slap causes a hollow, rapping noise that is most noticeable on acceleration when the engine is cold. Depending on the piston condition, the noise may disappear when the engine reaches normal operating temperature. A collapsed piston skirt, cracked piston, excessive piston-to-cylinder wall clearance, lack of lubrication, or a misaligned connecting rod are causes of piston slap. Correction of this problem requires piston or connecting rod replacement and possible reboring of the cylinder.

Piston Pin Noise

A loose, worn piston pin causes a sharp, metallic rapping noise that occurs with the engine idling. This noise may be caused by a worn piston pin, worn pin bores, cracked piston in the pin bore area, a worn rod bushing, or a lack of lubrication. An oversized pin may be installed to correct the problem. Severely worn or cracked pistons must be replaced.

Piston Ring Noise

If the piston rings are excessively loose in the piston ring grooves, a high-pitched clicking noise is noticeable in the upper cylinder area during acceleration. To correct this noise, the rings, and possibly the pistons, must be replaced. If the cylinders are severely worn, cylinder reboring is necessary.

Ring Ridge Noise

When the top piston rings are striking the ring ridge on the cylinder wall (Figure 3-3), a high-pitched clicking noise is heard. This noise intensifies when the engine is accelerated. Ring ridge noise may be caused by the installation of new rings without removing the ring ridge at the top of the cylinder. This noise may also be caused by a loose connecting rod bearing or piston pin, which allows the piston to move above its normal travel in the cylinder. To correct this problem, remove the ring ridge and check the connecting rod bearing and piston pin.

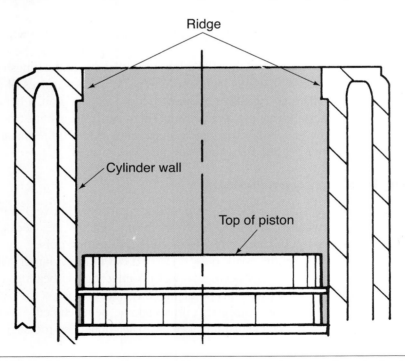

Figure 3-3 When the piston strikes the ridge at the top of the cylinder, a high-pitched rapping or clicking sound is heard.

Valve Train Noise

Valve train noise, such as valve lifters or rocker arms, appears as a light clicking noise when the engine is idling. Valve train noise is slower than piston or connecting rod noise because the camshaft is turning at one-half the crankshaft speed. This noise is less noticeable when the engine is accelerated. Sticking or worn valve lifters may provide an irregular clicking noise that is more likely to occur when the engine is first started. Valve train noise may be caused by a lack of lubrication or contaminated oil. To correct valve train noise, check the valve adjustment where applicable, and check the valve lifters, pushrods, and rocker arms.

Camshaft Noise

If one lobe on a camshaft is worn, a heavy clicking noise is heard with the engine running at 2,000 to 3,000 rpm. When several camshaft lobes and lifter bottoms are scored, a continuous, heavy clicking noise is evident at idle speed. To correct this problem, the camshaft must be machined or replaced, and the valve lifter condition should be checked. On occasion, it is possible for a worn camshaft to make little or no noise. The symptom will most likely be a cylinder miss and possibly a popping noise while accelerating. Depending on the valve system design and adjustment method, a worn camshaft lobe may no longer come in contact with the lifter or follower. Therefore, no audible valvetrain noise will be detected.

A worn, loose timing chain, sprockets, or chain tensioners causes a whirring and light rattling noise when the engine is accelerated and decelerated. Severely worn timing chains, sprockets, and tensioners may cause these noises at idle speed. The timing chain, sprockets, and tensioners must be checked and replaced as necessary to correct this noise.

Combustion Noises

The most common abnormal combustion noise is caused by detonation in the cylinders. Detonation occurs in a cylinder when the air-fuel mixture suddenly explodes rather than burning smoothly. This action drives the piston against the cylinder wall and suddenly forces the connecting rod insert against the crankshaft journal. Under this condition, a **pinging** noise occurs that is similar to marbles rattling inside a metal can. The pinging noise usually occurs when the engine is accelerated. Detonation may be caused by excessive ignition spark advance, low octane fuel, higher-than-normal engine temperature, and hot carbon spots on top of the piston. A lean air-fuel mixture or an inoperative exhaust gas recirculation (EGR) valve contributes to detonation. Detonation at a steady cruising speed may be caused by a defective EGR valve or related controls. To correct the detonation problem, check the ignition timing and spark advance, air-fuel mixture, engine compression, engine temperature, and EGR valve.

Flywheel and Vibration Damper Noise

A loose flywheel or cracked flexplate causes a thumping noise at the back of the engine. This noise varies depending on whether the engine is equipped with a flywheel or a flexplate. The noise also varies depending on the looseness of the flywheel. To correct this problem requires transmission removal and tightening of the flywheel-to-crankshaft bolts. If the bolt holes in the flywheel are damaged, flywheel replacement is necessary.

A loose vibration damper on the front of the crankshaft causes a rumbling or thumping noise at the front of the engine, and this noise may be accompanied by engine vibrations. When the engine is accelerated under load, the noise is more noticeable. The vibration damper must be replaced to correct this problem.

Engine Exhaust Diagnosis

> ⚠ **WARNING:** Vehicle exhaust contains carbon monoxide (CO), which is a poisonous gas. Breathing CO results in health problems, and a concentrated mixture of CO and air is fatal! Do not operate a vehicle in the shop unless a shop exhaust hose is connected to the tailpipe.

Some engine problems may be diagnosed by the color, smell, or sound of the exhaust. If the engine is operating normally, the exhaust should be colorless. In severely cold weather, it is normal to see a swirl of white vapor coming from the tailpipe, especially when the engine and exhaust system are cold. This vapor is moisture in the exhaust, a normal byproduct of the combustion process.

Blue exhaust indicates excessive amounts of engine oil entering the combustion chambers.

If the exhaust is blue, excessive amounts of oil are entering the combustion chamber and are being burned with the fuel. When the blue smoke in the exhaust is more noticeable on deceleration, the oil is likely getting past the rings into the cylinder. Vacuum in the cylinder is high on deceleration, and this high vacuum is pulling oil past the rings into the combustion chamber. If the blue smoke appears in the exhaust immediately after a hot engine is restarted, the oil is likely leaking down the valve guides. In this case, the seals, guides, or valve stems may be worn.

Black exhaust is caused by a rich air-fuel mixture.

If black smoke appears in the exhaust, the air-fuel mixture is too rich. This condition may be caused by a defect in the fuel injection system. In a fuel injection system, a defective pressure regulator, injectors, or input sensors may cause a rich air-fuel mixture. A restriction in the air intake, such as a plugged air filter, may be responsible for a rich air-fuel mixture. The evaporative emission controls may also be at fault.

If gray smoke is present in the exhaust immediately after the engine is started, coolant is probably entering the combustion chamber.

Gray smoke in the exhaust may be caused by a coolant leak into the combustion chambers. If a coolant leak is the cause of gray smoke in the exhaust, this smoke is usually most noticeable when the engine is first started after it has been shut off for over half an hour.

On catalytic converter-equipped vehicles, a strong sulphur smell in the exhaust indicates a rich air-fuel mixture. Some sulphur smell on these engines is normal, especially during engine warmup. When a vehicle is not equipped with a catalytic converter, excessive odor in the exhaust usually indicates a rich air-fuel mixture. You may also notice this odor causes your eyes to water.

Exhaust Noise

> ⚠ **WARNING:** Exhaust components may be extremely hot. If you must touch them, wear recommended protective gloves to avoid burns.

When the engine is idling, the exhaust at the tailpipe should have a smooth, even sound. If the exhaust in the tailpipe has a puff sound at regular intervals, a cylinder may be misfiring. When this sound is present, check the engine's ignition and fuel systems and the engine's compression.

If the vehicle has excessive exhaust noise, check the exhaust system for leaks. The noise from exhaust leaks is most noticeable when the engine is accelerated. A small exhaust leak may cause a whistling noise when the engine is accelerated.

If the exhaust system produces a rattling noise when the engine is accelerated, check the muffler and catalytic converter for loose internal components. When this problem is suspected, rap on the muffler and converter with the engine off. If a rattling noise is heard, replace the appropriate component. This rattling noise in the exhaust system may also be caused by loose exhaust system hangers or an exhaust system component hitting the chassis.

When the engine has a wheezing noise at idle or with the engine running at higher rpm, check for a restricted exhaust system. Connect a vacuum gauge to the intake manifold and observe the vacuum with the engine idling. Accelerate the engine to 2,500 rpm and hold this speed for 2 minutes. If the exhaust is restricted, the vacuum will slowly drop below the recorded vacuum at idle. When the car is driven under normal loads, this vacuum drop is even more noticeable.

Diagnosis of Oil Consumption

With advances in piston ring technology, modern engines consume very little oil. These engines may be driven for 1,000 to 3,000 miles (1,600 to 4,800 km) without adding oil to the engine. If an engine uses excessive oil, it may be leaking from the engine or burning in the combustion chambers. A plugged PCV system causes excessive pressure in the engine, and this pressure may force oil from some of the engine gaskets. If excessive amounts of oil are burned in the combustion chambers, the exhaust contains blue smoke and the spark plugs may be fouled with oil. Excessive oil burning in the combustion chambers may be caused by worn rings and cylinders or worn valve guides and valve seals.

● **CUSTOMER CARE:** Always inform the customer if you notice problems, such as engine noises, with his or her car. Most engine failures are evidenced by unusual noises for a considerable length of time prior to the actual failure. If you inform the customer about a noise that may indicate a serious engine problem, the customer will likely consent to further diagnosis of the problem. When the diagnosis is complete, the customer will likely approve the necessary repairs, and these repairs will likely be less expensive than the repairs of a complete engine failure. The alternative procedure is to let the customer drive the car until the engine noise develops into a serious engine failure. This problem may occur when the customer is miles from home. The customer will likely appreciate your diagnosis of the impending engine failure before it develops into a major problem at an inconvenient time.

Engine Oil Pressure Tests

If the low oil pressure warning light or pressure gauge indicates low oil pressure, the engine oil level should be checked immediately. The oil should also be checked for contamination with gasoline or coolant. If the oil is contaminated with gasoline, the oil is thinner and oil pressure is reduced. When the oil level and quality are satisfactory but the oil pressure warning light or gauge indicates low pressure, the oil pressure should be checked. If the engine runs for a few minutes without proper lubrication, the crankshaft bearings and cylinder walls may be severely scored. A vehicle should not be driven if the oil pressure warning light or gauge indicates low oil pressure.

A pressure gauge must be connected to the main oil gallery to check engine oil pressure. The oil pressure sending unit may be removed from the main oil gallery to install the pressure gauge (Figure 3-4). Photo Sequence 2 shows a typical procedure for testing engine oil pressure.

Special Tools

Oil pressure gauge and adapters

Low engine oil pressure may be caused by a worn oil pump, plugged pick-up screen, leaking pick-up pipe, loose main bearings, loose connecting rod bearings, worn camshaft bearings, and worn valve lifter bores.

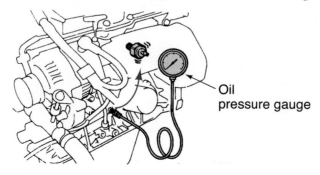

Figure 3-4 Oil pressure test gauge connected to the opening of the oil pressure gauge sending unit.

Oil pressure gauge

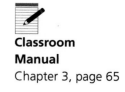

Classroom Manual
Chapter 3, page 65

Photo Sequence 2
Typical Procedure for Testing Engine Oil Pressure

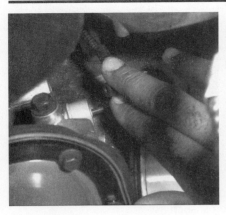

P2-1 Disconnect the wire from the oil sending unit, and remove the oil sending unit.

P2-2 Thread the fitting on the oil pressure gauge hose into the oil sending unit opening, and tighten the fitting.

P2-3 Place the oil pressure gauge where it will not contact rotating or hot components.

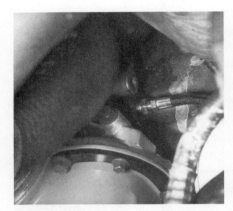

P2-4 Start the engine and check for leaks at the oil pressure gauge fittings.

P2-5 Operate the engine until it reaches normal operating temperature, and observe the oil pressure on the gauge at idle speed and 2,000 rpm.

P2-6 Compare the oil pressure gauge reading to the vehicle manufacturer's specifications.

P2-7 Remove the oil pressure gauge from the oil sending unit opening.

P2-8 Install the oil sending unit and tighten it to the specified torque, and install the sending unit wire.

P2-9 Start the engine and check for oil leaks at the oil sending unit.

The engine should be at normal operating temperature before checking the oil pressure. Engine oil pressure should be checked with the engine idling and with the engine operating at 2,000 rpm. If the oil pressure is equal to the vehicle manufacturer's specifications but the oil pressure warning light or gauge indicated low oil pressure, the sending unit is probably defective. When the oil pressure is lower than specified, one of these problems may exist:

1. Worn oil pump
2. Plugged oil pump pickup screen
3. Leaking oil pump pickup tube
4. Sticking pressure regulator valve in the pump
5. Broken oil pump drive
6. Loose main and connecting rod bearings
7. Worn camshaft bearings
8. Worn valve lifter bores
9. Missing or leaking oil gallery plugs

On many engines, when the oil pressure is lower than specified, the oil pan must be removed to check the cause of the low pressure. The oil pump and pressure regulator valve are accessible without removing the oil pan on some engines.

Engine Temperature Tests

Lower-than-normal coolant temperature reduces engine efficiency and fuel economy. If the engine coolant temperature is too low, insufficient heat is discharged from the vehicle heater in cold weather. Lower-than-normal coolant temperature affects some computer input sensor signals. Cooling system temperature is more critical today due to computerized engine controls. Many functions are determined in part by the engine's operating temperature. The correct temperature must be maintained precisely throughout the entire operating cycle, from startup to shutdown. Other functions or systems that can be affected are idle speed, mixture, timing, exhaust gas recirculation (EGR), transmission shift points, and other emission controls. If they are not working properly, degraded engine performance and fuel economy can result. If the coolant temperature is higher than the rated temperature of the thermostat, the engine may overheat.

The thermostat's operating temperature can be checked while installed in the engine or removed from it. You will need to monitor the temperature as part of the test, regardless. You can tape a thermometer to the upper radiator hose positioned so that the sensing bulb is in direct contact with the hose. Another choice is to install a radiator thermometer directly in the coolant at the radiator fill neck. Some digital multimeters (DMMs) have a thermometer accessory that connects to the unit that indicates the temperature on the DMM's readout display. There are also non-contact, infrared thermometers available that display an object's temperature by aiming the unit at the object and reading the unit's display. Start the engine and run it at a fast idle for approximately 15 minutes or until the coolant starts to move in the radiator. Observe the temperature reading when the coolant begins to flow in the radiator. This is the temperature the thermostat started to open. The temperature observed should be within ±5 degrees of the thermostat's rated temperature. The thermostat can also be checked when removed from the vehicle. Monitor the temperature and thermostat as described above. After removing the thermostat from the engine, place it in a container of water suspended so it does not come in contact with the container walls or bottom (Figure 3-5). A scan tool can also be used to read the output of the coolant temperature sensor value to determine coolant temperature.

Heat the water and note the temperature of the water when the thermostat begins to open. Depending on the rating, most thermostats will open from 185°F to 196°F and should be fully

Classroom Manual
Chapter 3, page 87

There are many causes of engine overheating. During the diagnosis of an overheating situation, all these causes should be checked until the source of the problem is found. Begin the diagnosis with the quickest tests or checks.

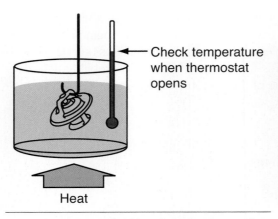

Check temperature when thermostat opens

Heat

Figure 3-5 Checking thermostat operation.

Recessed area

Figure 3-6 Be sure that the thermostat fits into the recess in the housing or manifold.

open at 212°F, the boiling point of water at sea level. If the thermostat opens much earlier or later than its rating, it should be replaced. Obviously, if the thermostat is stuck open or closed, regardless of the temperature, it needs to be replaced. When replacing or installing the thermostat, be certain to thoroughly scrape and clean the mating surfaces of the thermostat housing and block, and use a new gasket or sealant as recommended by the manufacturer. Also, look for a recess in the block where the thermostat must be seated (Figure 3-6).

If the thermostat is cocked or not seated during installation it will not seal, and you may break the thermostat housing while torque tightening the thermostat housing bolts.

When the engine operating temperature is higher than the rated temperature of the thermostat, check these causes of overheating:

1. Loose or slipping fan belt
2. Electric-drive cooling fan not operating at the proper temperature
3. Defective viscous-drive fan clutch
4. Restricted air passages in radiator core or other restricted airflow to the radiator
5. Partially plugged coolant passages in radiator core
6. Restricted or collapsed radiator hoses
7. Thermostat not opening at the proper temperature
8. Thermostat improperly installed
9. Leaking cylinder head gasket
10. External cooling system leaks causing low coolant level
11. Leaking radiator pressure cap
12. Defective water pump, loose impeller
13. Improper mixture of antifreeze and water
14. Excessive engine load
15. Late ignition timing and/or spark advance
16. Lean air-fuel mixture, engine vacuum leaks
17. Automatic transmission overheating
18. Dragging brakes
19. Broken, improperly positioned radiator shroud

Cooling System Inspection and Diagnosis

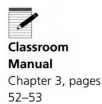

Classroom Manual
Chapter 3, pages 52–53

When a cooling system is suspected as being faulty, it should be inspected and tested. Keep in mind that an engine and its systems will only be efficient when the engine is at normal operating temperatures (Photo Sequence 3).

Photo Sequence 3
Typical Procedure for Testing a Cooling System

P3-1 After allowing the engine to cool, the technician starts his coolant service by doing a visual inspection of the belts, hoses, and any obvious signs of leakage. Also check for any blockages in front of the radiator.

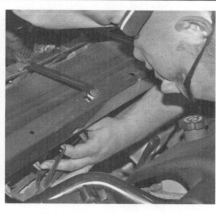

P3-2 If the vehicle has electric fans or a viscous fan clutch, check its condition.

P3-3 The technician opens the radiator cap and inspects the level and condition of the coolant. Is it rusty or contaminated in appearance? Check the coolant freezing and boiling point with a refractometer.

P3-4 Pressure test the cooling system using a tester. Make sure the system is able to hold pressure rated on the radiator cap.

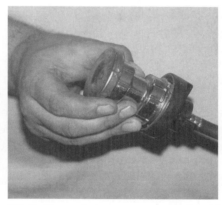

P3-5 The radiator pressure cap should also be checked. It must hold the pressure that it is rated.

P3-6 If during inspection any work was required, make sure you double check your repairs. Remember many vehicles have special fill procedure to prevent air locks.

Visual Inspection

⚠ WARNING: Use caution and wear protective gloves when touching cooling system components. If the engine has been running, these components may be very hot, and touching them could result in severe burns.

⚠ WARNING: Keep hands, tools, and clothing away from electric-drive cooling fans. On many vehicles, an electric-drive fan may start turning even with the ignition switch off.

All cooling system hoses should be inspected for soft spots, swelling, hardening, chafing, leaks, and collapsing. If any of these conditions are present, hose replacement is necessary. Hose clamps should be inspected to make sure they are tight. Some radiator hoses contain a wire coil inside them to prevent hose collapse as the coolant temperature decreases. Remember to include heater hoses and the bypass hose in the hose inspection.

✔ **SERVICE TIP:** If the upper radiator hose collapses when the coolant is cold, the vacuum valve may be sticking in the radiator pressure cap or the hose to the coolant recovery container may be restricted.

The radiator should be inspected for restrictions such as bugs and debris in the air passages through the core. An air hose and air gun may be used to blow bugs and debris from the core. The radiator should also be inspected for leaks and loose brackets. Radiator repairs are usually done in a radiator repair shop. When the radiator cap is removed, the openings in some of the radiator tubes are visible through the filler neck. These tube openings should be free from rust and corrosion. If the tubes are restricted, cooling system flushing may remove the restriction. When flushing does not clean the radiator tubes, the radiator must be cleaned at a radiator shop. Any restrictions in the radiator air passages or coolant tubes result in higher engine operating temperature and possible overheating.

Many radiators have a shroud behind the radiator to concentrate the flow of air through the radiator core. Inspect the **radiator shroud** for looseness and cracks or missing pieces. If the shroud is loose, improperly positioned, or broken, airflow through the radiator is reduced and engine overheating may result.

The radiator cap should be inspected for a damaged sealing gasket or vacuum valve (Figure 3-7). Check the pressure rating stamped on top of the cap, and be sure this is the same as the radiator cap specification in the vehicle manufacturer's service manual. Replace the cap if the pressure rating does not meet the vehicle manufacturer's specified pressure. If the sealing gasket or vacuum valve is damaged, replace the cap. Check the seat in the filler neck for damage or burrs, and remove any rough spots with fine emery paper. Visually inspect the filler neck for rust accumulation, which may indicate a contaminated cooling system. If the pressure cap sealing gasket or seat is damaged, the engine will overheat and coolant will be lost to the coolant recovery system. Under this condition, the coolant recovery container becomes overfilled with coolant.

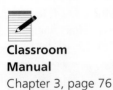

**Classroom
Manual**
Chapter 3, page 76

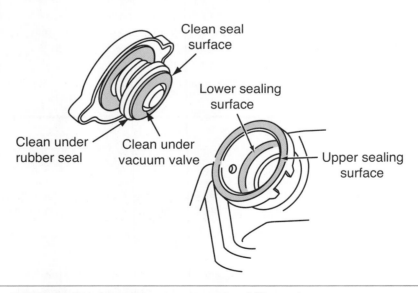

Figure 3-7 Inspecting the radiator pressure cap and filler neck.

If the cap vacuum valve is sticking, a vacuum may occur in the cooling system after the engine is shut off and the coolant temperature decreases. This vacuum may cause collapsed cooling system hoses.

The cooling system pressure tester is used to pressurize the entire cooling system to check the sealing integrity of the system, including the radiator pressure cap. Pressurizing the system will make it easier to detect any coolant leaks. After installing the pressure tester to the radiator filler neck, the technician manually pumps the tester to a specific pressure equal to the operating pressure of the vehicle's cooling system (see Figure 3-1). If the coolant is low in the radiator, top it off with water, and then perform the test. The pressure will hold if no leaks are present. If the pressure drops steadily over several minutes, there is enough pressure loss to indicate a cooling system leak. Inspect the entire cooling system for any evidence of coolant leaks. Be sure to include the radiator, hoses, water pump, heater core, and water jacket expansion plugs in the engine block and cylinder heads. Also, do not overlook the less obvious such as coolant in the engine's crankcase or in a cylinder. Do not forget to check the rear seat heating system components, if so equipped. If you added water to the radiator, be sure to add the proper amount of coolant to bring the mixture to the proper protection level after the repair of any leak.

The pressure tester is also used to test the radiator cap. An adapter is used to connect the cap to the pressure tester. The pressure tester should be pumped up to build enough pressure to overcome the pressure rating of the radiator cap. If the cap is good, it should hold pressure equal to its rating. Any pressure beyond the specified rating should force open the spring-loaded seal in the cap, allowing the excess pressure to escape. This is how it would function on the radiator in the event that the pressure is too high in the cooling system. The excess coolant under higher pressure is captured in the recovery tank. This is a cyclic occurrence and is normal, especially during hot operating conditions.

A coolant hydrometer is used to check the specific gravity of the coolant mixture in the cooling system. A sample of the coolant is drawn into the hydrometer (following tool manufacturer's directions) from the radiator. The coolant will work against a calibrated float device inside the hydrometer once the proper level is drawn in. The float device will indicate the temperature protection value of the coolant in degrees. When adding or replacing coolant, follow the manufacturer's instructions for the proper mixture of the coolant and water. Actually, pure coolant offers no protection. It needs to be mixed with water to be effective. The mix is usually 50 percent coolant and 50 percent water for most applications. Otherwise, check the label on the coolant container for specific protection mixtures. Usually the 50-50 mix of coolant and water gives freezing protection to approximately –34°F (–37°C) and boiling protection up to 265°F (129°C). Other tests that can be performed on the coolant include acidity content levels and rust protection levels. Follow the vehicle manufacturer's specific recommendations for cooling system servicing, which may include a system flush at certain mileage or time intervals. It should be noted that not all coolant/anti-freeze compounds are compatible with each other. For example, General Motors warns against mixing DEX-COOL® (silicate-free) coolant with other types of coolant. The special properties of DEX-COOL® are unique for its application and therefore not compatible with other types of coolant. The manufacturer states that this coolant is designed to remain in the vehicle for 5 years or 150,000 miles. Furthermore, it warns that if DEX-COOL® is mixed with other types of coolant, cooling system or engine damage may occur, jeopardizing the new vehicle warranty.

⚠ WARNING: Used engine coolant is a hazardous waste, and pollution laws prohibit the dumping of coolant in sewers. Coolant recycling equipment is available to clean contaminants from the coolant so it may be reused.

Defective radiator pressure caps may cause excessive coolant loss to the coolant recovery system, engine overheating, and collapsed hoses.

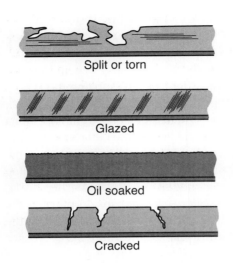

Split or torn

Glazed

Oil soaked

Cracked

Figure 3-8 Defective drive belt conditions.

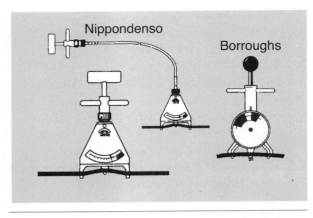

Nippondenso

Borroughs

Figure 3-9 Checking belt tension with a belt tension gauge.

Classroom Manual
Chapter 3, page 78

All the drive belts, including the water pump belt, should be inspected for cracks, oil contamination, glazing, and tears or splits (Figure 3-8). If any of these conditions are present, replace the belt or belts. Since the friction surfaces are the sides of a V-belt, the belt must be replaced if the sides are worn and the belt is contacting the bottom of the pulley.

The belt tension is checked with the engine off and a belt tension gauge placed over the belt at the center of the belt span (Figure 3-9). If the belt tension is less than specified, the belt must be tightened. A loose or worn belt may cause a squealing noise when the engine is accelerated. On many vehicles, the same belt drives the water pump and the alternator. If the engine has a V-belt, loosen the alternator bracket bolt and the alternator mounting bolt, and then pry on the alternator housing to tighten the belt. Ribbed V-belts usually have a spring-loaded belt tensioner. Tighten the alternator bolts and recheck the belt tension. Since misaligned pulleys cause premature belt failure, be sure the pulleys are properly aligned.

The belt tension may also be checked by measuring the amount of belt deflection with the engine shut off. Use your thumb to depress the belt at the center of the belt span. If the belt tension is correct, the belt should have 1/2 inch of deflection per foot of belt span.

With the engine off, grasp the fan blades or the water pump hub and try to move the blades from side to side. This action checks for looseness in the water pump bearing. If there is any side-to-side movement in the bearing, water pump replacement is required. Check the fan blade for cracks or a bent condition. A cracked fan blade causes a clicking noise at idle speed. If the fan blades are cracked or bent, replace the fan blade assembly. Bent fan blades cause an imbalance and vibration condition, which damages the water pump bearing.

WARNING: If a cracked or bent fan blade is discovered, replace the fan blade immediately. Do not start the engine. If a fan blade breaks while the engine is running, the blade becomes a very dangerous projectile that may cause serious personal injury or vehicle damage.

WARNING: Do not attempt to straighten bent fan blades. This action weakens the metal and may cause the blade to break suddenly, resulting in serious personal injury or vehicle damage.

Grasp the water pump hub to check the pump bearing for side-to-side movement. Check for coolant leaks at the water pump drain hole in the bottom of the pump and at the inlet hose connected to the pump. When coolant is dripping from the pump drain hole, replace the pump. If coolant is leaking around the inlet hose connection on the pump, check the hose and clamp.

Check the coolant recovery container and hose for leaks. Observe the level of coolant in the recovery container. In cooling systems with a coolant recovery system, coolant is added to the recovery container. Most of these containers have cold and hot coolant level marks. The coolant level should be at the appropriate mark on the recovery container, depending on engine temperature.

The thermostat must be removed to inspect it. Some of the coolant must be drained from the cooling system before the thermostat is removed. Loosen the radiator drain plug and drain the coolant into a clean coolant drain pan. Remove the thermostat housing bolts and remove the housing. Check the thermostat valve and replace the thermostat if the valve is stuck open. Visually inspect the thermostat housing area for rust and contaminants. If excessive rust and contaminants are present, flush the cooling system and replace or recycle the coolant.

If the opening temperature of the thermostat is higher than the rated temperature, the engine will overheat. When the thermostat opening temperature is lower than the rated temperature, the engine runs cooler than normal, providing a reduction in engine efficiency and performance. On engines with computer-controlled air-fuel mixture and spark advance, lower-than-normal coolant temperature changes some input sensor signals and reduces fuel economy.

Many thermostats are marked for proper installation. The pellet end of the thermostat faces toward the engine. Be sure the thermostat fits properly in the recess in the upper or lower part of the housing. Clean the mating surfaces carefully with a scraper to remove the old gasket. Install a new housing gasket and tighten the housing bolts to the specified torque.

If the radiator tubes and coolant passages in the block and cylinder head are restricted with rust and other contaminants, these components may be flushed. Cooling system flushing equipment is available for this purpose. Always operate the flushing equipment according to the equipment manufacturer's directions, and be sure that your service procedure conforms to pollution laws in your state. Coolant reconditioning machines are available to remove harmful particles and restore corrosion additives so the coolant can be returned to the cooling system (Figure 3-10).

Figure 3-10   An engine coolant (anti-freeze) recycling machine.

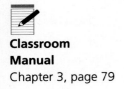
Special Tools

12-volt test light, jumper wire, DMM

P.S.I.A. = Pounds/in.² absolute. Absolute means the gauge reads atmospheric pressure when disconnected from a pressure source. At sea level, a PSIA gauge would read 14.7 pounds/in.²

The viscous-drive fan clutch should be visually inspected for leaks. If there are oily streaks radiating outward from the hub shaft, the fluid has leaked out of the clutch.

With the engine off, rotate the cooling fan by hand. When the engine is cold, the viscous clutch should offer a slight amount of resistance. The clutch should offer more resistance to turning when the engine is hot. If the viscous clutch allows the fan blades to rotate easily both hot and cold, the clutch should be replaced. A slipping viscous clutch results in engine overheating. Push one of the fan blades in and out and check for looseness in the clutch shaft bearing. If any looseness is present, replace the viscous clutch.

Electric-Drive Cooling Fan Circuit Diagnosis

Electric cooling fan circuits vary depending on the vehicle and the model year. Always follow the diagnostic procedure in the vehicle manufacturer's service manual. If the electric-drive cooling fan does not operate at the coolant temperature specified by the vehicle manufacturer, engine overheating will result, especially at idle and lower speeds when airflow through the radiator is reduced (Figure 3-11). Following is a typical electric cooling fan diagnostic procedure:

1. Check both the fusible link and the fuse.
2. Do a visual check of the relay and connections.
3. Make sure conditions are right for the fan to operate (or use a scan tool to command the coolant fan on). Check for power at the relay terminals 1, 2, and 4, and check terminal 3 for continuity to ground with a DMM. (Do not use a test lamp or self-powered test light.)
   - All terminals show correct voltage—Check the fan motor wiring and fan motor.
   - No power on terminal 1—Check the power feed from the fuse block to the fan relay.

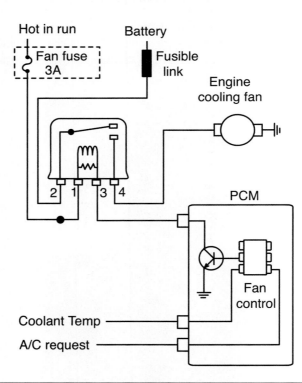

Figure 3-11 A computer-controlled cooling fan circuit.

- No power on terminal 2—Check the circuit from the fusible link to the fan relay.
- No ground on terminal 3—Use the scan tool to see if the fan is showing the correct coolant temperature. If not, run the diagnostics for the coolant temperature sensor. If the coolant temp sensor is okay, then check the wiring between the PCM and relay. If okay, replace the PCM.
- No power at terminal 4—Remove the relay and, using a fused jumper wire, jumper terminals 2 and 4. If the fan runs replace the relay or if the fan is still inoperative, repeat steps 1 and 2.

Vacuum Tests

A vacuum test on an engine is one of those quick and easy tests that will tell you quite a bit about an engine. By monitoring the vacuum of an engine, a technician is able to see how well each of the engine's cylinders seal and how well the engine's systems support the intake stroke of each cylinder.

Each cylinder of an engine is a vacuum pump. A vacuum is produced during each intake stroke. Ideally, all cylinders of the engine operate under the same conditions and produce the same amount of power. When this does not exist, the extra power produced by the stronger cylinders will be partially consumed by the weaker cylinders. The cylinders' function as a vacuum pump should be equal. Each cylinder should produce the same amount of vacuum during its intake stroke. In order to produce the same amount of power, each cylinder must first produce equal amounts of vacuum.

Vacuum is formed during each piston's intake stroke. The better the piston is sealed in the cylinder, the lower the pressure will be in the cylinder at the end of the intake stroke. A poor sealing cylinder will allow air to leak into the cylinder keeping the pressure high. Remember, intake air and fuel enter the cylinder because of the pressure differential between the pressure inside the cylinder and the pressure of the air outside the engine. When a cylinder leaks, vacuum is low and the incorrect amount of air and fuel will enter the cylinder.

The amount and quality of the air present in the cylinder after the exhaust stroke also determines how much vacuum will be formed on the intake stroke. If high pressure remains after the exhaust stroke, vacuum in the cylinder will be low.

To measure engine vacuum, connect a vacuum gauge (Figure 3-12) to the intake manifold. An ideal reading on the gauge, with the engine idling, is a steady reading of 16 or more inches of mercury (in. Hg). Any movement of the gauge's needle indicates the cylinders are not producing

P.S.I.G. = Pounds/in.² gauge. This means that when disconnected, the gauge reads zero regardless of atmospheric pressure.

Special Tools

Vacuum gauge, various lengths of vacuum hose, tee-fittings

Vacuum is any pressure lower than atmospheric pressure.

Classroom Manual Chapter 3, page 46

Figure 3-12 Typical vacuum gauge.

Figure 3-13 Various vacuum gauge readings and what the readings indicate.

the same amount of vacuum. The more the needle fluctuates, the greater the problem. If all of the cylinders are producing the same amount of vacuum but the amount is too low, the engine could be worn from many miles of use. Refer to Figure 3-13 for various vacuum gauge readings.

Exhaust Gas Analyzer

An exhaust gas analyzer is sometimes called an infra-red emissions analyzer.

An exhaust gas analyzer is another diagnostic tool that does not take too much time and is easy to use. The analyzer looks at the results of the combustion process. Since anything that affects the combustion process will affect the exhaust, the entire engine can be analyzed with the exhaust. The key to using an exhaust gas analyzer is the proper interpretation of the measured gases.

Higher-than-normal HC emissions may be caused by ignition system misfiring, improper ignition timing, excessively lean or rich air-fuel ratio, low cylinder compression, leaking head gasket, defective valves, guides, or lifters, and/or defective rings, pistons, or cylinders. In summary, HCs are affected by anything that affects the combustion process. Remember that HCs are present in the exhaust because of incomplete combustion. The better the combustion, the lower the percentage of HC in the exhaust.

Higher-than-normal CO emissions may be caused by a rich air-fuel ratio, dirty air filter, faulty injectors, higher-than-specified fuel pressure, and/or defective input sensors. CO is the result of combustion. Therefore, if combustion does not take place or if the combustion is very incomplete, CO levels will be very low. CO results from a lack of air (O_2) in the cylinder during combustion.

When both HC and CO emissions are higher than normal, a rich condition, suspect a plugged PCV system, an inoperative or disconnected air pump, inoperative catalytic converter, or the engine oil is diluted with gasoline.

Each condition listed in the following chart can affect one or more gases measured by the infrared exhaust gas analyzer. This chart is designed to be used as a quick reference guide to help understand the relationship between the five gases that are typically read by the analyzer. Depending on the problem or condition, the particular exhaust gas may increase, decrease, or stay about the same when comparing normal and abnormal readings. The information here is not intended to replace actual vehicle specifications. All tests are performed with the secondary downstream air blocked off (if equipped) with the engine at normal operating temperature. Some computer controls can override the root fault conditions, therefore causing no apparent difference in the gas readings. Since different vehicles use various computerized engine management strategies, it is possible for the conditions listed below to cause varying or subtle test results. Obviously, the more gases your shop equipment is capable of reading (two, three, four, or five gases), the more information you can obtain.

Lower-than-normal CO_2 levels normally accompany higher-than-normal CO levels. Therefore, all of the possible causes for high CO should be suspect for lower-than-normal CO_2 levels. This problem may also be caused by exhaust gas sample dilution because of leaking exhaust system and/or an excessively rich air-fuel ratio.

When the O_2 readings are lower than normal and CO readings are higher than normal, check for a rich air-fuel ratio, defective injectors (pintles not seating properly and dripping fuel), higher-than-specified fuel pressure, defective input sensors, restricted PCV system, and/or carbon canister purging at idle and low speeds.

When the O_2 readings are higher than normal and CO readings are lower than normal, suspect a lean air-fuel ratio, vacuum leak, lower-than-specified fuel pressure, defective injectors, defective input sensors, and/or the air pump or pulsed secondary air injection system connected during infrared exhaust gas analysis.

Higher-than-normal NO_x levels are caused by excessive combustion chamber temperatures. This can be caused by an overheated engine to an excessively lean air-fuel mixture. NO_x is formed by combustion, therefore NO_x levels will tend to be lower with increased amounts of HC in the exhaust.

To pinpoint the exact cause of the improper exhaust gas analyzer readings, perform detailed tests of the individual systems and components that could cause the abnormal readings. The cause for readings that could be caused by the engine—lack of sealing, improper operating temperatures, etc.—should be checked prior to testing other systems.

Five-Gas Emissions Interrelationship Chart	Exhaust Gas Components As Sampled From The Tailpipe				
Condition	**HC ppm**	**CO %**	**O_2 %**	**CO_2 %**	**NO_x ppm**
Mechanical (Compression related)	▲	▼	▲	▼	▼
Electrical (Ignition Misfire)	▲	▼	▲	▼	▼
Rich air/fuel mixture condition	▲	▲	▼	▼	▼
Lean air/fuel mixture condition	▲	▼	▲	▼	▲
Engine Ping; Detonation; High Combustion Temperatures	▲	▼	▲	▼	▲
Cooler than Normal Operating Temperature	▲	▲	▼	▼	▼
EGR Malfunction	=	=	=	=	▲
Catalytic Converter Inactive	▲	▲	▲	▼	▲
Typical average preferred readings at idle speed with a normally functioning catalytic converter* *(secondary air disabled, hoses to exhaust blocked off during testing)	<100	<1.0%	<1.0%	>12%	See NO_x note below

Key: ▲ = increase in number; ▼ = decrease in number; = = no significant change in number
NO_x Note: Readings should be <400 ppm at idle unloaded and <900 ppm at 2,000 rpm loaded. The readings should be baselined with several known good vehicles with your own shop equipment. Refer to equipment manufacturer's procedures. NO_x readings may vary with equipment used and method of test performed. Check with the specific vehicle manufacturer and any state or federal regulations regarding specifications and testing.

Exhaust gas analyzers have a warmup and calibration period that is usually about 15 minutes. Modern analyzers perform this calibration automatically. Always be sure the analyzer is calibrated properly so it provides accurate readings. Some older infrared analyzers had to be calibrated manually with calibration controls on the analyzer.

On most exhaust gas analyzers, a warning light is illuminated if the exhaust flow through the analyzer is restricted because of a plugged filter, probe, or hose. This warning light must be off before proceeding with the infrared exhaust analysis.

Most vehicle manufacturers and test equipment manufacturers recommend disconnecting the air pump or pulsed secondary air injection system during an infrared exhaust gas analysis. Always follow the recommended procedure in the vehicle manufacturer's service manual or the equipment operator's manual.

Quick Tests Using the Infrared Exhaust Gas Analyzer

The infrared exhaust gas analyzer can be used as a diagnostic tool as well as an emission-testing tool. Following is a list of quick tests using the four-gas infrared. Generally, these tests should be done with the engine at normal operating temperature with the transmission in park (neutral for manual), the parking brake set, and the wheels blocked. In addition, if the engine is equipped with any type of secondary air injection system, it should be disabled if the test calls for readings sampled from the tailpipe. The best way to do that is to pinch off the secondary air hose to the exhaust after the diverter valve with a pair of needle nose clamping pliers.

Look for the hose or tube that goes to the catalytic converter, if so equipped (Figure 3-14). The secondary air normally dilutes the readings of the true combustion results. Any problems

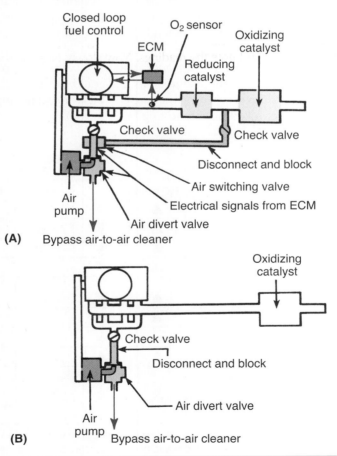

Figure 3-14 The procedure for conducting a cylinder power balance test differs, depending on whether the engine (A) has an air-fuel mixture feedback control or (B) does not have an air-fuel mixture feedback control.

may be masked by the extra air that is pumped into the exhaust. Do not forget to remove the pliers from the hose(s) when you are done testing. Please note that there are no hard and fast rules to these tests. They are intended as quick tests only and are not to be used in place of specific manufacturer's procedures where applicable.

1. *Engine Manifold Vacuum Leaks*. Place the sample probe of the analyzer in the tailpipe and observe the readings. In the case of a vacuum leak, you may notice an increased amount of O_2 and/or HC and a decrease of CO_2. To pinpoint the source of the vacuum leak, run the engine and carefully release a small metered amount of propane along the intake manifold and vacuum-operated accessories. When the propane is drawn into the area of the leak, the infrared will detect the change. Give the process several seconds to occur. At the time the propane reacts on the vacuum leak, it momentarily enriches the mixture. You will notice a decrease of O_2/HC and perhaps a momentary increase in CO_2/CO.

2. *Leaking Injectors*. If you suspect the fuel injectors are leaking fuel after the engine is shut down, use the infrared exhaust analyzer to confirm this. With the engine off, place the sample probe in the intake manifold past the throttle plates (fuel injected engines). Observe the hydrocarbon (HC) readings. You will notice the HC levels will rise as the infrared draws any excess fuel from the manifold at the start of the test. Gradually, the HC levels will decrease and should stay low. If the numbers start to increase again or otherwise remain high, fuel is continuing to enter the manifold. This could be due to a leaking injector.

3. *Fuel Combustion Efficiency Test*. Use the infrared to monitor exhaust gases while running the engine at various speeds. You should expect CO and HC levels to go lower as rpm increases. Start at idle speed and note the readings of CO, HC, O_2, and CO_2. Allow the readings to stabilize. Next, increase to and maintain engine speed of 1,500 rpm, allow the analyzer to stabilize, and note the readings. Now slowly increase the engine speed from 1,500 rpm to 2,500 rpm and maintain that speed allowing readings to stabilize and note the readings again. Compare the three test speed results. CO and HC should decrease with each increase of engine speed. CO_2 levels may increase, indicating an efficient combustion. If the CO and HC levels increase with higher engine rpm, there may be a fuel delivery or fuel control problem. This vehicle may have poor fuel economy and sluggish performance due to the mixture being too rich at city and highway speeds.

4. *Contaminated Motor Oil Test*. If the crankcase becomes contaminated with gasoline (HC), premature engine wear and performance problems will result. Gasoline in a large volume can be the result of a rich mixture, leaking injectors, an engine misfire, mechanical problems, or the engine becoming flooded. Suspect a contaminated crankcase any time these conditions exist. Of course, you will need to address the original problem first, then change the engine oil and oil filter afterward. To check for crankcase contamination, bring the engine to operating temperature and shut the engine off. Remove the oil filler cap and place the sample probe into the engine. Be sure to keep the probe from any oil in the engine. Allow the reading to stabilize and observe the hydrocarbons (HCs). An allowable amount will be anywhere from 0 to 300 parts per million (ppm) HC. If the reading is between 400 ppm to 500 ppm or more, the engine oil and oil filter need to be changed due to contamination. Please note that this is not intended to determine oil change intervals. It is merely a quick-test gauge to indicate crankcase contamination as a result of another problem.

5. *PCV Test*. To test the positive crankcase ventilation (PCV) system, place the sample probe in the tailpipe and run the engine at an idle. Remove the PCV valve from the engine and place your thumb over the end of the valve. The PCV valve is actually a calibrated vacuum leak. If you block off the valve, you should see a slight increase in carbon monoxide (CO). This quick test is checking airflow through the valve and hose. It does not confirm the valve is the correct one.

6. *Air Injection Reactor (AIR) System, Pulse Air, Secondary Air Test.* This test will simply confirm that secondary air is being delivered to the exhaust. Place the sample probe into the tailpipe and observe the oxygen (O_2) readings. The first reading can be made with the downstream air hose blocked off. The O_2 reading will be low, usually around 0.5 to 1.5 percent. Release the clamp(s) from the secondary air hose and compare the O_2 readings. It should increase by at least 3 percent.

7. *General Emissions Test.* Many states have an emission testing program of one type or another. This quick test will usually not take the place of the state test; however, it can help to identify a problem at a glance. This quick test can be performed with or without the secondary air disabled. In fact, if the readings are acceptable without secondary air (air blocked off), the results will be even better with secondary air enabled. Place the sample probe into the tailpipe and observe the readings while at an idle and at 2,000 rpm. Ideally, the reading with a catalytic converter should be as follows: CO = <1.0 percent, HC = <100 ppm, O_2 = 0.5 to 3.0 percent, and CO_2 = >12 percent. On a vehicle without a catalytic converter, the CO and HC readings will likely be higher. Depending on the model year and vehicle mileage, the CO will be approximately 2.5 percent to 3.5 percent and the HC should be in the range of 250 ppm to 350 ppm.

8. *Fuel Enrichment Test.* This test will verify proper fuel enrichment during a snap acceleration test. Place the sample probe in the tailpipe and snap the throttle to wide-open throttle (WOT). Watch for at least a 1 percent to 2 percent rise in CO without the O_2 rising prior to the CO increasing. The increasing CO indicates the extra fuel from the snap was burned. If you noticed a weak CO increase or a rise in O_2 before the CO increased, the engine could be hesitating on acceleration. The hesitation could be due to loose throttle linkage, dirty injectors, or a throttle plate out of adjustment, to name a few.

9. *Combustion Chamber Leaks.* As mentioned earlier, the exhaust analyzer can be used to check for coolant leaks in the combustion chamber. Use extreme caution when working on a hot engine. Because you will be opening the radiator cap, it may be a good idea to wait for the engine to cool down first. Remove the radiator cap and place the sample probe above the radiator filler neck. Do not allow the sample probe to come in contact with the coolant. Run the engine at a fast idle and observe the CO_2 reading (four-gas unit) or the HC reading if you are using a two-gas analyzer. If CO_2 is present in the cooling system, it means combustion gases are escaping into the cooling system, probably due to a leaking head gasket or similar failure. A cooling system pressure test can assist in locating the source of the leak.

10. *Locating a Fuel Leak.* Sometimes fuel leaks can be difficult to locate, especially when the fuel lines are routed in hard-to-see areas. The infrared can help locate the area of the fuel leak. With the engine off, guide the sample probe tip along the fuel line(s) from the fuel tank to the engine. Observe the HC as you move the sample probe along the route. A fuel leak will cause the HC reading to increase. This test can be particularly helpful if gasoline odor is detected but there are no visible signs of a leak. Be sure to check as far on top of the fuel tank as possible, and check the evaporative emission canister.

11. *Excessive Valve Guide Wear.* Worn internal engine components can cause compression-related problems and other symptoms such as excessive blowby or oil consumption. Although there are no firm numbers, CO_2 detected in the valve cover area while the engine is running is a good indication there is something worn internally. Test several known good engines of similar types to establish what may be considered a normal reading. The lower, the better. CO_2 is a byproduct of combustion, and if it is present in the valve cover area it is because of worn valve guides or blowby from worn piston rings. These conditions usually have other symptoms associated with them; therefore, use other available testing methods to determine the source of the actual problem.

Engine Power Balance Test

The engine power balance test checks the efficiency of individual cylinders. If the vacuum test revealed a cylinder that was an inefficient vacuum pump, that cylinder can be identified by the power balance test. The test is performed with an engine analyzer (Figure 3-15) and with the engine at normal operating temperature. Many engine analyzers have the ability to test cylinder power balance using one of several methods performed either automatically or manually by the technician. It requires the analyzer to be connected to the vehicle's primary and secondary ignition system. In addition, certain components may need to be disconnected or disabled while performing the various tests. Refer to the equipment manufacturer's operating instructions when making connections and disabling specific controls on the engine. On computerized engine systems, certain idle, timing, fuel mixture, and/or emission controls will have to be disabled during the tests so the vehicle's system will not attempt to compensate for what the engine analyzer is doing. Depending on the type of analyzer, one method of performing a power balance is to physically interrupt the cylinder firing so spark is not delivered to the spark plug. This creates a total shutdown of that cylinder, resulting in what should be a noticeable rpm drop. When an engine analyzer is performing the test, it will cancel the cylinders one at a time, recording the rpm drop for each one. Upon completion, it will compare the test results relative to that engine. Any rpm drop that is inconsistent with the other cylinders will be identified as being from a cylinder with a weaker power contribution. Some computerized engine analyzers incorporate additional functions, such as vacuum and exhaust change per cylinder, as they perform the cylinder power balance tests. This can be helpful when locating fuel delivery problems on a per-cylinder basis. The analyzer will track the difference in exhaust gas change before and after the spark is interrupted to help determine the volume of fuel being exchanged in the cylinder. Ideally, all cylinders will have the same effect on engine speed when they are cancelled. The amount of rpm drop per cylinder depends on the size and type of engine. A four-cylinder engine will have more of an rpm drop per cylinder as compared to a V-8 engine. If there is very little rpm drop

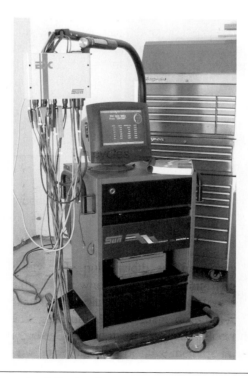

Figure 3-15 A computerized diagnostic engine analyzer.

when the cylinder is cancelled, that particular cylinder is not contributing to engine power. Some analyzers interrupt fuel delivery to the cylinder as another way to interrupt cylinder power for the power balance test. This is done with certain engine analyzers on only a few types of engines with sequential fuel injection. Either of the methods used to momentarily interrupt power to the cylinder, be it spark or fuel, is considered an intrusive test. It should be noted that the analyzers will interrupt power to the cylinder for only several seconds to avoid catalytic converter contamination or damage. If you are performing the test manually, be sure to limit the amount of time the cylinder is cancelled to only 5 to 10 seconds. Also, allow an equal amount of time before cancelling the next cylinder.

As with technology in general, advances in performing cylinder power balance tests have been made. Some computerized engine analyzers perform a nonintrusive test by calculating each cylinder's power contribution using a complex algorithm that looks at the crankshaft speed variance between cylinder firings. In other words, it will compute and display any momentary change of crankshaft speed due to a problem cylinder. It is expected that under normal operating conditions as each cylinder fires, the rotational speed of the crankshaft will stay relatively constant between cylinder firings. If a problem in a cylinder were present that would reduce power output, the crankshaft speed would slow momentarily at the point the problem cylinder was to fire. Obviously, the crankshaft speed will increase once the next normal firing cylinder provides the power. The analyzer detects this crankshaft speed change and will average it into an equation for the duration of the power balance test, thus identifying the problem cylinder without having to interrupt spark or fuel.

Regardless of the method used to measure cylinder power contribution, any cylinder that is below performance standards will need further inspection. A weak cylinder can be caused by an ignition problem, fuel delivery, or a mechanical condition. Use other tests to confirm the satisfactory operation of the other engine systems to help isolate the problem.

Compression Tests

Special Tools

Compression gauge, remote starter switch, squirt can of engine oil

Classroom Manual
Chapter 3, page 46

WARNING: If the injection system is not disabled during a compression test on a gasoline fuel injected engine, the injectors continue injecting fuel into the intake ports during the compression test. These fuel vapors are discharged from the spark plug openings and, if ignited, may cause a serious explosion resulting in personal injury and/or property damage.

CAUTION: Since a diesel engine has much higher compression than a gasoline engine, a special compression tester is required for a diesel engine.

CAUTION: If the injection system is not disabled during a compression test on a diesel engine, the injectors continue injecting fuel into the combustion chambers during the test. On a diesel engine, the compression is high enough to self-ignite the fuel in the cylinder in which the compression is tested. This self-ignition destroys the hose on the compression tester and possibly the tester.

The compression test checks the sealing qualities of the rings, valves, and combustion chamber. Operate the engine until it reaches normal operating temperature prior to conducting a compression test. Photo Sequence 4 shows the typical procedure for conducting a compression test. The following are some guidelines to follow when conducting this test:

1. Disable the ignition system by disconnecting the positive primary wire from the ignition coil. Insulate this wire with electrician's tape so it does not contact the vehicle ground. On distributorless ignition systems, disconnect the complete primary coil connector on the coil pack.

Photo Sequence 4
Conducting a Cylinder Compression Test

P4-1 Before conducting a compression test, disable the ignition and the fuel injection system (if the engine is so equipped).

P4-2 Prop the throttle plate into a wide-open position. This will allow an unrestricted amount of air to enter the cylinders during the test.

P4-3 Remove all of the engine's spark plugs.

P4-4 Connect a remote starter button to the starter system.

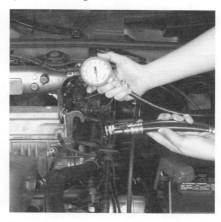

P4-5 Many types of compression gauges are available. The screw-in type tends to be the most accurate and easiest to use.

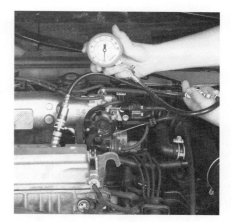

P4-6 Carefully install the gauge into the spark plug hole of the first cylinder.

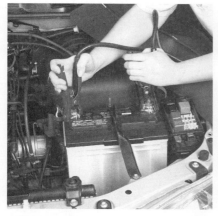

P4-7 Connect a battery charger to the car. This will allow the engine to crank at consistent and normal speeds that are needed for accurate test results.

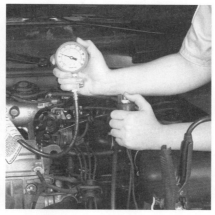

P4-8 Depress the remote starter button and observe the gauge's reading after the first engine revolution.

P4-9 Allow the engine to turn through four revolutions, and observe the reading after the fourth. The reading should increase with each revolution.

Photo Sequence 4
Conducting a Cylinder Compression Test (continued)

P4-10 Readings observed should be recorded. After all cylinders have been tested, a comparison of cylinders can be made.

P4-11 Before removing the gauge from the cylinder, release the pressure from it using the release valve on the gauge.

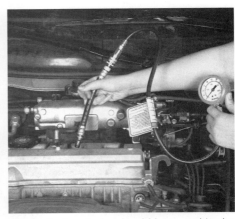

P4-12 Each cylinder should be tested in the same way.

P4-13 After completing the test on all cylinders, compare them. If one or more cylinders is much lower than the others, continue testing those cylinders with the wet test.

P4-14 Squirt a few drops of oil into the weak cylinder(s).

P4-15 Reinstall the compression gauge into that cylinder and conduct the test.

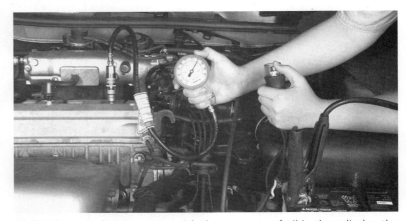

P4-16 If the reading increases with the presence of oil in the cylinder, the most likely cause of the original low readings was poor piston ring sealing. Using oil during a compression test is normally referred to as a wet test.

2. If the engine is fuel injected, disable the injection system by shutting off the fuel pump. On many vehicles, disconnect the wires from the fuel pump relay to shut off the fuel pump. On General Motors products, disconnect the fuel pump wire at the fuel tank. If the vehicle has a diesel engine, disconnect the fuel shut-off solenoid wire at the injection pump.

3. Loosen the spark plugs and blow any dirt from the plug recesses with an air blow-gun. Remove all plugs.

4. Place a screwdriver in the throttle linkage to hold the throttle open. If the engine has a carburetor, be sure the choke is open.

5. Install the compression tester in a spark plug hole. Some compression testers are threaded into the spark plug openings, whereas other testers lock into position when the compression pressure is applied to them. Other testers must be held in the spark plug openings by hand.

6. Crank each cylinder through at least four compression strokes noting the compression on the first puff and the final puff. Each of the four compression strokes can be observed as a definite increase on the gauge pointer reading.

7. Release the pressure from the compression tester, and follow the same procedure to obtain the compression reading on each cylinder. Record the reading obtained on each cylinder.

Compression Test Interpretation

Compare the compression readings to the vehicle manufacturer's compression specifications given in the service manual. If the compression readings on all the cylinders are equal to the specified compression, the readings are satisfactory. Higher-than-expected compression readings are typically caused by carbon buildup in the cylinders, which raises the compression ratio. If all of the cylinders have low readings, the engine is worn or the timing belt or chain is misaligned. When interpreting the results of this test, keep in mind the test shows the cylinders' ability to compress air. The better a cylinder is sealed, the more air it can compress. Likewise, the higher the compression ratio, the more the air in the cylinder will be compressed.

If the compression on one or more cylinders is lower than the specified compression, the valves or rings are likely worn. When the compression readings on a cylinder are low on the first stroke, increase to some extent on the next three strokes, but remain below the specifications, the rings are probably worn. If the reading is low on the first compression stroke and there is very little increase on the following three strokes, the valves are probably leaking. When all the compression readings are even but considerably lower than the specifications, the rings and cylinders are probably worn or camshaft timing is wrong.

When the compression readings on two adjacent cylinders are lower than specified, the cylinder head gasket is probably leaking between the cylinders. Higher-than-specified compression readings usually indicate excessive carbon in the combustion chamber, such as on top of the piston.

If the compression reading on a cylinder is zero, there may be a hole in the piston, or an exhaust valve may be severely burned. When the zero compression reading is caused by a hole in a piston, there is excessive blowby from the crankcase. This blowby is visible at the positive crankcase ventilation (PCV) valve opening in the rocker cover if this valve is removed. Zero compression can also be caused by an intake valve not opening.

Wet Compression Test

SERVICE TIP: If you are attempting to squirt engine oil through a spark plug opening into the cylinder to perform a **wet compression test** and this opening is difficult to access, place a length of vacuum hose on the end of the squirt can nozzle. Install the other end of this hose in the spark plug opening.

If a cylinder compression reading is below specifications, a wet test may be performed to determine if the valves or rings are the cause of the problem. Squirt about two or three teaspoons of engine oil through the spark plug opening into the cylinder with the low compression reading. Crank the engine for several revolutions to distribute the oil around the cylinder wall. Repeat the compression test on the cylinder with the oil added in the cylinder. If the compression reading improves considerably, the rings or cylinders are worn. When there is very little change in the compression reading, one of the valves is leaking.

Running Compression Test

The running compression test is a complement to the cranking compression test. The running compression test is performed on one cylinder at a time, with the compression gauge installed in the same manner as for the cranking compression test. The difference, of course, is that the engine is running at an idle. During a running compression test, the piston is moving approximately three times faster than cranking speed. This causes the compression test pressures to be lower compared to the cranking test. There are no firm specifications for running compression test results, and it is a relative test. The results of one cylinder reading need to be compared to the readings of the other cylinders. During a typical running compression test performed at idle, you could expect to see readings of 50 to 70 psi. If you slowly increase the rpm, the pressure will go down proportionately. The next step is to snap accelerate the throttle. The rapid increased throttle opening increases the amount of air into the cylinder, creating a sudden rise in compression pressures. This helps to reinforce the understanding of the principles of volumetric efficiency. The sudden increase in air intake results in higher compression pressures. The increased pressure with added fuel creates more speed. As mentioned, these tests are relative tests compared to the other cylinders. By reviewing test results, you can best determine a potential problem due to volumetric efficiency. For example, typical readings at idle may be at 60 psi and then drop to about 40 psi while holding the engine speed at 1,500 rpm. The snap acceleration reading should be approximately 80 percent of the cranking compression test results. If the snap acceleration readings are low, it could indicate an air intake restriction due to a valvetrain problem or carbon buildup on the back of the intake valve. If the snap acceleration reading is high, it may indicate an exhaust-related restriction or exhaust valve problem. The purpose of the cranking compression test is to check for cylinder sealing integrity. The running compression test is done to check volumetric efficiency.

Cylinder Leakage Test

Some engines have a tendency to rotate unless they are very close to TDC when the air is applied.

The leakage test may be used to determine if a low compression reading is caused by worn rings, burned valves, or other combustion chamber leaks. During this test, a regulated amount of air from the shop air supply is forced into the cylinder with both exhaust and intake valves closed. The gauge on the leakage tester indicates the percentage of leakage in the cylinder. A gauge reading of 0 percent indicates no cylinder leakage; if the reading is 100 percent, the cylinder is not holding any air.

The spark plug must be removed from the cylinder to be leak tested. Do not remove the other spark plugs, or the air forced into the cylinder will rotate the engine until one of the valves in the cylinder opens, which causes a leak and ruins the leakage tester reading. The engine must be at normal operating temperature prior to the cylinder leakage test. There are many differences in engines, ignition systems, and fuel systems at present. A high percentage of engines produced in recent years are equipped with distributorless ignition. Other engines have an electronic ignition system with computer-controlled spark advance. These variations in engines and systems complicate the diagnostic procedures, and in some cases many different procedures are recommended because of these system differences. The following cylinder leakage test procedure may be used on most engines regardless of the type of ignition system or fuel system:

1. With the pressure regulator valve in the tester in the off position, connect the shop air supply to the tester. Adjust the pressure regulator valve to obtain a zero gauge reading if necessary.

2. Using the same methods described previously in the compression test, disable the ignition system and the fuel injection system.

3. Loosen the spark plug in the cylinder to be tested and use an air gun to blow any dirt from the plug recess. Remove the spark plug from the cylinder to be tested.

4. Place your thumb over the spark plug hole in the cylinder to be tested. If the spark plug hole is not accessible, install a compression gauge in the spark plug hole of the cylinder to be tested.

5. Connect a remote control switch to the starter solenoid and crank the engine a very small amount at a time. When compression is available at the spark plug hole, stop cranking the engine.

6. Connect the spark plug removed from the cylinder to the end of the disconnected spark plug wire. Attach a jumper wire from the spark plug case to ground on the engine.

7. Reconnect the coil primary wires that were disconnected in step 2. Connect a timing pickup light to the spark plug wire on the cylinder being tested, and attach the positive and negative timing light leads to the battery terminals with the correct polarity.

> ⚠ **WARNING:** When a remote control switch is used to crank the engine slowly with the ignition system and the fuel system in operation, the engine may start. Therefore, keep tools and hands away from the cooling fan and belts to avoid personal injury.

> ⚠ **WARNING:** Never attempt to rotate the engine by installing a wrench on the crankshaft pulley bolt with the ignition system and fuel system in operation. Under this condition, the engine may start and rotate the wrench with the crankshaft. This action could result in serious personal injury and expensive vehicle damage.

8. Crank the engine a very small amount at a time with the remote control switch. When the timing light fires, stop cranking the engine. The piston is now at or near TDC on the compression stroke with both valves closed. As an alternative, you can place a TDC whistle in place of the spark plug. "Bump" the starter, causing the crankshaft to move small increments at a time. As the piston in the test cylinder begins to move to TDC on the compression stroke, the whistle will sound, indicating the piston is close to the top dead center position.

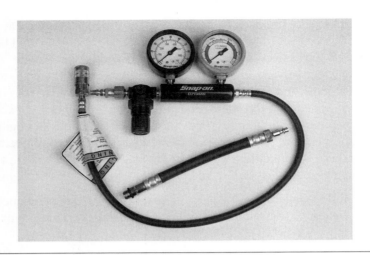

Figure 3-16 A cylinder leakage tester.

9. Connect the leakage tester air hose to the cylinder (Figure 3-16). Various adapters are available to fit different spark plug threads.
10. If the reading exceeds 20 percent, check for air escaping from the tailpipe, positive crankcase valve (PCV) opening, and the top of the throttle body or carburetor (Figure 3-17). Air escaping from the tailpipe indicates an exhaust valve leak. When the air is coming out of the PCV valve opening, the piston rings are leaking. An intake valve is leaking if air is escaping from the top of the throttle body or carburetor. In addition, check for air escaping into the cooling system or into an adjacent cylinder. This can indicate a head gasket or cylinder head leak.
11. Follow the same procedure in steps 2 through 10 to perform a leakage test on other cylinders.

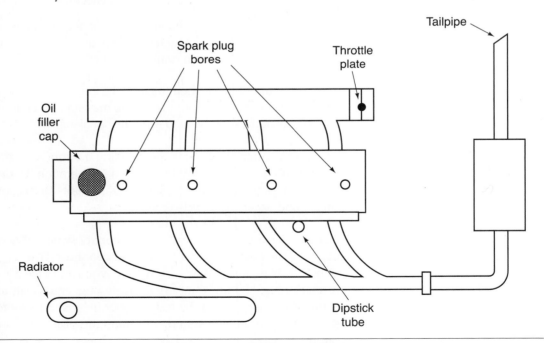

Figure 3-17 During a cylinder leakage test, air may be felt or heard leaking from these areas.

Valve Timing Checks

If the timing belt or chain has slipped on the camshaft sprocket, the engine may fail to start because the valves are not properly timed in relation to the crankshaft. When the timing belt or chain has only slipped a few cogs on the camshaft sprocket, the engine has a lack of power and fuel consumption is excessive. Follow these steps to check the valve timing:

1. Remove the spark plug from number one cylinder and place your thumb on top of the spark plug hole. If this hole is not accessible, place a compression gauge in the opening. Disable the ignition system.

2. Crank the engine until compression is felt at the spark plug hole.

3. Connect a remote control switch to the starter solenoid terminal and the battery terminal on the solenoid. Slowly crank the engine until the timing mark lines up with the zero-degree position on the timing indicator. The number one piston is now at TDC on the compression stroke. On many engines, the timing mark is on the crankshaft pulley and the timing indicator is mounted above the pulley.

4. Slowly crank the engine for one revolution until the timing mark lines up with the zero-degree position on the timing indicator. The number one piston is now at TDC on the exhaust stroke.

5. Remove the rocker arm cover and install a breaker bar and socket on the crankshaft pulley nut. Observe the valve action while rotating the crankshaft about 30 degrees before and after TDC on the exhaust stroke. In this crankshaft position, the exhaust valve should close a few degrees after TDC on the exhaust stroke and the intake valve should open a few degrees before TDC on the exhaust stroke. This valve position with the piston at TDC on the exhaust stroke is called valve overlap. If the valves do not open properly in relation to the crankshaft position, the valve timing is incorrect. Under this condition, the timing chain or belt cover should be removed to check the position of the camshaft sprocket in relation to the crankshaft sprocket position. When the valve timing is incorrect, the timing chain or belt and/or sprockets must be replaced.

Valve Adjustment

Some engines have **mechanical valve lifters** in place of hydraulic valve lifters, and these engines require a valve adjustment. Various car manufacturers recommend different valve adjustment procedures depending on the engine. For example, on some engines the manufacturer recommends adjusting the valves with the engine cold, whereas other manufacturers recommend performing this adjustment with the engine at normal operating temperature. Always follow the vehicle manufacturer's recommended valve adjustment procedure in the service manual.

The valves in each cylinder must be adjusted with the piston at TDC on the compression stroke and both valves closed. To position each piston at TDC on the compression stroke, connect a timing light to the spark plug wire and crank the engine slowly with the ignition switch on and a remote starter switch connected to the starter solenoid. When the timing light flashes, the piston is near TDC on the compression stroke.

Some rocker arms have an adjustment screw and a lock screw on the pushrod end of the rocker arm. With the piston at TDC on the compression stroke, place a feeler gauge of the

Classroom Manual
Chapter 3, page 47

It may be easier to remove the timing belt cover to check valve timing on some overhead cam engines.

Mechanical valve lifters are commonly called solid lifters or solid tappets.

Special Tools

Remote starter switch, feeler gauge

Adjusting pads, or shims, are located in some mechanical valve lifters to provide a valve adjustment.

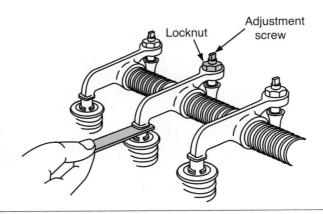

Figure 3-18 Adjusting valves with an adjusting screw and locknut on the pushrod end of the rocker arm.

specified thickness between the rocker arm and the valve stem (Figure 3-18). If the valve adjustment is correct, the feeler gauge slides between the rocker arm and the valve stem with a light push fit. If necessary, loosen the locknut and turn the adjustment screw to obtain the specified valve clearance. Hold the adjustment screw with a slotted screwdriver and tighten the locknut. On most engines, the specified valve clearance is different on the exhaust valves than the intake valves.

In some four-valve per cylinder (4V) engines, the rocker arms are mounted under the overhead camshafts and the camshaft lobes contact a friction pad on the top of the rocker arm. One end of the rocker arm contacts the valve stem, and the other end of the rocker arm is mounted in a pivot in the cylinder head. The manufacturer recommends measuring the valve clearance on these engines with the cylinder head temperature less than 100°F (38°C). Follow these steps for valve adjustment:

1. Remove the rocker arm cover.
2. Disable the ignition system by disconnecting the negative primary coil wire from the coil.
3. Crank the engine slowly until the number one piston is at TDC with the marks on the camshaft sprockets pointing upward (Figure 3-19).
4. Adjust the valves on the number one cylinder. Place the specified feeler gauge between the camshaft lobe and the upper side of the rocker arm (Figure 3-20). The

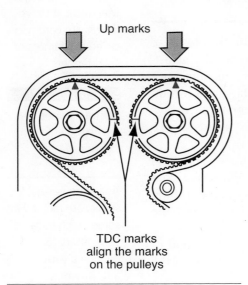

Figure 3-19 Number one piston at TDC and the camshaft sprocket marks facing upward.

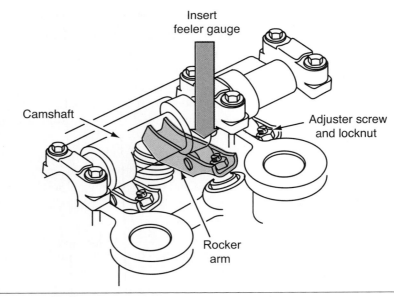

Figure 3-20 Measuring the valve clearance between the camshaft and the rocker arm.

feeler gauge should be a light push fit between these components. If the valve clearance requires adjusting, loosen the locknut and turn the adjusting screw in the rocker arm until the specified clearance is obtained. Tighten the locknut.

5. Rotate the crankshaft 180 degrees counterclockwise so the camshaft sprockets move 90 degrees. Under this condition, the marks on the camshaft sprockets face the exhaust side of the cylinder head (Figure 3-21). Adjust the valve clearance on the number three cylinder.

6. Rotate the crankshaft 180 degrees counterclockwise so the marks on the camshaft sprockets are facing downward (Figure 3-22). Adjust the valves on the number four cylinder.

7. Rotate the crankshaft 180 degrees counterclockwise so the marks on the camshaft gears are facing the intake side of the cylinder head (Figure 3-23). Adjust the valves on the number two cylinder.

8. Install a new rocker arm cover gasket and install the rocker arm cover. Tighten all bolts to the specified torque.

Some engines with hydraulic valve lifters do not have a valve adjustment. In these engines, if all the valvetrain components are in satisfactory condition, the hydraulic valve lifters maintain zero valve clearance. However, worn or improperly serviced valvetrain components may affect the valve lifter operation and result in valvetrain noise or improper engine performance.

A valve clearance measurement with the valve lifter bottomed is recommended on many valvetrains that do not have a valve adjustment. With the piston at TDC on the compression stroke, push downward on the pushrod end of the rocker arm until the valve lifter bottoms. Measure the clearance between the end of the rocker arm and the valve stem. If this clearance is more than specified, valvetrain components such as the rocker arm, pivot, or pushrod are worn, and this problem may cause a clicking noise at idle and low speed.

When this clearance is less than specified, the installed height of the valve in the cylinder head is too high. Excessive valve stem height is caused by grinding too much material from the valve seat or valve face without removing material from the end of the valve stem when the valves are reconditioned. If the valve stem height is excessively high, the valve lifter plunger may be bottomed in the lifter and the valve may not be allowed to close. Under this condition, cylinder misfiring occurs.

Some valvetrains have hydraulic valve lifters and individual rocker arm pivots retained with self-locking nuts. These valvetrains require an initial adjustment of the rocker arm nut to position the lifter plunger. With the valve closed, loosen the rocker arm nut until there is clearance between the end of the rocker arm and the valve stem. Slowly turn the rocker arm nut clockwise

Valve stem installed height is the distance from the valve spring seat to the top of the valve stem when the valve is closed.

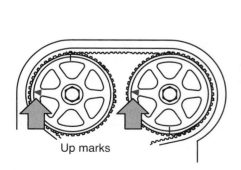

Figure 3-21 Camshaft sprockets positioned to adjust the valves in cylinder number three.

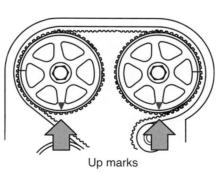

Figure 3-22 Camshaft sprockets positioned to adjust the valves in cylinder number four.

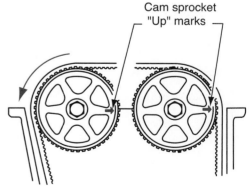

Figure 3-23 Camshaft sprockets positioned to adjust the valves in cylinder number two.

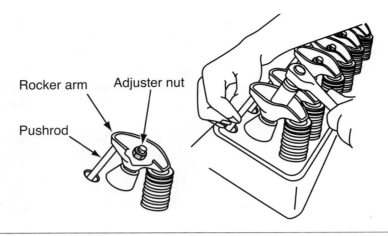

Rocker arm

Adjuster nut

Pushrod

Figure 3-24 Rocker arm adjustment with hydraulic lifters and individual rocker arms.

while rotating the pushrod (Figure 3-24). Continue rotating the rocker arm nut until the end of the rocker arm contacts the end of the valve stem and the pushrod becomes more difficult to turn. Continue turning the rocker arm nut clockwise the number of turns specified in the service manual. In some engines, this specification is one turn plus or minus one-quarter turn.

CASE STUDY

A police car was brought to the shop with the explanation that a local transmission shop had rebuilt the transmission three times and still has a shudder in fourth gear. The technician drove the car and discovered that it did indeed have a pronounced shutter from about 45-55 mph in high gear. If you placed you foot on the brake pedal just hard enough to release the torque converter clutch but not hard enough to apply the brakes, the shudder was unnoticeable. Several technicians have been mistaken in deciding whether the problem was engine or transmission related, especially with torque converter clutch. A torque converter clutch locks the engine and transmission together mechanically just like a manual transmission. Any vibration or irregularity in the operation of the transmission or engine can definitely be felt, especially in a low torque overdrive gear range. To make the problem worse, there were several cases of bad converters that caused the same problem. Another common cause of the shudder was fuel injector tip clogging. The technician carefully evaluated the vehicle. Although the engine idled just a bit rough, the customer did not mention anything about it, probably because it seemed so minor. The technician, however, was accustomed to this engine being very smooth since this engine had a balance shaft. The technician decided to perform a compression test. The results were 160 pounds compression on all but one cylinder in the back that had 125 pounds compression. This falls in the acceptable range of at least 75 percent of the highest compression, but the technician decided to perform a cylinder leakage test anyway. It was found that there was a 40 percent cylinder leakage from the exhaust valve on number four cylinder. High-speed driving had caused the exhaust valves to recess. The number four cylinder was the worst, but it was obvious that the others were also wearing. Both heads were sent to the machine shop and new exhaust seats were installed. After reassembling the engine, it test drove fine. You really need to be able to do some "detective" work on your own to be a technician. Basic tests made the difference.

Terms to Know

Mechanical valve lifters

Pinging

Radiator shroud

Wet compression test

ASE-Style Review Questions

1. Two technicians were discussing oil leak diagnosis:
   Technician A says that special dye is available to pinpoint the location of oil leaks.
   Technician B says that the customer may have to drive the vehicle for a few days for the dye to show up.
   Who is correct?
   - **A.** A only
   - **B.** B only
   - **C.** Both A and B
   - **D.** Neither A nor B

2. While discussing engine coolant leaks:
   Technician A says a leaking head gasket lowers engine operating temperature.
   Technician B says a leaking heater core may cause windshield fogging.
   Who is correct?
   - **A.** A only
   - **B.** B only
   - **C.** Both A and B
   - **D.** Neither A nor B

3. While discussing the diagnosis of engine exhaust color:
   Technician A says blue smoke in the exhaust during deceleration may be caused by a rich air-fuel ratio.
   Technician B says blue smoke in the exhaust during deceleration may be caused by loose camshaft bearings.
   Who is correct?
   - **A.** A only
   - **B.** B only
   - **C.** Both A and B
   - **D.** Neither A nor B

4. While discussing exhaust color and odor diagnosis:
   Technician A says black smoke in the exhaust during acceleration indicates a lean air-fuel mixture.
   Technician B says a strong sulphur smell from a catalytic converter-equipped vehicle may be caused by a rich air-fuel mixture.
   Who is correct?
   - **A.** A only
   - **B.** B only
   - **C.** Both A and B
   - **D.** Neither A nor B

5. A technician is performing a compression test. Which statement below is *least* likely to be true?
   - **A.** All cylinders with higher-than-normal readings could be caused by excessive carbon buildup.
   - **B.** All cylinders reading even but lower than normal may be caused by a slipped timing chain/belt.
   - **C.** Low readings on two adjacent cylinders may be caused by a blown head gasket.
   - **D.** A low reading on one cylinder may be caused by a vacuum leak in that cylinder.

6. While discussing an engine power balance test:
   Technician A says the engine may have an ignition problem on the weak cylinder.
   Technician B says the engine may have a mechanical condition on the weak cylinder.
   Who is correct?
   - **A.** A only
   - **B.** B only
   - **C.** Both A and B
   - **D.** Neither A nor B

7. While discussing an engine compression test:
   Technician A says the engine should be cranked until four compression strokes occur during the compression test on each cylinder.
   Technician B says the gasoline fuel injection system should be left in operation during a compression test.
   Who is correct?
   - **A.** A only
   - **B.** B only
   - **C.** Both A and B
   - **D.** Neither A nor B

8. While discussing the cylinder leakage test:
   Technician A says during a cylinder leakage test, if the air escapes from the tailpipe, an intake valve is leaking.
   Technician B says during a cylinder leakage test, the piston must be at TDC on the exhaust stroke.
   Who is correct?
   - **A.** A only
   - **B.** B only
   - **C.** Both A and B
   - **D.** Neither A nor B

9. While discussing engine noise diagnosis:
   Technician A says loose main bearings cause a light rapping noise during acceleration.
   Technician B says piston slap causes a hollow rapping noise that is most noticeable on acceleration with a cold engine.
   Who is correct?
   - **A.** A only
   - **B.** B only
   - **C.** Both A and B
   - **D.** Neither A nor B

10. While discussing engine oil pressure:
    Technician A says low oil pressure may be caused by a leaking oil pump pickup tube.
    Technician B says low oil pressure may be caused by loose camshaft bearings.
    Who is correct?
    - **A.** A only
    - **B.** B only
    - **C.** Both A and B
    - **D.** Neither A nor B

ASE Challenge Questions

1. A V-6 engine has oil leaking from the vicinity of the lower front engine covers. The most probable cause of the leak is:
   A. Worn front main seal
   B. Timing belt cover
   C. Valve cover(s)
   D. Oil pressure sending unit

2. The customer states the coolant level must be topped off every few days. There are no visible leaks.
   Technician A says this may be caused by a bad radiator cap.
   Technician B says a stuck open thermostat will cause this problem.
   Who is correct?
   A. A only
   B. B only
   C. Both A and B
   D. Neither A nor B

3. Main bearing wear is being discussed.
   Technician A says a heavy thumping knock at irregular intervals during acceleration may be caused by worn main thrust bearings.
   Technician B says this noise indicates loose/worn main bearings.
   Who is correct?
   A. A only
   B. B only
   C. Both A and B
   D. Neither A nor B

4. During an engine vacuum test, the gauge reading fluctuates between 11 and 16 in. Hg. This indicates service should be performed on:
   A. Improper idle mixture
   B. The valve and valve springs
   C. Fuel injectors
   D. Either A or C

5. A vacuum leak is suspected for a rough idle concern. Using a four-gas infrared exhaust analyzer, *Technician A* says O_2 will be higher than normal.
   Technician B says CO will be higher than normal.
   Who is correct?
   A. A only
   B. B only
   C. Both A and B
   D. Neither A nor B

Job Sheet 5

5

Name _____ Date _____

Cooling System Inspection and Diagnosis

Upon completion of this job sheet, you should be able to inspect the entire cooling system and determine the cause of abnormal conditions.

ASE Correlation

- Verify engine operating temperature, check coolant level and condition, perform cooling system pressure test; determine needed repairs.
- Inspect and test mechanical/electrical fans, fan clutch, fan shroud, ducting and fan control devices; determine needed repairs.

Tools and Materials

A vehicle
Clean cloth rag
Cooling system pressure tester
Service manual for the above vehicle

Describe the vehicle being worked on.

Year_____ Make _____ VIN_____

Model_____

Procedures

Task Completed

1. Inspect the following components and describe their condition:

   Radiator _____

   Radiator hoses _____

   Hose clamps _____

   Heater hoses _____

   Overflow tank _____

   Cooling fan _____

   Drive belts _____

   Any obvious defects should be corrected before continuing with testing the system.

2. Remove the radiator cap. Make sure the engine is cooled down.
   Do not do this if the engine is warm. □

3. With a clean rag, wipe off the radiator filler neck, then inspect it. Is the sealing area free of accumulated dirt, nicks, or anything that would prevent the radiator cap from sealing good?   □ Yes □ No

   If yes, describe the condition _____

   _____

4. Check the locking tabs and/or cams on the filler neck for damage. Describe their condition.

   _____

5. Inspect the overflow tank, hose, and connections. Look for cracks or other problems that may not allow the system to seal. Describe the condition of the overflow tank and connecting hose.

   _____

   _____

☐

6. Run a wire through the overflow tube or hose to clear any obstructions.

7. What is the pressure rating of the radiator cap? _____ psi

8. Using the proper adapter, attach the cooling system pressure tester to the radiator cap.

   Pump the tester to bring the pressure to the rating of the radiator cap.

   Are you able to bring the pressure to that amount?   ☐ Yes   ☐ No

   Once you are at that amount, does the pressure hold?   ☐ Yes   ☐ No

   If the cap does not hold the specified pressure, it should be replaced.

☐

   Slowly release the pressure from the tester and cap.

9. Attach the cooling system pressure tester to the radiator filler neck. Pump the tester to bring the pressure to the rating of the radiator cap.

   Are you able to bring the pressure to that amount?   ☐ Yes   ☐ No

   Once you are at that amount, does the pressure hold?   ☐ Yes   ☐ No

   If the system does not hold the specified pressure, the system leaks. Look over the cooling system to find the source of the leak. If the pressure drops quickly to zero, it may be necessary to keep pumping the tester to apply pressure to the system and to find the leak.

   If the system held pressure, there are no leaks. Then carefully relieve the pressure from the system and tester by following the procedure given by the tester's manufacturer.

☐

10. Top off the coolant to bring it to the proper level.

11. Install the radiator cap.

Instructor's Response _____

_____

_____

Job Sheet 6

Name _____ Date _____

Conduct a Cylinder Power Balance Test

Upon completion of this job sheet, you should be able to conduct a cylinder power balance test and accurately interpret the results.

ASE Correlation

Perform cylinder balance test; determine needed action.

Tools and Materials

Tune-up scope

Describe the vehicle being worked on.

Year _____ Make _____ VIN _____

Model_____

Engine type and size _____Firing order _____

Procedures

Task Completed

1. Describe the general running condition of the engine.

   _____

   _____

2. Connect the scope to the engine according to the instructions given with the equipment. ☐

3. Using proper testing procedures, conduct a cylinder power balance test on the engine. (Make sure not to short a cylinder for a long time if the vehicle is equipped with a catalytic converter.) Record the results below:

   Cylinder #   1   2   3   4   5   6   7   8

   RPM loss   ____ ____ ____ ____ ____ ____ ____ ____

4. Describe what is indicated by the results of this test:

   _____

   _____

5. Briefly explain what is actually being measured by a cylinder power balance test.

   _____

   _____

6. Based on the test results, describe the condition of the engine: (Does it agree with your original description of the engine?)

   _____

   _____

7. How and why would the readings be different if the camshaft intake lobes for number two cylinder were severely worn?

_____

_____

Instructor's Response_____

_____

_____

Job Sheet 7

Name _____ Date _____

Conduct a Cylinder Compression and Leakage Test

Upon completion of this job sheet, you should be able to conduct a cylinder compression and cylinder leakage test on an engine and interpret the results.

ASE Correlation

- Perform cylinder cranking compression test; determine needed repairs.
- Perform cylinder/leak-drain test; determine needed repairs.

Tools and Materials

Compression gauge
Remote starter
Cylinder leakage tester
Shop air supply
Safety glasses

Describe the vehicle being worked on.

Year _____ Make _____ VIN _____

Model_____

Engine size_____ # of cyl _____ compression ratio_____:_____

Expected compression readings: _____

Source of information: _____

Describe the general running condition of the engine:

_____

What does a compression test actually measure?

_____

Why would it be wise to connect a battery charger to the vehicle while conducting a compression test?

_____

_____

Procedures

Perform the following tasks exactly as they are written.

1. Using the proper procedures, conduct a dry compression test on the engine and record the results below:

    ____ ____ ____ ____ ____ ____ ____ ____
     1   2   3   4   5   6   7   8

    Summarize and explain the test results:

    _____

    _____

    _____

2. Using the proper procedures, conduct a wet compression test on the engine and record the results below:

____	____	____	____	____	____	____	____
1	2	3	4	5	6	7	8

Summarize and explain the test results and compare them to the dry results: (Why are they the same or different?)

_____

_____

3. Reinstall all of the spark plugs, except one. Repeat the compression test on the cylinder without a spark plug.

Your readings now:_____

Summarize the results and state why they are different or the same as the previous test.

_____

_____

4. What does a cylinder leakage test actually measure?

_____

_____

Using the proper procedures, conduct a cylinder leakage test on the engine, then record the results below:

Cylinder #	Reading	Point of leakage
1	_____	_____
2	_____	_____
3	_____	_____
4	_____	_____
5	_____	_____
6	_____	_____
7	_____	_____
8	_____	_____

Summarize and explain the results of the test below:

_____

_____

5. Based on the results from the compression and leakage tests, describe the general condition of the engine (be sure to explain why):

_____

_____

6. Does the above summary agree with your original description of the running condition of the engine? Why or why not?

_____

_____

7. How and why would the readings on the compression and leakage be different if the camshaft intake lobes for number two cylinder were severely worn?

_____

_____

Instructor's Response_____

_____

_____

Basic Electrical Tests and Service

Upon completion and review of this chapter, you should be able to:

❏ Diagnose electrical problems by logic and symptom description.

❏ Perform troubleshooting procedures using meters, test lights, and jumper wires.

❏ Inspect and repair wiring harness and connectors.

❏ Explain the proper use of a digital storage oscilloscope.

❏ Explain the proper use of a logic probe.

❏ Perform general battery service such as cleaning battery terminals and a battery case as well as removing and replacing a battery.

❏ Perform battery hydrometer test and determine battery condition.

❏ Perform battery capacity test and determine battery condition.

❏ Perform battery charging procedures.

❏ Connect booster battery cables properly.

❏ Measure battery drain accurately.

❏ Perform starter current draw test.

❏ Perform voltage drop tests on starter and starter control circuits.

❏ Inspect AC generator belt condition and check and adjust belt tension.

❏ Perform an alternator output test and determine alternator condition.

❏ Check alternator regulator voltage.

Basic Tools

Hand Tools, service manual, jumper wires, DMM

Basic Electrical Diagnosis

When electrons are able to flow along a path (wire) between two points, an electrical circuit is formed. An electrical circuit is considered complete when there is a path that connects the positive and negative terminals of the electrical power source. Somewhere in the circuit there must be a *load* or resistance to control the amount of current in the circuit. Most automotive electrical circuits use the chassis as the path to the negative side of the battery. Electrical components have a lead that connects them to the chassis. These are called the **chassis ground** connections. In a complete circuit, the flow of electricity can be controlled and applied to do useful work, such as light a headlamp or turn over a starter motor. Components that use electrical power put a load on the circuit and consume electrical energy.

The **chassis ground** is commonly referred to as the ground of the circuit.

Two basic types of wire are used in automobiles: solid and stranded. Solid wires are single-strand conductors. Stranded wires are made up of a number of small solid wires twisted together to form a single conductor. Stranded wires are the most commonly used type of wire in an automobile. Electronic units, such as computers, use specially shielded twisted cable for protection from unwanted induced voltages that can interfere with computer functions. In addition, some solid state components use printed circuits.

Battery cables need to be large gauge wires capable of carrying high current for the starter motor.

The current-carrying capacity and the amount of voltage drop in an electrical wire are determined by its length and gauge (size). The wire sizes are established by the SAE, which is the **American Wire Gauge (AWG)** system. Sizes are identified by a numbering system ranging from number 0000 to 50, with number 0000 having the largest cross-sectional area and number 50 the smallest. Most automotive wiring ranges from number 10 to 18 with battery cables that are at least number 4 gauge.

The **American Wire Gauge (AWG)** system designates wire sizes established by the SAE.

Metric Size (mm^2)	AWG (Gauge) Size	Ampere Capacity
0.5	20	4
0.8	18	6
1.0	16	8
2.0	14	15
3.0	12	20
5.0	10	30
8.0	8	40
13.0	6	50
19.0	4	60

Figure 4-1 Metric to AWG wire sizes.

In the metric system, wiring size is identified by the cross-sectional area of the wire. Metric wire size is expressed in square millimeters, so the larger the number, the larger the wire. A chart that compares metric and standard wire sizes and the amperage they are capable of carrying is shown in Figure 4-1.

Electrical Wiring Diagrams

Wiring diagrams are used to show how circuits are wired and how the components are connected. A typical service manual contains dozens of wiring diagrams vital to the diagnosis and repair of the vehicle.

A wiring diagram does not show the actual position of the parts on the vehicle nor their appearance, nor does it indicate the length of the wire that runs between components. It usually indicates the color of the wire's insulation and sometimes the wire gauge size. The first letter of the color coding is a combination of letters usually indicating the base color. The second letter usually refers to the strip color (if any). Tracing a circuit through a vehicle is basically a matter of following the colored wires.

Many different symbols are also used to represent components such as motors, batteries, switches, transistors, and diodes. Common symbols are shown in Figure 4-2. Part of a typical wiring diagram is shown in Figure 4-3 (page 132). Notice that the components are also labeled.

SERVICE TIP: Keep in mind that electrical symbols are not standardized throughout the automotive industry. Different manufacturers may have different methods of representing certain components, particularly the less common ones. Always refer to the symbol reference charts, wire color code charts, and abbreviation tables listed in the vehicle's service manual to avoid confusion when reading wiring diagrams.

Wiring diagrams can become quite complex. To avoid this, the vehicle's electrical system may be divided into many diagrams, each illustrating only one system, such as the backup light circuit, oil pressure indicator light circuit, or wiper motor circuit. In more complex ignition, electronic fuel injection, and computer control systems, one diagram may be used to illustrate only part of the entire circuit.

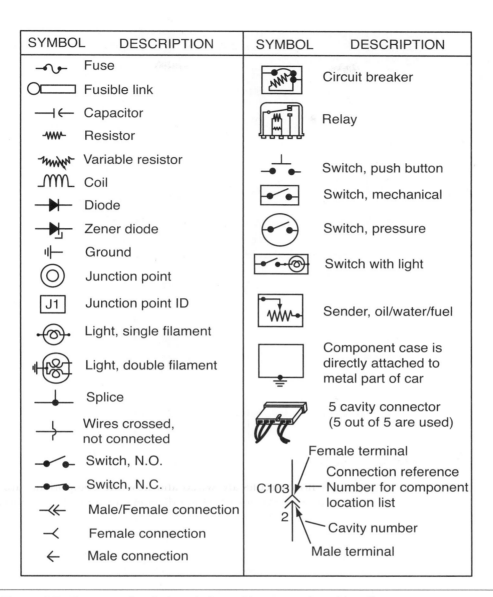

SYMBOL	DESCRIPTION	SYMBOL	DESCRIPTION
	Fuse		Circuit breaker
	Fusible link		Relay
	Capacitor		Switch, push button
	Resistor		Switch, mechanical
	Variable resistor		Switch, pressure
	Coil		Switch with light
	Diode		Sender, oil/water/fuel
	Zener diode		Component case is directly attached to metal part of car
	Ground		5 cavity connector (5 out of 5 are used)
	Junction point		
J1	Junction point ID		Female terminal
	Light, single filament	C103	Connection reference Number for component location list
	Light, double filament	2	Cavity number
	Splice		Male terminal
	Wires crossed, not connected		
	Switch, N.O.		
	Switch, N.C.		
	Male/Female connection		
	Female connection		
	Male connection		

Figure 4-2 Common electrical symbols used in automotive wiring diagrams.

Electrical Problems

All electrical problems can be classified into one of three categories: opens, shorts, or high-resistance problems. Identifying the type of problem allows you to identify the correct tests to perform when diagnosing an electrical problem.

An *open* occurs when a circuit has a break in the wire. Without a completed path, current cannot flow and the load or component cannot work. An open circuit can be caused by a disconnected wire, a broken wire, or a switch in the off position. Although voltage will be present up to the open point, there is no current flow. Without current flow, there are no voltage drops across the various loads.

An open circuit is often called a break in the circuit's continuity.

A short to ground will cause fuses to fail.

A circuit with a short to ground is sometimes called a grounded circuit.

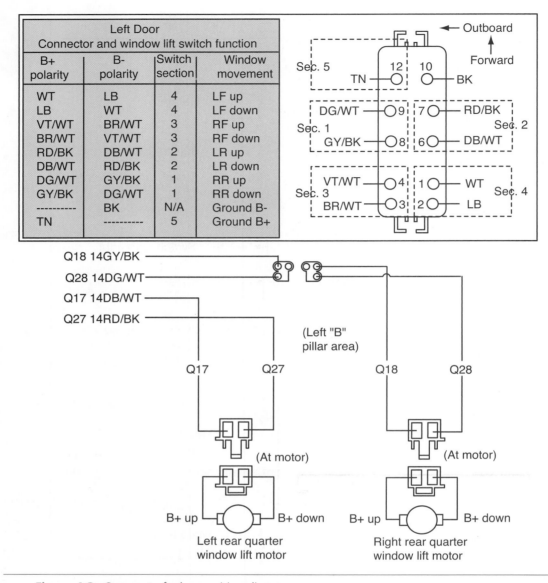

Left Door Connector and window lift switch function			
B+ polarity	B- polarity	Switch section	Window movement
WT	LB	4	LF up
LB	WT	4	LF down
VT/WT	BR/WT	3	RF up
BR/WT	VT/WT	3	RF down
RD/BK	DB/WT	2	LR up
DB/WT	RD/BK	2	LR down
DG/WT	GY/BK	1	RR up
GY/BK	DG/WT	1	RR down
----------	BK	N/A	Ground B-
TN	----------	5	Ground B+

Figure 4-3 One part of a large wiring diagram.

A *short* results from an unwanted path for current. Shorts cause an increase in current flow. This increased current flow can burn wires or components. Sometimes two circuits become shorted together. When this happens, one circuit powers another. This may result in strange happenings, such as the horn blasting every time the brake pedal is depressed. In this case, the brake light circuit is shorted to the horn circuit (Figure 4-4). Improper wiring or damaged insulation are the two major causes of short circuits.

A short can also be an unwanted short to ground (Figure 4-5). This problem provides a low resistance path to ground. This problem causes extremely high current flow that will damage wires and electrical components.

High resistance problems occur when there is unwanted resistance in the circuit. The higher-than-normal resistance causes the current flow to be lower than normal and the components in the circuit are unable to operate properly (Figure 4-6). A common cause of this type of problem is corrosion at a connector. The corrosion becomes an additional load in the circuit. This load not only decreases the circuit's current, but also uses some of the circuit's voltage, which prevents full voltage to the normal loads in the circuit. In this case, the bulb would be dim.

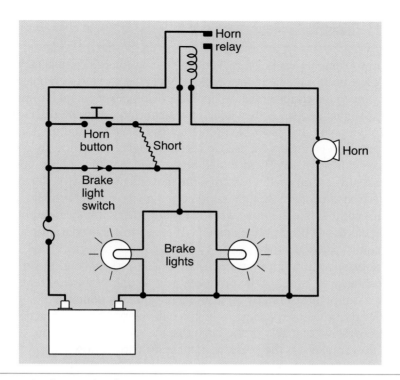

Figure 4-4 A wire-to-wire short.

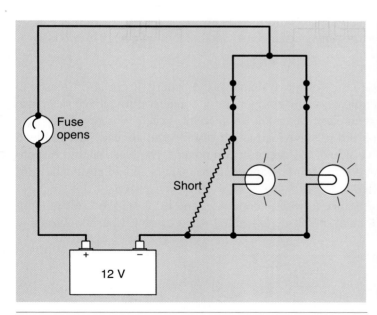

Figure 4-5 A short to ground.

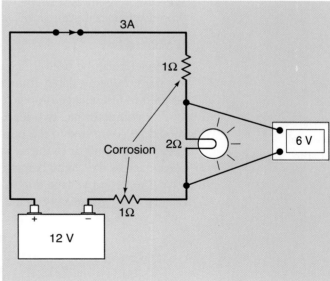

Figure 4-6 A simple light circuit with unwanted resistance. Notice the reduced voltage drop across the lamp and the reduced circuit current.

Electrical Test Equipment

To troubleshoot a problem, always begin by verifying the customer's complaint. Then operate the systems for a complete understanding of the problem. Often there are other problems that are not as evident or bothersome to the customer that will provide helpful information for diagnostics. Refer to the correct wiring diagram and study the circuit that is affected. From the diagram you should be able to identify testing points and probable problem areas. Then test and use logic to identify the cause of the problem. Several meters are used to test and diagnose electrical systems. These are the voltmeter, ohmmeter, and ammeter. These should be used along with test lights and jumper wires.

An ampere is usually called an amp.

Electrical current is a term used to describe the movement or flow of electricity. The greater the number of electrons flowing past a given point in a given amount of time, the more current the circuit has. This current, like the flow of water or any other substance, can be measured. The unit for measuring electrical current is the ampere. The instrument used to measure electrical current flow in a circuit is called an ammeter.

When any substance flows, it meets resistance. The resistance to electrical flow can be measured. The resistance to current flow produces heat. A unit of measured resistance is called an ohm. Resistance can be measured by an instrument called an ohmmeter.

Voltage is electrical pressure. Voltage is the force developed by the attraction of electrons to protons. The more positive one side of the circuit is, the more voltage is present in the circuit. Voltage does not flow, rather it is the pressure that causes current flow.

To have electricity, some force is needed to move the electrons between atoms. This **electromotive force (EMF)** is the pressure that exists between a positive and negative point within an electrical circuit. This force is measured in units called volts. One volt is the amount of pressure (force) required to move 1 ampere of current through a resistance of 1 ohm. Voltage is measured by an instrument called a voltmeter.

The amount of current that flows in a circuit is determined by the resistance in that circuit. As resistance goes up, the current goes down. The energy used by a load is measured in volts. Amperage stays constant in a circuit, but the voltage is dropped as it powers a load. Measuring voltage drop determines the amount of electrical energy changed to another form of energy by the load.

Voltmeter

A voltmeter can be used to measure the voltage available at the battery. It can also be used to test the voltage available at the terminals of any component or connector. A voltmeter can also be used to test voltage drop across an electrical circuit, component, switch, or connector.

A voltmeter has two leads: a red positive lead and a black negative lead. The red lead should be connected to the positive side of the circuit or component. The black should be connected to ground or to the negative side of the component. Voltmeters should always be connected across the circuit being tested.

The loss of voltage due to resistance in wires, connectors, and loads is called voltage drop (Figure 4-7). Voltage drop is the amount of voltage expended when current is pushed through a

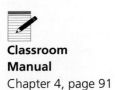

Classroom Manual
Chapter 4, page 91

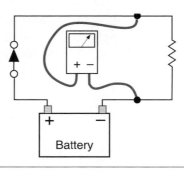

Figure 4-7   A voltmeter is connected in parallel or across the component or part of the circuit being tested for voltage drop.

resistance. Voltage drop that exceeds specifications for a given circuit indicates an undesired load in the circuit. Too much unwanted resistance or measurable voltage drop may reduce the amount of voltage, causing the component (load) to work improperly. The procedure for measuring a voltage drop is shown in Photo Sequence 5.

A voltmeter can also be used to check for proper circuit grounding. For example, if a voltmeter reading indicates full voltage at the lights but no lighting is seen, the bulbs or sockets could be bad or the ground connection is faulty.

An easy way to check for a defective bulb is to replace it with one known to be good. You can also use an ohmmeter to check for electrical continuity through the bulb.

If the bulbs are good, the problem lies in either the light sockets or ground wires. Connect the voltmeter to the ground wire and a good ground. If the light socket is defective, the voltmeter would read 0 volts. If the socket was not defective but the ground wire was broken or disconnected, the voltmeter would read very close to battery voltage. In fact, any voltage reading would indicate a bad or poor ground circuit. The higher the voltage, the greater the problem.

Photo Sequence 5
Performing a Voltage Drop Test

P5-1 The tools required to perform a voltage drop test are a voltmeter and fender covers.

P5-2 Set the voltmeter to its highest DC voltage scale; then move down in scale until you get the best reading.

P5-3 To test the voltage drop of an entire power feed circuit, connect the red (positive) lead to the positive terminal of the battery.

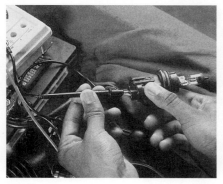

P5-4 Connect the black (negative) lead to the power wire at the connector for the load. A low beam headlight circuit is used in this photo. Turn the circuit on.

P5-5 Read the voltmeter. If there are no unwanted resistances in the circuit, the voltmeter will read less than 0.1 volt. If the reading is greater than 0.1 volt, wire or connector problems are indicated. To find the source of the problem, move the black test lead to another circuit connector closer to the battery.

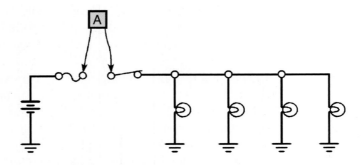

Figure 4-8 Measuring current with an ammeter. Notice this meter must be placed in series with the circuit.

Ammeters

An ammeter measures current flow in a circuit. An ammeter must be placed into or in series with the circuit being tested (Figure 4-8). Normally, this requires disconnecting a wire or connector from a component and connecting the ammeter between the wire or connector and the component. The red lead of the ammeter should always be connected to the side of the connector closest to the positive side of the battery, and the black lead should be connected to the other side.

> **CAUTION:** Never place the leads of an ammeter across the battery or a load. This puts the meter in parallel with the circuit and will blow the fuse in the ammeter or possibly destroy the meter.

It is much easier to test current using an ammeter with an inductive pickup. The pickup clamps around the wire or cable being tested. These ammeters measure amperage based on the magnetic field created by the current flowing through the wire. This type of pickup eliminates the need to separate the circuit to insert the meter.

Because ammeters are built with very low internal resistance, connecting them in series does not add any appreciable resistance to the circuit. Therefore, an accurate measurement of the current flow can be taken.

For example, assume that a circuit normally draws 5 amps and is protected by a 6-amp fuse. If the circuit constantly blows the 6-amp fuse, a short exists somewhere in the circuit. Mathematically, each light should draw 1.25 amperes ($5 \div 4 = 1.25$). To find the short, disconnect all lights by removing them from their sockets. Then, close the switch and read the ammeter. With the load disconnected, the meter should read 0 amperes. If there is any reading, the wire between the fuse block and the socket is shorted to ground.

If zero amps was measured, reconnect each light in sequence. The reading should increase 1.25 amperes with each bulb. If when making any connection the reading is higher than expected, the problem is in that part of the light circuit.

> **CAUTION:** When testing for a short, always use a fuse. Never bypass the fuse with a wire. The fuse should be rated at no more than 50 percent higher capacity than specifications. This offers circuit protection and provides enough amperage for testing. After the problem is found and corrected, be sure to install the specified rating of fuse for circuit protection.

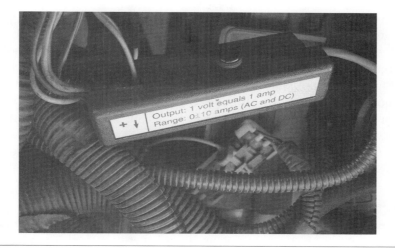

Figure 4-9 A low-amp probe.

Current Probes

Current probes have been used for many years, especially for larger amperages such as battery and starting. In recent years, the trend has been towards **low-amp probes** (Figure 4-9) that are small and are designed to measure small currents in the milliampere range. As stated earlier, the amp probe actually measures the magnetic field around the wire. This eliminates the need for back probing and possibly damaging the wiring and its connectors. A typical use for the low-amp probe is shown in Figure 4-10. The diagram is a rough drawing of a pair of transmission shift solenoids. Battery positive is supplied to the solenoids which are wired in parallel. In

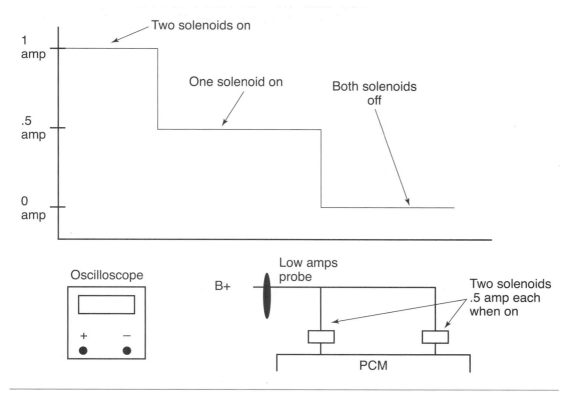

Figure 4-10 A typical use for the low-amp probe.

order to confirm that the solenoids were being activated by the PCM at the appropriate time (the solenoids are difficult to access), a low-amp probe was installed on the B+ wire into the transmission. It was possible to determine when both, one, or none of the solenoids were active when commanded on by measuring the current flow as shown in the diagram. Of course, it would also be possible to determine if the solenoids or the wiring were shorted or open by the amount of current flow. Also, the circuit is running at normal amperage which is a better check of component health than using an ohmmeter to check winding resistance. Current probes can also be used to check electric motors, coils, and solenoids for a correct **current ramp**, which is a characteristic waveform that uses amperage instead of voltage to examine the waveform of the component.

Ohmmeters

An ohmmeter measures resistance to current flow in a circuit. In contrast to the voltmeter, which uses the voltage available in the circuit, an ohmmeter is battery powered. The circuit being tested must be open. If the power is on in the circuit, the ohmmeter will be damaged.

The two leads of the ohmmeter are placed across or in parallel with the circuit or component being tested (Figure 4-11). The red lead is placed on the positive side of the circuit, and the black lead is placed on the negative side of the circuit. The meter sends current through the component and determines the amount of resistance based on the voltage dropped across the load. The scale of an ohmmeter reads from 0 to infinity (∞). A zero reading means there is no resistance in the circuit and may indicate a short in a component that should show a specific resistance. An infinity reading indicates a number higher than the meter can measure. This usually is an indication of an open circuit.

Ohmmeters are also used to trace and check wires or cables. Assume that one wire of a four-wire cable is to be found. Connect one probe of the ohmmeter to the known wire at one end of the cable, and touch the other probe to each wire at the other end of the cable. Any evidence of resistance, such as meter needle deflection, indicates the correct wire. You can use this same method to check a suspected defective wire. If resistance is shown on the meter, the wire is sound. If no resistance is measured, the wire is defective (open). If the wire is okay, continue checking by connecting the probe to other leads. Any indication of resistance indicates that the wire is shorted to one of the other wires and that the harness is defective.

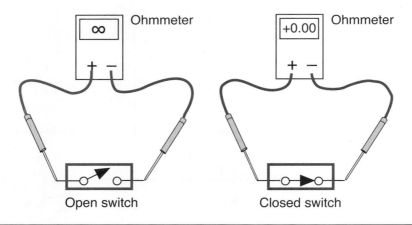

Figure 4-11 Measuring resistance with an ohmmeter. The meter is placed in parallel with the component after power is removed from the circuit.

AUTHOR'S NOTE

Ohmmeters to measure resistance or voltage drop?

Many service technicians have asked: "If I have an ohmmeter to measure resistance, why would I have to perform a voltage drop to find unwanted resistance in a circuit?" A DVOM calculates resistance based on Ohm's Law. A current is applied through the DVOM's internal circuitry from the meter battery to the circuit or component being tested. This current is usually very, very small compared to the amount of current normally running in the same circuit. This small current does not really load the circuit at enough amperage to determine if there is a problem at operating current. Since we cannot use the ohmmeter to measure current with the power applied, think of voltage drop as a dynamic test of circuit resistance. Consider this: A battery cable that has most of its strands cut through would still read "good" on a resistance test with an ohmmeter. The same cable under the considerable load of a starter motor would quickly heat up and show a significant "voltage drop." While the ohmmeter is an indispensable tool, the service technician must remember that it does not "dynamically" test a circuit or component. Current flow or voltage drop can often expose "it tested okay but still does not work" components or circuits.

Digital Multimeter (DMM) Usage

A multimeter is one of the most versatile tools used in diagnosing engine performance and electrical systems. Analog meters have low input impedance, but most digital meters have a high input impedance, usually at least 10 megohms. Metered voltage for resistance tests is well below 5 volts, reducing the risk of damaging sensitive components and delicate computer circuits.

Multimeters have either an auto range feature, in which the appropriate scale is automatically selected by the meter, or they must be set to a particular range. In either case, you should be familiar with the ranges and the different settings available on the meter you are using. To designate particular ranges and readings, meters display a prefix before the reading or range. If the meter has a setting for mAmps, this means the readings will be given in milliamps, or 1/1000th of an amp. Ohmmeter scales are expressed as a multiple of tens or use the prefix K or M. K stands for kilo, or one thousand. A reading of 10K ohms equals 10,000 ohms. An M stands for mega, or one million. A reading of 10M ohms equals 10,000,000 ohms. When using a meter with an auto range, make sure you note the range being used by the meter. There is a big difference between 10 ohms and 10,000,000 ohms. The common abbreviations and symbols used on multimeters are shown in Figure 4-12.

10 megohms equals 10,000,000 ohms.

PREFIX	SYMBOL	RELATION TO BASIC UNIT
Mega	M	1,000,000
Kilo	K	1.000
Milli	m	0.001 or $\frac{1}{1000}$
Micro	μ	0.000001 or $\frac{1}{1000000}$
Nano	n	0.000000001
Pico	p	0.000000000001

Figure 4-12 Common prefixes used on meters.

$$0.345 \text{ K}\Omega = 345 \, \Omega$$

$$1025 \text{ mAmps} = 1.025 \text{ Amps}$$

Figure 4-13 Placement of decimal and scale should be noticed when measuring with a meter with auto range.

CAUTION: Many digital multimeters with auto range display the measurement with a decimal point. Make sure you observe the decimal and the range being used by the meter (Figure 4-13). A reading of 0.972 K ohms equals 972 ohms. If you ignore the decimal point, you will read 972,000 ohms.

Root mean square (RMS) meters convert the AC signal to DC voltage signal.

Average responding meters show the average voltage peak.

There are two ways multimeters display AC voltage: **root mean square (RMS)** and **average responding**. When the AC voltage signal is a true sine wave, both methods will display the same reading. Since most automotive sensors do not produce pure sine wave signals, it is important to know how the meter will display the AC voltage reading when comparing measured voltage to specifications. RMS meters convert the AC signal to a comparable DC voltage signal. Average responding meters display the average voltage peak. Always check the voltage specification to see if the specification is for RMS voltage. If this is the case, use an RMS meter.

When using the ohmmeter function, the DMM will show zero or close to zero when there is good continuity. If the continuity is very poor, the meter will display an infinite reading. This reading is usually shown as a blinking "1.000", a blinking "1", or an "OL". Before taking any measurement, calibrate the meter. This is done by holding the two leads together and adjusting the meter reading to zero. Not all meters need to be calibrated; some digital meters automatically calibrate when a scale is selected. On meters that require calibration, it is recommended that the meter be zeroed after changing scales.

Some multimeters also feature a min/max function. This function displays the maximum, minimum, and average voltage the meter recorded during the time of the test. This feature is valuable when checking sensors or when looking for electrical noise. Noise is primarily caused by **radio frequency interference (RFI)** which may come from the ignition system. RFI is an unwanted voltage signal that rides on a signal. This noise can cause intermittent problems with unpredictable results. The noise causes slight increases and decreases in the voltage. When a computer receives a voltage signal with noise, it will try to react to the minute changes. As a result, the computer responds to the noise rather than the voltage signal.

Multimeters may also have the ability to measure duty cycle, pulse width, and frequency. All of these represent voltage pulses caused by the turning on and off of a circuit or the increase and decrease of voltage in a circuit. **Duty cycle** is a measurement of the amount of time something is on compared to the time of one cycle (Figure 4-14). Duty cycle is measured in a percentage. A 60-percent duty cycle means that a device is on 60 percent of the time and off 40 percent of one cycle. Duty cycle is similar to dwell, which is normally expressed in degrees. For example, a 30-degree dwell on a 60-degree scale equals 50 percent duty cycle.

Pulse width is similar to duty cycle except that it is the exact time something is turned on (Figure 4-15). Pulse width is normally measured in milliseconds. When measuring duty cycle, you are looking at the amount of time something is on during one cycle. When measuring pulse width, you are looking at the amount of time something is on.

The number of cycles that occur in one second is called the **frequency** (Figure 4-16). The higher the frequency, the more cycles occur in a second. Frequencies are measured in hertz. One hertz is equal to one cycle per second. The example in Figure 4-16 equals 10 hertz (10 Hz).

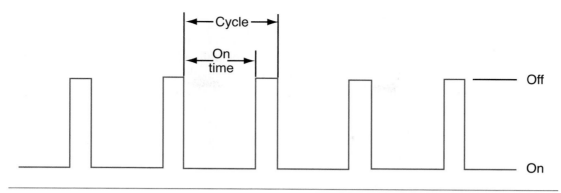

Figure 4-14 Duty cycle.

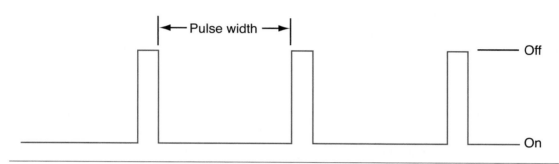

Figure 4-15 Pulse width.

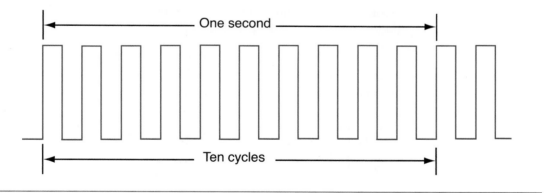

Figure 4-16 Signal frequency.

To accurately measure duty cycle, pulse width, and frequency, the meter's trigger level must be set. The trigger level tells the meter when to start counting. Trigger levels can be set at certain voltage levels or at a rise or fall in the voltage. Normally, meters have a built-in trigger level that corresponds with the voltage range setting. If the voltage does not reach the trigger level, the meter will not begin to recognize a cycle. On some meters you can select between a rise or fall in voltage to trigger the cycle count. A rise in voltage is a positive increase in voltage. This setting is used to monitor the activity of devices whose power feed is controlled by a computer. A fall in voltage is negative voltage. This setting is used to monitor ground-controlled devices.

Graphing Multimeter

A power-graphing multimeter has a history feature that can capture signals during an operator-selectable or preset timeframe. This can be particularly helpful in locating intermittent problems or performing dynamic tests on various components. The graphing multimeter is connected to the device or circuit you want to test and is set up to capture the signals. Various configurations can be set, including specific triggering voltage, minimum and maximum limits, and time scale. When you are done with the dynamic portion of the test, simply review the signal information recorded or stored in the graphing multimeter. The playback feature allows you to easily observe the results and spot glitches. Some units will have a library of signals, patterns, or other databases that can help you determine results.

Lab Scope Usage

An oscilloscope is a visual voltmeter. For many years, technicians have used scopes to diagnose ignition, fuel injection, and charging systems. These scopes, called tune-up scopes, were normally part of a large diagnostic machine, although some were stand-alone units. In recent years, an electronic scope, referred to as a lab scope, has become the diagnostic tool of choice for many good technicians.

A scope allows a technician to see voltage changes over time (Figure 4-17). Voltage is displayed across the screen of the scope as a waveform. By displaying the waveform, the scope shows the slightest changes in voltage. This is a valuable feature for a diagnostic tool. Measuring voltage is a common test for diagnostics. Precise measurement is possible with a scope. When measuring voltage with an analog voltmeter, the meter only displays the average values at the point being probed. Digital voltmeters simply sample the voltage several times each second and update the meter's reading at a particular rate. If the voltage is constant, good measurements can be made with both types of voltmeters. A scope will display any change in voltage as it occurs. This is especially important for diagnosing intermittent problems.

The screen of a lab scope is divided into small divisions of time and voltage (Figure 4-18). These divisions set up a grid pattern on the screen. Time is represented by the horizontal move-

The divisions on a scope are called graticules.

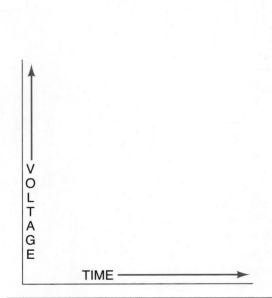

Figure 4-17 Voltage and time axis on a scope's screen.

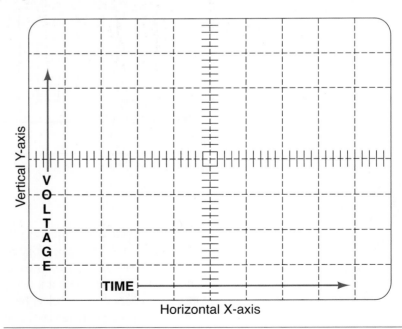

Figure 4-18 Grids on a scope screen that serve as a time and voltage reference.

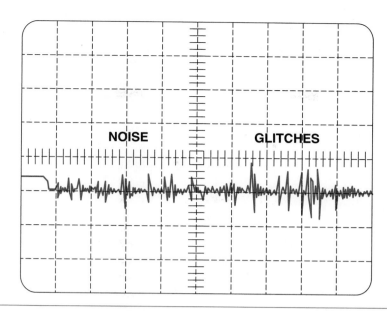

Figure 4-19 RFI noise and glitches may appear on a voltage signal.

ment of the waveform. Voltage is measured with the vertical position of the waveform. Since the scope displays voltage over time, the waveform moves from the left (the beginning of measured time) to the right (the end of measured time). The value of the divisions can be adjusted to improve the view of the voltage waveform. For example, the vertical scale can be adjusted so that each division represents 0.5 volts, and the horizontal scale can be adjusted so that each division equals 0.005 (5 milliseconds). This allows the technician to view small changes in voltage that occur in a very short period of time. The grid serves as a reference for measurements.

Since a scope displays actual voltage, it will display any electrical noise or disturbances that accompany the voltage signal (Figure 4-19). Noise is generally caused by RFI. Noise on a voltage signal causes the signal to change. These slight changes cause the computer to react, making changes to system operations when they are not necessary.

Electrical disturbances, or **glitches**, are momentary changes in the signal. These can be caused by intermittent shorts to ground, shorts to power, or opens in the circuit. These problems can occur for only a moment or may last for some time. A lab scope is handy for finding these and other causes of intermittent problems. By observing a voltage signal and wiggling or pulling a wiring harness, any looseness can be detected by a change in the voltage signal.

Analog versus Digital Scopes

Analog scopes show the actual activity of a circuit and are called real-time scopes. This simply means that what is taking place at the point being measured or probed is what you see on the screen. Analog scopes have a fast update rate that allows for the display of activity without delay.

A digital scope, commonly called a **digital storage oscilloscope (DSO)**, converts the voltage signal into digital information and stores it into its memory. Some DSOs send the signal directly to a computer or a printer or save it to a disk or memory card. To help in diagnostics, a technician can freeze the captured signal for close analysis. DSOs also have the ability to capture low-frequency signals. Low-frequency signals tend to flicker when displayed on an analog screen. To have a clean waveform on an analog scope, the signal must be repetitive and occurring in real time. The signal on a DSO is not quite real time. Rather, it displays the signal as it occurred a short time before.

This slight delay is actually very slight. Most DSOs have a sampling rate of one million samples per second. This is quick enough to serve as an excellent diagnostic tool. This fast sampling rate allows slight changes in voltage to be observed. Slight and quick voltage changes cannot be observed on an analog scope.

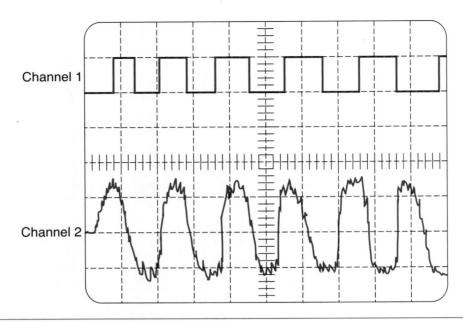

Figure 4-20 Some scopes have the feature of being able to display two traces at one time. These are called dual trace scopes.

A DSO uses an analog-to-digital (A/D) converter to digitize the input signal. Since digital signals are based on binary numbers, the trace appears slightly choppy when compared to an analog trace. However, the voltage signal is sampled more often, resulting in a more accurate waveform. The waveform is constantly being refreshed as the signal is pulled from the scope's memory. Remember, the sampling rate of a DSO can be as high as 1 million times per second.

Both an analog and a digital scope can be dual trace scopes (Figure 4-20). This means they both have the capability of displaying two traces at one time. By watching two traces simultaneously, you can watch the cause and effect of a sensor as well as compare a good or normal waveform to the one being displayed.

Waveforms

A waveform represents voltage over time. Any change in the amplitude of the trace indicates a change in the voltage. When the trace is a straight horizontal line, the voltage is constant (Figure 4-21). A diagonal line up or down represents a gradual increase or decrease in voltage. A sudden rise or fall in the trace indicates a sudden change in voltage. A **sinusoidal** wave represents an equal rise and fall from positive to negative. However, if the rise and fall are different, the wave is called **non-sinusoidal.**

Scopes can display AC and DC voltage, either one at a time or both together, as in the case of noise caused by RFI. Noise results from AC voltage riding on a DC voltage signal. The consistent change of polarity and amplitude of the AC signal causes slight changes in the DC voltage signal. A normal AC signal changes its polarity and amplitude over a period of time. The waveform created by AC voltage is typically called a sine wave (Figure 4-22). One complete sine wave shows the voltage moving from zero to its positive peak then moving down through zero to its negative peak and returning to zero.

One complete sine wave is a cycle. The number of cycles that occur per second is the frequency of the signal. Checking frequency or cycle time is one way of checking the operation of some electrical components. Input sensors are the most common components that produce AC voltage. Permanent magnet voltage generators produce an AC voltage that can be checked on a

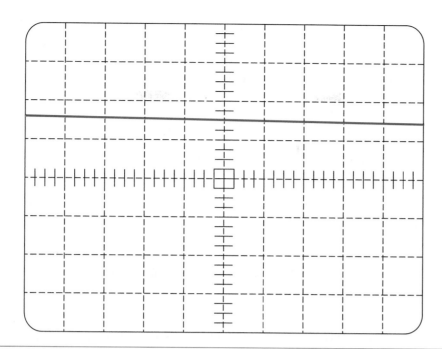

Figure 4-21 A constant voltage waveform.

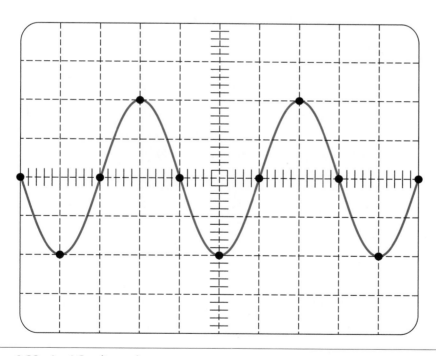

Figure 4-22 An AC voltage sine wave.

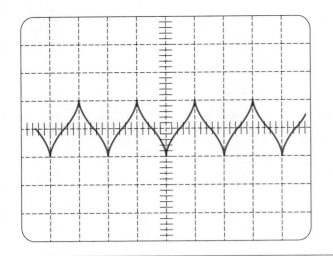

Figure 4-23 An AC voltage trace from a typical permanent magnetic generator-type pickup or sensor.

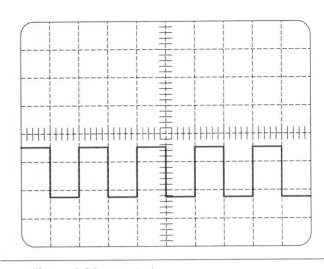

Figure 4-24 A typical square wave.

scope (Figure 4-23). AC voltage waveforms should also be checked for noise and glitches. These may send false information to the computer.

DC voltage waveforms may appear as a straight line or a line showing a change in voltage. Sometimes a DC voltage waveform will appear as a square wave, which shows voltage making an immediate change (Figure 4-24). Square waves are identified by having straight vertical sides and a flat top and bottom. This type of wave represents voltage being applied (circuit being turned on), voltage being maintained (circuit remaining on), and no voltage applied (circuit is turned off). A DC voltage waveform may also show gradual voltage changes.

Scope Controls

Depending on the manufacturer and model of the scope, the type and number of its controls will vary. However, nearly all scopes have these: intensity, vertical (Y-axis) adjustments, horizontal (X-axis) adjustments, and trigger adjustments. The intensity control is used to adjust the brightness of the trace. This allows for clear viewing regardless of the light around the scope screen.

The vertical adjustment controls the voltage displayed. The voltage setting of the scope is the voltage that will be shown per division (Figure 4-25). If the scope is set at 0.5 (500 milli)

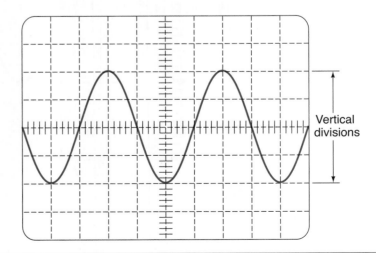

Figure 4-25 Vertical divisions represent voltage.

volts, that means a 5-volt signal will need ten divisions. Likewise, if the scope is set to one volt, 5 volts will need only five divisions. While using a scope, it is important to set the vertical so that voltage can be accurately read. Setting the voltage too low may cause the waveform to move off the screen, while setting it too high may cause the trace to be flat and unreadable. The vertical position control allows the vertical position of the trace to be moved anywhere on the screen.

The horizontal position control allows the horizontal position of the trace to be set on the screen. The horizontal control is the time control of the trace (Figure 4-26). Setting the horizontal control is setting the time base of the scope's sweep rate. If the time per division is set too low, the complete trace may not show across the screen. Also if the time per division is set too high, the trace may be too crowded for detailed observation. The time per division (time/div) can be set from very short periods of time (millionths of a second) to full seconds.

Trigger controls tell the scope when to begin a trace across the screen. Setting the trigger is important when trying to observe the timing of something. Proper triggering will allow the trace to repeatedly begin and end at the same points on the screen. There are typically numerous trigger controls on a scope. The trigger mode selector has a norm and auto position. In the norm setting, no trace will appear on the screen until a voltage signal occurs within the set time base. The auto setting will display a trace regardless of the time base.

Slope and level controls are used to define the actual trigger voltage. The slope switch determines whether the trace will begin on a rising or falling of the voltage signal

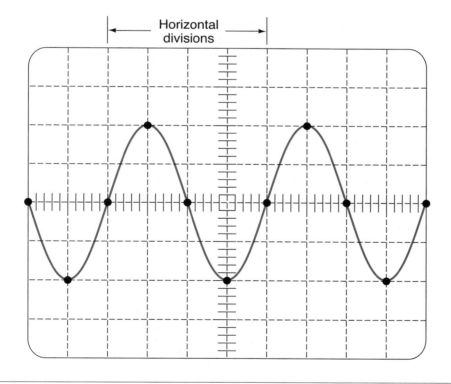

Figure 4-26 Horizontal divisions represent time.

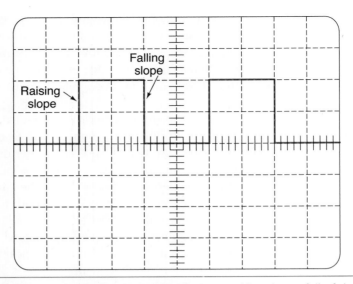

Figure 4-27 The trigger can be set to start the trace with a rise or fall of the voltage.

(Figure 4-27). The level control determines where the time base will be triggered according to a certain point on the slope.

A trigger source switch tells the scope which input signal to trigger on. This can be Channel 1, Channel 2, line voltage, or an external signal. External signal triggering is very useful when desiring to observe a trace of a component that may be affected by the operation of another component. An example of this would be observing fuel injector activity when changes in throttle position are made. The external trigger would be voltage change at the Throttle Position Sensor. The displayed trace would be the cycling of a fuel injector. Channel 1 and Channel 2 inputs are determined by the points of the circuit being probed. Some scopes have a switch that allows inputs from both channels to be observed at the same time or alternately.

Logic Probes

In some circuits, pulsed or digital signals pass through the wires. These on-off digital signals either carry information or provide power to drive a component. Many sensors used in a computer-control circuit send digital information back to the computer. To check the continuity of the wires that carry digital signals, a logic probe can be used (Figure 4-28).

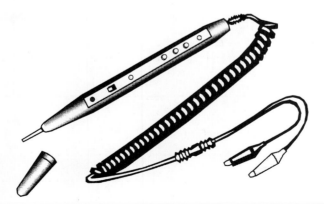

Figure 4-28 A typical logic probe.

A **logic probe** is similar in appearance to a test light. It contains three different colored light-emitting diodes (LEDs). A red LED lights when there is high voltage at the point being probed. A green LED lights to indicate low voltage. A yellow LED indicates the presence of a voltage pulse. The logic probe is powered by the circuit and reflects only the activity at the point being probed. When the probe's test leads are attached to a circuit, the LEDs display the activity.

If a digital signal is present, the yellow LED will turn on. When there is no signal, the LED is off. If voltage is present, the red or green LEDs will light depending on the amount of voltage. When there is a digital signal and the voltage cycles from low to high, the yellow LED will be lit and the red and green LEDs will cycle indicating a change in the voltage.

Basic Electrical Troubleshooting

Troubleshooting electrical problems involves the same tools and methods regardless of what circuit has the problem. All electrical circuits must have voltage, current, and resistance. Therefore, testing and measuring these in addition to comparing your measurements to specifications are the key to effective diagnosis.

A shorted circuit decreases the resistance of the circuit. This happens by shorting across to another circuit or by shorting to a ground. When there is a circuit-to-circuit short, one of the circuits is not controlled by its switch. The shorted-to circuit becomes a new parallel leg to the shorted circuit. With this type of problem, many strange things can happen. When a circuit is shorted to ground, a new parallel leg is present. This new leg has very low resistance and causes the current in the circuit to increase drastically. High-resistance problems can occur anywhere in the circuit. However, the effect of high-resistance is the same regardless of where it is. Additional or unwanted resistance in series with a circuit will always reduce the current in the circuit and will reduce the amount of voltage drop by the component in the circuit.

To troubleshoot a problem, always begin by verifying the customer's complaint. Then operate the affected system and others. There are often other problems that are not as evident nor bothersome to the customer that will provide helpful information for diagnostics. With the correct wiring diagram for the car, study the circuit that is affected. From the diagram you should be able to identify testing points and probable problem areas. Then test and use logic to identify the cause of the problem.

An ammeter and a voltmeter connected to a circuit at the different locations shown in Figure 4-29 should give readings as indicated when there are no problems in the circuit.

If there is an open anywhere in the circuit, the ammeter will read zero current. If the open is in the 1-ohm resistor, a voltmeter connected from C to ground will read zero. However, if the resistor is open and the voltmeter is connected to points B and C, the reading will be 12 volts.

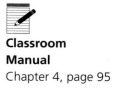

Classroom Manual Chapter 4, page 95

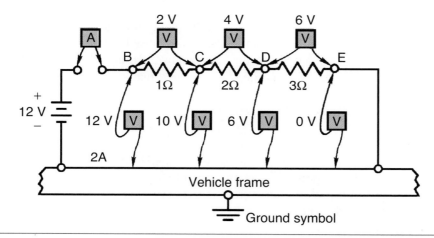

Figure 4-29 A basic circuit being tested with an ammeter and a voltmeter.

The reason is that the battery, ammeter, voltmeter, 2-ohm resistor, and 3-ohm resistor are all connected together to form a series circuit. Because of the open in the circuit, there is only current flow in the circuit through the meter, not the rest of the circuit. This current flow is very low because the meter has such high resistance. Therefore, the voltmeter will show a reading of 12 volts, indicating little, if any, voltage drop across the resistors.

To understand this concept, look at what happens if the 2-ohm resistor was open instead of the 1-ohm resistor. A voltmeter connected from point C to ground would indicate 12 volts. The 1-ohm resistor in series in the high resistance of the voltmeter would have little affect on the circuit. If an open should occur between point E and ground, a voltmeter connected from points B, C, D, or E to ground would read 12 volts. A voltmeter connected across any one of the resistors, from B to C, C to D, or D to E, would also read zero volts because there will be no voltage drops if there is no current flow.

A short would be indicated by excessive current and/or abnormal voltage drops. These examples illustrate how a voltmeter and ammeter may be used to check for problems in a circuit. An ohmmeter also may be used to measure the values of each component and compared to specifications. If there is no continuity across a part, it is open. If there is more resistance than called for, there is high internal resistance. If there is less resistance than specified, the part is shorted.

According to many manuals, the maximum allowable voltage drop for an entire circuit, except for the drop across the load, is 10 percent of the source voltage. Although 1.2 volts is the maximum acceptable amount, it is still too much. Many good technicians use 0.5 volts as the maximum allowable drop. However, there should be no more than 0.1 volts dropped across any one wire or connector. This is the most important specification to consider and remember.

Basic Electrical Repairs

Many automotive electrical problems can be traced to faulty wiring. Loose or corroded terminals, frayed, broken, or oil-soaked wires, and faulty insulation are the most common causes. Wires, fuses, and connections should be checked carefully during troubleshooting. Keep in mind that the insulation does not always appear to be damaged when the wire inside is broken. Also, a terminal may be tight but still may be corroded.

Wiring Harness and Terminal Diagnosis and Repair

CAUTION: Always follow the vehicle manufacturer's wiring and terminal repair procedure given in the service manual. On some components and circuits, manufacturers recommend component replacement rather than wiring repairs. For example, some manufacturers recommend replacing air bag system components such as sensors if the wiring or terminals are damaged on these components.

Wires should be checked for a burned or melted condition (Figure 4-30). Connector ring terminals should be checked for loose retaining nuts, which cause high resistance or intermittent open circuits (Figure 4-31). An open circuit may be caused by a terminal that is backed out of the connector (Figure 4-32). Terminals that are bent or damaged may cause shorts or open circuits (Figure 4-33). An open circuit occurs when the terminal is crimped over the insulation instead of the wire core (Figure 4-34). A greenish-white corrosion on terminals results in high resistance or an open circuit (Figure 4-35).

Wire end terminals are connecting devices. They are generally made of tin-plated copper and come in many shapes and sizes. They may be either soldered or crimped in place. When installing a terminal, select the appropriate size and type of terminal. Be sure it fits the unit's connecting post or prongs and it has enough current-carrying capacity for the circuit. Also, make sure it is heavy enough to endure normal wire flexing and vibration.

Remember Ohm's Law when trying to understand circuit defects: high resistance = low current flow; low resistance = high current flow.

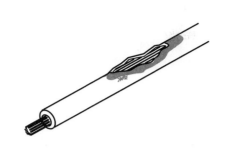

Figure 4-30 Damaged wire insulation.

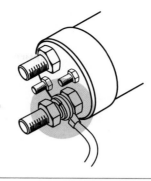

Figure 4-31 Loose retaining nuts on a ring-type terminal can cause high resistance or an intermittent open.

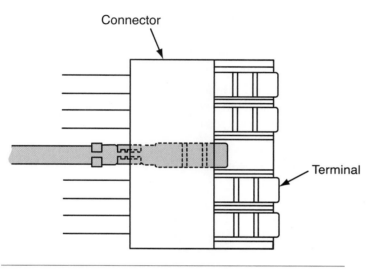

Connector

Terminal

Figure 4-32 Terminals that are backed out of their connector can cause an open circuit.

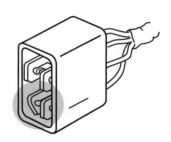

Figure 4-33 Bent or damaged component terminals may result in a shorted or open circuit.

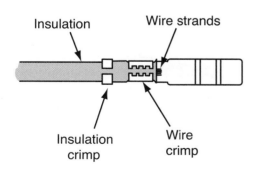

Insulation

Wire strands

Insulation crimp

Wire crimp

Figure 4-34 An open circuit occurs when a terminal is crimped over the insulation of the wire.

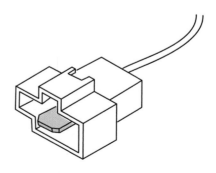

Figure 4-35 A greenish-white film of corrosion will cause high resistance and maybe an open in the circuit. It can also cause arcing across the terminals.

Some wiring harnesses contain a drain wire. When repairing wires in this type of harness, splice the wire or wires as explained previously. If more than one splice is required, stagger the wiring splices and tape the splices with mylar tape. Do not tape the drain wire onto the harness. Splice the drain wire separately (Figure 4-36) and then place mylar tape over the spliced area with a winding motion. Figure 4-37 shows how to crimp a terminal. Be sure to use the proper crimping tool and to follow the tool manufacturer's instructions.

CAUTION: Do not crimp a terminal with the cutting edge of a pair of pliers. While this method may crimp the terminal, it also weakens it.

General procedures for crimping on a connector follow:

1. Use the correct size stripping opening on the crimping tool and remove enough insulation to allow the wire to completely penetrate the connector.
2. Place the wire into the connector and crimp the connector. To get a proper crimp, place the open area of the connector facing toward the anvil. Make sure the wire is compressed under the crimp.
3. If connecting two wires, insert the stripped end of the other wire into the connector and crimp in the same manner.
4. Use electrical tape or heat-shrink tubing to tightly seal the connection. This will provide good protection for the wire and connector.

The preferred way to connect wires or to install a connector is by soldering. Some car manufacturers have used aluminum in their wiring. Aluminum cannot be soldered. Follow the manufacturer's guidelines and use the proper repair kits when repairing aluminum wiring.

CAUTION: Never use acid-core solder. It creates corrosion and can damage electronic components. Always use rosin-core solder.

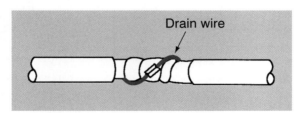

Figure 4-36 Separate splice for a drain wire in a wiring harness.

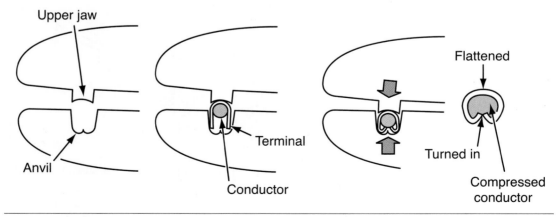

Figure 4-37 Properly crimping a connection with a crimping tool.

Splicing Copper Wire With Splice Clips

Follow this procedure to splice copper wire with splice clips:

1. Cut the tape off the harness if necessary. Be careful not to cut the wire insulation.
2. Cut the wire to be repaired as necessary. Do not cut more than necessary off the wire.
3. Use the appropriate stripper opening in a pair of wire strippers to strip the proper amount of insulation off the wire ends. Enough insulation must be stripped off the wire ends to allow the wire ends to fit into the splice clip. A small amount of wire should be visible on each side of the clip.
4. Place the wire ends in the splice clip so the clip is centered on the wire ends. Be sure all the wire strands are in the clip.
5. Place the proper size opening in the crimping tool over the splice clip so this tool is even with one edge of the clip. Apply steady pressure on the crimping tool to crimp the clip onto the wire.
6. Repeat step 5 on the opposite end of the clip.
7. Heat the splice clip with a soldering gun and melt rosin-core solder into the clip.
8. Use splicing tape to cover the splice, and do not flag the tape.
9. If the wire is installed in a harness, tape the complete harness in the area where the tape was removed. When the wire remains by itself, wrap splicing tape over the spliced area and the wire insulation near the spliced area.

Splicing Copper Wire With Crimp and Seal Splice Sleeves

Crimp and seal splice sleeves may be used to splice copper wire. These sleeves have a heat-shrink sleeve over the outside of the sleeve. The stripped wire ends are placed in the sleeve until they contact the stop in the center of the sleeve. A wire crimping tool is used to crimp the sleeve near each end. After the crimping procedure, the sleeve is heated slightly with a heat torch to shrink the insulating heat-shrink sleeve onto the terminal. When the shrinking process is complete, sealant should appear at each end of the sleeve (Figure 4-38).

When working with wiring and connectors, never pull on the wires to separate the connectors. This can loosen the connector and cause an intermittent problem that may be very difficult to find later. Always follow the correct procedures and use the tools designed for separating connectors.

Check all connectors for corrosion, dirt, and looseness. Nearly all connectors have push-down release type locks (Figure 4-39). Make sure these are not damaged when disconnecting the

Special Tools

Wire strippers, splice clip, crimping tool, soldering gun, rosin-core solder, splicing tape

The term tape flagging refers to a large piece of excess splicing tape hanging from one side of a wire.

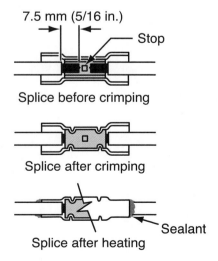

Figure 4-38 Crimp and seal splice sleeve after crimping and heating.

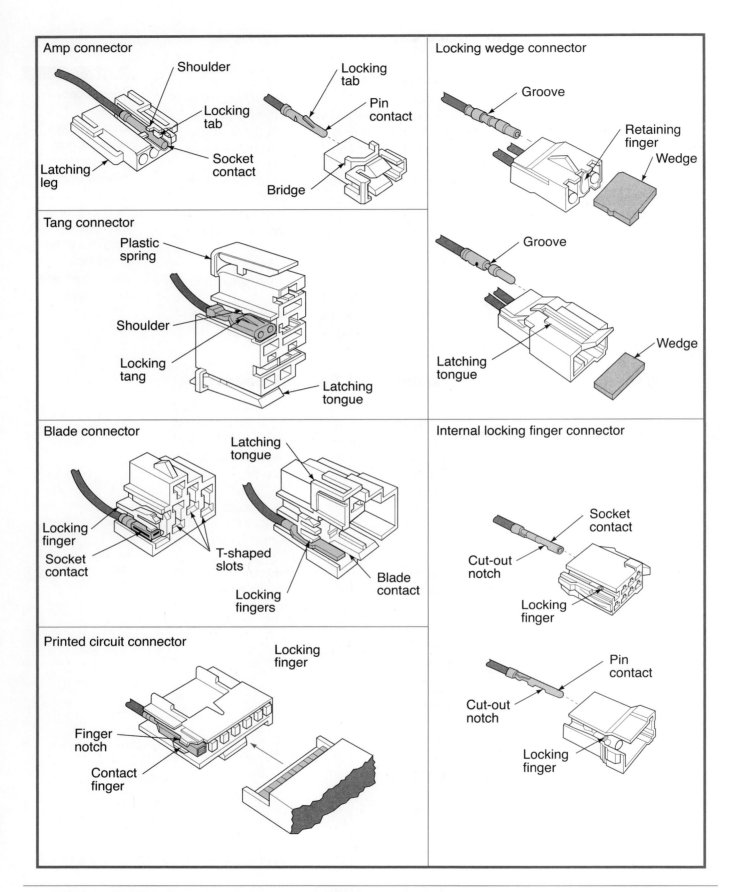

Figure 4-39 Different multiple-wire hard shell connectors and their locking mechanisms.

connectors. Many connectors have covers over them to protect them from dirt and moisture. Make sure these are properly installed to provide for that protection.

Never reroute wires when making repairs. Rerouting wires can result in induced voltages from nearby components. These stray voltages can interfere with the function of electronic circuits.

Dielectric grease should be used to moisture-proof and to protect connections from corrosion. If the manufacturer specifies that a connector be filled with grease, make sure it is. If the old grease is contaminated, replace it. Some car manufacturers suggest using petroleum jelly to protect connection points.

Testing Basic Electrical Components

The testing of individual electrical components requires a knowledge of the component's operation, the various diagnostic tools, and a thorough understanding of basic electricity. This section looks at some of the methods to test common electrical components.

> **CAUTION:** Fuses and other protection devices do not normally wear out. They go bad because something went wrong. Never replace a fuse or fusible link or reset a circuit breaker without finding out why it went bad.

Fuses

A protection device is designed to turn the system off whenever excessive current or an overload occurs (Figure 4-40). There are three basic types of fuses in automotive use: cartridge, blade, and ceramic. The cartridge fuse is found on most older domestic cars and a few imports. To check this type of fuse, look for a break in the internal metal strip. Discoloration of the glass cover or glue bubbling around the metal end caps is an indication of overheating. Late-model domestic vehicles and many imports use blade or spade fuses. To check the fuse, pull it from the fuse panel and look at the fuse element through the transparent plastic housing. Look for internal breaks and discoloration. The ceramic fuse is used on many European imports. To check this type of fuse, look for a break in the contact strip on the outside of the fuse. All types of fuses can be checked with an ohmmeter or test light. If the fuse is good, there will be continuity through it.

Fuses are rated by the current at which they are designed to blow. A three-letter code is used to indicate the type and size of fuses. Blade fuses have codes ATC or ATO. All glass SFE fuses have the same diameter, but the length varies with the current rating. Ceramic fuses are available in two sizes, code GBF (small) and the more common code GBC (large). The amperage rating is also embossed on the insulator. Codes such as AGA, AGW, and AGC indicate the length and diameter of the fuse. Fuse lengths in each of these series is the same, but the current rating can vary. The code and the current rating are usually stamped on the end cap.

A blade-type fuse is called a spade fuse.

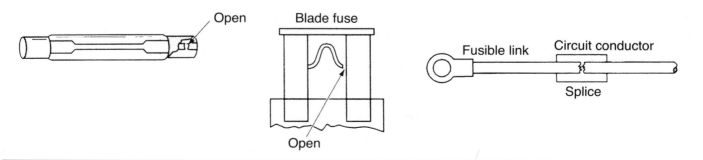

Figure 4-40 A protection device can have a fault that is not seen by a technician. They should always be tested with a volt or ohmmeter.

Blade Fuse Color Coding	
Ampere Rating	**Housing Color**
4	pink
5	tan
10	red
15	light blue
20	yellow
25	natural
30	light green
Fuse Link Color Coding	
Wire Link Size	**Insulation color**
20 GA	blue
18 GA	brown or red
16 GA	black or orange
14 GA	green
12 GA	gray
Maxi-fuse Color Coding	
Ampere Rating	**Housing color**
20	yellow
30	light green
40	amber
50	red
60	blue

Figure 4-41 Typical color coding of protection devices.

The current rating for blade fuses is indicated by the color of the plastic case (Figure 4-41). In addition, it is usually marked on the top. The insulator on ceramic fuses is color coded to indicate different current ratings.

Fusible Links

A fuse link is commonly called a fusible link.

Fuse (or fusible) links are used in circuits where limiting the maximum current is not extremely critical. They are often installed in the positive battery lead to the ignition switch and other circuits that have power with the key off. Fusible links are normally found in the engine compartment near the battery. Fusible links are also used when it would be awkward to run wiring from the battery to the fuse panel and back to the load.

Because a fuse link is a lighter gauge of wire than the main conductor, it melts and opens the circuit before damage can occur in the rest of the circuit. Fuse link wire is covered with a special insulation that bubbles when it overheats, indicating that the link has melted. If the insulation appears good, pull lightly on the wire. If the link stretches, the wire has melted. When it is hard to determine if the fuse link is burned out, check for continuity through the link with a test light or ohmmeter.

CAUTION: Do not mistake a resistor wire for a fuse link. A resistor wire is generally longer and is clearly marked "Resistor—do not cut or splice."

To replace a fuse link, cut the protected wire where it is connected to the fuse link. Then tightly crimp or solder a new fusible link of the same rating as the original link. Since the insulation on the manufacturer's fuse links is flameproof, never fabricate a fuse link from ordinary wire because the insulation may not be flameproof.

 WARNING: Always disconnect the battery ground cable prior to servicing any fuse link.

Maxi-Fuses

Many late-model vehicles use *maxi-fuses* instead of fusible links. Maxi-fuses look and operate like two-prong, blade, or spade fuses except they are much larger and can handle more current. Typically, a maxi-fuse is four to five times larger. Maxi-fuses are located in their own underhood fuse block.

Maxi-fuses are easier to inspect and replace than fuse links. To check a maxi-fuse, look at the fuse element through the transparent plastic housing. If there is a break in the element, the maxi-fuse has blown. To replace it, pull it from its fuse box or panel. Always replace a blown maxi-fuse with a new one having the same ampere rating.

Maxi-fuses allow the vehicle's electrical system to be broken down into smaller circuits that are easy to diagnose and repair. For example, in some vehicles a single fusible link controls one-half or more of all circuitry. If it burns out, many electrical systems are lost. By replacing this single fusible link with several maxi-fuses, the number of systems lost due to a problem in one circuit is drastically reduced. This makes it easy to pinpoint the source of trouble.

Circuit Breakers

Some circuits are protected by circuit breakers (Figure 4-42). Like fuses, they are rated in amperes. A circuit breaker can be cycling or must be manually reset. In the cycling type, the bimetal arm will begin to cool once the current to it is stopped. Once it returns to its original shape, the contacts are closed and power is restored. If the current is still too high, the cycle of breaking the circuit will be repeated.

Two types of non-cycling or resettable circuit breakers are used. One is reset by removing the power from the circuit. The other type is reset by depressing a reset button.

A visual inspection of a fuse or fusible link will not always determine if it is blown. To accurately test a circuit protection device, use an ohmmeter, voltmeter, or test light.

A circuit breaker is typically abbreviated c.b. in the fuse chart of a service manual.

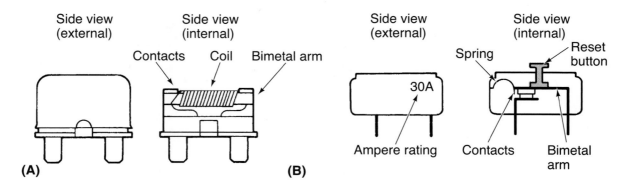

Figure 4-42 Two basic types of circuit breakers.

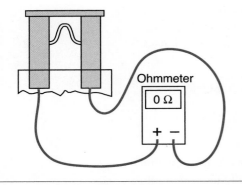

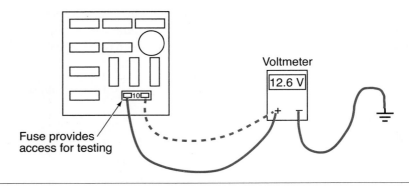

Figure 4-43 A good fuse will have zero resistance across it.

Figure 4-44 Circuit protection devices can be tested with a voltmeter. Make sure there is voltage present on both sides of the device.

With the fuse or circuit breaker removed from the vehicle, connect the ohmmeter's test leads across the protection device's terminals (Figure 4-43). On its lowest scale, the ohmmeter should read 0 ohms. If it reads infinity, the protection device is open. Test a fusible link in the same way. Before connecting the ohmmeter across the fusible link, make sure there is no current flow through the circuit. To be safe, disconnect the negative cable of the battery.

To test a circuit protection device with a voltmeter, check for available voltage at both terminals of the unit (Figure 4-44). If the device is good, voltage will be present on both sides. A test light can be used in place of a voltmeter.

☑ **SERVICE TIP:** Before using a test light, it is good practice to check the tester's lamp. To do this, simply connect the test light across the battery. The light should come on.

Measuring voltage drop across a fuse or other circuit protection device will tell you more about its condition than just if it is open. If a fuse, a fuse link, or circuit breaker is in good condition, a voltage drop of zero will be measured. If 12 volts is read, the fuse is open. Any reading between zero and 12 volts indicates some voltage drop. If there is voltage drop across the fuse, it has resistance and should be replaced. Make sure you check the fuse holder for resistance as well.

Switches

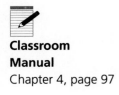

Classroom Manual
Chapter 4, page 97

To check a switch, disconnect the connector to the switch. With an ohmmeter, check for continuity between the terminals of the switch (Figure 4-45) with the switch moved to the on position and

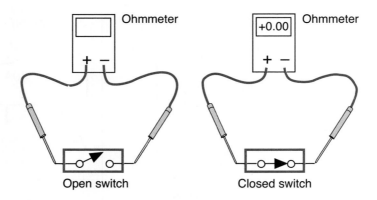

Figure 4-45 The continuity through a switch can be checked with an ohmmeter. With the switch closed, there should be zero resistance. With the switch open, there should be infinite resistance.

to the off position. With the switch in the off position, there should be no continuity between the terminals. With the switch on, there should be good continuity between the terminals. If the switch is activated by something mechanical and does not complete the circuit when it should, check the adjustment of the switch. (Some switches are not adjustable.) If the adjustment is correct, replace the switch. Another way to check the clutch switch is to simply bypass it with a jumper wire. If the component works when the switch is jumped, a bad switch is indicated.

> **CAUTION:** Before replacing a switch or any component connected to the vehicle's control computer, disconnect the negative cable from the battery.

Voltage drop across switches should also be checked. Ideally, when the switch is closed there should be no voltage drop. Any voltage drop indicates resistance and the switch should be replaced.

Relays

A relay can be checked with a jumper wire, voltmeter, ohmmeter, or test light. If the terminals are accessible, a jumper wire and test light will be the quickest method.

Check the wiring diagram for the relay being tested to determine if the control is through an insulated or ground switch. If the relay is controlled on the ground side, follow this procedure to test the relay.

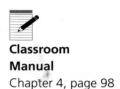

Classroom Manual
Chapter 4, page 98

1. Use the test light to check for voltage at the battery side of the relay. If voltage is not present, the fault is in the battery feed to the relay. If there is voltage, continue testing.
2. Probe for voltage at the control terminal. If voltage is not present, the relay coil is faulty. If voltage is present, continue testing.
3. Use a jumper wire to connect the control terminal to a good ground. If the relay works, the fault is in the control circuit. If the relay does not work, continue testing.
4. Connect the jumper wire from the battery to the output terminal of the relay. If the device operated by the relay works, the relay is bad. If the device does not work, the circuit between the relay and the device's ground is faulty.

If the relay is controlled by a computer, do not use a test light. Rather, use a high-impedance voltmeter set to the 20-volt DC scale, then:

1. Connect the negative lead of the voltmeter to a good ground.
2. Connect the positive lead to the output wire. If no voltage is present, continue testing. If there is voltage, disconnect the ground circuit of the relay. The voltmeter should now read 0 volts. If it does, the relay is good. If voltage is still present, the relay is faulty and should be replaced.
3. Connect the positive voltmeter lead to the power input terminal. Close to battery voltage should be there. If not, the relay is faulty. If the correct voltage is present, continue testing.
4. Connect the positive meter lead to the control terminal. Close to battery voltage should be there. If not, check the circuit from the battery to the relay. If the correct voltage is there, continue testing.
5. Connect the positive meter lead to the relay ground terminal. If more than 1 volt is present, the circuit has a poor ground.

If the relay terminals are not accessible, remove the relay from its mounting and test it on a bench. Use an ohmmeter to test for continuity between the relay coil terminals. If the meter indicates an infinity reading, replace the coil. If there is continuity, use a pair of jumper wires to energize the coil. Check for continuity through the relay contacts. If there is an infinity reading, the relay is faulty. If there is continuity, the relay is good and the circuits need to be tested carefully.

Be sure to check your service manual for resistance specifications and compare the relay to them. It is easy to check for an open coil; however, a shorted coil will also prevent the relay from working. Low resistance across a coil would indicate that is shorted. Too low of a resistance may also damage the transistors and/or driver circuits because of the excessive current that would result.

Stepped Resistors

Stepped resistors are often used to control heater blower fan speeds.

The best method for testing a stepped resistor is to use an ohmmeter. To do this, remove the resistor from its mounting. Connect the ohmmeter leads to the two ends of the resistor. Compare the results against specifications. Make sure the ohmmeter is set to the correct scale for the anticipated amount of resistance.

A stepped resistor can also be checked with a voltmeter. By measuring the voltage after each part of the resistor block and comparing the readings to specifications, you can tell if the resistor is good or not.

Variable Resistors

Classroom Manual
Chapter 4, page 96

Potentiometers are used as computer inputs. The most common example is a Throttle Position Sensor (TPS).

One way to test a variable resistor is with an ohmmeter. However, one can also be effectively checked dynamically by observing its output voltage.

To test a rheostat, identify the input and output terminals and connect the ohmmeter across them. Rotate the control while observing the meter. The resistance value should remain within the limits specified for the switch. If the resistance values do not match the specified amounts or if there is a sudden change in resistance as the control is moved, the unit is faulty. When using a voltmeter, the readings should be smooth and consistent as the control is moved.

To test a potentiometer, connect an ohmmeter across the resistor. The readings should be within the range listed in the specifications. If they are not, the resistor needs to be replaced. Then move the leads to the input and output of the resistor. The readings should sweep evenly and consistently within the specified resistance values. The condition of a potentiometer can also be checked with a voltmeter or a lab scope. The lab scope can be used to check signal integrity as well as voltage values. The DMM can show dynamic voltage changes and minimum and maximum voltages. As it is moved through its dynamic range, the lab scope can show any erratic voltage changes or "noise."

Diodes

Classroom Manual
Chapter 4, page 101

Regardless of the bias of the diode, it should allow current flow in one direction only. To test a diode, use an analog ohmmeter. Connect the meter's leads across the diode (Figure 4-46). Observe the reading on the meter, then reverse the meter's leads and observe the reading on the

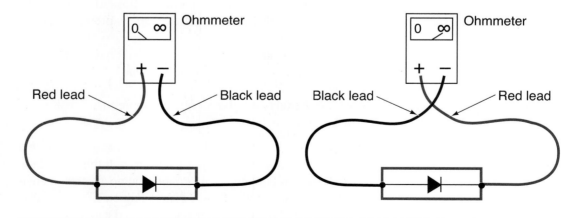

Figure 4-46 Use an ohmmeter to test a diode for an open and/or short.

meter. The resistance in one direction should be very high or infinity, and in the other direction, the resistance should be very low. If any other readings were observed, the diode is bad. A diode that has low resistance in both directions is shorted. A diode that has high resistance or an infinity reading in both directions is open.

Problems may be encountered when checking a diode with a high-impedance digital ohmmeter. Since many diodes will not allow current flow through them unless the voltage is at least 0.6 volts, a digital meter may not be able to forward bias the diode. This will result in readings that indicate the diode is open, when in fact it may not be. Because of this problem, many multimeters are equipped with a diode testing feature. This test allows for increased voltage at the test leads. Some meters will display the voltage required to forward bias the diode. If the diode is open, the meter will display "OL" or another reading to indicate infinity or out-of-range. Some meters during a diode check will make a beeping noise when there is continuity.

Diodes may also be tested with a voltmeter. Using the same logic as when testing with an ohmmeter, test the voltage drop across the diode. The meter should read low voltage in one direction and near source voltage in the other direction.

Battery Diagnosis and Service

CAUTION: Battery electrolyte contains sulfuric acid, a strong, corrosive acid. Electrolyte is very damaging to vehicle paint, upholstery, and clothing. Do not allow the electrolyte to contact these items.

WARNING: Battery electrolyte is very harmful to human skin and eyes. Always wear face and eye protection, protective gloves, and protective clothing when handling batteries or electrolyte. If electrolyte contacts your skin or eyes, flush with clean water immediately and obtain medical help.

WARNING: Never smoke or allow other sources of ignition near a battery. Hydrogen gas discharged while charging the battery is explosive (Figure 4-47)! A battery explosion may cause personal injury or property damage. Even when the battery has not been charged for several hours, hydrogen gas still may be present under the battery cover.

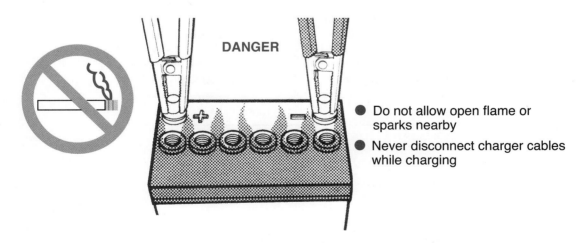

Figure 4-47 Never smoke or allow a spark or excessive heat near a battery.

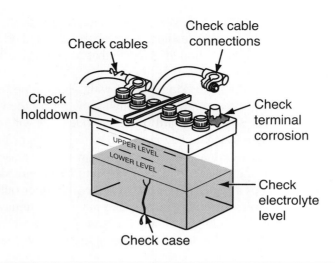

Figure 4-48 Guidelines for the inspection of a battery.

Classroom Manual
Chapter 4, page 106

A battery should be checked for terminal corrosion, loose or broken holddown, frayed or corroded cables, cracked case, and electrolyte level (Figure 4-48). If the battery has removable vent caps, the electrolyte level may be checked through the vent cap openings. When the vent caps are not removable, the electrolyte level may be checked by observing the level through the translucent plastic case. The battery may be removed from the vehicle, washed with a baking soda and water solution, and rinsed with clean water.

Open Circuit Voltage Test

The battery test can begin with an open circuit voltage (OCV) test by connecting a voltmeter to the battery terminals. This can be done in place of a battery hydrometer test. A static voltage reading of the battery can give an indication of the battery's condition and state of charge. The battery temperature should be between 60°F and 100°F (15.5°C and 37.7°C). Before performing the test, the battery voltage must stabilize for 10 minutes. If the battery is coming off a charge, apply a load for 15 seconds ($1/2$ cca), allow the battery to stabilize, and then test. An acceptable reading is 12.6 volts (see Figure 4-49). If the OCV indicates a charge below 75 percent of a full charge, the battery needs to be charged and load tested.

Open Circuit Voltage	State of Charge
12.6 or greater	100%
12.4 to 12.6	75–100%
12.2 to 12.4	50–75%
12.0 to 12.2	25–50%
11.7 to 12.0	0–25%
11.7 or less	0%

Figure 4-49 Battery open circuit voltage as an indicator of state of charge.

Figure 4-50 Performing a battery leakage test.

Battery Leakage Test

Connect the negative voltmeter lead to the battery terminal or post and touch the positive voltmeter lead at various points across the top and sides of the battery (Figure 4-50). If there is a reading on the voltmeter, voltage may be leaking from the battery. Over time, this can be a somewhat normal occurrence due to battery gassing and dirt collecting on top of the battery. Clean the battery with either battery cleaner or baking soda and water. Be sure to remove and clean the battery cable ends and battery posts.

WARNING: On air bag-equipped vehicles, the negative battery cable should be disconnected and the technician should wait 1 minute before working on the vehicle electrical system to avoid accidental air bag deployment. The wait time may vary depending on the vehicle manufacturer. Always follow the instructions in the vehicle manufacturer's service manual.

SERVICE TIP: Always disconnect the negative battery cable before the positive cable. If the positive cable is disconnected first, the wrench may slip and ground this cable to the chassis. This action may result in severe burns or a battery explosion.

Disconnecting a battery in a computer-equipped vehicle causes these problems:

1. Erases adaptive memories in various computers.
2. Erases memory in memory seats and memory mirror systems.
3. Erases station memory programmed in the radio.
4. On some cars, such as Chrysler products, the theft deterrent computer will not allow the vehicle to restart after a battery disconnection until one of the door lock cylinders is cycled from locked to unlocked.

To avoid these problems, some tool manufacturer's supply a dry cell voltage source that may be plugged into the cigarette lighter prior to disconnecting the battery.

CAUTION: Do not hammer or twist battery cable ends to loosen them. This action will loosen the terminal in the battery cover and cause electrolyte leaks.

Figure 4-51 Use a terminal cleaning tool to clean the battery's terminals and the cable ends.

Special Tools

Battery cable puller, battery terminal cleaner, hydrometer

After the battery terminal nuts have been loosened, the battery cables should be removed from the battery with a terminal puller. Never pry the cable off—the battery may be damaged this way. Battery terminals and cable ends may be cleaned with a battery terminal cleaner (Figure 4-51).

CAUTION: Always tighten the battery holddown bolts to the specified torque. Overtightening these bolts may damage the battery cover and case.

Hydrometer Testing

Although most batteries made today are maintenance-free, some may still have cell caps that are removable, allowing access for a hydrometer test. The hydrometer test measures the **specific gravity** of the electrolyte. Remove the vent caps and insert the hydrometer pickup tube into the electrolyte. Squeeze the hydrometer bulb and release it to draw electrolyte into the hydrometer until the float is floating freely (Figure 4-52).

Bend over and observe the electrolyte level on the hydrometer float. Read the electrolyte specific gravity on the float at the electrolyte level. A fully charged battery has a specific gravity of 1.270. If the battery-specific gravity is less than 1.270, the battery requires charging. A battery with a 1.155 specific gravity is 50 percent charged (Figure 4-53). A voltmeter may be connected across the battery terminals to check the open-circuit voltage.

Since the specific gravity is affected by temperature, some correction of the specific gravity reading in relation to temperature is required. A thermometer in the lower part of some hydrometers measures electrolyte temperature. For every 10°F below 80°, subtract 4 points from the hydrometer reading, and add 4 points to the reading for each 10° above 80°F.

Some maintenance-free batteries have a built-in hydrometer in the cover. When a green dot appears in the center of the hydrometer, the battery is ready for testing. If the hydrometer appears dark with no green dot, the battery and charging circuit should be tested. A light yellow or bright hydrometer indicates the electrolyte is below the bottom of the hydrometer (Figure 4-54). Under this condition, battery replacement is required and the battery should not be tested or charged.

The **specific gravity** of a liquid is the weight of a liquid in relation to the weight of an equal volume of water. Water has a specific gravity of 1.000, sulfuric acid has a specific gravity of 1.835, and the electrolyte in a fully charged battery has a specific gravity of 1.270.

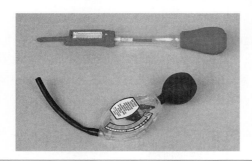

Figure 4-52 Two types of battery hydrometers.

STATE OF CHARGE	SPECIFIC GRAVITY
100%	1.265
75%	1.225
50%	1.190
25%	1.155
DEAD	1.120

Figure 4-53 The state of charge of a battery in relation to its specific gravity.

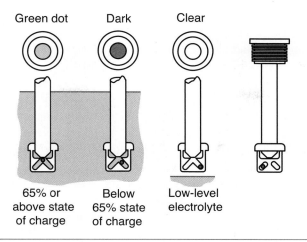

Green dot	Dark	Clear
65% or above state of charge	Below 65% state of charge	Low-level electrolyte

Figure 4-54 Design and operation of built-in hydrometers on maintenance-free sealed batteries.

Battery Capacity Test

A battery must have a minimum specific gravity of 1.190 prior to undergoing a capacity test. Prior to running this test, disconnect and remove the battery cables. The battery capacity test procedure may vary depending on the type of test equipment and whether the tester has digital or analog meters. Always follow the test equipment manufacturer's instructions.

Follow these steps for the capacity test with an analog-type capacity tester:

1. Rotate the capacity tester load control to the off position and set the voltmeter control to the 18-volt position.
2. Check the mechanical zero on each meter and adjust if necessary.

SERVICE TIP: On automotive electrical test equipment such as battery testers and chargers, the positive cable clamp is red and the negative cable clamp is black.

3. Connect the tester leads to the battery terminals with the proper polarity. The red lead is always connected to the positive terminal and the black lead is always connected to the negative terminal (Figure 4-55).
4. Observe the battery open circuit voltage, which should be above 12.2 volts.

Special Tools

Volt-ampere tester

Battery open circuit voltage is the voltage of the battery with no load on it.

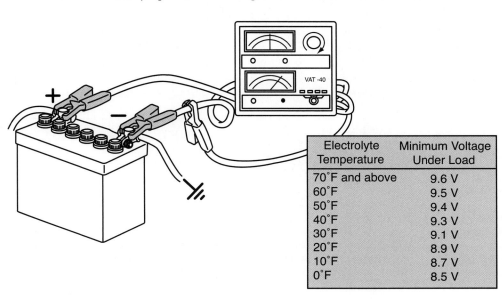

Electrolyte Temperature	Minimum Voltage Under Load
70°F and above	9.6 V
60°F	9.5 V
50°F	9.4 V
40°F	9.3 V
30°F	9.1 V
20°F	8.9 V
10°F	8.7 V
0°F	8.5 V

Figure 4-55 Volt-ampere tester connections for testing the battery.

5. Set the test selector to the number two charging position and adjust the ammeter pointer to the zero position with the electrical zero adjust control.
6. Connect the ammeter inductive clamp over the negative tester cable.
7. Set the test selector to the number one starting position.
8. Rotate the load control until the ammeter reads one-half the cold cranking rating or three times the ampere-hour rating.
9. Maintain this load position for 15 seconds. If necessary, adjust the load control slightly to maintain the proper ampere reading.
10. After the 15 seconds are completed, observe the voltmeter reading and immediately rotate the load control to the off position.
11. If the battery temperature is 70°F (21°C) and the voltage is above 9.6 volts after 15 seconds on the capacity test, the battery is satisfactory and may be returned to service. Since temperature affects battery capacity, the minimum voltage at the end of the capacity test varies depending on temperature.

If the battery voltage is less than the minimum load voltage at the end of the load test, the battery should be charged and the load test repeated. After charging and repeating a load test, if the battery voltage is less than the minimum load voltage, the battery should be replaced. As discussed in the steps above, the minimum voltage the battery must maintain during the load test is 9.6 volts. However, most healthy batteries will be able to maintain a higher voltage. Although the battery you are testing may be in borderline condition (age and performance) and may barely be able to maintain minimum voltage, consider discontinuing its service. As mileage and vehicle age continue to increase over time, the electrical demands of the vehicle do not decrease. Therefore, a marginal battery nearing the end of its life cycle should be a candidate for replacement. Extreme hot or cold temperatures can suddenly cause a weak battery to fail. Photo Sequence 6 shows a typical procedure for testing battery capacity.

Charging a Battery

There are two basic ways to recharge a battery: slow or fast. Knowing when to fast charge or slow charge a battery is important to your own safety and to the life of the battery.

Fast Charging

Special Tools

Battery charger

Before charging the battery, the battery case and terminals should be cleaned and the vent caps should be removed. Check the electrolyte level and add distilled water as required. If the battery is in the vehicle, disconnect both battery cables. Determine the charging rate and time (Figure 4-56), then follow these steps to properly and safely charge the battery:

Reserve Capacity Rating	20-Hour Rating	5 Amperes	10 Amperes	20 Amperes	30 Amperes	40 Amperes
75 minutes or less	50 ampere-hours or less	10 hours	5 hours	2 1/2 hours	2 hours	
Above 75 to 115 minutes	Above 50 to 75 ampere-hours	15 hours	7 1/2 hours	3 1/4 hours	2 1/2 hours	2 hours
Above 115 to 160 minutes	Above 75 to 100 ampere-hours	20 hours	10 hours	5 hours	3 hours	2 1/2 hours
Above 160 to 245 minutes	Above 100 to 150 ampere-hours	30 hours	15 hours	7 1/2 hours	5 hours	3 1/2 hours

Figure 4-56 Battery charging time in relation to the battery's rating.

Photo Sequence 6
Typical Procedure for Testing Battery Capacity

P6-1 Observe the hydrometer in the battery top. If the hydrometer is green, proceed with the load test. A black hydrometer indicates battery charging is necessary, and a yellow hydrometer indicates battery replacement is required.

P6-2 Remove the battery cables and install adapters in the battery terminals.

P6-3 Be sure the load control is in the off position and set the voltmeter control to 18 volts.

P6-4 Connect the volt-ampere tester cables to the battery terminals with the correct polarity.

P6-5 Connect the ammeter clamp over the negative tester cable.

P6-6 Set the tester control to the battery test position.

P6-7 Rotate the load control until the ammeter reads one-half the cold cranking ampere rating of the battery.

P6-8 Hold the load control in this position for 15 seconds and read the battery voltage with the load still applied. If the battery temperature is 70°F, the voltage should be above 9.6 volts.

P6-9 Immediately rotate the load control to the off position and disconnect the tester cables.

⚠ WARNING: Never connect or disconnect battery charger cables from the battery terminals with the charger in operation. This procedure causes sparks at the battery terminals, which may result in a battery explosion, personal injury, and/or property damage.

1. Be sure the charger timer switch and main switch are off.
2. Connect the negative charger cable to the negative battery terminal and the positive charger cable to the positive battery terminal.
3. Connect the charger power cable to an electrical outlet.
4. Set the charger voltage switch to the proper battery voltage then turn on the main switch.
5. Set the timer to the required battery charging time.
6. Adjust the current control to obtain the proper charging rate. Some battery chargers have a reverse polarity protector that prevents charger operation if the charger cables are connected to a battery with reversed polarity. If the charger cables are connected properly and the battery voltage is very low, the reverse polarity protector may not allow the charger to begin charging the battery. Many chargers have a bypass button that must be pressed to bypass the reverse polarity protector when this condition occurs.
7. Shut the charger off while testing the battery. Continue charging the battery until the specific gravity is above 1.250 and the open circuit voltage is above 12.6 volts.
8. When the battery is above 1.225 specific gravity, a high charging rate causes excessive gassing. Reduce the charging rate as required to prevent excessive gassing.
9. Shut the timer, current control, and main switch off, then disconnect the charger cables from the battery.

Slow Charging

A battery may be slow charged at one-tenth of the ampere-hour rating. For example, an 80 ampere-hour battery should be charged at 8 amperes. Some slow chargers are capable of charging several batteries connected in series. A current control on the charger is rotated until the desired charging rate is obtained on the charger ammeter. The battery should be charged until the specific gravity is above 1.250 or until there is no further increase in specific gravity or battery voltage in 1 hour.

Charging Maintenance-Free Batteries

Maintenance-free batteries require different charging rates compared to low-maintenance batteries. Some vehicle manufacturers recommend using the reserve capacity rating of the battery to determine the ampere-hours charge required by the battery. For example, a battery with a reserve capacity of 75 minutes requires 25 amperes × 3 hours charge = 75 ampere-hours. Continue charging the battery until the green dot appears in the hydrometer and the battery passes a capacity test. Always use the vehicle or battery manufacturer's recommended charging rate.

Special Tools

Jumper wire

⬛ CAUTION: When jump starting a vehicle, always turn off all electrical accessories in the vehicle being boosted and the boost vehicle. Electrical accessories left on during the boost procedure may be damaged.

CAUTION: While recharging a battery, watch for excessive gassing and observe the battery case temperature. Immediately slow the charge rate if the battery is boiling or the temperature goes over 125° F. If the open circuit voltage (OCV) is less than 11 volts, the battery will take several hours to charge. It may be preferable to begin with a slow rate of charge.

Battery Drain Testing

Most computers have a few milliamperes of current flow when they are not in operation. This current flow is called battery drain. A computer-equipped vehicle has a battery drain that may be called a **parasitic drain**. Since many vehicles today have several computers, this current flow may discharge the battery if the vehicle is not driven for several weeks. Many computers have a higher drain for a short period of time after the ignition switch is turned off and then the drain is reduced. It is very important for technicians to understand how to perform an accurate battery drain test and determine if the drain is normal or excessive. To do this:

Battery drain refers to the current draw on the battery when the ignition switch is off.

1. Be sure the ignition switch is off. Make sure the glove compartment and trunk lights are off and the doors are closed.
2. Disconnect the negative battery cable and connect a 12-gauge jumper wire between the cable end and the negative battery terminal. Do not attempt to start the engine once this jumper is connected.
3. Connect the ammeter on a multimeter in series between the negative battery cable end and the negative battery terminal (Figure 4-57). Set the ammeter scale on 20 amperes.
4. Turn the ignition switch on, then turn on computer-controlled accessories, such as the A/C and radio/stereo.
5. Turn off all electrical accessories and the ignition switch.
6. Disconnect the 12-gauge jumper wire.
7. If the ammeter reading is very low, switch the ammeter scale to the lowest milliampere scale and observe the reading.

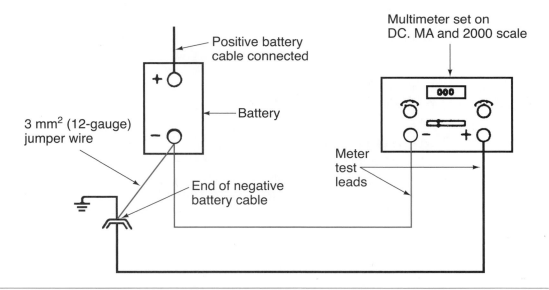

Figure 4-57   Jumper wire and ammeter connected to test battery drain.

COMPONENT OR MODULE	Current draw (mA)	
	Typical	Max
ELC (after 7 min. time out)		1.0
Generator	2.0	2.0
Heated windshield control module	0.3	0.4
Heated seat control modules		0.5
Instrument panel digital cluster	4.0	6.0
Light control module	0.5	1.0
Multi-function chime module (key removed from ignition)	1.0	1.0
Oil level module		0.1
Pass key decoder module	.75	1.0
PCM	5.0	7.0
Radio U1X	7.0	8.5
CD U1B	1.8	3.5
RAC (theft deterrent) (illuminated entry) (auto door locks)		3.8

Figure 4-58   Computer drain specifications for different common accessories (typical).

8. Check the vehicle manufacturer's specifications for the required time for the computers to reduce the amount of drain (Figure 4-58).
9. Compare the drain on the ammeter to the vehicle manufacturer's specifications. If the drain is excessive, disconnect the fuses or circuit breakers in various circuits to locate the source of the high drain.
10. Disconnect the multimeter, then tighten the negative battery cable.

Starter Diagnosis and Service

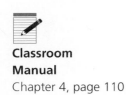

Classroom Manual
Chapter 4, page 110

Special Tools

Volt-ampere tester

The battery-specific gravity must be 1.190 or above before the starter current draw test is performed. Always follow the starter current draw test procedure in the vehicle manufacturer's service manual. The following is a typical starter current draw test procedure:

1. Rotate the load control on the tester to the off position.
2. If an analog tester is used, check the mechanical zero on each meter and adjust as necessary.
3. Set the voltmeter selector to the internal (INT) 18-volt position.
4. Connect the positive tester cable to the positive battery cable, and connect the negative tester cable to the negative battery cable.
5. Set the test selector to the number two charging position.
6. Adjust the ammeter to read zero with the zero adjust control.

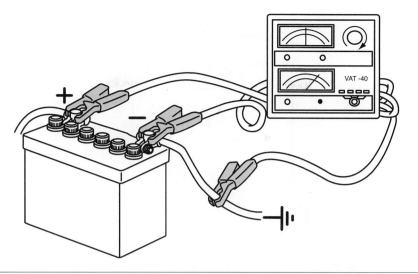

Figure 4-59 Volt-ampere tester connections for testing the starter and charging system.

7. Connect the ammeter inductive clamp on the negative battery cable or cables (Figure 4-59), including any chassis or body ground coming off the negative cable.
8. Be sure all the electrical accessories on the vehicle are off and the doors are closed.
9. Set the test selector to the number one starting position, if applicable.

■ **CAUTION:** Always follow the vehicle manufacturer's recommended procedure for disabling the ignition system. If these instructions are not followed, electronic components may be damaged.

10. Disable the ignition system by disconnecting the positive primary coil wire. Always follow the vehicle manufacturer's recommended procedure for disabling the ignition system.
11. Crank the engine and observe the ammeter and voltmeter readings. The starter current draw should equal the specifications, and the voltage should be above the minimum cranking voltage.
12. Disconnect the tester cables and reconnect the positive primary coil wire.

Starter Current Draw Test Results

■ **CAUTION:** The average starter should turn the engine over at about 200 rpm.

High current draw and low cranking speed usually indicate a defective starter. High current draw may also be caused by internal engine problems, such as partial bearing seizure. A low cranking speed and low current draw with high cranking voltage usually indicate excessive resistance in the starter circuit, such as cables and connections. Always remember that the battery and battery cable connections must be in satisfactory condition to obtain accurate starter current draw test results.

■ **CAUTION:** Perform voltage drop tests on suspect components, cables, and connections.

A starter that draws too much current and cranks the engine slowly may be called a dragging starter.

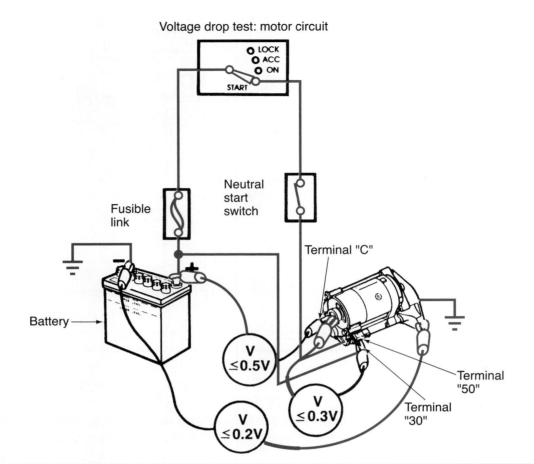

Figure 4-60 Voltmeter connections for measuring voltage drops in the starter and starter control circuits.

Voltage Drop Test, Starting Motor Circuit Insulated Side

The resistance in an electrical wire may be checked by measuring the voltage drop across the wire with normal current flow in the wire. To measure the voltage drop across the positive battery cable, connect the positive voltmeter lead to the positive battery cable at the battery and connect the negative voltmeter lead to the other end of the positive battery cable at the starter solenoid (Figure 4-60). Set the voltmeter selector switch to the lowest scale and disable the ignition system. Crank the engine. The voltage drop indicated on the meter should not exceed 0.5 volts. If the voltage reading is above this figure, the cable has excessive resistance. If the cable ends are clean and tight, replace the cable.

Connect the positive voltmeter lead to the positive battery cable on the starter solenoid and connect the negative voltmeter lead to the starting motor terminal on the other side of the solenoid. Leave the voltmeter on the lowest scale and crank the engine. If the voltage drop exceeds 0.3 volts, the solenoid disc and terminals have excessive resistance.

Voltage Drop Test, Starting Motor Ground Side

Connect the positive voltmeter lead to the starter case and connect the negative voltmeter lead to the negative battery cable on the battery. Leave the ignition system disabled and place the voltmeter selector on the lowest scale. If the voltage drop reading exceeds 0.2 volts while cranking the engine, the negative battery cable or ground return circuit has excessive resistance.

Voltage Drop Tests, Control Circuit

Connect the positive voltmeter lead to the positive battery cable at the battery, and connect the negative voltmeter lead to the solenoid winding terminal on the solenoid. Leave the ignition system disabled and place the voltmeter selector on the lowest scale. If the voltage drop across the control circuit exceeds 1.5 volts while cranking the engine, individual voltage drop tests on control circuit components are necessary to locate the high resistance problem.

Connect the voltmeter leads across individual control circuit components such as the ignition switch, neutral safety switch, and starter relay contacts to measure the voltage drop across these components while cranking the engine. In most starting motor control components, if the voltage drop exceeds 0.2 volts, the component is defective. The voltmeter leads may also be connected across individual wires in the starter control circuit to test voltage drop in the wires. Before any starting motor circuit component is replaced, always remove the negative battery cable. High resistance in starting motor control components and wires may cause the starting circuit to be inoperative. High resistance in the solenoid disc and terminals may cause a clicking action when the ignition switch is turned to the start position.

 SERVICE TIP: A lower reading is desirable in *any* voltage drop test.

AC Generator Diagnosis and Service

Classroom Manual Chapter 4, page 117

A ribbed V-belt may be called a serpentine belt.

AC generator belt condition and tension are extremely important for satisfactory AC generator operation. A loose belt causes low AC generator output and a discharged battery. A loose, dry, or worn belt may cause squealing and chirping noises, especially during engine acceleration and cornering.

The AC generator belt should be checked for cracks, oil soaking, worn or glazed edges, tears, and splits (Figure 4-61). If any of these conditions are present, belt replacement is necessary.

Since the friction surfaces are on the sides of a V-type belt, wear occurs in this area. If the belt edges are worn, the belt may be rubbing on the bottom of the pulley. This condition requires belt replacement. Belt tension may be checked by measuring the belt deflection. Press on the belt with the engine stopped to measure the belt deflection, which should be ½ inch per foot of free span. The belt tension may be checked with a belt tension gauge placed over the belt (Figure 4-62). The

Special Tools

Belt tension gauge

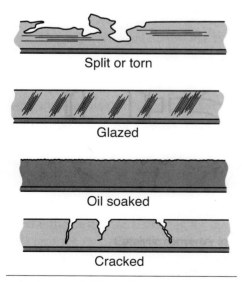

Figure 4-61 Defective drive belt conditions.

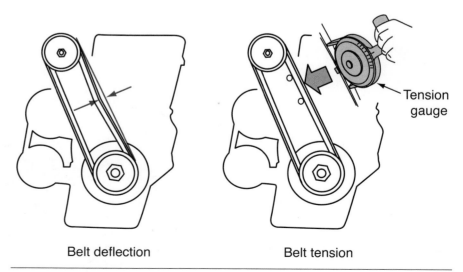

Figure 4-62 Methods for checking drive belt tension.

tension on the gauge should equal the vehicle specifications. If a V-belt requires tightening, follow this procedure:

1. Loosen the AC generator bracket bolt.
2. Loosen the AC generator mounting bolt.
3. Check the bracket and AC generator mounting bolts for wear. If these bolts or bolt openings in the bracket or AC generator housing are worn, replacement is necessary.
4. Pry against the AC generator housing with a pry bar to tighten the belt.
5. Hold the AC generator in the position described in step 4 and tighten the bracket bolt.
6. Recheck the belt tension with the tension gauge. If the belt does not have the specified tension, repeat steps 1 through 5.
7. Tighten the bracket bolt and the mounting bolt to the specified torque.

Some AC generators have a ribbed V-belt. Many ribbed V-belts have an automatic tensioning pulley. Therefore, a tension adjustment is not required. The ribbed V-belt should be checked to make sure it is installed properly on each pulley in the belt drive system. The tension on a ribbed V-belt may be checked with a belt tension gauge in the same way as the tension on a V-belt.

Many ribbed V-belts have a spring-loaded tensioner pulley that automatically maintains belt tension. As the belt wears or stretches, the spring moves the tensioner pulley to maintain the belt tension. Some of these tensioners have a belt length scale that indicates new belt range and used belt range. If the indicator on the tensioner is out of the used belt length range, belt replacement is required. Many belt tensioners have a one-half inch drive opening in which a ratchet or flex handle may be installed to move the tensioner pulley off the belt during belt replacement (Figure 4-63).

Belt pulleys must be properly aligned to minimize belt wear. The edges of the pulleys must be in line when a straightedge is placed on the pulleys.

AC Generator Output Test

Special Tools

Volt-ampere tester

CAUTION: Never allow the AC generator voltage to go above 15.5 volts. High voltage may damage computers and other electrical and electronic equipment on the vehicle.

The AC generator belt must be in satisfactory condition before the output test is performed. Always follow the vehicle recommended procedure for testing AC generator output. Photo Sequence 7 covers a typical AC generator output test procedure.

CAUTION: Never disconnect the circuit between the AC generator and the battery with the engine running. This action may cause extremely high voltage momentarily, which causes electronic component damage.

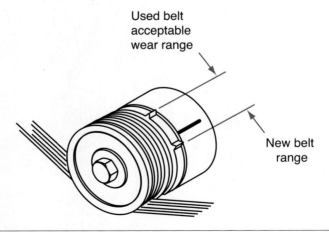

Used belt acceptable wear range

New belt range

Figure 4-63 Drive belt tensioner.

Photo Sequence 7
Using a Volt and Amp Tester to Test the Charging System

P7-1 Testing the charging system begins with a visual inspection of the battery and its cables. Make sure they are clean and free of corrosion. The alternator drive belt should also be inspected and its tension checked.

P7-2 To test the charging system, the tester's ampere pickup probe must be positioned around the battery's negative cable and its cables connected to the battery.

P7-3 To observe the tester's meters during this test, the tester should be placed where it can easily be seen from the driver's seat.

P7-4 It is recommended that engine speed be monitored. Therefore, connect a tachometer to the engine. Some testers, such as the one shown, are equipped with a tachometer that monitors ignition pulses at the battery.

P7-5 At about 2,000 rpm, the charging system will deliver a maximum amount of charge to the battery. This charging system is providing 14.6 volts at 52.5 amperes. Readings should be compared to specifications. Engine rpm does not show in the photo because the photo was taken after the test, not during it.

P7-6 If the readings are low, this may indicate that there is a fault in the charging system or that the battery is fully charged and not allowing the alternator to work at its peak. To determine the cause, the voltage regulator should be bypassed according to the procedures outlined in the service manual.

P7-7 With the regulator bypassed, this alternator increased its voltage output. This indicates that the regulator had been regulating voltage output. The amperage reading is also well within specifications. Therefore, the charging system is functioning properly.

Since the ammeter inductive pickup is installed on the negative battery cable, the ammeter only indicates the AC generator current flow through the battery during the output test. The current is supplied directly from the AC generator to the ignition system and AC generator field, so the current draw for these two systems must be added to the AC generator output to obtain an accurate output reading. If the AC generator output is not within 10 percent of the specified output, further diagnosis is required.

AC Generator Regulator Voltage Test

To check the AC generator for proper regulator voltage, leave the volt-ampere tester connected as in the output test. Operate the engine at 2,000 rpm until the battery is sufficiently charged to provide a charging rate of 10 amperes or less. Observe the voltmeter reading. The voltage should be at the specified regulator voltage.

If the AC generator output is zero, the AC generator field circuit may have an open circuit. The most likely place for an open circuit in the AC generator field is at the slip rings and brushes. When the AC generator voltage is normal but the amps output is zero, the fuse link between the AC generator battery terminal and the positive battery cable may be open. With the ignition switch off, battery voltage should be available on each end of the fuse link. If the voltage is normal on the battery side of the fuse link and zero on the AC generator side, the fuse link is open and must be replaced.

If the AC generator output is less than specified, always be sure that the belt and belt tension are satisfactory. When the belt and belt tension are satisfactory and the AC generator output is less than specified, the AC generator is defective.

Battery overcharging causes excessive gassing of the battery. Overcharging may be caused by a defective voltage regulator, which allows a higher-than-specified charging voltage.

Battery undercharging eventually results in a discharged battery and a no-start problem. Undercharging may be caused by a loose or worn belt, a defective AC generator, low regulator voltage, high resistance in the battery wire between the AC generator and the battery, or an open fuse link.

Alternator Output Requirement Test

AC generators are rated according to their output capacity in amperes. On occasion, it may be necessary to choose the proper alternator based on the vehicle's electrical load requirements. If an undersized alternator is installed on a vehicle containing many electrical components and accessories, the battery will likely fail under certain conditions because it is undercharged. In addition, the AC generator may be stressed to keep up with the electrical demands and may also experience problems. To determine the correct ampere output alternator, perform the following:

1. Connect the amp probe from the tester to the negative battery cable and zero the ammeter if necessary.
2. Switch the ignition to the "on" position. Turn on all the accessories in the vehicle. Be sure to include lights, rear window defroster, heater blower motor, brake lamps, etc.
3. If applicable, add an approximate estimated amperage value for the electric cooling fan(s) and electric fuel pump.
4. Total all live and calculated ampere values.
5. As a rule of thumb, the alternator should be rated approximately 5 amperes over the maximum calculated ampere requirement.

Diode Pattern Testing

✓ **SERVICE TIP:** It is good practice to check the output waveform of an AC genera—tor any time an electronic component fails. Because the electronics of the vehicle cannot accept AC current, the damage to the faulty component may have been caused by a bad diode allowing AC noise to be present in the charging system's output.

One of the final tests that should be performed on the alternator is the diode or diode/stator test. This test checks the AC component of the DC voltage after rectification. This is measured with an AC voltmeter. Some charging system testers show this as a separate step in the test sequence while other testers automatically perform it. There are several good reasons to check this part of the alternator. If one of the six diodes in the AC generator circuit fails, the alternator can lose at least 17 percent of its output. Often, this can take place without any instrument-warning lamp alert to the driver. Depending on the type of diode problem, results can include a loss of output, AC generator noise being heard through the entertainment system, or the generator-warning lamp being illuminated dimly while running *or* with the engine off. In addition, an excessive amount of AC voltage output from the AC generator due to a failed diode can cause false or phantom signals to be generated and may influence computer input and output signals or the ignition system. To perform the diode or diode/stator test, simply have the tester connected with the selector set and run the engine at 2,000 rpm. Load the system to about 12.5 volts while maintaining 2,000 rpm and observe the tester reading. It usually shows a go or no-go test result. Some computerized engine analyzers also perform this test during what is referred to as a high cruise test. During this test, there is usually an electrical load placed on the system by turning on the headlamps and high-speed blower. This works the alternator at a high cruise speed, causing it to put out a higher charge. The diode ripple voltage is displayed either digitally or as a waveform. The analyzer will usually flag an alternator ripple greater than 0.12 volts, just as you would if you were using any DMM. Most oscilloscopes can be set up to display an AC generator diode pattern. Connect the appropriate leads either on the positive battery terminal or preferably on the generator's output terminal. Set the scope controls for this test. Start the engine and hold it at about 2,000 rpm. Next, load the alternator and observe the pattern. A representation of both good and problem patterns is illustrated in Figure 4-64. To perform this test using a DMM, set the meter to low AC voltage, place the leads across the battery terminals, and run the engine as stated earlier. Again, the reading should not exceed 0.12 AC volts. If the readings or diode patterns are unsatisfactory, a teardown or replacement of the alternator is necessary.

Electrically heated windshield's circuit uses unrectified AC to heat the glass. The AC is picked up at the AC generator before the diodes.

Special Tools

Tune-up or lab scope, DMM

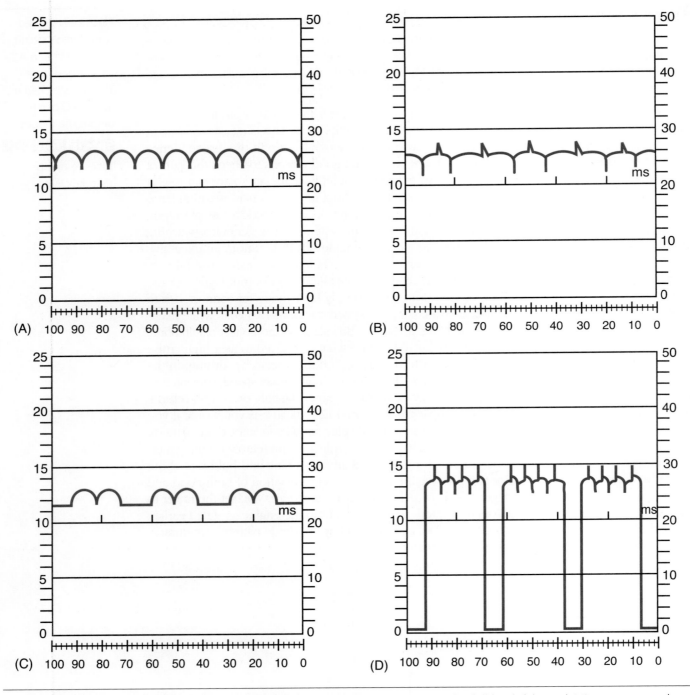

Figure 4-64 AC generator oscilloscope patterns: (A) good AC generator under full load, (B) good AC generator under no load, (C) shorted diode and/or stator winding under full load, and (D) open diode in diode trio.

A customer complained about a no-start problem on an Oldsmobile Alero. When the service writer questioned the customer about the problem, the customer indicated the battery seemed to be discharged and would not crank the engine. The customer also indicated this problem occurred about once a month.

When the technician lifted the hood, she noticed the battery, AC generator, and starting motor had been replaced, which indicated that someone else had likely been trying to correct this problem. The technician checked the belt and all wiring connections in the starting and charging systems without finding any problems. Next, the technician performed starter current draw and AC generator output tests, and these systems performed satisfactorily. The technician checked the regulator voltage and found it was within specifications. After the battery was charged, the technician performed a capacity test and the battery performance was satisfactory. Since the battery, starting system, and charging system tests were normal, the technician concluded there must be a battery drain problem. The technician connected an ammeter and jumper wire in series between the negative battery cable and terminal and proceeded with the drain test. When the jumper wire was disconnected, the ammeter indicated a continual 1 ampere drain. The technician began removing the fuses one at a time and checking the ammeter reading. When the courtesy lamp fuse was disconnected, the drain disappeared.

The technician checked the wiring diagram for the vehicle and found the vehicle had a remote keyless entry module. When the courtesy lamp fuse was installed and the remote keyless entry module disconnected, the drain disappeared on the ammeter. This module was draining current through it to ground, but the current flow was not high enough to illuminate the courtesy lights.

A new remote keyless entry module was installed and the drain was rechecked. The drain was now within the vehicle manufacturer's specifications.

Terms to Know

American Wire Gauge (AWG)	Electromotive force (EMF)	Parasitic drain
Average responding	Frequency	Pulse width
Chassis ground	Glitches	Radio frequency interference (RFI)
Current ramp	Logic probe	Root mean square (RMS)
Digital storage oscilloscope (DSO)	Low-amp probes	Sinusoidal
Duty cycle	Non-sinusoidal	Specific gravity

ASE-Style Review Questions

1. While discussing AWG wiring numbering systems:
   Technician A says that most wiring used in vehicles has an AWG number from 10 to 18 gauge.
   Technician B says that on the AWG scale, 0 is the smallest and 20 is the largest.
   Who is correct?
   - **A.** A only
   - **B.** B only
   - **C.** Both A and B
   - **D.** Neither A nor B

2. While discussing resistance:
   Technician A says current will increase with a decrease in resistance.
   Technician B says current will decrease with an increase in resistance.
   Who is correct?
   - **A.** A only
   - **B.** B only
   - **C.** Both A and B
   - **D.** Neither A nor B

3. While discussing oscilloscope patterns:
   Technician A says frequency is measured by counting the number of cycles in 1 second.
   Technician B says a frequency of 10 Hz means there are ten cycles per second.
   Who is correct?
   A. A only   **C.** Both A and B
   B. B only   **D.** Neither A nor B

4. While discussing electricity:
   Technician A says an open causes unwanted voltage drops.
   Technician B says high resistance problems cause increased current flow.
   Who is correct?
   A. A only   **C.** Both A and B
   B. B only   **D.** Neither A nor B

5. While discussing battery service:
   Technician A says if a battery is disconnected, computer adaptive memories are erased.
   Technician B says if a battery is disconnected, the memory seat and memory mirror positions are erased.
   Who is correct?
   A. A only   **C.** Both A and B
   B. B only   **D.** Neither A nor B

6. While discussing the battery capacity test:
   Technician A says the battery should be discharged at one-third of the cold cranking ampere rating.
   Technician B says a satisfactory battery at 70°F has at least 9.6 volts at the end of the 15-second capacity test.
   Who is correct?
   A. A only   **C.** Both A and B
   B. B only   **D.** Neither A nor B

7. While discussing battery drain testing:
   Technician A says computers have a low milliampere drain with the ignition off.
   Technician B says the fuses may be disconnected one at a time to locate the cause of a battery drain.
   Who is correct?
   A. A only   **C.** Both A and B
   B. B only   **D.** Neither A nor B

8. While discussing the starter current draw test:
   Technician A says the battery should be at least 50 percent charged before the starter draw test is performed.
   Technician B says the electrical accessories on the vehicle should be on during the starter draw test.
   Who is correct?
   A. A only   **C.** Both A and B
   B. B only   **D.** Neither A nor B

9. While discussing AC generator output:
   Technician A says zero AC generator output may be caused by an open field circuit.
   Technician B says zero AC generator amperage output may be caused by an open AC generator fuse link.
   Who is correct?
   A. A only   **C.** Both A and B
   B. B only   **D.** Neither A nor B

10. While discussing wiring harness and terminal repairs:
    Technician A says acid-core solder should be used for soldering electrical terminals.
    Technician B says sealant should appear at both ends of a crimp and seal sleeve after it is heated.
    Who is correct?
    A. A only   **C.** Both A and B
    B. B only   **D.** Neither A nor B

ASE Challenge Questions

1. A parasitic battery drain test is being performed.
   Technician A says the ammeter must be connected in series between the battery's negative post and the negative cable.
   Technician B says all circuits must be off (open) for this test to be valid.
   Who is correct?
   - **A.** A only
   - **B.** B only
   - **C.** Both A and B
   - **D.** Neither A nor B

2. A vehicle with a no-crank condition is being discussed.
   Technician A says corrosion on and in the battery's positive cable could cause this condition.
   Technician B says an open between the ignition switch and the ignition module would result in a no-crank condition.
   Who is correct?
   - **A.** A only
   - **B.** B only
   - **C.** Both A and B
   - **D.** Neither A nor B

3. A jumper wire has been used to energize the coil of a relay.
   Technician A says an infinity resistance reading across the contacts indicates a good relay.
   Technician B says a zero resistance reading across the contacts indicates the problem is elsewhere in the circuit.
   Who is correct?
   - **A.** A only
   - **B.** B only
   - **C.** Both A and B
   - **D.** Neither A nor B

4. A diode is being tested with an analog ohmmeter. The meter reads very high resistance in one direction and very low resistance in the opposite direction. This indicates a(n):
   - **A.** shorted condition.
   - **B.** open condition.
   - **C.** normal condition.
   - **D.** additional testing is needed.

5. Any of the following may cause low AC generator output *except*:
   - **A.** glazed slip rings and brushes.
   - **B.** loose AC generator drive belt.
   - **C.** shorted diode.
   - **D.** low regulator voltage.

Job Sheet 8

Name _____ Date _____

Using a DSO on Sensors and Switches

Upon completion of this job sheet, you should be able to connect a DSO and observe the activity of various sensors and switches.

ASE Correlation

Diagnose engine mechanical, electrical, electronic, fuel, and ignition problems with an oscilloscope and/or engine analyzer; determine needed action.

Tools and Materials

A vehicle with accessible sensors and switches
Service manual for the above vehicle
Component locator manual for the above vehicle
A DSO
A DMM

Describe the vehicle being worked on.

Year _____ Make _____ VIN _____

Model_____

Procedures

1. Connect the DSO across the battery. Make sure the scope is properly set. Observe the trace on the scope. Is there evidence of noise? Explain.

   _____

   _____

   _____

   _____

2. Locate the A/C compressor clutch control wires. Start and run the engine. Connect the DMM to read available voltage. Observe the meter, then turn the compressor on. What happened on the meter?

   _____

   _____

   Now connect the DSO to the same point with the compressor turned off. Observe the waveform, then turn the compressor on. What happened to the trace?

   _____

   _____

3. Turn the engine off but keep the ignition on. Locate the TP sensor and identify the purpose of each wire to it. List each wire and describe the purpose of each.

_____

_____

4. Connect the DMM to read reference voltage at the TP sensor. What do you read?

_____

_____

Now move the leads to read the output of the sensor. Starting with the throttle closed, slowly open the throttle until it is wide open. Watch the voltmeter while doing this. Describe your readings.

_____

_____

5. Now connect the DSO to read reference voltage at the TP sensor. What do you see on the trace?

_____

_____

Now move the leads to read the output of the sensor. Starting with the throttle closed, slowly open the throttle until it is wide open. Watch the trace while doing this. Describe your readings.

_____

_____

6. Now run the engine. Locate the oxygen sensor and identify the purpose of each wire to it. Connect the DMM to read voltage generated by the sensor. (To do this, you may use an electrical connector for the O_2 sensor that is positioned away from the hot exhaust manifold.) Watch the meter and describe below what happened.

_____

_____

_____

_____

7. Now connect the DSO to read voltage output from the sensor. Watch the trace and describe what happened.

_____

_____

_____

_____

8. Explain what you observed as the differences between testing with a DMM and a DSO.

_____

_____

_____

_____

Instructor's Response_____

_____

_____

Job Sheet 9

9

Name _____ Date _____

Inspecting a Battery

Upon completion of this job sheet, you should be able to visually inspect a battery.

Tools and Materials

A vehicle with a 12-volt battery
A DMM

ASE Correlation

Test and diagnose emissions or driveability problems caused by battery condition, connections, or excessive key-off battery drain; determine needed action.

Task Completed

Procedures

1. Describe the general appearance of the battery.

   _____
   _____
   _____

2. Describe the general appearance of the cable and terminals.

   _____
   _____
   _____

3. Check the tightness of the cables at both ends. Describe their condition.

   _____
   _____
   _____

4. Connect the positive lead of the meter (set on DC volts) to the positive terminal of the battery. ☐

5. Put the negative lead on the battery case and move it all around the top and sides of the case. What readings do you get on the voltmeter?

   _____
   _____
   _____

6. What is indicated by the readings?

   _____
   _____
   _____

7. Measure the voltage of the battery. Your reading was: _____ volts

8. What do you know about the condition of the battery based on the visual inspection and above tests?

_____

_____

_____

Instructor's Response_____

_____

_____

Job Sheet 10

Name _____ Date _____

Testing the Battery's Capacity

Upon completion of this job sheet, you should be able to test a battery's capacity.

Tools and Materials

A vehicle with a 12-volt battery
Service manual for the above vehicle
Starting/charging system tester (VAT-40 or similar)

ASE Correlation

Test and diagnose emissions or driveability problems caused by battery condition, connections, or excessive key-off battery drain; determine needed repairs.

Describe the vehicle being worked on.

Task Completed

Year _____ Make _____ VIN _____

Model _____

Procedures

1. Perform a battery state-of-charge test:
   a. Record the specific gravity readings for each cell:
      (1) _____ (2) _____ (3) _____ (4) _____ (5) _____ (6) _____
   b. If the battery is a maintenance-free type of battery, what is the open circuit voltage?
      _____ volts

2. Summarize the battery's state of charge from the above and indicate the percentage of charge.

   _____

   _____

   _____

3. Connect the starting/charging system tester to the battery. ☐

4. Locate the CCA rating of the battery. What is the rating? _____

5. Based on the CCA, how much load should be put on the battery during the capacity test?
   _____ amps

6. Conduct the battery load test.

   Battery voltage decreased to _____ volts after _____ seconds.

7. Describe the results of the battery load (capacity) test. Include in the results your service recommendations and the reasons for them.

_____
_____
_____
_____
_____
_____

Instructor's Response_____

_____

_____

Job Sheet 11

11

Name _____ Date _____

Testing Charging System Output

Upon completion of this job sheet, you should be able to measure the output of the charging system.

Tools and Materials

A vehicle
Service manual for the above vehicle
Starting/charging system tester (VAT-40 or similar)

ASE Correlation

Test and diagnose engine performance problems resulting from an undercharge, overcharge, or a no-charge condition; determine needed action.

Describe the vehicle being worked on.

Year _____ Make _____ VIN _____

Model _____

Task Completed

Procedure

1. Identify the type and model of AC generator. What type is it?

   _____

   What are the output specifications for this AC generator?

   _____ amps and _____ volts at _____ rpm

2. Connect the starting/charging system tester to the vehicle. ☐

3. Start the engine and run it at the specified engine speed. ☐

4. Apply the load to the electrical system in order to obtain the highest amp reading. ☐

5. Observe the output to the battery. The meter readings are:

   _____ amps and _____ volts (after load is removed and voltage stabilizes)

6. Compare the readings to specifications and give recommendations.

   _____

   _____

   _____

7. If readings are outside of the specifications, refer to the service manual to find the proper way to full-field the AC generator. Describe the method.

   _____

   _____

   _____

8. Full-field the generator and observe the output to the battery. The meter readings are:

   _____ amps and _____ volts

9. Compare readings to specifications and give recommendations.

_____

_____

_____

_____

Instructor's Response_____

_____

_____

Intake and Exhaust System Diagnosis and Service

CHAPTER 6

Upon completion and review of this chapter, you should be able to:

❏ Service and replace air filters.

❏ Inspect and troubleshoot vacuum and air induction systems.

❏ Visually inspect the exhaust system and determine needed repairs.

❏ Diagnose catalytic converters.

❏ Remove, replace, and service exhaust manifolds.

❏ Remove, replace, and service mufflers, pipes, and catalytic converters.

❏ Perform a turbocharger inspection.

❏ Diagnose intake and exhaust leaks that affect turbocharger operation.

❏ Test turbocharger boost pressure.

❏ Measure axial shaft movement in a turbocharger.

❏ Measure wastegate stroke.

❏ Inspect a turbocharger and describe some common turbocharger problems.

❏ Explain supercharger operation and identify common supercharger problems.

An internal combustion engine requires air to operate. This air supply is drawn into the engine by the vacuum created during the intake stroke of the pistons. The air is mixed with fuel and is delivered to the combustion chambers. Controlling the flow of air and the air-fuel mixture is the job of the induction system.

Prior to the introduction of emission control devices, the induction system was quite simple. It consisted of an air cleaner housing mounted on top of the engine with a filter inside the housing. Its function was to filter dust and grit from the air being drawn into the carburetor.

AUTHOR'S NOTE: If the MAF is mounted on a remote air cleaner box, any leak in the air intake hose from the MAF to the intake results in un-metered air entering the engine, resulting in a very poorly running engine. This is usually most noticeable on acceleration when a cracked hose tends to open with engine movement.

Modern air induction systems do much more than simply filter the air. The introduction of emission standards and fuel economy standards encouraged the development of intake air temperature controls. The air intake system on a modern fuel injected engine is much more complicated (Figure 5-1). Ducts channel cool air from outside the engine compartment to the throttle plate assembly. The air filter has been moved to a position below the top of the engine to allow

Basic Tools

Vacuum gauge, basic tool set, fender covers, safety glasses

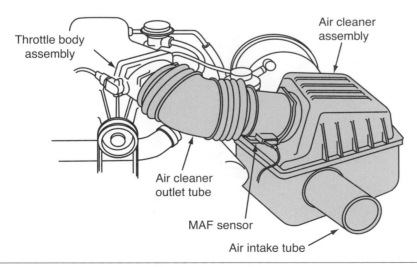

Figure 5-1 Typical air intake system for a late-model engine.

Throttle body assembly

Air cleaner assembly

Air cleaner outlet tube

MAF sensor

Air intake tube

for aerodynamic body designs. Electronic meters measure airflow, temperature, and density. These components allow the air induction system to perform the following functions:

❏ Provide the air that the engine needs to operate

❏ Filter the air to protect the engine from wear

❏ Monitor airflow temperature and density for more efficient combustion and a reduction of hydrocarbon (HC) and carbon monoxide (CO) emissions

❏ Operate with the positive crankcase ventilation (PCV) system to burn the crankcase fumes in the engine

❏ Provide air for some air injection systems

Air Filters

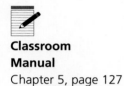

Classroom Manual
Chapter 5, page 127

When servicing an air filter, always service the positive crankcase ventilation (PCV) inlet filter.

If the air filter becomes very dirty, the dirt can block the flow of air into the fuel charging assembly. Without enough air, the engine will constantly receive a rich air-fuel mixture. The use of extra fuel means poor fuel economy and a lack of power.

The manufacturer's recommended air filter replacement interval is usually specified in the vehicle owner's manual. However, replacement of the air filter might be required on a more frequent schedule if the vehicle is subjected to continuous operation in an extremely dusty or severe off-the-road environment. A damaged air filter may cause increased wear on cylinder walls, pistons, and piston rings.

Certain types of heavy-duty air filters can be cleaned and reused. These filters have a heavy paper media encased in an element with metal end caps. Generally, reusable heavy-duty air filter elements should be replaced with a new element after six cleanings or once a year.

Follow these steps for air filter service or replacement:

1. Remove the wing nut or retaining bolts on the air cleaner cover and remove the cover to access the air filter element (Figure 5-2).

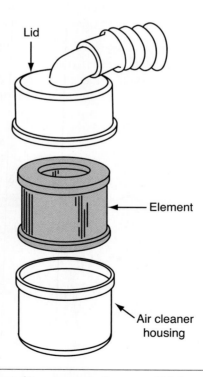

Lid

Element

Air cleaner housing

Figure 5-2 To remove the air filter, separate the cover (lid) from the housing.

2. Remove the air filter element from the air cleaner and be sure that no foreign material, such as small stones, drops into the throttle body while removing the element. If the air filter assembly is remote, remove all dirt, rocks, etc., from the air filter housing and ductwork.

3. Visually inspect the air filter for pin holes in the paper element and damage to the paper element, sealing surfaces, or metal screens on both sides of the element. If the element is damaged or contains pin holes, replace the element.

4. Place a trouble light on the inside surface of the air filter and look through the filter element to the light. The light should be visible through the paper element but no holes should appear in the element. If the paper element is plugged with dirt or oil, the light is not visible through the element. When the element is plugged with dirt or oil, replace the element. If the air filter element is contaminated with oil, excessive blowby or a defective PCV system is causing a pressure buildup in the engine. Under this condition, oil is forced up the clean air hose from the rocker arm cover into the air cleaner. Clean air should normally flow from the air cleaner through this hose to the engine.

5. If there is any dirt in the air cleaner body around the air filter, remove the air cleaner body and use a clean shop towel to remove the foreign material from the air cleaner body.

6. Check the PCV inlet filter for dirt accumulation (Figure 5-3). Many PCV inlet filters are made from foam plastic, and these filters may be washed in an approved solvent and reused. If the PCV inlet filter is damaged or plugged with dirt, replace the filter.

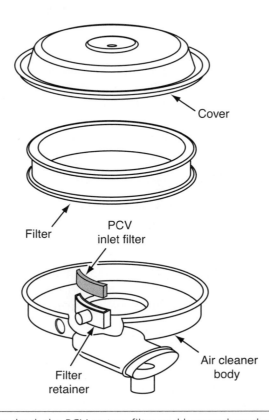

Cover

Filter

PCV inlet filter

Filter retainer

Air cleaner body

Figure 5-3 Always check the PCV system filter and hoses when changing air filters.

7. Be sure the gasket between the air cleaner and the throttle body is in satisfactory condition and install the air cleaner body.

8. Install the air filter element and be sure the seal on the lower side of the element fits evenly against the matching surface on the air cleaner body.

9. Install the air cleaner cover and be sure the cover sealing area fits against the element.

10. Install and tighten the wing nut or cover retaining bolts.

11. Be sure the PCV hose and any other hoses or sensors are properly connected to the air cleaner.

An air gun on a shop air hose may be used to blow the dirt and foreign material from the air filter element. Always keep the air gun 6 inches away from the air filter and direct the air against the inside of the air filter. While blowing dirt out of an air filter, do not allow the air pressure to exceed 30 psi (207 kPa). After blowing the dirt from the air filter, reinspect the air filter element for pin holes and remaining dirt with a shop light.

CAUTION: If air is directed from an air gun against the outside of an air filter element, the blast of air may blow dirt particles through the element and create pin holes in the element.

WARNING: Do not direct air from an air gun against any part of your body. If air penetrates the skin and enters the bloodstream, it will cause serious personal injury or death.

Heavy-Duty Air Cleaner Service

CAUTION: Do not wring or twist a polyurethane air cleaner element cover. This action stretches the cover and makes it useless.

Some heavy-duty air filter elements have a polyurethane cover over the top of the paper air filter element. This polyurethane cover may be removed and washed in an approved solvent. After washing the cover, the excess solvent should be squeezed from the element. A light coating of ordinary crankcase oil should be placed on the polyurethane cover, and the cover may be squeezed to distribute the oil evenly.

Some heavy-duty air filter elements have a heavy paper element encased between two metal end caps. This type of air filter element may be cleaned and reused. Before cleaning the element, it should be inspected for tears, punctures, pin holes, and bent end caps. If any of these conditions are present, replace the element.

Other heavy-duty air filter elements may be washed in an approved filter cleaning solution for 15 minutes or more. The filter may be rinsed with low-pressure water from a water hose and then allowed to dry. Always follow the vehicle manufacturer's air filter element cleaning instructions in the service manual.

A few heavy-duty air cleaners have a precleaner containing a series of tubes and fins. These precleaners should be inspected for dirt accumulation. A brush with stiff nylon or fiber bristles may be used to clean the tubes and fins in the precleaner.

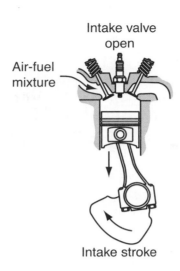

Figure 5-4 Vacuum is formed on the engine's intake stroke.

Engine Vacuum

Vacuum is measured in relation to atmospheric pressure. Atmospheric pressure is caused by the weight of the surrounding air. At sea level, the pressure exerted by the atmosphere is 14.7 psi. Most pressure gauges ignore atmospheric pressure and read zero under normal conditions. The normal measure of vacuum is in inches of mercury (in. Hg) instead of psi. Other units of measurement for vacuum are kilopascals and bars. Normal atmospheric pressure at sea level is about 1 bar or 100 kilopascals.

Vacuum in any four-stroke engine is created by the downward movement of the piston during the intake stroke (Figure 5-4). With the intake valve open and the piston moving downward, a partial vacuum is created within the cylinder and intake manifold. The air passing the intake valve does not move fast enough to fill the cylinder, thereby causing the lower pressure. This partial vacuum is continuous in a multi-cylinder engine since at least one cylinder is always at some stage of its intake stroke.

The amount of vacuum created is partially related to the positioning of the throttle plate. The throttle plate not only admits air or air-fuel into the intake manifold, but also helps control the amount of vacuum available during engine operation. At closed throttle idle, the vacuum available is usually between 15 to 22 inches. At a wide-open throttle acceleration, the vacuum can drop to zero.

Vacuum Schematic

An engine emissions vacuum schematic is required to be displayed on all domestic and import vehicles. It is located on an underhood decal. This schematic shows the vacuum hose routings and vacuum source for all emissions-related equipment. The vacuum schematic shown in Figure 5-5 shows the relationship and position of components as they are mounted on the engine. It is important to remember that these schematics only show the vacuum-controlled parts of the emission system. Hose routing and positions for vacuum devices not related to the emissions control system can be found in the service manual.

Classroom Manual
Chapter 5, page 135

An intake vacuum leak causes a low steady reading on a vacuum gauge connected to the intake manifold.

The vacuum in the intake manifold may operate many systems such as emission controls, brake boosters, heater/air conditioners, cruise controls, and more.

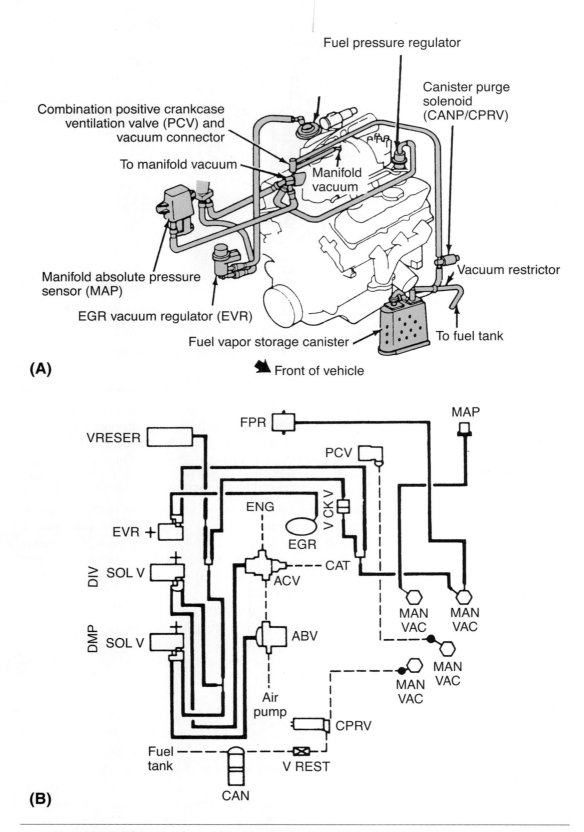

Figure 5-5 (A) New style and (B) old style vacuum schematics.

Diagnosis and Troubleshooting

Vacuum system problems can produce or contribute to the following driveability symptoms:

- ❑ Stalls
- ❑ No start (cold)
- ❑ Hard start (hot soak)
- ❑ Backfire (deceleration)
- ❑ Rough idle
- ❑ Poor acceleration

- ❑ Rich or lean stumble
- ❑ Overheating
- ❑ Detonation, or knock or pinging
- ❑ Rotten eggs exhaust odor
- ❑ Poor fuel economy

As a routine part of problem diagnosis, a technician suspecting a vacuum problem should first visually inspect all vacuum hoses and vacuum-operated components. The following are some guidelines for inspecting the vacuum systems:

1. Check all vacuum hoses to be sure they are properly routed and connected. The underhood vacuum decal indicates proper vacuum hose routing.

2. Inspect all vacuum hoses for proper tight fit between the hoses and nipples.

3. Inspect all vacuum hoses for kinks, breaks, and cuts.

4. Be sure vacuum hoses are not burned because they are positioned near hot components such as exhaust manifolds or exhaust gas recirculation (EGR) tubes.

5. Inspect all vacuum-operated components for damage, such as broken nipples on a **thermal vacuum switch (TVS)**.

6. Check for evidence of oil on vacuum hose connections. Oil in a vacuum hose may contaminate vacuum-operated components or plug the vacuum hose.

Broken or disconnected hoses allow vacuum leaks that admit more air into the intake manifold than the engine is calibrated for. The most common result is a rough running engine due to the leaner air-fuel mixture created by the excess air.

Kinked hoses can cut off vacuum to a component, thereby disabling it. For example, if the vacuum hose to the EGR valve is kinked, vacuum cannot be used to move the diaphragm and the valve will not open.

To check vacuum controls, refer to the service manual for the correct location and identification of the components. Typical locations of vacuum-controlled components are shown in Figure 5-6.

A **thermal vacuum switch (TVS)** controls vacuum signals based on coolant or air temperatures.

Many vacuum system components such as thermal vacuum switches may be tested with a hand vacuum pump and a vacuum gauge.

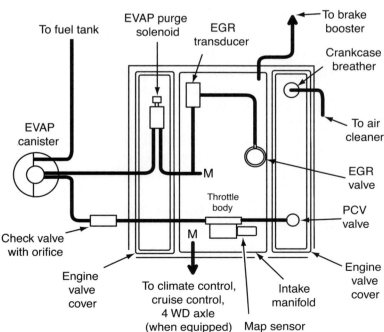

Figure 5-6 Typical vacuum devices and controls.

Tears and kinks in any vacuum line can affect engine operation. Any defective hoses should be replaced one at a time to avoid misrouting. OEM vacuum lines are installed in a harness consisting of 1/8-inch or larger outer diameter and 1/16-inch inner diameter nylon hose with bonded nylon or rubber connectors. Occasionally, a rubber hose might be connected to the harness. The nylon connectors have rubber inserts to provide a seal between the nylon connector and the component connection (nipple). In recent years, many domestic car manufacturers have been using ganged steel vacuum lines.

Vacuum Test Equipment

The vacuum gauge is one of the most important engine diagnostic tools used by technicians. With the gauge connected to the intake manifold and the engine warm and idling, watch the action of the gauge's needle. A healthy engine will give a steady, constant vacuum reading between 17 and 22 in. Hg on some four- and six-cylinder engines. However, a reading of 15 inches is considered acceptable. Figure 5-7 shows some of the common readings and what engine malfunctions they indicate. With high-performance engines, a slight flicker of the needle can also be expected. Keep in mind that a normal vacuum gauge reading will be 1 inch lower for each 1,000 feet above sea level. This is caused by the decrease in atmospheric pressure at higher altitudes (Figure 5-8).

Vacuum Leak Diagnosis

Broken or disconnected vacuum hoses may cause an air leak into the intake manifold, resulting in a lean air-fuel mixture. When the air-fuel mixture is leaner than normal, engine idle operation is erratic. On fuel injected engines, an intake manifold vacuum leak may cause the engine to idle faster than normal. An intake manifold vacuum leak may cause a hesitation on low-speed acceleration. If the vacuum leak is positioned in the intake manifold so it leans the air-fuel mixture on one cylinder more than the other cylinders, the cylinder may misfire at idle and lower engine speeds. Once the engine speed increases, the reduced intake manifold vacuum does not pull as much air through the vacuum leak and the cylinder stops misfiring.

A vacuum gauge connected to the intake manifold indicates a low steady reading if there is a vacuum leak into the intake manifold. Intake manifold mounting bolts should be checked periodically for proper torque. Loose intake manifold bolts cause leaks between the intake manifold and the cylinder head. Throttle body base gasket leaks are a common cause of uncontrollable high idle problems on many engines.

On some V-type engines, the intake manifold covers the valve lifter chamber. If the intake manifold gaskets are leaking on the underside of the intake manifold on these applications, oil splash from the valve lifter chamber is moved past the intake manifold gasket into the cylinders. This action results in oil consumption and blue smoke in the exhaust.

Kinked vacuum hoses shut off the vacuum to a component, making it inoperative. For example, if the vacuum hose to the EGR valve is kinked, vacuum is not supplied to this valve and the valve remains closed. The EGR valve recirculates some exhaust into the intake manifold. Since this exhaust gas contains very little oxygen, it does not burn in the combustion chambers and combustion temperature is reduced, which lowers oxides of nitrogen (NO_X) emissions. If the EGR valve does not open, combustion chamber temperatures are higher than normal and the engine may detonate. When the EGR valve is inoperative, NO_X emissions are high.

WARNING: When using a propane cylinder to check for intake system leaks, do not smoke and do not place the end of the hose near any source of ignition. Failure to observe this precaution may result in an explosion, causing personal injury and/or property damage.

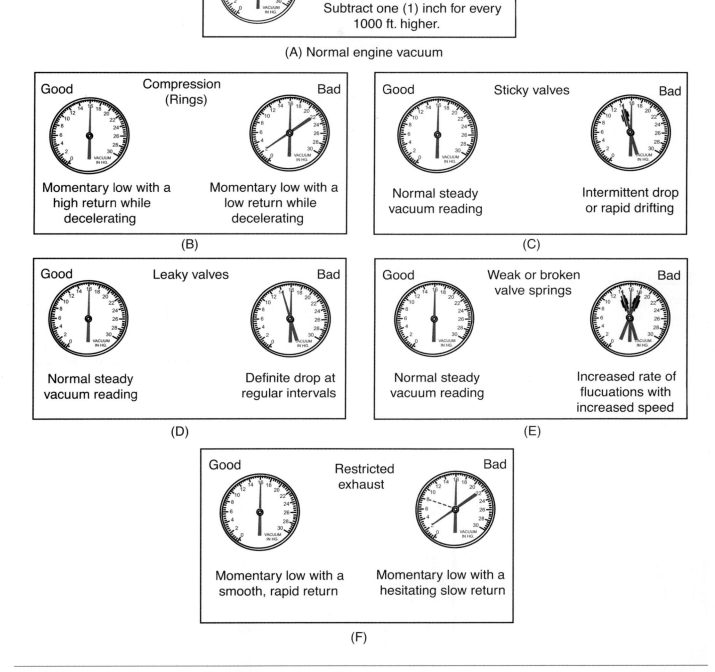

Figure 5-7 Common vacuum gauge readings and what they indicate.

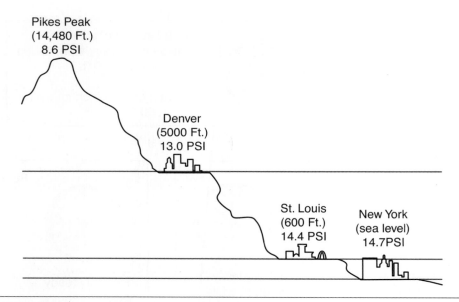

Pikes Peak
(14,480 Ft.)
8.6 PSI

Denver
(5000 Ft.)
13.0 PSI

St. Louis
(600 Ft.)
14.4 PSI

New York
(sea level)
14.7PSI

Figure 5-8 Atmospheric pressure changes with changes in elevation.

Special Tools

Propane bottle,
  metering valve

A propane cylinder with a metering valve and hose may be used to check the intake system for leaks. With the engine idling, open the metering valve a small amount and position the end of the hose near the locations of any suspected leaks in the intake manifold. When a leak is present, the engine speed increases (Figure 5-9). Close the metering valve when the test is completed.

⚠ **WARNING:** When using a propane cylinder to check for intake system leaks, maintain the cylinder in an upright position and keep the cylinder away from rotating components, hot components, or sources of ignition. Failure to observe this precaution may result in personal injury and/or property damage.

Another way to locate vacuum leaks is with an electronic leak detector (Figure 5-10). With the engine running, simply move the tester along suspected areas. The tester will emit an audible sound near the location of the leak. The point where the sound is the greatest is the point of the leak.

Figure 5-9 Using propane to locate a vacuum leak.

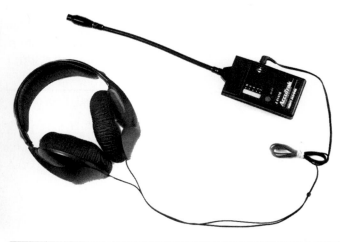

Figure 5-10 Electronic vacuum leak detector.

On engines with port-type fuel injection, vacuum leaks can cause a variety of driveability problems. This is especially true with late-model computer-controlled engines. These systems are consistently adjusting the air-fuel mixture according to the oxygen content in the exhaust. The PCM relies on the O_2 sensor for mixture information. The O_2 sensor could be biased rich or lean. This means the sensor tends to always send a signal indicating low oxygen or high oxygen instead of toggling between high and low. Since the PCM always adds fuel when exhaust shows lean and subtracts fuel when the exhaust shows rich, a biased O_2 sensor will cause the PCM to provide a constantly rich or lean mixture depending on the sensor's bias. If the sensor signals are out of range, the PCM will ignore the O_2 sensor signals and set a trouble code. However, if the signals are within the normal range, the PCM will always try to correct the mixture.

A vacuum leak will keep the O_2 sensor's signal voltage low. The PCM would then compensate by adding fuel. This would bias the adaptive fuel rich. To correct the problem, the vacuum leak must be found. A propane bottle or an electronic leak detector might have worked well on older carbureted cars. However, new port injected cars require new methods of vacuum leak testing that are actually easier to perform.

To locate the vacuum leak, you need compressed air, an air nozzle, and a squirt bottle full of bubble soap. Begin by finding a large vacuum hose, such as the PCV hose or power brake booster hose. Hook up the air nozzle to the source of compressed air and insert the nozzle into the hose. Allow air to enter the hose and fill the intake manifold. The positive pressure of the air will try to exit the manifold. Where the air escapes is where the vacuum leak is. Spray the manifold and vacuum lines with the soap. Look for bubbles. The location of the bubbles is the source of the vacuum leak.

Special Tools

Compressed air, air nozzle, soap

Exhaust System Service

Exhaust system components are subject to both physical and chemical damage. Any physical damage to an exhaust system part that causes a partially restricted or blocked exhaust system usually results in loss of power or backfire up through the throttle plate(s). In addition to improper engine operation, a blocked or restricted exhaust system causes increased noise and air pollution.

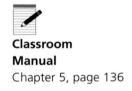

Classroom Manual
Chapter 5, page 136

> ⚠ **WARNING:** Exhaust gas contains poisonous carbon monoxide (CO) gas. This gas can cause illness and death by asphyxiation. Exhaust system leaks can be dangerous for customers and for technicians.

> ⚠ **WARNING:** Exhaust system components may be extremely hot if the engine has been running. Allow the engine and exhaust system to cool down prior to exhaust system service, and wear protective gloves.

Exhaust System Problems

Exhaust system components are subject to rust from inside and outside the components. Since exhaust components are exposed to road splash, they tend to rust on the outside. Stainless steel exhaust system components have excellent rust-resistant qualities. Exhaust system components are subject to large variations in temperature, which form condensation inside these components. Condensation inside the exhaust system components rusts them on the inside. A vehicle that is driven short distances and then shut off tends to have more condensation in the exhaust system than a vehicle that is driven continually for longer time periods. If a vehicle is driven for a short distance, the exhaust system heat does not have sufficient time to vaporize the condensation in the system. The rust erodes exhaust system components, causing exhaust leaks and excessive noise.

Exhaust components, particularly mufflers, catalytic converters, and pipes, may become restricted. Mufflers and catalytic converters may become restricted by loose internal components. Pipes may become restricted by physical damage or collapsed inner walls on double-walled pipes. This exhaust restriction causes a loss of engine power and increased fuel consumption.

Exhaust system components cause a rattling noise if they are touching, or almost touching, chassis components. Loose internal structure in mufflers and catalytic converters may cause a rattling noise, especially when the engine is accelerated.

Exhaust System Inspection

Most parts of the exhaust system, particularly the exhaust pipe, muffler, and tailpipe, are subject to rust, corrosion, and cracking. Broken or loose clamps and hangers can allow parts to separate or hit the road as the car moves.

WARNING: During all exhaust inspection and repair work, wear safety glasses or equivalent eye protection.

Any inspection of the exhaust system should include listening for hissing or rumbling that would result from a leak in the system. An on-lift inspection should pinpoint any damage. All of the exhaust system components should be inspected for physical damage, kinks, dents, holes, rust, loose internal components, loose clamps and hangers, proper clearance, and improperly positioned shields. Physical damage on exhaust system components includes flattened areas, abnormal bends, and scrapes. Flattened exhaust system components cause restrictions. Components with this type of damage should be replaced.

Loose or broken hangers (Figure 5-11) and clamps may allow exhaust system components to contact chassis components, resulting in a rattling noise. Broken hangers cause excessive movement of exhaust system components, and this action may result in premature breaking of these components.

A block of wood or a wrench may be used to tap mufflers and catalytic converters to check for loose internal components. If the internal structure is loose, a rattling noise is evident when the component is taped.

Push upward on the muffler and catalytic converter to check for weak, rusted inlet and outlet pipes. Check the entire exhaust system for leaks. With the engine running, hold a length of hose near your ear and run the other end of the hose around the exhaust system components. The hose amplifies the sound of an exhaust leak and helps to locate the source of a leak.

Inspect all exhaust system shields for proper position, and check for the correct clearance between the exhaust system components and the shields (Figure 5-12). Check the catalytic converter for overheating, which is indicated by bluish or brownish discoloration.

The system should also be tested for blockage and restrictions using a vacuum and/or pressure gauge and a tachometer. Before beginning work on the system, be sure it is cool to the touch. Some technicians disconnect the battery ground to avoid short-circuiting the electrical system. Soak all rusted nuts, bolts, etc., in a good quality penetrating oil. Finally, check the system for critical clearance points so they can be maintained when new components are installed.

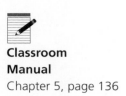

Classroom Manual
Chapter 5, page 136

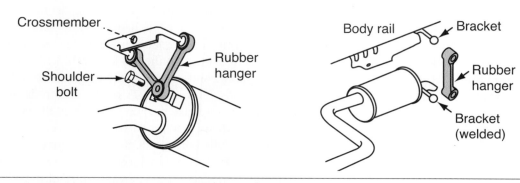

Figure 5-11 Various exhaust hangers used on today's vehicles.

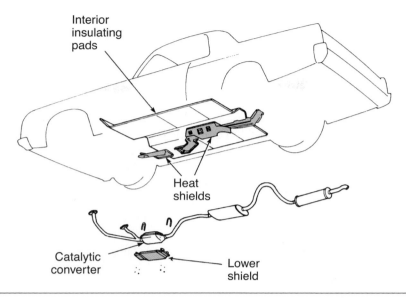

Interior
insulating
pads

Heat
shields

Catalytic
converter

Lower
shield

Figure 5-12 Location of exhaust system heat shields.

Catalytic Converters

■ **CAUTION:** Avoid prolonged engine compression measurement or spark jump testing. They can allow fuel to collect in the converter. When you need to conduct these tests, do them as rapidly as possible.

A plugged converter or any exhaust restriction can cause loss of power at high speeds, stalling after starting (if totally blocked), a drop in engine vacuum as engine rpm increases, or sometimes popping or backfiring. Catalytic converters can also be destroyeed by overheating, usually caused by a very rich fuel mixture due to a fuel delivery problem. Leaded fuel, although not widely available, will also ruin a catalytic converter.

There are many ways to test a catalytic converter. One of these is to simply strike the converter with a rubber mallet. If the converter rattles, it needs to be replaced and there is no need for further testing. A rattle indicates loose catalyst substrate, which will soon rattle into small pieces. This is one test and is not used to determine if the catalyst is good. Keep in mind a converter may also be bad if it is restricted or not working properly.

A vacuum gauge can be used to watch engine vacuum while the engine is accelerated. Another way to check for a restricted exhaust or catalyst is to insert a pressure gauge in the exhaust manifold's bore for the O_2 sensor (Figure 5-13). On some engines, it is not that easy. On

Classroom Manual Chapter 5, page 138

Special Tools

Vacuum gauge, pressure gauge, pyrometer, exhaust gas analyzer

The catalytic converter uses the catalysts to change CO, HC, and NO_x into water vapor, CO_2, N, and O_2.

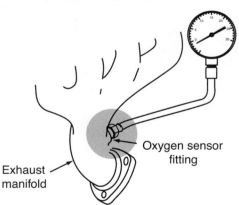

Oxygen sensor
fitting

Exhaust
manifold

Figure 5-13 Insert pressure gauge in oxygen sensor fitting.

Two approaches can be used to increase engine compression. One is to modify the internal configuration of the engine to increase the basic compression ratio. The less expensive way is to increase the quantity of the intake charge. The two processes of artificially increasing the amount of airflow into the engine are known as turbocharging and supercharging.

A PFE is a pressure feedback exhaust gas recirculation sensor. According to SAE J1930, this sensor should be called a feedback pressure EGR sensor.

A **pyrometer** is an electronic device that measures heat.

engines with a PFE, the port for the PFE can be used to install the gauge. On other engines, it may be necessary to fabricate an adapter for the pressure gauge from an old O_2 sensor or air check valve. Once the gauge is in place, hold the engine's speed at 2,000 rpm and watch the gauge. The desired pressure reading will be less than 1.25 psi. A very bad restriction will give a reading of over 2.75 psi.

The converter should be checked for its ability to convert CO and HC into CO_2 and water. There are three separate tests for doing this. The first method is the delta temperature test. As with all converter tests, make sure the catalyst is warmed up before conducting the test.

To conduct this test, use a hand-held digital **pyrometer** (Figure 5-14). By touching the pyrometer probe to the exhaust pipe just ahead of and just behind the converter, there should be an increase of at least 100°F, or 8 percent above the inlet temperature reading, as the exhaust gases pass through the converter. If the outlet temperature is the same or lower, nothing is happening inside the converter. Do not be quick to condemn the converter. To do its job efficiently, the converter needs a steady supply of oxygen from the air pump if so equipped. A bad pump, faulty diverter valve or control valve, leaky air connections, or faulty computer control over the air injection system could be preventing the needed oxygen from reaching the converter. If the converter fails this test, check those systems. If a problem is found, fix it, then retest the converter.

The next test is called the O_2 storage test and is based on the fact that a good converter stores oxygen. The following test is for closed-loop feedback systems. Non-feedback systems require a different procedure. Begin by disabling the air injection system. Turn on the gas analyzer and allow it to warm up. Start the engine and allow the converter to warm up. Once the analyzer and converter are ready, hold the engine at 2,000 rpm. Watch the readings on the exhaust analyzer. If the converter was cold, the readings will continue to drop until the converter reaches light-off temperature. Once the numbers stop dropping, check the oxygen level on the gas analyzers. The O_2 readings should be about 0.5 to 1 percent. This shows the converter is using most of the available oxygen. There is one exception to this, if there is no CO left, there may be a higher amount of oxygen in the exhaust. However, it still should be less than 2.5 percent.

It is important to observe the O_2 reading as soon as the CO begins to drop. Otherwise, good converters will fail this test. The O_2 will go way over 1.25 after CO starts to drop. If the O_2 reading is too high and there is no CO in the exhaust, stop the test and make sure the system has control of the air-fuel mixture. If the system is in control, use a propane enrichment tool to bring

Figure 5-14 A digital pyrometer used to measure inlet and outlet temperatures of a catalytic converter.

the CO level up to about 0.5 percent. Now the O_2 level should drop to zero. Once you have a solid oxygen reading, snap the throttle open, then let it drop back to idle. Check the rise in oxygen. It should not rise above 1.2 percent. If the converter passes these tests, it is working properly. If the converter fails the tests, chances are that it is working poorly or not at all.

This final converter test uses a principle that checks the converter's efficiency. Remember, the catalytic converter changes CO and HC into CO_2 and water. Before beginning this test, make sure the converter is warmed up. Calibrate a four-gas analyzer and insert its probe into the tailpipe. If the vehicle has dual exhaust with a crossover, plug the side that the probe is not in. If the vehicle has a true dual exhaust system, check both sides separately.

Disable the ignition, then crank the engine for 9 seconds while pumping the throttle. Watch the readings on the analyzer; the CO_2 on fuel-injected vehicles should be over 11 percent and carbureted vehicles should have a reading of over 10 percent. As soon as you have your readings, reconnect the ignition and start the engine. Do this as quickly as possible to cool off the catalytic converter. If while the engine is cranking the HC goes above 1,500 ppm, stop cranking; the converter is not working. Also, stop cranking once the CO_2 readings reach 10 or 11 percent; the converter is good. If the catalytic converter is bad, there will be high HC and low CO_2 at the tailpipe. Do not repeat this test more than one time without running the engine in between.

Typically, today's catalytic converters are not serviced or repaired; they are generally replaced with another converter.

If a catalytic converter is found to be bad, it is replaced. There are two types of replacement. Installation kits (Figure 5-15) include all necessary components for the installation. There are also direct-fit catalytic converters (Figure 5-16) (Photo Sequence 8).

Catalytic Converter Removal and Replacement

Since there are many different catalytic converter mountings, the converter removal and replacement procedure varies depending on the vehicle. The following is a typical converter removal-and-replacement procedure for a converter mounted directly to the exhaust manifold:

1. Allow the engine and exhaust system time to cool down before working on the vehicle. Disconnect the negative battery terminal, and wait 1 minute before working on the vehicle if the car has an air bag. This gives the air bag module time to power down and prevents accidental air bag deployment.

2. Lift the vehicle on a hoist and disconnect the two bolts and front exhaust pipe bracket. Remove the two bolts holding the front exhaust pipe to the center exhaust pipe.

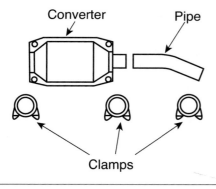

Figure 5-15 Catalytic converter installation kit.

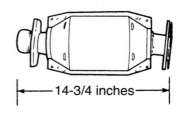

Figure 5-16 A direct-fit replacement catalytic converter.

Photo Sequence 8
Typical Procedure for Catalytic Converter Diagnosis

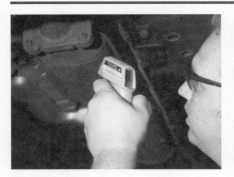

P8-1 The first method that is commonly used to check the operation of a converter is the temperature difference test, also called a Delta temperature test.

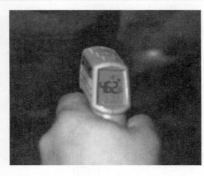

P8-2 The converter outlet should be at least 100°F hotter than the inlet with the vehicle at normal operating temperature.

P8-3 If the converter fails this test, make certain that the AIR system is performing as intended before replacing the converter.

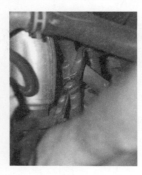

P8-4 The next test is the oxygen storage test. For this test, we will temporarily disable the AIR system (if equipped) and make sure the vehicle is at operating temperature.

P8-5 The technician holds the engine rpm at 2,000 rpm compared to idle readings, HC and CO will usually drop. As soon as CO readings start to drop, read the level of O_2. O_2 should be $1/2$ to 1 percent (with 0 percent CO, the O_2 level may be as high as 2.5 percent).

P8-6 If the level of O_2 is too high and there is no CO in the exhaust, make sure the vehicle is in control of the air-fuel mixture. If it is, add enough propane to bring the O_2 level to zero.

P8-7 Once the O_2 level is stable, snap the throttle open and allow it to return to idle. The O_2 level should not rise above 1.2 percent. This gives an indication of how well the converter stores oxygen. If the converter fails the test, it should be replaced.

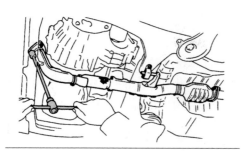

Figure 5-17 Removing the nuts holding the exhaust pipe to the exhaust manifold.

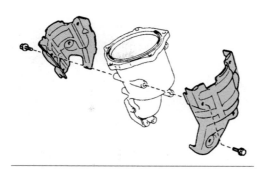

Figure 5-18 Separating the outer heat shields around the catalytic converter.

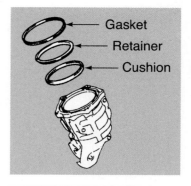

Figure 5-19 Installing a new cushion, retainer, and gasket on the catalytic converter.

3. Disconnect the three nuts holding the front exhaust pipe to the front catalytic converter (Figure 5-17).

4. Be sure the front catalytic converter is cool, and disconnect the suboxygen sensor connector.

5. Remove the bolt, nut, and number one manifold bracket.

6. Remove the bolt, nut, and number two manifold bracket.

7. Remove the two nuts and three bolts holding the catalytic converter to the manifold, and remove the converter. Remove the gasket, retainer, and cushion from the converter. Remove the eight bolts holding the outer heat insulators on the converter and remove the heat shields (Figure 5-18).

8. Install the heat shields on the new catalytic converter and tighten the retaining bolts to the specified torque.

9. Place a new cushion, retainer, and gasket on the catalytic converter (Figure 5-19).

10. Install the catalytic converter on the exhaust manifold, and install the three new bolts and two new nuts (Figure 5-20). Tighten the bolts and nuts to the specified torque.

11. Install the number one and number two exhaust manifold support brackets, and tighten the fasteners to the specified torque.

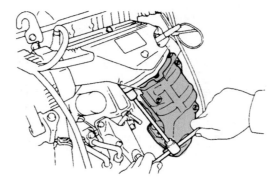

Figure 5-20 Installing the catalytic converter onto the exhaust manifold.

12. Place new gaskets on the front and rear of the front exhaust pipe, and install the front exhaust pipe in its proper position. Install the two bolts and nuts holding the front exhaust pipe to the center exhaust pipe.

13. Install the three nuts holding the front exhaust pipe to the converter, and tighten these nuts to the specified torque.

14. Tighten the two bolts and nuts holding the front exhaust pipe to the center exhaust pipe. Install the front exhaust pipe bracket, and tighten the fastener to the specified torque.

Replacing Exhaust System Components

Most exhaust system servicing involves the replacement of parts. When replacing exhaust system components, it is important that original equipment parts (or their equivalent) are used to ensure proper alignment with other parts in the system and to provide acceptable exhaust noise levels. When replacing only one component in an exhaust system, it is not always necessary to take the parts behind it off.

Exhaust Manifold

Classroom Manual
Chapter 5, page 136

The exhaust manifold gasket seals the joint between the head and exhaust manifold. Many new engines are assembled without exhaust manifold gaskets. This is possible because new manifolds are flat and fit tightly against the head without leaks. Exhaust manifolds go through many heating/cooling cycles. This causes stress and some corrosion in the exhaust manifold. Removing the manifold will usually distort the manifold slightly so it is no longer flat enough to seal without a gasket. Exhaust manifold gaskets are normally used to eliminate leaks when exhaust manifolds are reinstalled (Figure 5-21).

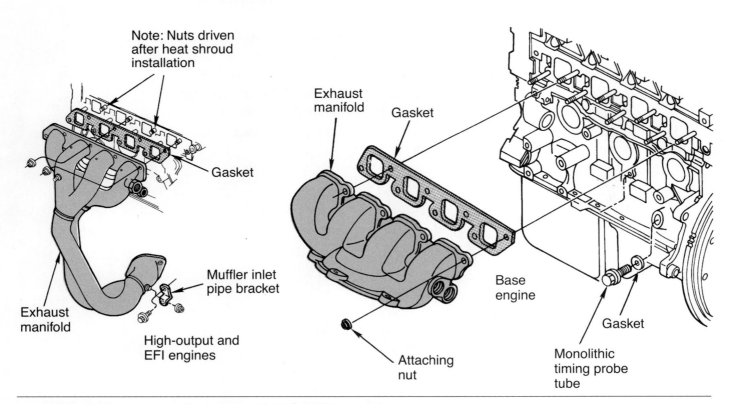

Figure 5-21 Examples of exhaust manifolds for in-line engines.

208

Exhaust manifolds can warp because of the heating and cooling cycles they go through. They can also crack because of the high temperatures generated by the engine. This usually occurs after the car passes through a large puddle and cold water splashes on the manifold's hot surface. If the manifold is warped beyond manufacturer's specifications or is cracked, it must be replaced.

CAUTION: Manifolds warp more easily if an attempt is made to remove them while still hot. Remember, heat expands metal, making assembly bolts more difficult to remove and easier to break.

SERVICE TIP: Since exhaust system fasteners are subject to extreme heat, they may be difficult to loosen. A generous application of penetrating solvent on these fasteners makes removal easier.

To replace an exhaust manifold, remove the exhaust pipe bolts at the manifold flange and disconnect any other components in the manifold, such as an oxygen (O_2) sensor. In some applications, it is easier to disconnect the sensor wires and leave the sensor in the manifold. However, if the exhaust manifold is being replaced, the sensor must be removed and installed in the new manifold. Remove the bolts retaining the manifold to the cylinder head and lift the manifold from the engine compartment. Remove the manifold heat shield.

Use a scraper to clean the matching surfaces on the exhaust manifold and cylinder head. Measure the exhaust manifold surface for warping with a straightedge and feeler gauge. Perform this measurement at three locations on the manifold surface. If the manifold is warped more than specified by the vehicle manufacturer, replace the manifold. Remove any gasket material from the manifold flange. If the flange has a ball connection, be sure the ball is not damaged.

If the manifold has an O_2 sensor, place some antiseize compound on the threads and install the sensor in the manifold. When this sensor is easy to access, it may be installed after the manifold is installed on the cylinder head. Install a new gasket between the cylinder head and the manifold, and install the manifold against the cylinder head. Install exhaust manifold-to-cylinder head mounting bolts, and tighten these bolts to the specified torque. Many exhaust manifold bolts, nuts, and studs are special heat-resistant fasteners. If exhaust manifold bolts or nuts must be replaced, do not substitute ordinary fasteners for heat-resistant fasteners.

Be sure the exhaust pipe mounting surface that fits against the exhaust manifold flange is in satisfactory condition. If an exhaust manifold flange gasket is required, install a new gasket and tighten the flange bolts to the specified torque (Figure 5-22).

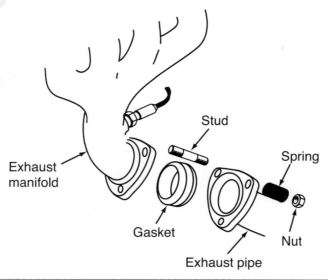

Figure 5-22 Exhaust pipe-to-manifold flange connections.

Exhaust Pipes and Mufflers

To replace a damaged exhaust pipe, begin by supporting the converter to keep it from falling. Carefully remove the oxygen sensor if there is one. Remove any hangers or clamps holding the exhaust pipe to the frame. Unbolt the flange holding the exhaust pipe to the exhaust manifold. When removing the exhaust pipe, check to see if there is a gasket. If so, discard it and replace it with a new one. Once the joint has been taken apart, the gasket loses its effectiveness. Disconnect the pipe from the converter and pull the front exhaust pipe loose and remove it.

✓ SERVICE TIP: An easy way to break off rusted nuts is to tighten them instead of loosening them. Sometimes a badly rusted clamp or hanger strap will snap off with ease. Sometimes the old exhaust system will not drop free of the body because a large part is in the way, such as the rear end or the transmission support. Use a large, cold chisel, pipe cutter, hacksaw, muffler cutter, or chain cutter to cut the old system at convenient points to make the exhaust assembly smaller.

Many original mufflers and catalytic converters are integral with the interconnecting pipes. When these components are replaced, they must be cut from the exhaust system with a cutting tool (Figure 5-23). The inlet and outlet pipes on the replacement muffler or converter must have a 1.5-in. (3.8-cm) overlap on the connecting pipes. Before cutting the pipes to remove the muffler or converter, measure the length of the new component, and always cut these pipes to provide the required overlap.

If the muffler or converter inlet and outlet pipes are clamped to the connecting pipes, remove the clamp. When the muffler or converter is tight on the connecting pipe, a slitting tool and hammer may by used to slit and loosen the muffler pipe. A slitting tool on an air chisel may be used for this job.

The old exhaust pipe might be rusted into the muffler or converter opening. Attempt to collapse the old pipe by using a cold chisel or slitting tool and a hammer (Figure 5-24). While freeing the pipe, try not to damage the muffler inlet. It must be perfectly round to accept the new pipe.

Slide the new pipe into the muffler (some lubricant might be helpful). Attach the front end to the manifold. The pipe must fit at least 1-1/2 inches into the converter or muffler. Before tightening the connectors, check the system for alignment. When it is properly aligned, tighten the clamps.

Figure 5-23 An exhaust pipe cutting tool.

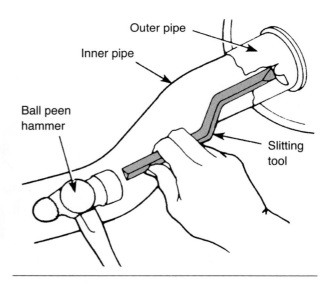

Figure 5-24 Using a slitting tool to separate a muffler from an exhaust pipe.

Figure 5-25 Hydraulically operated pipe expanding tool.

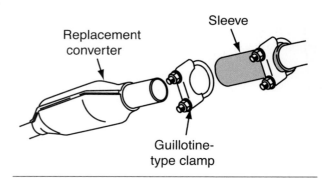

Figure 5-26 Exhaust pipe sleeve installation.

When a new muffler or converter is installed, the connecting pipes may not fit perfectly. A hydraulically operated expanding tool may be used to expand the pipe and provide the necessary fit (Figure 5-25). In some cases, adaptors may be used to provide the necessary pipe fit. A sleeve may be used to join two pieces of pipe (Figure 5-26). When the new exhaust system components are installed, there must be adequate clearance between all the exhaust components and the chassis. Install and tighten all clamps and hangers securely.

Turbocharger Diagnosis

⚠ WARNING: If the engine has been running, turbochargers and related components are extremely hot. Use caution, and wear protective gloves to avoid burns when servicing these components.

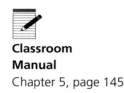

Classroom Manual
Chapter 5, page 145

The first step in turbocharger diagnosis is to check all linkages and hoses connected to the turbocharger (Figure 5-27). Inspect the wastegate diaphragm linkage for looseness and binding, and check the hose from the wastegate diaphragm to the intake manifold for cracks, kinks, and restrictions. Check the coolant hoses and oil line connected to the turbocharger for leaks.

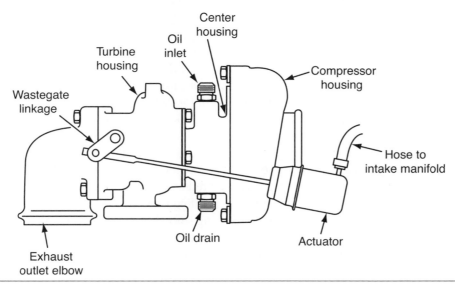

Figure 5-27 A turbocharger assembly.

Exhaust leaks
between the
cylinders and the
turbocharger
decrease
turbocharger
efficiency.

Excessive blue smoke in the exhaust may indicate worn turbocharger seals. The technician must remember that worn valve guide seals or piston rings also cause oil consumption and blue smoke in the exhaust. When oil leakage is noted at the turbine end of the turbocharger, always check the turbocharger oil drain tube and the engine crankcase breathers for restrictions. When sludged engine oil is found, the engine's oil and oil filter must be changed.

Check all turbocharger mounting bolts for looseness. A rattling noise may be caused by loose turbocharger mounting bolts. Some whirring noise is normal when the turbocharger shaft is spinning at high speed. Excessive internal turbocharger noise may be caused by too much end-play on the shaft, which allows the blades to strike the housings.

Check for exhaust leaks in the turbine housing and related pipe connections. If exhaust gas is escaping before it reaches the turbine wheel, turbocharger effectiveness is reduced. Check for intake system leaks. If there is a leak in the intake system before the compressor housing, dirt may enter the turbocharger and damage the compressor or turbine wheel blades. When a leak is present in the intake system between the compressor wheel housing and the cylinders, turbocharger pressure is reduced.

When turbochargers have computer-controlled boost pressure, a diagnostic trouble code (DTC) is stored in the PCM memory if a fault is present in the boost control solenoid or solenoid-to-PCM wiring.

Common Turbocharger Problems

The turbocharger, with proper care and servicing, will provide years of reliable service. Most turbocharger failures are caused by lack of lubricant, ingestion of foreign objects, or contamination of lubricant. Refer to Figure 5-28 and the service manual for the symptoms of turbocharger failures and a summary of causes and recommended remedies.

Turbocharger Inspection

Special Tools

Stethoscope, flash-
light

To inspect a turbocharger, start the engine and listen to the sound the turbo system makes. As a technician becomes more familiar with this characteristic sound, it will be easier to identify an air leak between the compressor outlet and engine or an exhaust leak between engine and turbo by the presence of a higher-pitched sound. If the turbo sound cycles or changes in intensity, the likely causes are a plugged air cleaner or loose material in the compressor inlet ducts or dirt buildup on the compressor wheel and housing.

After listening, check the air cleaner and remove the ducting from the air cleaner to turbo and look for dirt buildup or damage from foreign objects. Check for loose clamps on the compressor outlet connections and check the engine intake system for loose bolts or leaking gaskets. Then, disconnect the exhaust pipe and look for restrictions or loose material. Examine the exhaust system for cracks, loose nuts, or blown gaskets. Rotate the turbo shaft assembly. Does it rotate freely? Are there signs of rubbing or wheel impact damage?

Leaks in the air
intake system may
allow dirt particles to
enter the
turbocharger and
damage the blades.

Visually inspect all hoses, gaskets, and tubing for proper fit, damage, and wear. Check the low pressure, or air cleaner, side of the intake system for vacuum leaks.

On the pressure side of the system, you can check for leaks by using soapy water. After applying the soap mixture, look for bubbles to pinpoint the source of the leak.

Leakage in the exhaust system upstream from the turbine housing will also affect turbo operation. If exhaust gases are allowed to escape prior to entering the turbine housing, the reduced temperature and pressure will cause a proportionate reduction in boost and an accompanying loss of power. If the wastegate does not appear to be operating properly (too much or too little boost), check to make sure the connecting linkage is operating smoothly and is not binding. Also, check to make sure the pressure-sensing hose is clear and properly connected. Photo Sequence 9 shows a typical procedure for inspecting turbochargers and testing boost pressure.

TURBOCHARGER TROUBLESHOOTING GUIDE

Condition	Possible Causes Code Numbers	Remedy Description by Code Numbers
Engine lacks power	1, 4, 5, 6, 7, 8, 9, 10, 11, 18, 20, 21, 22, 25, 26, 27, 28, 29, 30, 37, 38, 39, 40, 41, 42, 43	1. Dirty air cleaner element 2. Plugged crankcase breathers 3. Air cleaner element missing, leaking, not sealing correctly; loose connections to turbocharger 4. Collapsed or restricted air tube before turbocharger
Black smoke	1, 4, 5, 6, 7, 8, 9, 10, 11, 18, 20, 21, 22, 25, 26, 27, 28, 29, 30, 37, 38, 39, 40, 41, 43	5. Restricted-damaged crossover pipe, turbocharger to inlet manifold 6. Foreign object between air cleaner and turbocharger 7. Foreign object in exhaust system (from engine, check engine)
Blue smoke	1, 2, 4, 6, 8, 9, 17, 19, 20, 21, 22, 32, 33, 34, 37, 45	8. Turbocharger flanges, clamps, or bolts loose. 9. Inlet manifold cracked; gaskets loose or missing; connections loose
Excessive oil consumption	2, 8, 15, 17, 19, 20, 29, 30, 31, 33, 34, 37, 45	10. Exhaust manifold cracked, burned; gaskets loose, blown, or missing 11. Restricted exhaust system
Excessive oil turbine end	2, 7, 8, 17, 19, 20, 22, 29, 30, 32, 33, 34, 45	12. Oil lag (oil delay to turbocharger at start-up) 13. Insufficient lubrication 14. Lubricating oil contaminated with dirt or other material 15. Improper type lubricating oil used
Excessive oil compressor end	1, 2, 4, 5, 6, 8, 19, 20, 21, 29, 30, 33, 34, 45	16. Restricted oil feed line 17. Restricted oil drain line 18. Turbine housing damaged or restricted 19. Turbocharger seal leakage
Insufficient lubrication	8, 12, 14, 15, 16, 23, 24, 31, 34, 35, 36, 44, 46	20. Worn journal bearings 21. Excessive carbon buildup in compressor housing 22. Excessive carbon buildup behind turbine wheel 23. Too fast acceleration at initial start (oil lag)
Oil in exhaust manifold	2, 17, 18, 19, 20, 22, 29, 30, 33, 34, 45	24. Too little warm-up time 25. Fuel pump malfunction 26. Worn or damaged injectors 27. Valve timing
Damaged compressor wheel	3, 4, 6, 8, 12, 15, 16, 20, 21, 23, 24, 31, 34, 35, 36, 44, 46	28. Burned valves 29. Worn piston rings 30. Burned pistons 31. Leaking oil feed lines
Damaged turbine wheel	7, 8, 12, 13, 14, 15, 16, 18, 20, 22, 23, 24, 25, 28, 30, 31, 34, 35, 36, 44, 46	32. Excessive engine pre-oil 33. Excessive engine idle 34. Coked or sludged center housing 35. Oil pump malfunction

Drag or bind in rotating assembly	3, 6, 7, 8, 12, 13, 14, 15, 16, 18, 20, 21, 22, 23, 24, 31, 34, 35, 36, 44, 46	36. Oil filter plugged
		37. Oil-bath-type air cleaner:
		  ◆ Air inlet screen restricted
		  ◆ Oil pullover
		  ◆ Dirty air cleaner
		  ◆ Oil viscosity low
		  ◆ Oil viscosity high

Worn bearings, journals, bearing bores	6, 7, 8, 12, 13, 14, 15, 16, 23, 24, 31, 35, 36, 44, 46	38. Actuator damaged or defective 39. Waste gate binding 40. Electronic control module or connector(s) defective 41. Waste gate actuator solenoid or connector defective
Noisy	1, 3, 4, 5, 6, 7, 8, 9, 10, 11, 12, 13, 14, 15, 16, 18, 20, 21, 22, 23, 24, 31, 34, 35, 36, 37, 44, 46	42. EGR valve defective 43. Alternator voltage incorrect 44. Engine shut off without adequate cool-down time
Sludged or coked center housing	2, 11, 13, 14, 15, 17, 18, 24, 31, 35, 36, 44, 46	45. Leaking valve guide seals 46. Low oil level

Figure 5-28 Turbocharger troubleshooting guide.

Photo Sequence 9
Typical Procedure for Inspecting Turbochargers and Testing Boost Pressure

P9-1 Check all turbocharger linkages for looseness, and check all turbocharger hoses for leaks, cracks, kinks, and restrictions.

P9-2 Check the level and condition of the engine oil on the dipstick.

P9-3 Check the exhaust for evidence of blue smoke when the engine is accelerated.

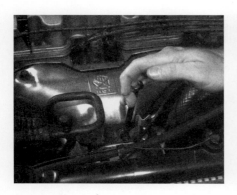

P9-4 Check all turbocharger mounting bolts for looseness.

P9-5 Check for exhaust leaks between the engine and the turbocharger, and use a stethoscope to listen for excessive turbocharger noise.

P9-6 Use a propane cylinder, metering valve, and hose to check for intake manifold vacuum leaks.

P9-7 Connect a pressure gauge to the intake manifold and locate the gauge in the passenger compartment where it can be easily seen by the driver.

P9-8 Road test the car and accelerate from 0 to 60 mph at wide-open throttle while observing the pressure gauge.

P9-9 Disconnect the pressure gauge from the intake manifold and remove the gauge from the passenger compartment.

Testing Boost Pressure

Special Tools

Pressure gauge, dial indicator

✓ **SERVICE TIP:** If the engine has low cylinder compression, there is reduced airflow through the cylinders, which results in lower turbocharger shaft speed and boost pressure.

✓ **SERVICE TIP:** Excessive boost pressure causes engine detonation and possible engine damage.

Connect a pressure gauge to the intake manifold to check the boost pressure. The pressure gauge hose should be long enough so the gauge may be positioned in the passenger compartment. One of the front windows may be left down enough to allow the gauge hose to extend into the passenger compartment. Road test the vehicle at the speed specified by the vehicle manufacturer, and observe the boost pressure. Some vehicle manufacturers recommend accelerating from a stop to 60 mph (96 kph) at wide-open throttle while observing the boost pressure.

Higher-than-specified boost pressure may be caused by a defective wastegate system. Low boost pressure may be caused by the wastegate system or turbocharger defects, such as damaged wheel blades or worn bearings. An engine with low cylinder compression will usually have low boost pressure.

Classroom Manual
Chapter 5, page 147

Wastegate Service

Wastegate malfunctions can usually be traced to carbon buildup that keeps the unit from closing or causes it to bind. A defective diaphragm or leaking vacuum hose can result in an inoperative wastegate (Figure 5-29). But before condemning the wastegate, check the ignition timing, the spark-retard system, vacuum hoses, knock sensor, oxygen sensor, and computer to be sure each is operating properly.

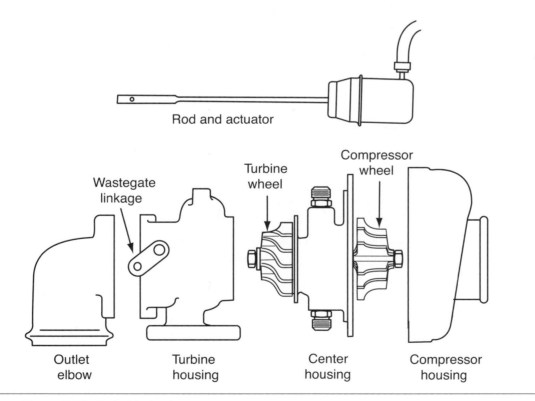

Figure 5-29 Typical turbocharger.

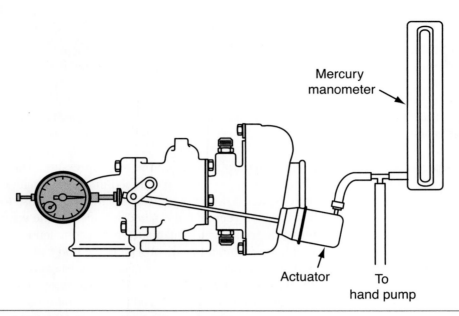

Figure 5-30 Measuring wastegate rod movement.

CAUTION: When removing carbon deposits from turbine and wastegate parts, never use a hard metal tool or sandpaper. Remember that any gouges or scratches on these metal parts can cause severe vibration or damage to the turbocharger. To clean these parts, use a soft brush and a solvent.

If the wastegate stroke is reduced, the wastegate valve opening is decreased and boost pressure is increased. Connect a hand pressure pump and a pressure gauge to the wastegate diaphragm. Position a dial indicator against the outer end of the wastegate diaphragm rod (Figure 5-30). Supply the specified pressure to the wastegate diaphragm and observe the dial indicator movement. If the wastegate rod movement is less than specified, disconnect the rod from the wastegate valve linkage and check the linkage for binding. If this linkage moves freely, replace the wastegate diaphragm. When the wastegate valve linkage is binding, turbocharger repair or replacement is required.

Turbocharger Removal

The turbocharger removal procedure varies depending on the engine. On some cars, such as a Nissan 300 ZX, the manufacturer recommends the engine be removed to gain access to the turbocharger. On other applications, the turbocharger may be removed with the engine in the vehicle. Always follow the turbocharger removal procedure in the vehicle manufacturer's service manual. The following is a typical turbocharger removal procedure:

1. Disconnect the negative battery cable and drain the cooling system.
2. Disconnect the exhaust pipe from the turbocharger.
3. Remove the support bracket between the turbocharger and the engine block.
4. Remove the bolts from the oil drain back housing on the turbocharger.
5. Disconnect the turbocharger coolant inlet tube nut at the block outlet and remove the tube-support bracket.
6. Remove the air cleaner element, air cleaner box, bracket, and related components.
7. Disconnect the accelerator linkage, throttle body electrical connector, and vacuum hoses.

8. Loosen the throttle body-to-turbocharger inlet hose clamps and remove the three throttle body-to-intake manifold attaching screws. Remove the throttle body.

9. Loosen the lower turbocharger discharge hose clamp on the compressor wheel housing.

10. Remove the fuel rail-to-intake manifold screws and the fuel line bracket screw. Remove the two fuel rail bracket-to-heat shield retaining clips and pull the fuel rail and injectors upward out of the way. Tie the fuel rail in this position with a piece of wire.

11. Disconnect the oil supply line from the turbocharger housing.

12. Remove the intake manifold heat shield.

13. Disconnect the coolant return line from the turbocharger and the water box. Remove the line support bracket from the cylinder head and remove the line.

14. Remove the four nuts retaining the turbocharger to the exhaust manifold and remove the turbocharger from the manifold studs. Move the turbocharger downward toward the passenger side of the vehicle, and then lift the unit up and out of the engine compartment.

Measuring Turbocharger Shaft Axial Movement

After the turbocharger is removed from the engine, remove the turbine outlet elbow from the turbine housing. Position a dial indicator against the shaft and move the shaft inward and outward while observing the dial indicator reading (Figure 5-31). The axial shaft movement should not exceed the vehicle manufacturer's specifications. On some turbochargers, the maximum axial shaft movement is 0.003 in. (0.91 mm). If the axial shaft movement exceeds the manufacturer's specifications, the turbocharger must be repaired or replaced. Some manufacturers recommend complete turbocharger replacement, whereas other manufacturers recommend replacing the center housing assembly as a unit. Some turbocharger manufacturers recommend replacement of individual components. Always follow the service procedures in the vehicle manufacturer's service manual.

Turbocharger Component Inspection

If the vehicle manufacturer recommends turbocharger disassembly, inspect the wheels and shaft after the end housings are removed. Lack of lubricant or lubrication with contaminated oil results in bearing failure, which leads to wheel rub on the end housings. A contaminated cooling system may provide reduced turbocharger bearing cooling and premature bearing failure. Bearing failure will likely lead to seal damage. Inspect the shaft and bearings for a burned condition (Figure 5-32). If the shaft and bearings are burned, replace the complete center housing assembly or individual parts as recommended by the manufacturer.

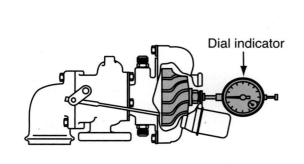

Figure 5-31 Measuring axial movement.

Figure 5-32 Inspecting turbocharger shaft, bearings, and

Figure 5-33 Inspecting turbocharger end housings.

If the shaft and bearings are in satisfactory condition but the blades are damaged, check the air intake system for leaks or a faulty air cleaner element. When the blades or shaft and bearings must be replaced, always check the end housings for damage (Figure 5-33). When these housings are marked or scored, replacement is necessary. Since turbocharger components are subjected to extreme heat, use a straightedge to check all mating surfaces for a warped condition. Replace warped components as necessary.

Turbocharger Installation

CAUTION: Failure to prelubricate turbocharger bearings may result in premature bearing failure.

Prior to reinstalling the turbocharger, be sure the engine oil and filter are in satisfactory condition. Change the oil and filter as required and be sure the proper oil level is indicated on the dipstick. Check the coolant for contamination. Flush the cooling system if contamination is present. Reverse the turbocharger removal procedure explained previously in this chapter to install the turbocharger. Replace all gaskets and be sure all fasteners are tightened to the specified torque. Follow the vehicle manufacturer's recommended procedure for filling the cooling system. On some Chrysler engines, this involves removing a plug on top of the water box and pouring coolant into the radiator filler neck until coolant runs out the hole in the top of the water box (Figure 5-34). Install the plug in the top of the water box and tighten the plug to the specified torque. Continue filling the cooling system to the maximum level mark on the reserve tank (Figure 5-35).

The turbocharger bearings must be prelubricated before starting the engine to prevent bearing damage. Some vehicle manufacturers recommend removing the turbocharger oil supply pipe and pouring a half pint of the specified engine oil into the turbocharger to prelubricate the bearings (Figure 5-36).

Other vehicle manufacturers recommend disabling the ignition system by disconnecting the positive primary coil wire and cranking the engine for 10 to 15 seconds to allow the engine lubrication system to lubricate the turbocharger bearings. Always follow the turbocharger prelubrication instructions in the vehicle manufacturer's service manual.

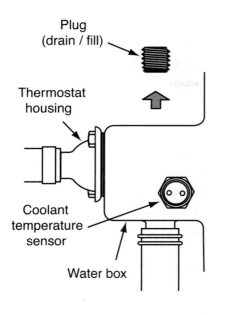

Plug
(drain / fill)

Thermostat
housing

Coolant
temperature
sensor

Water box

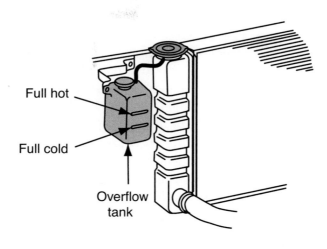

Full hot

Full cold

Overflow
tank

Figure 5-34 Removing the plug in the top of the water box while filling the cooling system.

Figure 5-35 Coolant level marks on the reserve tank.

Figure 5-36 Pouring engine oil into the turbocharger oil inlet to prelubricate the turbocharger's bearings.

Turbo Startup and Shutdown

After replacement of a turbocharger or after an engine has been unused or stored, there can be a considerable lag after engine startup before the oil pressure is sufficient to deliver oil to the turbo's bearings. To prevent this problem, follow these simple steps:

1. When installing a new or remanufactured turbocharger, make certain that the oil inlet and drain lines are clean before connecting them.
2. Be sure the engine oil is clean and at the proper level.

3. Fill the oil filter with clean oil.

4. Leave the oil drain line disconnected at the turbo and crank the engine without starting it until oil flows out of the turbo drain port.

5. Connect the drain line, start the engine, and operate it at low idle for a few minutes before running it at higher speeds.

● **CUSTOMER CARE:** Remind your customer to not immediately stop a turbocharged engine after pulling a trailer, or driving at a high speed, or driving uphill. Idle the engine for 20 to 120 seconds. The turbocharger will continue to rotate after the engine oil pressure has dropped to zero, which can cause bearing damage. Also remind the owner of the proper starting procedures and the importance of proper maintenance.

Contaminated oil can cause sludge buildups within the turbo. Check the drain outlet for sludge buildup with the oil drain line removed. Failure to follow these steps result in bearing failure on the turbo's main shaft. Remember, shaft rotation speed in modern superchargers can easily exceed 2,000 revolutions per second.

A turbocharger should never be operated under load if the engine has less than 30 psi oil pressure. A turbocharger is much more sensitive to a limited oil supply than an engine due to the high rotational speed of the shaft and the relatively small area of the bearing surfaces. Low oil pressure and slow oil delivery during engine starting can destroy the bearings in a turbocharger. During normal engine starting, this should not be a problem. There are, of course, abnormal starting conditions. Oil lag conditions will most often occur during the first engine start after an engine oil and filter change. Before the engine is put under load and the turbo activated, the engine should be run for 3 to 5 minutes at idle to prevent oil starvation to the turbo. Similar conditions can also exist if an engine has not been operated for a long period of time. Engine lube systems have a tendency to bleed down. Before allowing the engine to start, the engine should be cranked over until a steady oil pressure reading is observed. This is called priming the lubricating system. The same starting procedure should be followed in cold weather. The thick engine oil will take a longer period of time to flow.

Supercharger Diagnosis and Service

Classroom Manual
Chapter 5, page 149

A supercharger is often called a blower.

A BIT OF HISTORY

The 1906 American Chadwick had a supercharger. Since then, many manufacturers have equipped engines with superchargers. Supercharged Dusenbergs, Hispano-Suizas, and Mercedes-Benzes were giants among luxury-car marques, as well as winners on the race tracks in the 1920s and 1930s. After World War II, supercharging started to fade, although both Ford and American Motors sold supercharged passenger cars into the late 1950s. However, after being displaced first by larger V8 engines, then by turbochargers, the supercharger started to make a comeback with the 1989 models.

A supercharger (Figure 5-37) should be trouble-free for many miles. Since it does not operate in the same high heat as a turbocharger, it tends to be more reliable. However, it does operate at high speeds and at close tolerances (Figure 5-38). Proper lubrication is essential to sustained reliability.

The effectiveness of a supercharger is quickly seen at the dragstrip watching funny cars and top fuel dragsters.

The fluid in the front supercharger housing lubricates the rotor drive gears. This fluid does not require changing for the life of the vehicle. However, this fluid level should be checked at 30,000-mi. (48,000-km) intervals. To check the fluid level in the front supercharger housing,

Figure 5-37 A supercharger installed on a late-model V6 engine.

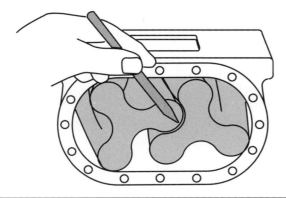

Figure 5-38 Notice the close tolerances between the lobes of the supercharger.

remove the Allen head plug in the top of this housing. The fluid should be level with the bottom of the threads in the plug opening. If the fluid level is low, add the required amount of synthetic fluid that meets the vehicle manufacturer's specifications.

Supercharger Diagnosis

The supercharger on most vehicles is serviced only as an assembly. Only specially equipped shops rebuild them. Therefore, a technician must diagnose supercharger problems and replace the supercharger if necessary. Supercharger problems include low boost, high boost, reduced vehicle response and/or fuel economy, noise, and oil leaks.

Supercharger Removal

Follow these steps for supercharger removal:

1. Disconnect the negative battery cable.
2. Remove the air inlet tube from the throttle body.
3. Remove the cowl vent screens.

4. Drain the coolant from the radiator.

5. Disconnect the spark plug wires from the spark plugs in the right cylinder head and position them out of the way.

6. Remove the electrical connections from the air charge temperature sensor, throttle position sensor, and idle air control valve.

7. Disconnect the vacuum hoses from the supercharger air inlet plenum.

8. Remove the EGR transducer from the bracket and disconnect the vacuum hose to this component (Figure 5-39).

9. Disconnect the PCV tube from the supercharger air inlet plenum.

10. Disconnect the throttle linkage and remove the throttle linkage bracket. Position this linkage and bracket out of the way. Remove the cruise control linkage if equipped.

11. Remove the two EGR valve attaching bolts, and place this valve out of the way.

12. Remove the coolant hoses from the throttle body.

13. Remove the supercharger drive belt.

14. Remove the inlet and outlet tubes from the intercooler (Figure 5-40).

15. Remove three intercooler adapter attaching bolts from the intake manifold.

16. Remove three supercharger mounting bolts (Figure 5-41).

17. Remove the supercharger and air intake plenum as an assembly.

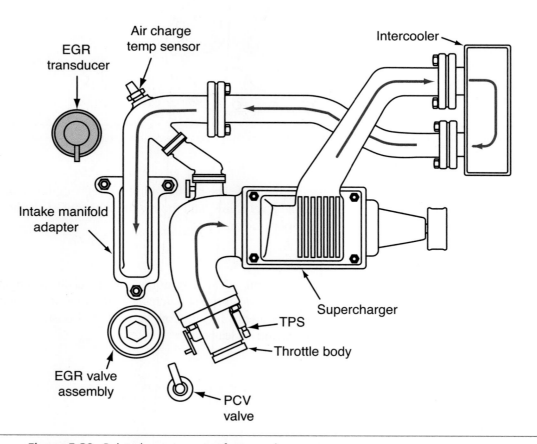

Figure 5-39 Related components of a supercharger.

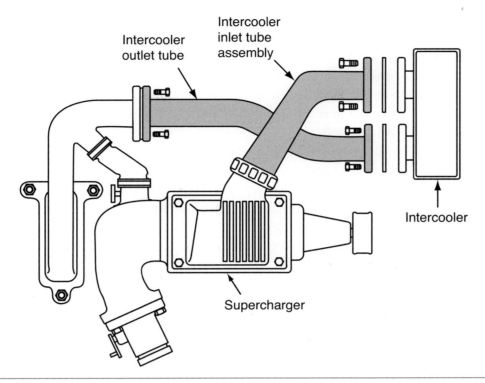

Figure 5-40 Removing the inlet and outlet tubes.

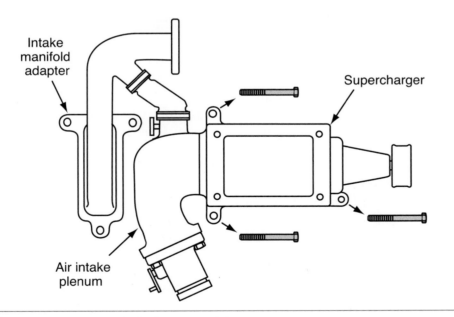

Figure 5-41 Mounting bolts for a supercharger.

This procedure should be followed for supercharger installation:

1. Clean all gasket surfaces and inspect these surfaces for scratches and metal burrs. Remove metal burrs as required.

2. Place a new gasket on the intake manifold surface that mates with the supercharger intercooler adapter.

3. Install the supercharger, throttle body, and air intake plenum as an assembly.

4. Install three supercharger mounting bolts and tighten these bolts to the specified torque.

5. Install three bolts that retain the intercooler adapter to the intake manifold and tighten these bolts to the specified torque.

6. Install the intercooler inlet and outlet tubes and tighten all mounting bolts to the specified torque.

7. Install the coolant hoses on the throttle body and tighten these hose clamps.

8. Install a new EGR valve gasket and install the EGR valve. Tighten the EGR valve mounting bolts to the specified torque.

9. Install the throttle linkage and bracket and tighten the bracket bolts to the specified torque.

10. Connect all the vacuum hoses to the original locations on the air intake plenum, EGR valve, and EGR transducer.

11. Install the spark plug wires on the spark plugs in the right cylinder head and connect the electrical connectors to the throttle position sensor, air charge temperature sensor, and idle air control valve.

12. Install the cowl covers.

13. Install and tighten the air inlet tube on the throttle body.

14. Install the supercharger drive belt.

15. Refill the cooling system to the proper level.

16. Connect the battery ground cable.

CASE STUDY

A customer complained about lack of power on a Dodge Spirit with a 2.5L turbocharged engine. The customer said the engine had a much slower acceleration compared to what it had previously.

The technician performed a basic turbocharger inspection and found that all linkages, hoses, and lines were in normal condition. A check for intake and exhaust leaks did not reveal any problems and the turbocharger did not have an excessive noise level. The technician checked the turbocharger boost pressure during a road test and discovered it was 3 psi below the specified boost pressure. There was some evidence of blue smoke in the exhaust during the road test.

The technician decided to check the engine compression before any further turbocharger service. The compression test revealed that all four cylinders had much lower-than-specified compression. The technician placed a small amount of oil in each cylinder and repeated the compression test. With the oil in the cylinders, the compression pressure improved considerably on each cylinder, indicating worn piston rings and cylinders. The customer was advised that the low engine compression was the cause of the reduced power and that the engine required rebuilding.

Terms to Know

Pyrometer   Thermal vacuum switch (TVS)

ASE-Style Review Questions

1. While discussing intake manifold vacuum leaks:
   Technician A says an intake manifold vacuum leak may result in faster-than-normal idle speed on fuel injected engines.
   Technician B says an intake manifold vacuum leak causes a low steady reading on a vacuum gauge connected to the intake manifold.
   Who is correct?
   A. A only   **C.** Both A and B
   B. B only   **D.** Neither A nor B

2. While discussing exhaust restrictions:
   Technician A says that an exhaust restriction can be detected by measuring backpressure.
   Technician B says that the pressure should be no more than 5.00 psi on a good system.
   Who is correct?
   A. A only   **C.** Both A and B
   B. B only   **D.** Neither A nor B

3. While discussing exhaust system diagnosis:
   Technician A says a rattling noise when the engine is accelerated may be caused by loose internal components in the muffler.
   Technician B says a lack of engine power may be caused by a restricted exhaust system.
   Who is correct?
   A. A only   **C.** Both A and B
   B. B only   **D.** Neither A nor B

4. While discussing catalytic converter diagnosis:
   Technician A says the temperature should be the same at the catalytic converter inlet and outlet if the converter is working normally.
   Technician B says a defective belt-driven air pump may cause improper converter operation.
   Who is correct?
   A. A only   **C.** Both A and B
   B. B only   **D.** Neither A nor B

5. *Technician A* says a vacuum leak results in less air entering the engine, which causes a richer air-fuel mixture.
   Technician B says a vacuum leak anywhere in the system can cause the engine to run poorly.
   Who is correct?
   A. A only   **C.** Both A and B
   B. B only   **D.** Neither A nor B

6. *Technician A* says that a catalytic converter breaks down HC and CO to relatively harmless byproducts.
   Technician B says that using leaded gasoline or allowing the converter to overheat can destroy its usefulness.
   Who is correct?
   A. A only   **C.** Both A and B
   B. B only   **D.** Neither A nor B

7. While discussing turbocharger service:
   Technician A says if the end housings are scored, they must be replaced.
   Technician B says a straightedge should be used to check all turbocharger mating surfaces for a warped condition.
   Who is correct?
   A. A only   **C.** Both A and B
   B. B only   **D.** Neither A nor B

8. While discussing turbocharger inspection and diagnosis:
   Technician A says a turbocharger should not be operated until the engine has idled for 3 to 5 minutes.
   Technician B says an intake system air leak upstream from the compressor wheel may allow dirt particles to enter the turbocharger.
   Who is correct?
   A. A only   **C.** Both A and B
   B. B only   **D.** Neither A nor B

9. While discussing turbocharger boost pressure:
   Technician A says low cylinder compression does not affect turbocharger operation.
   Technician B says a wastegate sticking in the closed position decreases boost pressure.
   Who is correct?
   A. A only   **C.** Both A and B
   B. B only   **D.** Neither A nor B

10. While discussing turbocharger service:
    Technician A says turbocharger bearing failure may be caused by a contaminated cooling system.
    Technician B says turbocharger bearing failure may be caused by contaminated engine oil.
    Who is correct?
    A. A only   **C.** Both A and B
    B. B only   **D.** Neither A nor B

ASE Challenge Questions

1. A restricted exhaust may cause any of the following *except*:
   A. poor upshift qualities in an automatic transmission.
   B. engine stumble at high rpm.
   C. poor- or no-start condition.
   D. low intake vacuum at cruise.

2. *Technician A* says that catalytic converter operation can be confirmed by the delta temperature test. *Technician B* says that the converter operation can be confirmed by a 4- or 5-gas analyzer.
   Who is correct?
   A. A only
   B. B only
   C. Both A and B
   D. Neither A nor B

3. Turbochargers are being discussed.
   Technician A says worn shaft seals may be indicated by blue smoke in the exhaust.
   Technician B says a clogged PCV valve may cause leakage at the shaft seals.
   Who is correct?
   A. A only
   B. B only
   C. Both A and B
   D. Neither A nor B

4. A turbocharger with too little boost may be a result of:
   A. binding linkage.
   B. leaking engine exhaust.
   C. worn engine piston rings.
   D. all of the above.

5. An engine has a piston with a hole in it. This may be caused by:
   A. excessive turbocharger boost pressure.
   B. a leaking engine exhaust system.
   C. damaged turbocharger shaft bearings.
   D. either A or C.

Job Sheet 12

Name _____ Date _____

Check Engine Manifold Vacuum

Upon completion of this job sheet, you should be able to measure engine manifold vacuum and determine the condition of the engine.

ASE Correlation

This job sheet is related to the ASE Engine Performance Test's content area: General engine diagnosis; task: perform engine absolute manifold pressure tests; determine needed repairs.

Tools and Materials

Vacuum gauge
Clean rag

Describe the vehicle being worked on.

Year_____ Make _____ VIN _____

Model_____

Engine size_____# of cyls_____ compression ratio_____:_____

Expected manifold vacuum readings at idle: _____

Source of information: _____

Describe the general running condition of the engine:

_____

_____

Procedures

1. What does manifold vacuum represent?

   _____

   _____

2. What is the difference between manifold and ported vacuum?

   _____

   _____

3. Connect a vacuum gauge to a manifold vacuum source. Where is the source you used?

   _____

   _____

4. What is the vacuum reading with the engine at idle? _____

5. What is the vacuum reading with the engine at 1,500 rpm? _____

6. In what unit of measurement is vacuum measured? _____

7. Conduct the following tests using the guidelines included in this job sheet, record the readings and explain what you just found out!

Cranking vacuum test

Results and conclusions:

_____

_____

PCV valve test

Results and conclusions:

_____

Vacuum leak test

Results and conclusions:

_____

Valve action test

Results and conclusions:

_____

_____

Exhaust restriction test

Results and conclusions:

_____

_____

Piston ring test

Results and conclusions:

_____

_____

Idle mixture test

Results and conclusions:

_____

_____

8. Based on these tests, how would you summarize the condition of this engine?

_____

_____

_____

_____

_____

_____

_____

_____

_____

GUIDELINES FOR VACUUM TESTS:

Cranking Vacuum

Completely close the throttle plate, disable the ignition, connect the vacuum gauge to the manifold and crank the engine.

5 inches or more = a good engine

Less than 5 inches = cylinders leak or there is a leak in the intake system

Less than 1 inch = possible timing belt or chain problem

PCV System Test

With the engine idling and the gauge connected, either pinch the PCV hose shut or hold your finger over the end of the valve. The reading should increase with the valve plugged off.

Vacuum Leak Test

Gauge connected, engine idling. Look for a steady reading of 16 to 21 inches.

Rhythmic drop of 1 to 2 inches = valves not sealing

A few inches below normal with needle flutter may indicate worn intake valve guides or a mixture problem.

A few inches below normal but with a steady needle would indicate a common small leak.

3 to 9 inches below normal indicates an intake system leak—use oil or carburetor cleaner to identify the location. This can also be caused by low compression.

Valve Action Test

Run engine

If the reading fluctuates between 5 and 7 inches = possible late ignition timing

If the reading drops irregularly from a normal reading = a sticking valve

Raise engine speed to 2,000 rpm

If the readings fluctuate wildly between 12 and 24 inches = weak valve springs

If the readings fluctuate wildly but irregularly between 12 and 24 inches = broken valve springs

Ignition Timing Test

Engine idling

Readings below normal = late timing

Readings above normal = early timing

Exhaust Restriction Test

Increase engine speed slowly to 2,000 rpm. Needle should increase over normal. Close throttle quickly, vacuum should return to normal as quickly as it rose.

Vacuum does not increase with speed = plugged exhaust

Vacuum returns slowly to normal = plugged exhaust

Piston Ring Test

All other test results must be satisfactory and engine oil level okay.

Increase speed to 2,000 rpm and quickly release the throttle.

Increase of 5 or more inches of vacuum = rings in good shape

Increase of less than 5 inches of vacuum = suspect worn rings

Idle Mixture Test

Idle engine. NOTE: All other test results must be satisfactory and engine oil level okay.

Increase speed to 2,000 rpm and quickly release the throttle.

Steady reading = mixture is fine

Floating needle = incorrect mixture

Instructor's Response _____

_____

_____

_____

Job Sheet 13

Name _____ Date _____

Test a Catalytic Converter for Efficiency

Upon completion of this job sheet, you should be able to check the condition and efficiency of a catalytic converter.

ASE Correlation

This job sheet is related to the ASE Engine Performance Test's content areas: fuel, air induction, and exhaust system diagnosis and repair and emissions control systems diagnosis and repair. Tasks: perform exhaust system backpressure test; and determine needed action as well as inspect and test components of the catalytic converter system and replace as needed.

Tools and Materials

Rubber mallet   Exhaust gas analyzer
Pyrometer   A vehicle
Pressure gauge   Hoist
Propane enrichment tool

Describe the vehicle being worked on:

Year _____ Make_____ VIN_____

Model _____

Procedures

Task Completed

1. Securely raise the vehicle on a hoist.   ☐

   Make sure you have easy access to the catalytic converter and that it is not hot.

2. Smack the exhaust pipe by the converter with a rubber mallet. Did it rattle?

   _____

   If it did, it needs to be replaced and there is no need to do other testing. A rattle indicates loose substrate that will soon rattle into small pieces.

3. If the converter passed this test it does not mean it is in good shape. It should be checked for plugging or restrictions.

4. Lower the vehicle and open the hood.   ☐

5. Remove the O_2 sensor.   ☐

6. Install a pressure gauge into the sensor's bore. On some engines, it is not that easy. On   ☐
   engines with a PFE, we can use its port to install our gauge. On other engines, we may
   need to fabricate a tester from an old O_2 sensor or air check valve.

7. After the gauge is in place, start the engine and hold the engine's speed at 2,000 rpm.
   Record the reading on the pressure gauge. _____psi

   We are looking for exhaust pressure under 1.25 psi. A very bad restriction would give us
   over 2.75 psi.

8. What does your reading tell you?

_____

_____

Newer cars should have pressures well under 1.25. Some older ones can be as high as 1.75 and still be good. You will notice if you quickly rev up the engine that the pressure goes up. This is normal. Remember, do this test at 2,000 rpm, not wide-open throttle.

☐ **9.** Remove the pressure gauge, turn off the engine, allow the exhaust to cool, then install the O_2 sensor.

☐ **10.** If the converter passed this test, we can now check its efficiency. There are three ways to do this. The first way is the delta temperature test. Start the engine and allow it to warm up. With the engine running, carefully raise the vehicle on the hoist or lift.

11. With a pyrometer, measure and record the inlet temperature of the converter. The reading is: _____

12. Now measure the temperature of the converter's outlet. The reading is: _____

13. What is the percentage increase of the temperature at the outlet compared to the temperature of the inlet? _____ There should be a temperature increase of about 8 percent or 100 degrees at the cat's outlet. If the temperature does not increase by 8 percent, replace the converter. What are your conclusions from this test?

_____

_____

14. How does temperature show the efficiency of a catalytic converter?

_____

_____

☐ **15.** Now we will do the O_2 storage test. This is based on the fact that a good converter stores oxygen. The following test is for closed-loop feedback systems. Non-feedback systems require a different procedure. Begin by disabling the air injection system.

☐ **16.** Turn your gas analyzer on and allow it to warm up. Start the engine and warm the cat up as well.

17. When everything is ready, hold the engine at 2,000 rpm. Watch the exhaust readings. Record the readings:

_____	_____	_____	_____	_____
HC	CO	O_2	CO_2	NO_X

If the converter was cold, the readings should continue to drop until the converter reaches light-off temperature.

18. When the numbers stop dropping, check the oxygen levels. Check and record the oxygen level. The reading is _____. O_2 should be about 0.5 to 1 percent. This shows the converter is using most of the available oxygen. There is one exception to this. If there is no CO left, there can be more oxygen in the exhaust. However, it still should be less than 2.5 percent. It is important that you reach your O_2 reading as soon as the CO begins to drop. Otherwise good converters will fail this test. The O_2 will go way over 1.25 after CO starts to drop.

☐ **19.** If there is too much oxygen left and no CO in the exhaust, stop the test and make sure the system has control of the air-fuel mixture. If the system is in control, use your propane enrichment tool to bring the CO level up to about 0.5 percent. Now the O_2 level should drop to zero.

20. Once you have a solid oxygen reading, snap the throttle open, then let it drop back to idle. Check the oxygen. The reading is _____. It should not rise above 1.2 percent.

21. If the converter passes these tests, it is working properly. If the converter fails the tests, chances are that it is working poorly or not at all. The final converter test uses a principle that checks the converter in its actual job, converting CO and HC into CO_2 and water.

22. Allow the converter to warm up by running the engine. ☐

23. Calibrate the gas analyzer and insert its probe into the exhaust pipe. If the vehicle has dual exhaust with a crossover, plug the side that the probe is not in. If the vehicle has a true dual exhaust system, check both sides separately. ☐

24. Turn off the engine and disable the ignition. ☐

25. Crank the engine for 9 seconds while pumping the throttle. Look at the gas analyzer and record the CO_2 reading _____. The CO_2 for injected cars should be over 11 percent and over 10 percent for engines with a carburetor. If you are cranking the engine and the HC goes above 1,500 ppm, stop cranking. The converter is not working. Also stop cranking once you hit your 10 or 11 percent CO_2 mark; the converter is good. If the converter is bad, you should see high HC and low CO_2 at the tailpipe. What are your conclusions from this test?

_____

_____

26. Do not repeat this test more than one time without running the engine in between. ☐

27. Reconnect the ignition and start the engine. Do this as quickly as possible to cool off the converter. ☐

Instructor's Response_____

_____

_____

_____

Engine Control System Diagnosis and Service

Upon completion and review of this chapter, you should be able to:

❏ Perform a flash code diagnosis on various vehicles.

❏ Obtain fault codes with an analog voltmeter.

❏ Perform a continuous self-test.

❏ Erase fault codes.

❏ Perform a scan tester diagnosis on various vehicles.

❏ Perform a cylinder output test.

❏ Remove and replace computer chips.

❏ Diagnose computer voltage supply and ground wires.

❏ Test and diagnose switch-type input sensors.

❏ Test and diagnose engine coolant temperature sensors.

❏ Test and diagnose air charge temperature sensors.

❏ Test, diagnose, and adjust throttle position sensors.

❏ Test and diagnose exhaust gas recirculation valve position sensors.

❏ Test and diagnose oxygen sensors.

❏ Test and diagnose knock sensors.

❏ Test and diagnose vehicle speed sensors.

❏ Test and diagnose different types of manifold absolute pressure sensors.

❏ Test and diagnose various types of mass air flow sensors.

Basic Tools

Basic tool set, DMM, lab scope, scanner, appropriate service manuals

---

A computer is an electronic device that processes and stores data. It is also capable of operating other devices.

A program is a set of instructions the computer must follow to achieve the desired results.

Failsoft action is commonly known as the computer's limp-in mode.

This chapter will look at basic computer operation, some manufacturer computer systems, and finally sensor diagnosis. The manufacturer systems described in this chapter are OBD I with some notes about significant differences in the OBD II systems. OBD I and OBD II use practically the same sensors with some differences noted. OBD II systems are discussed in detail later in the book.

For more than two decades, a computer has played an important role in the way an engine runs. The role of the computer has evolved from the control of a single system to the control of nearly all of the engine's systems. Understanding what a computer does and how it works is extremely important to effective diagnosis of driveability problems.

The operation of a computer can be divided into four basic functions:

1. Input: A voltage signal that is sent from an input device. This device can be a sensor or a switch activated by the driver, the technician, or another device.

2. Processing: The computer uses the input information and compares it to programmed instructions. The logic circuits process the input signals into output commands.

3. Storage: The program instructions are stored into an electronic memory. Some of the input signals are also stored for processing later.

4. Output: After the computer has processed the sensor input and checked its programmed instructions, it will put out control commands to various output devices. These output devices may be a system actuator or an indicator. The output of one computer can be used as an input to another computer.

Understanding these four computer functions will help you organize the troubleshooting process. When a system is tested, you are attempting to isolate a problem with one of these functions.

In the process of controlling the various engine systems, the PCM continuously monitors operating conditions for possible system malfunctions. The computer compares system conditions against programmed parameters. If conditions fall outside the limits of these parameters, the computer detects a malfunction. A DTC is set to indicate the portion of the system that is at fault. A technician can access the code as an aid in troubleshooting.

Failsoft means the computer will substitute a fixed input value if a sensor circuit should fail. This provides system operation, but at a limited functioning level.

Impedance is the combined opposition to current created by the resistance, capacitance, and inductance of a meter.

If a malfunction results in improper system operation, the computer may minimize the effects by using failsoft action. In other words, the computer may substitute a fixed value in place of the real value from a sensor to avoid shutting down the entire system. This fixed value can be programmed into the computer's memory or it can be the last received signal from the sensor prior to failure. This allows the system to operate on a limited basis instead of shutting down completely.

There are several things you need to know prior to learning how to access trouble codes in a computer's memory. You need to become familiar with what you are looking at, and you must follow proper precautions when servicing these systems.

Electronic Service Precautions

A technician must take some precautions before servicing a computer or its circuit. The PCM is designed to withstand normal current draws associated with normal operation. However, overloading the system will destroy the computer. To prevent damage to the PCM and its related components, follow these service precautions:

1. Never ground or apply voltage to any controlled circuit unless the service manual instructs you to do so.

2. Use only a high impedance multimeter (10 megaohms or higher) to test the circuits. Never use a test light unless instructed to do so in the manufacturer's procedures.

3. Make sure the ignition switch is turned off before disconnecting or connecting electrical terminals at the PCM.

4. Unless instructed otherwise, turn the ignition switch off before disconnecting or connecting any electrical connections to sensors or actuators.

5. Turn the ignition switch off whenever disconnecting or connecting the battery terminals. Also, turn it off when replacing a fuse.

6. Do not connect any electrical accessories to the insulated or ground circuits of computer-controlled systems.

7. Use only manufacturer's specific test and replacement procedures for the year and model of the vehicle being serviced.

Electrostatic Discharge

Some manufacturers mark certain components and circuits with a code or symbol to warn technicians that they are sensitive to electrostatic discharge (Figure 6-1). Static electricity can destroy or render a component useless.

When handling any electronic part, especially those that are static sensitive, follow the guidelines below to reduce the possibility of electrostatic buildup on your body and the inadvertent discharge to the electronic part. If you are not sure if a part is sensitive to static, treat it as if it is.

Static electricity can be 25,000 volts or higher.

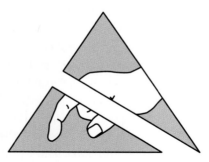

Figure 6-1 GM's electrostatic discharge (ESD) symbol that warns a technician that the component or circuit is sensitive to static.

1. Always touch a known good ground before handling the part. This should be repeated while handling the part and more frequently after sliding across a seat, sitting down from a standing position, or walking a distance.

2. Avoid touching the electrical terminals of the part unless you are instructed to do so in the written service procedures. It is good practice to keep your fingers off all electrical terminals since the oil from your skin can cause corrosion.

3. When you are using a voltmeter, always connect the negative meter lead first.

4. Do not remove a part from its protective package until it is time to install the part.

5. Before removing the part from its package, ground yourself and the package to a known good ground on the vehicle.

6. When replacing a PROM, ground your body by putting a metal wire around your wrist and connect the wire to a good ground.

⬤**CUSTOMER CARE:** It is always an excellent customer service practice to return the vehicle to the customer with no evidence or side effects of being repaired. This applies to seat position, mirror adjustment, cleaning of smudges, and perhaps even cleaning the windshield. Refrain from changing radio stations and volume or tone controls as well as eating, drinking, or smoking in the car. This also includes resetting electronic devices and accessories any time computer-related work is performed or if an interruption of battery power occurs. Items that will probably need to be reset include the clock, radio (if the stations are known), seat memories, and mirrors. At the very least, offer assistance to the customer at the time of pickup, if possible. He or she may not recall how to reset certain accessories because it is usually not done regularly. Additionally, be sure all DTCs are cleared from the computer's memory.

Basic Diagnosis of Electronic Engine Control Systems

Diagnosing a computer-controlled system is much more than accessing the DTCs in the computer's memory. When diagnosing any system, you need to know what to test, when to test it, and how to test it. Because the capabilities of the engine control computer have evolved from simple to complex, it is important to know the capabilities of the system you are working with before attempting to diagnose a problem. Refer to the service manual for this information. After you understand the system and its capabilities, begin your diagnosis using your knowledge and logic.

The importance of logical troubleshooting cannot be overemphasized. The ability to diagnose a problem (to find its cause and its solution) is what separates an automotive technician from a parts changer.

There are two logics used to diagnose and service electronic engine controls: computer logic and a technician's logical diagnosis.

Computer Logic Flow

In order to control an engine system, the computer makes a series of decisions. Decisions are made in a step-by-step fashion until a conclusion is reached. Generally, the first decision is to determine the engine mode. For example, to control air-fuel mixture, the computer first determines whether the engine is cranking, idling, cruising, or accelerating. Then, the computer can choose the best system strategy for the present engine mode. In a typical example, sensor input indicates the engine is warm, rpm is high, manifold absolute pressure is high, and the throttle plate is wide open. The computer determines the vehicle is under heavy acceleration or wide-open throttle. Next, the computer determines the goal to be reached. For example, with heavy acceleration, the goal is to create a rich air-fuel mixture. When operating in open-loop fuel

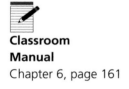

Classroom Manual
Chapter 6, page 161

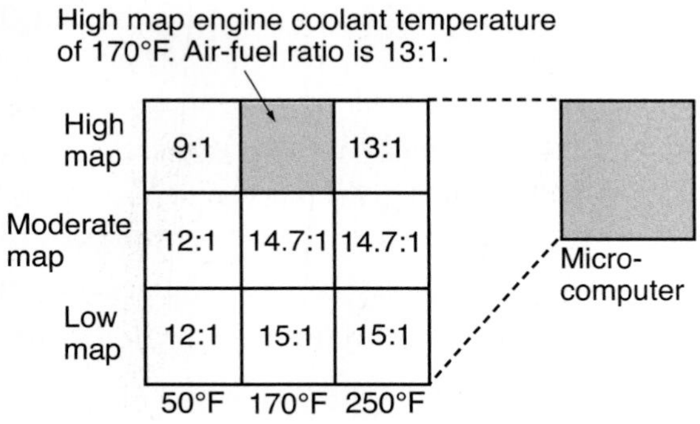

High map engine coolant temperature
of 170°F. Air-fuel ratio is 13:1.

	50°F	170°F	250°F	
High map	9:1		13:1	
Moderate map	12:1	14.7:1	14.7:1	
Low map	12:1	15:1	15:1	

Micro-computer

Figure 6-2 A typical computer lookup table.

control, the computer uses a lookup table similar to that shown in Figure 6-2. At wide-open throttle with high manifold absolute pressure and coolant temperature of 170°F, the table indicates the air-fuel ratio should be 13:1. That is, 13 pounds of air for every 1 pound of fuel. An air-fuel ratio of 13:1 creates the rich air-fuel mixture needed for heavy acceleration.

In a final series of decisions, the computer determines how the goal can be achieved. In our example, a rich air-fuel mixture is achieved by increasing fuel injector pulse width. The injector nozzle remains open longer and more fuel is drawn into the cylinder, providing the additional power needed.

Technician's Logical Diagnosis

The best automotive technicians use this logical process to diagnose engine problems. When faced with an abnormal engine condition, they compare clues (such as meter readings, oscilloscope readings, visible problems) with their knowledge of proper conditions and discover a logical reason for the way the engine is performing. Logical diagnosis means following a simple basic procedure. Start with the most likely cause and work to the most unlikely. In other words, check out the easiest, most obvious solutions first before proceeding to the less likely and more difficult solutions. Do not guess at the problem nor jump to a conclusion before considering all of the factors.

The logical approach has a special application to troubleshooting electronic engine controls. Check all traditional non-electronic engine control possibilities before attempting to diagnose the electronic engine control itself. For example, low battery voltage might result in faulty sensor readings. The distributor could also be sending faulty signals to the computer, resulting in poor ignition timing.

An additional problem that occurs when diagnosing any electronic part is that the part itself cannot be accurately tested. Most electronic parts are completely sealed due to complex, interlocking circuitry and cannot be checked at their input and output connections.

This problem can be overcome by using a logical procedure called the process of elimination. In other words, if every related part checks out okay, it must be the electronic part that is bad. This means thoroughly checking every component in a system as well as checking for such basic factors as proper current supply and good grounds.

AUTHOR'S NOTE: When beginning a diagnostic procedure, do not skip steps nor attempt shortcuts such as assuming all the basic items will check out fine. It is critical to have proper power supply voltage and system ground circuits that are sound. Multiple symptoms and driveability problems that appear to be complex may require minor repairs such as cleaning and tightening a loose ground wire. A recent report from an OEM indicated that less than 2 percent of the PCMs returned for warranty claims were actually verified to be faulty. That means 98 percent of returns were misdiagnosed and had no problems. A detailed systematic approach to effective problem solving and diagnosis was discussed in the beginning of Chapter 2 of this manual.

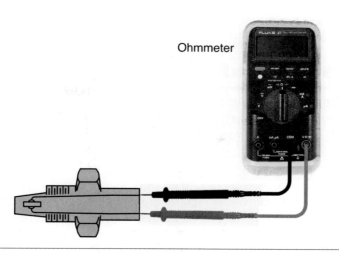

Ohmmeter

Figure 6-3 Checking a temperature sensor with an ohmmeter.

Isolating Computerized Engine Control Problems

Determining which part or area of a computerized engine control system is defective requires a thorough knowledge of how the system works and following the logical troubleshooting process previously explained.

Electronic engine control problems are usually caused by defective sensors, wiring, connections, and output devices. The logical procedure in most cases is to check the input sensors and wiring first, then the output devices and their wiring, and, finally, the computer.

Most late-model computerized engine controls have self-diagnosis capabilities. A malfunction in any sensor, output device, or in the computer itself may be stored in the computer's memory as a trouble code. Stored codes can be retrieved and the indicated problem areas checked further.

Some malfunctions may cause driveability problems without stored codes. These methods can be used to check individual system components:

1. Visual Checks. This means looking for obvious problems. Any part that is burned, broken, cracked, corroded, or has any other visible problem must be replaced before continuing the diagnosis. Examples of visible problems include disconnected sensor vacuum hoses and broken or disconnected wiring.

2. Ohmmeter Checks. Most sensors and output devices can be checked with an ohmmeter. For example, Figure 6-3 shows an ohmmeter used to check a temperature sensor. The ohmmeter reading is normally low on a cold engine and high or infinity on a hot engine if the sensor is a PTC. If the sensor is an NTC, the opposite readings would be expected. Output devices such as coils or motors can also be checked with an ohmmeter.

3. Voltmeter Checks. Many sensors, output devices, and their wiring can be diagnosed by checking the voltage flowing to them and, in some cases, from them. Even some oxygen sensors can be checked in this manner.

4. Lab Scope Checks. The activity of sensors and actuators can be monitored with a lab scope. By watching their activity, you are doing more than testing them. Problems elsewhere in the system will often cause a device to behave abnormally. These situations are identified by the trace on a scope and by the technician's understanding of a scope and the device being monitored.

In some cases, a final check on the computer or sensor can be made only by substitution. Substitution is not the most desirable way to diagnose problems. To substitute, replace the suspected part with a known good unit and recheck the system. If the system now operates normally, the original part is defective.

Service Bulletin Information

When diagnosing engine control system problems, service bulletin information is absolutely essential. If a technician does not have service bulletin information, many hours of diagnostic time may be wasted. This information is available from different suppliers of CD-ROM and paper technical service bulletins (TSBs), as well as Internet-based products. We will discuss the solutions to some difficult problems that are only quickly discovered by using a TSB or by having much experience with similar problems and their causes.

Many fuel-injected General Motors engines are equipped with Multec-I injectors. Some of these injectors may be prone to shorting problems in the windings, especially if the fuel contained some alcohol. If the injectors become shorted, they draw excessive current. General Motors P4 PCMs have a sense line connected to the quad driver that operates the injectors. When this sense line experiences excessive current flow from the shorted injectors, the quad driver shuts off and stops operating the injectors. This action protects the quad driver, but also causes the engine to stall. After a few minutes, the engine will usually restart. If a technician does not have this information available in a service bulletin, he or she may waste a great deal of time locating the problem.

General Motors has a number of programmable read only memory (PROM) changes for their PCMs to correct various performance problems such as engine detonation and stalling. General Motors' service bulletins provide the information regarding PROM changes. Several automotive test equipment suppliers and publishers also provide General Motors PROM identification (ID) and service bulletin books. This PROM ID and service bulletin information is absolutely necessary when diagnosing fuel injection problems on General Motors vehicles. Technicians must not have the idea they should immediately change a PROM to correct a problem. Most of the service bulletins relating to PROM changes provide a diagnostic procedure that directs the technician to test and eliminate all other causes of the problem before changing the PROM.

If a service bulletin or a fault code indicates that a PROM, calibration package (CALPAK) or memory calibrator (MEM-CAL) chip replacement is necessary, these chips are serviced separately from the PCM. A fault code representing one of these chips may indicate that the chip is defective or improperly installed. The PROM and CALPAK chips are used in 160-baud PCMs and the MEM-CAL chip is found in P4 PCMs. The MEM-CAL chip is a combined PROM, CALPAK, and electronic spark control (ESC) module. If PCM replacement is required, these chips are not supplied with the replacement PCM. Therefore, the chip or chips from the old PCM have to be installed in the new PCM. Always disconnect the negative battery cable before any chip replacement is attempted.

Follow these steps for PROM replacement (see Photo Sequence 10):

1. Remove the PROM access cover from the PCM.

2. Attach the PROM removal tool to the PROM and use a rocking action on alternate ends of the PROM carrier to remove the PROM.

3. Align the notch on the PROM carrier with the notch in the PROM socket in the PCM. Press on the PROM carrier only and push the PROM and carrier into the socket (Figure 6-4). Never remove the PROM from the carrier.

To remove and replace a MEM-CAL chip, the process is similar:

1. Remove the MEM-CAL access cover on the PCM.

2. Use two fingers to push the retaining clips back away from the MEM-CAL, then grasp the MEM-CAL at both ends and lift it upward out of the socket.

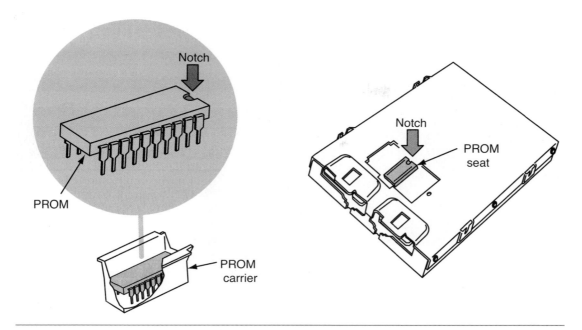

Figure 6-4 PROM installation.

Installing a PROM backwards or exposure to electrostatic discharge will destroy the PROM.

3. Align the small notches on the MEM-CAL with the notches in the MEM-CAL socket in the PCM.

4. Press on the ends of the MEM-CAL and push it into the socket until the retaining clips push into the ends of the MEM-CAL (Figure 6-5).

On the early 160-baud General Motors computers, the pins on the internal components extended through the circuit board tracks and soldering was done on the opposite side of the board from where the components were located. On P4 PCMs, a surface mount technology was developed in which the component pins were bent at a 90-degree angle and then soldered on

The baud rate of a computer is related to the computer's processing speed. The higher the baud, the faster it works.

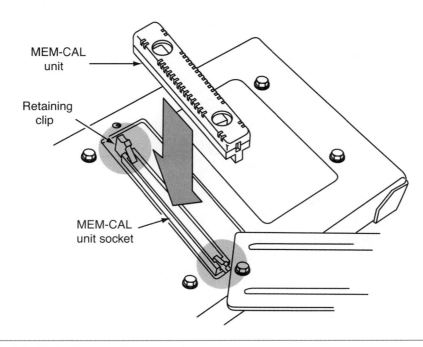

Figure 6-5 MEM-CAL installation.

Photo Sequence 10
Replacing a PROM

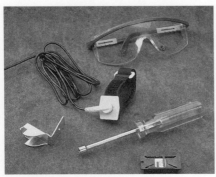

P10-1 Tools required to remove and replace the PROM: rocker-type PROM removal tool, ESD strap, safety glasses, and replacement PROM.

P10-2 Place the PCM onto the workbench with the PROM access cover facing up. Be careful not to touch the electrical connectors with your fingers.

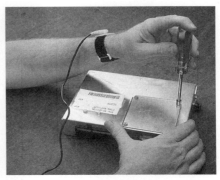

P10-3 Remove the PROM access cover.

P10-4 Using the rocker-type PROM removal tool, engage one end of the PROM *carrier* with hook end of tool. Grasp the PROM carrier with the tool only at the narrow ends of the carrier.

P10-5 Press on the vertical bar end of the tool. Rock the end of the PROM carrier up as far as possible.

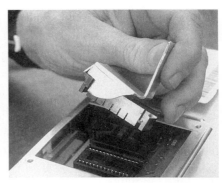

P10-6 Repeat the process on the other end of the carrier until the PROM carrier is removed from the socket.

P10-7 Inspect the replacement PROM part number for proper calibration.

P10-8 Check for proper orientation of the PROM in the carrier. The notch in the PROM should be referenced to the smaller notch in the carrier. If the replacement PROM does not come in its own carrier, it will be necessary to remove the old PROM and install the replacement PROM into the carrier. Be careful not to bend the pins.

P10-9 Align the PROM carrier with the socket. The small notch of the carrier must be aligned with the small notch in the socket.

Replacing a PROM (continued)

P10-10 Press the PROM carrier until it is firmly seated into the socket. Do not press on the PROM.

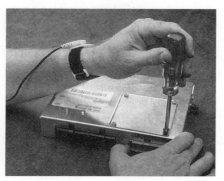

P10-11 Replace the PROM cover. Reinstall the PCM into the vehicle.

top of the tracks on the circuit boards. In some cases, loose connections developed in the computers that were surface mounted. These loose connections usually caused the engine to quit. If a technician suspects this problem, the PCM may be removed with the wiring harness connected. Start the engine and give the PCM a slap with the palm of your hand. If the engine stalls or the engine operation changes, a loose connection is present on the circuit board. When a technician does not have this information available in service bulletins, much diagnostic time may be wasted.

EEPROMS in OBD II Vehicles

OBD II computers have the EEPROM soldered in to prevent the installation of aftermarket "hot" chips. The computer is "re-flashed" to perform updates for driveability problems. The information is sent to the dealer monthly in case a PCM should have to be replaced. Aftermarket technicians can access the information needed over the Internet if they have the necessary equipment.

Chrysler experienced some low-speed surging during engine warmup on 3.3L and 3.8L engines. On these engines, the port fuel injectors sprayed against a hump in the intake port. As a result, fuel puddled behind this hump, especially while the engine was cold. When the engine temperature increased, this fuel evaporated and caused a rich air-fuel ratio and engine surging. Chrysler corrected this problem by introducing angled injectors with the orifices positioned at an angle so the fuel sprayed over the hump in the intake. When angled injectors are installed, the wiring connector must be positioned vertically. Angled injectors have beige exterior bodies. Technicians must have service bulletin information regarding problems like this.

Self-Diagnostic Systems

Today's computer-controlled systems are complicated. It would take endless amounts of time to diagnose these systems using pre-computer control methods. For this reason, most computerized engine controls have self-diagnostic capabilities. By entering a self-test mode, the computer is able to evaluate the condition of the entire electronic engine control system, including itself. If problems are

EXAMPLE: P0137 LOW VOLTAGE BANK 1 SENSOR 2

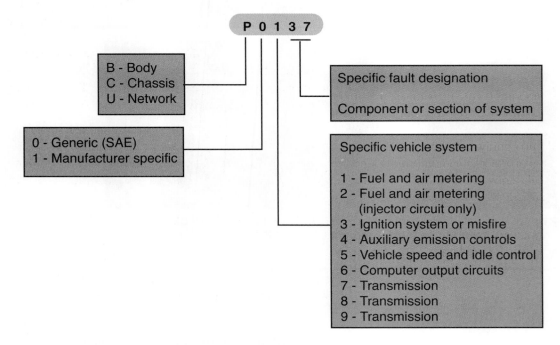

Typical generic code examples:

Mass Airflow	P0102, P0103
Intake Air Temperature	P0112, P0113, P0127
Barometric Pressure	P0106, P0107, P0108, P0109
Engine Coolant Temperature	P0117, P0118
Oxygen Sensor (one of several)	P0131, P0133, P0135, P0136, P0141
Throttle Position	P0121, P0122, P0123
EGR / EVP	P0400, P0401, P0402
Vehicle Speed	P0500, P0501, P0503

Note: Manufacturers will also use specific codes that apply only to their systems.

Figure 6-6 Different systems use different codes to identify problem areas.

found, they are identified as either hard faults (on-demand) or intermittent failures. Each type of fault or failure is assigned a numerical trouble code that is stored in computer memory (Figure 6-6).

A *hard fault* means a problem has been found somewhere in the system at the time of the self-test. An intermittent problem indicates a malfunction occurred (for example, a poor connection causing an intermittent open or short), but is not present at the time of the self-test. Nonvolatile RAM allows intermittent faults to be stored for up to a specific number of ignition key on/off cycles. If the trouble does not reappear during that period, it is erased from the computer's memory.

There are various methods of assessing the trouble codes generated by the computer. Most manufacturers have diagnostic equipment designed to monitor and test the electronic components of their vehicles. Aftermarket companies also manufacture scan tools that have the capability to read and record the input and output signals passing to and from the computer.

It was mandated that OBD I vehicle's computer systems be serviced/accessed with equipment that was commonly found in the shop. That is why OBD I vehicles had flash codes that could be read without special tools. Chrysler products were accessed by turning the ignition key

OBD I vehicles were manufactured prior to 1996.

on and off three times rapidly; General Motors cars were accessed by jumping two terminals of the ALDL; Fords were accessed with an analog voltmeter. Hondas and Nissans had LEDs that blinked codes built right into the PCM. Modern vehicles are equipped with OBD II systems that are many times faster than the early 1980's system. Scan tools have been in use for some time and are very advanced in their own right. For OBD II, a scan tool is required to read trouble codes; flash codes are no longer available. Since there is so much information to help the technician on the scan tool, a technician would want to use the scan tool anyway. The following section of the text book will talk about OBD I system trouble codes and how to access them.

Visual Inspection

Before reading self-diagnostic or trouble codes, do a visual check of the engine and its systems. This quick inspection can save much time during diagnosis. While visually inspecting the vehicle, make sure to include the following:

1. Inspect the condition of the air filter and related hardware around the filter.
2. Inspect the entire PCV system.
3. Check to make sure the vapor canister is neither saturated nor flooded.
4. Check the battery and its cables, the vehicle's wiring harnesses, connectors, and the charging system for loose or damaged connections. Also, check the connectors for signs of corrosion.
5. Check the condition of the battery and its terminals and cables.
6. Make sure all vacuum hoses are connected and are not pinched or cut.
7. Check all sensors and actuators for signs of physical damage.

Accessing Trouble Codes

Although the parts in any computerized system are amazingly reliable, they do occasionally fail. Diagnostic charts in service manuals (Figure 6-7) help you through troubleshooting procedures in

Trouble codes are referred to as Diagnostic Trouble Codes, or DTCs. DTCs became standardized among manufacturers when OBD II was mandated in 1996.

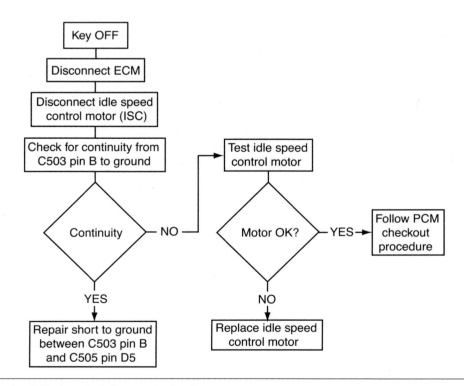

Figure 6-7 A typical diagnostic chart.

the proper order. Start at the top and follow the sequence down. There will be branches of the tree—yes or no, on or off, okay or not—to follow after making the check required in each step.

Tools to use when checking circuits, connections, sensors, and signals include the following:

❏ A digital multimeter (DMM) with an impedance of at least 10 megaohms

❏ Scan tool

❏ Lab scope (dual trace is preferred)

❏ Dwell meter

❏ Tachometer

❏ Vacuum gauge and pump

❏ Analog voltmeter or multimeter (used with some computer systems)

❏ Service manuals with trouble code charts and diagnostic procedures

A spark tester, fuel pressure gauge, a jumper wire with two alligator clips, and a non-powered test light can also come in handy.

There are two things that would tell the driver there is a problem with some part of the computer system: the MIL (check engine, power loss, or service engine soon light) or the car simply does not start.

The malfunction indicator lamp (MIL) comes on normally for a few seconds when the key is turned on and during cranking. It should go off when the engine starts. The check engine light should not come back on unless the computer finds a problem. If it does come on, there is probably a trouble code stored in memory.

Self-diagnostic procedures and diagnostic or trouble codes differ with vehicle make and year. Each time the key is turned to the on position, the system does a self-check. The self-check makes sure that all of the bulbs, fuses, and electronic modules are working. If the self-test finds a problem, it might store a code for later servicing. It may also instruct the computer to turn on the MIL to show that service is needed.

Using a Scanner

Scanners are available to diagnose nearly all engine control systems. The exact tester buttons and test procedures vary on these testers, but many of the same basic diagnostic functions are completed regardless of the tester make. When test procedures are performed with a scan tester, these precautions must be observed:

1. Always follow the directions in the manual supplied by the scan tester manufacturer.

2. Do not connect or disconnect any connectors or components with the ignition switch on. This includes the scan tester power wires and the connection from the tester to the vehicle diagnostic connector.

3. Never short across or ground any terminals in the electronic system except those recommended by the vehicle manufacturer.

4. If the computer terminals must be removed, disconnect the scan tester diagnostic connector first.

Scan Tester Features

Scan testers vary depending on the manufacturer, but many of these testers have the following features:

1. Display window that displays data and messages to the technician. Messages are displayed from left to right. Most scan testers display at least four readings on the display at the same time.

2. Memory cartridge that plugs into the scan tester. These memory cartridges are designed for specific vehicles and electronic systems. For example, a different cartridge may be required for the transmission computer and the engine computer. Most scan tester manufacturers supply memory cartridges for domestic and imported vehicles.

3. Power cord connected from the scan tester to the battery terminals or cigarette lighter socket.

4. Adaptor cord that plugs into the scan tester and connects to the DLC on the vehicle. A special adaptor cord is supplied with the tester for the diagnostic connector on each make of vehicle.

5. Serial interface for optional devices such as a printer, terminal, or personal computer may be connected to this terminal.

6. Keypad that allows the technician to enter data and reply to tester messages. Depending on the model, a typical scan tester keypad may contain similar buttons (Figure 6-8): numbered keys—digits 0 through 9; up and down arrow keys that allow the technician to move back and forward through test modes and menus; enter key for entering information into the tester; mode key that allows the technician to interrupt the current procedure and go back to the previous modes; and F1 and F2 keys that allow the technician to perform special functions described in the scan tester manufacturer's manuals.

Most later model vehicles will automatically identify certain information to the scanner, once it is connected.

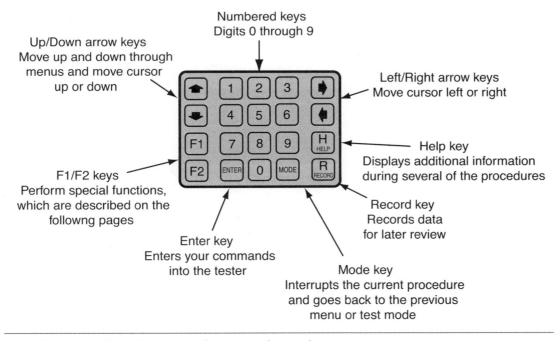

Figure 6-8 Typical scan tester buttons and controls.

Scan Tester Initial Entries

Photo Sequence 11 shows a typical procedure for using a scanner in diagnostics. Scan tester operation varies depending on the make of the tester, but a typical example of initial entries follows:

1. Be sure the engine is at normal operating temperature and the ignition switch is off. With the correct module in the scan tester (Figure 6-9), connect the power cord to the vehicle battery.

2. Enter the vehicle year. The technician is prompted for the model year and enters it on the keypad.

3. Enter the VIN code. This is usually a two-digit code based on the model year and engine type. These codes are listed in the scan tester operator's manual. The technician enters the appropriate two-digit code and presses the enter key.

4. Connect the scan tester adaptor cord to the diagnostic connector on the vehicle being tested.

After the scanner has been programmed by performing the initial entries, some entry options appear on the screen. These entry options vary depending on the scan tester and the vehicle being tested. The following is a typical list of initial entry options:

1. Engine

2. Antilock brake system (ABS)

3. Suspension

4. Transmission

5. Data line

6. Deluxe CDR

7. Test mode

The technician presses the number beside the desired selection to proceed with the test procedure. In the first four selections, the tester is asking the technician to select the computer system to be tested. If data line is selected, the scan tester provides a voltage reading from each input sensor in the system.

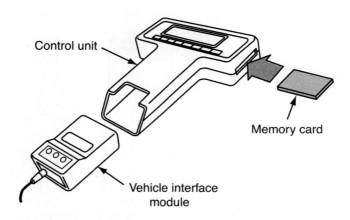

Figure 6-9 Some scanners use modules or cartridges for specific years, makes, and models of vehicles that are inserted into the scan tool.

Photo Sequence 11
Diagnosis with a Scan Tester

P11-1 Be sure the engine is at normal operating temperature and the ignition switch is off.

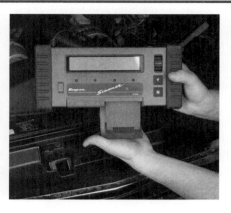

P11-2 Install the proper module for the vehicle and system into the scan tool.

P11-3 Connect the scan tool power leads to the battery or cigar lighter (depending on the design of the scan tool).

P11-4 Enter the vehicle's model year and VIN code into the scan tool.

P11-5 Select the proper scan tool adapter for the vehicle's DLC.

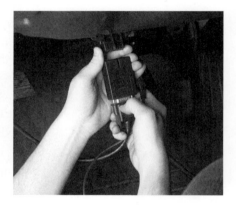

P11-6 Connect the scan tool to the DLC.

P11-7 Retrieve the DTCs with the scan tool. Interpret the codes by using the service manual.

P11-8 Start the engine and obtain the input sensor and output actuator data on the scan tool. If a printer for the tool is available, print out the data report.

P11-9 Compare the input sensor and output actuator data to the specifications given in the service manual. Mark all data that is not within specifications.

After the test has been selected, the scanner moves on to the actual test selections. These selections vary depending on the scan tester and the vehicle being tested. The following list includes many of the possible test selections and a brief explanation:

1. Fault codes. Displays fault codes on the scan tester display.

2. Switch tests. Allows the technician to operate switch inputs such as the brake switch to the PCM. Each switch input should change the reading on the tester.

3. ATM tests. Forces the PCM to cycle all the solenoids and relays in the system for 5 minutes or until the ignition switch is turned off.

4. Sensor tests. Provides a voltage reading from each sensor.

5. Automatic idle speed (AIS) motor. Forces the PCM to operate the AIS motor when the up and down arrows are pressed. The engine speed should increase 100 rpm each time the up arrow is touched. RPM is limited to 1,500 or 2,000 rpm.

6. Solenoid state tests or output state tests. Displays the on or off status of each solenoid in the system.

7. Emission maintenance reminder (EMR) tests. Allows the technician to reset the EMR module. The EMR light reminds the driver when emission maintenance is required.

8. Wiggle test. Allows the technician to wiggle solenoid and relay connections. An audible beep is heard from the scan tester if a loose connection is present.

9. Key on, engine off (KOEO) test. Allows the technician to perform this test with the scan tester on Ford products.

10. Computed timing check. Forces the PCM to move the spark advance 20 degrees ahead of the initial timing setting on Ford products.

11. Key on, engine running (KOER) test. Allows the technician to perform this test with the scan tester on Ford products.

12. Clear memory or erase codes. Quickly erases fault codes in the PCM memory.

13. Code library. Reviews fault codes.

14. Basic test. Allows the technician to perform a faster test procedure without prompts.

15. Cruise control test. Allows the technician to test the cruise control switch inputs if the cruise module is in the PCM.

Snapshot Testing

Many scan testers have snapshot capabilities on some vehicles that allow the technician to operate the vehicle under the exact condition when a certain problem occurs and freeze the sensor voltage readings into the tester memory. The vehicle may be driven back to the shop and the technician can play back the recorded sensor readings. During the playback, the technician watches closely for a momentary change in any sensor reading that indicates a defective sensor or wire. This action is similar to taking a series of sensor reading "snapshots" and then reviewing the pictures later. The snapshot test procedure may be performed on most vehicles with a data line from the computer to the DLC.

Retrieving Troubles Codes

Prior to the implementation of the mandates of OBD II, each car manufacturer required a different method for retrieving stored DTCs from the computer's memory. The diagnostic connector looked different and was located in different places. Also, the diagnostic code represented different things. Although all of the computer systems basically had the same functions, diagnostic methods were as different as day and night. What follows are some of the typical procedures for retrieving DTCs from some popular model vehicles. Always refer to the appropriate system within a service manual when following these procedures.

General Motors' Vehicles (pre-OBD II)

The main components of General Motors' diagnostic system include an electronic control module or powertrain control module (PCM), a MIL—check engine or service engine soon warning light, and an assembly line communications link (ALCL) or DLC. The check engine light serves two purposes (Figure 6-10). First, it is a signal to the vehicle operator that a detectable failure has occurred. Second, it can also be used by a technician to aid in the location of system malfunctions. Systems that are in compliance with OBD II do not have this flashout feature. All codes must be retrieved with the use of a scan tool.

The preliminary diagnostic procedure mentioned previously in this chapter must be completed before the diagnostic trouble code diagnosis. If a defect occurs in the system and a DTC is set in the PCM memory, the malfunction indicator lamp (MIL) on the instrument panel is illuminated. If the fault disappears, the MIL goes out but a DTC is likely set in the computer's memory. If a system defect occurs and the MIL is illuminated, the system is usually in a limp-in mode. In this mode, the PCM provides a rich air-fuel ratio and a fixed spark advance. Therefore, engine performance and economy decrease and emission levels increase, but the vehicle may be driven to a service center.

To find the problem, you must perform a diagnostic circuit check to be sure the diagnostic system is working. Turn the ignition key on but do not start the engine. If the check engine light does not come on, follow the diagnostic chart in the service manual. If the light comes on, ground the diagnostic test terminal. This terminal is located on the **data link connector (DLC)** (Figure 6-11). Connect pin A (A ground) to pin B (test terminal). Watch the check engine light to see if it displays code 12. Code 12 (think of it as 1,2) is one flash, a short pause and two flashes (Figure 6-12). If the computer's diagnostic program is working properly, it flashes code 12 three times. If code 12 does not appear, follow the diagnostic chart in the manual. The system then displays any fault codes in the same manner.

GM's computer command control system was commonly called the 3C system.

Special Tools

Jumper wire with male spade connectors at each end

The **data link connector (DLC)** is a group of terminals connected to a diagnostic scan tool.

The DLC was also called the ALDL.

Prior to OBD II, manufacturers' codes were uniquely numbered for their vehicles.

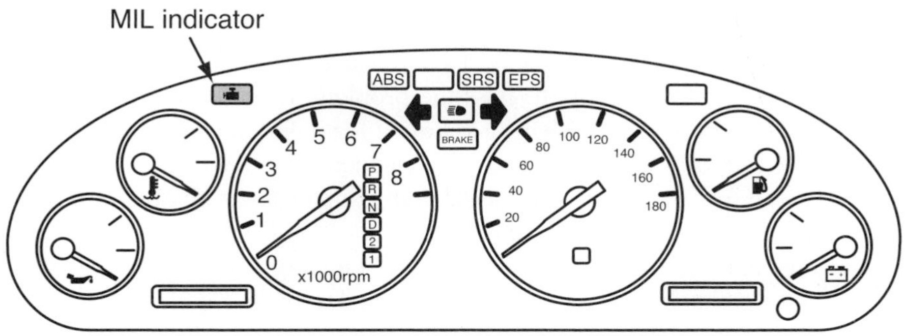

Figure 6-10 A standard MIL.

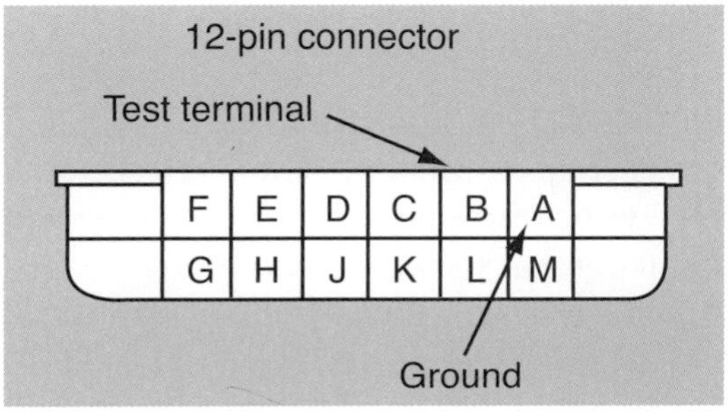

Figure 6-11 To initiate self-diagnosis, the self-test terminal is connected to the ground terminal at the DLC.

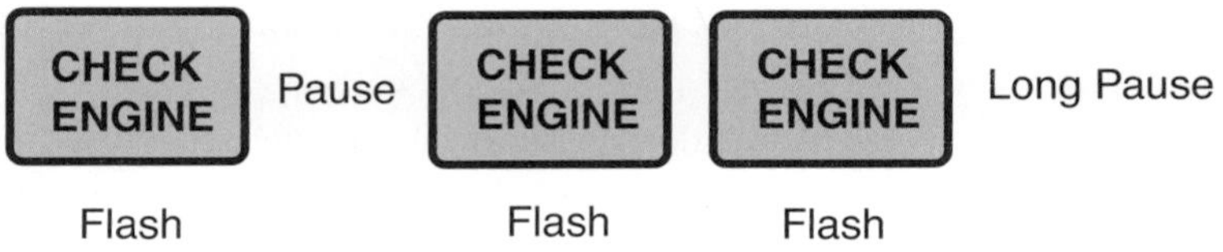

Figure 6-12 The MIL signals the codes.

Connect a jumper wire from terminals A to B in the DLC and turn the ignition switch on to obtain the DTCs. The DTCs are flashed out by the MIL. Two lamp flashes followed by a brief pause and two more flashes indicate code 22. Each code is repeated three times. When more than one code is present, the codes are given in numerical order. Code 12, which indicates that the PCM is capable of diagnosis, is given first. The DTC sequence continues to repeat until the ignition switch is turned off. A DTC indicates a problem in a specific area. Some voltmeter or ohmmeter tests may be necessary to locate the exact defect.

If the A and B terminals are connected in the DLC and the engine is started, the PCM enters the field service mode. In this mode, the speed of the MIL flashes indicate whether the system is in open loop or closed loop. If the system is in open loop, the MIL flashes quickly. When the system enters closed loop, the MIL flashes at half the speed of the open-loop flashes.

Codes are identified through the flashing MIL as follows. The number of times the light flashes on before it pauses tells the number to which it refers. In other words, if the MIL flashes like this: "on-off-on-off-on-off, pause," that means number 3 because it was on three times before it paused. If it flashed "on-off-on-off, pause," that means number 2. If it flashes "on-off, pause," that means number 1.

Because each code has two numbers to it, such as 12, 23, or 32, the flashes are read as a set. This means you always see two series of flashes separated by a pause: first number, pause, second number. The sequence repeats itself three times.

If the PCM has stored more than one trouble code, the codes flash starting with the lowest number code first and then work their way to the highest number code. For example, if the codes were 13 and 24, 13 would flash first then 24 would flash. Keep in mind that you have to watch closely while you do this.

If the PCM is good, the first code that flashes when the connector is grounded and the ignition switch turned on will be twelve (12). This is the tachometer code. It is a normal condition when the key is on but the engine is not running. If a code 12 is not received, there is something wrong in the DLC or related circuits.

Once the codes have been identified, check the service manual (Figure 6-13). Each code directs you to a specific troubleshooting tree. If you have a code, follow the tree exactly. Never try to skip a step or you will be certain to miss the problem. Even though the codes may mean the same thing, how you trace the problems is different for different cars. This is one of the main reasons a manual for the car you are working on is needed.

A hand-held scan tool can be used by itself or in conjunction with MIL. When using a scan tool, follow the manufacturer's directions to the letter.

After correcting all faults or problems, clear the PCM memory of any current codes by pulling the PCM fuse at the fuse panel for 10 seconds. Then, to make sure the codes have cleared, remove the test terminal ground and set the parking brake. Put the transmission in park and run the engine for 2 to 5 minutes. Watch the check engine light. If the light comes on again, ground the test lead and note the flashing trouble code. If no light comes on, check the codes recorded earlier, if any.

Erasing Fault Codes

A quick-disconnect wire connected to the battery positive cable may be disconnected for 10 seconds with the ignition switch off to erase DTCs. If the vehicle does not have a quick-disconnect wire, remove the PCM battery fuse to erase codes. On later-model General Motors vehicles with P4 PCMs, the quick-disconnect or PCM fuse may need to be disconnected for a longer time to erase the codes. Depending on the model year, a scan tool may be used to clear codes.

It is a good idea to give the vehicle a road test. Some parts, such as the vehicle speed sensor, do not show any problems with the engine just idling.

If DTCs are left in a computer after the defect is corrected, the codes are erased automatically when the engine is stopped and started thirty to fifty times. This applies to most computer-equipped vehicles.

DIAGNOSTIC CODE IDENTIFICATION	
The "Service Engine Soon" light will be "ON" if the malfunction exists under the conditions listed below. If the malfunction clears, the light will go out and the code will be stored in the ECM. Any codes stored will be erased if no problem reoccurs within fifty engine starts.	
Code and Circuit	**Probable cause**
Code 13 - Oxygen sensor circuit (Open)	Indicates that the oxygen sensor circuit or sensor was open for 1 minute while off idle
Code 14 - Coolant temperature sensor circuit (High)	Sets if the sensor or signal line becomes grounded for 3 seconds
Code 15 - Coolant temperature sensor circuit (Low)	Sets if the sensor, wires, or connector is open for more than 3 seconds
Code 21 - Throttle position sensor circuit (High)	TPS voltage greater than 2.5 volts for 3 seconds with less than 1200 rpm
Code 22 - Throttle position sensor circuit (Low)	A short to ground or open signal circuit will set code in 3 seconds
Code 23 - Intake air temperature sensor circuit (Low)	Sets if the sensor, wires, or connector is open for more than 3 seconds
Code 24 - Vehicle speed sensor	No vehicle speed present during a road load decel
Code 25 - Intake air temperature sensor circuit (High)	Sets if the sensor or signal line becomes grounded for more than 3 seconds
Code 32 - Exhaust gas recirculation system	Vacuum switch shorted to ground on startup OR Switch not closed after the ECM has commanded EGR for a specific period of time OR EGR solenoid circuit open for a specific period of time

Figure 6-13 These are typical GM trouble codes with their probable causes (pre-OBD II) *(continues on next page).*

DIAGNOSTIC CODE IDENTIFICATION (CONTINUED)	
Code and Circuit	Probable Cause
Code 33 - Manifold absolute pressure (High)	MAP sensor or circuit output high for 5 seconds
Code 34 - Manifold absolute pressure (Low)	Low or no output from sensor with engine running
Code 35 - Idle air control system	IAC error
Code 42 - Electronic spark timing	ECM has seen an open or grounded EST or bypass circuit
Code 43 - Electronic spark control circuit	Signal to the ECM has remained low for too long or the system has failed a functional check
Code 44 - Oxygen sensor circuit (Lean exhaust)	Sets if sensor voltage remains below 0.2 volts for about 20 seconds
Code 45 - Oxygen sensor circuit (Rich exhaust)	Sets if sensor voltage remains above 0.7 volts for about 60 seconds
Code 51 - Faulty MEM-CAL or PROM	Faulty MEM-CAL, PROM or ECM
Code 52 - Fuel CALPAK missing	Fuel CAL-PAK missing or faulty
Code 53 - System over voltage	System overvoltage indicates generator problem
Code 54 - Fuel pump circuit (Low voltage)	Sets when the fuel pump voltage is less than 2 volts when ignition reference pulses are being received
Code 55 - Faulty ECM	Faulty ECM

Figure 6-13 *(continued)* These are typical GM trouble codes with their probable causes (pre-OBD II).

Complete DLC Terminal Explanation

The actual number of wires in the DLC varies depending on the vehicle make and year. The purpose of each DLC terminal may be explained as follows:

A. Ground terminal.

B. Diagnostic request.

C. Air injection reactor. When this terminal is grounded, the air injection reactor (AIR) pump air is directed upstream to the exhaust ports continually because this connection grounds the AIR system port solenoid. This applies to AIR systems with a converter solenoid and a port solenoid. However, this action does not apply to newer AIR systems with an electric diverter valve (EDV) solenoid.

D. MIL. When this terminal is grounded on some systems, the check engine light is illuminated continually.

E. Serial data slow speed 160-baud PCM. This terminal supplies input sensor data to the scan tester on 160-baud PCM systems.

F. Torque converter clutch (TCC). If the vehicle is lifted and the engine accelerated until the transmission shifts through all the gears, a 12-volt test light may be connected from this terminal to ground to diagnose the TCC system. The light is on when the TCC is not locked up and the light goes out when TCC lockup occurs.

G. Fuel pump test. On some models, when 12 volts are supplied to this terminal with the ignition off, current flows through the top fuel pump relay contacts to the fuel pump, and the pump should run. Other models have a separate fuel pump test lead located under the hood.

H. Antilock brake system (ABS) cars and trucks. When a jumper wire is connected from this terminal to terminal A, the ABS computer flashes the ABS warning light to provide fault codes.

I. Not used.

J. Air bag supplemental inflatable restraint (SIR) system. When this terminal is connected to terminal A, the SIR computer flashes fault codes on the SIR warning light.

K. Not used.

L. High-speed serial data P4 PCM. This terminal supplies sensor data to the scan tester on P4 PCM systems.

Ford Motor Company Vehicles (pre-OBD II)

At the center of Ford's EEC-IV system is a computer unit called an electronic control assembly (ECA) or PCM. Like the other computer-controlled systems discussed, the PCM has self-diagnostic capabilities. By entering a mode known as self-test, the computer is able to evaluate the condition of the entire electronic system, including itself. If problems are found, they show up as either hard faults (on demand) or intermittent failures.

Many Ford products with computer-controlled carburetors do not have a MIL in the instrument panel. On these systems, an analog voltmeter may be connected to the DLC to obtain fault codes. The positive voltmeter lead must be connected to the positive battery terminal and the negative voltmeter lead must be connected to the proper DLC terminal. With the ignition switch off, connect a jumper wire from the separate self-test input wire to the proper DLC terminal (Figure 6-14). If the vehicle does not have a separate self-test input wire, connect the jumper wire to the proper self-test connector terminals (Figure 6-15).

The DLC is located under the hood on Ford vehicles with EEC-IV systems. A separate self-test input wire is located near the DLC on many of these systems (Figure 6-16). However, on some Ford systems, the self-test input wire is an integral part of the DLC (Figure 6-17).

On 1988 and newer model vehicles, the diagnostic trouble codes are displayed by the MIL on the instrument panel. Since 1994, Ford has equipped some models with the EEC-V system, which is an OBD II-compliant system. This system does not display codes on the MIL. The Ford self-test diagnostic procedure can be broken down into four parts.

❏ Key on/engine off: checks system inputs for hard faults (malfunctions that occur during the self-test) and intermittent faults (malfunctions that occurred sometime prior to the self-test and were stored in memory).

❏ Computed ignition timing check: determines the PCM's ability to advance or retard ignition timing. This check is made while the self-test is activated and the engine is running.

❏ Engine running segment: checks system output for hard faults only.

❏ Continuous monitoring test (wiggle test): allows the technician to look for and set intermittent faults while the engine is running.

Special Tools

Jumper wire with
  male spade
  connectors at
  each end, analog
  voltmeter

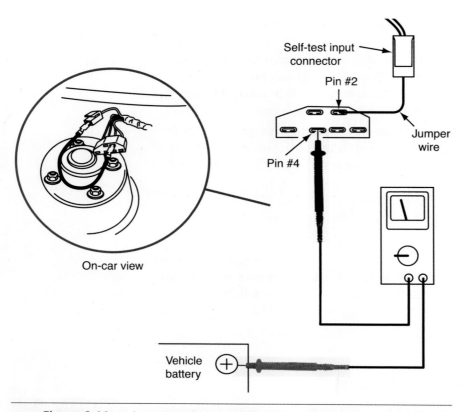

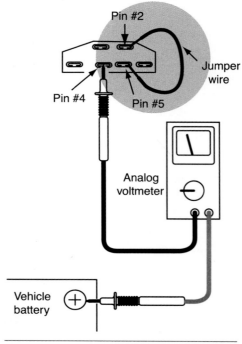

Figure 6-14 Voltmeter and jumper wire connections to the DLC and self-test input wire.

Figure 6-15 Voltmeter and jumper connections to the DLC.

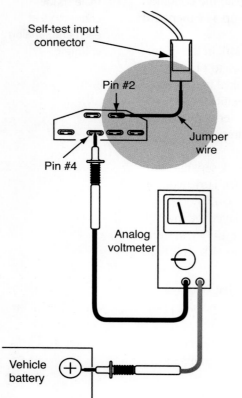

Figure 6-16 DLC with a separate self-test input wire.

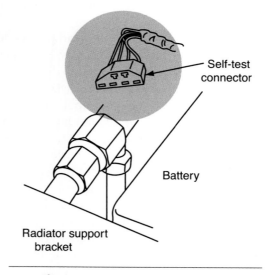

Figure 6-17 DLC with an integral self-test input wire.

Prior to any fault code diagnosis, the engine must be at normal operating temperature and the preliminary diagnostic procedure mentioned previously in the chapter must be completed. Photo Sequence 12 goes through a flash code procedure for a Ford product with a MIL. A jumper wire must be connected from the self-test input wire to the appropriate DLC terminal to enter the self-test mode. When the ignition switch is turned on after this jumper wire connection, the MIL begins to flash any DTCs in the PCM memory.

Within these four tests there are six types of service codes: on-demand codes, keep alive memory codes, separator codes, dynamic response codes, fast codes, and engine ID codes. To understand what you are dealing with when the codes start to display, a brief explanation of each type is necessary.

On-demand codes are used to identify hard faults. As mentioned earlier, a hard fault means that a system failure has been detected and is still present at the time of testing. The term "on-demand" simply means a technician is asking the computer if a problem exists right now. That is, the computer locates them while it is running its own self-test.

Keep alive memory codes mean that a malfunction was noted sometime during the last twenty vehicle warmups but is not present now (if it were, it would be recorded as a hard fault). The continuous memory code comes on after an approximate 10-second delay. Make the on-demand code repairs first. Once you have completed repairs, repeat the key on, engine off test. If all the parts are repaired correctly, a pass code of 11 should be received.

A separator code (10) indicates that the on-demand codes are over and the memory codes are about to begin. The separator code occurs as part of the key-on, engine-off segment of the self-test only.

When a code 10 appears during the engine running segment of the self-test, it is referred to as a dynamic response code. The dynamic response code is a signal to the technician to goose the throttle momentarily so that the PCM can verify the operation of the throttle position (TP) and MAP sensors. On some models, the technician must goose the throttle, step on the brake pedal, and turn the steering wheel 180 degrees. Failure to respond to the dynamic code within 15 seconds after it appears sets a code 77 (operator did not respond).

Fast codes are of no value to the service technician unless he happens to have a scan tool that is capable of reading them. They are designed for factory use and are transmitted about one hundred times faster than a scan tester can read. On the voltmeter, fast codes cause the needle to rapidly pulse between 0 and 3 volts. On the scan tester, the LED light flickers. Although fast codes have no practical use in the service bay, pay attention to when they occur so you know what is coming next. Fast codes appear twice during the entire self-test: once at the very beginning of the key-on, engine off test (right before the on-demand codes) and again after the dynamic response code.

Engine identification codes are used to tell automated assembly line equipment how many cylinders the engine has. Two needle pulses indicate a four-cylinder, three pulses a six-cylinder, and four pulses identify the engine as an eight-cylinder model. Engine ID codes appear at the beginning of the engine-running segment only.

Key On, Engine Off (KOEO) Test

Follow these steps for the **key on, engine off (KOEO)** fault code diagnostic procedure:

1. With the ignition switch off, connect the jumper wire to the self-test input wire and the appropriate terminal in the DLC.
2. If the vehicle does not have a MIL, connect the voltmeter to the positive battery terminal and the appropriate DLC terminal.

Key on, engine off (KOEO) signifies performing a test with the key in the "on" position but without the engine running.

Photo Sequence 12
Performing a Ford Flash Code Diagnosis, Key On, Engine Off (KOEO), and Key On, Engine Running (KOER) Tests

P12-1 Be sure the engine is at normal operating temperature and the ignition switch is off.

P12-2 Connect a jumper wire to the proper terminals in the DLC.

P12-3 Turn the ignition switch on.

P12-4 Observe the MIL flashes. The sequence of codes in the KOEO test is hard faults, separator code (10), and intermittent fault codes. Record all codes.

P12-5 Turn the ignition switch off and wait 10 seconds.

P12-6 Start the engine and observe the MIL flashes. The sequence of codes in the KOER test is engine ID code, separator code (10), and hard faults. Record all codes.

P12-7 Compare all codes to the fault interpretation chart in the service manual. Determine what tests need to be conducted to find the cause of the problem(s).

P12-8 Turn the ignition switch off and wait 10 seconds.

P12-9 Erase the faults by turning the ignition on and removing the jumper wire at the DLC while the MIL is flashing in the KOEO mode.

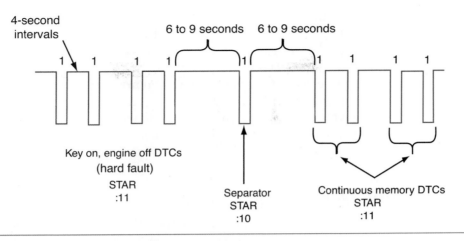

Figure 6-18 Key on, engine off DTC display.

3. Turn on the ignition switch and observe the MIL or voltmeter. Hard fault DTCs are displayed, followed by a separator code 10 and continuous memory DTCs (Figure 6-18).

Hard fault DTCs are present at the time of testing, whereas memory DTCs represent intermittent faults that occurred sometime ago and are set in the computer memory. Separator code 10 is displayed as one flash of the MIL or one sweep of the voltmeter pointer. Each fault DTC is displayed twice and are provided in numerical order. If there are no DTCs, system pass code 11 is displayed. If the technician wants to repeat the test or proceed to another test, the ignition switch must be turned off for 10 seconds.

Key On, Engine Running (KOER) Test

Follow these steps to obtain the fault codes in the **key on, engine running (KOER)** test sequence:

1. Connect the jumper wire and the voltmeter as explained in step 1 and step 2 of the KOEO test.
2. Start the engine and observe the MIL or the voltmeter. The engine identification code is followed by the separator code 10 and hard fault codes (Figure 6-19).

Key on, engine running (KOER) indicates a test conducted while the engine is running.

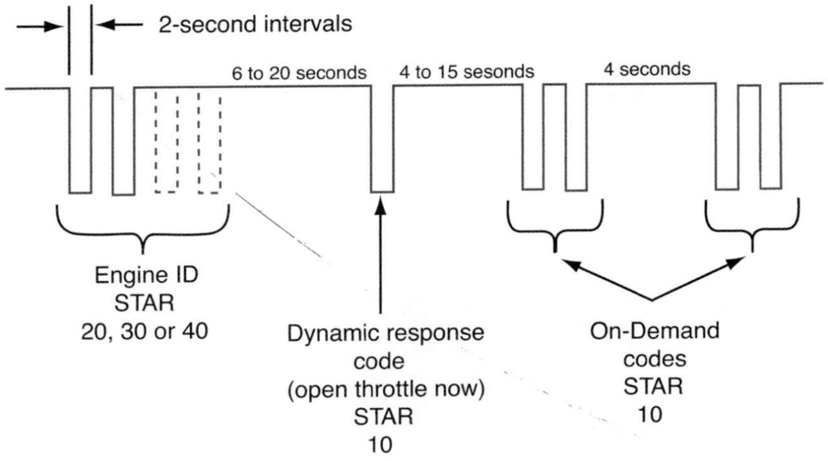

Figure 6-19 Key on, engine running DTC display.

On some Ford products, the brake on/off (BOO) switch and the power steering pressure switch (PSPS) must be activated after the engine ID code or DTCs 52 and 74, representing these switches, are displayed. Step on the brake pedal and turn the steering wheel to activate these switches immediately after the engine ID display.

Separator code 10 is presented during the KOER test on many Ford products. When this code is displayed, the throttle must be pushed momentarily to the wide-open position. The best way to provide a wide-open throttle is to push the gas pedal to the floor momentarily. On some Ford products, the separator code 10 is not displayed during the KOER test and this throttle action is not required.

Continuous Self-Test

During the **wiggle test** the PCM "sees" and indicates a problem with wires or connections while the technician wiggles or moves the wiring.

The continuous self-test may be referred to as a **wiggle test**. This test allows the technician to wiggle suspected wiring harness connectors while the system is being monitored. The continuous monitor test may be performed at the end of the KOER test. Leave the jumper wire connected to the self-test connector. Approximately 2 minutes after the last code is displayed, the continuous monitor test is started. If suspected wiring connectors are wiggled, the voltmeter pointer will deflect when a loose connection is present.

Cylinder Output Test

A cylinder output test may be performed on Electronic Engine Control IV (EEC IV) SFI engines at the end of the KOER test. The cylinder output test is also available on many SFI Ford products regardless of whether the KOER test was completed with a scan tester or with the flash code method.

When the KOER test is completed, momentarily push the throttle wide open to start the cylinder output test. After this throttle action, the PCM may require up to 2 minutes to enter into the cylinder output test. In the cylinder output test, the PCM stops grounding each injector for about 20 seconds. This action causes each cylinder to misfire.

While a cylinder is misfiring, the PCM looks at how much the engine slows down. If the engine does not slow down, there is a problem in the injector, ignition system, engine compression, or a vacuum leak. If the engine does not slow down as much on one cylinder, a fault code is set in the PCM memory. For example, code 50 indicates a problem in number 5 cylinder. The correct DTC list must be used for each model year.

Fault Code Erasing

DTCs may be erased by entering the KOEO test procedure and disconnecting the jumper wire between the self-test input wire and the DLC during the code display.

Breakout Box Testing

On Chrysler and General Motors products, a data line is connected from the computer to the DLC. Sensor voltage signals are transmitted on this data line and are read on the scan tester. If these sensor voltage signals are normal on the scan tester, the technician knows that these signals are reaching the computer.

Figure 6-20 A breakbox is connected in series with the PCM to allow meter readings of individual circuits.

In 1989, Ford introduced data links on some Lincoln Continental models. Since that time, Ford has gradually installed data links on their engine computers. If a Ford vehicle has data links, there are two extra pairs of wires in the DLC. Ford has introduced a New Generation Star (NGS) tester that has the capability to read data on some Ford engine computers. This technology is now available on other scan testers.

Since most Ford products prior to 1990 do not have a data line from the computer to the DLC, the sensor voltage signals cannot be displayed on the scan tester. Therefore, some other method must be used to prove that the sensor voltage signals are received by the computer. A breakout box is available from Ford Motor Company and some other suppliers (Figure 6-20).

Two large wiring connectors on the breakout box allow the box to be connected in series with the PCM wiring. Once the breakout box is connected, each PCM terminal is connected to a corresponding numbered breakout box terminal. These terminals match the terminals listed in Ford PCM wiring diagrams. On many Ford EEC IV systems, PCM terminals 37 and 57 are 12-volt supply terminals from the power relay to the PCM. Therefore, with the ignition switch on and the power relay closed, a digital voltmeter may be connected from breakout box terminals 37 and 57 to ground; 12 volts should be available at these terminals. The wiring diagram for the model year and system being tested must be used to identify the breakout box terminals. Ford issues templates for the breakout box that correspond with the wiring of the different models.

Chrysler Corporation Vehicles (pre-OBD II)

As far as basic system operation is concerned, Chrysler's engine control system is typical of most computer-controlled designs. At the heart of Chrysler's computer-controlled system is a digital preprogrammed microprocessor known as the logic module (LM). From its location behind the right front kick pad (in the passenger compartment), the LM issues commands affecting fuel delivery, ignition timing, idle speed, and the operation of various emission control devices.

Chrysler Corporation is now a division of DaimlerChrysler.

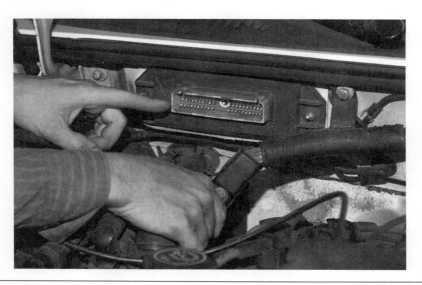

Figure 6-21 Power module location.

To accomplish all these tasks, the logic module operates in conjunction with a subordinate control unit called the power module (PM). The power module, located inside the engine compartment (Figure 6-21), is given the responsibility of controlling the injector and ignition coil ground circuit (based on the LM's commands). It is also in charge of supplying the ground to the automatic shutdown relay (ASD). The ASD relay controls the voltage supply to the fuel pump, logic module, injector, and coil drive circuits and is energized through the PM when the ignition switch is turned on.

Late-model Chrysler products have a single unit that actually consists of the logic and power modules. This unit is called the Single Board Electronic Controller (SBEC). On these systems, the LM and PM function as a single unit. However, each of these still have the same primary purposes as they did when they were in separate units.

Under normal operating conditions, the LM bases its decisions on information from several input devices (Figure 6-22).

From the information it receives, the logic module calculates the injector's pulse width, determines precise spark advance, maintains proper idle speed, governs turbo boost, controls the EGR and purge solenoids, and directs the operation of the electric cooling fan (among other things). When a computer-related failure occurs, the LM switches to a limp-in mode.

Chrysler has a simple method for checking trouble codes. Without starting the engine, turn the ignition key on and off three times within 5 seconds, ending with it on. The check engine or power loss light glows a short time to test the bulb, then starts flashing. Count the first set of flashes as tens. There is a half second pause before the light starts flashing again; this time count by ones. Add the two sets of flashes together to obtain the trouble code. For example, three flashes, a half second pause, followed by five flashes would be read as code 35. Watch carefully, because each trouble code is displayed only once. Look the code up in the service manual. This tells you which circuit to check. Once the trouble codes are flashed, the computer signals a code 55. If the light does not flash at all, there are no trouble codes stored.

The same diagnostic trouble code tests can be done using a scan tool. Connect the tool to the diagnostic connector on the left fender apron. Follow the same sequence with the ignition key. The trouble codes are displayed on the scan tool readout. You can also check the circuits, switches, and relays with the scan tool. The service manual contains the troubleshooting trees to continue your diagnosis of the problem.

A scan tester may be used to diagnose different types of computer-controlled systems. Scan testers are supplied by the vehicle manufacturer and several independent suppliers. Many scan

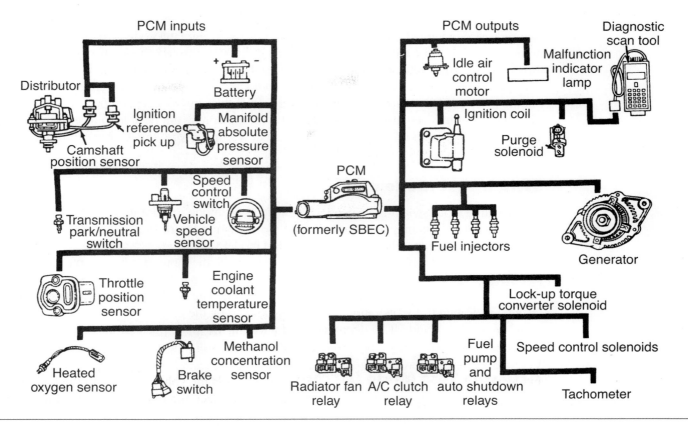

Figure 6-22 Chrysler's electronic engine control system.

testers have a removable module plugged into the tester. This module is designed for a specific vehicle make and model year. Always be sure the proper module is installed in the tester for the vehicle being diagnosed.

Always use the diagnostic procedure recommended in the vehicle manufacturer's appropriate service manual. Some Chrysler products with early O_2 feedback carbureted systems and self-diagnostic capabilities do not have a MIL. These systems are diagnosed with a scan tester (Figure 6-23). The scan tester must be connected to the diagnostic connector in the engine compartment (Figure 6-24).

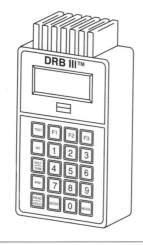

Figure 6-23 Chrysler's DRB III scan tool.

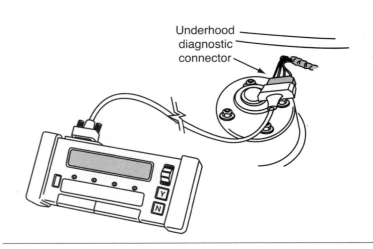

Figure 6-24 Chrysler's DLC connected to a scanner.

After the scan tester is connected to the diagnostic connector, follow the scan tester manufacturer's recommended procedure to obtain the DTCs. These fault codes are displayed once in numerical order. If the technician wants to repeat the DTC display, the ignition switch must be turned off and the fault code procedure repeated. DTCs in any computer system indicate a fault in a specific area, not necessarily in a specific component. For example, if a code representing a throttle position (TP) sensor is present, the wires from the sensor to the computer may be defective, the sensor may require replacement, or the computer may not be able to receive this sensor signal. The technician may have to perform some voltmeter or ohmmeter tests on the sensor wiring to locate the exact cause of the code.

Switch Test Mode

If a defective input switch is suspected, the switch test mode may be used to check the switch inputs. Switch inputs may include the brake switch, neutral park switch, and A/C switch. The switch inputs vary depending on the vehicle and model year.

When the fault code display is completed, be sure all the input switches are off and then follow the scan tester manufacturer's recommended procedure to enter the switch test mode. During the switch test mode, each switch is turned on and off. The reading on the scan tester should change, indicating this switch input signal was received by the computer. For example, the brake switch is turned on and off by depressing and releasing the brake pedal. If the scan tester display does not change when any input switch is activated, the switch or connecting wires are defective.

Actuator Test Mode (ATM)

If a fault code is obtained representing a solenoid or relay, the ATM may be used to cycle the component on and off to locate the exact cause of the problem. When code 55 is displayed at the end of the fault code diagnosis, follow the scan tester manufacturer's recommended procedure to enter the ATM mode.

Typical components that may be cycled during the ATM are the canister purge solenoid, cooling fan relay, and throttle kicker solenoid. The actual components that may be cycled in the ATM vary depending on the vehicle and model year. Each relay or solenoid may be cycled for 5 minutes or until the ignition switch is turned off. When the appropriate scan tester button is pressed, the tester begins cycling the next relay or solenoid.

On logic module and power module systems, disconnect the quick-disconnect connector at the positive battery cable for 10 seconds with the ignition switch off to erase DTCs. On later-model PCMs, this connector must be disconnected for 30 minutes to erase fault codes.

Toyota Vehicles

Special Tools

Jumper wires

To retrieve flash out codes from most non-OBD II-compliant Toyota vehicles, follow these steps:

1. Turn on the ignition switch and connect a jumper wire between terminals E1 and TE1 in the data link connector (DLC). Some round DLCs are located under the instrument panel (Figure 6-25) while the rectangular-shaped DLCs are positioned in the engine compartment (Figure 6-26).

2. Observe the MIL flashes. If the light flashes on and off at 0.26-second intervals, there are no DTCs in the computer memory (Figure 6-27).

3. If there are DTCs in the computer memory, the MIL flashes out the DTCs in numerical order. For example, one flash followed by a pause and three flashes is code 13, and three flashes followed by a pause and one flash represents code 31 (Figure 6-28). The codes will be repeated as long as terminals E1 and TE1 are connected and the ignition switch is on.

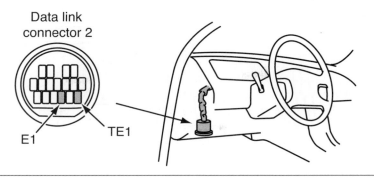

Figure 6-25 Toyota's round DLC.

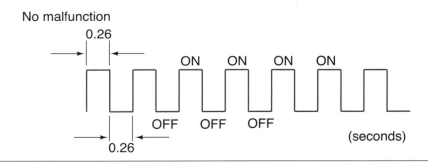

Figure 6-26 Toyota's rectangular DLC.

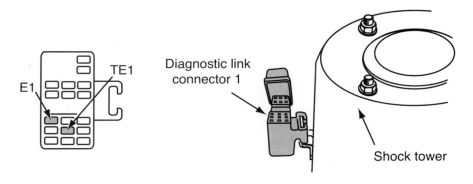

Figure 6-27 Sequence of the MIL when no DTCs are present.

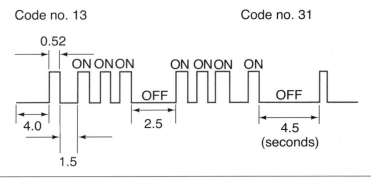

Figure 6-28 Examples of Toyota's codes.

Figure 6-29 Drive test mode terminals and their identification.

4. Remove the jumper wire from the DLC to end testing.

To identify the cause of intermittent faults, this system can be set into a drive test mode. This mode should only be used after any problems causing a hard code have been identified and repaired. To diagnose intermittent problems:

1. Turn the ignition switch on and connect terminals E1 and TE2 in the DLC (Figure 6-29).
2. Start the engine and drive the vehicle at speeds above 6 mph (10 km/h). Simulate the conditions when the problem occurs.
3. Stop the engine and connect a jumper wire between terminals E1 and TE1 on the DLC.
4. Observe the flashes of the MIL to read the DTCs.
5. Remove the jumper wire from the DLC to end the testing.

Nissan Vehicles

On some Nissan electronic concentrated engine control systems (ECCS), the PCM has two light-emitting diodes (LEDs) that flash a fault code if a defect occurs in the system. One of these LEDs is red, the other LED is green. The technician observes the flashing pattern of the two LEDs to determine the DTC. If there are no DTCs in the ECCS, the LEDs flash a system pass code. The flash code procedure varies depending on the year and model of the vehicle, and the procedure in the manufacturer's service manual must be followed. Later-model Nissan engine computers have a five-mode diagnostic procedure. Be sure the engine is at normal operating temperature and complete the preliminary diagnostic procedure explained previously in this chapter. Turn the diagnosis mode selector in the PCM clockwise to enter into the diagnostic modes (Figure 6-30).

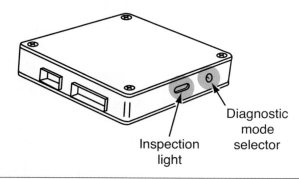

Figure 6-30 Diagnosis mode selector on a Nissan PCM.

The diagnostic modes available on some Nissan products are as follows: Mode 1 checks the oxygen sensor signal. With the system in closed loop and the engine idling, the green light should flash on each time the oxygen sensor detects a lean condition. This light goes out when the oxygen sensor detects a rich condition. After 5 to 10 seconds, the PCM "clamps" on the ideal air-fuel ratio and pulse width and the green light may be on or off. This PCM clamping of the pulse width only occurs at idle speed.

In Mode 2, the green light comes on each time the oxygen sensor detects a lean mixture, and the red light comes on when the PCM receives this signal and makes the necessary correction in pulse width. Mode 3 provides DTCs representing various defects in the system. Switch inputs to the PCM are tested in Mode 4. This mode cancels codes available in Mode 3. Mode 5 increases the diagnostic sensitivity of the PCM for diagnosing intermittent faults while the vehicle is driven on the road.

After the defect is corrected, turn the ignition switch off, rotate the diagnosis mode selector counterclockwise, and install the PCM securely in the original position.

Diagnosis of Computer Voltage Supply and Ground Wires

SERVICE TIP: Never replace a computer unless the ground wires and voltage supply wires are proven to be in satisfactory condition.

CAUTION: Always observe the correct meter polarity when connecting test meters. Since nearly all cars and light trucks have negative battery ground, the negative meter lead is connected to ground for many tests. Connecting a meter with incorrect polarity will damage the meter.

This goes along with a common computer saying: garbage in, garbage out.

A computer cannot operate properly unless it has proper ground connections and a satisfactory voltage supply at the required terminals. A computer wiring diagram for the vehicle being tested must be available for these tests. Backprobe the bat terminal on the computer and connect a pair of digital voltmeter leads from this terminal to ground (Figure 6-31). Always ground the black meter lead.

The voltage at this terminal should be 12 volts with the ignition switch off. If 12 volts are not available at this terminal, check the computer fuse and related circuit. Turn the ignition switch on and connect the red voltmeter lead to the other battery terminals at the PCM with the black lead still grounded. The voltage measured at these terminals should be 12 volts with the ignition switch on. When the specified voltage is not available, test the voltage supply wires to these terminals. These terminals may be connected through fuses, fuse links, or relays. Always refer to the vehicle manufacturer's wiring diagram for the vehicle being tested.

SERVICE TIP: When diagnosing computer problems, it is usually helpful to ask the customer about service work that has been performed lately on his or her vehicle. If service work has been performed in the engine compartment, it is possible that a computer harness or connector may have been disturbed, causing a problem.

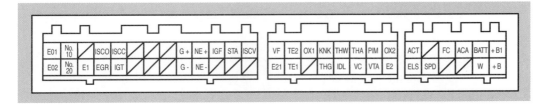

Figure 6-31 Typical computer terminals and their identification.

Computer ground wires usually extend from the computer to a ground connection on the engine or battery. With the ignition switch on, connect a pair of digital voltmeter leads from the battery ground to the computer ground. The voltage drop across the ground wires should be 30 millivolts or less. If the voltage reading is greater than or more than specified by the manufacturer, repair the ground wires or connection.

Not only should the computer ground be checked, but so should the ground (and positive) connection at the battery. Checking the condition of the battery and its cables should always be part of the initial visual inspection before beginning diagnosis of an engine control system.

A voltage drop test is a quick way of checking the condition of any wire. To do this, connect a voltmeter across the wire or device being tested. Then turn the circuit on. Ideally there should be a 0-volt reading across any wire, unless it is a resistance wire that is designed to drop voltage. Even then, check the drop against specifications to see if it is dropping too much.

A good ground is especially critical for all reference voltage sensors. The problem here is not obvious until it is thought about. A bad ground will cause the reference voltage (normally 5 volts) to be higher than normal. Normally in a circuit, the added resistance of a bad ground would cause less voltage at a load. Because of the way reference voltage sensors are wired, the opposite is true. If the reference voltage to a sensor is too high, the output signal from the sensor to the computer will also be too high. As a result, the computer will be making decisions based on the wrong information. If the output signal is within the normal range for that sensor, the computer will not notice the wrong information and will not set a DTC.

The **reference voltage** is a regulated value provided by the PCM that operates vehicle sensors.

To explain why the **reference voltage** increases with a bad ground, let's look at a voltage divider circuit (Figure 6-32). This circuit is designed to provide a 5-volt reference signal off the tap. A vehicle's computer feeds a regulated 12 volts to a similar circuit to ensure the reference voltage to the sensors is very close to 5 volts. The voltage divider circuit consists of two resistors connected in series with a total resistance of 12 ohms. The reference voltage tap is between the two resistors. The first resistor drops 7 volts (Figure 6-33) which leaves 5 volts for the second resistor and for the reference voltage tap. This 5-volt reference signal will be always available at the tap as long as 12 volts are available for the circuit.

If the circuit has a poor ground—one that has resistance—the voltage drop across the first resistor will be decreased. This will cause the reference voltage to increase. In Figure 6-34, to simulate a bad ground, a 4-ohm resistor was added into the circuit at the ground connection at the battery. This increases the total resistance of the circuit to 16 ohms and decreases the current flowing throughout the circuit. With less current flow through the circuit, the voltage drop across the first resistor decreases to 5.25 volts (Figure 6-35). This means the voltage available at the tap will be higher than 5 volts. It will be 6.75 volts.

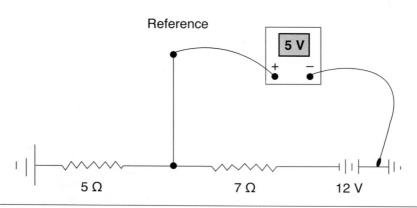

Figure 6-32 Basic voltage divider circuit.

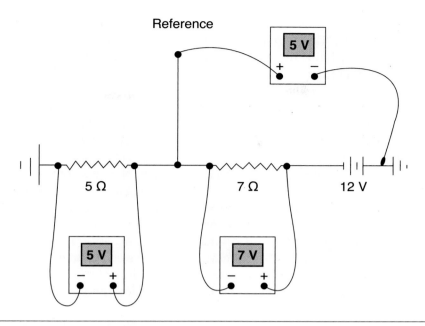

Figure 6-33 Divider circuit with voltage values.

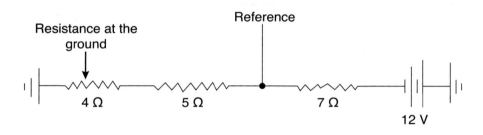

Figure 6-34 Voltage divider circuit with a bad ground.

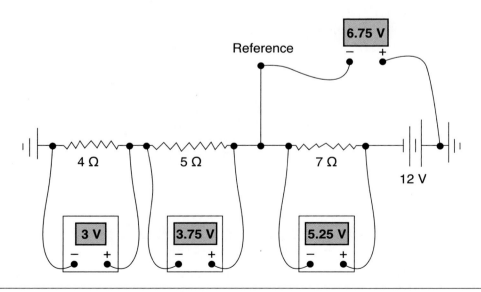

Figure 6-35 Figure 6-34 with voltage readings.

Poor grounds can also allow electromagnetic interference (EMI) or noise to be present on the reference voltage signal. This noise causes minute changes in the voltage going to the sensor. Therefore, the output signal from the sensor will also have these voltage changes. The computer will try to respond to these changes, which can cause a driveability problem. The best way to check for noise is to use a lab scope.

Connect the lab scope between the 5-volt reference signal into the sensor and the ground. The trace on the scope should be flat (Figure 6-36). If noise is present, move the scope's negative probe to a known good ground. If the noise disappears, the sensor's ground circuit is bad or has resistance. If the noise is still present, the voltage feed circuit is bad or there is EMI in the circuit from another source such as the AC generator. Find and repair the cause of the noise.

Circuit noise may be present in the positive side or the negative side of a circuit. It may also be evident by a flickering MIL, a popping noise on the radio, or by an intermittent engine miss. However, noise can cause a variety of problems in any electrical circuit. The most common sources of noise are electric motors, relays and solenoids, AC generators, ignition systems, switches, and A/C compressor clutches. Typically, noise is the result of an electrical device being turned on and off. Sometimes the source of the noise is a defective suppression device. Manufacturers include these devices to minimize or eliminate electrical noise. Some of the commonly used noise suppression devices are resistor-type secondary cables and spark plugs, shielded

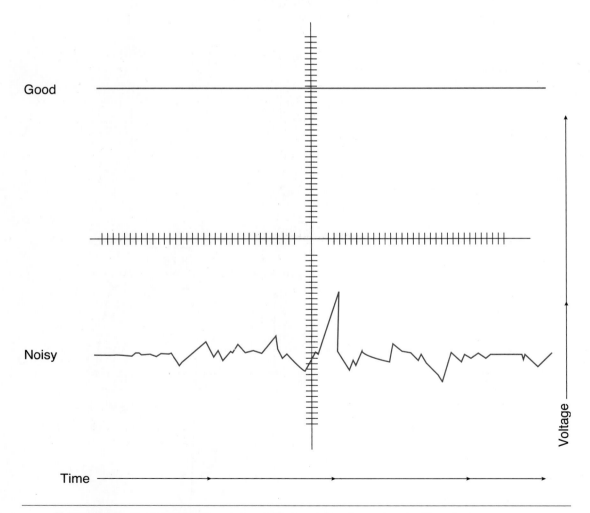

Figure 6-36 (Top) Good voltage signal. (Bottom) A voltage signal with noise.

cables, capacitors, diodes, and resistors. If the source of the noise is not a poor ground or a defective component, check the suppression devices.

Diodes and resistors are the most commonly used noise suppression devices. Resistors do not eliminate the spikes but limit their intensity. If a voltage trace has a large spike and the circuit is fitted with a resistor to limit noise, the resistor may be bad. Clamping diodes are used on devices like A/C compressor clutches to eliminate voltage spikes. If the diode is bad, a negative spike will result (Figure 6-37). Capacitors or noise filters are used to control noise from a motor or generator (Figure 6-38 and Figure 6-39).

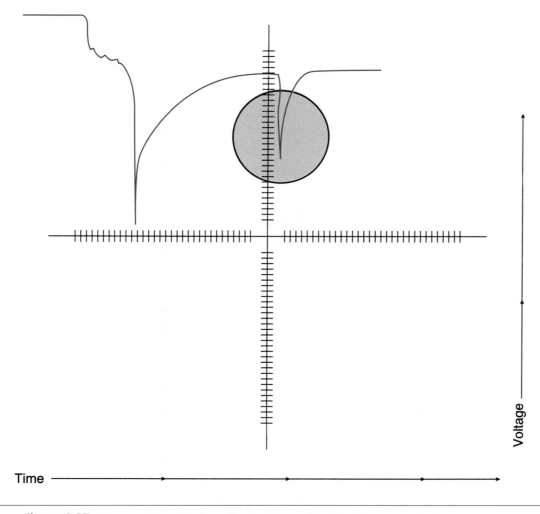

Figure 6-37 A trace of an A/C compressor clutch with a bad clamping diode.

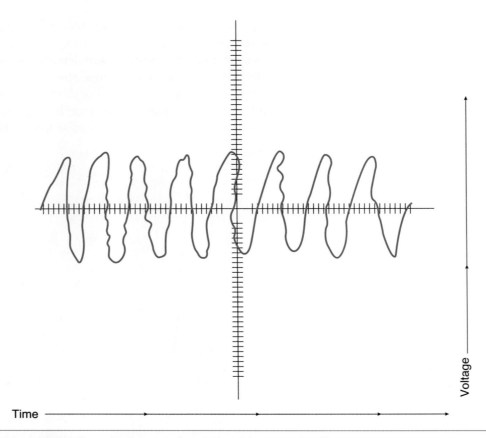

Figure 6-38 The voltage trace of a motor without a noise filter.

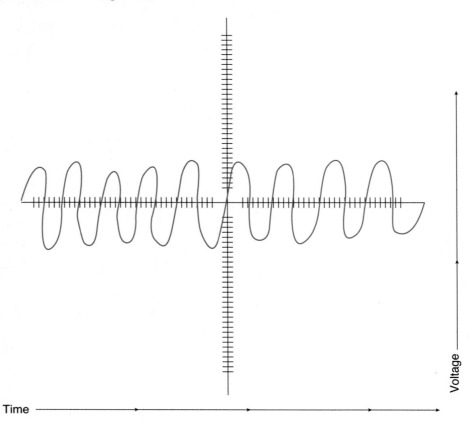

Figure 6-39 The voltage trace of a motor with a noise filter.

To avoid added frustration when diagnosing a computer system, especially the inputs, check the integrity of the ground before doing anything else. The test does not take much time and is quite simple. Overlooking the ground circuit will make system diagnosis much more difficult, if not impossible.

Testing Input Sensors

If a DTC directs you to a faulty sensor or sensor circuit or if you suspect that a sensor is faulty, it should be tested. Testing sensors is included here to orient you to the basic procedures. The recommended procedures given in a service manual for testing individual sensors may be different than those described here. Always follow the manufacturer's recommendations. Sensors are tested with a DMM, scanner, lab scope, and/or break-out box.

Since the controls are different on the various types of lab scopes that can be used for automotive diagnostic work, the connections and settings for a lab scope are loosely defined in this discussion. Make sure you follow the instructions of the scope's manufacturer when using a lab scope. These are not simple test instruments! You will become more familiar and comfortable with a scope as you use it. If the scope is set wrong, the scope will not break. It just will not show you what you want to be shown. To help with understanding how to set the controls on a scope, keep the following things in mind. The vertical voltage scale must be adjusted in relation to the voltage expected in the signal being displayed. The horizontal time base or milliseconds per division must be adjusted so the waveform appears properly on the screen. Many waveforms are clearly displayed when the horizontal time base is adjusted so three waveforms are displayed on the screen.

The trigger is the signal that tells the lab scope to start drawing a waveform. A marker indicates the trigger line on the screen, and minor adjustments of the trigger line may be necessary to position the waveform in the desired vertical position. Trigger slope indicates the direction in which the voltage signal is moving when it crosses the trigger line. A positive trigger slope means the voltage signal is moving upward as it crosses the trigger line, whereas a negative trigger slope indicates the voltage signal is moving downward when it crosses the trigger line.

There are many different types of sensors, and their design depends upon what they are monitoring. Some sensors are simple on-off switches. Others are some form of variable resistor that changes resistance according to temperature changes. Some sensors are voltage or frequency generators, while others send varying signals according to the rotational speed of another device. Knowing what they are measuring and how they respond to changes are the keys to being able to accurately test an input sensor. What follows are typical testing procedures for common input sensors. The sensors are grouped according to the type of sensor they are.

Switches

Switches are turned on and off through an action of a device or by the actions of the driver. Some of these switches are grounding switches that complete a circuit when they are closed. Others send a 5- or 12-volt signal to the PCM when they are closed. Switches inform the PCM of certain conditions or when another device is being operated.

Grounding switches always send a digital signal to the PCM. The switch's circuit is either on or off. These circuits contain a fixed resistor to limit the circuit's current and prevent voltage

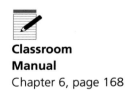

Classroom Manual
Chapter 6, page 168

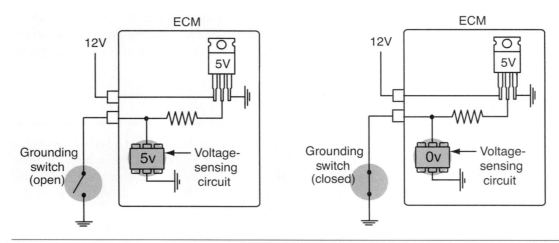

Figure 6-40 Changes in voltage signals as a grounding-type switch opens and closes. The voltage sensing circuit is somewhat functioning as a voltmeter measuring voltage drop.

spikes in the circuit as the switch opens and closes. When the switch is closed, the voltage signal to the PCM is low or zero. When the switch is open, there is a high voltage signal (Figure 6-40).

A typical grounding switch is an idle tracking switch. This switch lets the PCM know when the throttle plates are open. The switch is open when the throttle is closed and closes as soon as the throttle is opened. When the throttle is closed, a low voltage signal is sent to the PCM. This tells the PCM that it can control idle speed if necessary.

The switch can easily be tested with an ohmmeter. Disconnect the connector to the idle tracking switch, then connect the ohmmeter across the switch's terminals. When the throttle is opened wide, there should be zero-ohm resistance across the terminals. Now slowly allow the throttle to close. The switch should remain closed until the throttle is closed. At that point, the ohmmeter should give an infinity reading. If the switch reacts in any other way, the switch is bad and should be replaced. Some systems have the idle tracking switch as part of the throttle position sensor. The idle tracking function of the TP sensor can be tested in the same way as other idle tracking switches.

Some switches are adjustable and must be set so they can close and open at the correct time. An example of this is the clutch engaged switch. This switch is used to inform the computer when there is no load (clutch engaged) on the engine. The switch is also connected into the starting circuit. The switch prevents the engine from starting unless the clutch pedal is fully depressed. The switch is normally open when the clutch pedal is released. When the clutch pedal is depressed, the switch closes and completes the circuit between the ignition switch and the starter solenoid. It also sends a signal of no-load to the PCM.

This switch can be the cause of no-start problems. Because the clutch pedal switch is wired in series between the starter relay coil and the ignition switch, the engine will not crank if the switch is faulty. Some clutch switches are part of the clutch pedal assembly, while others are part of the clutch master cylinder's pushrod.

To check a clutch switch, disconnect the connector to the switch at the clutch pedal assembly. With an ohmmeter, check for continuity between the terminals of the switch with the clutch released and with it engaged. With the clutch released there should be no continuity between the terminals. With the clutch engaged, there should be good continuity between the terminals. If the switch does not complete the circuit when the clutch is engaged, check the adjustment of the switch (some switches are not adjustable). If the adjustment is correct, replace the switch. Another way to check the clutch switch is to simply bypass it with a jumper wire. If the engine starts when the switch is jumped, a bad switch is indicated.

If the switch is adjustable, make sure the clutch pedal height and pushrod (free) play are correct. Then check the release point of the clutch. Begin by engaging the parking brake and installing wheel chocks. Start the engine and allow it to idle. Without depressing the clutch pedal, slowly shift the shift lever into reverse until the gears contact. Gradually depress the clutch pedal and measure the stroke distance from the point the gear noise stops (this is the release point) up to the full stroke end position. If the distance is not within specifications, check the pedal height, pushrod play, and pedal freeplay. If the distance is correct, check the clearance between the switch and the pedal assembly when the clutch is fully depressed. Loosen the switch and adjust its position to provide for the specified clearance.

Most grounding switches react to some mechanical action to open or close. However, there are some that respond to changes of condition. These may respond to changes in pressure or temperature. An example of this type of switch is the power steering pressure switch. This switch informs the PCM when power steering pressures reach a particular point. When the power steering pressure exceeds that point, the PCM knows there is an additional load on the engine and will increase idle speed.

To test this type of switch, monitor its activity with a DMM or lab scope. With the engine running at idle speed, turn the steering wheel to its maximum position on one side. The voltage signal should drop as soon as the pressure in the power steering unit has reached a high level. If the voltage does not drop, either the power steering assembly is incapable of producing high pressures or the switch is bad.

Temperature responding switches operate in the same way. When a particular temperature is reached, the switch opens. This type switch is best measured by removing it and submerging it in heated water. Watch the ohmmeter as the temperature increases. A good temperature responding switch will open (have an infinity reading) when the water temperature reaches the specified amount. If the switch fails this test, it should be replaced.

Voltage input signal switches send a high voltage signal to the PCM when they are closed. An example of this type switch is the A/C compressor switch. This switch lets the PCM know when the extra load on the engine is caused by the air conditioning compressor's clutch engaging. Some A/C switches close when the driver selects A/C. Always check the service manual to determine what action closes this type of switch.

These switches can be tested with a voltmeter. Connect the meter to the output of the switch and to a good ground. When the A/C is turned on, the meter should read 12 volts. If it does not, check the wiring to the switch and the switch itself. Make sure voltage is available to the input side of the switch before condemning the switch. The switch can also be checked with an ohmmeter. Disconnect the wiring to the switch and connect the meter across the input and output of the switch. The switch should open and close with the cycling of the A/C compressor.

Variable Resistor-Type Sensors

Many sensors send a voltage signal to the PCM in direct response to changes in operating conditions. The voltage signal changes as the resistance in the sensor changes. Most often these variable resistor-type sensors are thermistors and potentiometers.

> A thermistor is a variable resistor made of semiconductor material.

Engine Coolant Temperature (ECT) Sensor

CAUTION: Never apply an open flame to an engine coolant temperature (ECT) sensor or intake air temperature (IAT) sensor for test purposes. This action will damage the sensor.

A defective **engine coolant temperature (ECT) sensor** may cause some of the following problems:

1. Hard engine starting
2. Rich or lean air-fuel ratio
3. Improper operation of emission devices

> An engine coolant temperature sensor (ECT) is a negative temperature coefficient (NTC) device that is affected by temperature.

Special Tools

Container, heat
  source, thermometer

**Classroom
Manual**
Chapter 6, page 176

4. Reduced fuel economy

5. Improper converter clutch lockup

6. Hesitation on acceleration

7. Engine stalling

The ECT sensor may be removed and placed in a container of water with an ohmmeter connected across the sensor terminals (Figure 6-41). A thermometer is also placed in the water. When the water is heated, the sensor should have the specified resistance at any temperature (Figure 6-42). Always use the vehicle manufacturer's specifications. If the sensor does not have the specified resistance, replace the sensor.

The wiring to the sensor can also be checked with an ohmmeter. With the wiring connectors disconnected from the ECT sensor and the computer, connect an ohmmeter from each sensor terminal to the computer terminal to which the wire is connected. Both sensor wires should indicate less resistance than specified by the vehicle manufacturer. If the wires have higher resistance than specified, the wires or wiring connectors must be repaired.

CAUTION: Before disconnecting any computer system component, be sure the ignition switch is turned off. Disconnecting components may cause high induced voltages and computer damage.

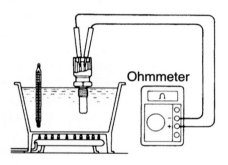

Figure 6-41 Testing an ECT sensor.

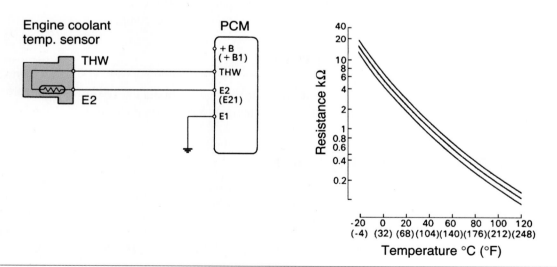

Figure 6-42 Specifications for an ECT sensor.

Cold—10,000-ohm resistor	Hot—909-ohm resistor
−20°F 4.7 V	110°F 4.2 V
0°F 4.4 V	130°F 3.7 V
20°F 4.1 V	150°F 3.4 V
40°F 3.6 V	170°F 3.0 V
60°F 3.0 V	180°F 2.8 V
80°F 2.4 V	200°F 2.4 V
100°F 1.8 V	220°F 2.0 V
120°F 1.25 V	240°F 1.62 V

Figure 6-43 Voltage drop specifications for an ECT sensor. A dual-range sensor specification is shown.

With the sensor installed in the engine, the sensor terminals may be backprobed to connect a digital voltmeter to the sensor terminals. The sensor should provide the specified voltage drop at any coolant temperature (Figure 6-43).

Some computers have internal resistors connected in series with the ECT sensor. The computer switches these resistors at approximately 120°F (49°C). This resistance change inside the computer causes a significant change in voltage drop across the sensor as indicated in the specifications. This is a normal condition on any computer with this feature. This change in voltage drop is always evident in the vehicle manufacturer's specifications.

Intake Air Temperature Sensors

A defective intake air temperature (IAT) sensor may cause these problems:

1. Rich or lean air-fuel ratio
2. Hard engine starting
3. Engine stalling or surging
4. Acceleration stumbles
5. Excessive fuel consumption

The IAT sensor may be removed from the engine and placed in a container of water with a thermometer. When a pair of ohmmeter leads is connected to the sensor terminals and the water in the container is heated, the sensor should have the specified resistance at any temperature. If the sensor does not have the specified resistance, sensor replacement is required.

Classroom Manual
Chapter 6, page 176

Prior to J1930, some IAT sensors were called air charge temperature (ACT) sensors.

CHARGED TEMPERATURE SENSOR TEMPERATURE VS. VOLTAGE CURVE	
Temperature	Voltage
−20°F	4.81 V
0°F	4.70 V
20°F	4.47 V
60°F	3.67 V
100°F	2.51 V
140°F	1.52 V
180°F	0.86 V
220°F	0.48 V
260°F	0.28 V

Figure 6-44 Intake air temperature sensor specifications.

With the IAT sensor installed in the engine, the sensor terminals may be backprobed and a voltmeter may be connected across the sensor terminals. The sensor should have the specified voltage drop at any temperature (Figure 6-44). The wires between the air charge temperature sensor and the computer may be tested in the same way as the ECT wires.

Throttle Position (TP) Sensor

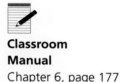

Classroom Manual
Chapter 6, page 177

Prior to J1930, a TP sensor was called a TPS by almost everyone.

A malfunctioning throttle position sensor (TP sensor) may cause acceleration stumbles, engine stalling, and improper idle speed. Backprobe the sensor terminals to complete the meter connections. With the ignition switch on, connect a voltmeter from the 5-volt reference wire to ground (Figure 6-45). The voltage reading on this wire should be approximately 5 volts. Always refer to the vehicle manufacturer's specifications.

If the reference wire is not supplying the specified voltage, check the voltage on this wire at the computer terminal. If the voltage is within specifications at the computer but low at the sensor, repair the reference wire. When this voltage is low at the computer check the voltage supply wires and ground wires on the computer. If these wires are satisfactory, replace the computer.

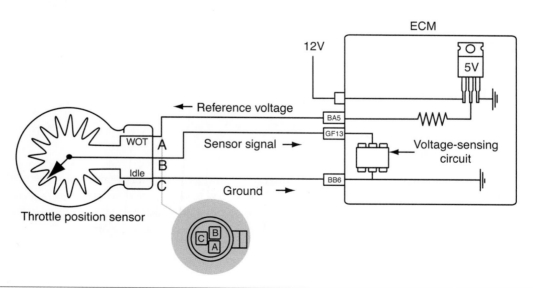

Figure 6-45 Wiring diagram for a TP sensor.

With the ignition switch on, connect the voltmeter from the sensor ground wire to the battery ground. If the voltage drop across this circuit exceeds specifications, repair the ground wire from the sensor to the computer.

> ✓ **SERVICE TIP:** When testing the throttle position sensor voltage signal, use a DSO or graphing DMM because the gradual voltage increase on this wire is easier to monitor. If the sensor voltage increase is erratic, the reading fluctuates.

> ✓ **SERVICE TIP:** When the throttle is opened gradually to check the throttle position sensor voltage signal, tap the sensor lightly and watch for fluctuations on the voltmeter or scope, indicating a defective sensor.

With the ignition switch on, connect a voltmeter from the sensor signal wire to ground. Slowly open the throttle and observe the voltmeter. The voltmeter reading should increase smoothly and gradually. Typical TP sensor voltage readings would be 0.5 volt to 1 volt with the throttle in the idle position and 4 to 5 volts at wide-open throttle. Always refer to the vehicle manufacturer's specifications. If the TP sensor does not have the specified voltage or if the voltage signal is erratic, replace the sensor.

Some TP sensors contain an idle switch that is connected to the computer. These sensors are similar to other TP sensors but have an additional wire for the idle switch (Figure 6-46). The four-wire TP sensor is tested with an ohmmeter connected from the sensor ground terminal to each of the other terminals (Figure 6-47). A specified feeler gauge must be placed between the throttle lever and the stop for some of the ohmmeter tests.

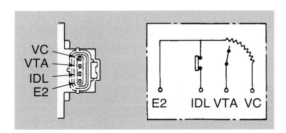

Figure 6-46 A TP sensor with an idle switch.

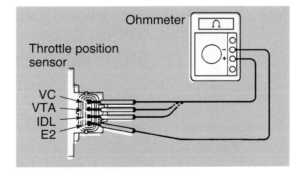

Figure 6-47 Ohmmeter tests on a TP sensor.

Many four-wire TP sensors contain an idle switch that informs the computer when the throttle is in the idle position.

On some applications, the TP sensor mounting bolts may be loosened and the sensor housing rotated to adjust the voltage signal with the throttle in the idle position.

While checking the TP voltage signal as the throttle is opened, the sensor should be tapped lightly to check for sensor defects.

The EVP sensor voltage signal should change from about 0.8 volts with the EGR valve closed to 4.5 volts with the EGR valve wide open.

Classroom Manual
Chapter 6, page 183

Special Tools

Hand-operated
    vacuum pump

The EGR position sensor is a linear potentiometer.

A TP sensor adjustment may be performed on some vehicles, but this adjustment is not possible on other applications. Check the vehicle manufacturer's service manual for the TP sensor adjustment procedure. An improper TP sensor adjustment may cause inaccurate idle speed, engine stalling, and acceleration stumbles. Follow these steps for a typical TP sensor adjustment:

1. Backprobe the TP sensor signal wire and connect a voltmeter from this wire to ground.

2. Turn on the ignition switch and observe the voltmeter reading with the throttle in the idle position.

3. If the TP sensor does not provide the specified signal voltage, loosen the TP sensor mounting bolts and rotate the sensor housing until the specified voltage is indicated on the voltmeter (Figure 6-48).

4. Hold the sensor in this position and tighten the mounting bolts to the specified torque.

Either type of TP sensor can be tested with a lab scope. Connect the scope to the sensor's output and a good ground and watch the trace as the throttle is opened and closed. The resulting trace should look smooth and clean, without any sharp breaks or spikes in the signal (Figure 6-49). A bad sensor will typically have a glitch (a downward spike) somewhere in the trace (Figure 6-50) or will not have a smooth transition from high to low. These glitches are an indication of an open or short in the sensor. On some carbureted engines, an open in the TP sensor will show up as an upward spike. This is because the computer supplies a 5-volt signal to the sensor's output wire if the TP sensor has an open.

Exhaust Gas Recirculation Valve Position Sensor

Many exhaust gas recirculation valve position (EVP) sensors have a 5-volt reference wire, a voltage signal wire, and a ground wire. The reference wire and the ground wire may be checked using the same procedure explained previously on TP sensors. Connect a pair of voltmeter leads from the voltage signal wire to ground and turn the ignition switch on. The voltage signal should be approximately 0.8 volt. Connect a vacuum hand pump to the vacuum fitting on the EGR valve and slowly increase the vacuum from zero to 20 in. Hg. The EVP sensor voltage signal should gradually increase to 4.5 volts at 20 in. Hg. Always use the EVP test procedure and specifications supplied by the vehicle manufacturer. If the EVP sensor does not have the specified voltage, replace the sensor.

It is good practice to check all variable resistor sensors with a lab scope. Any defects in the sensor will show up as glitches in the waveform. These are often unnoticeable on a DMM. The trace from all variable resistors should be clean and smooth.

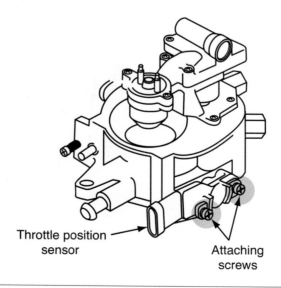

Throttle position sensor

Attaching screws

Figure 6-48 A TP sensor with elongated slots for sensor adjustments (some models).

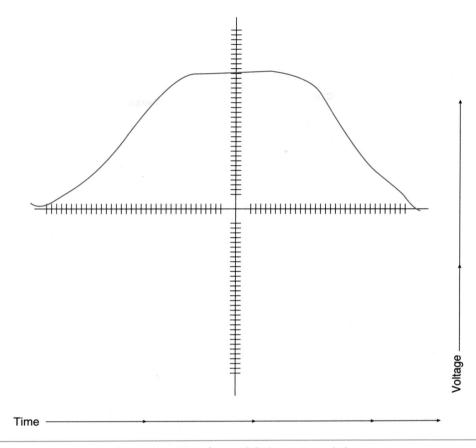

Time ───────────────────────▶

Voltage ───

Figure 6-49 A normal TP sensor waveform while it opens and closes.

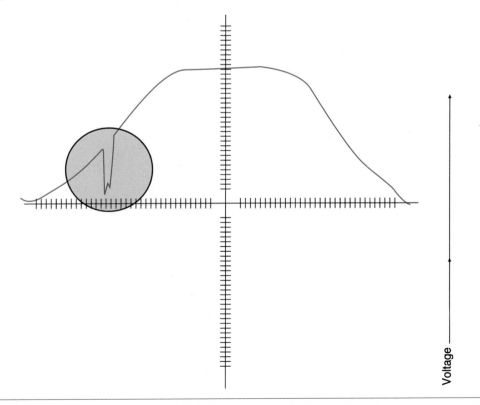

Voltage ───

Figure 6-50 The waveform of a malfunctioning TP sensor. Notice the glitch while the throttle opens.

**Classroom
Manual**
Chapter 6, page 172

Most O₂ sensors are
zinconium-type
sensors. A zinconium
battery produces
between 0 and
1 volt depending
on the amount of
oxygen it is exposed
to.

A defective O₂ sensor
heater may cause
extended open loop
time and reduced
fuel economy.

Special Tools

Propane bottle,
  bottle valve, hose

If the O₂ sensor
voltage signal is
higher than
specified, the air-fuel
ration may be rich or
the sensor may be
contaminated.

**Room temperature
vulcanizing (RTV)
sealant** is a sealant
that cures at room
temperature (also
known as silicone
sealant).

Generating Sensors

Some engine sensors generate a voltage or frequency signal in response to changing conditions. The most common voltage-generating sensors are the oxygen sensor and crankshaft position sensors. Common frequency generator-type sensors are some MAP sensors, some MAF sensors, and reference signals.

Oxygen (O₂) Sensors

Oxygen sensors produce a voltage based on the amount of oxygen in the exhaust. Large amounts of oxygen result from lean mixtures and result in low voltage output from the O₂ sensor. Rich mixtures have released lower amounts of oxygen in the exhaust—therefore, the O₂ sensor voltage is high. The engine must be at normal operating temperature before the oxygen sensor is tested. Always follow the test procedure in the vehicle manufacturer's service manual and use the specifications supplied by the manufacturer.

 CAUTION: An oxygen sensor must be tested with a digital voltmeter. If an analog meter is used for this purpose, the sensor may be damaged.

 SERVICE TIP: A contaminated oxygen sensor may provide a continually high voltage reading because the oxygen in the exhaust stream does not contact the sensor.

Testing with a DMM

Connect the voltmeter between the O₂ sensor wire and ground to test this sensor (Figure 6-51). Backprobe the connector near the O₂ sensor to connect the voltmeter to the sensor signal wire. If possible, avoid probing through the insulation to connect a meter to the wire. With the engine idling, if the O₂ sensor and the computer system are working properly, the sensor voltage should be cycling from low voltage to high voltage. The signal from most O₂ sensors varies between 0 and 1 volt.

If the voltage is continually high, the air-fuel ratio may be rich or the sensor may be contaminated. The O₂ sensor may be contaminated with **room temperature vulcanizing (RTV) sealant**, antifreeze or lead from leaded gasoline. When the O₂ sensor voltage is continually low, the air-fuel ratio may be lean, the sensor may be defective, or the wire between the sensor and the computer may have a high-resistance problem. If the O₂ sensor voltage signal remains in a mid-range position, the computer may be in open loop or the sensor may be defective.

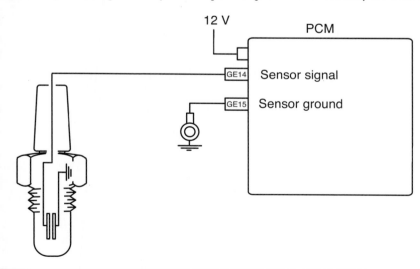

Figure 6-51 Wiring diagram for a single-wire O₂ sensor.

The sensor can also be tested after it is removed from the exhaust manifold. Connect the voltmeter between the sensor wire and the case of the sensor. Using a propane torch, heat the sensor element. The propane flame keeps the oxygen in the air away from the sensor element, causing the sensor to produce voltage. While the sensor element is in the flame, the voltage should be nearly 1 volt. The voltage should drop to zero immediately when the flame is removed from the sensor. If the sensor does not produce the specified voltage or if the sensor does not quickly respond to the change, it should be replaced.

If a defect in the O_2 sensor signal wire is suspected, backprobe the sensor signal wire at the computer and connect a digital voltmeter from the signal wire to ground with the engine idling. The difference between the voltage readings at the sensor and at the computer should not exceed the vehicle manufacturer's specifications. A typical specification for voltage drop across the average sensor wire is 0.2 volt.

Now check the sensor's ground. With the engine idling, connect the voltmeter from the sensor case to the sensor ground wire on the computer. Typically, the maximum allowable voltage drop across the sensor ground circuit is 0.2 volt. Always use the vehicle manufacturer's specifications. If the voltage drop across the sensor ground exceeds specifications, repair the ground wire or the sensor ground in the exhaust manifold.

Most late-model engines are fitted with heated O_2 sensors (HO$_2$S). If the O_2 sensor heater is not working, the sensor warm-up time is extended and the computer stays in open loop longer. In this mode, the computer supplies a richer air-fuel ratio. As a result, the engine's emissions are high and its fuel economy is reduced. To test the heater circuit, disconnect the O_2 sensor connector and connect a voltmeter between the heater voltage supply wire and ground. With the ignition switch on, 12 volts should be supplied on this wire. If the voltage is less than 12 volts, repair the fuse in this voltage supply wire or the wire itself.

With the O_2 sensor wire disconnected, connect an ohmmeter across the heater terminals in the sensor connector (Figure 6-52). If the heater does not have the specified resistance, replace the sensor.

When the O_2 sensor voltage signal is lower than specified, the air-fuel ratio may be lean or the sensor may be defective.

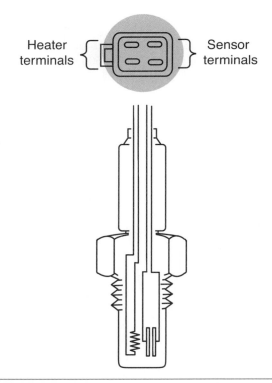

Figure 6-52 Electrical terminals for a heated O_2 sensor.

Testing with a Scanner

The output from an O_2 sensor should constantly cycle between high and low voltages as the engine is running in closed loop. This cycling is the result of the computer constantly correcting the air-fuel ratio in response to feedback from the O_2 sensor. When the O_2 sensor reads lean, the computer will richen the mixture. When the O_2 sensor reads rich, the computer will lean the mixture. This enables the computer to control the air-fuel mixture. Many things can occur to take that control away from the computer. One of these is a faulty O_2 sensor.

The activity of the sensor can be monitored on a scanner. By watching the scanner while the engine is running, the O_2 voltage should move to nearly 1 volt then drop back to close to 0 volt. Immediately after it drops, the voltage signal should move back up. This immediate cycling is an important function of an O_2 sensor. If the response is slow, the sensor is lazy and should be replaced. With the engine at about 2,500 rpm, the O_2 sensor should cycle from high to low ten to forty times in 10 seconds. The voltage readings shown on the scanner are also an indicator of how well the sensor works. When testing the O_2 sensor, make sure the sensor is heated and the system is in closed loop.

Testing with a Lab Scope (including a graphing multimeter or digital storage oscilloscope)

A faulty O_2 sensor can cause many different types of problems. It can cause excessively high HC and CO emissions and all sorts of driveability problems. Most computer systems monitor the activity of the O_2 sensor and store a code when the sensor's output is not within the desired range. Again, the normal range is between 0 and 1 volt, and the sensor should constantly toggle from close to 0.2 volt to 0.8 volt then back to 0.2 volt (Figure 6-53). If the sensor toggles within the specifications, the computer will think everything is normal and respond accordingly. This does not mean the sensor is working properly.

The voltage signal from an O_2 sensor should have two to three cross counts with the engine without a load at 2,000 rpm. O_2 signal cross counts (Figure 6-54) are the number of times the O_2 voltage signal changes above or below 0.45 volt in a second. If there are not enough cross counts, the sensor is contaminated or lazy. It should be replaced.

If the sensor's voltage toggles between 0 volt and 500 millivolts, it is toggling within its normal range but is not operating normally. It is biased low or lean. As a result, the computer will be constantly adding fuel to try to reach the upper limit of the sensor. Something is causing the sensor to be biased lean. If the toggling only occurs at the higher limits of the voltage range, the sensor is biased rich. In either case, the computer does not have true control of the air-fuel mixture because of the faulty O_2 signals.

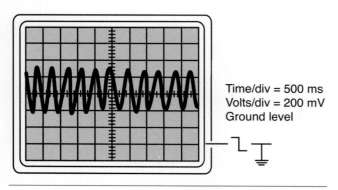

Figure 6-53 A good O_2 sensor trace.

Time/div = 500 ms
Volts/div = 200 mV
Ground level

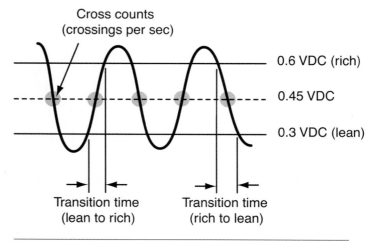

Cross counts
(crossings per sec)

0.6 VDC (rich)

0.45 VDC

0.3 VDC (lean)

Transition time
(lean to rich)

Transition time
(rich to lean)

Figure 6-54 O_2 sensor signal cross counts.

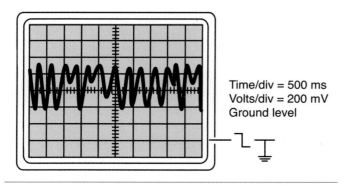

Figure 6-55 O_2 sensor signal caused by a shorted spark plug wire.

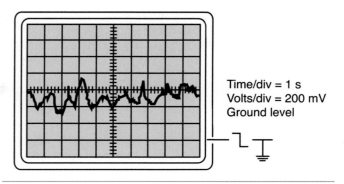

Figure 6-56 O_2 sensor signal caused by an open secondary cable.

The O_2 can be biased rich or lean, not work at all, or work too slowly to ensure good emissions and fuel economy. To test the O_2 sensor for all of these concerns, use a lab scope. Begin by allowing the engine and O_2 sensor to warm up. Insert the hose of a propane enrichment tool into the power brake booster vacuum hose or simply install it into the nozzle of the air cleaner assembly. This will drive the mixture rich. Most good O_2 sensors will produce almost 1 volt when driven full rich. The typical specification is at least 800 millivolts.

Connect the lab scope to the sensor and a good ground. Set the scope to display the trace at 200 millivolts per division and 500 milliseconds per division. Inject some propane into the air cleaner assembly. Observe the O_2 signal's trace. The O_2 sensor should show over 800 millivolts. If the voltage does not go high, the O_2 sensor is bad and should be replaced. Now, remove the propane bottle and cause a vacuum leak by pulling off an intake vacuum hose. Watch the scope to see how the O_2 sensor reacts. It should drop to under 175 millivolts. If it does not, replace the sensor. These tests check the O_2 sensor, not the system, therefore they are reliable O_2 sensor checks.

Also keep in mind that on an air pump-equipped car, it is a good idea to disable the air pump before doing this test. Unwanted air may bias the results.

Observing the trace of an O_2 sensor can also help in the diagnosis of other engine performance problems. Figure 6-55 and Figure 6-56 show how ignition problems affect the signal from the O_2 sensor. Keep in mind that during complete combustion, nearly all of the oxygen in the combustion chamber is combined with the fuel. This means there will be little O_2 in the exhaust of a very efficient engine. As combustion becomes more incomplete, the levels of oxygen increase. Ignition problems cause incomplete combustion and there is much oxygen in the exhaust. This is also true of lean mixtures, overadvanced ignition timing, or anything else that causes incomplete combustion.

When the mixture is rich, combustion has a better chance of being complete. Therefore, the oxygen levels in the exhaust decrease. The O_2 sensor output will respond to the low oxygen with a high voltage signal (Figure 6-57 and Figure 6-58). Remember that the PCM will always try to do

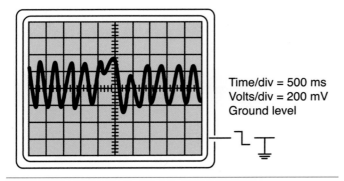

Figure 6-57 O_2 sensor signal caused by a leaking injector.

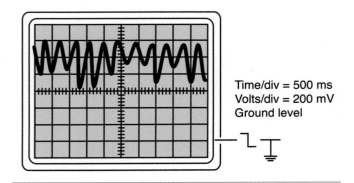

Figure 6-58 O_2 sensor signal caused by a fuel injector problem.

the opposite of what it receives from the O_2 sensor. When the O_2 shows lean, the PCM goes rich and vice versa. When a lean exhaust signal is not caused by an air-fuel problem, the PCM does not know what the true cause is and will richen the mixture in response to the signal. This may make the engine run worse than it did.

Hall Effect Switches

To test a Hall effect sensor, disconnect its wiring harness. Connect a voltage source of the correct low voltage level across the positive and negative terminals of the Hall layer. Then connect a voltmeter across the negative and signal voltage terminals.

Insert a metal feeler gauge between the Hall layer and the magnet. Make sure the feeler gauge is touching the Hall element. If the sensor is operating properly, the meter will read close to battery voltage. When the feeler gauge blade is removed, the voltage should decrease. On some units, the voltage will drop to near zero. Check the service manual to see what voltage you should observe when installing and removing the feeler gauge.

PIP (profile ignition pickup) and SPOUT (spark out) are terms specific to Ford vehicles—however, they are similar to GM's EST (electronic spark timing) and ignition reference signals. The PIP signal is an ignition reference signal from the crankshaft position sensor, which is a Hall effect sensor. The PCM uses the signal to determine crankshaft position and engine speed. The computer uses PIP to build the SPOUT which it sends to the ignition module. The SPOUT signal is a ground controlled signal. It appears in a trace as a downward pulse, which also represents the beginning of dwell (Figure 6-59).

When observing a Hall effect sensor on a lab scope, pay attention to the downward and upward pulses. These should be straight (Figure 6-60). If they appear at an angle (Figure 6-61),

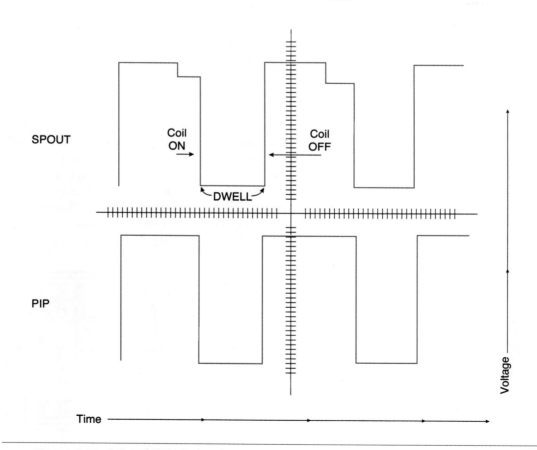

Figure 6-59 PIP and SPOUT signals.

Classroom Manual
Chapter 6, page 185

According to J1930, the PIP sensor should be called the crankshaft position (CKP) sensor.

Special Tools

Metal feeler gauge

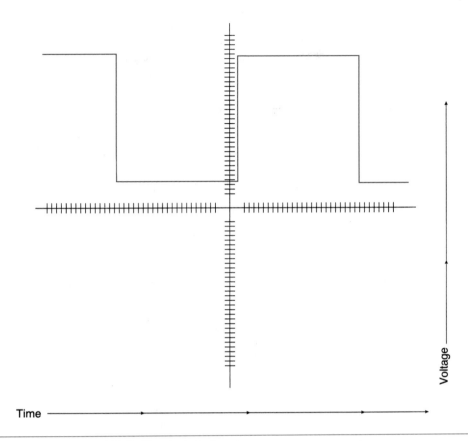

Voltage

Time

Figure 6-60   A good Hall effect switch signal.

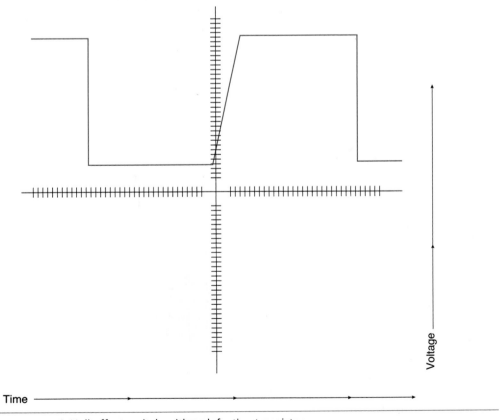

Voltage

Time

Figure 6-61   A Hall effect switch with a defective transistor.

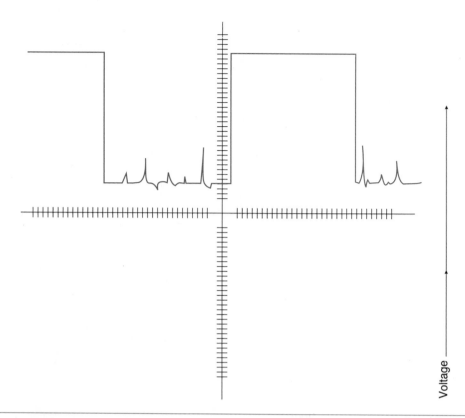

Voltage

Figure 6-62 Check the square wave for any and all glitches.

**Classroom
Manual**
Chapter 6, page 164

Magnetic pulse generators use the principles of magnetic induction to produce a voltage signal.

this indicates the transistor is faulty causing the voltage to rise slowly. This can cause a no-start condition. The entire square wave from a Hall effect unit should be flat. Any change from a normal trace means the sensor should be replaced (Figure 6-62).

Magnetic Pulse Generators

Ignition pickup coils, wheel speed sensors, and vehicle speed sensors are permanent magnet voltage producing sensors. The procedures for testing each of these is similar.

A pickup coil is a common AC voltage source. The frequency and amplitude of the output signal from an ignition pickup coil will depend on the numbers of cylinders and the speed of the engine. A good pickup coil will have even peaks that reach at least 300 millivolts when the engine is cranking (Figure 6-63).

The waveforms from most crankshaft position sensors will have a number of equally spaced pulses and one double pulse or sync signal, as shown in Figure 6-64. The number of

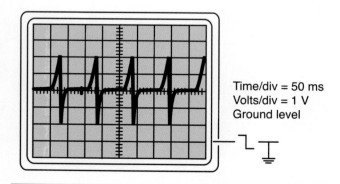

Time/div = 50 ms
Volts/div = 1 V
Ground level

Figure 6-63 The trace of a good PM generator.

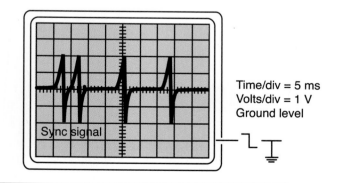

Sync signal

Time/div = 5 ms
Volts/div = 1 V
Ground level

Figure 6-64 The trace of a good crankshaft position sensor.

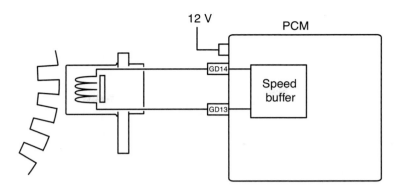

Figure 6-65 Wiring diagram for a VSS.

evenly spaced pulses equals the number of cylinders that the engine has. Carefully examine the trace. Any glitches indicate a problem with the sensor or sensor circuit.

Vehicle Speed Sensor

A defective vehicle speed sensor may cause different problems depending on the computer output control functions. A defective **vehicle speed sensor (VSS)** may cause these problems:

1. Improper converter clutch lockup
2. Improper cruise control operation
3. Inaccurate speedometer operation

Prior to VSS diagnosis, the vehicle should be lifted on a hoist so the drive wheels are free to rotate. Backprobe the VSS output wire (Figure 6-65) and connect the voltmeter leads from this wire to ground. Select the 20-volt AC scale on the voltmeter, then start the engine.

Place the transaxle in drive and allow the drive wheels to rotate. If the VSS voltage signal is not 0.5 volt or more, replace the sensor. When the VSS provides the specified voltage signal, backprobe the VSS terminal at the PCM and repeat the voltage signal test with the drive wheels rotating. If 0.5 volt is available at this terminal, the trouble may be in the PCM.

When 0.5 volt is not available at this terminal, turn the ignition switch off and disconnect the wire from the VSS to the PCM. Connect the ohmmeter leads across the wire. The meter should read zero ohm. Repeat the test with the ohmmeter leads connected to the VSS ground terminal and the PCM ground terminal. This wire should also have zero ohm resistance. If the resistance in these wires is more than specified, repair the wires.

The condition of a speed sensor can be checked by going through the diagnostic routines for the system. If this test indicates that a sensor is faulty, the sensor should be replaced. Speed sensors can also be checked with an ohmmeter. Most manufacturers list a resistance specification. The resistance of the sensor is measured across the sensor's terminals. The typical range for a good sensor is 800 to 1,400 ohms of resistance.

Knock Sensors

A defective knock sensor may cause engine detonation or reduced spark advance and fuel economy. When a knock sensor is removed and replaced, the sensor torque is critical. The procedure for checking a knock sensor varies depending upon the vehicle make and year. Always follow the vehicle manufacturer's recommended test procedure and specifications. Follow these steps for a typical knock sensor diagnosis:

1. Disconnect the knock sensor wiring connector and turn the ignition switch on.
2. Connect a voltmeter from the disconnected knock sensor wire to ground. The voltage should be 4 to 6 volts. If the specified voltage is not available at this wire, backprobe

Classroom Manual
Chapter 6, page 184

The **vehicle speed sensor (VSS)** is used by the computer to control the torque converter clutch lockup, cruise control, and the electronic speedometer.

The VSS produces an analog A/C voltage.

Classroom Manual
Chapter 6, page 182

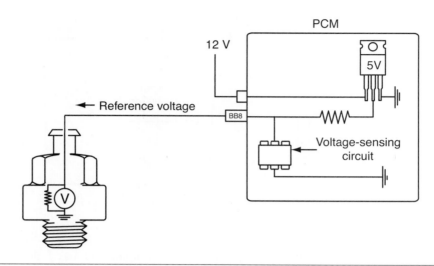

Figure 6-66 Wiring diagram for a knock sensor.

the knock sensor wire at the computer and read the voltage at this terminal (Figure 6-66). If the voltage is satisfactory at this terminal, repair the knock sensor wire. When the voltage is not within specifications at the computer terminal, replace the computer.

3. Connect an ohmmeter from the knock sensor terminal to ground. Some knock sensors should have 3,300 ohms to 4,500 ohms. If the knock sensor does not have the specified resistance, replace the sensor.

Manifold Absolute Pressure (MAP) Sensors

Classroom Manual
Chapter 6, page 178

The **manifold absolute pressure (MAP) sensors** monitor engine loads. They are connected to the engine vacuum.

A defective **manifold absolute pressure (MAP) sensor** may cause a rich or lean air-fuel ratio, excessive fuel consumption, and engine surging. This diagnosis applies to MAP sensors that produce an analog voltage signal. With the ignition switch on, backprobe the 5-volt reference wire and connect a voltmeter from the reference wire to ground (Figure 6-67).

✔ **SERVICE TIP:** Manifold absolute pressure sensors have a much different calibration on turbocharged engines than on non-turbocharged engines. Be sure you are using the proper specifications for the sensor being tested.

If the reference wire is not supplying the specified voltage, check the voltage on this wire

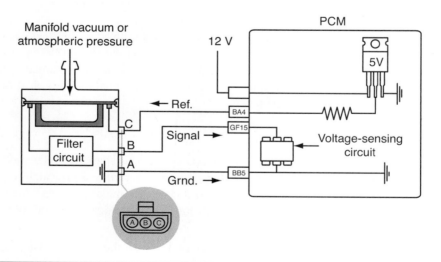

Figure 6-67 Wiring diagram for a MAP sensor.

at the computer. If the voltage is within specifications at the computer but low at the sensor, repair the reference wire. When this voltage is low at the computer, check the voltage supply wires and ground wires on the computer. If these wires are satisfactory, replace the computer.

With the ignition switch on, connect the voltmeter from the sensor ground wire to the battery ground. If the voltage drop across this circuit exceeds specifications, repair the ground wire from the sensor to the computer.

Backprobe the MAP sensor signal wire and connect a voltmeter from this wire to ground with the ignition switch on. The voltage reading indicates the barometric pressure signal from the MAP sensor to the computer. Many MAP sensors send a barometric pressure signal to the computer each time the ignition switch is turned on and each time the throttle is in the wide-open position. If the voltage supplied by the barometric pressure signal in the MAP sensor does not equal the vehicle manufacturer's specifications, replace the MAP sensor.

The barometric pressure voltage signal varies depending on altitude and atmospheric conditions. Follow this calculation to obtain an accurate barometric pressure reading:

1. Phone your local weather or TV station and obtain the present barometric pressure reading; for example, 29.85 inches. The pressure they quote is usually corrected to sea level.

2. Multiply your altitude by 0.001; for example, 600 feet × 0.001 = 0.6.

3. Subtract the altitude correction from the present barometric pressure reading: 29.85 − 0.6 = 29.79.

4. Check the vehicle manufacturer's specifications to obtain the proper barometric pressure voltage signal in relation to the present barometric pressure (Figure 6-68).

To check the voltage signal of a MAP, turn the ignition switch on and connect a voltmeter to the MAP sensor signal wire. Connect a vacuum hand pump to the MAP sensor vacuum connection and apply 5 inches of vacuum to the sensor. On some MAP sensors, the sensor voltage signal should change 0.7 to 1.0 volt for every 5 inches of vacuum change applied to the sensor. Always use the vehicle manufacturer's specifications. If the barometric pressure voltage signal was 4.5 volts with 5 inches of vacuum applied to the MAP sensor, the voltage should be 3.5 volts to 3.8 volts. When 10 inches of vacuum is applied to the sensor, the voltage signal should be 2.5 volts to 3.1 volts. Check the MAP sensor voltage at 5-inch intervals from 0 to 25 inches. If the MAP sensor voltage is not within specifications at any vacuum, replace the sensor.

To check a MAP sensor with a lab scope, connect the scope to the MAP output and a good ground. When the engine is accelerated and returned to idle, the output voltage should increase

Special Tools

Hand-operated vacuum pump, frequency tester

Absolute BARO reading	Lowest allowable voltage at −40°F	Lowest allowable voltage at 257°F	Lowest allowable voltage at 77°F	Designed output voltage	Highest allowable voltage at 77°F	Highest allowable voltage at 257°F	Highest allowable voltage at −40°F
31.0"	4.548 V	4.632 V	4.716 V	4.800 V	4.884 V	4.968 V	5.052 V
30.9"	4.531 V	4.615 V	4.699 V	4.783 V	4.867 V	4.951 V	5.035 V
30.8"	4.514 V	4.598 V	4.682 V	4.766 V	4.850 V	4.934 V	5.018 V
30.7"	4.497 V	4.581 V	4.665 V	4.749 V	4.833 V	4.917 V	5.001 V
30.6"	4.480 V	4.564 V	4.648 V	4.732 V	4.816 V	4.900 V	4.984 V
30.5"	4.463 V	4.547 V	4.631 V	4.715 V	4.799 V	4.883 V	4.967 V
30.4"	4.446 V	4.530 V	4.614 V	4.698 V	4.782 V	4.866 V	4.950 V
30.3"	4.430 V	4.514 V	4.598 V	4.682 V	4.766 V	4.850 V	4.934 V
30.2"	4.413 V	4.497 V	4.581 V	4.665 V	4.749 V	4.833 V	4.917 V
30.1"	4.396 V	4.480 V	4.564 V	4.648 V	4.732 V	4.816 V	4.900 V
30.0"	4.379 V	4.463 V	4.547 V	4.631 V	4.715 V	4.799 V	4.883V

Figure 6-68 Barometric pressure voltage signal specifications at different barometric pressures.

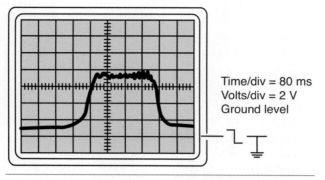

Time/div = 80 ms
Volts/div = 2 V
Ground level

Figure 6-69 The trace of a normal MAP sensor.

Figure 6-70 A MAP sensor tester.

and decrease (Figure 6-69). If the engine is accelerated and the MAP sensor voltage does not rise and fall or if the signal is erratic, the sensor or sensor wires are defective.

If the MAP sensor produces a digital voltage signal of varying frequency, check the 5-volt reference wire and the ground wire with the same procedure used on other MAP sensors. This sensor diagnosis is based on the use of a MAP sensor tester that changes the MAP sensor varying frequency voltage to an analog voltage. Photo Sequence 13 shows a typical procedure for testing a Ford MAP sensor, which has a varying frequency. Follow these steps to test the MAP sensor voltage signal:

1. Turn the ignition switch off and disconnect the wiring connector from the MAP sensor.
2. Connect the connector on the MAP sensor tester to the MAP sensor (Figure 6-70).
3. Connect the MAP sensor tester battery leads to a 12-volt battery.
4. Connect a pair of digital voltmeter leads to the MAP tester signal wire and ground.
5. Turn the ignition switch on and observe the barometric pressure voltage signal on the meter. If this voltage signal does not equal the manufacturer's specifications, replace the sensor.
6. Supply the specified vacuum to the MAP sensor with a hand vacuum pump.
7. Observe the voltmeter reading at each specified vacuum. If the MAP sensor voltage signal does not equal the manufacturer's specifications at any vacuum, replace the sensor.

If a special MAP tester is not available, the sensor can be checked with a DMM that measures frequency. Connect the meter to the MAP sensor. Measure the voltage, duty cycle, and frequency at the sensor with no vacuum applied. Then apply about 18 in. Hg of vacuum to the MAP. Observe and record the same readings. A good MAP will have about the same amount of voltage and duty cycle with or without the vacuum. However, the frequency should decrease (Figure 6-71). Normally, a frequency of about 155 hertz is expected at sea level with no vacuum

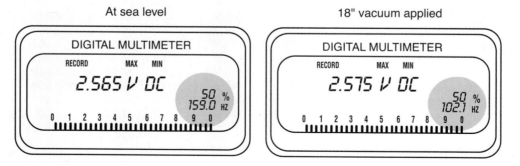

Figure 6-71 Reaction of a good MAP sensor with vacuum applied.

Photo Sequence 13
Testing a Ford MAP Sensor

P13-1 Remove the MAP sensor wiring connector and vacuum hose.

P13-2 Connect the MAP sensor tester to the MAP sensor.

P13-3 Connect the digital voltmeter leads from the proper MAP sensor tester lead to ground.

P13-4 Connect the MAP sensor tester leads to the battery terminals with the proper polarity.

P13-5 Observe the MAP sensor barometric pressure (Baro) voltage reading on the voltmeter and compare this reading to specifications.

P13-6 Connect a vacuum hand pump hose to the MAP sensor and apply 5 in. Hg to the MAP sensor. Observe the MAP sensor voltage signal on the voltmeter. Compare this voltmeter reading to specifications.

P13-7 Apply 10 in. Hg to the MAP sensor with the hand pump and observe the voltmeter reading. Compare this reading to specifications.

P13-8 Apply 15 in. Hg to the MAP sensor with the hand pump and observe the voltmeter reading. Compare this reading to specifications.

P13-9 Apply 20 in. Hg to the MAP sensor with the hand pump and observe the voltmeter reading. Compare this reading to specifications. If any of the MAP sensor readings do not meet specifications, replace the MAP sensor.

Vacuum Applied	Output Frequency
0 in. Hg	152-155 Hz
5 in. Hg	138-140 Hz
10 in. Hg	124-127 Hz
15 in. Hg	111-114 Hz
20 in. Hg	93-98 Hz

Figure 6-72 Typical MAP frequency outputs with different amounts of vacuum applied.

applied to the MAP. When vacuum is applied, the frequency should decrease to around 95 hertz (Figure 6-72).

A lab scope can be used to check a Ford MAP sensor. The upper horizontal line of the trace should be at 5 volts and the lower horizontal line should be close to zero (Figure 6-73). Check the waveform for unusual movements of the trace. If the waveform is anything but normal, replace the sensor.

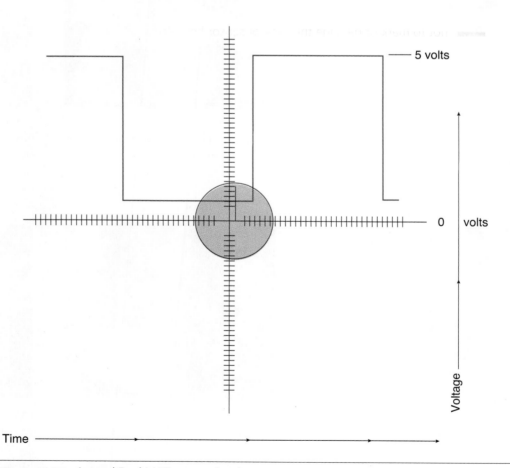

Figure 6-73 A good Ford MAP sensor signal.

Mass Air Flow Sensors

The **mass air flow (MAF) sensor** measures the mass of the air being drawn into the engine. The signal from the MAF is used by the PCM to calculate injector pulse width. Two types of MAF sensors are commonly used: grid and hot-wire types. A faulty MAF will cause driveability problems resulting from incorrect ignition timing and improper air-fuel ratios.

Volume Air Flow Meters

Begin checking a vane-type sensor by checking the voltage supply wire and the ground wire to the sensor's module before checking the sensor voltage signal. Always follow the recommended test procedure in the manufacturer's service manual and use the specifications supplied by the manufacturer. The following procedure is based on the use of a Fluke multimeter. Follow these steps to measure the sensor voltage signal:

1. Set the multimeter on the VDC scale and connect the black meter lead to the COM terminal in the meter while the red meter lead is installed in the V/RPM meter connection.

2. Connect the red meter lead to the sensor's signal wire with a special piercing probe and connect the black meter lead to ground (Figure 6-74).

3. Turn the ignition switch on and press the min/max button to activate the min/max feature (Figure 6-75).

> **CAUTION:** While pushing the air flow sensor vane open and closed, be careful not to mark or damage the vane or sensor housing.

Classroom Manual
Chapter 6, page 180

The **mass air flow (MAF) sensor** monitors incoming air flow, including speed, temperature, and pressure.

There are two styles of vane style air flow sensors. In the latest type, the voltage should decrease as the vane is opened. In the earlier style, the voltage increased.

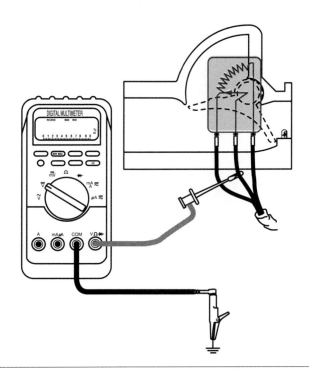

Figure 6-74 A voltmeter connected to measure the signal from a MAF sensor.

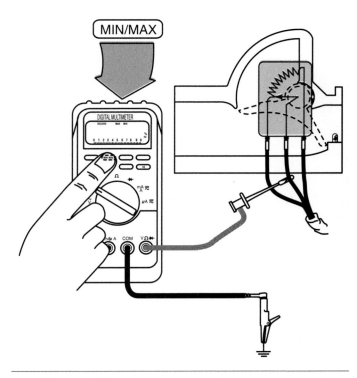

Figure 6-75 Press the min/max button to engage the min/max test mode.

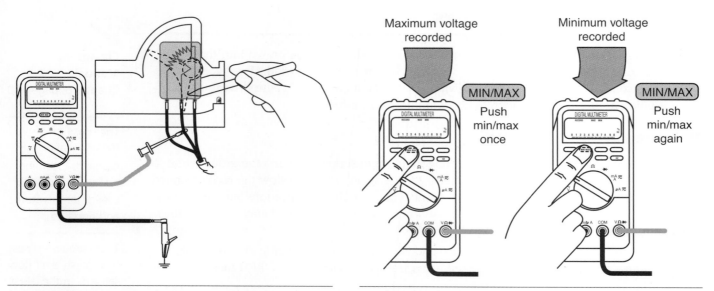

Figure 6-76 Watch the meter as the sensor's flap (vane) is pushed open.

Figure 6-77 Use the min/max function to record minimum and maximum sensor voltage signals.

4. Slowly push the vane from the closed to the wide-open position and allow the vane to slowly return to the closed position (Figure 6-76).

5. Touch the min/max button once to read the maximum voltage signal recorded and press this button again to read the minimum voltage signal (Figure 6-77). If the minimum voltage signal is zero, there may be an open circuit in the sensor's variable resistor. When the voltage signal is not within the manufacturer's specifications, replace the sensor.

Some vehicle manufacturers specify ohmmeter tests for the air flow sensor. With the sensor removed, connect the ohmmeter across the sensor's output and input terminals (Figure 6-78). The resistance at these terminals is normally 200 to 600 ohms.

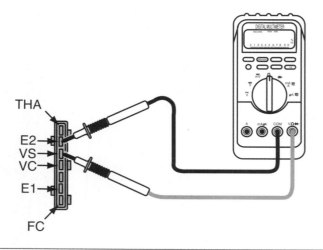

Figure 6-78 Ohmmeter connections to an air flow sensor.

BETWEEN TERMINALS	RESISTANCE (Ω)	TEMPERATURE °C (°F)
VS – E2	200 – 600	–
VC – E2	200 – 400	–
THA – E2	10,000 – 20,000	–20 (–4)
	4,000 – 7,000	0 (32)
FC – E1	Infinity	–

Figure 6-79 Resistance specifications for a typical air flow sensor.

Connect the ohmmeter leads to the other recommended air flow sensor terminals. Compare the resistance readings to the specifications (Figure 6-79). In this example, since the THA and E2 sensor terminals are connected internally to the thermistor, temperature affects the ohm reading at these terminals as indicated in the specifications. If the specified resistance is not available in any of the test connections, replace the sensor.

Connect the ohmmeter leads to the specified air flow sensor terminals and move the vane from the fully closed to the fully open position. With each specified meter connection and vane position, the ohmmeter should indicate the specified resistance (Figure 6-80). When the ohmmeter leads are connected to the sensor's input and output terminals, the ohm reading should transition smoothly as the sensor vane is opened and closed.

To check a vane-type sensor with a lab scope, connect the positive lead to the output signal terminal and the negative scope lead to a good ground. This type air flow sensor should display an analog voltage signal when the engine is accelerated. A defective air flow sensor will have sudden and erratic voltage changes (Figure 6-81).

Some heated resistor-type or hot wire-type MAF sensors produce a frequency voltage signal that may be checked with a multimeter with frequency capabilities.

Hot Wire-Type MAF Sensors

The test procedure for heated resistor and **hot wire MAF sensors** varies depending on the vehicle make and year. Always follow the test procedure in the vehicle manufacturer's service manual. A frequency test may be performed on some MAF sensors, such as the AC Delco MAF on

The resistance of the heated wire in a **hot wire MAF sensor** changes depending on air flow.

BETWEEN TERMINALS	RESISTANCE (Ω)	MEASURING PLATE OPENING
FC – E1	Infinity	Fully closed
FC – E1	Zero	Other than closed
VS – E2	200 – 600	Fully closed
VS – E2	20 – 1,200	Fully open

Figure 6-80 Resistance specifications for a typical sensor with door open and closed.

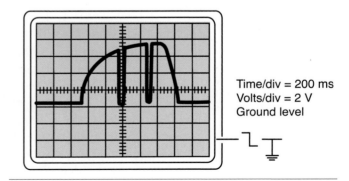

Time/div = 200 ms
Volts/div = 2 V
Ground level

Figure 6-81 The trace of a malfunctioning vane-type sensor.

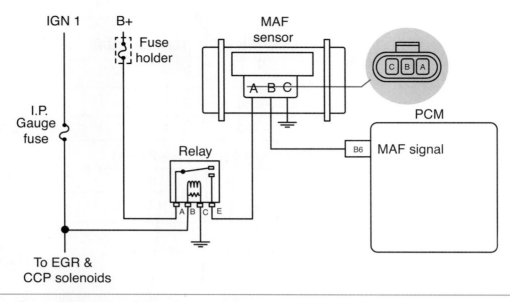

Figure 6-82 Wiring diagram for a MAF sensor.

some General Motors products. The following test procedure is based on the use of a Fluke multimeter. Follow these steps to check the MAF sensor voltage signal and frequency:

1. Place the multimeter on the V/RPM scale and connect the meter leads from the MAF voltage signal wire to the ground wire (Figure 6-82).

2. Start the engine and observe the voltmeter reading. On some MAF sensors, this reading should be 2.5 volts. Always refer to the manufacturer's specifications.

3. Lightly tap the MAF sensor housing with a screwdriver handle and watch the voltmeter pointer. If the pointer fluctuates or the engine misfires, replace the MAF sensor. Some MAF sensors have experienced loose internal connections, which cause erratic voltage signals and engine misfiring and surging.

4. Be sure the meter dial is on DC volts and press the rpm button three times so the meter displays voltage frequency. The meter should indicate about 30 hertz with the engine idling.

5. Increase the engine speed, and record the meter reading at various speeds.

6. Graph the frequency readings. The MAF sensor frequency should increase smoothly and gradually in relation to engine speed. If the MAF sensor frequency reading is erratic, replace the sensor (Figure 6-83).

While diagnosing a General Motors vehicle with a scanner, one test mode displays grams per second from the MAF sensor. This mode provides an accurate test of the MAF sensor. The grams per second reading should be 4 to 7 with the engine idling. This reading should gradually increase as the engine speed increases. When the engine speed is constant, the grams-per-second reading should remain constant. If the grams-per-second reading is erratic at a constant engine speed or if this reading varies when the sensor is tapped lightly, the sensor is defective. A MAF sensor fault code may not be present with an erratic grams-per-second reading, but the erratic reading indicates a defective sensor.

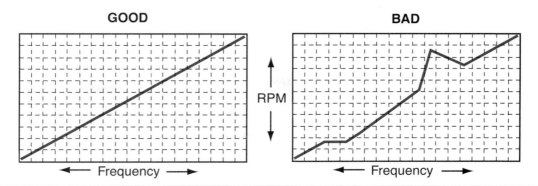

Figure 6-83   Satisfactory and unsatisfactory MAF sensor frequency readings.

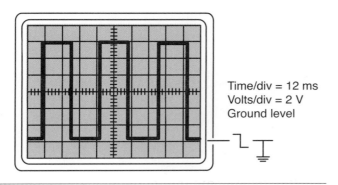

Time/div = 12 ms
Volts/div = 2 V
Ground level

Figure 6-84   A normal trace for a frequency varying MAF sensor.

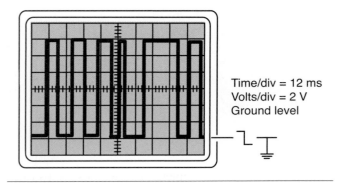

Time/div = 12 ms
Volts/div = 2 V
Ground level

Figure 6-85   The trace of a defective frequency varying MAF sensor.

Frequency varying types of MAF sensors can be tested with a lab scope. The waveform should appear as a series of square waves (Figure 6-84). When the engine speed and intake airflow increases, the frequency of the MAF sensor signals should increase smoothly and proportionately to the change in engine speed. If the MAF or connecting wires are defective, the trace will show an erratic change in frequency (Figure 6-85).

Accelerator Pedal Position Sensor (OBD II)

The accelerator pedal position (APP) sensor is comprised of three individual sensors in one case. Each of these sensors has a dedicated ground, power, and sensor return wiring. The APP system is only used on those vehicles with "fly by wire" throttle control systems which use an electric motor instead of a throttle cable to control the accelerator.

Each sensor has a unique voltage curve. Sensor 1 goes from below 1 volt at closed throttle to about 2 volts at wide-open throttle. Sensor 2 uses a signal that decreases from about 4 volts at closed throttle to 2.9 volts at wide-open throttle. Sensor 3 voltage is from 3.8 volts at closed throttle to about 3.1 volts at wide-open throttle. See Figure 6-86.

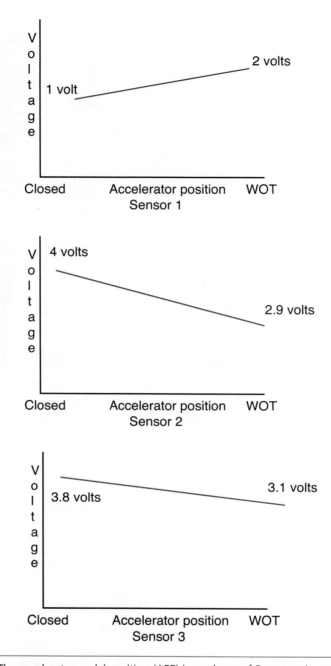

Figure 6-86 The accelerator pedal position (APP) is made up of 3 sensors in one housing.

Fuel Tank Pressure Sensor (OBD II)

The fuel tank pressure sensor is used by the enhanced emission system to detect fuel tank pressure and vacuum. The system is used by the EVAP system to monitor for leaks in the sealed system. The fuel tank pressure sensor voltage varies between 0 and 5 volts depending on the amount of pressure in the tank. The voltage decreases as fuel tank pressure increases on most vehicles. The MIL is armed the first time a failure is detected and is illuminated on the second consecutive failure. See Figure 6-87.

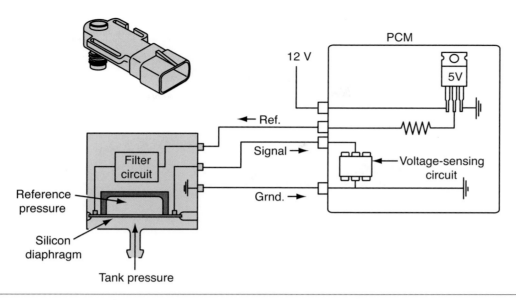

Figure 6-87 The fuel tank pressure sensor circuit.

CASE STUDY

A customer brought their car in for a severe hesitation on acceleration. The customer had already had the fuel injectors cleaned at another shop, but the problem still existed. After confirming the problem, the technician checked the fuel pressure (which was also in specification), and a quick visual check did not give any clues to the cause of the problem.

The technician decided to take the car for a test drive with the scan tool. Using the scan tool's snapshot function to avoid trying to read the scan tool and drive, the technician duplicated the condition again and stored the event for reading in the shop.

On returning to the shop, the technician began to analyze the information he had gathered on the test drive. He watched the TP sensor reading rise as the vehicle was accelerated, and the readings looked normal. Next, he watched the TP sensor and oxygen sensors together and noticed that on acceleration, the mixture was running very lean. The oxygen sensor was suspect, but the technician noticed that it seemed to have a good range of activity and switched normally at idle.

Next, the technician checked the MAF sensor against the TP sensor and noticed that there was very little rise in air flow shown under hard acceleration. The technician had already checked the air filter and the inlet hose during his visual inspection but decided to check the MAF sensor more closely. When the MAF sensor was removed from the vehicle, it was noticed that the hot wire was contaminated. When the hot wire was carefully cleaned, the vehicle was test driven and the problem was corrected.

The technician explained to the customer that the contamination had insulated the MAF sensor hot wire and thus the MAF sensor was slow to pick up on a sudden increase in airflow. Without the correct input, the computer did not add enough fuel to support a heavy acceleration. The PCM did not set a trouble code because the MAF readings did not go beyond the specified range.

The technician did an efficient job of repair by using the large amount of information available on a modern vehicle. More importantly, the technician was able to use the information to repair the vehicle.

Terms to Know

Data link connector (DLC)

Engine coolant temperature (ECT) sensor

Hot wire MAF sensor

Key on, engine off (KOEO)

Key on, engine running (KOER)

Manifold absolute pressure (MAP) sensor

Mass air flow (MAF) sensor

Reference voltage

Room temperature vulcanizing (RTV) sealant

Vehicle speed sensor (VSS)

Wiggle test

ASE-Style Review Questions

1. While discussing O$_2$ sensor diagnostics:
   Technician A says the speed of switching rich to lean is important.
   Technician B says the voltage levels attained is also important.
   Who is correct?
   A. A only
   B. B only
   C. Both A and B
   D. Neither A nor B

2. While discussing ECT sensor diagnosis:
   Technician A says a defective ECT sensor may cause hard cold engine starting.
   Technician B says a defective ECT sensor may cause improper operation of emission control devices.
   Who is correct?
   A. A only
   B. B only
   C. Both A and B
   D. Neither A nor B

3. While discussing trouble codes:
   Technician A says hard code failures are those that have occurred in the past, but were not present during the last test of the PCM.
   Technican B says intermittent codes are those that were detected the last time the PCM tested the circuit.
   Who is correct?
   A. A only
   B. B only
   C. Both A and B
   D. Neither A nor B

4. While discussing the testing of sensors:
   Technician A says sensors can be tested with a scan tool.
   Technician B says a breakout box can be used to test sensor outputs.
   Who is correct?
   A. A only
   B. B only
   C. Both A and B
   D. Neither A nor B

5. While discussing magnetic pulse generators tests:
   Technician A says use an ohmmeter to test the resistance of the coil.
   Technician B says the voltage generated by the sensor can be measured by connecting a voltmeter across the sensor's terminals.
   Who is correct?
   A. A only
   B. B only
   C. Both A and B
   D. Neither A nor B

6. *Technician A* says installing a PROM chip in backwards will immediately destroy the chip.
   Technican B says electrostatic discharge will destroy the chip.
   Who is correct?
   A. A only
   B. B only
   C. Both A and B
   D. Neither A nor B

7. While discussing TP sensor diagnosis:
   Technician A says a four-wire TP sensor contains an idle switch.
   Technician B says in some applications, the TP sensor mounting bolts may be loosened and the TP sensor housing rotated to adjust the voltage signal with the throttle in the idle position.
   Who is correct?
   A. A only
   B. B only
   C. Both A and B
   D. Neither A nor B

8. While discussing airflow sensor diagnosis:
   Technician A says on a vane-type airflow sensor, the voltage signal should be checked as the vane is moved from fully closed to fully open.
   Technician B says on a vane-type sensor, the voltage signal should show a smooth transition as the vane is opened.
   Who is correct?
   A. A only
   B. B only
   C. Both A and B
   D. Neither A nor B

9. While discussing flash code diagnosis:
Technician A says in a General Motors MFI system, the check engine light flashes each fault code four times.
Technician B says in a General Motors MFI system, terminals A and D must be connected in the DLC to obtain the fault codes.
Who is correct?
A. A only   C. Both A and B
B. B only   D. Neither A nor B

10. While discussing flash code diagnosis:
Technician A says the check engine light flashes fault codes in numerical order.
Technician B says if a fault code is set in the computer memory, the system may be in the limp-in mode.
Who is correct?
A. A only   C. Both A and B
B. B only   D. Neither A nor B

ASE Challenge Questions

1. The code or symbol shown above depicts:
   A. danger of electrical shock.
   B. non-conductive footwear and gloves should be worn.
   C. anti-static measures and devices should be utilized.
   D. high voltage is present.

2. Electronic engine controls are being discussed.
   Technician A says problems in these systems are more likely caused by the mechanical portion of output devices.
   Technician B says a mechanical malfunction may result in a DTC being stored for an unrelated sensor.
   Who is correct?
   A. A only   C. Both A and B
   B. B only   D. Neither A nor B

3. Electronic engine control system diagnosis is being discussed.
   Technician A says a breakout box is needed to make pinpoint tests on most vehicles.
   Technician B says DTCs can be retrieved without a scan tool on most vehicles.
   Who is correct?
   A. A only   C. Both A and B
   B. B only   D. Neither A nor B

4. The idle tracking switch is being discussed.
   Technician A says a poor idle may be caused by a high voltage signal present when the switch is open.
   Technician B says an infinity ohmmeter reading at part throttle indicates a defective switch.
   Who is correct?
   A. A only   C. Both A and B
   B. B only   D. Neither A nor B

5. A low voltage on a sensor's reference wire could be caused by any of the following *except*:
   A. poor computer ground.
   B. shorted reference wire.
   C. high resistance in the sensor.
   D. excessive voltage drop on sensor ground.

Job Sheet 14

Name _____ Date _____

Retrieve Codes from the Computer of an OBD I Engine Control System

Upon completion of this job sheet, you should be able to retrieve diagnostic trouble codes from a typical computerized engine control system.

ASE Correlation

This job sheet is related to the ASE Engine Performance Test's content area: computerized engine controls diagnosis and repair; task: retrieve and record stored diagnostic codes.

Tools and Materials

A Chrysler product with flash code capability
Appropriate service manual

Describe the vehicle being worked on.

Year _____ Make _____ VIN_____
Model _____

Procedures

                                                                    Task Completed

1. Block the drive wheels.   ☐
2. Apply the parking brake and start the engine.   ☐
3. Press the brake pedal. Move the shift lever through all positions and return it to park.   ☐
4. Turn the A/C on and off if the vehicle is so equipped.   ☐
5. Turn the engine off.   ☐
6. Turn the ignition switch on-off, on-off, and on.   ☐
7. Count the flashes of the MIL and record the DTCs.

   _____
   _____

8. What is indicated by the flash codes?

   _____
   _____
   _____

Instructor's Response_____

_____
_____
_____

Job Sheet 15

15

Name _____ Date _____

Test an ECT Sensor

Upon completion of this job sheet, you should be able to check the operation of an engine coolant temperature sensor.

ASE Correlation

This job sheet is related to the ASE Engine Performance Test's content area: computerized engine controls diagnosis and repair; task: inspect, test, adjust, and replace computerized engine control system sensors, powertrain control module, actuators, and circuits.

Tools and Materials

DMM

Describe the vehicle being worked on.

Year _____ Make _____ VIN _____

Model _____

Procedures

Task Completed

1. Describe the location of the ECT sensor.

   _____

   _____

   _____

2. What color of wires are connected to the sensor?

   _____

   _____

3. Record the resistance specifications for a normal ECT sensor for this vehicle.

   _____

   _____

   _____

   _____

4. Disconnect the electrical connector to the sensor. ☐

5. Measure the resistance of the sensor _____ ohms at approximately _____°F.

6. Conclusions: _____
_____
_____
_____
_____
_____

Instructor's Response_____
_____
_____
_____

Job Sheet 16

16

Name _____ Date _____

Check the Operation of a TP Sensor

Upon completion of this job sheet, you should be able to test the operation of a throttle position sensor with a variety of test instruments.

ASE Correlation

This job sheet is related to the ASE Engine Performance Test's content area: computerized engine controls diagnosis and repair; task: inspect, test, adjust, and replace computerized engine control system sensors, powertrain control module, actuators, and circuits.

Tools and Materials

DMM   Lab scope

Describe the vehicle being worked on.

Year_____ Make _____ VIN_____

Model_____

Procedures

Task Completed

(Perform test with the engine off.)

1. Connect the lab scope across the TP sensor. ☐

2. With the ignition on, move the throttle from closed to fully open and then allow it to close slowly. ☐

3. Observe the trace on the scope while moving the throttle. Describe what the trace looked like.

   _____

   _____

   _____

4. Based on the waveform of the TP sensor, what can you tell about the sensor?

   _____

   _____

5. With a voltmeter, measure the reference voltage to the TP sensor. The reading should be _____ volts. The reading is _____ volts.

6. What is the output voltage from the sensor when the throttle is closed? _____ volts

7. What is the output voltage of the sensor when the throttle is opened? _____ volts

8. Move the throttle from closed to fully open and then allow it to close slowly. Describe the action of the voltmeter.

   _____

   _____

9. Conclusions:

_____
_____
_____
_____

Instructor's Response_____

_____
_____
_____

Job Sheet 17

17

Name _____ Date _____

Test an O₂ Sensor

Upon completion of this job sheet, you should be able to test an oxygen sensor with a voltmeter and a lab scope.

ASE Correlation

This job sheet is related to the ASE Engine Performance Test's content area: computerized engine controls diagnosis and repair; task: inspect, test, adjust, and replace computerized engine control system sensors, powertrain control module, actuators, and circuits.

Tools and Materials

DMM   Scan tool
Propane torch

Describe the vehicle being worked on.

Year _____ Make _____ VIN _____

Model _____

Procedures

Task Completed

1. Connect the voltmeter between the O_2 sensor wire and ground. Backprobe the connector near the O_2 sensor to connect the voltmeter to the sensor signal wire. ☐

2. With the engine idling, record and describe the voltmeter readings:

   _____

   _____

3. What does this test tell you about the sensor?

   _____

   _____

4. Remove the sensor from the exhaust manifold. ☐

5. Connect the voltmeter between the sensor wire and the case of the sensor. ☐

6. Using a propane torch, heat the sensor element. ☐

7. Observe and record the voltmeter reading:

   _____

   _____

8. Your conclusions from this test:

   _____

   _____

9. Backprobe the sensor signal wire at the computer and connect a digital voltmeter from the signal wire to ground with the engine idling.

10. Record the voltmeter readings:

_____

_____

11. Connect the voltmeter from the sensor case to the sensor ground wire on the computer.

12. Record the voltmeter readings:

_____

_____

13. Your conclusions from these two tests:

_____

_____

14. Connect the scan tool to the DLC.

15. Observe and record what happens to the voltage reading from the sensor.

_____

_____

16. How many cross counts were there? _____

17. Your explanation of the above test:

_____

_____

18. Connect a lab scope to the sensor and observe the trace.

19. Observe and record what happens to the voltage reading from the sensor.

_____

_____

20. How many cross counts were there? _____

21. Your explanation of the above test:

_____

_____

Instructor's Response _____

_____

_____

Job Sheet 18

18

Name _____ Date _____

Testing a MAP Sensor

Upon completion of this job sheet, you should be able to test a manifold absolute pressure sensor in a variety of ways.

ASE Correlation

This job sheet is related to the ASE Engine Performance Test's content area: computerized engine controls diagnosis and repair; task: inspect, test, adjust, and replace computerized engine control system sensors, powertrain control module, actuators, and circuits.

Tools and Materials

Hand-operated vacuum pump
Lab Scope
DMM

Describe the vehicle being worked on.

Year _____ Make _____ VIN _____

Model _____

Procedures

Task Completed

1. If the MAP sensor produces an analog voltage signal, follow this procedure. ☐

2. With the ignition switch on, backprobe the 5-volt reference wire. ☐

3. Connect a voltmeter from the reference wire to ground. The reading is _____ volts.

4. If the reference wire is not supplying the specified voltage, what should be checked next?
   _____

5. With the ignition switch on, connect the voltmeter from the sensor ground wire to the battery ground.

6. What is the measured voltage drop? _____ volts

7. What does this indicate?
   _____
   _____

8. Backprobe the MAP sensor signal wire and connect a voltmeter from this wire to ground with the ignition switch on. ☐

9. What is the measured voltage? _____ volts

10. What does this indicate?
    _____
    _____

11. How do you determine the barometric pressure based on these voltage readings?
    _____
    _____

☐

12. Turn the ignition switch on and connect a voltmeter to the MAP sensor signal wire.

13. Connect a vacuum hand pump to the MAP sensor vacuum connection and apply 5 inches of vacuum to the sensor. Record the voltage reading: _____ volts

Note: On some MAP sensors, the sensor voltage signal should change 0.7 to 1.0 volt for every 5 inches of vacuum change applied to the sensor. Always use the vehicle manufacturer's specifications. If the barometric pressure voltage signal was 4.5 volts with 5 inches of vacuum applied to the MAP sensor, the voltage should be 3.5 to 3.8 volts. When 10 inches of vacuum is applied to the sensor, the voltage signal should be 2.5 to 3.1 volts. Check the MAP sensor voltage at 5-inch intervals from 0 to 25 inches.

If the MAP sensor voltage is not within specifications at any vacuum, replace the sensor.

14. Record the results of all vacuum checks:

_____

_____

15. What did these tests indicate?

_____

_____

☐

16. Connect the scope to the MAP output and a good ground.

17. Accelerate the engine and allow it to return to idle. Observe and describe the trace:

_____

_____

18. What did the trace show about the sensor?

_____

_____

Note: If the MAP sensor produces a digital voltage signal of varying frequency, check the 5-volt reference wire and the ground wire with the same procedure used on other MAP sensors. This sensor diagnosis is based on the use of a MAP sensor tester that changes the MAP sensor varying frequency voltage to an analog voltage. Follow these steps to test the MAP sensor voltage signal:

☐ 1. Turn the ignition switch off and disconnect the wiring connector from the MAP sensor.

☐ 2. Connect the connector on the MAP sensor tester to the MAP sensor.

☐ 3. Connect the MAP sensor tester battery leads to a 12-volt battery.

☐ 4. Connect a pair of digital voltmeter leads to the MAP tester signal wire and ground.

5. Turn the ignition switch on and observe the barometric pressure voltage signal on the meter. Observe and record the voltmeter readings:

_____

_____

☐ 6. Supply the specified vacuum to the MAP sensor with a hand vacuum pump.

7. Observe the voltmeter reading at each specified vacuum. Record the readings:

_____

_____

8. What do these readings indicate?

_____

_____

Instructor's Response_____

_____

_____

Servicing Computer Outputs and Networks

Upon completion and review of this chapter, you should be able to:

❏ Diagnose problems with high-side and low-side drivers.

❏ Troubleshoot computer output drivers.

❏ Diagnose computer output actuators.

❏ Diagnose output drivers in the operation of fuel injectors.

❏ Describe computer control of the IAC system.

❏ Diagnose computer-operated EGR valves.

❏ Diagnose vehicle network module function.

Introduction

In this chapter, we will look at diagnosing computer outputs and vehicle networks. As computers continue to become more prevalent in vehicle design, the service technician should not feel overwhelmed. Understanding computer operation is an essential skill. We actually have more information and diagnostic capability than we would have imagined a short time ago. Always make sure that you receive an accurate description of the problem from the customer.

Diagnosing Computer Outputs

Computer output problems can be of three different sources: wiring problems, actuator problems, or computer problems. **Computer output circuits** are just as vulnerable to defects as computer inputs. Make sure you have an accurate wiring diagram, check for technical bulletins, and as always, double-check power and ground circuits. The **actuator** can, of course, be at fault, and you should be able to test the component. In the case of a computer problem, your fault might be caused by a bad input that is preventing the PCM from activating a component, or the computer itself could be defective. In this chapter we will concentrate on actuator and computer problems.

The most common computer-controlled devices are:

❏ EGR solenoid

❏ Canister purse solenoid

❏ Fuel injectors (which are also specialized solenoids)

❏ Fuel pump and A/C relays

❏ Idle speed control motor

❏ A/C controller

❏ Cooling fan module or relay

❏ MIL and shift indicator

❏ Automatic transmission solenoids and pressure regulators

Most systems allow for testing of the actuator through a scan tool. Actuators that are duty cycled by the computer are more accurately diagnosed through this method. Prior to diagnosing an actuator, make sure the engine's compression, ignition system, and intake system is in good condition. Using a scanner, serial data can be used to diagnose outputs. The displayed data should be compared against specifications to determine the condition of any actuator. Also, when an actuator is suspected as being faulty, make sure the inputs related to the control of that actuator

Classroom Manual
Chapter 7, page 196

Actuators perform the actual work commanded by the PCM. They can be in the form of a motor, relay, switch, or solenoid.

Remember an actuator can be electrically or mechanically bad.

are within normal range. Faulty inputs will cause an actuator to appear faulty. Some examples of this are:

1. A TP sensor signal that is lower than specifications at idle speed may cause the IAC motor counts to be higher than normal at idle, resulting in a higher than desired idle speed.

2. A TP sensor signal that is higher than specifications at idle speed may cause the IAC motor counts to be lower than normal at idle, resulting in a lower than desired idle speed.

3. A low or lean biased O_2 signal will cause the PCM to richen the mixture. This will decrease fuel economy.

4. A high or rich biased O_2 signal will cause the PCM to lean out the mixture. This will result in engine surging and a hesitation during acceleration.

5. An ECT sensor signal that indicates a coolant temperature well below the actual temperature may increase injector pulse width and spark advance. It may also have a negative effect on emissions by delaying the operation of the EGR, radiator cooling fan, and/or canister purge.

If the actuator is tested by other means than a scanner, always follow the manufacturer's recommended procedures. Because many actuators operate with 5 to 7 volts, never connect a jumper wire from a 12-volt source unless directed to do so by the appropriate service procedure. Some actuators are easily tested with a voltmeter by testing for input voltage to the actuator. If there is the correct amount of input voltage, check the condition of the ground. If both are in good condition, the actuator is faulty. If an ohmmeter needs to be used to measure the resistance of an actuator, disconnect it from the circuit first.

When checking components with an ohmmeter, logic can dictate good and bad readings. If the meter reads infinity, there is an open. Based on what is being measured across, an open could be good or bad. The same is true for very low resistance readings. Across some components, this would indicate a short. For example, there should not be an infinity reading across the windings of a solenoid. It should be low resistance. However, an infinity reading from one winding terminal to the case of the solenoid is normal. If the resistance is low, the winding is shorted to the case.

On late-model vehicles, more high-side drivers are being used.

Testing Actuators with a Lab Scope

Most computer-controlled circuits are ground controlled circuit. The PCM energizes the actuator by providing the ground. On a scope trace, the on-time pulse is the downward pulse. On positive-feed circuits, where the computer is supplying the voltage to turn a circuit on, the on-time pulse is the upward pulse (Figure 7-1). One complete cycle is measured from one on-time pulse to the beginning of the next on-time pulse.

Actuators are electromechanical devices, meaning they are electrical devices that cause some mechanical action. Actuators are faulty because they are either electrically or mechanically

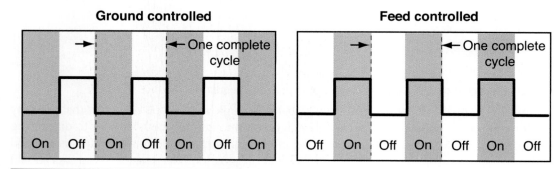

Figure 7-1 The difference between ground-controlled (low-side) and power controlled (high-side) actuator traces.

faulty. By observing the action of an actuator on a lab scope, you will be able to watch its electrical activity. Normally, if there is a mechanical fault, this will affect its electrical activity as well. Therefore, you have a good sense of the actuator's condition by watching it on a lab scope.

To test an actuator, you need to know what it is. Most actuators are solenoids. The computer controls the action of the solenoid by controlling the pulse width of the control signal. By watching the control signal, you can see the turning on and off of the solenoid (Figure 7-2). The voltage spikes are caused by the discharge of the coil in the solenoid.

Some actuators are controlled pulse width modulated signals (Figure 7-3). These signals show a changing pulse width. These devices are controlled by varying the pulse width, signal frequency, and voltage levels.

Both waveforms should be checked for amplitude, time, and shape. You should also observe changes to the pulse width as operating conditions change. A bad waveform will have noise, glitches, or rounded corners. You should be able to see evidence that the actuator immediately turns off and on according to the commands of the computer.

A fuel injector is actually a solenoid. The PCM's signals to an injector vary in frequency and pulse width. Frequency varies with engine speed, and the pulse width varies with fuel control. Increasing an injector's on time increases the amount of fuel delivered to the cylinders. The trace of a normally operating fuel injector is shown in Figure 7-4.

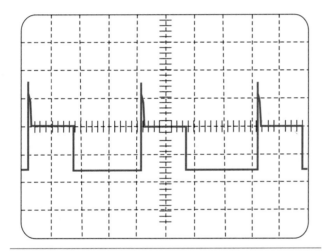

Figure 7-2 A typical solenoid control signal.

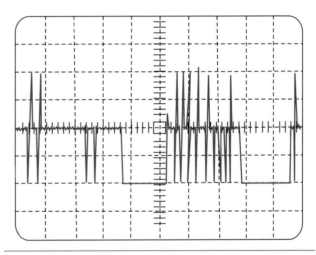

Figure 7-3 A typical pulse width modulated solenoid control signal.

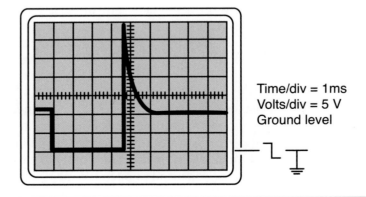

Time/div = 1ms
Volts/div = 5 V
Ground level

Figure 7-4 The trace of a normally operating fuel injector.

317

Testing an Actuator With a Scan Tool

The scan tool has many capabilities, but one of the best diagnostic uses of a scan tool comes from its **bidirectional control.** Bidirectional control means that the technician can tell a computer to exercise an output without the regular required input. Some bidirectional controls are not available on all scan tools. The most capable bidirectional scan tools are the manufacture-dedicated tools; it seems to take awhile for some of these functions to find their way into popular scan tools used in the aftermarket. For example, we will look at a Jeep's output control in Photo Sequence 14. Of course, output controls are different for each make and model vehicle, so do not assume that every vehicle has all these options.

Serial Data

Serial data is a term used to describe module to module or computer to computer communication. Basically scan tools "speak the language" of the computer system installed on the vehicle. Just like spoken language, which is often termed a protocol, there are many different types of languages in use depending on the manufacturer's preference or model year. These different protocols are the reason non-manufacturer specific scan tools need Asian, Domestic, or European cartridges to decode data.

There were several types of serial data used for OBD II communication, such as Class B, ISO9141, SCP, Class 2, and KWP2000. The communication types vary somewhat between and within manufacturers, so there are no hard and fast rules until CAN or (controller area network) becomes the mandatory standard.

❏ The earliest standard was called UART for universal asynchronous receive and transmit or Class A. UART was very slow but still used for some communications.

❏ Class B, which is also called J1850 is used by most manufacturers. GM called their Class B communication Class 2. Ford has a special faster version of Class B that they call SCP for standard corporate protocol.

❏ ISO 9141 is a Class B communication that is only used for a scan tool to communicate with the modules and is not used for module to module communication on the vehicle. ISO9141 is usually associated with the Japanese but is also used by Ford, Chrysler and GM.

❏ KWP2000 is usually associated with European manufacturers, but once again is used by GM and others as well.

Idle Air Control

The computer is in control of idle speed by varying the amount of air past the throttle plate with a stepper motor called the IAC (Figure 7-5). The stepper motor can be positioned in 'steps' by the PCM. Naturally, the idle is based on coolant temperature, TPS, and other sensors depending on make and model. If the IAC is suspected of causing a problem or being inoperable, the valve can usually be commanded open and closed by the scan tool. This would allow the technician to check the IAC through the range of operation. If the IAC valve responds properly, then the technician would need to look for a problem with a PCM input that would cause an idle problem, a minimum air plug that had been tampered with, or a vacuum leak.

EGR

Vehicles with digital or linear EGR valves can be commanded open with the scan tool allowing the technician to perform a functional check of the system very quickly. Usually, the EGR valve is opened while the vehicle is at idle. This results in an immediate rough idle if the system is operating normally. Additionally, if the system uses a pintle position sensor, then the technician can see the results of the command to open the valve. The pintle position sensor tells the PCM that its command for EGR is being met. If the valve does not open, then the technician would check the EGR valve wiring for response when commanded by the PCM. If the harness check showed the EGR valve was receiving commands but not opening, then the EGR valve is faulty. If

Photo Sequence 14
Testing Computer Outputs Using a Scan Tool

P14-1 Connect the scan tool, enter the vehicle information, then go to the main menu and select 'functional tests.'

P14-2 For this test, we selected the auto shutdown relay.

P14-3 The ASD (auto shut down) relay can be heard clicking in the fuse block.

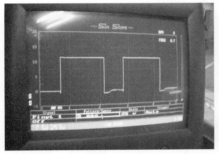

P14-4 The output of the relay can be graphed on the oscilloscope.

P14-5 Multimeter reading of relay output.

P14-6 The AIS (IAC) motor can be commanded to go fully closed (minimum air setting) operated in steps open or closed, or cycled fully in and out.

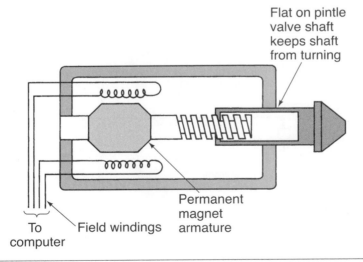

Figure 7-5 The IAC controls idle by varying the amount of airflow past the throttle plate and is controlled by PCM outputs.

the commands are not received, then the PCM or wiring is faulty. If the valve shows opening as commanded, then the valve would be removed for inspection of the passages for blockage.

Computer Networks

Vehicle networking has become common and will continue to be as technological advances are made. The use of many modules all connected to a common data bus allows for the use of smart modules that have diagnostic capabilities of their own. The modules have to use a common communication protocol or language. The most commonly used today is **J1850**. Ford calls their communication method **SCP** (for **Standard Corporate Protocol**) but does not use it for all of their modules. GM calls their J1850 communication **Class 2. ISO 9141** is used by Ford for some modules and in Japanese vehicles and KWP 2000 for European imports. Figure 7-6 shows

If network wiring is repaired, make certain to maintain the length as close as possible to the original.

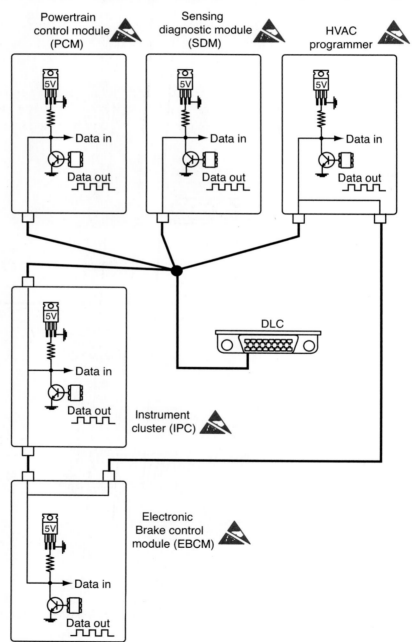

Figure 7-6 GM's (Class 2) communication system.

a GM Class 2 network. The new CAN protocol is mandatory for communication beginning in the year 2008. Figure 7-7 diagrams the pins used in the DLC. A diagram of the Ford ISO 9141 modules is shown in Figure 7-8, and the SCP or J1850 is shown in Figure 7-9. A vehicle's communication method or methods could be determined by observation of the pins in the DLC. Compare Figure 7-7 to Figure 7-8 and Figure 7-9.

Ford PATS

For one example of a networked module function, we will examine the function of Ford's **Passive Anti-Theft System,** or **(PATS)**, from a Crown Victoria. Figure 7-10 shows a wiring diagram for the system, which consists of a control module, transceiver, and PATS indicator.

Each key for the PATS-equipped vehicle must be digitally encoded into the PATS module. There is a transceiver with an antenna located in the steering column that reads the key in the ignition. The PATS module and the PCM communicate with each other using sophisticated commands. Diagnosis of the system begins with ensuring the key is properly coded and not damaged. Of course, if one of the modules or PCM has a concern or thee is a wiring problem, the vehicle will not start. A **New Generation Star (NGS) tester** is required to test the PATS system. The trouble codes for the PATS system are shown in Figure 7-11. Note the U codes are for network problems. It has been found that certain fast-pay keychain devices might have an adverse effect on the PATS system and that customers should hold the rest of the keys away from the ignition cylinder when starting the vehicle to help prevent interference.

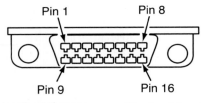

Pin 1: Manufacturer discretionary
Pin 2: J1850 bus positive
Pin 3: Manufacturer discretionary
Pin 4: Chassis ground
Pin 5: Signal ground
Pin 6: Can high
Pin 7: ISO 1941-2 "K" line
Pin 8: Manufacturer discretionary

Pin 9: Manufacturer discretionary
Pin 10: J1850 bus negative
Pin 11: Manufacturer discretionary
Pin 12: Manufacturer discretionary
Pin 13: Manufacturer discretionary
Pin 14: Can low
Pin 15: ISO 9141-2 "L" line
Pin 16: Battery power

Figure 7-7 Data links to the 16-pin DLC.

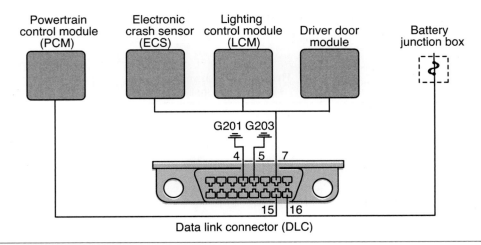

Figure 7-8 Ford ISO 9141 modules.

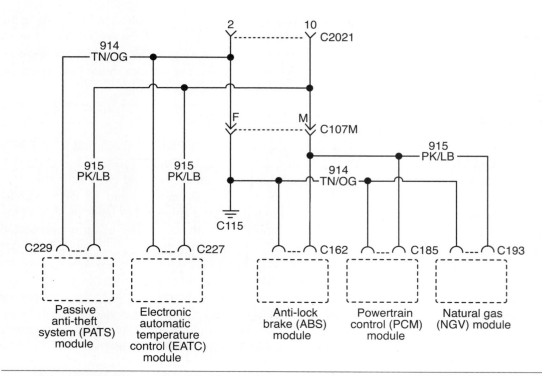

Figure 7-9 J1850 module diagram for a Ford product.

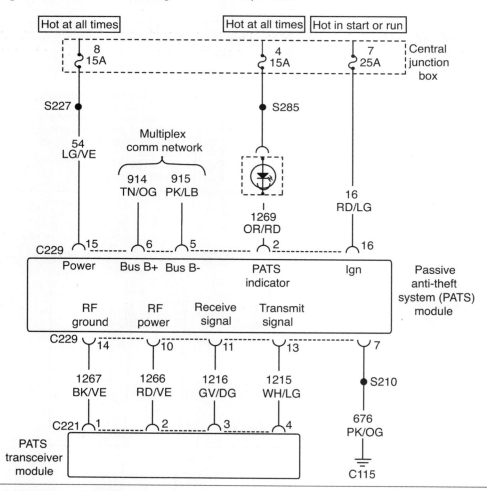

Figure 7-10 PATS wiring diagram.

PASSIVE ANTI-THEFT SYSTEM

DTC	Description
B1213	The number of programmed encoded keys is less than 2
B2103 or B1232	The antenna is not connected to the transceiver or the transceiver is defective
B1600	The key code was not received. The key is damaged or non-PATS
B1601	Encoded key—but not programmed
B1602	Only part of the key code was received
B1681	Signal not received from PATS transceiver. Possibly not connected, damaged, or wiring
B2139	PCM identification does not match between the PATS module and PCM
B2141	PATS and PCM did not exchange security identification
U1147	Faulty Standard Corporate Protocol link/PCM calibration incorrect
U1262	Standard Corporate Protocol message missing

Figure 7-11 PATS codes.

Electronic Automatic Temperature Control

The Ford **electronic automatic temperature control (EATC)** system is part of the SCP network. The PCM and the EATC module communicate with each other, and both these modules set codes for the A/C system. The system can perform a self-test as well as detect intermittent conditions and set them as run-time faults. The self-test can be run only with the A/C off; the run-time codes are set with the system active. The PCM can set codes that will shut down the A/C system. The EATC module will not record the PCM codes; to receive all related codes, it is necessary to use the NGS. Refer to Figure 7-12 for trouble codes from both the PCM and EATC module. A block diagram for the EATC system is shown in Figure 7-13.

New Generation Star Tester Codes	Module tests		Description
	Self-Test Faults	Intermittent Faults	
B1249	024	022 025	Blend door short Blend door failure
B1251	031	N/A	Open in car temperature sensor circuit
B1253	030	N/A	Grounded in car temperature sensor
B1255	041	043	Open ambient temperature sensor circuit
B1257	040	042	Ambient temperature sensor shorted to ground
B1261	050	052	Grounded sun radiation sensor
U1041	N/A	N/A	Vehicle speed data missing or invalid (SCP)
U1073	N/A	N/A	Engine coolant data missing or invalid (SCP)
U1222	N/A	N/A	Interior lamps data missing or invalid (SCP)

(PCM Diagnostic Trouble Code Index	
DTC	Description
P1460	Internal driver malfunction in wide-open throttle cut-out of A/C
P1469	A/C cycling too often
P1474	PCM internal driver for low-speed fan failed
P1479	PCM internal driver for high-speed fan failed
P1464	A/C demand switch detected on during self-test

Figure 7-12 EATC and PCM codes related to climate control.

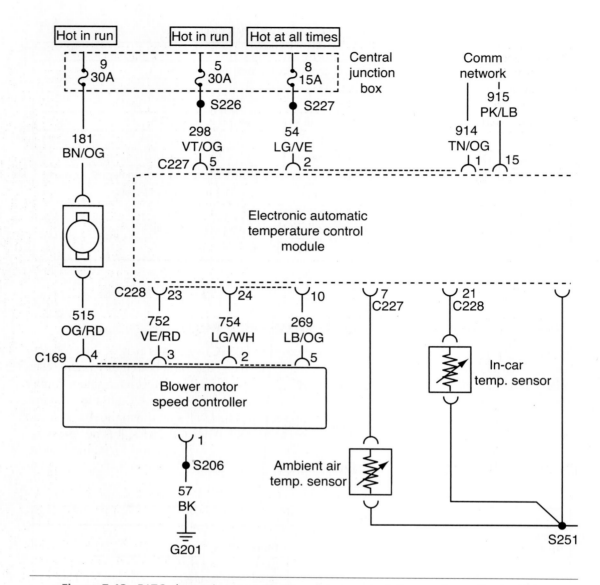

Figure 7-13 EATC electronic automatic temperature control diagram from a Ford vehicle.

Replacing Modules

Many modules have to be reprogrammed for the vehicle if they are to be replaced, just like the PCM. Make sure you have the proper tools to replace and reprogram a module before you begin. Another factor is the **customization** of certain features that the owner has put into the systems. One example of customization is the lighting modules; some customers like the dome light to stay on after the front door is closed, some do not. In some localities, the alarm cannot chirp the horn when it is set. When a module is replaced, make sure that you know how the customer had his system programmed if possible. Since it takes a scan tool to reset the customization features on some vehicles, the customer will have to make an extra trip to the shop if it is not done.

SHIFT SOLENOID OPERATION CHART

Transaxle range selector lever position	Powertrain Control Module Gear commanded	Eng braking	AX4N solenoids		
			SS 1	SS 2	SS 3
P/N[a]	P/N	No	Off[b]	On[b]	Off
R (Reverse)	R	Yes	Off	On	Off
	1	No	Off	On	Off
	2	No	Off	Off	Off
Overdrive	3	No	On	Off	On
	4	Yes	On	On	On
	1	No	Off	On	Off
D (Drive)	2	No	Off	Off	Off
	3	Yes	On	Off	Off
	2[c]	Yes	Off	On	Off
Manual 1	3[c]	Yes	Off	Off	Off
		Yes	On	Off	Off

a When transmission fluid temperature is below 50ß then SS 1 = Off, SS 2 = On, SS 3 = On to prevent cold creep.

b Not contributing to powerflow

c When a manual pull-in occurs above calibrated speed, the transaxle will downshift from the higher gear until the vehicle speed drops below this calibrated speed.

Figure 7-14 An example of solenoid activity during different gear ranges.

Automatic Transmission Controls

While this text is primarily concerned with engine performance issues we did feel the need to mention automatic transmissions in the discussion of electronically controlled components. Automatic transmissions are some of the heaviest users of electronic actuators on a late model vehicle. Most electronically shifted transmissions have the capability to display scan tool data and trouble codes just as the engine management system. Many transmissions can be shifted manually by commands from a scan tool or special transmission testers. This gives the technician the capability to determine whether a fault is in the transmission control system or the transmission. Additionally, the technician can see which gear ratio the computer has commanded and compare the command to the actual gear ratio. Refer to Figure 7-14 for a chart of solenoid usage for a Ford AX4N transmission.

Terms to Know

Actuators

Bidirectional control

Class 2

Computer output circuits

Customization

Electronic Automatic Temperature Control (EATC)

ISO 9141

J1850

New Generation Star (NGS) tester

Passive Anti-Theft System (PATS)

Standard Corporate Protocol (SCP)

Vehicle networking

ASE-Style Review Questions

1. *Technician A* says some scan tools have bidirectional controls to help diagnose computer problems. *Technician B* says the best scan tools for bidirectional control are aftermarket scan tools. Who is correct?
   - **A.** A only
   - **B.** B only
   - **C.** Both A and B
   - **D.** Neither A nor B

2. *Technician A* says that serial data is what the scan tool uses to communicate with the PCM and modules. *Technician B* says serial data can be used to diagnose outputs. Who is correct?
   - **A.** A only
   - **B.** B only
   - **C.** Both A and B
   - **D.** Neither A nor B

3. *Technician A* says the IAC valve controls the idle speed by varying the amount of air past the throttle plate. *Technician B* says the IAC valve is a stepper motor. Who is correct?
   - **A.** A only
   - **B.** B only
   - **C.** Both A and B
   - **D.** Neither A nor B

4. *Technician A* says that the passages for the EGR might be blocked if the engine runs rough when the EGR is commanded open at idle. *Technician B* says the pintle position sensor measures the location of the EGR pintle. Who is correct?
   - **A.** A only
   - **B.** B only
   - **C.** Both A and B
   - **D.** Neither A nor B

5. *Technician A* says the PATS system can be affected by certain fast-pay modules a customer may have on their keychain. *Technician B* says the transceiver and antenna is located in the steering column. Who is correct?
   - **A.** A only
   - **B.** B only
   - **C.** Both A and B
   - **D.** Neither A nor B

6. *Technician A* says that 'U' codes are undefined codes. *Technician B* says that 'U' codes are network codes. Who is correct?
   - **A.** A only
   - **B.** B only
   - **C.** Both A and B
   - **D.** Neither A nor B

7. *Technician A* says that Ford calls their J1850 communication language Class 2. *Technician B* says GM calls their J1850 communication system SCP. Who is correct?
   - **A.** A only
   - **B.** B only
   - **C.** Both A and B
   - **D.** Neither A nor B

8. *Technician A* says that actuators can be electrically or mechanically faulty. *Technician B* says that actuators are usually solenoids. Who is correct?
   - **A.** A only
   - **B.** B only
   - **C.** Both A and B
   - **D.** Neither A nor B

9. *Technician A* says that the EATC can set codes during a self-test for the A/C system. *Technician B* says the EATC codes take the place of any PCM codes for the A/C system. Who is correct?
   - **A.** A only
   - **B.** B only
   - **C.** Both A and B
   - **D.** Neither A nor B

10. *Technician A* says that the EATC can set intermittent codes with the system running. *Technician B* says that the EATC must have the A/C on to perform a self-test. Who is correct?
    - **A.** A only
    - **B.** B only
    - **C.** Both A and B
    - **D.** Neither A nor B

ASE Challenge Questions

1. *Technician A* says that when a ground-controlled (low-side) driver is turned off, the voltage trace on the oscilloscope at the actuator goes high.
   Technician B says that when a high-side driver-controlled actuator is turned off, the oscilloscope at the actuator goes low.
   Who is correct?
   A. A only   **C.** Both A and B
   B. B only   **D.** Neither A nor B

2. *Technician A* says that each key for the PATS system has to be coded into the PATS system.
   Technician B says the PATS and PCM share information.
   Who is correct?
   A. A only   **C.** Both A and B
   B. B only   **D.** Neither A nor B

3. *Technician A* says that the EATC can set intermittent codes with the system running.
   Technician B says that the EATC must have the A/C on to perform a self-test.
   Who is correct?
   A. A only   **C.** Both A and B
   B. B only   **D.** Neither A nor B

4. *Technician A* says that a linear or digital EGR valve can be commanded open with a scan tool.
   Technician B says that if the linear or digital EGR valve is commanded open at idle, the engine will run very rough or die, which verifies the operation of the EGR valve.
   Who is correct?
   A. A only   **C.** Both A and B
   B. B only   **D.** Neither A nor B

5. *Technician A* says that ISO 9141 is used only by the Japanese.
   Technician B says that SCP is used by European manufacturers.
   Who is correct?
   A. A only   **C.** Both A and B
   B. B only   **D.** Neither A nor B

Job Sheet 19

19

Name _____ Date _____

EGR Actuation

Upon completion of this job sheet, you should be able to diagnose emissions and driveability problems caused by malfunctions in the exhaust gas recirculation (EGR) system; determine necessary action. You must know how to perform this task to pass the ASE engine performance test.

Tools and Materials

A vehicle with an electronic EGR valve
Scan tool with bi-directional capability

Protective Clothing

Safety glasses

Describe the vehicle being worked on.

Year _____ Make _____ VIN _____

Model _____ Engine type and size _____

Task Completed

Procedures

1. Install the scan tool in the vehicle and go to the override selection. This may be called by several different names according to the scan tool manufacturer. Make sure to be familiar with your scan tool before performing this test. ☐

2. If possible, pull up the EGR percentage on the scan tool and activate the EGR valve and look for a change in the EGR valve opening. If this is not possible, with your equipment, look for a change in engine rpm as the valve is opened. ☐

3. If you were unable to command the EGR valve open with the scan tool, what cold be a possible cause?

   _____

   _____

4. If the scan tool showed a change in EGR position, but the engine did not show the effects (rough idle), what is a possible cause?

   _____

   _____

5. How could a technician use these tests with a scan tool to save time in diagnosis?

   _____

   _____

6. List problems encountered.

_____

_____

Instructor's Response_____

_____

_____

_____

Job Sheet 20

20

Name _____ Date _____

Computer Control of the Fuel Injector

Upon completion of this job sheet, you should be able to perform NATEF task of diagnosing enine mechanical, electrical, electronic, fuel, and ignition concerns with an oscilloscope and/or engine diagnostic equipment; determine necessary action. You must know how to perform this task to pass the ASE engine performance test.

Tools and Materials

Oscilloscope and test leads, fuel injected vehicle, service information for vehicle being tested

Protective Clothing

Safety glasses

Describe the vehicle being worked on.

Year _____ Make _____ VIN _____

Model _____ Engine type and size _____

Task Completed

Procedures

1. Bring vehicle into the shop, apply parking brake and wheel chocks. ☐

2. Using electronic service information or the appropriate service manual find the wiring diagram for the fuel injection circuit, in particular the injectors themselves. ☐

3. Connect the positive lead of the oscilloscope to the negative lead of the injector. The negative lead should be the one activated by the computer. Attach the negative lead of the oscilloscope to a good ground. ☐

4. Start the engine and adjust the oscilloscope to obtain the best view possible. Note: refer to the manufacturer's recommendations on setting up the oscilloscope and triggers. ☐

   Hint: Sometimes the best patterns are possible with the scope synchronized to cylinder one. Ask your instructor for help with the set up as needed. Using the oscilloscope will become easier with practice.

5. Once you have established a good signal, answer the following questions:

   Looking at the pattern, see what the voltage on the ground is during the "on" time of the injector.

   _____ mS

   Looking at the pattern, see what the voltage on the injector ground is during the "off" time of the injector. How do you explain the voltage you see?

   _____ V

   What is the approximate injector on time in milliseconds?

   _____ mS

Create a vacuum leak (large enough to force the engine to run rough) what is the injector "on time"? How do you explain what you see?

_____

_____

6. Reconnect the vacuum hose. Snap accelerate the engine from idle. Explain what happened to the injector on time on acceleration and then as the engine slowed down.

_____

_____

7. List problems encountered.

_____

_____

Instructor's Response_____

_____

_____

On-Board Diagnostic (OBD II) System Diagnosis and Service

Upon completion and review of this chapter, you should be able to:

❏ Explain how to logically approach diagnosing a problem in an OBD II system.

❏ Conduct preliminary checks on an OBD II system.

❏ Describe how a scan tool can be used in diagnostics.

❏ Use a symptom chart to set up a strategic approach to troubleshooting a problem.

❏ Define the terms associated with OBD II diagnostics.

❏ Identify the cause of an illuminated MIL.

❏ Explain the basic format of OBD II diagnostic trouble codes (DTCs).

❏ Monitor the activity of OBD II system components.

❏ Explain how to diagnose intermittent problems.

❏ Properly repair OBD II circuits.

OBD II regulations require that the powertrain control module (PCM) (Figure 8-1) monitor and perform some continuous tests on the emission control system and components. Some OBD II tests are completed at random, at specific intervals, or in response to a detected fault.

To perform the new strategies and tests on the emission control system, OBD II PCMs have diagnostic management software. The many diagnostic steps and tests required of OBD II systems must be performed under specific operating conditions referred to as **enable criteria.** The PCM's software organizes and prioritizes the diagnostic routines. The software determines if the conditions for running a test are present. Then it monitors the system for each test and records the results of these tests.

The PCM supplies a buffered low voltage to various sensors and switches. The input and output devices in the PCM include analog-to-digital converters, signal buffers, counters, and special drivers. The PCM controls most components with electronic switches that complete a ground circuit when turned on. These switches are arranged in groups of four and seven and are called one of the following: quad driver module or output driver module. The quad driver module can independently control up to four output terminals. The output driver module can independently control up to seven outputs.

Basic Tools

Scan tool, service manuals

The continuous tests by the PCM are also called active tests. Pre-OBD II tests are called passive tests because they wait for something to go wrong instead of looking for it.

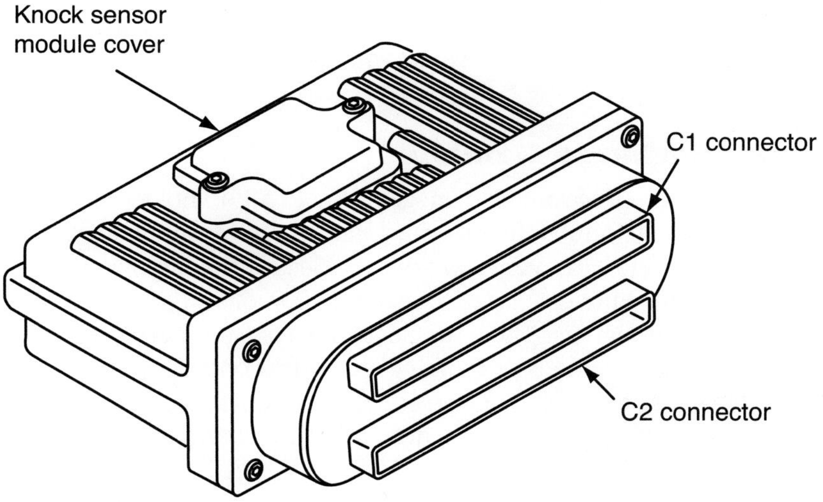

Knock sensor module cover

C1 connector

C2 connector

Figure 8-1 A typical OBD II PCM.

The PCM has a learning ability that allows the module to make corrections for minor variations in the fuel system in order to improve driveability. Whenever the battery cable is disconnected, the learning process resets. The driver may note a change in the vehicle's performance. In order to allow the PCM to relearn, drive the vehicle at part throttle with moderate acceleration. Some manufacturers have a specific idle relearn procedure that can speed the process. If available, this can be an important customer satisfaction tool.

Electronically erasable programmable read only memory (EEPROM) modules are soldered into the PCM. EEPROMs allow the manufacturer to update what is held in the PROM without replacing it. The PCM checks its internal circuits continuously for integrity. Using a check sum value unique to the program, it checks its EEPROM for accuracy of its data. It checks the actual values against what they are supposed to be and sets a code if they are different. Besides the hard-wired memory chip, the PCM also monitors its volatile keep alive memory. If that has been improperly changed or deleted, the PCM sets a code. This type of code will also be set if the vehicle's battery has been disconnected.

There is a continuous self-diagnosis on certain control functions. This diagnostic capability is complemented by the diagnostic procedures contained in the service manual. The language of communicating the source of the malfunction is a system of DTCs. When a malfunction is detected by the control module, a DTC will set and the MIL will illuminate.

The system monitoring diagnostic sequence is a unique segment of the software that is designed to coordinate and prioritize the diagnostic procedures as well as define the protocol for recording and displaying their results. The main diagnostic responsibilities of the PCM are:

❏ Monitoring the diagnostic test enabling conditions.
❏ Requesting the MIL.
❏ Illuminating the MIL.
❏ Recording pending, current, and history DTCs.
❏ Storing and erasing freeze-frame data.
❏ Monitoring and recording test status information.

The diagnostic tables and functional checks given in service manuals are designed to locate a faulty circuit or component through a process of logical decisions. The tables are prepared with the assumption that the vehicle functioned correctly at the time of assembly and that there are no multiple faults present.

CAUTION: The PCM is designed to withstand normal current draws associated with vehicle operation. Avoid overloading any circuit. When testing for opens or shorts, do not ground any of the PCM circuits unless instructed to do so. When testing for opens or shorts, do not apply voltage to any of the control module circuits unless instructed. Test these circuits with a lab scope or digital voltmeter only (Figure 8-2) while the PCM connectors remain connected to the PCM.

Electrostatic Discharge Damage

In order to prevent possible electrostatic damage to the PCM, do not touch the connector pins or the soldered components on the circuit board. Electronic components used in the control systems are often designed to carry very low voltage. Electronic components are susceptible to damage caused by electrostatic discharge. Less than 100 volts of static electricity can cause damage to electronic components. There are several ways for a person to become statically charged. The most common methods of charging are by friction and by induction. An example of charging by friction is a person sliding across a car seat. Charging by induction occurs when a person with well-insulated shoes stands near a highly charged object and momentarily

Classroom Manual
Chapter 8, page 218

GM calls the diagnostic function of the PCM the diagnostic executive because it is in charge of the diagnostic tests.

Special Tools

Lab scope, DMM

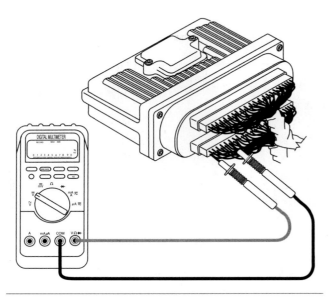

Figure 8-2 Using a digital voltmeter to check the PCM's circuit.

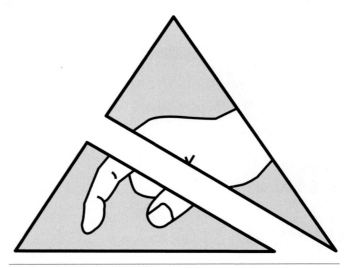

Figure 8-3 General Motors' electrostatic discharge (ESD) symbol warns technicians that a part or circuit is sensitive to static electricity.

touches ground. Charges of the same polarity are drained off leaving the person highly charged with the opposite polarity. Static charges can cause damage. Therefore, it is important to use care when handling and testing electronic components. Service manuals and wiring diagrams also carry the electrostatic discharge warning (Figure 8-3).

Aftermarket Electrical and Vacuum Equipment

In the design of a new car, there was no allowance made for add-on equipment. Therefore, do not install any add-on vacuum-operated equipment to new vehicles. Also, only connect the add-on electrically operated equipment to the vehicle's electrical system at the battery. Add-on equipment, even when installed under these guidelines, may still cause the powertrain system to malfunction. This may also include equipment not connected to the vehicle's electrical system such as portable telephones and radios. The first step in diagnosing any powertrain problem is to eliminate all aftermarket electrical equipment from the vehicle. If the problem still exists after this is done, diagnose the problem in the normal manner.

OBD II PCM

The OBD II PCM is similar in many respects to earlier PCMs but usually has more connectors and does not have a replaceable PROM. The PROM in the PCM is an EEPROM, or electrically erasable programmable read only memory chip, that can be reprogrammed without removing the PCM from the vehicle. This allows the updating of the PCM without replacing the PROM and means that a new PCM will have to be reprogrammed before it can be used in the vehicle. Some OBD II computers do have a chip that needs to be transferred to the new PCM, but this chip is a knock senor module as shown in Figure 8-1. The procedure to reprogram a PCM is explained in Photo Sequence 15. This procedure is not meant to replace the service information from vehicle manufacturers, but it is meant to be a guide only. Always consult vehicle manufacturer information before programming a PCM or any other procedure.

Aftermarket equipment is most commonly referred to as add-on equipment and includes anything installed on a vehicle after it has left the factory.

Only those procedures that are unique to OBD II systems are covered in this chapter. General testing of the engine and its major systems is done in the same way on OBD II systems as on non-OBD systems. Testing of individual components is important in diagnosing OBD II systems. Although the DTCs in these systems give much more detail on the problems, the PCM does not know the exact cause of the problem. That is the technician's job.

Photo Sequence 15
Reprogramming an OBD II PCM

P15-1 Technician is directed by a factory procedure or bulletin that the PCM must be reprogrammed or replaced.

P15-2 Before beginning the programming procedure, the battery voltage must be more than 12 volts. Charge the battery as needed, but never program a PCM with a battery charger on the vehicle.

P15-3 Make certain that unnecessary battery drains, such as blower motors, daytime running lights, A/C, etc., are switched off before beginning.

P15-4 Make sure the ignition switch is in the position directed by the scan tool. Do not turn the key on or off unless prompted to do so.

P15-5 Make certain that all connections are secure. A disconnection during programming may damage the PCM.

P15-6 Place the scan tool into the programming mode and enter the necessary information when requested.

P15-7 Make sure that the vehicle identification number (VIN) matches the VIN number of the vehicle. The VIN number gathered from the vehicle by the scan tool will ensure the proper calibration is downloaded into the scan tool at the computer terminal.

P15-8 The technician takes the scan tool back to the computer terminal to download calibration information.

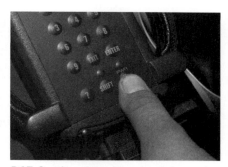

P15-9 After the calibration information is finished downloading to the PCM, turn off the scan tool BEFORE disconnecting it from the computer terminal.

Reprogramming an OBD II PCM (continued)

P15-10 The technician reinstalls the scan tool making certain all connections are secure; then turns the scan tool on and enters PCM programming mode.

P15-11 The technician then downloads the calibration to the vehicle's PCM following appropriate service procedures prescribed by the manufacturer of the scan tool and vehicle.

P15-12 After the download is complete, the technician should perform any necessary relearn procedures (such as idle learn) and drive the vehicle to confirm the repair.

OBD II Diagnostics

Diagnosing and servicing OBD II systems is much the same as diagnosing and servicing any other electronic engine control system. However, OBD II systems give more useful information than the previous designs, and, with OBD II, each make and model vehicle can have its own basic testing and servicing procedures.

OBD II systems note the deterioration of certain components before they fail, which allows owners to bring their vehicles in at their convenience and before it is too late.

OBD II monitors the vehicle's exhaust gas recirculation and fuel systems, oxygen sensors, catalytic converter performance, and other miscellaneous components. OBD II diagnosis is best done with a strategy based on the flow charts and other information given in the service manuals. Before beginning to diagnose a problem, verify the customer's complaint. In order to verify the complaint, a technician must know the normal operation of the system. Compare what the vehicle is doing to what it should be doing. Also, pay close attention to the vehicle's driveability. Sometimes there is another problem or symptom that the customer is not aware of or disregarded. This symptom may be the key to properly diagnosing the system. When diagnosing a vehicle with multiple symptoms, look for a problem that may be common or related to all complaints. By identifying the circuit of seemingly unrelated codes, common power and ground circuits may be discovered. This can be an excellent clue leading to a quick diagnosis.

Visual Inspection

After verifying the complaint (Figure 8-4), the next step is a careful visual inspection. Make sure all of the grounds, including the battery and computer ground, have clean and tight connections. Do a voltage drop test across all related ground circuits. The PCM and its systems cannot function correctly with a bad ground. A voltage drop of as little as 0.2 volts across a ground circuit can cause problems.

Check all vacuum lines and hoses, as well as the tightness of all attaching and mounting bolts in the induction system. Check for damaged air ducts. Check the ignition circuit, especially the secondary cables, for signs of deterioration, insulation cracks, corrosion, and looseness. Pay attention to any unusual noises or odors. These can be the hints needed to locate the fault.

Special Tools
DMM

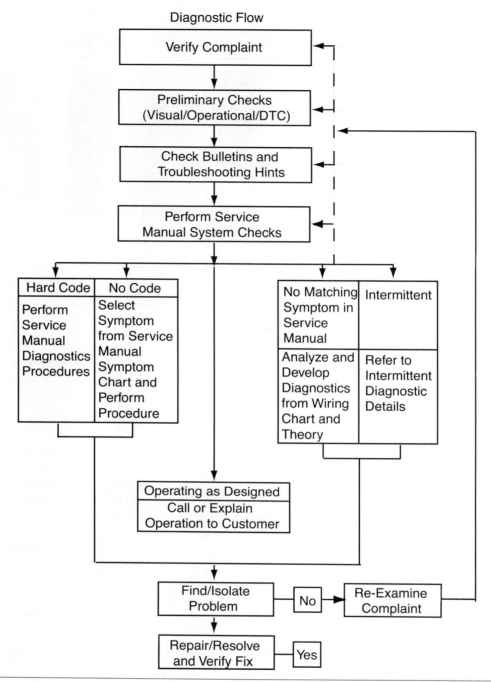

Figure 8-4 A strategy-based diagnostic tree.

Inspect all related wiring and connections at the PCM. These may cause an intermittent fault. Any circuitry that is suspected as causing an intermittent complaint should be thoroughly checked for the following conditions: backed-out terminals, improper mating, broken locks, improperly formed or damaged terminals, poor terminals to wiring connections, physical damage to the wiring harness, and corrosion.

▲ **WARNING:** Always block the drive wheels and set the parking brake while checking the system.

If the visual inspection did not identify the source of the customer's complaint, gather as much information as you can about the symptom. This should include a review of the vehicle's service history. Check for any published troubleshooting service bulletins (TSBs) relating to the exhibited symptoms. This should include videos, newsletters, and any electronically transmitted media. Do not depend solely on the diagnostic tests run by the PCM. A particular system may not be supported by one or more DTCs. System checks verify the proper operation of the system. This will lead the technician in an organized approach to diagnostics.

When a complaint cannot be isolated or the cause found, a reevaluation is necessary. The complaint should be reverified and could be intermittent or normal. Then check the following:

❑ Engine coolant temperature (ECT) sensor for initial coolant temperature reading close to ambient, then observe the rise in temperature while the engine is warming up.

❑ Throttle position (TP) sensor for proper sweep from 0 percent to 100 percent.

❑ Manifold absolute pressure (MAP) sensor for quick changes during changes in various engine loads.

❑ Oxygen sensor for proper rich-lean and lean-rich sweeps operation (Figure 8-5).

❑ Idle air control (IAC) valve for proper operation during warmup, which can be a crucial step in correctly diagnosing any driveability concern.

Careful observation of these sensors during engine warmup may reveal a slow responding sensor or a sensor that malfunctions only within a small portion of its range.

⚠️ **WARNING:** OBD II vehicles are probably equipped with air bags, possibly including driver and passenger-side air bags in addition to side curtain and rear seat systems. Before working around or on the circuits related to the air bags (Figure 8-6), make sure you refer to the precautions given in the service manual. Failure to follow these precautions can cause unexpected air bag deployment, which can cause injury and/or damage to the vehicle.

⬛ **CAUTION:** Never use a test light to diagnose powertrain electrical systems unless specifically instructed by the diagnostic procedures.

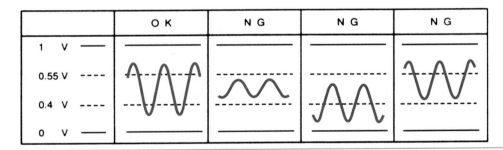

Figure 8-5 O₂ sensor activity should be checked with a lab scope.

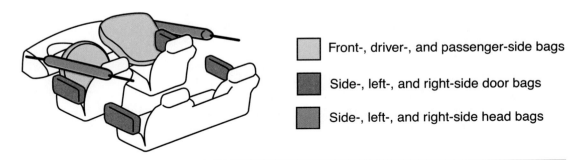

Front-, driver-, and passenger-side bags

Side-, left-, and right-side door bags

Side-, left-, and right-side head bags

Figure 8-6 The locations of various air bags.

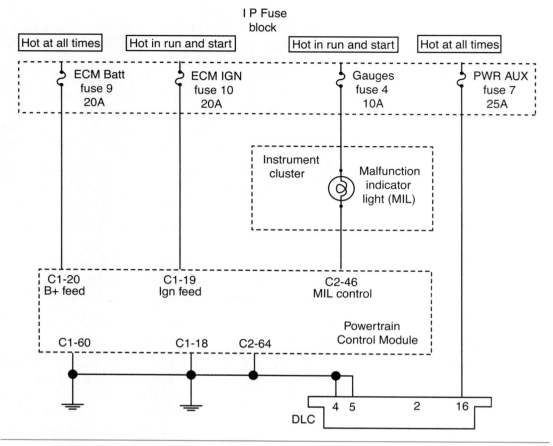

Figure 8-7 Wiring diagram for a powertrain OBD system check.

Diagnosis should continue with preliminary system checks. Each manufacturer will recommend that certain tests be completed at this stage. The purpose of this stage in diagnostics is to make sure the basic engine and its systems are in good shape. This includes checking for vacuum leaks, checking the engine's compression, ignition system, and air-fuel system.

GM recommends that a system check be conducted before proceeding with diagnosis. The powertrain OBD system check (Figure 8-7) is an organized approach to identifying a problem created by an electronic engine control system malfunction. The powertrain OBD system check is the starting point for any driveability complaint diagnosis. This directs the technician to the next logical step in diagnosing a complaint (Figure 8-8). Never perform this check if a driveability complaint does not exist. Understanding the system check table and using it correctly will reduce diagnostic time and prevent the replacement of good parts.

● **CUSTOMER CARE:** As simple as it may seem, advise your customers to tighten the fuel cap at least three to four clicks when reinstalling it after filling the fuel tank. Otherwise, the OBD II system may assume a problem exists within the system resulting from a poorly sealed fuel tank and turn on the MIL.

Scan Tester Diagnosis

When the ignition is initially turned on, the MIL will momentarily flash on then off and remain on until the engine is running or there are no DTCs stored in memory. Now connect the scan tool. If the scanner does not automatically detect the vehicle's VIN, enter the vehicle identifica-

Step	Action	Values	Yes	No
1	1. Turn the ignition switch ON with the engine lamp OFF. 2. Observe the Malfunction Indicator Lamp (MIL). Is the MIL on?	–	Go to Step 2	Go to No Malfunction Indicator Lamp
2	1. Turn the ignition switch OFF. 2. Install a scan tool. 3. Turn the ignition switch ON. Does the scan tool display PCM data?	–	Go to Step 3	Go to Data Link Connector Diagnosis
3	Using the scan tool, command the MIL OFF. Does the MIL turn OFF?	–	Go to Step 4	Go to Malfunction Indicator Lamp On Steady
4	Check for DTCs with the scan tool. Were any last test fail, history, or MIL request DTCs set?	–	Go to Step 5	Go to Step 6
5	Using the scan tool, record the freeze frame and failure records information. Is the action complete?	–	Go to applicable DTC table	–
6	Does the engine start and continue to run?	–	Go to Step 7	Go to engine cranks but does not run
7	1. Turn the ignition switch ON with the engine OFF. 2. Check the ECT and TP sensors for proper operation. 3. Start the engine. 4. Allow the running engine to reach operating temperature. 5. Allow the running engine to reach operating temperature, check the ECT, MAP, and O2S 1 sensors and IAC valve for proper operation. 6. Compare the scan tool data with the typical values shown in the Scan Tool Data. Are the display values normal or within typical ranges?	–	Go to symptoms	Go to the applicable diagnostic section

Figure 8-8 Powertrain OBD system check.

tion information (Figure 8-9), then retrieve the DTCs with the scan tool. The scan tool should be used to do more than simply retrieve codes. The information that is available, whether or not a DTC is present, can reduce diagnostic time. Some of the common uses of the scan tool are: identifying stored DTCs, clearing DTCs, performing output control tests, and reading serial data (Figure 8-10).

Depending on the scan tool being used, OEM-enhanced data and special diagnostic routines may be available in addition to generic OBD II data. The scan tester must have the appropriate connector to fit the DLC on an OBD II system (Figure 8-11) and the proper software for the vehicle being tested. Different inserts are available for the OBD II scan tester cable

Vehicle and engine selection
Diagnostic data link
View recorder areas
Digital measurement system
Generic OBD II functions

Figure 8-9 The main menu on a typical scan tool.

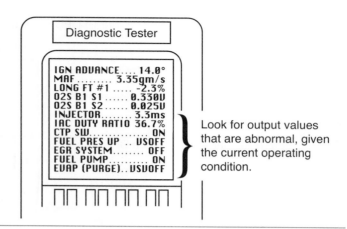

Figure 8-10 Use serial data to identify abnormal conditions.

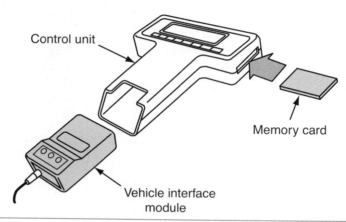

Figure 8-11 A scan tool with the necessary insert and adapter for use in an OBD II system.

depending on the vehicle being tested. These inserts are used to adapt a standard scan tool to the OBD system. Make sure the recommended insert for that make and model of vehicle is used in the scan tool. Always follow the instructions in the manuals supplied by the scan tester manufacturer. Many similar tests, such as the key on, engine off (KOEO) and key on, engine running (KOER) tests, are performed on OBD II systems as on previous systems.

Some OBD II systems flash DTCs with the MIL and OBD DTCs on the scan tool. On other systems, flash DTCs are not possible. Flash DTCs and OBD DTCs, with the conditions required to set the DTCs, are shown in Figure 8-12. When the option DTCs are selected on the scan tool, the scan tool can be used to read and erase those DTCs. On many vehicles, OBD II codes are more extensive and better defined than standard DTCs. This provides additional assistance for technicians.

Check the DTCs with the scan tool. If multiple DTCs are stored, refer to the DTC interpretation table in the following order:

1. PCM error DTCs

2. System voltage DTCs

3. Component level DTCs

4. System level DTCs

INPUTS

Component	Malfunction	MIL	J2012 DTC	Diagnostic Criteria to Set Fault	Mature Time
Sync. (cam/crank)	Rationality	11	P1390	Cam/Crank reference Angle>threshold	25 cts.
MAP Sensor	Rationality	13	P1297	Baro MAP<4 in. Hg at idle	2 sec.
	Short	14	P0170	Voltage signal <0.02 volts	2 sec.
	Open	14	P1080	Voltage signal >4.667 volts	2 sec.
Vehicle Speed Sensor	Rationality	15	P0500	No pulses during driving conditions	11 sec.
Coolant Temp. Sensor	Rationality	17	P0125	After start engine temperature <35°F	10 min.
	Short	22	P0117	Voltage signal <0.51 volts	3 sec.
	Open	22	P0118	Voltage signal >4.96 volts	3 sec.
02 Sensor Upstream	Shortened Low	21	P0131	Voltage stays below 0.15 volts during O_2 heater test	7 min. 3 sec.
02 Sensor Upstream	Shortened High	21	P0132	Voltage signal <1.22 volts	2 min.
02 Sensor Upstream	Open	21	P0134	Voltage stays >0.35 volts or <0.59 volts	7 min.
02 Sensor Downstream	Shortened Low	21	P0137	Voltage signal <1.22 volts	3 sec.
02 Sensor Downstream	Shortened High	21	P0138	A WOT voltage signal >0.61 volts and	15 sec.
	Rationality	21	P0139	At decel fuel shutoff voltage signal <0.29 volts	
Charge Temp. Sensor	Short	23	O0112	Voltage signal <0.51 volts	3 sec.
	Open	23	P0113	Voltage signal >4.96 volts	3 sec.
Throttle Position Sensor	Rationality	24	P0121	At idle voltage signal >4 volts OR at cruise voltage signal <0.51 volts	3 sec.
	Short	24	P0122	Voltage signal <0.157 volts	0.7 sec.
	Open	24	P0123	Voltage signal >4.706 volts	0.7 sec.
PCM	ROM sum	53	P0605	At power down but sum not equal to calibration	Immediate
	SPI	53	P0605	No communication between devices	
Cam Position Sensor	Rationality	54	P0340	No pulses seen	5 sec.
P/S Pressure Switch	Rationality	65	P0551	At vehicle speed >40 mph, high pressure	15 sec.

OUTPUTS

Component	Malfunction	MIL	J2012 DTC	Diagnostic Criteria to Set Fault	Mature Time
IAC motor	Open/Short	25	P0505	Voltage signal read ≠ to expected state	3 sec.
	Functionality	25	P1294	At idle, RPM not within ±300 of target	20 sec.
Injector 1	Open/Short	27	P0201	Voltage signal read ≠ to expected state	3 sec.
Injector 2	Open/Short	27	P1202	Voltage signal read ≠ to expected state	3 sec.
Injector 3	Open/Short	27	P0203	Voltage signal read ≠ to expected state	3 sec.
Injector 4	Open/Short	27	P0204	Voltage signal read ≠ to expected state	3 sec.
EVAP solenoid	Functionality	31	P0441	Shift in O_2 feedback or IAC or RPM from solenoid off to full on	25 sec.
	Open/Short	31	P0443	Voltage signal read ≠ to expected state	3 sec.
EGR solenoid	Open/Short	32	P0403	Voltage signal read ≠ to expected state	3 sec.
High speed fan relay	Open/Short	35	P1489	Voltage signal read ≠ to expected state	3 sec.
Low speed fan relay	Open/Short	35	P1490	Voltage signal read ≠ to expected state	3 sec.
Ignition coil 1 (DIS)	Open/Short	43	P0351	Max. dwell current ≠ to expected state	3 sec.
Ignition coil 2 (DIS)	Open/Short	43	P0352	Max. dwell current ≠ to expected state	3 sec.

Figure 8-12 Flash DTCs and OBD DTCs with the conditions required to set them.

CAUTION: Do not clear DTCs unless directed by a diagnostic procedure. Clearing DTCs will also clear valuable freeze-frame and failure records data.

Lower fuel trim numbers indicate less fuel required, while higher trim requires more fuel.

Using a scan tool's advanced function, command the MIL to turn off. If it turns off, turn it back on and continue. If it stays on, follow the manufacturer's diagnostic procedure for this problem.

The next step in diagnostics depends on the outcome of the DTC display. If there were DTCs stored in the PCM's memory, proceed by following the designated DTC table to make an effective diagnosis and repair. If no DTCs were displayed, match the symptom to the symptoms listed in the manufacturer's symptom tables (Figure 8-13) and follow the diagnostic paths or

Hesitation, Sag and Stumble

Step	Action	Values	Yes	No
	Definition: Momentary lack of response as the accelerator is pushed down. Can occur at all vehicle speeds. Usually most severe when first trying to make the vehicle move, as from a stop sign. May cause engine to stall if severe enough.			
1	1. Was the Powertrain On-Board Diagnostic (OBD) System Check performed?	–	Go to Step 2	Go to Powertrain OBD System Check 2.4 L Powertrain OBD System Check 2.2 L
2	1. Perform a bulletin search. 2. If a bulletin that addresses the symptom is found, correct the condition per bulletin instructions. Was a bulletin found that addresses the symptom?	–	Go to Step 16	Go to Step 3
3	1. Perform the careful visual/physical checks as described at the beginning of the Symptoms section. Is the action complete?	–	Go to Step 4	–
4	1. Check the TP sensor for binding or sticking. Voltage should increase at a steady rate as the throttle is moved toward wide-open throttle. 2. Repair if found faulty. Was a problem found?	–	Go to Step 16	Go to Step 5
5	1. Check the MAP sensor for proper operation. Refer to MAP Sensor Output. Check 2.4L or MAP Sensor Output. Check 2.2L. 2. Repair if found faulty. Was a problem found?	–	Go to Step 16	Go to Step 6
6	1. Check for fouled spark plugs. 2. Replace the spark plugs if found faulty. Was a problem found?	–	Go to Step 16	Go to Step 7
7	1. Check the ignition control module for proper grounds. 2. Repair if grounds are found faulty. Was a problem found?	–	Go to Step 16	Go to Step 8

Figure 8-13 A diagnostic sequence based on a symptom. *(continued on next page)*

Step	Action	Values	Yes	No
8	1. Check for poor fuel pressure. Refer to fuel system diagnosis 2.4L or 2.2L. 2. If a problem is found, repair as necessary. Was a problem found?	–	Go to Step 16	Go to Step 9
9	1. Check for contaminated fuel. 2. If a problem is found, repair as necessary. Was a problem found?	–	Go to Step 16	Go to Step 10
10	1. Perform a fuel injector coil test 2.4L or 2.2L and fuel injector balance test 2.4L or 2.2L. 2. If a problem is found, repair as necessary. Was a problem found?	–	Go to Step 16	Go to Step 11
11	1. Check for proper system performance. Use table EVAP control system diagnosis 2.4L or 2.2L. 2. If a problem is found, repair as necessary. Was a problem found?	–	Go to Step 16	Go to Step 12
12	1. Check for proper function of the thermostat. Also check for proper heat range. 2. If a problem is found, repair as necessary. Was a problem found?	–	Go to Step 16	Go to Step 13
13	1. Check the generator for proper output voltage. The generator output voltage should be between the specified values. 2. If a problem is found, repair as necessary. Was a problem found?	11V to 16V	Go to Step 16	Go to Step 14
14	1. Check the intake valves for valve deposits. 2. If deposits are found, remove as necessary. Were deposits found on valves?	–	Go to Step 16	Go to Step 15
15	1. Review all diagnostic procedures within this table. 2. If all procedures have been completed and no malfunctions have been found, review/inspect the following items: *Visual/physical inspection *Scan tool data *All electrical connections within a suspected circuit and/or system. 3. If a problem is found, repair as necessary. Was a problem found?	–	Go to Step 16	Check for service bulletin
16	1. Operate vehicle within the conditions under which the original symptom was noted. Does the system now operate properly?	–	System OK	Go to step 2

Figure 8-13 A diagnostic sequence based on a symptom. *(continued)*

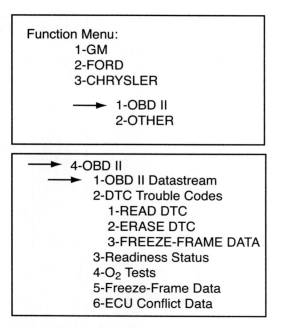

```
Function Menu:
    1-GM
    2-FORD
    3-CHRYSLER

        ──→  1-OBD II
             2-OTHER
```

```
    ──→  4-OBD II
         ──→  1-OBD II Datastream
              2-DTC Trouble Codes
                  1-READ DTC
                  2-ERASE DTC
                  3-FREEZE-FRAME DATA
              3-Readiness Status
              4-O₂ Tests
              5-Freeze-Frame Data
              6-ECU Conflict Data
```

Figure 8-14   OBD II scan tool menu.

suggestions to complete the repair or refer to the applicable component/system check. If there is not a matching symptom, analyze the complaint, then develop a plan for diagnostics. Utilize the wiring diagrams and theory of component and system operation. Call technical assistance for similar cases where repair history may be available. It is possible that the customer's complaint is a normal operating condition for that vehicle. If this is suspected, compare the complaint to the driveability of a known good vehicle.

After the DTCs and the MIL have been checked, the OBD II data stream can be selected (Figure 8-14). In this mode, the scan tool displays all of the data from the inputs. This information may include engine data (Figure 8-15), EGR data (Figure 8-16, page 349), three-way catalytic converter data (Figure 8-17, page 350), oxygen sensor data (Figure 8-18, page 351), and misfire data (Figure 8-19, page 352). The actual available data will depend on the vehicle and the scan tool. However, OBD codes dictate a minimum amount of data. Always refer to the appropriate service manual and the scan tool's operating instructions for proper identification of normal or expected data.

CAUTION: A scan tool that displays faulty data should not be used and the problem should be reported to the manufacturer. The use of a faulty scan tool can result in misdiagnosis and unnecessary parts replacement.

One of the selections on a scan tool is the DTC data. This selection contains two important categories of information: freeze frame, which displays the conditions that existed when a DTC was set, and failure records. The latter is valuable when the MIL was not turned on but a fault was detected. The failure records display the conditions present when the system failed the diagnostic test.

With the scan tool, record the freeze frame and failure records information. By storing the freeze-frame and failure records data (Figure 8-20, page 353) on the scan tool, an electronic copy of the data is created when the fault occurred and stored on the scan tool which can be referred to later. Using the scan tool with a four- or five-gas exhaust analyzer allows for a comparison of computer data with tailpipe emissions (Figure 8-21, page 353). A lab scope is also extremely handy for watching the activity of sensors and actuators. Both of these testers will aid in the diagnosis of a problem.

Parameter	Units Displayed	Typical Data Value
Engine Speed	rpm	+/- 100 rpm from desired idle
Desired Idle	rpm	PCM-commanded idle speed (varies with temperature)
ECT	°C, °F	85°C–105°C/ 185°F–220°F (varies with temperature)
IAT	°C, °F	Varies with ambient air temperature
IAC Position	Counts	10–40
TP Angle	Percent	0%
TP Sensor	Volts	0.20–0.74
Throttle at Idle	Yes/No	Yes
BARO	kPa	65–110 (Depends on engine load and barometric pressure)
MAP	kPa/Volts	29–48 kPa /1–2 volts (Depends on engine load and baro. pressure)
MAF Input Frequency	Hz	1200–3000 (Depends on engine load and barometric pressure)
MAF	gm/s	3–6 (Depends on engine load and barometric pressure)
Injection Pulse Width	millisecond	Varies with engine load
Air Fuel Ratio	Ratio	14.2–14.7
Fuel Pump	On/Off	On
VTD Fuel Disable	Active/Inactive	Inactive
ECT	°C, °F	85°C–105°C/ 185°F–220°F (varies with temperature)
Engine Run Time	Hr: Min: Sec	Depends on time since startup
Loop Status	Open/Closed	Closed
Fuel Trim Learn	Disabled/Enabled	Enabled
H02S Bn1 Sen. 1	Not Ready/Ready	Ready
H02S Bn2 Sen. 1	Not Ready/Ready	Ready
H02S Bn1 Sen. 1	millivolts	0 - 1000, constantly varying
H02S Bn2 Sen. 1	millivolts	0 - 1000, constantly varying

Figure 8-15 Engine data. *(continued on next page)*

Parameter	Units Displayed	Typical Data Value
Rich/Lean Bn 1	Rich/Lean	Constantly changing
Rich/Lean Bn 2	Rich/Lean	Constantly changing
HO₂S Bn 1 Sen. 2	millivolts	0–1000 constantly varying
HO₂S Bn 1 Sen. 3	millivolts	0–1000 varying
Short-Term FT Bn 1	Percent	+10%-10%
Long-Term FT Bn 1	Percent	+10%-10%
Short-Term FT Bn 2	Percent	+10%-10%
Long-Term FT Bn 1	Percent	+10%-10%
Fuel Trim Cell	Cell #	0
Engine load	Percent	2%–5%
Desired ERG Pos.	Percent	0%
Actual ERG Pos.	Percent	0%
Vehicle Speed	MPH, Km/h	0
Engine Speed	rpm	100 rpm from desired idle
Commaned TCC	Engaged/ Disengaged	Disengaged
Current Gear	1/2/3/4	1
Ignition 1	Volts	13 (varies)
Commanded GEN	On/Off	On
Cruise Engaged	Yes/No	No
Cruise Inhibited	Yes/No	Yes
Engine Load	Percent	2%–5% (varies)
TWC Protection	Active/Inactive	Inactive
Hot Open Loop	Active/Inactive	Inactive
ECT	°C, °F	85°C–105°C/ 185°F–220°F (varies with temperature)
Decel Fuel Mode	Active/Inactive	Inactive
Injector Pulse Width	millisecond	Varies with engine load
Power Enrichment	Active/Inactive	Inactive
TP Angle	Percent	0%

Figure 8-15 Engine data. *(continued)*

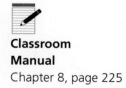

Classroom Manual
Chapter 8, page 225

Special Tools

Exhaust analyzer, lab scope

If the PCM reacts to a misfire by shutting down the injector of a cylinder, that bad cylinder will still have an effect on the engine and emissions. The piston of that cylinder is still moving, the valves are still opening, and the spark plug is still firing. All of this leads to excessive amounts of oxygen in the exhaust stream. Since the O_2 sensor doesn't know where the oxygen is coming from, it sends a very lean reading to the PCM. In turn, the PCM will increase the richness of the mixture. On most engines, this amount of enriching will be beyond the capacity of the system because no amount of added fuel will burn the pass-through oxygen of the dead cylinder. As a result, the PCM will set a code and turn on the MIL. When a cylinder misfire is present, check to see if two cylinders are involved. If there are, look for something shared by both cylinders such as the common coil or a common vacuum leak.

Parameter	Units Displayed	Typical Data Value
Desired EGR position	Percent	0%
Actual EGR position	Percent	0%
EGR test count	Counts	0
EGR feedback	Volts	0.14–1.0 volt
EGR duty cycle	Percent	0%
EGR pos. error	Percent	0%–9%
EGR close valve pintle position	Volts	0.14–1.0 volt
Engine speed	RPM	± 100 rpm from desired idle
Desired idle	RPM	PCM commanded idle speed
ECT	°C,°F	85°C–105°C / 85°F–220°F
IAT	°C,°F	Varies to ambient
IAC position	Counts	10–40
TP angle	Percent	0%
Baro	kPa	65–110
MAP	kPa/Volts	29–48 kPa, 1–2 volts
Loop status	Open/Closed	Closed
Fuel trim learn	Dis/enabled	Enabled
Rich/lean Bn1	Rich/lean	Constantly changing
Rich/lean Bn2	Rich/lean	Constantly changing
Fuel trim cell	Cell #	0
Engine load	Percent	2%–5% varies
Hot open loop	Active/inactive	Inactive
Ignition 1	Volts	13 varies
Knock retard	Degrees	0%
KS activity	Yes / No	No
Currrent gear	1/2/3/4	1
Trans. range	P/N, Reverse, Low, Drive 2, Drive 3 & 4	P/N
Commanded TCC	En/disengage	Disengaged
Trans hot mode	Active/inactive	Inactive
Engine load	Percent	2%–5% varies
TWC protection	Active/inactive	Inactive
Decel fuel mode	Active/inactive	Inactive
Power enrichment	Active/inactive	Inactive

Figure 8-16 EGR data.

Parameter	Units Displayed	Typical Data Value
TWC Monitor Test Counter	Counts	0
TWC Diagnostic	Enabled/disabled	Enabled
Engine Speed	RPM	± 100 rpm from desired idle
ECT	°C, °F	85°C–105°C/ 185°F–220°F (varies with temperature)
IAT	°C, °F	Varies with ambient temperature
IAC Position	Counts	10–40
TP Angle	Percent	0%
TP Sensor	Volts	0.20–0.74
Throttle at Idle	Yes/No	Yes
Engine Speed	RPM	± 100 rpm from desired idle
MAF	gm/s	3–6 (Depends on engine load and barometric pressure)
Engine Load	Percent	2%–5% (varies)
Air-Fuel Ratio	Ratio	14.2–14.7
ECT	°C, °F	85°C–105°C/ 185°F–220°F (varies with temperature)
Engine Run Time	HR: Min: Sec	Depends on time since start up
Loop Status	Open/Closed	Closed
Fuel Trim Learn	Disabled/enabled	Enabled
HO_2S Bn1 Sen. 2	Millivolts	0–1000, constantly varying
HO_2S Bn1 Sen. 3	Millivolts	0–1000 varying
Vehicle Speed	MPH, Km/h	0
MIL	On/Off	Off
Engine Load	Percent	2%–5% (varies)
TWC Protection	Active/inactive	Inactive
Hot Open Loop	Active/inactive	Inactive
ECT	°C, °F	85°C–105°C/ 185°F–220°F (varies with temperature)
Decel Fuel Mode	Active/inactive	Inactive
Inj. Pulse Width	Millisecond	Varies with eng. load
Power Enrichment	Active/inactive	Inactive
TP Angle	Percent	0%

Figure 8-17 Three-way catalytic converter data.

Parameter	Units Displayed	Typical Data Value
Start-up IAT	°C, °F	Depends on intake air temperature at time of startup
Start-up ECT	°C, °F	Depends on engine coolant temperature at time of startup
HO_2S Bn1 Sen. 1	Millivolts	0–1000, constantly varying
HO_2S Bn2 Sen. 1	Millivolts	0–1000, constantly varying
HO_2S Bn1 Sen.2	Millivolts	0–1000, constantly varying
HO_2S Bn 1Sen. 3	Millivolts	0–1000 varying
HO_2SX Counts Bn1	Counts	Varies
HO_2S X Counts Bn2	Counts	Varies
HO_2S Warm-Up Time Bn1 Sen. 1	Hr: Min: Sec	Depends on startup intake air temperature, startup engine coolant temperature, and time to HO_2S activity.
HO_2S Warm Up Bn2 SEN 1	HR: Min: Sec	Depends on startup intake air temperature, startup engine coolant temperature, and time to HO_2S activity.
HO_2S Warm Up Bn1 SEN 2	HR: Min: Sec	Depends on startup intake air temperature, startup engine coolant temperature, and time to HO_2S activity.
HO_2S Warm Up Bn1 SEN 3	HR: Min: Sec	Depends on startup intake air temperature, startup engine coolant temperature, and time to HO_2S activity.
Engine Speed	RPM	±100 rpm from desired idle
ECT	°C, °F	85°C–105°C/ 185°F–220°F (varies with temperature
MAF	gm/s	3–6 (Depends in engine load and barometric pressure).
EVAP Purge PWM	Percent	0%–25% (varies)
TP Angle	Percent	0%
Ignition 1	Volts	13 (varies)

Figure 8-18 Heated oxygen sensor data.

Parameter	Units Displayed	Typical Data Value
Misfire Current #1	Counts	0
Misfire History #1	Counts	0
Misfire Current #2	Counts	0
Misfire History #2	Counts	0
Misfire Current #3	Counts	0
Misfire History #3	Counts	0
Misfire Current #4	Counts	0
Misfire History #4	Counts	0
Misfire Current #5	Counts	0
Misfire History #5	Counts	0
Misfire Current #6	Counts	0
Misfire History #6	Counts	0
Misfire Failures Since First Fail	Counts	0
Misfire Passes Since First fail	Counts	0
Total Misfire Current Count	Counts	0
ECT	°C, °F	85°C–105°C/ 185°F–220°F (varies with temperature)
Hot Open Loop	Active/Inactive	Inactive
Loop Status	Open/Closed	Closed
Fuel Trim Learn	Disabled/Enabled	Enabled
HO$_2$S Bn 1 Sen. 1	Not Ready/Ready	Ready
HO$_2$S Bn 2 Sen. 1	Not Ready/Ready	Ready
Rich/Lean Sen. 1	Rich/Lean	Constantly changing
Rich/Lean Sen. 2	Rich/Lean	Constantly changing
HO$_2$S X Counts Bn 1	Counts	Varies
HO$_2$S X Counts Bn 2	Counts	Varies
Short-Term FT Bn 1	Percent	-10%–10%
Long-Term FT Bn 1	Percent	-10%–10%
Short-Term FT Bn 2	Percent	-10%–10%
Long-Term FT Bn 2	Percent	-10%–10%
Fuel Trim Cell	Cell #	0
Engine Load	Percent	2%–5% (varies)
Engine Speed	RPM	± 100 rpm from desired idle
Ignition Mode	IC/Bypass	IC
Spark	Degrees	16
Knock Retard	Degrees	0
Desired ERG Position	Percent	0%
Actual ERG Position	Percent	0%
Vehicle Speed	mph, Km/h	0
Current Gear	1/2/3/4	1
Engine Speed	RPM	± 100 rpm from desired idle
TCC Enable	On/Off	Off
IAC Position	Counts	10–40
Current Gear	1/2/3/4	1
Engine Load	Percent	2%–5%
TWC Protection	Active/Inactive	Inactive
Decel Fuel Mode	Active/Inactive	Inactive
Power Enrichment	Active/Inactive	Inactive
A/C Request	Yes/No	No (Yes with A/C "ON")
Commanded A/C	On/Off	Off (On with A/C compressor engaged)

Figure 8-19 Misfire data.

At Idle/Upper Radiator/Closed Throttle/Park or Neutral Closed Loop/Acc. OFF		
Scan Tool Parameter	**Units Displayed**	**Typical Data Value**
DTC Freeze Frame	#	DTC #
Scan Tool Parameter	Units Displayed	Typical Data Value
Air-Fuel Ratio	Ratio	Varies
Calc. Air Flow	g/S	Varies
ECT	°C–°F	Varies
BARO	kPa, V	Varies
Base PWM Cyl. 1	mS	Varies
MAP	kPa,V	Varies
Short-Term FT	Counts, %	Varies
Long-Term FT	Counts, %	Varies

At Idle/Upper Radiator/Closed Throttle/Park or Neutral Closed Loop/Acc. OFF		
Scan Tool Parameter	**Units Displayed**	**Typical Data Value**
Loop Status	Open/Closed	Varies
MPH	MPH and km/h	Varies
Load (2.4L Engine)	%	15–21% (16–29% @2500 rpm)
Load (2.2L Engine)	%	22–26% (19–25% @2500 rpm)
TP Angle	%	Varies
Engine Speed	RPM	Varies

Figure 8-20 Freeze-frame and failure record data.

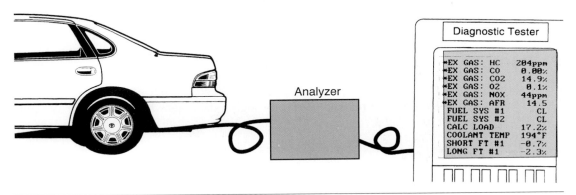

Figure 8-21 Combining the readings of a scan tool with those of an exhaust gas analyzer are very valuable in diagnostics.

OBD II Terminology

Drive Cycle

An OBD II drive cycle is a method of driving that consists of an engine start and vehicle operation that brings the vehicle into closed loop and includes whatever specific operating conditions are necessary either to initiate and complete a specific monitoring sequence or to verify a symptom or verify a repair. A monitoring sequence is an operational strategy designed to test the operation of a specific system, function, or component.

Enable Criteria

Enabling criteria defines the criteria that must be met in order for a monitor test to be completed. The enable criteria varies with the make and model of the vehicle. Therefore, refer to the service manual to identify the manufacturer's requirements. If the enable criteria is not met, that particular monitor test is not completed and the condition of the tested system is unknown.

On-Board Diagnostics

A diagnostic test is a series of steps that has a beginning and an end, the result of which is a pass or fail reported to the PCM. When a diagnostic test reports a pass result, the PCM records the following data: the diagnostic test has completed since the last ignition cycle, the diagnostic test has passed during the current ignition cycle, and the fault identified by the diagnostic test is not currently active. When a diagnostic test reports a fail result, the PCM records the following data: the diagnostic test has completed since the last ignition, the fault identified by the diagnostic test is currently active, the fault has been active during the ignition cycle, and the operating conditions at the time of the failure.

Pending Situation

A **pending situation** prevents a test from being performed due to an uncorrected fault.

Under certain circumstances, the PCM will postpone and not run a particular test if the MIL is on and a DTC stored. This creates a **pending situation** in that a test will not run or be completed until the fault is corrected or goes undetected for the required number of driving cycles. Pending situations occur when a sensor is malfunctioning but its signal is important to the testing of another component or system. An example of this is the oxygen sensor. If the O$_2$ sensor signal is out of acceptable range or if it not responding according to the guidelines set in the PCM, a DTC will be stored. However, the O$_2$ signal is needed to measure the efficiency of the catalytic converter. This prevents the catalyst test from being conducted and the test is pending.

Serial Data

Serial data is the PCM information sent to the scanner and other control modules.

Serial data is a term that refers to a method of data transmission. Serial data is transmitted, one bit at a time, over a single wire. The data travels through a data bus where it is made available to the PCM and other associated control modules.

Similar Conditions

This term defines the operating conditions that must be present during a second trip to set a code when there is a defect and to erase a code after a repair has been made. For the conditions to be similar, these conditions must be present: the load conditions must be within 10 percent of the vehicle load present when the diagnostic test reported the malfunction, the engine speed must be within 375 rpm of the engine speed present when the DTC was set, and the engine coolant temperature must have been in the same range present when the PCM turned the MIL on.

Trip

The ability for a diagnostic test to run depends upon whether or not a trip has been completed. A trip for a particular diagnostic test is defined as a key on and key off cycle in which all the enabling criteria for a given diagnostic test has been met allowing the PCM to run vehicle operation, followed by an engine off period of duration and driving mode such that any particular diagnostic test has had sufficient time to complete at least once. The requirements for trips vary since they may involve items of an unrelated nature such as driving style, length of trip, ambient temperature, etc. Some diagnostic tests run only once per trip, such as the catalyst monitor. Others run continuously, such as the misfire and fuel system monitor systems. If the proper enabling conditions are not met during that ignition cycle, the tests may be incomplete or the test may not have run.

The OBD II trip consists of an engine start following an engine off period, with enough vehicle travel to allow the following monitoring sequences to complete their tests:

Misfire, fuel system, and comprehensive system components. These are checked continuously throughout the trip.

EGR. This test requires a series of idle speed operations, acceleration, and deceleration to satisfy the conditions needed for completion. This is performed once per trip.

HO$_2$S. This test requires a steady speed for about 20 seconds at speeds between 20 and 45 mph after warmup to be complete. This test is performed once per trip.

Two Trip Monitors

The first time the OBD II monitor detects a fault during any drive cycle, it sets a pending code in the memory of the PCM. Before a pending code becomes a DTC and turns the MIL on, the fault must be repeated under similar conditions. A pending code can remain in memory for some time while it waits for the conditions to repeat themselves. When the same conditions exist, the PCM checks the fault. If the fault is still present, a DTC is stored and the MIL is illuminated. Two trip monitors are those monitors that require a second trip to set a code. Some misfire and catalyst faults can cause the PCM to turn the MIL on after only one trip.

A maturing code is one that has not been repeated nor has it caused the MIL to illuminate.

Warm-up Cycle

A warm-up cycle consists of engine startup and vehicle operation such that the coolant temperature has risen greater than 40°F from startup temperature and reached at least 160°F. If this condition is not met during the ignition cycle, the diagnostic tests may not run.

Malfunction Indicator Lamp (MIL)

The MIL is on the instrument panel. The MIL informs the driver that a fault that affects the vehicle's emission levels has occurred. The owner should take the vehicle in for service as soon as possible. The MIL will illuminate if a component or system that has an impact on vehicle emissions indicates a malfunction or fails to pass an emissions-related diagnostic test (Figure 8-22). The MIL will stay illuminated until the system or component passes the same test for three consecutive trips with no emissions-related faults. After making the repair, technicians may need to take the vehicle for three trips to ensure the MIL does not illuminate again.

As a bulb and system check, the MIL comes on with the ignition switch on and the engine off. When the engine is started, the MIL turns off if there are no DTCs set. When the MIL remains on while the engine is running or a malfunction is suspected due to a driveability or emissions problem, perform an OBD system check. These checks expose faults a technician may not detect if other diagnostics are performed first.

If the vehicle is experiencing a malfunction that may cause damage to the catalytic converter, the MIL will flash once per second. If the MIL flashes, the driver needs to get the vehicle to the service department as soon as possible. If the driver reduces speed or load and the MIL stops flashing, a code is set and the MIL stays on. This means the conditions that presented potential problems to the converter have passed with the changing operating conditions.

On some systems, the MIL may also light if the gas cap is off or is loose. This is due to a loss of pressure in the fuel vapor system and will be identified with an evaporative system DTC. Some Ford vehicles will illuminate a "check fuel cap" lamp if a large EVAP leak is detected.

Classroom Manual
Chapter 8, page 238

Components Intended to Illuminate MIL
Automatic Transmission Temperature Sensor
Engine Coolant Temperature (ECT) Sensor
Evaporative Emission Canister Purge (EVAP Canister Purge)
Evaporative Emission Purge Vacuum Switch
Idle Air Control (IAC) Coil
Intake Air Temperature (IAT) Sensor
Ignition Control (IC) System
Ignition Sensor (Cam Sync, Diag)
Ignition Sensor High Resolution 7x
Knock Sensor (KS)
Manifold Absolute Pressure (MAP) Sensor
Mass Air Flow (MAF) Sensor
Throttle Position (TP) Sensor A, B
Transmission Range (TR) Mode Pressure Switch
Transmission Shift Solenoid A
Transmission Shift Solenoid B
Transmission TCC Enable Solenoid
Transmission 3/4 Shift Solenoid
Transmission Torque Converter Clutch (TCC) Control Solenoid
Transmission Turbine Sensor (HI/LO)
Transmission Vehicle Speed Sensor (HI/LO)
Transmission Vehicle Speed Sensor (HI/LO)

Figure 8-22 A list of components that will cause the MIL to turn on when the PCM detects they are defective or are malfunctioning.

Given the comprehensive nature of these diagnostic tests and readiness monitors, many state I/M programs are implementing an OBD II test for vehicles produced in 1996 or later. A simple DCL connection allows immediate access to vehicle data that will reveal conditions affecting emissions and is likely more accurate than conventional tailpipe testing.

MIL History Codes

The PCM must acknowledge when all the emissions-related diagnostic tests have reported a pass or fail condition since the last ignition cycle. Each diagnostic test is separated into four types: emissions-related that turns the MIL on the first time the PCM reports a fault; emissions-related that turns the MIL on if the fault is active for two consecutive driving cycles; nonemissions-related that does not turn the MIL on but will turn the service light on; and nonemissions-related that does not turn the MIL on or the service light on.

The PCM has the option of turning the MIL off when the diagnostic test that caused the MIL to light passes for three consecutive trips. In the case of misfire or fuel trim malfunctions, there are additional requirements as follows:

❏ The load conditions must be within 10 percent of the vehicle load present when the diagnostic test reported the malfunction.

❏ The engine speed must be within 375 rpm of the engine speed present when the DTC was set.

❏ The engine coolant temperature must have been in the same range present when the PCM turned on the MIL.

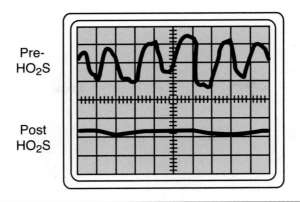

Figure 8-23 O₂ sensor signals from before and after a good converter.

When the PCM requests the service light to be turned on, a history DTC is also recorded for the diagnostic test. The provision for clearing a history DTC for any diagnostic test requires forty subsequent warm-up cycles during which no diagnostic tests have reported a fail, a battery disconnect, or a scan tool clear info command.

Unique to the misfire diagnostic, the PCM has the capability of alerting the driver of potentially damaging levels of misfire. If a misfire condition exists that could potentially damage the catalytic converter, the PCM will command the MIL to flash at a rate of once per second during the times that the catalyst-damaging misfire condition is present. The PCM may also shut down injectors for the misfiring cylinders to prevent raw fuel from entering the converter.

With OBD II, there are multiple oxygen sensors. One sensor is located before the catalyst and detects rich-lean swings in the exhaust stream. Another oxygen sensor is located after the catalytic converter. It also looks at the rich-lean swings of the exhaust stream. If the catalytic converter is working properly, there should be no swings! Large swings mean that the converter is not working properly (Figure 8-23). The PCM responds by setting a fault code. If the downstream O₂ sensor signal is the same as the upstream sensor, this means the catalytic converter is not doing anything.

Each time a fuel trim malfunction is detected, the engine load, engine speed, and engine coolant temperature is recorded. When the ignition is turned off, the last reported set of conditions remains stored. During subsequent ignition cycles, the stored conditions are used as a reference for similar conditions. If a fuel trim malfunction occurs during two consecutive trips, the PCM treats the failure as a normal malfunction and does not use the stored data. However, if a fuel trim malfunction occurs on two non-consecutive trips, the stored conditions are compared with the current conditions. The MIL will illuminate if there are similar conditions to those present at the time the fault was detected.

In the case of an intermittent fault, the MIL may illuminate and turn off after three trips. However, the corresponding DTC will be stored in memory. When unexpected DTCs appear, check for an intermittent malfunction (Photo Sequence 16).

Photo Sequence 16
Comparing O₂ Signals

P16-1 A two-channel oscilloscope can read both front and rear O₂ sensors simultaneously. The technician will compare the two signals.

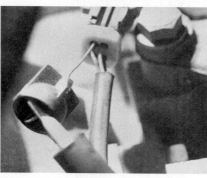

P16-2 The technician connects to the O₂ sensor using a jumper harness to avoid piercing the wire. Both sensors will be connected in the same way. The red lead of the scope will be connected to the sensor output wire and the black to ground.

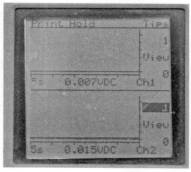

P16-3 Turn the key on and look for about 0.5 volts on both channels.

P16-4 The engine is started and allowed to warm up. Take note of the cold O₂ sensor activity.

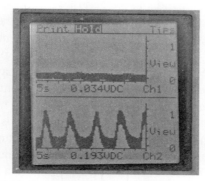

P16-5 After the engine is warmed up, raise the engine speed to 2,000 rpm and watch the patterns.

P16-6 Note the difference in the upstream and downstream sensors. Compare the upstream and downstream average voltage.

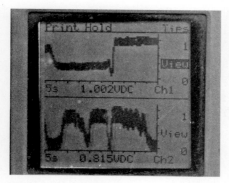

P16-7 If the throttle is raised quickly, the rear O₂ sensor may pick up on the additional fuel flow and show a small bit of activity for a short time.

Freeze Frame

Government regulations require that engine operating conditions be captured whenever the MIL is illuminated. The data captured is called freeze-frame data. Whenever the MIL is illuminated, the corresponding record of operating conditions is recorded to the freeze-frame buffer. Each time a diagnostic test reports a failure, the current engine operating conditions are recorded in the failure records buffer. A subsequent failure will update the recorded operating conditions. The following operating conditions for the diagnostic test that failed typically include the following parameters:

- ❏ Air-fuel ratio
- ❏ Airflow rate
- ❏ Fuel trim
- ❏ Engine speed
- ❏ Engine load
- ❏ Engine coolant temperature
- ❏ Vehicle speed
- ❏ TP angle
- ❏ MAP/BARO
- ❏ Injector base pulse width
- ❏ Loop status

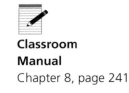

Classroom Manual Chapter 8, page 241

Freeze-frame data can only be overwritten with data associated with a misfire or fuel trim malfunction. Data from these faults take precedence over data associated with any other fault. The freeze-frame data will not be erased unless the associated history DTC is cleared. Technicians should utilize freeze frame to help duplicate intermittent conditions.

Diagnostic Trouble Codes

OBD II standards call for standardized diagnostics and DTCs. The standardized diagnostic codes give somewhat detailed descriptions of the faults detected by the PCM (Figure 8-24). In

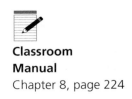

Classroom Manual Chapter 8, page 224

- • DTC No
  Indicates the diagostic trouble code.

- • Detection Item
  Indicates the system of the problem or contents of the problem.

DTC Chart (SAE Controlled)

HINT: Parameters listed in the chart may not be exactly the same as your reading due to the type of instrument or other factors.

If a malfunction code is displayed during the DTC check in check mode, check the circuit for the code listed in the table below. For details of each code, turn to the page referred to under the "See page" for the respective "DTC No." in the DTC chart.

DTC No. (See page)	Detection Item	Trouble Area	MIL	Memory
P0100 (EG-244)	Mass Air Flow Circuit Malfunction	• Open or short in mass air flow meter circuit • Mass air flow meter • ECM	O	O
P0101 (EG-247)	Mass Air Flow Circuit Range/Performance Problem	• Mass air flow meter	O	O
P0110 (EG-248)	Intake Air Temp. Circuit Malfunction	• Open or short in intake air temp. sensor circuit • Intake air temp. sensor • ECM	O	O
P0115 (EG-241)	Engine Coolant Temp. Circuit Malfunction	• Open or short in engine coolant temp. sensor circuit • Engine coolant temp. sensor • ECM	O	O

...olant temp. sensor

- • Page or Instructions
  Indicates the page where the inspection procedure for each circuit is to be found, or gives instructions for checking and repairs.

- • Trouble Area
  Indicates the suspect area of the problem.

Figure 8-24 Explanation of a typical DTC chart.

the standardized list there are gaps in the numbering that will allow for codes to be added in the future. Also, the list of codes includes some equipment not included on all engines. This means some of the codes are not available on some engines. The importance of the standardized codes is simply every car make and model will use the same code to define a fault.

All DTCs are displayed as a five-character alphanumeric code in which each character has a specific meaning. The first character is the prefix letter that indicates the area of the fault.

B = body   C = chassis   P = powertrain

The second character is a number that indicates whether the DTC that follows is a standard, common, or manufacturer-specific code. A 0 indicates that it is a generic or common code, and a 1 means the code is manufacturer specific. The latter means the same code may indicate something different on another vehicle make or model.

The third character indicates the subsystem of the area that the fault lies in. Engine performance problems will set a DTC that begins with a P. The third character defines where in the powertrain the fault resides. The following list includes the possible third characters and their associated subgroups:

0 = entire system   5 = idle speed control
1 = air-fuel control   6 = PCM and its inputs and outputs
2 = air-fuel control   7 = transmission
3 = ignition system   8 = non-OBD II system powertrain components
4 = auxiliary emissions controls

The fourth and fifth characters identify the specific detected fault. These characters not only indicate the component or circuit, but also give a description of the type of fault. As an example of this, consider DTC P0108. This is a generic powertrain DTC. The fault is in the air-fuel system. The last two characters indicate that the condition that triggered the fault is the input voltage to the MAP sensor is above the acceptable standard (Figure 8-25). With this sort of information, a technician knows what to check (Figure 8-26).

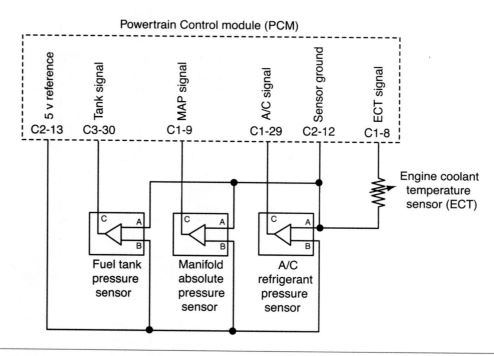

Figure 8-25 Circuit pertaining to PO 108 DTC.

DTC PO108 MAP sensor circuit high voltage

Step	Action	Values	Yes	No
1	Was the Powertrain On-Board Diagnostic (OBD) System Check performed?	–	Go to Step 2	Go to Powertrain OBD System Check 2.4 L Powertrain OBD System Check 2.2 L
2	1. Install a scan tool 2. Engine at idle. Does the scan tool display the MAP voltage specified?	4.0V	Go to Step 3	Go to Step 4
3	1. Turn the ignition swith OFF. 2. Disconnect the MAP sensor electrical connector. 3. Turn the ignition switch ON. Does the scan tool display the MAP voltage specified?	1.0V	Go to Step 5	Go to Step 6
4	1. Turn the ignition switch ON, with the engine OFF, review freeze frame data. 2. Operate the vehicle within the freeze frame conditions. Does the scan tool display the MAP voltage specified?	4.0V	Go to Step 3	Go to diagnostic aids
5	Probe the MAP sensor signal ground circuit with a test light connected to battery voltage. Does the test light illuminate?	–	Go to Step 7	Go to Step 11
6	Check the MAP sensor signal circuit for a short to voltage and repair as necessary. Was a repair necessary?	–	Go to Step 14	Go to Step 12
7	With a DVM connected to ground, probe the 5 volt reference circuit. Does the DVM display the specified voltage?	5V	Go to Step 8	Go to Step 9
8	Check the MAP sensor vacuum source for being plugged or leaking. Was a problem found?	–	Go to Step 10	Go to Step 13
9	Check the 5-volt reference circuit for a short to voltage and repair as necessary.	–	Go to Step 14	Go to Step 12
10	Repair the vacuum hose as necessary. Is the action complete?	–	Go to Step 14	–
11	Check for an open in the MAP sensor ground circuit and repair as necessary. Was a repair necessary?	–	Go to Step 14	Go to Step 12
12	Replace the PCM. Is the action complete?	–	Go to Step 14	–
13	Replace the MAP sensor. Is the action complete?	–	Go to Step 14	–
14	1. Using the scan tool, clear the DTCs. 2. Start the engine and idle at normal temperature. 3. Operate the vehicle within the conditions for setting the DTC. Does the scan tool indicate that the diagnostic ran and passed?	–	Go to Step 15	Go to Step 2
15	Check for any additional DTC's. Are any DTCs displayed that have not been diagnosed?	–	Go to specific DTC charts	System OK

Figure 8-26 Diagnostic chart for a DTC — PO108.

OBD II mandates that all DTCs be stored according to a priority. DTCs with a higher priority take precedence over a DTC with a lower priority. The higher priority codes are set the first time the fault occurs and turn on the MIL immediately. These are Type A code. The next level of code priorities are those faults that set a code the first time they are detected but the MIL is not lit until they occur a second time. These codes are called Type B codes. The lower level codes are faults that relate to non-emissions systems and are called C and D-type codes.

Data Link Connector

Classroom Manual
Chapter 8, page 236

All vehicles sold since the beginning of 1996 are equipped with a standard DLC. This 16-pin connector (Figure 8-27) is used to connect the scan tool to the computer circuit. There the scan tool can read serial data, retrieve and erase DTCs, and perform system diagnostic tests.

Federal regulations require that all automobile manufacturers establish a common communications system. GM began using a 5-volt Universal Asynchronous Receive and Transmit (UART) serial data bus before OBD II was mandated. UART is transmitted on pin 1 or 9 of the DLC (Figure 8-28). When data is not being transmitted on the UART line, 5 volts are present. When the PCM or scan tool is transmitting information to the MIL or the scan tool, the voltage approaches 0 volts. The result was a digital signal that changed with the toggling of components.

Because of the OBD II mandates, improved data communication was necessary. GM uses the Class 2 communication system to meet these requirements. Each bit of information can have one of two lengths: long or short. The pulse width of a long signal is 125 microseconds, and the short pulse is 75 microseconds long. This allows vehicle wiring to be reduced by the transmission and reception of multiple signals over a single wire. The voltage on the line toggles from 0 to 7 volts. The messages carried on Class 2 data streams are also prioritized. If two messages attempt to establish communications on the data line at the same time, only the message with the higher priority will continue. The lower priority message waits until the high priority message is complete. The most significant result of this regulation is that scan tool manufacturers now have the capability of accessing data from any make or model vehicle that is sold in the United States.

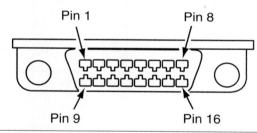

Figure 8-27 Typical DLC.

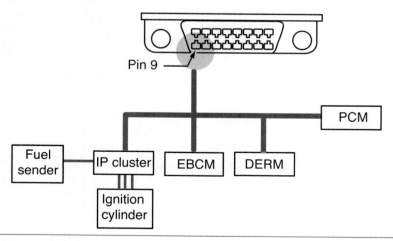

Figure 8-28 Serial data from the PCM is available at Pin 9 of the DLC.

Diagnostic Software

The diagnostic management software in the PCM is designed to organize and prioritize all of the monitor tests and their procedures, to record and display DTCs and freeze-frame information, and control the operation of the MIL. All OBD II systems have the same basic test modes. These test modes must be accessible with an OBD II scan tool. Mode 1 is the **parameter identification (PID) mode**. It allows access to certain data values, analog and digital inputs and outputs, calculated values, and system status information. Some of the PID values will be manufacturer specific; others are common to all OBD II vehicles.

In service manuals, there will be mention of PID values. Some of the PID references are from a list of generic PID values that all scan tools must be able to reference. If a referenced non-generic PID is not on the list, it can be accessed with the manufacturer-specific scan tool or its equivalent. If a generic scan tool is used for non-generic OBD II PIDs, a string of characters (in some cases, a hexadecimal number) may need to be entered. The necessary numbers will be supplied by the manufacturer of the scan tool. Newer generic scan tools have the capability to directly access all or some selected manufacturer codes.

Mode 2 is the Freeze-Frame Data Access mode. This mode permits access to emission-related data values from specific generic PIDs. These values represent the operating conditions at the time the fault was recognized and logged into memory as a DTC. Once a DTC and a set of freeze-frame data are stored in memory, they will stay in memory even if other emission-related DTCs are stored. The number of these sets of freeze frames that can be stored are limited. On 1996 GM vehicles, the possible number of stored sets is five.

There is one type of failure that is an exception to this rule: misfire. Fuel system misfires will overwrite any other type of data except for other fuel system misfire data. This data can only be removed with a scan tool. When a scan tool is used to erase a DTC, it automatically erases all freeze-frame data associated with the events that lead to that DTC.

Mode 3 permits scan tools to obtain stored DTCs. The information is transmitted from the PCM to the scan tool following an OBD II Mode 3 request. Either the DTC, its descriptive text, or both will be displayed on the scan tool.

The PCM reset mode, Mode 4, allows the scan tool to clear all emission-related diagnostic information from its memory. Once the PCM has been reset, the PCM stores an inspection maintenance readiness code until all OBD II system monitors or components have been tested to satisfy an OBD trip cycle without any other faults occurring. Specific conditions must be met before the requirements for a trip are satisfied.

Mode 5 is the oxygen sensor monitoring test. This mode gives the oxygen sensor fault limits and the actual oxygen sensor outputs during the test cycle. The test cycle includes specific operating conditions that must be met to complete the test. This information helps determine the effectiveness of the catalytic converter.

Mode 6 is the **output state test mode (OTM),** which allows a technician to activate or deactivate the system's actuators through using a scan tool that utilizes OE specific software. When the OTM is engaged, the actuators can be controlled without affecting the radiator fans. The fans are controlled separately. This gives a pure look at the effectiveness and action of the outputs.

As an example, GM systems have three available sections: controls, switches, and modes. The *controls* section allows the technician to place the vehicle's engine at a specific idle speed, IAC position, or transmission gear. The technician can also selectively deactivate fuel injectors while the vehicle is running in order to perform a cylinder power balance test. The *switches* section allows the activation or deactivation of various systems such as the MIL, secondary air system, or fuel pump. The *modes* section can be used to perform evaporative emission system tests, reset fuel trims, or conduct a crankshaft sensor tooth error learn mode.

Adaptive Fuel Control Strategy

The **short-term fuel trim (SFT)** and **long-term fuel trim (LFT)** refer to the PCM's temporary or permanent fuel control correction as part of its adaptive memory.

OBD II's **short-term fuel trim (SFT)** and **long-term fuel trim (LFT)** strategies monitor the oxygen sensor signal. The information gathered by the PCM is used to make adjustments to the fuel control calculations. The adaptive fuel control strategy allows for changes in the amount of fuel delivered to the cylinders according to operating conditions.

During open loop, the PCM changes pulse width without any feedback from the O_2 sensor, and the short-term adaptive memory value is one. The one represents a 0 percent change. Once the engine warms up, the PCM moves into closed loop and begins to recognize the signals from the O_2. The system remains in closed loop until the engine stops unless the throttle is fully opened or the engine's temperature drops below a specified limit. In both of these cases, the system goes into open loop.

The PCM controls the pulse width of the injectors in response to the amount of oxygen in the exhaust. When the system is in open loop, the injectors operate at a fixed-base pulse width, reinforcing the airflow calculations of a MAF sensor or load calculations of a MAP sensor. During closed-loop operation, the pulse width is either lengthened or shortened. It is positively or negatively corrected to ensure the proper air-fuel mixture for the operating conditions. As the voltage from the oxygen sensor increases in response to a rich mixture, the short-term fuel trim decreases, which means the pulse width is shortened. Decreases in short-term fuel trim are indicated on a scan tool as a number below 1. For example, the short-term adaptive value of 0.75 means the pulse width was shortened by 25 percent, and the percent of change on the scan tool will be –25. A short-term adaptive value of 1.25 means the pulse width was lengthened by 25 percent. The latter will be displayed as +25.

Once the engine reaches a specified temperature (normally 180°F), the PCM begins to update the long-term fuel trim. The adaptive setting is based on engine speed and the short-term fuel trim. If the short-term fuel trim moves 3 percent and stays there for a period of time, the PCM adjusts the long-term fuel trim. The long-term fuel trim becomes a new but temporary base. In other words, the long-term fuel trim changes the length of the pulse width that is being changed by the short-term fuel trim. Long-term fuel trim works to bring short-term fuel trim close to zero percent correction.

If a lean condition exists because of a vacuum leak or restricted fuel injectors, the long-term fuel trim will have a plus (+) number on the scan tool. If the injectors leak or the fuel pressure regulator is faulty, there will be a rich condition. The rich condition will be evident by minus (–) numbers in the long-term adaptive on the scan tool.

Should the exhaust from a vehicle indicate a rich condition, the O_2 sensor signal prompts the computer to subtract fuel. The reverse condition is an engine running lean in which the PCM adds fuel.

On an OBD II vehicle, zero is the midpoint of the fuel strategy when the computer is in closed loop. Ford's fuel cells are illustrated as a percentage: numbers without a minus sign indicate fuel is being added, and numbers with a minus sign indicate fuel is being subtracted. The constant change or crossing above and below the zero line indicate proper system operation. If the SFT readings are constantly on either side of the zero line, the engine is not operating efficiently.

The GM long-term fuel trim strategy is displayed on the scan tool in the same way as their SFT. Long-term fuel trim represents the PCM learning driver habits, engine variables, and road conditions. If the number displayed on the scan tool is above 0 percent, the computer has learned to compensate for a lean exhaust. If the number is below 0 percent, the computer has learned to compensate for a rich exhaust.

The OBD II LFT strategy is displayed as a percentage on the scan tool in the same way as the SFT. LFT is the computer learning the driver's habits, engine variables, and road conditions. Numbers without a minus sign indicate the computer has compensated for a lean exhaust. Numbers with a minus sign indicate the PCM has compensated for a rich exhaust. As the LFT learns to compensate for an exhaust that is rich or lean, the SFT returns to a value that crosses the zero reference point. If the engine's condition is too far toward either the lean or rich side, the LFT will not compensate and a DTC will be set.

Some intermittent fuel-related driveability problems may be diagnosed by making a recording of the computer's data stream. Before you begin recording with a scan tool, test your understanding of LFT and SFT.

To diagnose this system, connect the scan tool, start the engine, and pull up the PCM's data stream display. Observe the PID called SFT and note the value. With the engine running, pull off a large vacuum hose and watch for the SFT value to rise above zero to as high as 33 percent. Now look at the LFT value, which should have risen above zero, to learn the engine condition. Be aware that a smaller vacuum leak must be made to watch the SFT return to zero as the result of LFT compensation. Since these systems use two oxygen sensors, a pair of SFT PIDs will appear on the scan tool. Use SFT to isolate the side of the engine that is running rich or lean by making two separate recordings. For example, let's say the engine bucks and jerks at highway speeds. After performing a self-test, we know whether the driveability problem did or did not set a fuel-related code. If the fuel injector's wiring or connector opened for a moment causing the engine to run lean, the recording corresponding to that injector bank would reveal the problem.

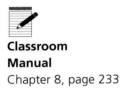

Classroom Manual
Chapter 8, page 233

OBD II Monitor Test Results

All OBD II scan tools include a readiness function showing all of the monitoring sequences on the vehicle and the status of each, complete or incomplete. If vehicle travel time, operating conditions, or other parameters were insufficient for a monitoring sequence to complete a test, the scanner will indicate which monitoring sequence is not yet complete.

The specific set of driving conditions that will set the requirements for all OBD II monitoring sequences follows:

1. Start the engine. Do not turn the engine off for the remainder of the drive cycle.
2. Drive the vehicle for at least 4 minutes.
3. Continue to drive until the engine's temperature reaches 180° or more.
4. Idle the engine for 45 seconds.
5. Open the throttle to about 25 percent to accelerate from a standstill to 45 mph in about 10 seconds.
6. Drive between 30 and 40 mph with a steady throttle for at least 1 minute.
7. Drive for at least 4 minutes at speeds between 20 and 45 mph. If during this phase you must slow down below 20 mph, repeat this phase of the drive cycle. During this time, do not fully open the throttle.
8. Decelerate and idle for at least 10 seconds.
9. Accelerate to 55 mph with about half throttle.
10. Cruise at a constant speed of 40 to 65 mph for at least 80 seconds with a steady throttle.

11. Decelerate and allow the engine to idle.

12. Using the scan tool, check for diagnostic system readiness and retrieve any DTCs.

OBD II readiness status indicates if the various monitors in the OBD II system have completed or not. This mode does not indicate whether the emissions levels were excessive or not during each monitor.

When most monitor tests are run and a system or component fails a test, a pending code is set. When the fault is detected a second time, a DTC is set and the MIL is lit. The DTC will define the fault but will not give the exact location of the fault. The PCM does not know that. It only knows the description of the problem. A technician must use the available information (DTC and freeze frame) from the scan tool and the DTC charts in the service manual to locate the source of the problem.

Once the MIL has been turned on for misfire or a fuel system fault, the vehicle must go through three consecutive drive cycles, including the operating conditions similar to those that were present when the code was set.

It is possible that a DTC for a monitored circuit may not be entered into the PCM memory even though a malfunction has occurred. This may happen when the monitoring criteria has not been met.

Comprehensive Component Monitor

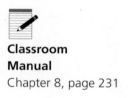

Classroom Manual
Chapter 8, page 231

The PCM's output signals are constantly being monitored by the diagnostic management software. These signals are observed to detect opens and shorts in the circuits. Some of the output devices are also monitored for their effectiveness and functionality. The PCM is programmed to expect certain things to happen when it commands a device to do something. Perhaps the simplest example of this is idle speed. The PCM has control over the engine's idle speed. This is accomplished in many different ways, one of these is through the stepper motor. The PCM sends voltage signals to the motor which changes the idle speed. When the PCM commands the motor to extend, the engine's speed should increase. The PCM watches engine speed and continues to adjust the motor until the desired speed is reached. If the PCM commands the motor to extend and the engine's speed does not increase, the PCM knows the motor is not doing what it is supposed to be doing. As a result, the PCM sets a DTC and illuminates the MIL.

☑ **SERVICE TIP:** On certain vehicles, the oxygen sensor may cool down after only a short period of operation at idle. This will cause the system to go into open loop. To restore closed-loop operation, run the engine at part throttle several minutes and accelerate from idle to part throttle a few times.

Intermittent Faults

An intermittent fault is a fault that is not always present. This type of fault may not activate the MIL or cause a DTC to be set. Therefore, intermittent problems can be difficult to diagnose. To use the DTC charts and test procedures in a service manual, the fault must be present. Prior to testing for the source of an intermittent problem, make sure the preliminary diagnostic checks are completed first.

By studying the system and the relationship of each component to another, you should be able to create a list of suspect sources of the intermittent problem. To help identify the cause of an intermittent problem, follow these steps:

1. Observe the history DTCs, the DTC modes, and the freeze-frame data.
2. Call technical assistance for similar cases where repair history may be available. Combine your knowledge and understanding of the system with the service information that is available.

3. Evaluate the symptoms and conditions described by the customer.
4. Use a check sheet or other method to identify the circuit or electrical system component that may have the problem.
5. Follow the suggestions for intermittent diagnosis found in service material and manuals.
6. Visually inspect the suspected circuit or system.
7. Use the data capturing capabilities of the scan tool.
8. Test the circuit's wiring for shorts, opens, and high resistance. This should be done with a DMM in a typical manner unless instructed differently in the service manual.

Most intermittent problems are caused by faulty electrical connections or wiring. Refer to a wiring diagram for each of the suspected circuits or components. This will help identify all of the connections and components in that circuit. The entire electrical system of the suspected circuit should be carefully and thoroughly inspected. Check for the following problems:

❑ Burnt or damaged wire insulation
❑ Damaged terminals at the connectors
❑ Corrosion at the connectors
❑ Loose connectors
❑ Wire terminals loose in the connector
❑ Disconnected or loose ground wire or strap

To locate the source of the problem, a voltmeter can be connected to the suspected circuit and the wiring harness wiggled (Figure 8-29). As a guideline for what voltage should be expected in the circuit, refer to the reference value table in the service manual. If the voltage reading changes with the wiggles, the problem is in that circuit. The vehicle can also be taken for a test drive with the voltmeter connected. If the voltmeter readings become abnormal with the changing operating conditions, the circuit being observed is probably the circuit with the defect.

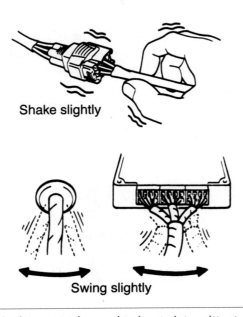

Shake slightly

Swing slightly

Figure 8-29 The wiggle test can be used to locate intermittent problems.

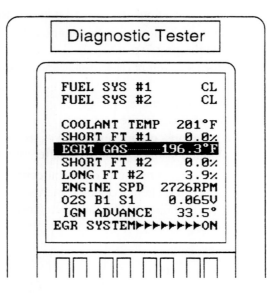

Figure 8-30 OBD II scan tools allow you to observe the results of an active test and to watch the effects of an actuator.

The vehicle can also be taken for a test drive with the scan tool connected to it. Scan tools have several features that make intermittent problem identification easier. The scan tool can be used to monitor the activity of an input or circuit while it is being driven. This allows for the observation of that circuit's response to changing operating conditions. The snapshot or freeze-frame feature stores engine conditions and operating parameters at command or when the PCM set a DTC. If the snapshot can be taken when the intermittent problem is occurring, the problem will be easier to diagnose.

With an OBD II scan tool, actuators can be activated and their functionality tested. The consequences of the change in output operation can be monitored (Figure 8-30). Also, the activity of the outputs can be monitored as they respond to changes in sensor signals to the PCM. This can be helpful in locating an intermittent problem.

When an actuator is activated, watch the response on the scan tool. Listen for the clicking of the relay that controls that output. If no clicking is heard, measure the voltage at the relay's control circuit. There should be a change of more than 4 volts when the output is activated.

To monitor how the PCM and an output responds to a change in sensor signal, use the scan tool. Select the mode that relates to the suspected circuit and view and record the scan data for that circuit. Compare the reading to specifications, then create a condition that would cause the related inputs to change. Observe the scan data to see if the change was appropriate.

☑ SERVICE TIP: If necessary, some systems will unlock the catalytic converter to perform more accurate misfire diagnostics. An unlocked converter allows the engine to be better isolated for testing.

Repairing the System

After isolating the cause, repairs should be made. Then the system should be rechecked to validate proper operation and to verify that the symptom has been corrected. This may involve road testing the vehicle in order to verify that the complaint has been resolved.

Parts Replacement

The components in an OBD II system are designed to perform within specifications for 100,000 miles. The parts in these systems may not be interchangeable with parts from previous systems even though they have the same exterior appearance. For example, if an electronic ignition (EI) module from an EEC IV system is installed on an EEC V system, inaccurate profile ignition pickup (PIP) sensor signals are sent to the PCM, which activates the engine misfire monitor and illuminates the MIL.

OBD II Circuit Repairs

When servicing or repairing OBD II circuits, the following guidelines are important:

- ❏ Do not connect aftermarket accessories into an OBD II circuit.
- ❏ Do not move or alter grounds from their manufactured locations.
- ❏ Only repair OBD II circuits in accordance with their manufactured configuration.
- ❏ Always replace a relay in an OBD II circuit with the same replacement part. Damaged relays should be thrown away, not repaired.
- ❏ Make sure all connector locks are in good condition and are in place.
- ❏ After repair of connectors or connector terminals, make sure to achieve proper terminal retention.
- ❏ When installing an electrical ground fastener, be sure to apply the specified torque.
- ❏ After repair of connectors, make sure to reinstall the connector seals.

HO₂S Repair

If the HO$_2$S wiring, connector or terminal is damaged, the entire oxygen sensor assembly should be replaced. Do not attempt to repair the wiring, connector, or terminals. In order for this sensor to work properly, it must have a clean air reference. This clean air reference is obtained by way of the oxygen sensor signal and heater wires. Any attempt to repair the wires, connectors, or terminals could result in the obstruction of the air reference and degraded oxygen sensor performance.

The following guidelines should be followed when servicing the heated oxygen sensor:

- ❏ Do not apply contact cleaner or other materials to the sensor or wiring harness connectors. These materials may get into the sensor, causing poor performance.
- ❏ The sensor pigtail and harness wires must not be damaged in such a way that the wires inside are exposed. This could provide a path for foreign materials to enter the sensor and cause performance problems.
- ❏ Neither the sensor nor vehicle lead wires should be bent sharply or kinked. Sharp bends, kinks, etc., could block the reference air path through the lead wire.
- ❏ Do not remove or defeat the oxygen sensor ground wire. Vehicles that utilize the ground wire sensor may rely on this ground as the only ground contact to the sensor. Removal of the ground wire will cause poor engine performance.
- ❏ To prevent damage due to water intrusion, be sure the peripheral seal remains intact on the vehicle harness.

Wire Repair

Special Tools

Crimp and seal splice sleeves, crimping tool, electrician's tape

When a damaged wire is found during the visual inspection or if it is the cause of a problem, it should be replaced or repaired. Proper repair to the wires and/or connectors is critical to the proper operation of an OBD II system. Improper repairs can cause excessive resistance immediately after the repair or in the future. Always follow the manufacturer's guidelines for repairing wire and connectors.

Most often, manufacturers recommend that damaged sensor pigtails should be replaced rather than repaired. Since the pigtail comes with the sensor, replacing the pigtail means replacing the sensor. To repair a wire, cut out the damaged part, then strip the insulation off the ends of the wire and splice them together. Normally, splice clips or solder are the recommended ways to join two wire ends together. After the wire has been spliced, tightly wrap electrical tape around the splice. Make sure the tape covers the entire splice. The tape should not be flagged; rather, it should be tightly rolled onto the spliced area (Figure 8-31). If the spliced wire does not belong in some conduit or other harness covering, tape the wire again. Wrap the tape to cover the entire first splice (Figure 8-32).

Crimp and seal splice sleeves may be used on all types of insulation, except those designated by the manufacturer, which normally includes coaxial cable used for networking. They are also used where there are special requirements such as moisture sealing. There are different splice sleeves and different sizes of wire. Always use the proper splice sleeve. The splice sleeves are color coded according to applicable wire size:

Splice Sleeve Color	Wire Gauge
Salmon	18, 20
Blue	14, 16
Yellow	10, 12

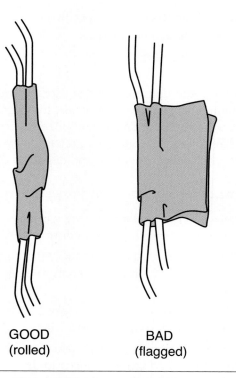

GOOD
(rolled)

BAD
(flagged)

TAPE AGAIN
(if needed)

Figure 8-31 A properly and improperly taped splice.

Figure 8-32 The appearance of the second taping on a splice.

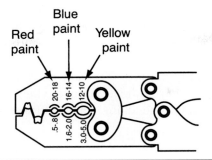

Figure 8-33 A hand crimping tool.

To install a splice sleeve, strip away about 5/16 inch of the insulation on both wire ends. Check the stripped wire for nicks or cut strands. If the wire is damaged, cut the damaged section and restrip. Select the proper splice sleeve according to wire size. Using the splice crimp tool, position the splice sleeve in the proper color nest of the crimp tool (Figure 8-33). The nests of the crimp tool are color coded and should be matched to the crimp sleeve color as follows:

Splice Sleeve Color	Crimp Tool Nest Color
Salmon	Red
Blue	Blue
Yellow	Yellow

Place the splice sleeve in the nest so that the crimp falls midway between the end of the barrel and the stop. The sleeve has a stop in the middle of the barrel to prevent the wire from going further (Figure 8-34). Lightly squeeze the crimper handles to hold the splice sleeve firmly

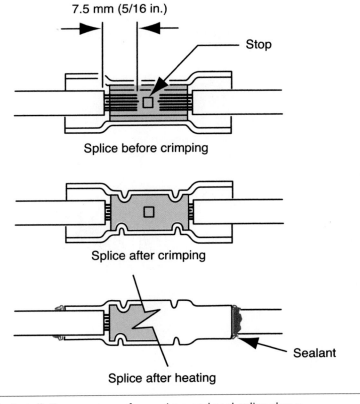

Figure 8-34 Installation sequence for a crimp and seal splice sleeve.

in the nest. Insert the wire into the sleeve until it hits the barrel stop and squeeze the crimper handles until the handles open when released. The crimper handles will not open until the proper amount of pressure is applied to the crimp sleeve. Repeat this same procedure to the other wire end.

Apply heat where the barrel is crimped. Gradually move the heat to the open end of the tubing, shrinking the tubing completely as the heat moves along the insulation. A small amount of sealant will come out of the end of the tubing when sufficient shrinking is achieved.

Connector Repair

Follow these steps to repair push-to-seat (Figure 8-35) and pull-to-seat (Figure 8-36) connectors:

GM refers to these retaining locks as Connector Position Assurance (CPA) Locks.

GM refers to these terminal locks as Terminal Position Assurance (TPA) Locks.

1. Remove any external locks used to hold the connector's halves together.
2. Remove any external locks used to keep the terminals from backing out of the connector.
3. Open any secondary locks on the connector. Secondary locks aid in the retention of the terminals in the connector, and they are normally molded to the connector shell.
4. Separate the connector halves and back out the seals.
5. Grasp the lead and push the terminal to its forwardmost position. Hold the lead in this position.
6. Locate the terminal lock in the connector canal.
7. Insert a pick directly into the canal.
8. Depress the locking tang to unseat the terminal. On push-to-seat connectors, gently pull on the lead to remove the terminal through the back of the connector. On pull-to-seat connectors, gently push the lead through the front of the connector.

 WARNING: Never force a terminal out of a connector.

9. Inspect the terminal and connector for damage or corrosion. Repair or replace the terminal as required.

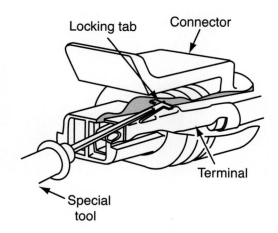

Figure 8-35 Typical push-to-seat connector and terminal.

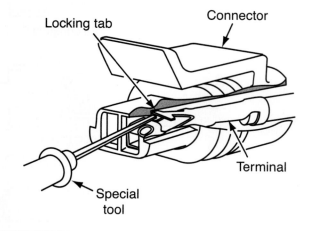

Figure 8-36 Typical pull-to-seat connector and terminal.

10. Reform the lock tang and reseat the terminal in the connector. Apply grease to the terminal if it was originally coated with grease.
11. Install the terminal and connector locks, close the secondary locks, and join the connector halves together.

To repair Weather Pack connectors (Figure 8-37):

1. Separate the connector halves.
2. Open the secondary lock.
3. Grasp the lead and push the terminal to its forwardmost position. Hold the lead in that position.
4. Insert the Weather Pack terminal removal tool into the front of the connector cavity until it rests on the cavity shoulder.
5. Gently pull on the lead to remove the terminal through the back of the connector.
6. Inspect the terminal and connector for damage. Repair or replace as required.
7. Reform the lock tang and reseat the terminal in the connector.
8. Close the secondary locks and join the connector halves.

1. Open secondary lock on connector.

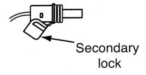

Secondary lock

2. Remove terminals using special tool.

3. Cut wire immediately behind cable seal.

4. Repair as shown below.

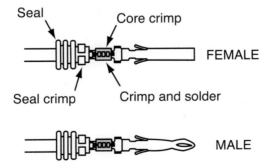

Figure 8-37 Typical Weather Pack connector and terminal.

Terminal Repair

The following procedure can be used to repair most types of terminal connectors, including pull-to-seat, push-to-seat, and Weather Pack terminals (Figure 8-38).

1. Cut off the terminal between the core and the insulation crimp and remove any seals.
2. Apply the correct seal according to the wire size and slide it far enough on the wire to allow for insulation stripping.
3. Strip the insulation the amount required for the terminal.
4. Align the seal with the end of the wire's insulation.
5. Position the stripped wire into the terminal.
6. Hand crimp the terminal's core wings.
7. Hand crimp the insulation wings. If the terminal is from a Weather Pack connector, make sure the seal is under the insulation wing before crimping.

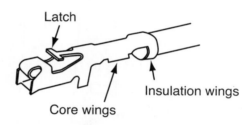

Typical push-to-seat terminal

Typical pull-to-seat terminal

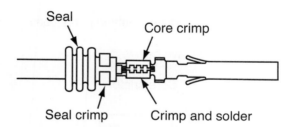

Typical weather pack® terminal

Figure 8-38 Typical connector terminals.

Verifying Vehicle Repair

Verification of repair is more comprehensive for vehicles with OBD II system diagnostics than earlier vehicles. After a repair, a technician should perform the following steps:

1. Review the fail records and the freeze-frame data for the DTC that was diagnosed. Record the fail records or freeze-frame data.

2. Clear the DTCs.

3. Operate the vehicle within the conditions noted in the fail records or the freeze-frame data.

4. Monitor the DTC status information for the specific DTC that has been diagnosed until the diagnostic test associated with that DTC runs. Some OEM scan tools can tell the PCM to run a monitor without waiting for enable criteria.

Clearing DTCs

In order to clear DTCs, use the diagnostic scan tool to clear DTCs or clear info function. When clearing DTCs, follow the instructions supplied by the tool manufacturer.

> **SERVICE TIP:** Never replace a computer until the ground wires and voltage supply wires to the computer are checked and proven to be in satisfactory condition. High resistance in computer ground wires may cause unusual problems.

CASE STUDY

A customer brought in his OBD II-equipped car into the dealership complaining of poor fuel economy. The customer stated that the gas mileage has been declining since the car was new. This is not normal. Gas mileage normally improves slightly as the engine is broken in. The customer had no other complaints. Because this is a very difficult problem to verify, the technician began the diagnostic process with a visual inspection and found nothing out of the ordinary. The car's MIL was not lit.

He then connected a scan tool to the DLC and reviewed the data. Comparing the input data being displayed to the normal range of values listed in the service manual, he discovered the upstream O_2 sensor was biased rich. That meant the PCM was seeing a lean condition and adding fuel to correct for this problem. To verify this, the technician watched the fuel trim. Sure enough, the long-term fuel trim had moved to add more fuel. Normally this is caused by a vacuum leak or restricted fuel injectors. The latter probable cause seemed unlikely since the O_2 sensor showed that added fuel was being delivered. Therefore, the technician began to look for a possible cause of a vacuum leak.

The process continued for quite some time as he checked all of the vacuum hoses and components. Nothing appeared to be leaking. As he leaned over to check something in the back of the engine, he heard a slight "pffft." The noise had somewhat of a rhythm to it, and he focused his attention to it. It did not sound like a vacuum leak. As he increased the engine's speed, the noise pulses became closer together and soon became a constant noise. The noise appeared to be coming from the lower part of the engine.

Using a stethoscope, he was able to identify the source of the noise. It appeared that the gasket joining the exhaust manifold to the exhaust pipe was not seated properly. To verify this, he raised the car on a hoist and took a look. He found the retaining bolts were loose. He took a quick look at the gasket and found it to be in reasonable shape, then tightened the bolts to specifications.

Not sure the noise or the exhaust leak was related to the problem but suspected that it could be, he connected a lab scope to the oxygen sensor and watched its activity. The sensor's signal no longer showed a bias. It appears that the leak in the exhaust was pulling air into the exhaust between each pulse. This was adding oxygen to the exhaust stream causing the computer to think the mixture was lean.

Terms to Know

Enable criteria	Parameter identification (PID) mode	Serial data
Long-term fuel trim (LTF)	Pending situation	Short-term fuel trim (SFT)
Output state test mode (OTM)		

ASE-Style Review Questions

1. While discussing the adaptive learning of a PCM:
   Technician A says when the battery is disconnected from the vehicle, the learning process resets.
   Technician B says the vehicle must be driven under part throttle with moderate acceleration in order for the PCM to relearn the system.
   Who is correct?
   A. A only   **C.** Both A and B
   B. B only   **D.** Neither A nor B

2. *Technician A* says that generic codes are designated P0 codes.
   Technician B says that manufacturer-specific codes are P1 codes.
   Who is correct?
   A. A only   **C.** Both A and B
   B. B only   **D.** Neither A nor B

3. *Technician A* says that if the PCM is correcting for a lean condition, the long-term fuel trim will indicate a negative number.
   Technician B says a rich condition causes the long-term fuel trim to show a positive number.
   Who is correct?
   A. A only   **C.** Both A and B
   B. B only   **D.** Neither A nor B

4. While discussing the MIL on OBD II systems:
   Technician A says the MIL will flash if the PCM detects a fault that would damage the catalytic converter.
   Technician B says whenever the PCM has detected a fault, it will set a code but not illuminate the MIL on the first failure if it is a B-type code.
   Who is correct?
   A. A only   **C.** Both A and B
   B. B only   **D.** Neither A nor B

5. While discussing diagnostic procedures:
   Technician A says after preliminary system checks have been made, the DTCs should be cleared from the memory of the PCM.
   Technician B says the data stream can be helpful when there are no DTCs but there is a fault in the system.
   Who is correct?
   A. A only   **C.** Both A and B
   B. B only   **D.** Neither A nor B

6. While discussing freeze frame-data:
   Technician A says this feature displays the conditions that existed when the PCM set a DTC.
   Technician B says the freeze-frame feature stores data even when a fault does not turn on the MIL.
   Who is correct?
   A. A only   **C.** Both A and B
   B. B only   **D.** Neither A nor B

7. *Technician A* says the enable criteria is the criteria that must be met before the PCM completes a monitor test.
   Technician B says a drive cycle includes operating the vehicle under specific conditions so that a monitor test can be completed.
   Who is correct?
   A. A only   **C.** Both A and B
   B. B only   **D.** Neither A nor B

8. While discussing DTCs:
   Technician A says OBD II regulations mandate that DTCs be stored according to a priority.
   Technician B says some of the DTCs listed in the OBD II regulations will not apply to some vehicles.
   Who is correct?
   A. A only   **C.** Both A and B
   B. B only   **D.** Neither A nor B

9. *Technician A* says the freeze-frame data can only be overriden with data associated with a misfire or fuel trim malfunction.
   Technician B says that all DTCs may not apply to some engines.
   Who is correct?
   A. A only   **C.** Both A and B
   B. B only   **D.** Neither A nor B

10. While discussing adaptive fuel control:
   Technician A says if there is a lean condition, the STFT will show a minus value on the scan tool.
   Technician B says the PCM will extend the injector pulse width if there is excess oxygen in the exhaust.
   Who is correct?
   A. A only   **C.** Both A and B
   B. B only   **D.** Neither A nor B

ASE Challenge Questions

1. A vehicle surges at cruise and erratic idle. The first operation the technician should perform after verifying the customer's complaint is:
   A. Check for stored DTCs.
   B. Check the fuel pressure.
   C. Test the battery and charging system.
   D. Disconnect any add-on equipment and repeat the road test.

2. *Technician A* says failure of a PCM to update the long-term fuel trim may be caused by the thermostat not being installed in the cooling system.
   Technician B says a thermostat rated too high may cause this condition.
   Who is correct?
   A. A only   **C.** Both A and B
   B. B only   **D.** Neither A nor B

3. An HO$_2$S DTC is retrieved and the fuel trim is incorrect.
   Technician A says this can be the result of a blocked referenced air path.
   Technician B says the sensor's wiring connectors may have been exposed to an engine cleaning agent.
   Who is correct?
   A. A only   **C.** Both A and B
   B. B only   **D.** Neither A nor B

4. *Technician A* says a wire nick caused by the stripping tool is acceptable and the splice can be completed.
   Technician B says splice sleeve may be squeezed closed with any clamping tool that will not cut the sleeve.
   Who is correct?
   A. A only   **C.** Both A and B
   B. B only   **D.** Neither A nor B

5. *Technician A* says a ground wire fastener must be torqued to specifications.
   Technician B says to completely clean the contact areas of the ground wire before fastening to the chassis.
   Who is correct?
   A. A only   **C.** Both A and B
   B. B only   **D.** Neither A nor B

Job Sheet 21

21

Name _____ Date _____

Conduct a Diagnostic Check on an Engine Equipped with OBD II

Upon completion of this job sheet, you should be able to conduct a system inspection and retrieve codes from the PCM of an OBD II-equipped engine.

ASE Correlation

This job sheet is related to the ASE Engine Performance Test's content area: computerized engine controls diagnosis and repair; task: retrieve and record stored diagnostic codes.

Tools and Materials

A vehicle equipped with OBD II   Service manual
Scan tool

Describe the vehicle being worked on.

Year _____ Make _____ VIN _____

Model_____

Engine size and type _____

Procedures

1. Check all vehicle grounds, including the battery and computer ground, for clean and tight connections. Comments:

   _____

2. Perform a voltage drop test across all related ground circuits. State where you tested and what your findings were:

   _____

   _____

3. Check all vacuum lines and hoses, as well as the tightness of all attaching and mounting bolts in the induction system. Comments:

   _____

4. Check for damaged air ducts. Comments:

   _____

5. Check the ignition circuit, especially the secondary cables, for signs of deterioration, insulation cracks, corrosion, and looseness. Comments:

   _____

   _____

6. Are there any unusual noises or odors? If there are, describe them and tell what may be causing them.

_____

_____

7. Inspect all related wiring and connections at the PCM. Comments:

_____

_____

8. Gather all pertinent information about the vehicle and the customer's complaint. This should include detailed information about the symptom from the customer, a review of the vehicle's service history, published TSBs, and the information in the service manual.

9. Are there any vacuum leaks? ☐ Yes ☐ No

10. Is the engine's compression normal? ☐ Yes ☐ No

11. Is the ignition system operating normally? ☐ Yes ☐ No

12. Are there any obvious problems with the air-fuel system? ☐ Yes ☐ No

13. Your conclusions from the above:

_____

_____

14. Check the operation of the MIL by turning the ignition on. Describe what happened and what this means.

_____

_____

☐ 15. Connect the scan tool to the DLC.

☐ 16. Enter the vehicle identification information into the scan tool.

☐ 17. Retrieve the DTCs with the scan tool.

18. List all codes retrieved by the scan tool.

_____

_____

_____

_____

19. Conclusions from these tests and checks:

_____

_____

_____

_____

Instructor's Response_____

_____

_____

Job Sheet 22

22

Name _____ Date _____

Monitor the Adaptive Fuel Strategy on an OBD II-Equipped Engine

Upon completion of this job sheet, you should be able to monitor the short-term and long-term fuel strategies of an OBD II system and determine if the trim is indicative of an abnormal condition.

ASE Correlation

This job sheet is related to the ASE Engine Performance Test's content area: computerized engine controls diagnosis and repair. Task: diagnose emissions or driveability problems resulting from failure of computerized engine controls with no diagnostic trouble codes stored and determine needed repairs.

Tools and Materials

A vehicle equipped with OBD II
Scan tool

Describe the vehicle being worked on.

Year _____ Make _____ VIN _____

Model _____

Procedures

Task Completed

1. Connect the scan tool. ☐
2. Start the engine. ☐
3. Pull up the PCM's data stream display. ☐
4. Observe the PID called SFT and note and record the value.
   _____

5. With the engine running, pull off a large vacuum hose and watch the SFT value. What did it do?
   _____

6. Look at the LFT value. What did it do?
   _____

7. State your conclusions from this test.
   _____

8. Reconnect the large vacuum hose. ☐
9. Observe the SFT and LFT values. Record what happened.
   _____

10. What was happening to the fuel trim and why did it happen?
    _____

Instructor's Response _____

_____

_____

_____

Diagnosing Related Systems

Upon completion and review of this chapter, you should be able to:

❏ Diagnose related systems for driveability problems.

❏ Name the common problems that occur with clutches that can directly affect driveability.

❏ Describe how to accurately perform a road test analysis.

❏ Explain when and how pressure tests are performed.

❏ Recognize conditions that adversely affect the performance of drums, shoes, linings, and related hardware.

❏ Describe how problems in the hydraulic brake system may seem like a driveability problem.

❏ Explain the importance of correct wheel alignment angles.

❏ Describe why certain factors affect tire performance, including inflation pressure, tire rotation, and tread wear.

Introduction

The purpose of this chapter is to make you aware of the relationship the engine has with the rest of the vehicle. This understanding is especially important when diagnosing driveability problems. Whenever there is a customer complaint that relates to how the car operates, always check the systems (Figure 9-1) that could cause the problem before diving into the engine and its systems. Often, diagnosis of these systems is much easier and less time consuming than those in the engine systems.

There is no attempt made in this chapter to cover all of the possible problems that will affect the way a vehicle operates. What is covered are the basic systems and components that are common causes of driveability problems. Further, the repair techniques for these problems are not covered; those should be covered in courses and textbooks that focus on these systems.

A driveability complaint is often centered on asking the vehicle to do something it was not designed to do. Carrying extremely heavy objects will put any vehicle to a test, especially

Basic Tools

Basic mechanic's tool set, appropriate service manual.

Look at the complete vehicle

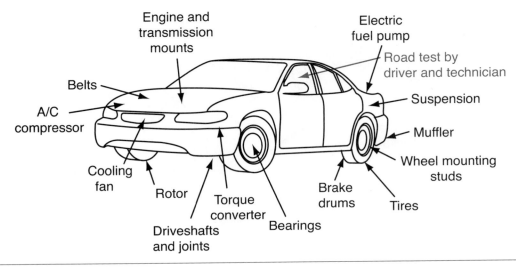

Figure 9-1 Possible sources of driveability symptoms.

if it is not powered by a larger engine or if the transmission's and final drive gear ratios are not designed for the heavy load. When this is the complaint, the owner should be advised not to use the vehicle with those kinds of loads. Heavy loads will cause poor acceleration, a general lack of power, and poor fuel economy. They can also create a very unsafe condition because the vehicle will have a hard time stopping and will not handle correctly.

⬤ **CUSTOMER CARE:** As a technician, it is necessary to become familiar with all aspects and systems of the vehicles being serviced. Always be aware of or inspect for other problems while the vehicle is being serviced. Regardless of the service being performed, take a look at other items such as abnormal tire wear, fluid or oil leaks, and fluid condition. Observe how the engine was running when it was driven into the service bay or listen for any abnormal brake noise. In other words, just because the vehicle is being serviced for perhaps only an oil and filter change, do not ignore other services that may be necessary even though they are unknown to the customer.

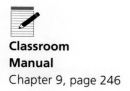

Classroom Manual
Chapter 9, page 246

Clutches

Clutches are used on vehicles with manual transmissions/transaxles. The clutch is used to mechanically connect the engine's flywheel to the transmission/transaxle input shaft (Figure 9-2). It does this through the use of a special friction plate that is splined to the input shaft. When the clutch is engaged, the friction plate contacts the flywheel and transfers power through the plate to the input shaft.

A bucking or jerking sensation can be felt if the clutch slips or if its splines to the transmission shaft are worn. Slipping can be caused by a contaminated clutch disc or incorrect clutch pedal freeplay (Figure 9-3). An unbalanced clutch assembly may cause the engine to run rough. This same problem can be caused by a loose flywheel or pressure plate. If the clutch does not release, the engine may be hard to start or perhaps will not crank at all unless the transmission is in neutral.

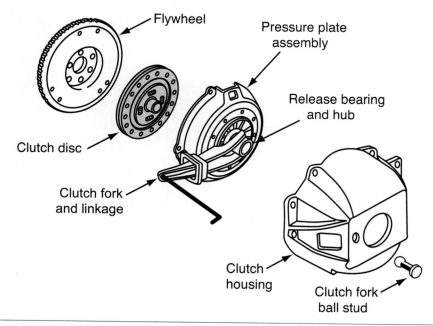

Figure 9-2 Parts of a clutch assembly.

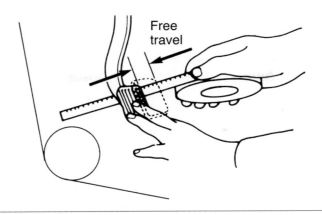

Free travel

Figure 9-3 Checking clutch pedal freeplay.

Slippage

Clutch slippage is a condition in which the engine overspeeds without generating any increase in torque to the driving wheels. It occurs when the clutch disc is not gripped firmly between the flywheel and the pressure plate. Instead, the clutch disc slips between these driving members. Slippage can occur during initial acceleration or subsequent shifts but is usually most noticeable in higher gears.

One way to check for slippage is by driving the vehicle. Normal acceleration from a stop and several gear changes indicate whether the clutch is slipping.

Slippage can also be checked in the shop. Check the service manual for correct procedures. A general procedure for checking clutch slippage follows. Be sure to follow the safety precautions stated earlier.

With the parking brake on, disengage the clutch. Shift the transmission into third gear and increase the engine speed to about 2,000 rpm. Slowly release the clutch pedal until the clutch engages. The engine should stall immediately.

If it does not stall within a few seconds, the clutch is slipping. Safely raise the vehicle and check the clutch linkage for binding or broken or bent parts. If no linkage problems are found, the transmission and the clutch assembly must be removed so that the clutch parts can be examined.

Clutch slippage can be caused by an oil-soaked or worn disc facing, warped pressure plate, weak diaphragm spring, or the release bearing contacting and applying pressure to the release levers.

Drag and Binding

If the clutch disc is not completely released when the clutch pedal is fully depressed, clutch drag occurs. Clutch drag causes gear clash, especially when shifting into reverse. It can also cause hard starting because the engine attempts to turn the transmission input shaft.

To check for clutch drag, start the engine, depress the clutch pedal completely, and shift the transmission into first gear. Do not release the clutch. Next, shift the transmission into neutral and wait 5 seconds before attempting to shift smoothly into reverse.

It should take no more than 5 seconds for the clutch disc, input shaft, and transmission gears to come to a complete stop after disengagement. This period, called the clutch spindown time, is normal and should not be mistaken for clutch drag.

If the shift into reverse causes gear clash, raise the vehicle safely and check the clutch linkage for binding or broken or bent parts. If no problems are found in the linkage, the transmission and clutch assembly must be removed so that the clutch parts can be examined.

Clutch drag can occur as a result of a warped disc or pressure plate, a loose disc facing, a defective release lever, or incorrect clutch pedal adjustment that results in excessive pedal play.

Binding can result when the splines in the clutch disc hub or on the transmission input shaft are damaged or when there are problems with the release levers.

Vibration

Clutch vibrations, unlike pedal pulsations, can be felt throughout the vehicle and can occur at any clutch pedal position. These vibrations usually occur at normal engine operating speeds (more than 1,500 rpm).

There are other possible sources of vibration that should be checked before disassembling the clutch to inspect it. Check the engine mounts and the crankshaft damper pulley. Look for any indication that engine parts are rubbing against the body or frame.

Accessories can also be a source of vibration. To check them, remove the drive belts one at a time. Set the transmission in neutral and securely set the emergency brake. Start the engine and check for vibrations. Do not run the engine for more than 1 minute with the belts removed.

If the source of vibration is not discovered through these checks, the clutch parts should be examined. Be sure to check for loose flywheel bolts, excessive flywheel runout, and pressure plate cover balance problems.

Manual Transmissions

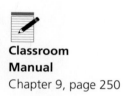

**Classroom
Manual**
Chapter 9, page 250

The purpose of the transmission or transaxle is to use gears of various sizes to give the engine a mechanical advantage over the driving wheels. During normal operating conditions, power from the engine is transferred through the engaged clutch to the input shaft of the transmission or transaxle. Gears in the transmission or transaxle housing alter the torque and speed of this power input before passing it on to other components in the powertrain. Without the mechanical advantage the gearing provides, an engine can generate only limited torque at low speeds. Without sufficient torque, moving a vehicle from a standing start would be impossible.

Unfortunately, nearly all driveability complaints caused by a manual transmission are actually the fault of the driver. The driver has simply selected the wrong gear for the conditions. Shifting quickly through the gears will cause the engine to seem sluggish. Without the benefit of gear reduction, the engine has a very hard time overcoming a load.

Automatic Transmissions

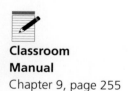

**Classroom
Manual**
Chapter 9, page 255

Until recently, all automatic transmissions were controlled by hydraulics. However, many new systems now feature computer-controlled operation of the torque converter and transmission. Based on input data supplied by electronic sensors and switches, the computer sets the torque converter's operating mode, controls the transmission's shifting sequence (Figure 9-4), and, in some cases, regulates transmission oil pressure.

A transmission that shifts too early or too late will affect fuel consumption and overall performance. When the transmission shifts too early, the available power to the wheels decreases, causing poor acceleration. When the gears change later than normal, the customer may complain about noise or may notice that fuel consumption has increased.

Shift quality is also important to good driveability. If the transmission slips, there will be a lack or power, poor acceleration, and poor fuel economy. Transmission slipping is normally caused by bad bands or clutches. Sometimes when there is slippage, the engine will seem to lurch or buck once the gear is finally fully engaged. There is a power loss and a sudden engagement of the gear that causes a surge of power.

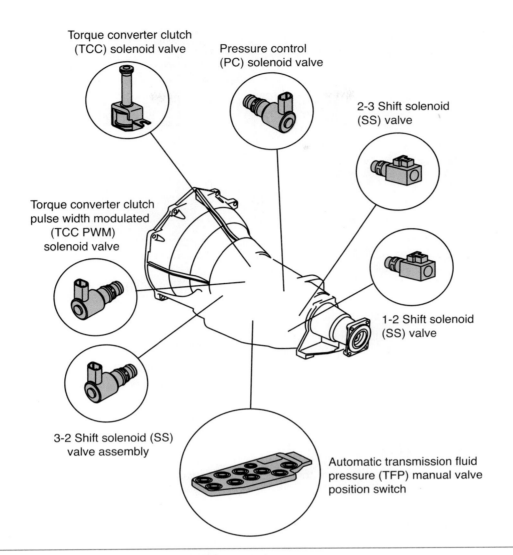

Torque converter clutch (TCC) solenoid valve

Pressure control (PC) solenoid valve

2-3 Shift solenoid (SS) valve

Torque converter clutch pulse width modulated (TCC PWM) solenoid valve

1-2 Shift solenoid (SS) valve

3-2 Shift solenoid (SS) valve assembly

Automatic transmission fluid pressure (TFP) manual valve position switch

Figure 9-4 An electronic automatic transmission.

Slippage can normally be traced to a leaking hydraulic circuit, low fluid level, or misadjusted band.

The engine can seem to run rough if the transmission—automatic or manual—mounts are loose or faulty. This is also true of bad engine mounts. The normal vibration of an engine is dampened by the mounts. If the mounts are bad, no or little dampening takes place.

Automatic transmission diagnosis and service is a highly specialized field of automotive work. Entire books cannot cover all the variables offered by the different manufacturers. But half of all automatic transmission problems result from improper fluid levels, fluid leaks, vacuum leaks, or electrical malfunctions.

Typical automatic transmission diagnostic steps may include the use of manufacturers' diagnostic guides, input from the customer, fluid and linkage checks, electrical checks, stall testing, road testing, and pressure testing.

✓ SERVICE TIP: A range reference chart is shown in Figure 9-5. The chart is useful in diagnosing problems with transmissions based on clutch and band application. For instance, the chart shows the high-reverse clutch on in both reverse and third gear. If this transmission had neither third gear nor reverse, then the high-reverse clutch is probably inoperative. Always use a range reference chart along with the road test to help accurately diagnose transmission problems.

Range		Gear ratio	Clutch		Low and reverse brake	Lockup	Band servo		One way clutch	Parking pawl
			High-reverse clutch (front)	Forward clutch (rear)			Operation	Release		
Park										X
Reverse		2.364	X		X					
Drive	D1	2.826		X					X	
	D2	1.543		X			X			
	D3	1.000	X	X		X	X	X		
2	2_1	2.865		X					X	
	2_2	1.543		X			X			
1	1_1	2.826		X	X				X	
	1_2	1.543		X			X			

Figure 9-5 A band and clutch application chart can be helpful when diagnosing transmission problems.

Diagnostic sheets, such as the one shown in Figure 9-6, are extremely helpful in accurately recording all useful information during testing and troubleshooting.

Road testing is exceedingly important when troubleshooting automatic transaxle and transmission problems. The objective of the road test is to confirm the customer's complaint. It is always wise to have the customer accompany the diagnostician on a road test to confirm the malfunction. The road test allows for the opportunity to check the transaxle or transmission operation for slipping, harshness, incorrect upshift speeds, and incorrect downshifts.

Torque Converters

Classroom Manual
Chapter 9, page 252

An automatic transmission eliminates the use of a mechanical clutch and shift lever. In place of a clutch, it uses a fluid coupling called a torque converter to transfer power from the engine's flywheel to the transmission input shaft. The torque converter allows for smooth transfer of power at all engine speeds.

A malfunctioning torque converter may provide less torque multiplication and a customer's complaint of a lack of power. An unbalanced or loose torque converter and/or flexplate will cause a vibration, which will make the engine seem to run rough or have a misfire (Figure 9-7).

Lockup Torque Converters

A lockup torque converter eliminates the 10 percent slip that takes place between the impeller and turbine at the coupling stage of operation. The engagement of a clutch between the engine crankshaft and the turbine assembly has the advantage of improving fuel economy and reducing torque converter operational heat and engine speed.

The lockup torque converter clutch assembly is controlled by the PCM. When the computer receives electronic signals from the different sensors confirming the requirements for lockup have been met, lockup clutch engagement begins. These sensors include an engine coolant sensor, vehicle speed sensor, engine vacuum sensor, and throttle position sensor.

If the converter clutch does not lock during cruising speeds, the customer will notice a decrease in fuel economy. Check the lockup circuit and the converter control circuit to determine why the clutch is not locking.

If the clutch locks too soon or remains locked throughout all engine speeds, many different complaints are possible. There may be a lack of power and poor acceleration. This is because while locked, the torque converter does not provide any torque multiplication. If the clutch is locked while the engine is idling, the engine will stall. If this situation exists, the engine may also stall during deceleration, especially when the wheels stop turning. Since the engine is mechani-

Date ____/____/____

TRANSMISSION/TRANSAXLE-CONCERN CHECK SHEET
(*information required for technical assistance)

*VIN _____ * Mileage_____ R.O.# ____

*Model year _____ *Vehicle Model_____ *Engine ____

* Trans. Model _____ *Trans. Serial _____

S E R V I C E A D V I S O R

CUSTOMER'S CONCERN
_____
_____
_____
_____

Check the items that describe the concern:

WHAT:	WHEN:	OCCURS:	USUALLY NOTICED:
_no power	_vehicle warm	_always	_idling
_shifting	_vehicle cold	_intermittent	_accelerating
_slips	_always	_seldom	_coasting
_noise	_not sure	_first time	_braking
_shudder			at ____MPH

T E C H N I C I A N

Preliminary Check Procedures   NOTE FINDINGS

Inspect
*fluid level and condition
*engine performance-vacuum & EMC codes
*TV cable and/or modulator vacuum
*manual linkage adjustment   *Simulate the conditions
road test to verify concern   under which the customer's
                                       concern was observed

* PROPOSED OR COMPLETED REPAIRS
_____
_____
_____
_____

TRANS. TEMPERATURE ____HOT ____COLD

R O A D   T E S T
P R E S S U R E   T E S T

PRND021

*VACUUM READINGS AT MODULATOR
(180C, 250C, 350C, 400, 410-T4)
Readings at Modulator_____IN. HG.(Engine at Hot Idle, Trans. in Drive)
Check for Vacuum Response During Accelerator Movement

	MINIMUM		MAXIMUM	
	SPEC. (from manual)	ACTUAL	SPEC. (from manual)	ACTUAL

*FINDINGS BASED ON ROAD TEST
Check items on road test that describe customer's comments about:

Garage Shift Feel	Upshifts	Downshifts	Torque Converter Clutch
_engine stops	_early	_busyness	_busyness
_harsh	_harsh	_harsh	_harsh
_delayed	_delayed	_delayed	_no release
_no drive	_slips	_slips	_shudder
	_no upshift	_no downshift	_early apply
			_no apply
			_late apply

Concerns Occur When/During:

Check	Gear	Range
		P-N
	1st	
	2nd	D
	3rd	
	4th	
	1st	
	2nd	D
	3rd	
	1st	2
	2nd	
	1st	1
	Reverse	R

During
_1-2 upshift
_2-3 upshift
_3-4 upshift
_4-3 downshift
_3-2 downshift
_2-1 downshift

Throttle Position
_light
_medium
_heavy
_W.O.T.

Noise
Type	When Noticed
_buzz	_always
_whine	_load sensitive
_clunk	_steering sensitive
	at _____MPH
	in _____gear

Pitch	Level
_low	_light
_medium	_medium
_high	_heavy

Figure 9-6 Typical transmission diagnostic checklist.

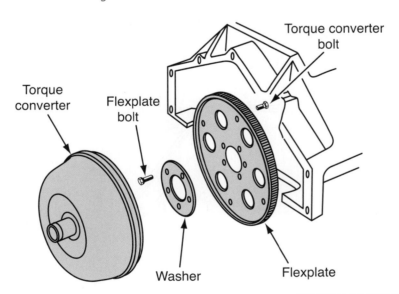

Figure 9-7 The torque converter mounted to a flexplate that is bolted to the engine's crankshaft.

cally connected to the drivetrain when the clutch is locked, if the wheels cannot turn, or if the engine's power is too low to make them turn, the engine will stop running.

If the converter locks too soon, there may be a surge or bucking as the clutch locks. The additional load of the drivetrain on the engine causes it to slow down until it produces enough power to overcome the load.

Diagnosing Transmission Fluid

If the fluid level is low and off the crosshatch section of the dipstick, the problem could be external fluid leaks. Low fluid levels cause a variety of problems. The condition allows air to be drawn into the pump inlet circuit. Aeration of the fluid causes slow pressure buildup and low fluid pressures, contributing to slipping automatic shifts.

High fluid levels cause the rotating planetary gears and parts to churn up the fluid and admit air developing foam. This condition is very similar to low fluid level. Aerated fluid can cause overheating, fluid oxidation—which contributes to varnish buildup—and interference with normal spool valve, clutch, and servo operation.

Uncontaminated automatic transmission fluid (ATF) is pinkish or red in color. A dark brownish or blackish color or a burnt odor indicates overheating. If the fluid has overheated, the ATF and the filter must be changed and the transmission inspected. A milky color can indicate that engine coolant is leaking into the transmission cooler in the radiator outlet tank. Bubbles on the dipstick indicate the presence of air. The bubbles are usually caused by a high-pressure leak.

Wipe the dipstick on absorbent white paper. Look at the fluid stain. Dark particles in the fluid indicate band or clutch material. Silvery metal particles indicate excessive wear on metal parts or housings. Varnish or gum deposits on a dipstick indicate the need to change the ATF and the transmission filter.

Band Adjustment

Slippage during shifting can indicate the need for band adjustment. Adjusting transmission bands may or may not be part of a scheduled maintenance program for the vehicle. Many transmissions have no provisions for adjusting bands. In other cases, the adjustment procedure involves draining the transmission fluid and removing the oil pan. Figure 9-8 shows a front kickdown band adjustment being made.

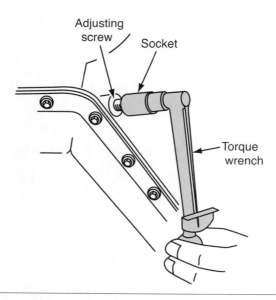

Figure 9-8 Adjusting a kickdown band.

Some late-model transmissions are sealed and require special procedures to check the transmission fluid level.

Overheated fluid has been described as smelling like burnt popcorn. It is also very dark and often contains particles of clutch material.

Electronically Shifted Automatic Transmissions

In modern automatic transmissions, the modulator has been replaced by the MAP sensor, the throttle valve cable by the throttle position (TP) sensor, and the governor by the vehicle speed sensor (VSS). Most transmissions have input and output speed sensors that the PCM uses to determine the actual gear ratio and compare it to the desired gear ratio. Some transmissions have line pressure controlled by pulse width modulated valves, and line pressure is determined by the PCM. Since these sensors also govern engine performance, a fault in these sensors can lead to complaints in both the engine and transmission. For instance, a faulty throttle position (TP) sensor can cause erratic torque converter clutch operation, high or low line pressure, erratic or fixed shift points, and improper engine operation, all of which could be interpreted as a major transmission failure (Figure 9-9). The technician must also be capable of determining if a transmission fault is electrical or hydraulic. Many transmissions have been replaced or disassembled only to find that the problem was electrical. Solid electrical diagnosis is a must on these systems. Some transmissions can be shifted by the scan tool or special diagnostic tools to aid in determining the fault (see Photo Sequence 17). For instance, if the transmission can be externally shifted by the use of a break-out box, then obviously the transmission is not at fault.

Components	Pass thru pins	Resistance at 20°C	Resistance at 100°C	Resistance to ground (case)
1-2 shift solenoid valve	A, E	19–24	24–31	Greater than 250M
2-3 shift solenoid valve	B, E	19–24	24–31	Greater than 250M
TCC solenoid valve	T, E	21–24	26–33	Greater than 250M
TCC PWM solenoid valve	U, E	10–11	13–15	Greater than 250M
3-2 shift solenoid valve assembly	S, E	20–24	29–32	Greater than 250M
Pressure control solenoid valve	C, D	3–5	4–7	Greater than 250M
Transmission fluid temp. (TFT) sensor	M, L	3088–3942	159–198	Greater than 10M
Vehicle speed sensor	A, B Vss conn	1420 @ 25°C	2140 @ 150°C	Greater than 10M

IMPORTANT: The resistance of this device is necessarily temperature dependent and will therefore vary far more than that of any other device. Refer to transmission fluid temp (TFT) sensor specifications.

Figure 9-9 Service manual page showing resistance values of a particular transmission.

Photo Sequence 17
Diagnosing an Electronic Automatic Transmission

P17-1 A vehicle is brought into the shop with a complaint of irregular transmission shifting and the MIL illuminated.

P17-2 After a preliminary quick check (fluid level, fuses), and a test drive that confirmed the transmission was not shifting, the technician installs a scan tool to read the trouble codes stored.

P17-3 The trouble code is P0751 "Shift Solenoid A Performance" (stuck off).

P17-4 The technician gathers service information on the shift solenoid action and DTC diagnostic information for the vehicle. The A solenoid is normally closed and is opened when activated.

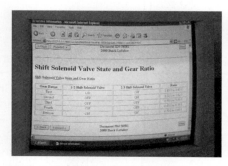

P17-5 The technician finds that according to the chart, if shift solenoid A is not able to turn on that the vehicle would start from first, be unable to shift into second or third, and finally shift into fourth. (The technician should not attempt to achieve fourth during a road test due to engine rpm.)

P17-6 As directed in the diagnostic chart for P0751, the technician moves the shifter through its range D1, D2, D3, D4, N, R, and Park while watching the scan tool "TR Switch" for a matching reading. The gear switch was determined to be okay.

P17-7 The technician raises the drive wheels off the ground and, with the scan tool, commands first, second, third, and fourth. Only first and fourth gears were present regardless of the commanded range.

P17-8 With the engine off and key on, use the scan tool to turn the solenoid on and off while listening with a stethoscope for the clicking sound of the solenoid as it activates. No click is heard.

P17-9 The technician checks for power and ground to the shift solenoid at the harness connector end when commanded by the scan tool. There is power to the connector but the PCM cannot or does not ground the solenoid when commanded.

Diagnosing an Electronic Automatic Transmission (continued)

P17-10 On close inspection, the technician finds that the terminal for the A solenoid is deformed. The technician replaces the terminal.

P17-11 The technician clears the codes and test drives the vehicle to verify the repair.

Computer and Solenoid Valve Tests

Refer to the manufacturer's service manual procedures for specific transmission computer tests. Be aware that improper test procedures can damage or destroy computer units immediately. Specific tests for input and output signals are made. Special hand-held computer testers might be required. If input signals are correct and output signals are incorrect, the computer unit must be replaced.

Solenoid valves (Figure 9-10) are tested by removing the electrical connector. Then the resistance of the solenoid coil can be checked between the connector and ground. Another way of testing a solenoid is to apply current with a small jumper wire to the disconnected solenoid valve. Correct operation can be heard or felt when the solenoid moves sharply. Refer to the manufacturer's service manuals and service bulletins for specific tests.

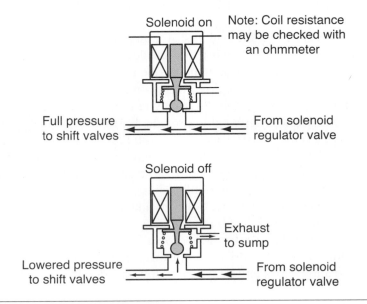

Figure 9-10 A converter clutch solenoid is checked the same way as any other solenoid.

Driveline

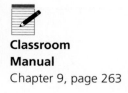

Classroom Manual
Chapter 9, page 263

Drivelines are used on rear-wheel drive vehicles and four-wheel drive vehicles. They connect the output shaft of the transmission with the gearing in the rear axle housing on rear-wheel drive vehicles. They are also used to connect the output shaft to the front and rear drive axles on a four-wheel drive vehicle.

A driveline consists of a hollow drive or propeller shaft that is connected to the transmission and drive axle differential by universal joints. These U-joints allow the drive shaft to move up and down with the rear suspension, preventing damage to the shaft. Driveline problems can set up a vibration that may appear as a rough running engine, especially at cruising or steady vehicle speeds.

Diagnosis of Drive Shaft and U-Joint Problems

A failed U-joint or damaged drive shaft can exhibit a variety of symptoms. A clunk that is heard when the transmission is shifted into gear is the most obvious. You can also encounter unusual noise, roughness, or vibration.

To help differentiate a potential drivetrain problem from other common sources of noise or vibration, it is important to note the speed and driving conditions at which the problem occurs. As a general guide, a worn U-joint is most noticeable during acceleration or deceleration and is less speed sensitive than an unbalanced tire (commonly occurring in the 30- to 60-mph range) or a bad wheel bearing (more noticeable at higher speeds). Unfortunately, it is often very difficult to accurately pinpoint drivetrain problems with only a road test. Therefore, expand the undercar investigation by putting the vehicle up on the lift where it is possible to achieve a good view of what is going on underneath.

Differential

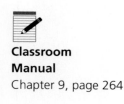

Classroom Manual
Chapter 9, page 264

All vehicles use a differential to provide an additional gear reduction (torque increase) above and beyond what the transmission or transaxle gearing can produce. This is known as the final drive gear.

In a transmission-equipped vehicle, the differential gearing is located in the rear axle housing. In a transaxle, the final reduction is produced by the final drive gears housed in the transaxle case.

The final drive gears consist of the pinion shaft pinion gear and the large differential ring gear. The fact that the driving pinion gear is much smaller than the driven ring gear leaves no doubt that there is substantial gear reduction and torque multiplication in the final drive gears.

Differentials are seldom the cause of driveability complaints unless the owner or previous owner has changed the ratio of the gearset. The new ratio may not offer the performance or fuel economy that the vehicle was designed to provide. If this is suspected, check the ratio of the final drive. If it is not a standard ratio, advise the customer. Also, check the accuracy of the speedometer. If the gearset was installed without installing the appropriate speedometer drive gear, the speedometer will read wrong. This alone can make the customer believe the vehicle is using too much fuel or is not accelerating quick enough (depending on the ratio installed). Engineers determine the ideal final drive ratio by considering the engine's power, the desired vehicle speed range, and the typical load on the vehicle.

A driveability complaint that can be caused by a final drive gearset is one of poor fuel economy. If the bearings in the final drive or transmission are bad, the shafts and gears will have a difficult time rotating. Extra engine power will be needed to keep them rotating. This wastes fuel.

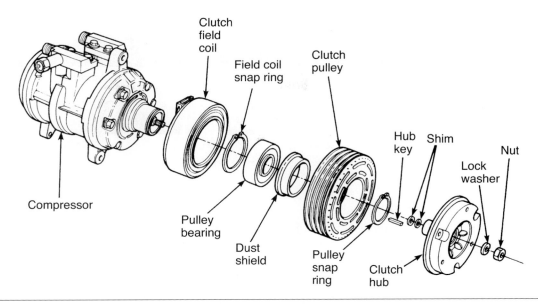

Figure 9-11 The air-conditioning compressor clutch and drive assembly.

Heating and Air Conditioning

An automotive air-conditioning system is a closed pressurized system. It consists of a compressor, condenser, receiver/dryer or accumulator, expansion valve or orifice tube, and an evaporator. A considerable amount of power from the engine is required to turn the compressor. Power loss when the air conditioning is running is normal. However, too much power loss suggests further diagnosis of the compressor and/or compressor drive (Figure 9-11).

Older vehicles were more prone to idle fluctuation as the A/C compressor and cooling fans operated and was considered normal operation. Although the operation of the A/C is still a heavy load for some engines, modern vehicles generally use "anticipatory" systems. Anticipatory means that since the PCM has control of both the idle speed and the air conditioning it "anticipates" the load of adding air conditioning. The idle can be adjusted and alternator output increased by the PCM to minimize the effect of a cycling air conditioning compressor before it comes on.

A normal condition may end up as a customer complaint—that is, a reduction in fuel economy. The load of the compressor uses power from the engine. The lost power requires a larger throttle opening to move the vehicle at a desired speed. This complaint is more common during very hot weather and when the vehicle is used in stop-and-go traffic.

Engine Cooling Fans

With the advent of transverse-mounted engines, electrical cooling fans found their way under the hood of the modern automobile. Today, most new cars are equipped with an electric cooling fan.

Air conditioned vehicles, especially those with small engines, usually have additional circuitry to ensure that the cooling fan comes on when the compressor cycles on. Airflow through the condenser must be present for A/C to function correctly. With the condenser mounted in front of the radiator, the logical method of ensuring airflow is to turn the cooling fan on as the A/C compressor cycles on.

On some engines, the turning on of the electric cooling fan will cause the engine to slow down. This is especially true if the air-conditioning compressor was engaged at the same time. This is a normal condition and cannot be corrected.

Classroom Manual
Chapter 9, page 265

Even though using an air conditioner will consume more fuel, it is more efficient than opening the windows. This is due to the added aerodynamic drag on the vehicle.

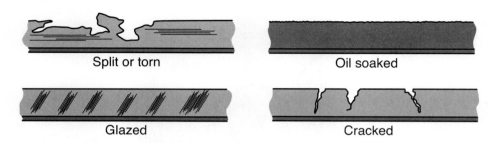

Split or torn   Oil soaked

Glazed   Cracked

Figure 9-12 Belt defects.

Although most newer vehicles have an electric cooling fan, some new vehicles have a cooling fan driven by the engine. These fans can cause a rough running engine if the fan is not balanced properly, if its drive hub is damaged, or if it is loose on its mounting. A vibration may also result from a bad drive belt (Figure 9-12). Of course, anything that is out of round or imbalanced and rotates with the engine will cause the engine to vibrate or run rough.

Speed Control Systems

**Classroom
Manual**
Chapter 9, page 268

Cruise or speed control systems are designed to allow the driver to maintain a constant speed (usually about 30 mph) without having to apply continual foot pressure on the accelerator pedal. Selected cruise speeds are easily maintained and can be easily changed. Several override systems also allow the vehicle to be accelerated, slowed, or stopped. Because of the constant changes and improvements in technology, each cruise control system may be considerably different.

If the cruise control unit has a hard time maintaining a fixed speed, the engine speed will surge up and down. To verify this problem as the cause of an engine surge at cruising speeds, simply allow the vehicle to cruise at highway speeds with and without the cruise control engaged. If the cruise control is at fault, the surging will only occur when it is engaged.

Brake Systems

**Classroom
Manual**
Chapter 9, page 269

The *caliper* of a disc brake assembly is a housing containing the pistons and related seals, springs, and boots, as well as the cylinders and fluid passages necessary to force the friction linings or pads against the rotor. The caliper resembles a hand in the way it wraps around the edge of the rotor. It is attached to the steering knuckle. Some models employ light spring pressure to keep the pads close against the rotor. In other caliper designs, this is achieved by a unique type of seal that allows the piston to be pushed out the necessary amount, then retracts it just enough to pull the pad off the rotor (Figure 9-13).

Unlike shoes in a drum brake, the pads act perpendicular to the rotation of the disc when the brakes are applied. This effect is different from that produced in a brake drum, where frictional drag actually pulls the shoe into the drum. Disc brakes are said to be non-energized, and so require more force to achieve the same braking effort. For this reason, they are ordinarily used in conjunction with a power brake unit.

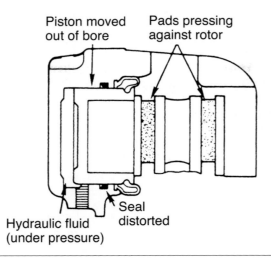

Piston moved out of bore

Pads pressing against rotor

Hydraulic fluid (under pressure)

Seal distorted

Figure 9-13 Cross section of a brake caliper. Notice the seal around the piston. When it relaxes, it helps pull the piston back into the bore.

Failure of the brakes to release is often caused by a tight or misaligned connection between the power unit and the brake linkage. Broken pistons, diaphragms, bellows, or return springs can also cause this problem.

To help pinpoint the problem, loosen the connection between the master cylinder and the brake booster. If the brakes release, the problem is caused by internal binding in the vacuum unit. If the brakes do not release, look for a crimped or restricted brake line or similar problem in the hydraulic system.

Proper adjustment of the master cylinder pushrod is necessary to ensure proper operation of the power brake system (Figure 9-14). A pushrod that is too long causes the master cylinder

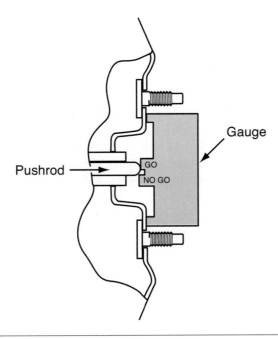

Gauge

Pushrod

GO
NO GO

Figure 9-14 A method for correctly adjusting the master cylinder's pushrod.

piston to close off the replenishing port, preventing hydraulic pressure from being released and resulting in brake drag. A pushrod that is too short causes excessive brake pedal travel and causes groaning noises to come from the booster when the brakes are applied. A properly adjusted pushrod that remains assembled to the booster with which it was matched during production should not require service adjustment. However, if the booster, master cylinder, or pushrod are replaced, the pushrod might require adjustment.

Inspection and Service

Road testing allows the brake technician to evaluate brake performance under actual driving conditions. Whenever practical, perform the road test before beginning any work on the brake system. In every case, road test the vehicle after any brake work to make sure the brake system is working safely and properly.

> ⚠️ **WARNING:** Before test driving any car, first check the fluid level in the master cylinder. Depress the brake pedal to be sure there is adequate pedal reserve. Make a series of low-speed stops to be sure that the brakes are safe for road testing. Always make a preliminary inspection of the brake system in the shop before taking the vehicle on the road.

Brakes should be road tested on a dry, clean, reasonably smooth, and level roadway. A true test of brake performance cannot be made if the roadway is wet, greasy, or covered with loose dirt. Not all tires grip the road equally. Testing is also adversely affected if the roadway is crowned so as to throw the weight of the vehicle toward the wheels on one side or if the roadway is so rough that wheels tend to bounce.

Test brakes at different speeds with both light and heavy pedal pressure. Avoid locking the wheels and sliding the tires on the roadway. There are external conditions that affect brake road test performance. Tires having unequal contact and grip on the road cause unequal braking. Tires must be equally inflated, and the tread pattern of right and left tires must be approximately equal. When the vehicle has unequal loading, the most heavily loaded wheels require more braking power than others. A heavily loaded vehicle requires more braking effort. A loose front-wheel bearing permits the drum and wheel to tilt and have spotty contact with the brake linings, causing erratic brake action. Misalignment of the front end causes the brakes to pull to one side. Also, a loose front-wheel bearing could permit the disc to tilt and have spotty contact with brake shoe linings, causing pulsations when the brakes are applied. Faulty shock absorbers that do not prevent the car from bouncing on quick stops can give the erroneous impression that the brakes are too severe.

Cylinder binding can be caused by rust deposits, swollen cups due to fluid contamination, or by a cup wedged into an excessive piston clearance. If the clearance between the pistons and the bore wall exceeds allowable values, a condition called heel drag might exist. It can result in rapid cup wear and can cause the piston to retract very slowly when the brakes are released. Brake caliper binding can be caused by sticking caliper slides.

Parking brakes can also be a cause of poor performance and bad fuel economy. If the parking brakes do not release, the rear brakes will drag. Even if the brakes are only slightly dragging, driveability will be affected.

Suspension and Steering Systems

Wheel alignment allows the wheels to roll without scuffing, dragging, or slipping on different types of road conditions. This gives greater safety in driving, easier steering, longer tire life, reduction in fuel consumption, and less strain on the parts that make up the front end of the vehicle.

Proper alignment of both the front and the rear wheels ensures easy steering, comfortable ride, long tire life, and reduced road vibrations.

There is a multitude of angles and specifications that the automotive manufacturers must consider when designing a car. The multiple functions of the suspension system complicate things a great deal for design engineers. They must take into account more than basic geometry. Durability, maintenance, tire wear, available space, and production cost are all critical elements. Most elements contain a degree of compromise in order to satisfy the minimum requirements of each.

Most technicians do not need to be concerned with all of this. All they need to do is restore the vehicle to the condition the design engineer specified. To do this, the technician must be totally familiar with the purpose of basic alignment angles.

The alignment angles are designed in the vehicle to properly locate the vehicle's weight on moving parts and to facilitate steering. If these angles are incorrect, the vehicle is misaligned. Before making adjustments, conduct the following prealignment checks.

Begin the alignment with a road test. While driving the car, check to see that the steering wheel is straight. Feel for vibration in the steering wheel as well as in the floor or seats. Notice any pulling or abnormal handling problems such as hard steering, tire squeal while cornering, or mechanical pops or clunks. This helps find problems that must be corrected before proceeding with the alignment.

Conduct a visual inspection. This includes tire wear and mismatched tire sizes or types. Look for the results of collision damage and towing damage. It should also include a ride height measurement. Every car is designed to ride at a specific curb height. Curb height specifications and the specific measuring points are given in service manuals.

With the vehicle raised, inspect all steering components such as control arm bushings, upper strut mounts, pitman arm, idler arm, center link, tie-rod ends, ball joints, and shock absorbers. Check the CV joints (if equipped) for looseness, popping sounds, binding, and broken boots. Damaged components must be repaired before adjusting alignment angles.

Caster is designed to provide steering stability. The caster angle for each wheel on an axle should be equal. Unequal caster angles cause the vehicle to steer toward the side with less caster. Too much negative caster can cause the vehicle to have sensitive steering at high speeds. The vehicle might wander as a result of negative caster. Caster is not a wear angle.

Caster is affected by worn or loose strut rod and control arm bushings. Caster adjustments are not possible on some strut suspension systems. Where they are provided, they can be made at the top or bottom mount of the strut assembly.

Camber angle changes, through the travel of the suspension system, are controlled by pivots. Camber is affected by worn or loose ball joints, control arm bushings, and wheel bearings. Anything that changes chassis height also affects camber. Camber is adjustable on most vehicles

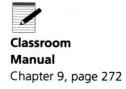

Classroom Manual
Chapter 9, page 272

Ride height is also called the curb or trim height.

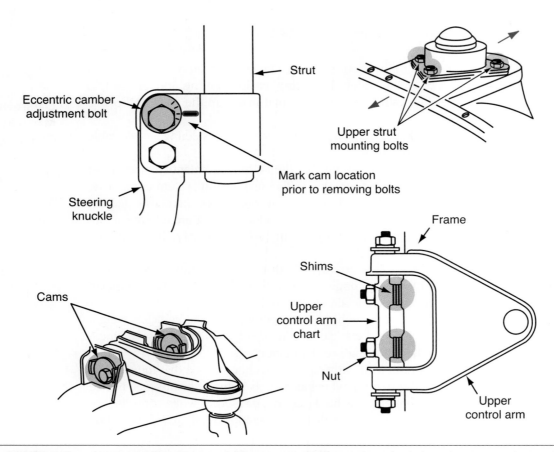

Figure 9-15 Different suspension designs and the various locations and methods for adjusting camber and caster.

(Figure 9-15). Some manufacturers prefer to include a camber adjustment at the spindle assembly. Camber adjustments are also provided on some strut suspension systems at the top mounting of the strut. Very little adjustment of camber (or caster) is required on strut suspensions if the tower and lower control arm positions are in their proper place. If serious camber error has occurred and the suspension mounting positions have not been damaged, it is an indication of bent suspension parts. In this case, diagnostic angle and dimensional checks should be made on the suspension parts. Damaged parts should be replaced.

Incorrect toe increases the rolling resistance of the tires. As the vehicle moves straight, the tires are dragged on their sides with excessive toe in and toe out. If this is suspected, check the tires for feathering.

Toe adjustments are made at the tie rod (Figure 9-16). They must be made evenly on both sides of the car. If the toe settings are not equal, the car may tend to pull due to the steering

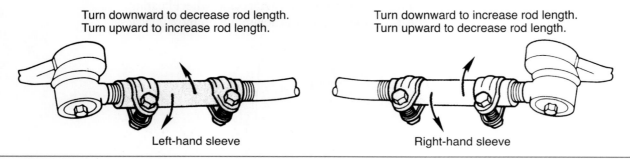

Figure 9-16 Toe is adjusted at the tie-rod sleeves.

wheel being off-center. This pull condition is especially common with power-assisted rack-and-pinion gears. It can be corrected by making the toe adjustments equal on both sides of the car. Toe is the last adjustment made in an alignment.

If rear toe does not parallel the vehicle centerline, a thrust direction to the left or right is created. This difference of rear toe from the geometric centerline is called the thrust angle. The vehicle tends to travel in the direction of the thrust line rather than straight ahead.

To correct this problem, begin by setting individual rear-wheel toe equally in reference to the geometric centerline. Four-wheel-alignment machines check individual toe on each wheel. Once the rear wheels are in alignment with the geometric centerline, set the individual front toe in reference to the thrust angle. Following this procedure assures that the steering wheel is straight ahead for straight-ahead travel. If you set the front toe to the vehicle geometric center-line ignoring the rear toe angle, a cocked steering wheel results.

Checking the SAI angle can help locate various problems that affect wheel alignment. For example, an SAI angle or SAI angle that varies from side to side may indicate an out-of-position upper strut tower, a bowed lower control arm, or a shifted center cross member.

On a short-long arm suspension, SAI is the angle between true vertical and a line drawn from the upper ball joint through the lower ball joint. In strut-equipped vehicles, this line is drawn through the center of the strut's upper mount down through the center of the lower ball joint.

When the camber angle is added to the SAI angle, the sum of the two is called the included angle. Comparing SAI included and camber angles can also help identify damaged or worn components. For example, if the SAI reading is correct but the camber and included angles are less than specifications, the steering knuckle or strut tower may be bent.

When a car has a steering problem, the first diagnostic check should be a visual inspection of the entire vehicle for anything obvious: bent wheels, misalignment of the cradle, and so on. If there is nothing obviously wrong with the car, make a series of diagnostic checks without disassembling the vehicle. One of the most useful checks that can be made with a minimum of equipment is a jounce-rebound check.

This jounce-rebound check determines if there is misalignment in the rack-and-pinion gear. For a quick check, unlock the steering wheel and see if it moves during the jounce or rebound. For a more careful check, use a pointer and a piece of chalk. Use the chalk to make a reference mark on the tire tread and place the pointer on the same line as the chalk mark. Jounce and rebound the suspension system a few times while someone watches the chalk mark and the pointer. If the mark on the wheel moves unequally in and out on both sides of the car, chances are there is a steering arm or gear out of alignment. If the mark does not move or moves equally in and out on both sides of the car, the steering arm and gear are probably all right. Each wheel or side should be checked.

Like front camber, rear camber affects both tire wear and handling. The ideal situation is to have zero running camber on all four wheels to keep the tread in full contact with the road for optimum traction and handling.

Camber is not a static angle. It changes as the suspension moves up and down. Camber also changes as the vehicle is loaded and the suspension sags under the weight.

CAUTION: Never jack up or lift a FWD vehicle on its rear axle. The weight of the vehicle may cause the axle to bend and result in misalignment of the rear wheels. Always lift the vehicle at the recommended lifting points.

Besides wearing the tires unevenly across the tread, uneven side-to-side camber (as when one wheel leans in and the other does not) creates a steering pull just like it does when the camber readings on the front wheels do not match. It is like leaning on a bicycle. A vehicle always pulls toward a wheel with the most positive camber. If the mismatch is at the rear wheels, the rear axle pulls toward the side with the greatest amount of positive camber. If the rear axle pulls to the right, the front of the car drifts to the left resulting in a steering pull even though the front wheels may be perfectly aligned.

Rear toe, like front toe, is a critical tire rear angle. If toed in or toed out, the rear tires scuff just like the front ones. Either condition can also contribute to steering instability as well as reduced braking effectiveness. (Keep this in mind with antilock brake systems.)

If rear toe is not within specifications, it affects tire wear and steering stability just as much as front toe. A total toe reading that is within specifications does not necessarily mean the wheels are properly aligned—especially when it comes to rear toe measurements. If one rear wheel is toed in while the other is toed out by an equal amount, total toe would be within specifications. However, the vehicle would have a steering pull because the rear wheels would not be parallel to center.

Remember, the ideal situation is to have all four wheels at zero running toe when the car is traveling down the road. This is especially true with antilock brakes where improper toe can affect brakes. Such a condition can affect brake balance when braking on slick or wet surfaces, causing the antilock brakes to cycle on and off to prevent a skid. Without antilock brakes, this condition may upset traction enough to cause an uncontrollable skid.

If rear toe cannot be easily changed, the next best alternative is to align the front wheels to the rear axle thrust line rather than the vehicle centerline. Doing this puts the steering wheel back on center and eliminates the steering pull—but it does not eliminate dog tracking.

Four-Wheel Alignment

The primary objective of **four-wheel alignment** (or total wheel alignment, as it is frequently called), whether front or rear drive, solid axle, or independent rear suspension, is to align all four wheels so the vehicle drives and tracks straight with the steering wheel centered. To accomplish this, the wheels must be parallel to one another and perpendicular to a common centerline.

Total toe for all four wheels must be determined and rear toe adjusted where possible to bring the rear axle or wheels into square with the chassis. The front toe setting can then be adjusted to compensate for any rear alignment deviation that might persist.

Four-wheel alignment also includes checking and adjusting rear-wheel camber as well as toe and performing all the traditional checks of front camber, toe, caster, toe out on turns, and steering axis inclination. The most important thing a four-wheel alignment job tells the technician is whether or not the rear axle or rear wheels are square with respect to the front wheels and chassis.

Wheels and Tires

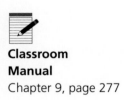

Classroom Manual
Chapter 9, page 277

The only contact a vehicle has with the road is through its tires and wheels. Tires are filled with air and made of rubber and other materials to give them strength. The air in the tires cushions the ride of the vehicle. Wheels are made of metal and are bolted to the axles or spindles. Wheels hold the tires in place. Wheels and tires come in many different sizes. Their sizes must be matched to one another and to the automobile. Different wheel widths with the same sized tires will cause pulling and will change the circumference of the tires.

Tires should be inflated to the recommended amount of pressure. The ideal amount of pressure is listed on the identification decal of the vehicle (Figure 9-17) and is given in the vehicle's owner's manual. Underinflated tires have more rolling resistance and, therefore, will cause fuel economy, as well as available power to decrease. Tires that are only a little low will still have a negative effect on fuel economy. Check the pressure in the tires and look at the tires for outside wear. If the tires are severely worn, tell the customer that they should be replaced and explain why.

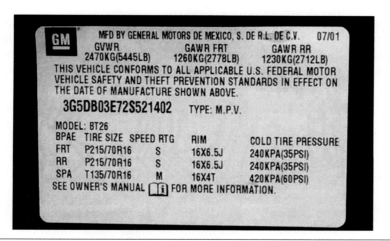

MFD BY GENERAL MOTORS DE MEXICO, S. DE R.L. DE C.V. 07/01
GVWR   GAWR FRT   GAWR RR
2470KG(5445LB)   1260KG(2778LB)   1230KG(2712LB)
THIS VEHICLE CONFORMS TO ALL APPLICABLE U.S. FEDERAL MOTOR
VEHICLE SAFETY AND THEFT PREVENTION STANDARDS IN EFFECT ON
THE DATE OF MANUFACTURE SHOWN ABOVE.

3G5DB03E72S521402   TYPE: M.P.V.

MODEL: BT26
BPAE TIRE SIZE SPEED RTG   RIM   COLD TIRE PRESSURE
FRT   P215/70R16   S   16X6.5J   240KPA(35PSI)
RR   P215/70R16   S   16X6.5J   240KPA(35PSI)
SPA   T135/70R16   M   16X4T   420KPA(60PSI)
SEE OWNER'S MANUAL [i] FOR MORE INFORMATION.

Figure 9-17 Typical tire placard on a vehicle.

Tire pressure monitoring systems will soon be installed on all new vehicles.

If replacement tires have been installed on the vehicle and they are not the same size as original equipment, fuel economy and engine performance will be affected. The change in tire diameter changes the overall gear ratio of the drivetrain. The overall drive gear ratio is determined by the gear ratios in the drivetrain and the circumference of the tire. The circumference of the tire determines how many times it will rotate in a mile. "Reflashing" the PCM is a common procedure outlined in Photo Sequence 15.

CASE STUDY

A regular customer brings a recently purchased 2003 Mustang GT into the shop. The used car was bought just two months ago. The customer's main complaint is that the car gets extremely poor gas mileage. The customer states that the car runs well and has plenty of power, but it seems to use much more fuel than he had anticipated it would.

The technician takes the car out for a test drive. The car runs very well. In fact, the car accelerates much quicker than the technician experienced in similar cars. Knowing the customer, the technician knows the complaint is legitimate. Therefore, she begins to think of the possible causes for poor fuel mileage. Because the engine runs well, it is unlikely that an engine problem is the cause unless the engine has been modified. Another possible cause is a fuel leak. After a detailed visual inspection, the technician finds no evidence of modifications or fuel leaks.

The technician decides to road test the car again. This time she notices a problem with the speedometer. It seems to be reading much higher speeds than the car is actually traveling. To verify this she drives the car onto the freeway and observes the odometer and the mileage markers along the side of the road. The speedometer is in error. It records speeds that are higher than actual speeds and measures a mile well before completing an actual mile.

Odometer error was causing the owner to perceive that the gas mileage was very poor. But why is the odometer in error? Another careful visual inspection is conducted. This time attention is paid to the transmission and the rear axle. Sealing compound is present around the outside of the differential housing. This indicates that the unit has been disassembled. She rotates the rear wheels while observing the drive shaft. Doing so, she is able to closely determine the final drive gear ratio. Comparing her findings to the specifications listed in the service manual, she finds that the gear ratio is much higher numerically than the standard gear ratio. She concludes that the gears had been changed for better performance and that is what caused the error in the speedometer.

The technician contacts the owner and explains what she has found. She gives the owner two options to correct the problem. Either he should replace the gears with gears of the standard ratio or have the PCM recalibrated for the different gear ratio with one that matches the final drive gear. Because the latter is less costly, the owner elects to recalibrate the speedometer.

Terms to Know

Clutch slippage   Four-wheel alignment

ASE-Style Review Questions

1. *Technician A* says tire inflation pressure greatly affects the amount of rolling resistance the tire has. *Technician B* says the recommended tire pressure is given on the side of the tire.
   Who is correct?
   A. A only   **C.** Both A and B
   B. B only   **D.** Neither A nor B

2. *Technician A* says that clutch slippage is most noticeable in higher gears.
   Technician B says that clutch slippage may not be as noticeable in lower gears.
   Who is correct?
   A. A only   **C.** Both A and B
   B. B only   **D.** Neither A nor B

3. *Technician A* says that clutch drag can cause hard starting.
   Technician B says that clutch drag can cause poor acceleration.
   Who is correct?
   A. A only   **C.** Both A and B
   B. B only   **D.** Neither A nor B

4. *Technician A* says throttle position is an important input in most electronic shift control systems.
   Technician B says vehicle speed is an important input for most electronic shift control systems.
   Who is correct?
   A. A only   **C.** Both A and B
   B. B only   **D.** Neither A nor B

5. *Technician A* says that automatic transmission fluid smells burned if the transmission has been overheated.
   Technician B says that a slipping automatic transmission can indicate the need for band adjustment.
   Who is correct?
   A. A only   **C.** Both A and B
   B. B only   **D.** Neither A nor B

6. *Technician A* says improper shifting can be caused by a hydraulic problem.
   Technician B says that improper shifting can be caused by an electrical manfunction.
   Who is correct?
   A. A only   **C.** Both A and B
   B. B only   **D.** Neither A nor B

7. *Technician A* says that a prerequisite to accurate road testing analysis is knowing what clutch or bands are applied in a particular gear range.
   Technician B says that all slipping conditions can be traced to a leaking hydraulic circuit.
   Who is correct?
   A. A only   **C.** Both A and B
   B. B only   **D.** Neither A nor B

8. *Technician A* says that if input signals are correct and output signals are incorrect, the computer unit must be replaced.
   Technician B tests a solenoid by checking the voltage between the solenoid's connector and ground.
   Who is correct?
   A. A only   **C.** Both A and B
   B. B only   **D.** Neither A nor B

9. *Technician A* says binding brakes may be caused by brake fluid contamination.
   Technician B says binding brakes may be caused by a worn caliper piston or cylinder.
   Who is correct?
   A. A only
   B. B only
   C. Both A and B
   D. Neither A nor B

10. *Technician A* says it is normal for the engine's idle to decrease when the air-conditioning compressor is turned on.
    Technician B says if the torque converter clutch does not disengage, the engine will not idle.
    Who is correct?
    A. A only
    B. B only
    C. Both A and B
    D. Neither A nor B

ASE Challenge Questions

1. A vehicle appears to have a power surge while traveling between 45 and 55 mph and up slight inclines.
   Technician A says this may be caused by a transmission torque converter "hunting."
   Technician B says this is caused by the A/C compressor cycling.
   Who is correct?
   A. A only
   B. B only
   C. Both A and B
   D. Neither A nor B

2. An engine has an above specified idle speed. This may be the result of a:
   A. leaking brake booster diaphragm.
   B. leaking transmission vacuum modulator diaphragm.
   C. restriction in the brake booster vacuum hose.
   D. either A or B.

3. *Technician A* says an electric cooling fan can slow the engine's idle speed.
   Technician B says the wrong size speedometer drive gear will cause a drop in fuel mileage.
   Who is correct?
   A. A only
   B. B only
   C. Both A and B
   D. Neither A nor B

4. A customer complains of a rough idle and engine stalling. This could be the result of:
   A. a leaking brake booster diaphragm.
   B. an improperly adjusted pushrod between the master cylinder and the booster.
   C. a leak in the cruise control diaphragm.
   D. any of the above.

5. It has been determined that the A/C was overcharged. The customer complaint was:
   A. belt squealing.
   B. intermittent engine stall at idle.
   C. a drop in fuel mileage.
   D. all or any of the above.

Job Sheet 23

23

Name _____ Date _____

Road Test a Vehicle to Check the Operation of the Automatic Transmission

Upon completion of this job sheet, you should be able to determine if the cause of a driveability complaint is the transmission or torque converter.

ASE Correlation

This job sheet is related to the ASE Engine Performance Test's content areas: general engine diagnosis and computerized engine controls diagnosis and repair. Tasks: interpret and verify complaint and determine needed repairs; diagnose driveability and emissions problems resulting from failures of interrelated systems and determine needed repairs.

Tools and Materials

A vehicle with an automatic transmission
Service manual
Clean shop rag
Pad of paper and pencil

Describe the vehicle being worked on.

Year _____ Make _____ VIN _____

Model_____

Model and type of transmission _____

Procedures

Task Completed

1. Park the vehicle on a level surface. ☐

2. Wipe all dirt off of the protective disc and the dipstick handle. ☐

3. Start the engine and allow it to reach operating temperature. ☐

4. Remove the dipstick and wipe it clean with a lint-free cloth or paper towel. ☐

5. Reinsert the dipstick, remove it again, and note and record the reading.

   _____

6. Describe the condition of the fluid (color, condition, and smell):

   _____

   _____

7. What is indicated by the fluid's condition?

   _____

   _____

8. Find and duplicate from a service manual the chart which shows the band and clutch application for different gear selector positions. Using these charts will greatly simplify your diagnosis of automatic transmission problems. It is also wise to have a notebook or piece of paper to jot down notes about the operation of the transmission. ☐

9. Inspect the transmission for signs of fluid leakage. Comments:

_____

_____

10. Drive the vehicle at normal speeds to warm the engine and transmission. Describe the behavior of the transmission and torque converter:

_____

_____

11. Place the shift selector into the drive position and allow the transmission to shift through all of its normal shifts. Describe the operation of the transmission and torque converter:

_____

_____

12. Check for proper operation in all forward ranges, especially the 1-2, 2-3, 3-4 upshifts and converter lockup during light throttle operation. Describe the operation:

_____

_____

13. Force the transmission to "kickdown" and record the quality of this shift and the speed at which it downshifts.

_____

_____

14. Manually cause the transmission to downshift. How did it react?

_____

_____

15. Record any vibrations or noises that occur during the test drive.

_____

_____

16. Park the vehicle and move the shifter into each gear range. Pay attention to shift quality as each range (including park) is selected. Record the results:

_____

_____

17. Compare your notes with the specifications and shifting chart. What are your conclusions about the transmission and torque converter? Could the transmission and/or torque converter be the cause of a driveability complaint?

_____

_____

_____

Instructor's Response_____

_____

_____

Job Sheet 24

24

Name _____ Date _____

Inspecting Drive Belts

Upon completion of this job sheet, you should be able to visually inspect drive belts and check their tightness.

ASE Correlation

This job sheet is related to the ASE Engine Performance Test's content areas: general engine diagnosis. Task: interpret and verify complaint and determine needed repairs; diagnose drive-ability and emissions problems resulting from failures of interrelated systems and determine needed repairs.

Tools and Materials

Two vehicles, one with a serpentine belt and the other with V-belts
Service manuals for the above vehicles
Belt tension gauge

Procedures

1. On the vehicle with a serpentine belt, carefully inspect the belt and describe the general condition of the belt.

   _____

   _____

   _____

   _____

2. With the proper belt tension gauge, check the tension of the belt. Belt tension should be _____. You found _____.

3. Based on the above, what are your recommendations?

   _____

   _____

4. Describe the procedure for adjusting the tension of the belt:

   _____

   _____

   _____

   _____

5. On the vehicle with V-belts, you will find more than one drive belt. List the different belts by their purpose.

   _____

   _____

   _____

   _____

6. Carefully inspect the belts and describe the general condition of each one:

_____

_____

_____

_____

7. Check the tension of the AC generator drive belt. Belt tension should be _____. You found _____.

8. Based on the above, what are your recommendations?

_____

_____

9. Describe the procedure for adjusting the tension of the belt:

_____

_____

_____

_____

Instructor's Response_____

_____

_____

Fuel System Diagnosis and Service

Upon completion and review of this chapter, you should be able to:

❏ Conduct a visual inspection on a fuel system.

❏ Test alcohol content in the fuel.

❏ Relieve fuel system pressure.

❏ Inspect and service fuel tanks.

❏ Remove, inspect, service, and replace electric fuel pumps and gauge sending units.

❏ Inspect and service fuel lines and tubing.

❏ Remove and replace fuel filters.

❏ Conduct a pressure and volume output test on an electric fuel pump.

❏ Service and test electric fuel pumps.

❏ Service and diagnose fuel injection systems.

Introduction

One of the key requirements for an efficient running engine is the correct amount of fuel. Although the carburetor or fuel injection system is responsible for the fuel delivery into the cylinders, many other components are responsible for the proper delivery of fuel. The fuel must be stored, pumped out of storage, piped to the engine, and filtered. All of this must be accomplished in an efficient and safe manner.

The fuel system should be checked whenever there is evidence of a fuel leak or fuel smell. Leaks are not only costly to the customer, but are also very dangerous. The fuel system should also be checked whenever basic tests suggest there is too little or too much fuel being delivered to the cylinders. Lean mixtures are often caused by insufficient amounts of fuel being drawn out of the fuel tank. Lean mixtures are suggested by many different test results including high hydrocarbon (HC) readings on an exhaust analyzer and high firing lines on a scope.

No-start conditions are also caused by a lack of fuel. When no fuel is delivered to the engine, the engine will not run. Remember, there must be four factors to have combustion: air, fuel, compression, and spark. On throttle body injected engines, it is easy to determine if fuel is being delivered. Simply look down the throttle body. If the surfaces are wet or you see fuel being sprayed while cranking the engine, fuel is there. With port injection it is a little more difficult. Connect a fuel pressure gauge to the fuel line or rail and observe the fuel pressure while cranking the engine. Testing fuel pressure is described later in this chapter. However, if there is no fuel pressure while cranking, there is no fuel being delivered to the engine.

There are many components in the fuel system. These components can be grouped into two categories: fuel delivery or fuel injection. Diagnosis and basic service to both of these are covered in this chapter. All of the tests in this chapter assume that the fuel is good and not severely contaminated. Obviously, water does not burn as well as gasoline. Therefore, water in the fuel tank can cause a driveability problem. If water is mixed with the gasoline, drain the tank and refill it with fresh clean gasoline. Also keep in mind that after gasoline has sat for awhile, it becomes less volatile. There are additives available to revitalize the fuel. However, if the fuel is so stale that an engine will not run, drain and refill the tank with fresh gasoline.

Basic Tools

Basic tool set, fender covers, clean rags, safety glasses, appropriate service manual

Alcohol in Fuel Test

Special Tools

100-milliliter graduated cylinder, approved gasoline containers

Some gasoline may contain a small quantity of alcohol. The percentage of alcohol mixed with the fuel does not usually exceed 10 percent. However, an excessive amount of alcohol mixed with gasoline may result in fuel system corrosion, fuel filter plugging, deterioration of rubber fuel system components, and a lean air-fuel ratio. These fuel system problems caused by excessive alcohol in the fuel may cause driveability complaints such as lack of power, acceleration stumbles, engine stalling, and no-start. If the correct amount of fuel is being delivered to the engine and there is evidence of a lean mixture, check for air leaks in the intake, then check the gasoline's alcohol content.

Some vehicle manufacturers supply test equipment to check the level of alcohol in the gasoline. The following alcohol-in-fuel test procedure requires only the use of a calibrated cylinder:

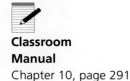

Classroom Manual
Chapter 10, page 291

1. Obtain a 100-milliliter (mL) cylinder graduated in 1-mL divisions.
2. Fill the cylinder to the 90-mL mark with gasoline.
3. Add 10 mL of water to the cylinder so it is filled to the 100-mL mark.
4. Install a stopper in the cylinder and shake it vigorously for 10 to 15 seconds.
5. Carefully loosen the stopper to relieve any pressure.
6. Install the stopper and shake vigorously for another 10 to 15 seconds.
7. Carefully loosen the stopper to relieve any pressure.
8. Place the cylinder on a level surface for 5 minutes to allow liquid separation.
9. Any alcohol in the fuel is absorbed by the water and settles to the bottom. If the water content in the bottom of the cylinder exceeds 10 mL, there is alcohol in the fuel. For example, if the water content is now 15 mL, there was 5 percent alcohol in the fuel. Since this procedure does not extract 100 percent of the alcohol from the fuel, the percentage of alcohol in the fuel may be higher than indicated.

Fuel System Pressure Relief

Special Tools

Fuel pressure gauge

Since electronic fuel injection (EFI) systems have a residual fuel pressure, this pressure must be relieved before disconnecting any fuel system component. Most port fuel injection (PFI) systems have a fuel pressure test port on the fuel rail (Figure 10-1). Follow this procedure for fuel system pressure relief:

WARNING: Failure to relieve the fuel pressure on electronic fuel injection (EFI) systems prior to fuel system service may result in gasoline spills, serious personal injury, and expensive property damage.

1. Disconnect the negative battery cable to avoid fuel discharge if an accidental attempt is made to start the engine.
2. Loosen the fuel tank filler cap to relieve any fuel tank vapor pressure.
3. Wrap a shop towel around the fuel pressure test port on the fuel rail and remove the dust cap from this valve.
4. Connect the fuel pressure gauge to the fuel pressure test port on the fuel rail.
5. Install the bleed hose on the gauge in an approved gasoline container and open the gauge bleed valve to relieve fuel pressure from the system into the gasoline container. Be sure all the fuel in the bleed hose is drained into the gasoline container.

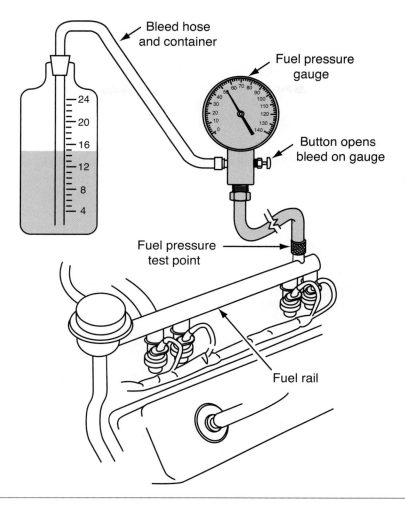

Figure 10-1 A port injection fuel rail with a Schrader valve.

On EFI systems that do not have a fuel pressure test port, such as most throttle body injection (TBI) systems, follow these steps for fuel system pressure relief:

1. Loosen the fuel tank filler cap to relieve any tank vapor pressure.
2. Remove the fuel pump fuse.
3. Start and run the engine until the fuel is used up in the fuel system and the engine stops.
4. Engage the starter for 3 seconds to relieve any remaining fuel pressure.
5. Disconnect the negative battery terminal to avoid possible fuel discharge if an accidental attempt is made to start the engine.

WARNING: Always wear eye protection and observe all other safety rules to avoid personal injury when servicing fuel system components.

Fuel Tanks

The fuel tank should be inspected for leaks, road damage, corrosion, and rust on metal tanks; loose, damaged, or defective seams; or loose mounting bolts and damaged mounting straps. Leaks in the fuel tank, lines, or filter may cause gasoline odor in and around the vehicle, especially during low-speed driving and idling. In most cases, the fuel tank must be removed for servicing.

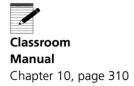

Classroom Manual
Chapter 10, page 310

Leaks in the metal fuel tank can be caused by a weak seam, rust, or road damage. The best method of permanently solving this problem is to replace the tank. Another method is to remove the tank and steam clean or boil it in a caustic solution to remove the gasoline residue. After this has been done, the leak can be soldered or brazed by a specialty shop that is equipped to do this.

Holes in a plastic tank can sometimes be repaired by using a special tank repair kit. Be sure to follow manufacturer's instructions when doing the repair.

When a fuel tank is leaking dirty water or has water in it, the tank must be cleaned and repaired or replaced.

SERVICE TIP: When a fuel tank must be removed, if possible, ask the customer to bring the vehicle to the shop with a minimal amount of fuel in the tank.

Fuel Tank Draining

CAUTION: Always drain gasoline into an approved container, and use a funnel to avoid gasoline spills.

The fuel tank must be drained prior to tank removal. If the tank has a drain bolt, this bolt may be removed and the fuel drained into an approved container. If the fuel tank does not have a drain bolt, follow these steps to drain the fuel tank:

1. Remove the negative battery cable.
2. Raise the vehicle on a hoist.
3. Locate the fuel tank drain pipe, and remove the drain pipe plug.
4. Install the appropriate adaptor in the fuel tank drain pipe and connect the intake hose from a hand-operated or air-operated pump to this adaptor (Figure 10-2). If the fuel tank does not have a drain pipe, install the pump hose through the filler pipe into the fuel tank.

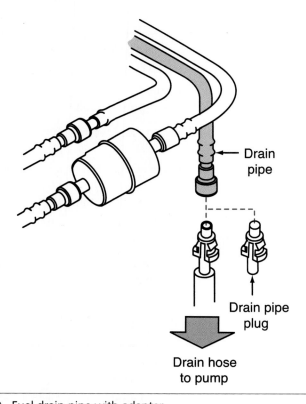

Drain pipe

Drain pipe plug

Drain hose to pump

Figure 10-2 Fuel drain pipe with adapter.

5. Install the discharge hose from the hand-operated or air-operated pump into an approved gasoline container, and operate the pump until all the fuel is removed from the tank.

 WARNING: When servicing fuel system components, always place a Class B fire extinguisher near the work area.

Fuel Tank Service

The fuel tank removal procedure varies depending on the vehicle make and year. Always follow the procedure in the vehicle manufacturer's service manual. The following is a typical procedure:

1. Disconnect the negative terminal from the battery.

2. Relieve the fuel system pressure, and drain the fuel tank.

3. Raise the vehicle on a hoist or lift the vehicle with a floor jack and lower the chassis onto jack stands.

4. Use compressed air to blow dirt from the fuel line fittings and wiring connectors.

5. Remove the fuel tank wiring harness connector from the body harness connector.

6. Remove the ground wire retaining screw from the chassis if used.

7. Disconnect the fuel lines from the fuel tank. If these lines have quick-disconnect fittings, follow the manufacturer's recommended removal procedure in the service manual. Some quick-disconnect fittings are hand releasable and others require the use of a special tool (Figure 10-3).

 WARNING: Abide by local laws for the disposal of contaminated fuels. Be sure to wear eye protection when working under the vehicle.

8. Wipe the filler pipe and vent pipe hose connections with a shop towel and then disconnect the hoses from the filler pipe and vent pipe to the fuel tank.

9. Unfasten the filler from the tank. If it is a rigid one-piece tube, remove the screws around the outside of the filler neck near the filler cap. If it is a three-piece unit, remove the neoprene hoses after the clamp has been loosened.

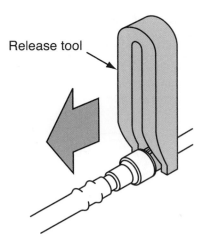

Release tool

Figure 10-3 Quick-disconnect fuel line tool.

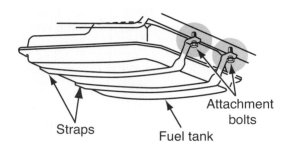

Straps   Fuel tank

Attachment bolts

Figure 10-4 Fuel tank strap mounting bolts.

10. Loosen the bolts holding the fuel tank straps to the vehicle (Figure 10-4) until they are about two threads from the end.

 WARNING: Do not heat the bolts on the fuel tank straps in order to loosen them. The heat could ignite the fumes.

11. Holding the tank securely against the underchassis with one hand, remove the strap bolts and lower the tank to the ground. When lowering the tank, make sure all wires and tubes are unhooked. Keep in mind that small amounts of fuel might still remain in the tank.

 CAUTION: Use a drain pan to catch any spilled fuel or be sure to clean it up immediately.

Make certain that the tank insulators are in place to prevent excessive fuel pump noise being transmitted into the vehicle.

12. To reinstall the new or repaired fuel tank, reverse the removal procedure. Be sure that all the rubber or felt tank insulators are in place. Then, with the tank straps in place, position the tank. Loosely fit the tank straps around the tank, but do not tighten them. Make sure that the hoses, wires, and vent tubes are connected properly. Check the filler neck for alignment and for insertion into the tank. Tighten the strap bolts and secure the tank to the car. Install all of the tank accessories (vent line, sending unit wires, ground wire, and filler tube). Fill the tank with fuel and check it for leaks, especially around the filler neck and the pickup assembly. Reconnect the battery and check the fuel gauge for proper operation.

Special Tools

Fuel pump testing tool

Classroom Manual
Chapter 10, page 317

Electric Fuel Pump Removal and Replacement

1. Remove the fuel tank from the vehicle.

2. Follow the vehicle manufacturer's recommended procedure to remove the fuel pump and gauge sending unit from the fuel tank. In many cases, a special tool must be used to remove this assembly (Figure 10-5).

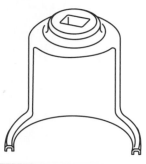

Figure 10-5 Special tool for removing a fuel pump and gauge sending unit assembly from a fuel tank.

3. Check the filter on the fuel pump inlet. If the filter is contaminated or damaged, replace the filter.

4. Inspect the fuel pump inlet for dirt and debris. Replace the fuel pump if these foreign particles are found in the pump inlet.

5. If the pump inlet filter is contaminated, it will be necessary to have the tank cleaned.

6. Check all fuel hoses and tubing on the fuel pump assembly. Replace fuel hoses that are cracked, deteriorated, or kinked. When fuel tubing on the pump assembly is damaged, replace the tubing or the pump.

7. Be sure the sound insulator sleeve is in place on the electric fuel pump, and check the position of the sound insulator on the bottom of the pump.

8. Clean the pump and sending unit mounting area in the fuel tank with a shop towel and install a new gasket or O-ring on the pump and sending unit. On some tanks, the gauge sending unit and fuel pump are mounted separately (Figure 10-6).

9. Install the fuel pump and gauge sending unit assembly in the fuel tank, and secure this assembly in the tank using the vehicle manufacturer's recommended procedure.

Some vehicles have a separate fuel pump and gauge sending unit. These units are held in the tank by either a retaining ring or screws. The easiest way to remove a sending unit retaining ring is to use a special tool designed for this purpose. This tool fits over the metal tabs on the retaining ring, and after about a quarter turn, the ring comes loose and the sender unit can be removed. If the special tool is not available, a drift punch and ball peen hammer usually do the job (Figure 10-7).

When removing the sending unit from the tank, be very careful not to damage the float arm, the float, or the fuel gauge sender. Check the unit carefully for any damaged components. Shake the float and if fuel can be heard inside, replace it. Make sure the float arm is not bent. It

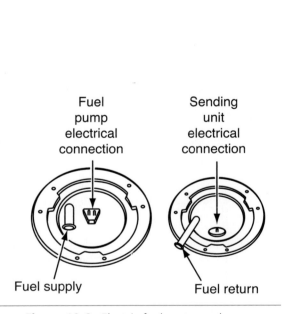

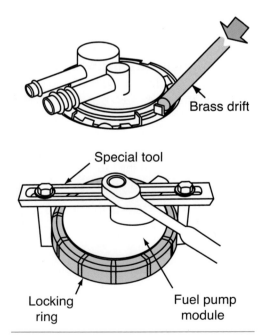

Figure 10-6 Electric fuel pump and gauge sending unit mounted separately in the fuel tank.

Figure 10-7 Tightening the fuel pump locking ring with a special tool or brass drift.

is necessary to replace the filter and O-ring before replacing the unit. Check the fuel gauge as described in the service manual. When reinstalling the pickup pipe-sending unit, be very careful not to damage any of the components.

Fuel Lines

Classroom Manual
Chapter 10, page 314

EVAP is the evaporative emissions control system for containment of gasoline vapors.

Fuel lines can be made of either metal tubing or flexible nylon or synthetic rubber hose. The latter must be able to resist gasoline. It must also be nonpermeable, so gas and gas vapors cannot evaporate through the hose. Ordinary rubber hose, such as that used for vacuum lines, deteriorates when exposed to gasoline. Only hoses made for fuel systems should be used for replacement. Similarly, vapor vent lines must be made of material that resists attack by fuel vapors. Replacement vent hoses are usually marked with the designation **EVAP** to indicate their intended use. The inside diameter of a fuel delivery hose is generally larger (5/16 to 3/8 inch) than that of a fuel return hose (1/4 inch).

Many fuel tanks have vent hoses to allow air in the fuel tank to escape when the tank is being filled with fuel. Vent hoses are usually installed alongside the filler neck.

The fuel lines carry fuel from the fuel tank to the fuel pump, fuel filter, and carburetor or fuel injection assembly. These lines are usually made of rigid metal, although some sections are constructed of rubber hose to allow for car vibrations. This fuel line, unlike filler neck or vent hoses, must work under pressure or vacuum. Because of this, the flexible synthetic hoses must be stronger. This is especially true for the hoses on fuel injection systems, where pressures reach 60 psi or more. For this reason, flexible fuel line hoses must also have special resistance properties. Many auto manufacturers recommend that flexible hose only be used as a delivery hose to the fuel metering unit in fuel injection systems. It should not be used on the pressure side of the injector systems. This application requires a special high-pressure hose.

All fuel lines should occasionally be inspected for holes, cracks, leaks, kinks, or dents. Many fuel system troubles that occur in the lines are blamed on the fuel pump or injectors. For instance, a small hole in the fuel line admits air but does not necessarily show any drip marks under the car. Air can then enter the fuel line, allowing the fuel to gravitate back into the tank. Then, instead of drawing fuel from the tank, the fuel pump sucks only air through the hole in the fuel line. When this condition exists, the fuel pump is frequently tested and, if there is insufficient fuel, it is considered faulty when in fact there is nothing wrong with it. If a hole is suspected, remove the coupling at the tank and the pump and pressurize the line with air. The leaking air is easily spotted.

Since the fuel is under pressure, leaks in the line between the pump and carburetor or injectors are relatively easy to recognize. When a damaged fuel line is found, replace it with one of similar construction—steel with steel or the flexible with nylon or synthetic rubber. When installing flexible tubing, always use new clamps. The old ones lose some of their tension when they are removed and do not provide an effective seal when used on the new line.

CAUTION: Do not substitute aluminum or copper tubing for steel tubing. Never use hose within 4 inches of any hot engine or exhaust system component. A metal line must be installed.

Fuel supply lines from the tank to the carburetor or injectors are routed to follow the frame along the underchassis of vehicles. Generally, rigid lines are used extending from near the tank to a point near the fuel pump. To absorb engine vibrations, the gaps between the frame and tank or fuel pump are joined by short lengths of flexible hose (Figure 10-8).

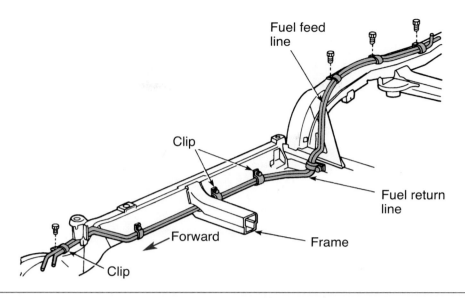

Figure 10-8 Gaps between the frame and tank are joined with flexible hose.

Steel tubing should be inspected for leaks, kinks, and deformation (Figure 10-9). This tubing should also be checked for loose connections and proper clamping to the chassis. If the fuel tubing threaded connections are loose, they must be tightened to the specified torque. Some threaded fuel line fittings contain an O-ring. If the fitting is removed, the O-ring should be replaced.

CAUTION: O-rings in fuel line fittings are usually made from fuel-resistant Viton. Other types of O-rings must not be substituted for fuel fitting O-rings.

Any damaged or leaking fuel line—either a portion or the entire length—must be replaced. To fabricate a new fuel line, select the correct tube and fitting dimension and start with a length that is slightly longer than the old line. With the old line as a reference, use a tubing bender to form the same bends in the new line as those that exist in the old. Although steel tubing can be bent by hand to obtain a gentle curve, any attempt to bend a tight curve by hand usually kinks the tubing. To avoid kinking, always use a bending tool like the ones shown in Figure 10-10.

Special Tools

Double-flare tool kit

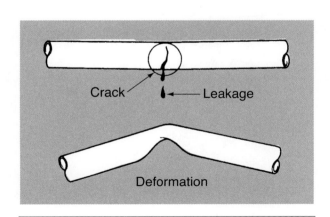

Figure 10-9 Steel fuel tubing should be inspected for leaks, kinks, and other damage.

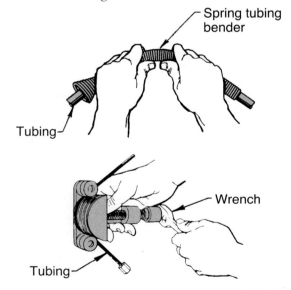

Figure 10-10 Two types of steel tubing bending tools.

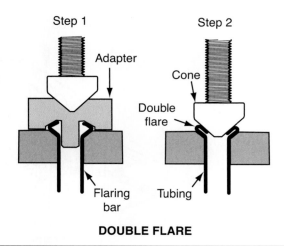

DOUBLE FLARE

Figure 10-11 Forming a double-flare fitting.

Double flare refers to the type of mating surfaces in fittings.

The most-used tubing fittings is the **double flare**. The double flare is made with a special tool that has an anvil and a cone (Figure 10-11). The double-flaring process is performed in two steps. First, the anvil begins to fold over the end of the tubing. Then, the cone is used to finish the flare by folding the tubing back on itself, doubling the thickness and creating two sealing surfaces.

The angle and size of the flare are determined by the tool. Careful use of the double flaring helps to produce strong, leakproof connections. Figure 10-12 shows other metal fuel line connections that are used by vehicle manufacturers.

The flare tool can also be used to make sure that nylon and synthetic rubber hoses stay in place. That is, to make sure the connection is secure, put a partial double-lip flare on the end of the tubing over which the hose is installed. This can be done quickly with the proper flaring tool by starting out as if it was going to be a double flare but stopping halfway through the procedure (Figure 10-13). This provides an excellent sealing ridge that does not cut into the hose. A clamp should be placed directly behind the ridge on the hose caused by the raised section on the metal line.

There are a variety of clamps used on fuel system lines, including the spring and screw types (Figure 10-14). The crimp clamps shown in Figure 10-15 are used most for metal tubing, but they require a special tool to install.

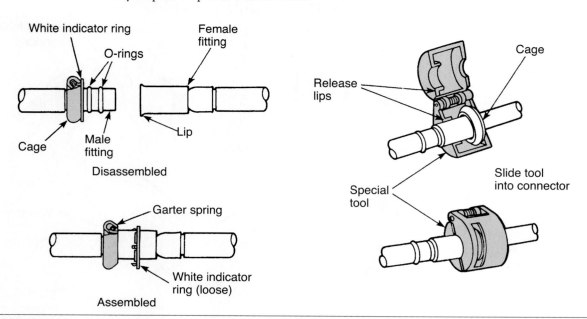

Figure 10-12 Common metal fuel line connections.

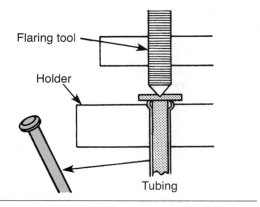

Figure 10-13 Using a flaring tool to secure hose connections.

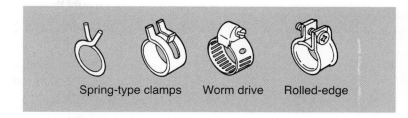

Spring-type clamps   Worm drive   Rolled-edge

Figure 10-14 Various clamps used on fuel lines.

To control the rate of vapor flow from the fuel tank to the vapor storage tank, a plastic or metal restrictor may be placed in either the end of the vent pipe or in the vapor-vent hose itself. When the latter hose must be replaced, the restrictor must be removed from the old vent hose and installed in the new one.

▲ **WARNING:** Always cover a nylon fuel pipe with a wet shop towel before using a torch or other source of heat near the line. Failure to observe this precaution may result in fuel leaks, personal injury, and property damage.

■ **CAUTION:** If a vehicle has nylon fuel pipes, do not expose the vehicle to temperatures above 194°F (90°C) for any extended period to avoid damage to the nylon fuel pipes.

Nylon fuel pipes should be inspected for leaks, nicks, scratches and cuts, kinks, melting, or loose fittings. If these fuel pipes are kinked or damaged in any way, they must be replaced. Nylon fuel pipes must be secured to the chassis at regular intervals to prevent fuel pipe wear and vibration.

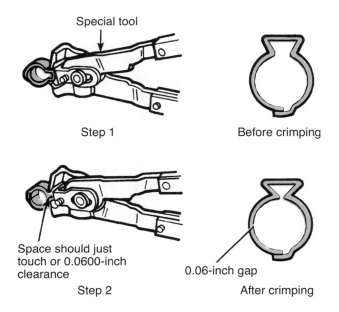

Special tool

Step 1   Before crimping

Space should just touch or 0.0600-inch clearance

0.06-inch gap

Step 2   After crimping

Figure 10-15 Crimp clamps require a special tool for installation.

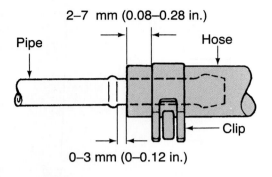

2–7 mm (0.08–0.28 in.)

Pipe

Hose

Clip

0–3 mm (0–0.12 in.)

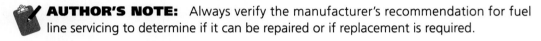

Figure 10-16 Rubber fuel hose installation on a steel fitting or tube.

 AUTHOR'S NOTE: Always verify the manufacturer's recommendation for fuel line servicing to determine if it can be repaired or if replacement is required.

Nylon fuel pipes provide a certain amount of flexibility and can be formed around gradual curves under the vehicle. Do not force a nylon fuel pipe into a sharp bend because this action may kink the pipe and restrict the flow of fuel. When nylon fuel pipes are exposed to gasoline, they may become stiffer, making them more susceptible to kinking. Be careful not to nick or scratch nylon fuel pipes.

Rubber fuel hose should be inspected for leaks, cracks, cuts, kinks, oil soaking, and soft spots or deterioration. If any of these conditions are found, the fuel hose should be replaced. When rubber fuel hose is installed, the hose should be installed to the proper depth on the metal fitting or line (Figure 10-16).

The rubber fuel hose clamp must be properly positioned on the hose in relation to the steel fitting or line as illustrated in the figure. Fuel hose clamps may be spring type or screw type. Screw-type fuel hose clamps must be tightened to the specified torque.

Fuel Filters

Classroom Manual Chapter 10, page 316

Automobiles and light trucks usually have an in-tank strainer and a gasoline filter (Figure 10-17). The strainer, located in the gasoline tank, is made of a finely woven fabric. The purpose of this

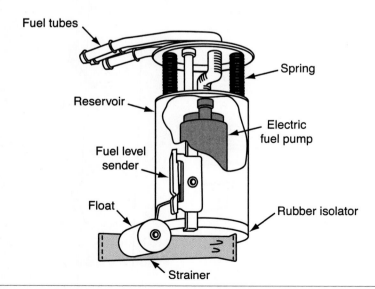

Fuel tubes

Spring

Reservoir

Electric fuel pump

Fuel level sender

Float

Rubber isolator

Strainer

Figure 10-17 A fuel strainer at the fuel pump pickup.

strainer is to prevent large contaminant particles from entering the fuel system where they could cause excessive fuel pump wear or plug fuel metering devices. It also helps to prevent passage of any water that might be present in the tank. Servicing of the fuel tank strainer is required when replacing the fuel pump.

The gasoline filter is usually located in the engine compartment and is the one this section examined because it is replaceable and might require service on a regular basis. The most common types of gasoline filters are in-line filters and out-pump filters.

Servicing Filters

Fuel filters and elements are serviced by replacement only. Some vehicle manufacturers recommend fuel filter replacement at 30,000 miles (48,000 km). Always replace the fuel filter at the vehicle manufacturer's recommended mileage. If dirty or contaminated fuel is placed in the fuel tank, the filter may require replacing before the recommended mileage. A plugged fuel filter may cause the engine to surge and cut out at high speed or hesitate on acceleration. A restricted fuel filter causes low fuel pump pressure and volume.

Fuel filter replacement procedure varies depending on the make and year of the vehicle and the type of fuel system. Always follow the filter replacement procedure in the appropriate service manual. Photo Sequence 18 shows a typical procedure for relieving fuel pressure and removing a fuel filter. The following is a typical filter replacement procedure on a vehicle with EFI:

1. Relieve fuel system pressure as mentioned previously.
2. Raise the vehicle on a hoist.
3. Flush the quick connectors on the filter with water, and use compressed air to blow debris from the connectors.
4. Disconnect the inlet connector first. Grasp the large connector collar, twist in both directions, and pull the connector off the filter.
5. Disconnect the outlet connector using the same procedure used on the inlet connector (Figure 10-18).
6. Loosen and remove the filter mounting bolts, and remove the filter from the vehicle.
7. Use a clean shop towel to wipe the male tube ends of the new filter.
8. Apply a few drops of clean engine oil to the male tube ends on the filter.
9. Check the quick connectors to be sure the large collar on each connector has rotated back to the original position. The springs must be visible on the inside diameter of each quick connector.

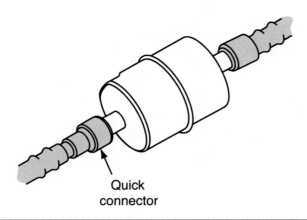

Quick connector

Figure 10-18 A fuel filter with quick-disconnect fittings.

Photo Sequence 18
Typical Procedure for Relieving Fuel Pressure and Removing Fuel Filter

P18-1 Disconnect the negative battery cable.

P18-2 Loosen the fuel tank filler cap to relieve any fuel tank vapor pressure.

P18-3 Wrap a shop towel around the Schrader valve on the fuel rail and remove the dust cap from this valve.

P18-4 Connect a fuel pressure gauge to the Schrader valve.

P18-5 Install the gauge bleed hose into an approved gasoline container and open the gauge bleed valve to relieve the fuel pressure.

P18-6 Place the lift arms under the manufacturer's specified lift points on the vehicle and lift the vehicle.

P18-7 Flush the fuel filter line connectors with water and use compressed air to blow debris from the connectors.

P18-8 Follow the vehicle manufacturer's recommended procedure to remove the inlet connector.

P18-9 Follow the vehicle manufacturer's recommended procedure to remove the outlet connector and the fuel filter.

10. Install the filter on the vehicle in the proper direction, and leave the mounting bolt slightly loose.

11. Install the outlet connector onto the filter outlet tube and press the connector firmly in place until the spring snaps into position. Grasp the fuel line and try to pull this line from the filter to be sure the quick connector is locked in place.

12. Do the same with the inlet connector.

13. Tighten the filter retaining bolt to the specified torque.

14. Lower the vehicle, start the engine, and check for fuel leaks at the filter.

Fuel Pumps

Electric Fuel Pumps

Electric fuel pumps offer important advantages over mechanical fuel pumps. Because electric fuel pumps maintain constant fuel pressure, they aid in starting and reduce vapor lock problems. The electric fuel pump can be located inside or outside the fuel tank. It may also be combined with the fuel gauge's sending unit. A typical wiring diagram for an electric fuel pump is shown in Figure 10-19.

Classroom Manual
Chapter 10, page 317

Some electric fuel pumps in EFI systems are computer-controlled, therefore testing these pumps must be done according to the manufacturer's recommendations.

● **CUSTOMER CARE:** Advise your customers to avoid leaving the level in the fuel tank extremely low. Vehicles with in-tank electric fuel pumps can actually overheat due to the lack of liquid in the tank because it helps dissipate the heat from the pump. Additionally, sediment or debris is more likely to be drawn into the pump as a result of repeated low fuel levels. Recommend that the fuel level not dip below a quarter tank of fuel.

Troubleshooting. Problems in fuel systems using electric fuel pumps are usually indicated by improper fuel system pressure or a dead or inoperative fuel pump.

■ **CAUTION:** The fuel supply lines can remain pressurized for long periods of time after the engine is shut down. This pressure must be relieved before servicing of the fuel system begins. A valve is normally provided on the throttle body or fuel rail for this purpose.

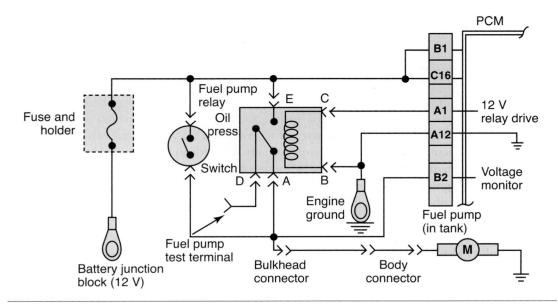

Figure 10-19 Typical wiring diagram for an electric fuel pump.

Even though there is a check valve in the pump, it is normal for some of the pressure to bleed down over a long period of time.

On vehicles that use an inertia switch, make sure that the switch has not been tripped (opened). These switches can sometimes be activated by hitting large bumps or railroad tracks.

Low pressure can be due to a clogged fuel filter, restricted fuel line, weak pump, leaky pump check valve, defective fuel pressure regulator, or dirty filter sock in the tank. It is possible to rule out filter and line restrictions as a cause of the problem by making a pressure check at the pump outlet. A higher reading at the pump outlet (at least 5 psi) means there is a restriction in the filter or line. If the reading at the pump outlet is unchanged, then either the pump is weak or is having trouble picking up fuel (clogged filter sock in the tank). Either way it is necessary to get inside the fuel tank. If the filter sock is gummed up with dirt or debris, it is also wise to clean out the tank when the filter sock is cleaned or replaced.

Another possible source of trouble is the pump check valve. Some pumps have one, others have two (positive displacement roller vane pumps). The purpose of the check valve is to prevent fuel movement through the pump when the pump is off so residual pressure remains at the injectors. This can be checked by watching the fuel pressure gauge after the engine is shut off.

Depending on the type of pump used, the check valve can also prevent the reverse flow of fuel and relieve internal pressure to regulate maximum pump output. Check valves can stick and leak. So, if a pump runs but does not pump fuel, a bad check valve is to blame. Unfortunately, the check valve is usually an integral part of the pump assembly that is sealed at the factory. Therefore, if the check valve is causing trouble, the entire pump must be replaced.

When an engine fails to start because there is no fuel delivery, the first check is the fuel gauge. A gauge that reads higher than a half tank probably means there is fuel in the tank, but not always. A defective sending unit or miscalibrated gauge might be giving a false indication. Sticking a wire or dowel rod down the fuel tank filler pipe tells whether or not there is really fuel in the tank. If the gauge is faulty, repair or replace it.

Listen for pump noise. When the key is turned on, the pump should buzz for a couple of seconds to build system pressure. On most late-model cars with computerized engine controls, the computer energizes a pump relay when it receives a cranking signal from the distributor pickup or crankshaft sensor (Figure 10-20). An oil pressure switch might still be included in the circuitry for safety purposes and to serve as a backup in case the relay or computer signal fails.

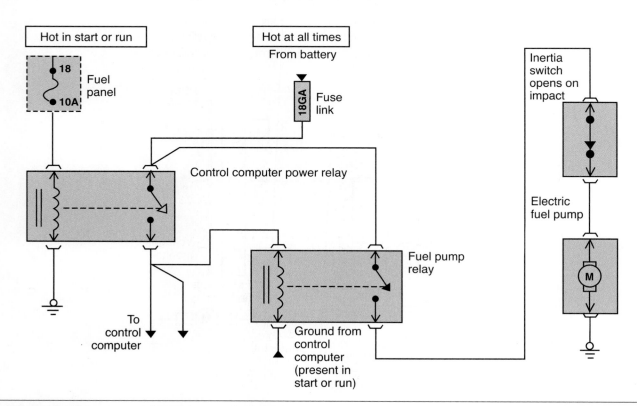

Figure 10-20 Electrical fuel delivery system wiring diagram.

Failure of the pump relay or computer driver signal can cause slow starting because the fuel pump does not come on until the engine cranks long enough to build up sufficient oil pressure to trip the oil pressure switch.

If a buzzing sound is not heard when the key is on or while the engine is being cranked, check for the presence of voltage at the pump electrical connectors. The pump might be good, but if it does not receive voltage and have a good ground, it does not run. To check the ground, connect a test light across the ground and feed wires at the pump to check for voltage, or use a voltmeter to read actual voltage and an ohmmeter to check ground resistance. The latter is the better test technique because a poor ground connection or low voltage can reduce pump operating speed and output. If the electrical circuit checks out but the pump does not run, the pump is probably bad and should be replaced.

No voltage at the pump terminal when the key is on and the engine is cranking indicates a faulty oil pressure switch, pump relay, relay drive circuit in the computer, or a wiring problem. Check the pump fuse to see if it is blown. Replacing the fuse might restore power to the pump, but until you have found out what caused the fuse to blow, the problem is not solved. The most likely cause of a blown fuse would be a short in the wiring between the relay and pump or a short inside the oil pressure switch or relay.

A faulty oil pressure switch can be checked by bypassing it with a jumper wire. If this restores power to the pump and the engine starts, replace the switch. If an oil pressure switch or relay sticks in the closed position, the pump can run continuously whether the key is on or off, depending on how the circuit is wired.

To check a pump relay, use a test light to check across the relays and ground terminals. This tells if the relay is receiving battery voltage and ground. Next, turn off the ignition, wait about 10 seconds, then turn it on. The relay should click and you should see battery voltage at the relay's pump terminal. If nothing happens, repeat the test checking for voltage at the relay terminal that is wired to the computer. The presence of a voltage signal means the computer is doing its job but the relay is failing to close and should be replaced. No voltage signal from the computer indicates an opening in that wiring circuit or a fault in the computer itself.

Fuel pumps can also be checked for amperage draw and current waveform. Check the amperage required for the fuel pump at the fuse block if possible. Make certain that you use a fused multimeter that is capable of measuring at least as much amperage as the vehicle's fuel pump fuse. Measure the amperage draw with the vehicle running and compare to vehicle specifications. Another method is to use a low amps probe (Figure 10-21) and look at the

Figure 10-21 A low amps probe.

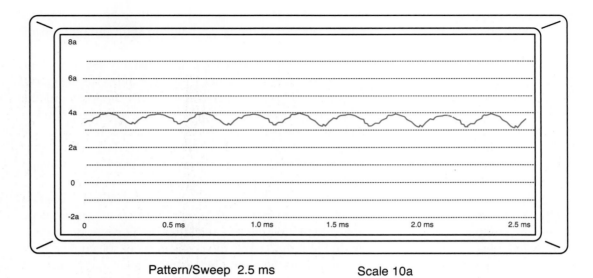

8a					
6a					
4a					
2a					
0					
-2a					

0   0.5 ms   1.0 ms   1.5 ms   2.0 ms   2.5 ms

Pattern/Sweep 2.5 ms   Scale 10a

Figure 10-22 Waveform from a low amps probe of a fuel pump. Notice even transitions caused by brushes on commutator bars.

amperage and/or the waveform the fuel pump produces (Figure 10-22). The low amps probe is popular with technicians because it is not necessary to backprobe or otherwise disturb vehicle wiring, just use a clamp around the wire.

⚠️ **WARNING:** When testing an electric fuel pump, do not let fuel contact any electrical wiring. The smallest spark (electric arc) could ignite the fuel.

The proper procedure for testing fuel pump pressure is shown in Photo Sequence 19. These photos outline the steps to follow while performing the test on an engine with fuel injection. To conduct this test on specific fuel injection systems, refer to the service manual for instructions. However, most systems have a *Schrader valve* on the fuel rail that can be used to connect the fuel pressure gauge. If the system does not have a Schrader valve, the fuel rail should be relieved of any pressure before loosening a fitting to install the pressure gauge. Special adapters may be needed.

Tips on fuel pressure checking:

❑ Always verify correct fuel pressure from service information. Do not rely on memory and past experience. Fuel pressures vary greatly between engines, injection types, and makes and model years. Special adaptors are sometimes required to check fuel pressure such as the "banjo" style shown in Figure 10-23.

A Schrader valve is very similar to a tire valve. The center of the valve must be depressed before anything can enter or leave.

A **"banjo" fitting** is named after the shape of the line attachment.

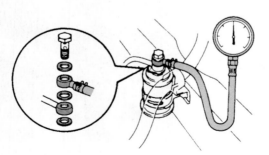

Figure 10-23 A fuel pressure gauge used with a "banjo-style" fuel fitting.

Photo Sequence 19
Checking the Fuel Pressure on a PFI System

P19-1 Many problems on today's cars can be caused by incorrect fuel pressure. Therefore, checking fuel pressure is an important step in diagnosing driveability problems.

P19-2 Prior to testing the fuel pump, a careful visual inspection of the injectors, fuel rail, and fuel lines and hoses is necessary. Any sign of a fuel leak should be noted and the cause corrected immediately.

P19-3 The supply line into the fuel rail is a likely point of leakage. Check the area around the fitting to make sure no leaks have occurred.

P19-4 Most fuel rails are equipped with a test fitting that can be used to relieve pressure and to test pressure.

P19-5 To test fuel pressure, connect the appropriate pressure gauge to the fuel rail test fitting (Schrader valve).

P19-6 Connect a hand-held vacuum pump to the fuel pressure regulator.

P19-7 Turn the ignition switch to the RUN position and observe the fuel pressure gauge. Compare the reading to specifications. A reading lower than normal indicates a faulty fuel pump or fuel delivery system.

P19-8 To test the fuel pressure regulator, create a vacuum at the regulator with the vacuum pump. Fuel pressure should decrease as vacuum increases. If pressure remains the same, the regulator is faulty.

❏ When checking fuel pressure, make sure the specifications are either for the engine running or engine off. If the specifications are for key on, engine off and the fuel system has a vacuum-operated pressure regulator, the key-on pressure will be higher than the idle pressure due to the vacuum at the pressure regulator. On most TBI and returnless fuel systems, the key on, engine off and engine running pressures will be about the same. Always check service information to be sure.

❏ If pressures are okay at idle, accelerate the engine while observing fuel pressure. The pressure should go up momentarily if the system has a vacuum-operated pressure regulator because of the drop in vacuum. Most returnless systems should remain steady because the fuel goes through the regulator at the fuel pump and is not affected by vacuum. A returnless style fuel pump module is shown in Figure 10-24.

❏ If fuel pressure is too high, either the fuel pressure regulator is defective or the return line is restricted. If the return line is bypassed with a hose placed into an approved container and the fuel pressure is normal, then the problem is the return line. If the pressure is still too high, then the regulator is defective. Of course, high fuel pressure in a returnless system is the fault of the regulator.

❏ If the gauge goes down on acceleration or is too low at any time, there is not enough volume, the pump is faulty, or the pressure regulator is defective. There is also a chance that a fuel injector could be stuck open. There are special adaptors available that may contain a shut-off valve or are able to be crimped to do this test. If the return line is blocked with the adaptor and the pressure rises (usually well above specs) the pressure regulator is faulty. The fuel feed line can be checked in the same manner. If the fuel feed line is restricted after pressure is allowed to build (if possible) and then the pressure holds after key off, the fuel pump check ball or a leak in the system (which could be located in the tank itself) is at fault. If the pressure still falls with the feed line restricted, there is a possibility of an injector sticking on. With a mechanic's stethoscope, listen for fuel escaping through an injector. If you do find an injector bleeding off, unplug the electrical connection at the injector and see if the pressure will hold. If it does, then the ground connection to the injector may be permanently grounded or the PCM is at fault. If the pressure still falls, then the injector itself is at fault. NOTE: The fuel pressure on most fuel-injected vehicles will bleed down after several minutes. It is important to remember that the fuel quantity in the lines is present at

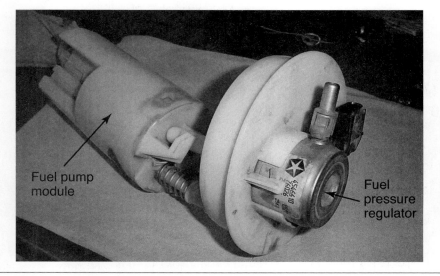

Figure 10-24 A returnless fuel pump module. Note the fuel pressure regulator location and single outlet.

startup if not full pressure. If the pressure is still too low, inspect the fuel filter and lines for restrictions, and check for proper power and ground feeds at the fuel pump. If none of these seem to be a problem, you will have to remove the tank and inspect the fuel pump, fuel strainer, and condition of the fuel in the tank.

Replacement. When replacing an electric pump, be sure that the new or rebuilt replacement unit meets the minimum requirements of pressure and volume for that particular vehicle. This information can be found in the service manual. To replace a typical electric fuel pump:

1. Disconnect the ground terminal at the battery.

2. Disconnect the electrical connectors on the electric fuel pump. Label the wires to aid in connecting it to the new pump. Reversing polarity on most pumps destroys the unit.

3. Fuel and vapor lines should be removed from the pump. Label the lines to aid in connecting them to the new pump.

4. The inside tank pump can usually be taken out of the tank by removing the fuel sending unit retaining or lock ring. However, it may be necessary to remove the fuel tank to reach the retaining ring.

5. On a fuel pump that is outside the tank, remove the bolts holding it in place. On in-tank models, loosen the retaining ring. Pull the pump and sending unit out of the tank (if they are combined in one unit) and discard the tank O-ring seal.

6. To remove the pump, twist off the filter sock, then push the pump up until the bottom is clear of the bracket. Swing the pump out to the side and pull it down to free it from the rubber fuel line coupler. The rubber sound insulator between the bottom of the pump and bracket and the rubber coupler on the fuel line are normally discarded because new ones are included with the replacement pump. Some pumps have a rubber jacket around them to quiet the pump. If this is the case, slip the jacket off and put it on the new pump.

7. Compare the replacement pump with the old one. If necessary, transfer any fuel line fittings from the old pump to the new one. Note the position of the filter sock on the pump so you can install a new one in the same relative position.

8. When inserting the new pump back into the sending unit bracket, be careful not to bend the bracket. Make sure the rubber sound insulator under the bottom of the pump is in place. Install a new filter sock on the pump inlet and reconnect the pump wires. Be absolutely certain you have correct polarity. Replace the O-ring seal on the fuel tank opening, then put the pump and sender assembly back in the tank and tighten the locking ring by rotating it clockwise. Some pump/sender assembly units are secured by bolts.

9. If the fuel was removed from the vehicle, replace it.

CAUTION: Avoid the temptation to test the new pump before reinstalling the fuel tank by energizing it with a couple of jumper wires. Running the pump dry can damage it because the pump relies on fuel for lubrication and cooling.

10. Reconnect the electrical connectors.

11. Reconnect the ground terminal at the battery.

12. Start the engine and check all connections for fuel leaks.

CAUTION: Never turn the ignition switch on or crank the engine with a disconnected fuel line. This action will result in gasoline discharge from the disconnected line, which may result in a fire, causing personal injury and/or property damage.

Some older fuel injection systems had rubber lines that could be carefully crimped with block-off pliers. Many later systems have nylon lines that should never be crimped. Make sure to follow service manual recommendations before crimping lines.

Fuel Injection Systems

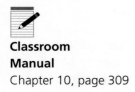

Classroom Manual
Chapter 10, page 309

Troubleshooting fuel injection systems requires systematic step-by-step test procedures. With so many interrelated components and sensors controlling fuel injection performance, a hit-or-miss approach to diagnosing problems can quickly become frustrating, time-consuming, and costly.

Most fuel injection systems are integrated into engine control systems. The self-test modes of these systems are designed to help in engine diagnosis. Unfortunately, when a problem upsets the smooth operation of the engine, many service technicians automatically assume that the computer (PCM) is at fault. But in the vast majority of cases, complaints about driveability, performance, fuel mileage, roughness, or hard starting or no-starting are due to something other than the computer itself, although many problems are caused by sensor malfunctions that can be traced using the self-test mode.

Before condemning sensors as bad, remember that weak or poorly operating engine components can often affect sensor readings and result in poor performance. For example, a sloppy timing chain, bad rings, or valves reduce vacuum and cylinder pressure, resulting in a lower exhaust temperature. This can affect the operation of a perfectly good oxygen or lambda sensor, which must heat up to approximately 600°F before functioning in its closed-loop mode.

A problem like an intake manifold leak can cause the manifold absolute pressure (MAP) sensor to adjust engine operation to less-than-ideal conditions.

One of the basic rules of electronic fuel injection servicing is that EFI cannot be adjusted to match the engine; you have to make the engine match EFI. In other words, make sure the rest of the engine is sound before condemning the fuel injection and engine control components.

Preliminary Checks

The best way to approach a problem on a vehicle with electronic fuel injection is to treat it as though it had no electronic controls at all. As the previous examples illustrate, any engine is susceptible to problems that are unrelated to the fuel system itself. Unless all engine support systems are operating correctly, the control system does not operate as designed.

Before proceeding with specific fuel injection checks and electronic control testing, be certain of the following:

❏ The battery is in good condition, fully charged, with clean terminals and connections.
❏ The charging and starting systems are operating properly.
❏ All fuses and fusible links are intact.
❏ All wiring harnesses are properly routed with connections free of corrosion and are tightly attached.
❏ All vacuum lines are in sound condition, properly routed, and are tightly attached.
❏ The PCV system is working properly and maintaining a sealed crankcase.
❏ All emission control systems are in place, hooked up, and are operating properly.
❏ The level and condition of the coolant/antifreeze is good and the thermostat is opening at the proper temperature.
❏ The secondary spark delivery components are in good shape with no signs of crossfiring, carbon tracking, corrosion, or wear.
❏ The base timing and idle speed are set to specifications.
❏ The engine is in good mechanical condition.
❏ The gasoline in the tank is of good quality and has not been substantially cut with alcohol or contaminated with water.

EFI System Component Checks

In any electronic throttle body or port injection system, three things must occur for the system to operate.

1. An adequate air supply must be supplied for the air-fuel mixture.
2. Fuel at the proper pressure must be delivered correctly to operating injectors.
3. The injectors must receive a trigger signal from the control computer.

If all of these preliminary checks do not reveal a problem, proceed to test the electronic control system and fuel injection components. Some older control systems require involved test procedures and special test equipment, but most newer designs have a self-test program designed to help diagnose the problem. These self-tests perform a number of checks on components within the system. Input sensors, output devices, wiring harnesses, and even the electronic control computer itself may be among the items tested.

If the **malfunction indicator lamp (MIL)** is on or there are trouble codes stored in the PCM, these can help a technician in diagnosing a problem. The meaning of trouble codes vary from manufacturer to manufacturer, year to year, and model to model, so it is important to have the appropriate service manuals.

Always remember that trouble codes only indicate the particular circuit in which a problem has been detected. They do not pinpoint individual components. So if a code indicates a defective lambda or oxygen sensor, the problem could be the sensor itself, the wiring to it, or its connector. Trouble codes are not a signal to replace components. They signal that a more thorough diagnosis is needed in that area.

The following sections outline general troubleshooting procedures for the most popular EFI designs in use today.

Air System Checks

In an injection system, particularly designs that rely on airflow meters or mass airflow sensors, all the air entering the engine must be accounted for by the air measuring device. If it is not, the air-fuel ratio becomes overly lean. For this reason, cracks or tears in the plumbing between the airflow sensor and throttle body are potential air leak sources that can affect the air-fuel ratio.

During a visual inspection of the air control system, pay close attention to these areas, looking for cracked or deteriorated ductwork. Also make sure all induction hose clamps are tight and are properly sealed. Look for possible air leaks in the crankcase, for example, dipstick tube and oil filter cap. Any extra air entering the intake manifold through the PCV system is not measured either and can upset the delicately balanced air-fuel mixture at idle.

Airflow Sensors

When looking for the cause of a performance complaint that relates to poor fuel economy, erratic performance/hesitation, or hard starting, make the following checks to determine if the airflow sensor is at fault (Figure 10-25). Start by removing the air intake duct from the airflow sensor to gain access to the sensor flap. Check for binding, sticking, or scraping by rotating the sensor flap (evenly and carefully) through its operating range. It should move freely, make no noise, and feel smooth.

On some systems, the movement of the sensor flap turns the fuel pump on. If the system is so equipped, turn the ignition on. Do not start the engine. Move the flap toward the open position. The electric fuel pump should come on as the flap is opened. If it does not, turn the ignition off, remove the sensor harness, and check for specific resistance values with an ohmmeter at each of the sensor's terminals.

Prior to the J1930 standards, the MIL was most commonly called the Check Engine Light.

The **malfunction indicator lamp (MIL)** is controlled by the PCM and is turned on to alert the driver to a problem.

Carefully inspect the intake hose on a MAF-equipped vehicle if the vehicle stumbles on acceleration. The movement of the engine will cause the hose to open up on acceleration.

If the seal on the dampening chamber is defective, the vane moves too freely causing idle and engine performance problems.

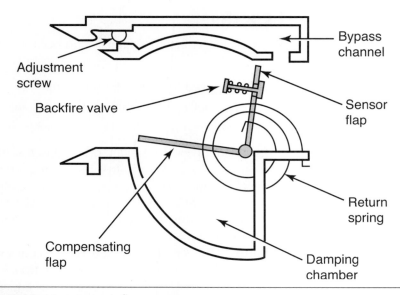

Figure 10-25 Vane-type air flow sensor.

On other models, it is possible to check the resistance values of the potentiometer by moving the air flap, but in either case a service manual is needed to identify the various terminals and to look up resistance specifications.

Vane-type airflow sensors measure the volume of air entering the engine. Even though later vane-type sensors included temperature sensors, the vane-type sensor is not as accurate at measuring airflow as the MAF sensor.

On systems that use a mass airflow meter or manifold pressure sensor to measure airflow, check for good electrical connections. However, a hand-held scan tool can be plugged into the diagnostic connector on some models to check for proper voltage values.

SERVICE TIP: If the engine does not start or idles poorly, unplug the MAF. If the engine starts or runs better with the sensor unplugged, the sensor is likely malfunctioning and should be replaced.

MAF sensors measure the **mass** of air entering the engine. Cold air has more oxygen content than warm air. Moist air burns differently and is heavier than dry air. The MAF sensor is a very accurate way to measure airflow into the engine. The more accurately the PCM can match the air with the fuel mixture, the more efficient the engine. MAF sensors can be located either on the throttle body or near the air cleaner (Figure 10-26). Locating the MAF on the throttle body helps reduce the chance of a split intake hose causing driveability problems, but these defects must still be repaired so that unfiltered air is not entering the engine. Most MAF sensors output a frequency signal that increases with airflow amount. An example of a MAF sensor waveform is shown in Figure 10-27. Look for consistency in the waveform. The higher the rpm, the higher the frequency read. The square waves will be smaller and closer together as the rpm increases. A voltmeter will not give you accurate readings on a MAF sensor because the computer is not looking for voltage variation. In fact, the voltage will vary very little from low to high rpm because most digital voltmeters average their voltage readings. The frequency of the MAF is important and changes in relation to airflow entering the engine. If the technician does not have an oscilloscope available, higher level digital multimeters do have a Hertz (Hz) or frequency measurement. This reading can be watched for smooth transitions during rpm changes.

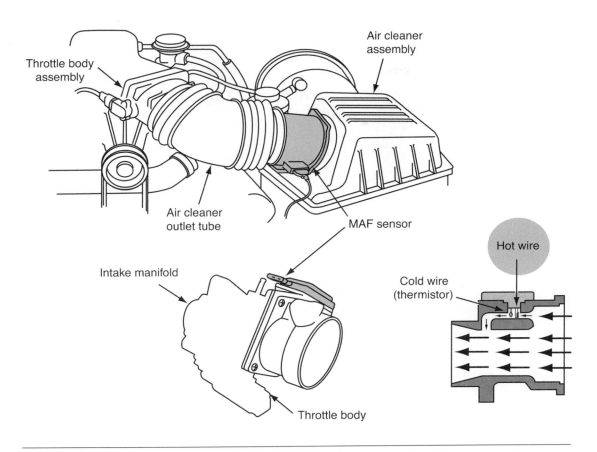

Figure 10-26 Mass Air Flow (MAF) sensors can be mounted on the air cleaner or intake manifold.

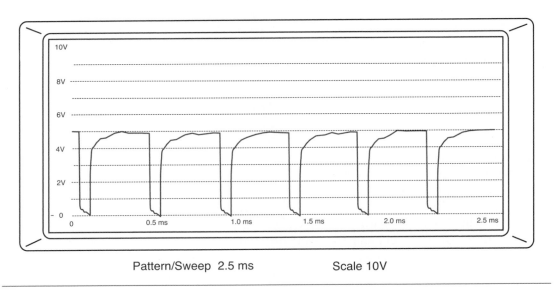

Pattern/Sweep 2.5 ms   Scale 10V

Figure 10-27 Voltage waveform from a hotwire MAF sensor.

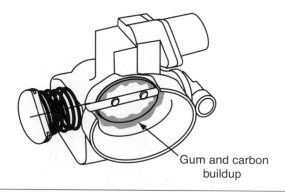

Gum and carbon
buildup

Figure 10-28 Gum and carbon can build up on the throttle bore and cause stalling or rough idle.

The hotwire can be cleaned with contact cleaner.

A hotwire MAF sensor can be checked for contamination. Just a small amount of contamination will cause the MAF sensor reading to be inaccurate. A contaminated MAF sensor wire will measure less airflow than is actually entering the engine, causing the mixture to go very lean on acceleration and results in a hesitation on acceleration.

The MAF sensor can be checked with a scan tool; usually readings are obtained in grams/second of airflow. Look for consistent readings under different operating conditions.

Throttle Body

Remove the air duct from the throttle assembly and check for carbon buildup inside the throttle bore and on the throttle plate. Soak a cloth with throttle body cleaning solution and wipe the bore and throttle plate to remove light to moderate amounts of carbon residue. Also, clean the backside of the throttle plate as shown in Figure 10-28). Then, remove the idle air control valve from the throttle body (if so equipped) and clean any carbon deposits from the pintle tip and the idle air control (IAC) air passage. Some throttle bodies should not be cleaned, so check the service information first.

Fuel System Checks

If the air control system is in working order, move on to the fuel delivery system. It is important to always remember that fuel injection systems operate at high fuel pressure levels. This pressure must be relieved before any fuel line connections can be broken. Spraying gasoline (under a pressure of 35 psi or more) on a hot engine creates a real hazard when dealing with a liquid that has a flash point of –45°F.

Follow the specific procedures given in the service manual when relieving the pressure in the fuel lines. If the procedures are not available, a safe alternative procedure is to apply 20 to 25 inches of vacuum to the externally mounted fuel pressure regulator (with a hand vacuum pump connected to the manifold control line of the regulator), which bleeds fuel pressure back into the tank (Figure 10-29). The fuel pump fuse can be pulled and the engine started and let run until it dies. The fuel pump can also be unplugged if necessary. (Sometimes the fuel pump and ignition are on the same fuse.) Also, some fuel pressure gauges are equipped with a fuel pressure bleed-off valve.

 CAUTION: Dispose of the fuel-soaked rag in a fireproof container.

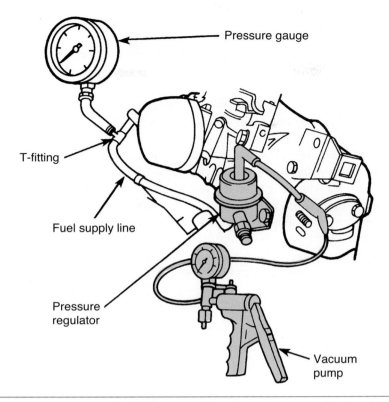

Pressure gauge

T-fitting

Fuel supply line

Pressure regulator

Vacuum pump

Figure 10-29 With an externally mounted fuel pressure regulator, it is possible to bleed off system pressure into the tank using a hand vacuum pump.

Fuel Delivery

When dealing with a fuel complaint that is preventing the vehicle from starting, the first step (after spark, compression, etc., have been verified) is to determine if fuel is reaching the cylinders (assuming there is gasoline in the tank). Checking for fuel delivery is a simple operation on throttle body systems. Remove the air cleaner, crank the engine, and watch the injector for signs of a spray pattern. If a better view of the injector's operation is required, an ordinary timing light does a great job of highlighting the spray pattern. Fuel pressure tests are shown in Figure 10-30.

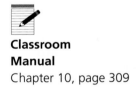

Classroom Manual
Chapter 10, page 309

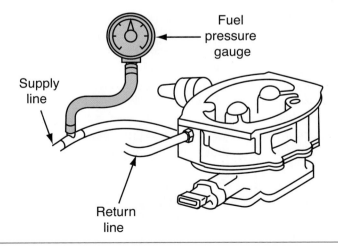

Fuel pressure gauge

Supply line

Return line

Figure 10-30 Connecting the fuel pressure gauge in series with the fuel inlet at a throttle body assembly.

It is impossible to visually inspect the spray pattern and volume of port system injectors. However, an accurate indication of their performance can be obtained by performing simple fuel pressure and fuel volume delivery tests.

Low fuel pressure can cause a no-start or poor-run problem. It can be caused by a clogged fuel filter, a faulty pressure regulator, or a restricted fuel line anywhere from the fuel tank to the fuel filter connection.

If a fuel volume test shows low fuel volume, it can indicate a bad fuel pump or blocked or restricted fuel line. When performing the test, visually inspect the fuel for signs of dirt or moisture. These indicate the fuel filter needs replacement.

High fuel pressure readings will result in a rich-running engine. A restricted fuel return line to the tank or a bad fuel regulator may be the problem. To isolate the cause of high pressure, relieve system pressure and connect a tap hose to the fuel return line. Direct the hose into a container and energize the fuel pump. If fuel pressure is now within specifications, the fuel return line is blocked. If pressure is still high, the pressure regulator is faulty.

If the first fuel pressure reading is within specs but the pressure slowly bleeds down, there may be a leak in the fuel pressure regulator, fuel pump check valve, or the injectors themselves. Remember, hard starting is a common symptom of system leaks.

Injector Checks

A fuel injector is nothing more than a solenoid-actuated fuel valve. Its operation is quite basic in that as long as it is held open and the fuel pressure remains steady, it delivers fuel until it is told to stop.

Because all fuel injectors operate in a similar manner, fuel injector problems tend to exhibit the same failure characteristics. The main difference is that in a TBI design, generally all cylinders suffer if the injectors malfunction, whereas in port systems the loss of one injector is not as crucial.

An injector that does not open causes hard starts on port-type systems and an obvious no-start on single-point TBI designs. An injector that is stuck partially open causes loss of fuel pressure (most noticeably after the engine is stopped and restarted within a short time period) and flooding due to raw fuel dribbling into the engine. In addition to a rich-running engine, a leaking injector also causes the engine to diesel or run on when the ignition is turned off.

Checking Voltage Signals

When an injector is suspected as the cause of a problem, the first step is to determine if the injector is receiving a signal from the control computer to fire. Fortunately, determining if the injector is receiving a voltage signal is easy and requires simple test equipment. Unfortunately, the location of the injector's electrical connector can make this simple voltage check somewhat difficult.

Once the injector's electrical connector has been removed, check for voltage at the injector using an ordinary test light or a convenient noid light that plugs into the connector (Figure 10-31). After making the test connections, crank the engine. A series of rapidly flickering lights indicates the computer is doing its job and supplying voltage or a ground to open the injector.

When performing this test, make sure to keep off the accelerator pedal. On some models, fully depressing the accelerator pedal activates the clear flood mode in which the voltage signal to the injectors is automatically cut off. Technicians unaware of this waste time tracing a phantom problem.

If sufficient voltage is present after checking each injector, check the electrical integrity of the injectors themselves. Use an ohmmeter to check each injector winding for shorts, opens, or excessive resistance. Compare resistance readings to the specifications found in the service manual.

An ohmmeter can be used to test the electrical soundness of an injector. Connect the ohmmeter across the injector terminals (Figure 10-32) after the wires to the injector have been disconnected. If the meter reading is infinity, the injector winding is open. If the meter shows more

Figure 10-31 Checking for voltage at the injector using a noid light.

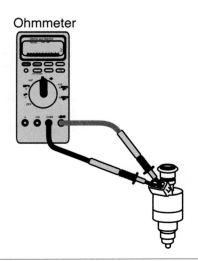

Figure 10-32 An ohmmeter can be connected across the injector terminals to test for shorted or open windings.

resistance than the specifications call for, there is high resistance in the winding. A reading that is lower than the specifications indicates that the winding is shorted. If the injector is even a little bit out of specifications, it must be replaced.

If the injector's electrical leads are difficult to access, an injector power balance test is hard to perform. As an alternative, start the engine and use a technician's stethoscope to listen for correct injector operation (Figure 10-33). A good injector makes a rhythmic clicking sound as the solenoid is energized and de-energized several times each second. If a clunk-clunk instead of a steady click-click is heard, chances are the problem injector has been found. Cleaning or replacement is in order. If a stethoscope is not handy, use a thin steel rod, wooden dowel, or fingers to feel for a steady on/off pulsing of the injector solenoid.

Another way to isolate an offending cylinder when injector access is limited is to perform the more traditional cylinder power balance test by momentarily grounding each spark plug wire instead of disabling the injectors. Remember to bypass the idle air control. Following the same warm-up and idle stabilizing procedures, watch the rpm drop and note any change in idle quality as each plug is shorted. If a lazy or dead cylinder is located, concentrate efforts on the portion of the fuel or ignition system pertaining to that cylinder, assuming no mechanical problems are present.

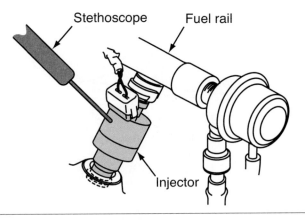

Figure 10-33 Checking the sound of an injector with a sound scope (stethoscope).

CAUTION: Any time cylinders are shorted during a power balance test, make the readings as quickly as possible. Prolonged operation of a shorted cylinder causes excessive amounts of unburned fuel to accumulate inside the catalytic converter and increase the risk of premature converter failure.

AUTHOR'S NOTE: With OBD II and even some earlier vehicles, the technician can actually do a balance test with the scan tool. The scan tool can also be checked for cylinder misfire with the weak cylinder identified by the PCM. Sometimes the injectors can be shut down while watching the rpm. Get to know the capability of your scan tool. If this is not possible on your vehicle and the injectors are accessible, the injectors can be unplugged one at a time instead of the spark plug wires. The advantage of these methods is that the unburned fuel from a missing spark does not go through the converter, and the chance of electric shock is also eliminated.

Oscilloscope Checks

An oscilloscope can be used to monitor the injector's pulse width and duty cycle when an injector-related problem is suspected. As covered earlier in the chapter, the pulse width is the time in milliseconds that the injector is energized. The duty cycle is the percentage of on time to total cycle time.

To check the injector's firing voltage on the scope, a typical hookup involves connecting the scope's positive lead to the injector supply wire and the scope's negative lead to an engine ground. Even though these connections are considered typical, it is still a good idea to read the instruction manual provided with the test equipment before making connections.

With the scope set on the low voltage scale and the pattern adjusted to fill the screen, a square-shaped voltage signal should be present with the engine running or cranking (Figure 10-34). If the voltage pattern reads higher than normal, excessive resistance in the injector circuit is indicated. Conversely, a low-voltage trace indicates low-circuit resistance. If the pattern forms a continuous straight line, it means the injector is not functioning due to an open circuit somewhere in the injector's electrical circuit.

The PCM controls fuel mixture by varying injector on time as needed.

Injector Replacement

Consult the vehicle's service manual for instructions on removing and installing injectors. Before installing the new one, always check to make sure the sealing O-ring is in place (Figure 10-35). Also, prior to installation, lightly lubricate the sealing ring with engine oil or automatic transmission fluid (avoid using silicone grease, which tends to clog the injectors) to prevent seal distortion or damage.

Photo Sequence 20 outlines a typical procedure for removing and installing an injector. Always refer to the service manual for the exact procedure for the engine being serviced.

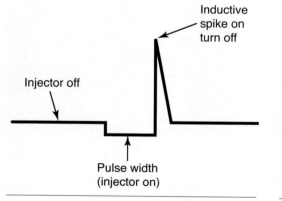

Figure 10-34 Injector waveform.

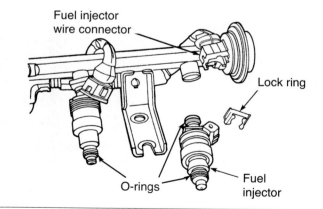

Figure 10-35 Replace the O-rings when replacing an injector.

Photo Sequence 20
Removing and Replacing a Fuel Injector on a PFI System

P20-1 Often, an individual injector needs to be replaced. Random disassembly of the components and improper procedures can result in damage to one of the various systems located near the injectors.

P20-2 The injectors are normally attached directly to a fuel rail and inserted into the intake manifold or cylinder head. They must be positively sealed because high-pressure fuel leaks can cause a serious safety hazard.

P20-3 Prior to loosening any fitting in the fuel system, the fuel pump fuse should be removed.

P20-4 As an extra precaution, many technicians disconnect the negative cable of the battery.

P20-5 To remove an injector, the fuel rail must be able to move away from the engine. The rail-holding brackets should be unbolted and the vacuum line to the pressure regulator disconnected.

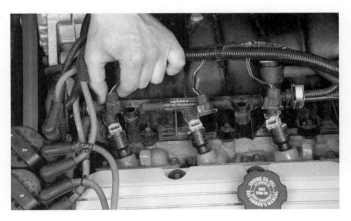

P20-6 Disconnect the wiring harness to the injectors by depressing the center of the attaching wire clip.

Removing and Replacing a Fuel Injector on a PFI System (Continued)

P20-7 The injectors are held to the fuel rail by a clip that fits over the top of the injector. O-rings at the top and at the bottom of the injector seal the injector.

P20-8 Pull up on the fuel rail assembly. The bottoms of the injectors will pull out of the manifold while the tops are secured to the rail by clips.

P20-9 Remove the clip from the top of the injector and remove the injector unit. Install new O-rings onto the new injector. Be careful not to damage the seals while installing them, and make sure they are in their proper locations.

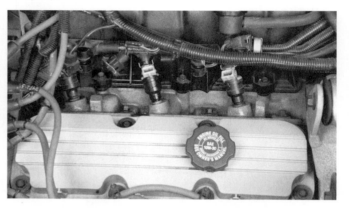

P20-10 Install the injector into the fuel rail and set the rail assembly into place.

P20-11 Tighten the fuel rail hold-down bolts according to manufacturer's specifications.

P20-12 Reconnect all parts that were disconnected. Install the fuel pump fuse and reconnect the battery. Turn the ignition switch to the run position and check the entire system for leaks. After a visual inspection has been completed, conduct a fuel pressure test on the system.

Idle Adjustment

In a fuel injection system, idle speed is regulated by controlling the amount of air that is allowed to bypass the airflow sensor or throttle plates. When presented with a car that tends to stall, especially when coming to a stop, or idles too fast, look for obvious problems like binding linkage and vacuum leaks first. If no problems are found, go through the minimum idle checking/setting procedure described on the underhood decal. The instructions listed on the decal spell out the necessary conditions that must be met prior to attempting an idle adjustment. These adjustment procedures can range from a simple twist of a throttle stop screw to more involved procedures requiring circumvention of idle air control devices, removal of casting plugs, or recalibration of the throttle position sensor. Specific idle adjustment procedures can also be found in the service manual.

Evaporative Emissions

Evaporative emissions, as we have seen previously, are important to reduce hydrocarbon (HC) emissions due to the evaporation of fuel stored in the vehicle. Evaporative emission systems were implemented in 1968. Early systems, even after the advent of computers on vehicles, did not set codes. Since 1996, evaporative emissions systems have to be checked for integrity by the PCM (Figure 10-36). Several different strategies have been employed. We will try to reproduce a generic system here, not meant to be exactly like any one system but a combination of designs. Make sure to check the latest service information for the specific vehicle you are working on before doing any work.

Canister

The canister, often referred to as the charcoal canister, is filled with activated charcoal. Instead of fuel vapors escaping through a tank vent into the air as HC emission, the vapor is stored and later burned in the engine. The canister is purged during driving when conditions are such that the richer mixture produced will not be detrimental to engine operation.

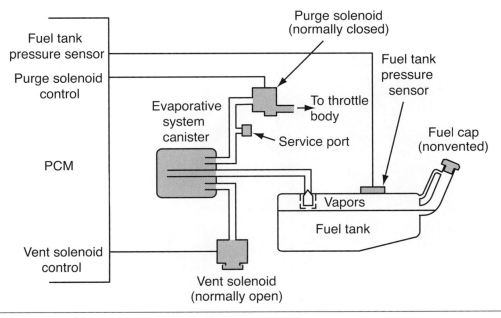

Figure 10-36 Evaporative emission diagram.

Purge Solenoid

The purge solenoid is used to purge the canister of vapors. The PCM supplies a duty cycle, or pulse width modulated signal, to the purge valve, thereby controlling the amount and timing of purge. Trouble codes PO441 through PO445 (generic) can be set if the purge valve is malfunctioning; PO458 and PO459 cover diagnosis of the purge valve electrical circuit. Manufacturers can also add manufacturer-specific codes for use on their vehicles.

Purge Flow Sensor

Some vehicles have a sensor to detect purge flow. The flow sensor is not used on all vehicles because the job of detecting purge flow is handled by the pressure sensor using software programs in the PCM. PO465 to PO469 pertain to purge flow sensor operation if so equipped.

Vent Solenoid

The canister vent solenoid not only allows fresh air into the canister during the purge cycle, but the solenoid allows the PCM to close off the evaporative emissions system to check for leaks. Leaks are determined by using engine vacuum that builds with a closed vent solenoid and an open purge solenoid. Generic trouble codes PO446 through PO449 cover the operation of the vent solenoid, its related wiring, and hose connections as applicable.

Vapor Pressure Sensor

The fuel pressure sensor is much like a MAP sensor, but more accurate. Some manufacturers call this sensor a fuel tank pressure transducer (FTPT), other manufacturers call this sensor the vapor pressure (VP) sensor, and locate the VP sensor in the fuel tank, canister, or in a line. This sensor is capable of measuring the amount of pressure (vacuum) inside the tank. This sensor is a vital part of the PCM's ability to check the system for leaks that are no larger than 0.020 inch in diameter. PO450 to PO454 concern the operation of the vapor pressure operation, hoses, and wiring.

Service Port

Evaporative emission control systems have connection fittings that can be used by the technician to check the system for leaks in the shop. These connections have a green cap and a Schrader valve to attach special tools to service the system (Figure 10-37).

Figure 10-37 Evaporative service port.

Fuel Cap

The fuel cap must be able to seal the evaporative system against leaks or the MIL will come on for a large leak. Some fuel caps have been designed to make it easier for consumers to tell when the cap is installed properly. PO457 is a code set for a loose or missing fuel cap on late-model vehicles.

Fuel Level Sensor

The fuel level of a vehicle is important to the PCM because the evaporative emissions check will not be performed on a tank with more than 85 percent and less than 15 percent of a full tank. The fuel level sensor can also be used to determine if the fuel tank has been filled recently, and some systems check periodically during driving to see if the fuel tank has been filled with the engine running. This is not recommended for safety reasons, but manufacturers have to take this into account because this will affect the way the PCM runs testing. NOTE: Some systems will set a trouble code if the tank is filled with the engine running and the PCM is performing its diagnostic test at the same time. PO460 to PO464 are trouble codes set for a fuel level sensor malfunction.

PCM Action

Depending on the model year and manufacturer, there are several ways to check the system for leaks as mandated by the federal government.

Many systems pull a vacuum on the system during cold engine operation, while some check for vacuum or pressure with the vent valve open to check for large leaks. If the system passes a large leak, then the system will check for smaller leaks by pulling a vacuum and measuring the rate of vacuum leak down. There are also systems that use a small pump to pressurize the system while a pressure sensor checks the system for leaks.

Some systems use the natural volatility of the fuel in the tank, looking for a rise in pressure with the system off.

Make certain that you follow the correct procedures when servicing these systems; including checking for any applicable service bulletins that may apply.

System Service

OBD II systems are required to check system integrity. The main concern for the service technician will be leaks in the system. Evaporative emission systems have to be able to detect a leak as small as 0.020 inch in diameter, so leaks can be difficult to find. Remember also that a kinked hose can also cause a trouble code if vacuum were unable to reach the pressure sensor or if a vent line did not allow the system to vent properly. Faults in the sensors and solenoids required for the PCM to check system integrity also have trouble codes to help the technician determine the point of failure. PO455 is set for a large leak, and code PO456 is set for a small leak.

Onboard Refueling Vapor Recovery (ORVR) System

The **Onboard Refueling Vapor Recovery (ORVR)** system was implemented in 1998 to control the release of hydrocarbons (HC) during vehicle refueling. It was determined that a greater amount of HC was released while refueling than was released during the burning of the tank

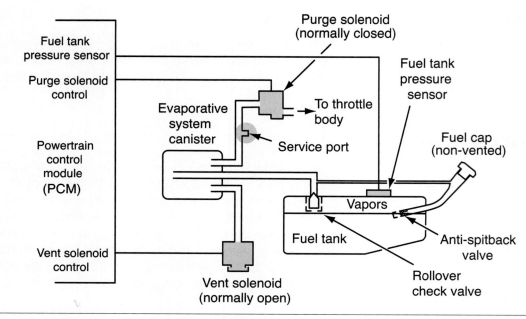

Figure 10-38 Onboard Refueling Recovery (ORVR) system.

of gasoline. The vapor is stored in the charcoal canister just like the normal vent vapors, although the canister is larger than systems without ORVR (Figure 10-38).

The fill pipe on ORVR-equipped vehicles is much smaller than was previously used, about 1 inch in diameter. The small hose is filled with liquid, and vapor is not blown back out of the tank while refueling. The fill pipe is also equipped with a check valve to prevent fuel from coming back out of the pipe while refueling.

The shutoff assembly actually controls the vent to the fuel tank. The valve floats on the fuel in the tank. When the tank level becomes high enough, the vent is closed. Since there is no vent, no more fuel can be added to the tank and the gas station fill nozzle shuts off. There are no codes for malfunctions in the ORVR system. Obviously, a problem in the vent system will make the fuel tank hard to fill. The shutoff valve also acts as a rollover check valve for the fuel tank vapor line.

Terms to Know

"Banjo" fitting

Double flare

EVAP

Malfunction indicator lamp (MIL)

Mass

Onboard Refueling Vapor Recovery (ORVR)

ASE-Style Review Questions

1. *Technician A* says to relieve pressure on an EFI system, connect a pressure gauge to the fuel pressure test port, and open the bleed valve on the gauge with the bleed hose installed in an approved container.
   Technician B says before fuel system components on an EFI system can be removed, the fuel system pressure must be relieved.
   Who is correct?
   A. A only   **C.** Both A and B
   B. B only   **D.** Neither A nor B

2. While discussing alcohol content in gasoline:
   Technician A says excessive quantities of alcohol in gasoline may cause fuel filter plugging.
   Technician B says excessive quantities of alcohol in gasoline may cause lack of engine power.
   Who is correct?
   A. A only   **C.** Both A and B
   B. B only   **D.** Neither A nor B

3. While discussing quick-disconnect fuel line fittings:
   Technician A says some quick-disconnect fittings may be disconnected with a pair of snap ring pliers.
   Technician B says some quick-disconnect fittings are hand releasable.
   Who is correct?
   A. A only   **C.** Both A and B
   B. B only   **D.** Neither A nor B

4. *Technician A* says that a trouble code identifies the exact location of a fuel system fault.
   Technician B says the trouble code only identifies the area of the fault and that the sensor wiring and PCM still have to be checked.
   Who is correct?
   A. A only   **C.** Both A and B
   B. B only   **D.** Neither A nor B

5. *Technician A* says the powertrain control module in TBI, MFI, and SFI systems provides the proper air-fuel ratio by controlling injector pulse width.
   Technician B says the powertrain control module controls the fuel pressure to provide the proper air-fuel ratio.
   Who is correct?
   A. A only   **C.** Both A and B
   B. B only   **D.** Neither A nor B

6. *Technician A* says the sending unit O-ring must be replaced when the sending unit is replaced.
   Technician B says a special tool may be required to remove the pump and sender from the tank.
   Who is correct?
   A. A only   **C.** Both A and B
   B. B only   **D.** Neither A nor B

7. *Technician A* says most fuel tanks contain an in-tank strainer to filter large particles of contaminate from the fuel filter.
   Technician B says the strainer also blocks water from entering the fuel supply.
   Who is correct?
   A. A only   **C.** Both A and B
   B. B only   **D.** Neither A nor B

8. *Technician A* says the evaporative emission system controls hydrocarbon emissions.
   Technician B says the evaporative emission system controls carbon monoxide emissions.
   Who is correct?
   A. A only   **C.** Both A and B
   B. B only   **D.** Neither A nor B

9. *Technician A* says the fuel pump can be checked for fuel volume.
   Technician B says the fuel pump can be checked for fuel pressure output.
   Who is correct?
   A. A only   **C.** Both A and B
   B. B only   **D.** Neither A nor B

10. *Technician A* says the MAF sensor sends a varying voltage signal back to the PCM.
    Technician B says that later-model vane-style airflow meters had a temperature sensor incorporated into the sensor.
    Who is correct?
    A. A only   **C.** Both A and B
    B. B only   **D.** Neither A nor B

ASE Challenge Questions

1. *Technician A* says gasoline contains 10 percent alcohol if the water content is 20 ml at the end of an alcohol-in-fuel test.
   Technician B says that during an alcohol-in-fuel test, the alcohol and water remain separated.
   Who is correct?
   A. A only
   B. B only
   C. Both A and B
   D. Neither A nor B

2. *Technician A* says a rich air-fuel mixture in TBI and MFI systems may be the result of higher-than-normal fuel pressure.
   Technician B says lower-than-normal fuel pressure in a TBI or MFI system causes a lean air-fuel ratio.
   Who is correct?
   A. A only
   B. B only
   C. Both A and B
   D. Neither A nor B

3. Which of the following should be checked first if there is no fuel pump pressure?
   A. Fusible link
   B. Inertia switch
   C. Fuse
   D. Electrical connections

4. *Technician A* says PO441 through PO445 cover purge valve operation.
   Technician B says that OBD II systems have no codes for the evaporative emissions system.
   Who is correct?
   A. A only
   B. B only
   C. Both A and B
   D. Neither A nor B

5. *Technician A* says engine stalling during vehicle driving may be caused by a vacuum leak.
   Technician B says engine stalling may be caused by a plugged fuel filter.
   Who is correct?
   A. A only
   B. B only
   C. Both A and B
   D. Neither A nor B

Job Sheet 25

25

Name _____ Date _____

Relieving Pressure in an EFI System

Upon completion of this job sheet, you should be able to relieve pressure from the fuel lines and system on a vehicle equipped with electronic fuel injection.

ASE Correlation

This job sheet is related to the ASE Engine Performance Test's content area: fuel, air induction, and exhaust system diagnosis and repair; tasks; inspect fuel tank and fuel cap; inspect and replace fuel lines, fittings, and hoses; and inspect and test mechanical and electrical fuel pumps and pump control systems; replace as needed.

Tools and Materials

Clean shop rags   Bleed hose
Pressure gauge with adapters   Approved gasoline container

Describe the vehicle being worked on.

Year _____ Make _____ VIN _____

Model_____

NOTE: Since electronic fuel injection systems have a residual fuel pressure, this pressure must be relieved before disconnecting any fuel system component. Failure to relieve the fuel pressure on electronic fuel injection (EFI) systems prior to fuel system service may result in gasoline spills, serious personal injury, and expensive property damage.

Procedures

Task Completed

1. Disconnect the negative battery cable to avoid fuel discharge if an accidental attempt is made to start the engine. ☐

2. Loosen the fuel tank filler cap to relieve any fuel tank vapor pressure. ☐

3. Wrap a shop towel around the fuel pressure test port on the fuel rail and remove the dust cap from this valve. ☐

4. Connect the fuel pressure gauge to the fuel pressure test port on the fuel rail. ☐

5. Install the bleed hose on the gauge in an approved gasoline container and open the gauge bleed valve to relieve fuel pressure from the system into the gasoline container. Be sure all the fuel in the bleed hose is drained into the gasoline container. ☐

6. Describe any problems encountered while following this procedure:

   _____

On EFI systems that do not have a fuel pressure test port, such as most throttle body injection (TBI) systems, follow these steps for fuel system pressure relief:

☐ 1. Loosen the fuel tank filler cap to relieve any tank vapor pressure.

☐ 2. Remove the fuel pump fuse.

☐ 3. Start and run the engine until the fuel is used up in the fuel system and the engine stops.

☐ 4. Engage the starter for 3 seconds to relieve any remaining fuel pressure.

☐ 5. Disconnect the negative battery terminal to avoid possible fuel discharge if an accidental attempt is made to start the engine.

6. Describe any problems encountered while following this procedure:

_____

Instructor's Response_____

_____

_____

Job Sheet 26

26

Name _____ Date _____

Testing Fuel Pressure on an EFI System

Upon completion of this job sheet, you should be able to test fuel pressure on a vehicle equipped with electronic fuel injection.

ASE Correlation

This job sheet is related to the ASE Engine Performance Test's content area: fuel, air induction, and exhaust system diagnosis and repair; tasks: inspect fuel tank and fuel cap; inspect and replace fuel lines, fittings, and hoses; and inspect and test mechanical and electrical fuel pumps and pump control systems; replace as needed.

Tools and Materials

Clean shop rags   Approved gasoline container
Pressure gauge with adapters   Hand-operated vacuum pump

Describe the vehicle being worked on.

Year_____ Make _____ VIN _____

Model_____

Engine size and type _____

Fuel pump pressure specifications _____ psi

Procedures

Task Completed

1. Carefully inspect the fuel rail and injectors for signs of leaks. Record findings:

   _____

   _____

2. Connect the fuel pressure tester to the Schrader valve on the fuel rail. ☐
3. Connect a hand-operated vacuum pump to the fuel pressure regulator. ☐
4. Turn the ignition on and observe the fuel pressure readings. Your readings are: _____ psi. ☐
5. Compare the readings to specifications. What is indicated by the readings?

   _____

   _____

6. Create a vacuum at the pressure regulator with the vacuum pump. ☐
7. What happened to the fuel pressure?

   _____

   _____

8. Conclusions:

   _____

   _____

Instructor's Response_____

_____

_____

Electronic Fuel Injection Diagnosis and Service

Upon completion and review of this chapter, you should be able to:

❏ Perform a preliminary diagnostic procedure on a fuel injection system.

❏ Remove, clean, inspect, and install throttle body assemblies.

❏ Explain the results of incorrect fuel pressure in a throttle body injection (TBI), multiport fuel injection (MFI), or sequential fuel injection (SFI) system.

❏ Perform an injector balance test and determine the injector condition.

❏ Clean injectors on an MFI or SFI system.

❏ Perform an injector sound test.

❏ Perform an injector ohmmeter test.

❏ Perform an injector noid light test.

❏ Perform an injector flow test and determine injector condition.

❏ Perform an injector leakage test.

❏ Check for leakage in the fuel pump check valve and the pressure regulator valve.

❏ Remove and replace the fuel rail, injectors, and pressure regulator.

❏ Perform a minimum idle speed adjustment.

❏ Diagnose causes on improper idle speed on vehicles with an idle air control motor.

❏ Diagnose idle contact switches and related circuits.

❏ Diagnose idle air control motors and idle air control bypass air motors.

❏ Remove, replace, and clean idle air control bypass air motors and related throttle body passages.

❏ Diagnose idle air control bypass air valves.

❏ Diagnose fast idle thermo valves.

❏ Diagnose starting air valves.

Introduction

Although fuel injection technology has been around since the 1920s, it was not until the 1980s that manufacturers began to replace carburetors with electronic fuel injection (EFI) systems. Many of the early EFI systems were throttle body injection (TBI) systems in which the fuel was injected above the throttle plates. Engines equipped with TBI have gradually become equipped with port fuel injection (PFI) which have injectors located in the intake ports of the cylinders. Since the 1995 model year, all new cars are equipped with an EFI system.

Troubleshooting fuel injection systems requires systematic step-by-step test procedures. With so many interrelated components and sensors controlling fuel injection performance (Figure 11-1), a hit-or-miss approach to diagnosing problems can quickly become frustrating, time-consuming, and costly.

Nearly all fuel injection systems are integrated into engine control systems. The self-test modes of these systems are designed to help in engine diagnosis. Unfortunately, when a problem upsets the smooth operation of the engine, many service technicians automatically assume that the computer (PCM) is at fault. But in the vast majority of cases, complaints about driveability, performance, fuel mileage, roughness, or hard starting or no-starting are due to factors other than the computer itself (although many problems are caused by sensor malfunctions that can be traced using the self-test mode).

Before condemning sensors as bad, remember that weak or poorly operating engine components can often affect sensor readings and result in poor performance. For example, a sloppy timing chain, bad rings, or valves reduce vacuum and cylinder pressure, resulting in a lower exhaust temperature. This can affect the operation of a perfectly good oxygen or lambda sensor, which must heat up to approximately 600°F before it works properly and is recognized

Basic Tools

Basic tool set, DMM, fuel pressure gauge

Classroom Manual
Chapter 11, page 327

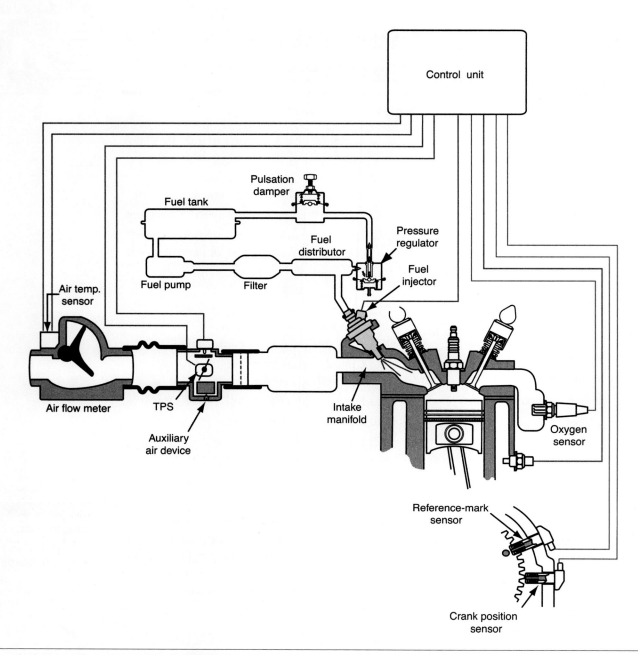

Figure 11-1 Layout of a late-model EFI system.

by the PCM. A problem like an intake manifold leak can cause a sensor, the MAP sensor in this case, to adjust engine operation to less than ideal conditions.

One of the basic rules of electronic fuel injection servicing is that EFI cannot be adjusted to match the engine; you have to make the engine match EFI. In other words, make sure the rest of the engine is sound before condemning the fuel injection and engine control components.

Preliminary Checks

The best way to approach a problem on a vehicle with electronic fuel injection is to treat it as though it had no electronic controls at all. Any engine is susceptible to problems that are unrelated to the fuel system itself. Unless all engine support systems are operating correctly, the control system does not operate as designed.

Before proceeding with specific fuel injection checks and electronic control testing, be certain of the following:

❑ Inspect the air cleaner and related components of the air intake system.

❑ The battery is in good condition, fully charged, with clean terminals and connections.

❑ The charging and starting systems are operating properly.

❑ All fuses and fusible links are intact.

❑ All wiring harnesses are properly routed with connections free of corrosion and are tightly attached.

❑ All vacuum lines are in sound condition, properly routed, and are tightly attached.

❑ The PCV system is working properly and maintaining a sealed crankcase.

❑ All emission control systems are in place, hooked up, and operating properly.

❑ The level and condition of the coolant/antifreeze is good and the thermostat is opening at the proper temperature.

❑ The secondary spark delivery components are in good shape with no signs of crossfiring, carbon tracking, corrosion, or wear.

❑ The base timing and idle speed are set to specifications.

❑ The engine is in good mechanical condition and has acceptable compression.

❑ The gasoline in the tank is of good quality and has not been substantially cut with alcohol or contaminated with water.

❑ The engine is at its normal operating temperature and the cooling system works normally.

❑ Prior to testing, make sure all accessories are turned off.

Service Precautions

These precautions must be observed when electronic fuel injection systems are diagnosed and serviced:

1. Always relieve the fuel pressure before disconnecting any component in the fuel system (Figure 11-2).

2. Never turn the ignition switch on when any fuel system component is disconnected.

3. Use only the test equipment recommended by the vehicle manufacturer.

4. Always turn the ignition switch off before connecting or disconnecting any system component or test equipment.

5. When arc welding is necessary on a computer-equipped vehicle, disconnect both battery cables before welding is started. Always disconnect the negative cable first.

6. Never allow electrical system voltage to exceed 16 volts. This could be done by disconnecting the circuit between the alternator and the battery with the engine running.

7. Avoid static electric discharges when handling computers, modules, and computer chips.

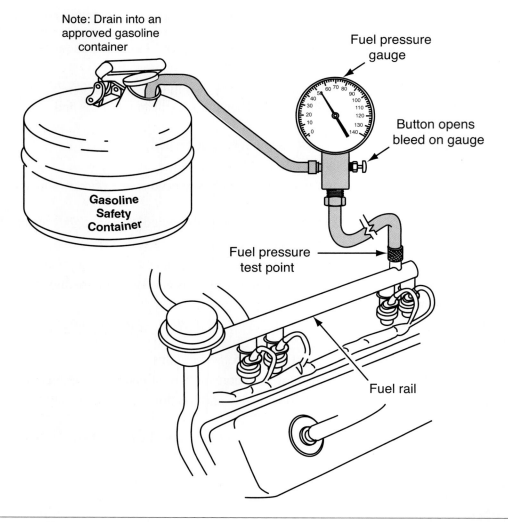

Note: Drain into an approved gasoline container

Fuel pressure gauge

Button opens bleed on gauge

Gasoline Safety Container

Fuel pressure test point

Fuel rail

Figure 11-2 Relieving fuel pressure from the system.

Disconnecting Battery Cables

While servicing electronic fuel injection systems, many procedures indicate the removal of the negative battery cable or both battery cables. If the negative battery cable is disconnected during diagnostic and service procedures, it has these effects:

1. Deprograms the radio
2. Deprograms other convenience items such as memory seats or mirrors
3. Erases the trip odometer if the vehicle has digital instrumentation
4. Erases the adaptive strategy in the computer

If the adaptive strategy in the computer is erased, engine operation may be rough at low speeds when the engine is restarted simply because the computer must relearn the computer system defects. Under this condition, the vehicle should be driven for 5 minutes with the engine at normal operating temperature. Some manufacturers recommend that a 12-volt dry cell battery be connected from the positive battery cable to ground if the battery is disconnected. The voltage supplied by the dry cell prevents deprogramming and memory erasing. Some voltage sources designed for this purpose plug into the cigarette lighter socket.

Basic EFI System Checks

Some older control systems require involved test procedures and special test equipment, but most newer designs have a self-test program designed to help diagnose the problem. These self-tests perform a number of checks on components within the system. Input sensors, output devices, wiring harnesses, and even the electronic control computer itself may be among the items tested.

Always remember that trouble codes only indicate the particular circuit in which a problem has been detected. They do not pinpoint individual components. Therefore, if a code indicates a defective lambda or oxygen sensor, the problem could be a fuel control problem, engine problem, the sensor itself, the wiring to it, or its connector. Trouble codes are not a signal to replace components. They signal that a more thorough diagnosis is needed in that area.

In any electronic throttle body or port injection system three things must occur for the system to operate:

1. An adequate air supply must be supplied for the air-fuel mixture.

2. A pressurized fuel supply must be delivered to properly operating injectors.

3. The injectors must receive a trigger signal from the control computer.

The following sections cover the testing of common EFI components that are involved with all of these things.

⬤ **CUSTOMER CARE:** Electric in-tank fuel pumps can be labor intensive to replace. When replacing a fuel pump, the fuel pump strainer should be replaced as well. The customer may also need to be advised of any debris or contamination within the fuel tank, which warrants cleaning and replacing the fuel filter. Otherwise, there is the risk that the vehicle will return with the same problem.

Oxygen Sensor Diagnosis

Oxygen sensors produce a voltage based on the amount of oxygen (O_2) in the exhaust. Large amounts of oxygen result from lean mixtures and produce a low voltage output from the O_2 sensor. Rich mixtures have released lower amounts of oxygen in the exhaust. Therefore, the O_2 sensor voltage is high. The engine must be at normal operating temperature before the O_2 sensor is tested. An O_2 sensor can be checked with a voltmeter.

Connect the voltmeter between the O_2 sensor wire and ground. The sensor's voltage should be cycling from low voltage to high voltage. The signal from most O_2 sensors varies between 0 and 1 volt. If the voltage is continually high, the air-fuel ratio may be rich or the sensor may be contaminated. When the O_2 sensor voltage is continually low, the air-fuel ratio may be lean, the sensor may be defective, or the wire between the sensor and the computer may have a high-resistance problem. If the O_2 sensor voltage signal remains in a mid-range position, the computer may be in open loop or the sensor may be defective.

The activity of the sensor can be monitored on a scanner. By watching the scanner while the engine is running, the O_2 voltage should move to nearly 1 volt than drop back to close to 0 volt. Immediately after it drops, the voltage signal should move back up. This immediate cycling is an important function of an O_2 sensor. If the response is slow, the sensor is lazy and should be replaced. With the engine at about 2,500 rpm, the O_2 sensor should cycle from high

---

On many vehicles, two terminals in the DLC must be connected to obtain flash codes from the check engine light.

OBD I vehicles had flash codes that could be accessed without a scan tool.

OBD II vehicles require a scan tool to access codes.

A breakout box may be connected in series with the PCM terminals on Ford products to obtain voltage readings from the input sensors.

Special Tools

Lab scope, scan tool

Classroom Manual
Chapter 11, page 334

Short-term and long-term fuel trims are values the PCM uses for fuel connection.

Normal operation

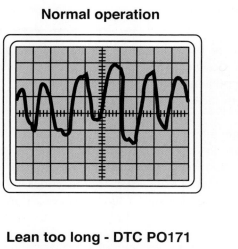

Lack of activity - DTC PO134

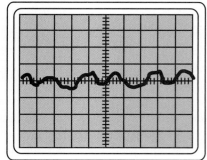
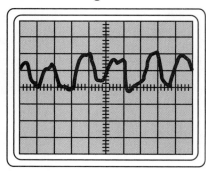

Lean too long - DTC PO171

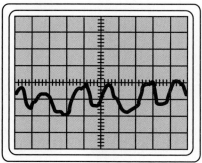

Rich too long - DTC PO172

Figure 11-3 Normal and abnormal O₂ sensor waveforms.

to low ten to forty times within a few seconds. When testing the O_2 sensor, make sure the sensor is heated and the system is in closed loop. The activity of an O_2 sensor is best monitored with a lab scope (Figure 11-3).

A scan tool is also an excellent way to monitor the fuel control of a fuel injected engine. There are many factors that determine the pulse width of the injectors, but it should always respond to the O_2 readings. Fuel correction for fuel injection systems is shown on a scan tool.

Short-term fuel trim (abbreviated STFT) is the short-term fuel correction. STFT numbers will vary with engine operating conditions. When there is no correction, the STFT number will be near zero. If fuel is being added, the numbers climb positively. If fuel is being subtracted, the numbers will go negative. On acceleration, for instance, the STFT numbers will raise quickly, reflecting the addition of fuel. On deceleration, the STFT numbers will fall, reflecting the change in this temporary condition.

Long-term fuel trim, or LTFT, is a long-term number for fuel trim. The numbers will rise and fall, but not immediately. Instead, these numbers reflect long-term trends. For instance, if a vacuum line was leaking or there was any condition that would cause a long-term lean condition, the PCM would see lower oxygen sensor readings. The PCM would add fuel to try and bring the oxygen sensor voltage back into **stoichiometry**, or **14.7:1 air-fuel ratio**. This shift would be reflected as positive numbers on the LTFT. On the other hand, if for instance there was a condition that would cause a long-term rich condition, the numbers would go negative, showing that the PCM was subtracting fuel from the engine in an effort to bring the air-fuel mixture into stoichiometry. The LTFT will change in an attempt to keep the STFT near 0.

The PCM will add and subtract fuel from its programmed delivery during its regular operation. This adding and subtracting of fuel is called **fuel control**. The PCM can only add or subtract so much fuel without setting a DTC. If the PCM reaches its programmed limit and is still lean, then a lean mixture code will set. If the PCM has subtracted its programmed limit and the oxygen sensor still shows a rich condition, then a rich code will set. When this condition has occurred, the PCM is said to be out of fuel control.

A technician can benefit from watching the fuel trim numbers. For instance, if the engine hesitates on acceleration, watch the STFT for changes. If it does not change, watch for a change in the throttle position, MAF, or MAP numbers. If a customer complains of poor fuel mileage, look at the LTFT numbers to see if the PCM has been adding or subtracting an excessive amount of fuel.

Modern oxygen sensors use a heater circuit to quickly place the fuel control system in closed loop. The PCM also monitors this circuit for failures.

Air-Fuel Ratio (A/F) Sensor

One of the later developments in exhaust oxygen detection is the air-fuel ratio sensor, also called the wide ratio sensor. The normal oxygen sensor only reads in a narrow range around 14.7:1. The wide-range sensor does a better job of determining just how rich or lean the engine is running. This allows for better fuel control. The mixture can be adjusted for optimal fuel mixture under a wider range of conditions not possible with the normal oxygen sensor.

The air-fuel ratio sensor is similar in appearance to the oxygen sensor, with some important differences. The A/F sensor runs at a much hotter 1200°F. The A/F sensor also alters current output as oxygen content in the exhaust varies. At 14.7:1, there is no current flow from the A/F sensor. A rich mixture produces a negative current flow which measures below 3.3 volts, and a lean mixture produces a positive current flow of over 3.3 volts. The only recommended way to check the A/F sensor is with the aid of a scan tool. The A/F sensor does not swing rich and lean as the oxygen sensor does.

Air System Checks

In an injection system, particularly designs that rely on airflow meters or mass airflow sensors, all the air entering the engine must be accounted for by the air measuring device. If it is not, the air-fuel ratio cannot be correct. For this reason, cracks or tears in the plumbing between the airflow sensor and throttle body and other potential air leak sources can affect the air-fuel ratio.

During a visual inspection of the air control system, pay close attention to these areas, looking for cracked or deteriorated ductwork. Also, make sure all induction hose clamps are tight and properly sealed. Look for possible air leaks in the crankcase, gaskets, dipstick tube, and oil filter cap. Any extra air entering the intake manifold through the PCV system can upset the delicately balanced air-fuel mixture at idle. What follows are brief discussions on the diagnosis and repair of EFI components related to the air delivery system.

Airflow Sensors

When looking for the cause of a performance complaint that relates to poor fuel economy, erratic performance/hesitation, or hard starting, make the following checks to determine if the airflow sensor is at fault. Begin the visual inspection by removing the air intake duct from the airflow sensor to gain access to the sensor flap. Check for binding, sticking, or scraping by rotating the sensor flap (evenly and carefully) through its operating range. It should move freely, make no noise, and feel smooth.

Classroom Manual
Chapter 11, page 332

Detailed instructions for checking airflow sensors and other sensors are included in Chapter 12.

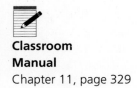

**Classroom
Manual**

Chapter 11, page 329

On some systems, the movement of the sensor flap turns the fuel pump on. If the system is so equipped, turn the ignition on. Do not start the engine. Move the flap toward the open position. The electric fuel pump should come on as the flap is opened. If it does not, turn the ignition off, remove the sensor harness, and check for specific resistance values with an ohmmeter at each of the sensor's terminals.

On other models it is possible to check the resistance values of the potentiometer by moving the air flap, but in either case a service manual is needed to identify the various terminals and to look up resistance specifications.

On systems that use a mass airflow meter or manifold pressure sensor to measure airflow, there are usually no actual physical checks other than checking for good electrical connections. However, a hand-held scan tool can be plugged into the data link connector (DLC) (Figure 11-4) to check for proper voltage values.

The MAF sensor converts the incoming airflow to grams per second (gps) or a frequency in hertz that the PCM uses for computing fuel and timing requirements. The MAF sensor can be accurately tested by observing the grams-per-second values using a scan tool at various engine speeds. Check the readings at idle through 2,500 rpm in 500-rpm increments. The gps value should increase proportionately to the engine rpm without any drastic or sudden jumps in the number. Some manufacturers have MAF grams-per-second or frequency values specified in their service manuals.

Throttle Body

The throttle body (Figure 11-5) allows the driver to control the amount of air that enters the engine, thereby controlling the speed of the engine. Each type of throttle body assembly is designed to allow a certain amount of air to pass through it at a particular amount of throttle opening. If anything accumulates on the throttle plates or in the throttle bore, especially with PFI systems, the amount of air that can pass through is reduced. This will cause stalling on deceleration or idle.

These deposits can be cleaned off the throttle assembly and the airflow through them restored. Begin by removing the air duct from the throttle assembly. This gives access to the plate and bore. The deposits can be cleaned with a spray cleaner or wiped off with a cloth. If

Fuel pressure should be relieved prior to disconnecting fuel system components.

Special Tools

Throttle body (carburetor) cleaner, clean rags, hand-operated vacuum pump, tachometer, 12-volt power source

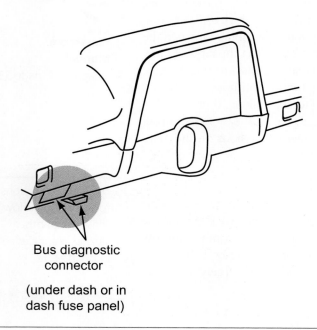

Bus diagnostic
connector

(under dash or in
dash fuse panel)

Figure 11-4 Data Link Connector (DLC) locations.

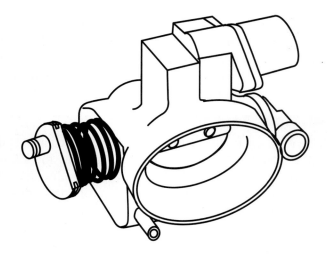

Figure 11-5 A throttle body assembly.

either of these cleaning methods do not remove the deposits, the throttle body should be removed, disassembled, and placed in an approved cleaning solution.

A pressurized can of throttle body cleaner may be used to spray around the throttle area without removing and disassembling the throttle body. The throttle assembly can also be cleaned by soaking a cloth in carburetor solvent and wiping the bore and throttle plate to remove light to moderate amounts of carbon residue. Also, clean the backside of the throttle plate. Then, remove the idle air control valve from the throttle body (if so equipped) and clean any carbon deposits from the pintle tip and the **idle air control** (**IAC**) air passage.

Throttle Body Inspection

Throttle body inspection and service procedures vary widely depending on the year and make of the vehicle. However, some components, such as the TP sensor, are found on nearly all throttle bodies. Since throttle bodies have some common components, inspection procedures often involve checking common components. The following throttle body service procedure example is typical:

1. Check for smooth movement of the throttle linkage from the idle position to the wide-open position. Check the throttle linkage and cable for wear and looseness.

2. Check the vacuum at each vacuum port on the throttle body while the engine is idling and running at a higher speed (Figure 11-6).

3. Apply vacuum from a hand vacuum pump to the throttle opener. Disconnect the TP sensor connector. Test the TP sensor with an ohmmeter connected across the appropriate terminals.

4. Loosen the two TP sensor mounting screws and rotate the TP sensor as required to obtain the specified ohmmeter readings (Figure 11-7). Retighten the mounting screws. If the TP sensor cannot be adjusted to obtain the proper ohmmeter readings, replace the TP sensor.

Some throttle bodies have a coating that resists carbon buildup. These should not be cleaned with solvent. Check service information.

Classroom Manual
Chapter 11, page 336

Idle air control (**IAC**) systems meter the amount of air bypassing the throttle plates to maintain proper idle speed.

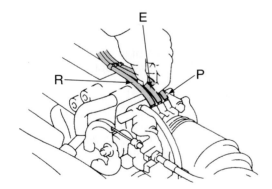

Port name	At idle	Other than idle
P	No vacuum	Vacuum
E	No vacuum	Vacuum
R	No vacuum	No vacuum

Figure 11-6 Throttle body vacuum ports and the specified vacuum at different throttle positions.

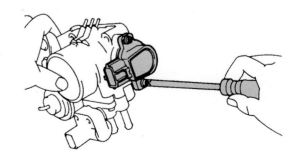

Figure 11-7 Loosen the TP sensor's mounting screws to adjust the sensor.

Use caution if a backup power supply is used to retain battery memory. Some have enough power to supply the air bag circuit.

5. Operate the engine until it reaches normal operating temperature and check the idle speed on a tachometer. Typical idle speed should be 700 to 800 rpm.

6. Disconnect and plug the vacuum hose from the throttle opener. Maintain 2,500 engine rpm.

7. Be sure the cooling fan is off. Release the throttle valve and observe the tachometer reading. When the throttle linkage strikes the throttle opener stem, the engine speed should be 1,300 to 1,500 rpm.

8. Adjust the throttle opener as necessary and reconnect the throttle opener vacuum hose.

Throttle Body Removal and Cleaning

Follow these steps for throttle body removal:

1. Connect a 12-volt power supply to the cigarette lighter socket and disconnect the negative battery cable. If the vehicle is equipped with an air bag, wait 1 minute.

2. Drain the engine coolant from the radiator.

3. Disconnect the accelerator cable from the throttle linkage. If the vehicle has an automatic transmission, disconnect the throttle cable from the throttle linkage.

4. Disconnect the air intake temperature sensor connector.

5. Remove the cruise control cable from the clamp on the air cleaner resonator.

6. Loosen the air cleaner hose clamp bolt at the throttle body and disconnect the air cleaner cap clips.

7. Disconnect the air cleaner hose from the throttle body and remove the air cleaner cap, air hose, and resonator (Figure 11-8).

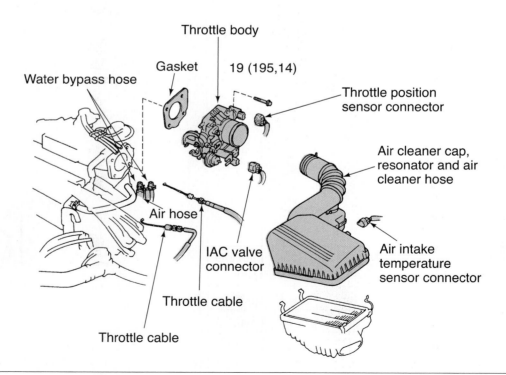

Figure 11-8 Throttle body and related components.

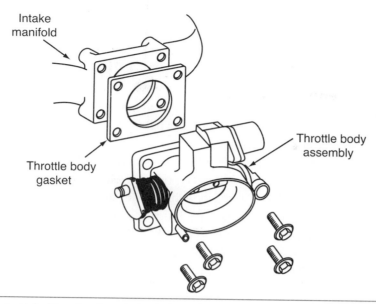

Intake manifold

Throttle body assembly

Throttle body gasket

Figure 11-9 Unbolting the throttle body from the manifold.

8. Disconnect the TP sensor wiring connector.

9. Disconnect the idle air control connector.

10. Remove the vacuum hoses from the throttle body and note the position of each hose so they may be installed in the same location.

11. Remove the throttle body mounting bolts, then remove the throttle body and gasket from the intake manifold (Figure 11-9).

12. Disconnect the water bypass hoses and air hose from the throttle body. Note the position of each hose so they can be reconnected properly.

13. Remove all nonmetallic parts such as the TP sensor, IAC valve, throttle opener, and the throttle body gasket from the throttle body.

14. Clean the throttle body assembly in the recommended throttle body cleaner and blow dry with compressed air. Blow out all passages in the throttle body assembly.

To reinstall the throttle body assembly:

1. Make sure all metal mating surfaces are clean and free from metal burrs and scratches. Install a new IAC valve gasket and install the IAC valve (Figure 11-10). Tighten the valve mounting screws to the proper torque (Figure 11-11).

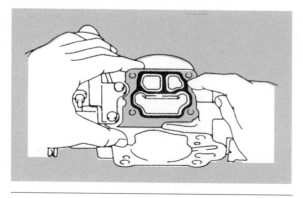

Figure 11-10 Installing a new IAC valve gasket.

Figure 11-11 IAC valve retaining screws must be tightened evenly.

2. With the TP sensor screws loose, connect the ohmmeter leads across the TP sensor. Apply vacuum to the throttle opener with a hand vacuum pump and place a 0.024-in. (0.60-mm) gauge between the throttle stop screw and the stop lever. Slowly rotate the TP sensor clockwise until the ohmmeter deflects. Then tighten the TP sensor screws.

3. Adjust the TP sensor as necessary.

4. Install the water bypass hoses and the air hose in their original locations on the throttle body. Be sure the hose clamps are tight.

5. Install a new throttle body gasket.

6. Install the throttle body attaching bolts, and tighten these bolts to the specified torque.

7. Install the vacuum hoses on the throttle body in their original location.

8. Connect the IAC valve and TP sensor wiring connectors.

9. Check the air cleaner element. Replace it if necessary. Inspect the air cleaner box, cap, hose, and resonator for cracks and distortion. Remove any debris from the air cleaner box. Connect the air cleaner hose and tighten the hose clamp. Install the air cleaner cap and the retaining clamps.

10. Connect the air intake temperature sensor connector. Install the cruise control cable in the clamp on the air cleaner resonator.

11. Connect the throttle cable and accelerator cable. Replace the engine coolant.

12. Connect the negative battery cable. Disconnect the 12-volt power supply.

 SERVICE TIP: Be sure to consult the service manual for the vehicle being serviced. Not all procedures are the same.

Fuel System Checks

Special Tools

Hand-operated vacuum pump, exhaust gas analyzer

If the air control system is in working order, move on to the fuel delivery system. It is important to always remember that fuel injection systems operate at high fuel pressure levels. This pressure must be relieved before any fuel line connections can be broken. Spraying gasoline (under a pressure of 35 psi or more) on a hot engine creates a real hazard. You are dealing with a liquid that has a flash point of –45°F.

Follow the specific procedures given in the service manual when relieving the pressure in the fuel lines. If the procedures are not available, a safe alternative procedure is to apply 20 to 25 inches of vacuum to the externally mounted fuel pressure regulator (with a hand vacuum pump connected to the manifold control line of the regulator), which bleeds fuel pressure back into the tank (Figure 11-12). If the vehicle does not have an externally mounted regulator or a service valve or injector to energize, use a rag to catch excess fuel while slowly breaking a fuel line connection. This method should only be used when all other possibilities have been tried.

 WARNING: Dispose of the fuel-soaked rag in a fireproof container.

Classroom Manual
Chapter 11, page 347

Fuel Delivery

When dealing with an alleged fuel complaint that is preventing the vehicle from starting, the first step (after spark, compression, etc., have been verified) is to determine if fuel is reaching the cylinders (assuming there is gasoline in the tank). Checking for fuel delivery is a simple

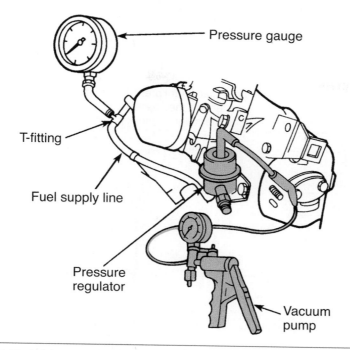

Figure 11-12 With an externally mounted fuel pressure regulator, it is possible to bleed the system's pressure into the fuel tank with a hand-operated vacuum pump.

operation on throttle body injection systems. Remove the air cleaner, crank the engine, and watch the injector for signs of a spray pattern. If a better view of the injector's operation is required, an ordinary timing light does a great job of highlighting the spray pattern.

It is impossible to visually inspect the spray pattern and volume of port system injectors. However, an accurate indication of fuel delivery to the fuel rail can be obtained by performing simple fuel pressure and fuel volume delivery tests (Figure 11-13).

If there is no fuel pump pressure, always check the inertia switch, fuse, or fuse link first.

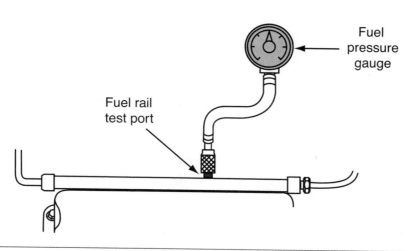

Figure 11-13 Checking the fuel pressure at the fuel rail.

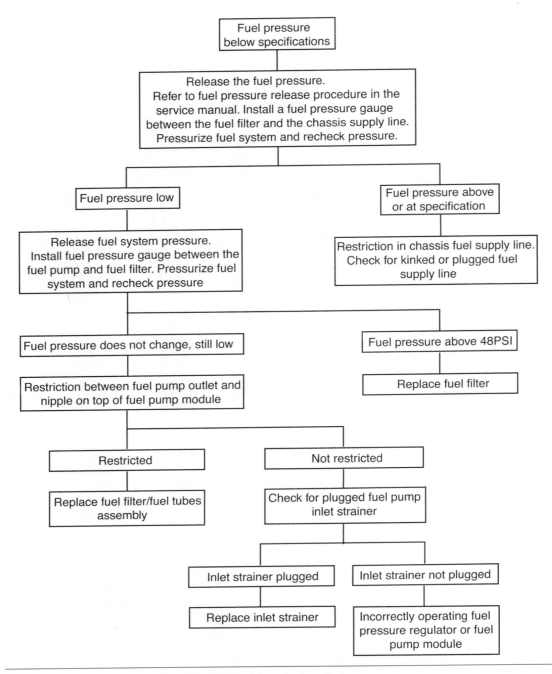

Figure 11-14 A typical troubleshooting chart for low fuel pressure.

Low fuel pressure can cause a no-start or poor-run problem (Figure 11-14). It can be caused by a clogged fuel filter, a faulty pressure regulator, or a restricted fuel line anywhere from the fuel tank to the fuel filter connection.

If a fuel volume test shows low fuel volume, it can indicate a bad fuel pump or blocked or restricted fuel line or pickup screen in the fuel tank. When performing the test, visually inspect the fuel for signs of dirt or moisture. These indicate if the fuel filter needs replacement.

High fuel pressure readings will result in a rich-running engine. A restricted fuel return line to the tank or a bad fuel regulator may be the problem. To isolate the cause of high pressure, relieve system pressure and connect a tap hose to the fuel return line. Direct the hose into a container and energize the fuel pump. If fuel pressure is now within specifications, the fuel return line is blocked. If pressure is still high, the pressure regulator is faulty.

If the first fuel pressure reading is within specs but the pressure slowly bleeds down, there may be a leak in the fuel pressure regulator, fuel pump check valve, or the injectors themselves. Remember, hard starting is a common symptom of system leaks.

When testing the fuel system, be sure to perform electrical and amperage tests on the fuel pump.

Injector Checks

Classroom Manual Chapter 11, page 334

A fuel injector is nothing more than a solenoid-actuated fuel valve. Its operation is quite basic in that as long as it is held open and the fuel pressure remains steady, it delivers fuel until it is told to stop.

Because all fuel injectors operate in a similar manner, fuel injector problems tend to exhibit the same failure characteristics. The main difference is that, in a TBI design, generally all cylinders suffer if the injectors malfunction, whereas in port systems the loss of one injector is not as crucial.

An injector that does not open causes hard starts on port-type systems and an obvious no-start on single-point TBI designs. An injector that is struck partially open causes loss of fuel pressure (most noticeably after the engine is stopped and restarted within a short time period) and flooding due to raw fuel dribbling into the engine. In addition to a rich-running engine, a leaking injector also causes the engine to diesel or run on when the ignition is turned off. Buildups of gum and other deposits on the tip of an injector can reduce the amount of fuel sprayed by the injector or they can prevent the injector from totally sealing, allowing it to leak. Since injectors on MFI and SFI systems are subject to more heat than TBI injectors, port injectors have more problems with tip deposits.

Because an injector adds the fuel part to the air-fuel mixture, any defect in the fuel injection system will cause the mixture to go rich or lean. If the mixture is too rich and the PCM is in control of the air-fuel ratio, a common cause is that one or more injectors are leaking. An easy way to verify this on port-injected engines is to use an exhaust gas analyzer.

With the engine warmed up, turned on but not running, remove the air duct from the airflow sensor. Insert the gas analyzer's probe into the intake plenum area. Be careful not to damage the airflow sensor or throttle plates while doing this. Look at the hydrocarbon (HC) readings on the analyzer. They should be low and drop as time passes. If an injector is leaking, the HC reading will be high and will not drop. This test does not locate the bad injector but does verify that one or more are leaking.

Another cause of a rich mixture is a leaking fuel pressure regulator. If the diaphragm of the regulator is ruptured, fuel will move intake manifold through the diaphragm, causing a rich mixture. The regulator can be checked by using two simple tests. After the engine has been run, disconnect the vacuum line to the fuel pressure regulator (Figure 11-15). If there are signs

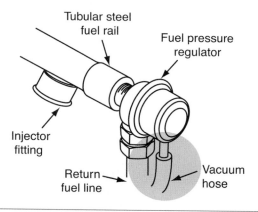

Figure 11-15 Location of a fuel pressure regulator on a fuel rail.

of fuel inside the hose or if fuel comes out of the hose, the regulator's diaphragm is leaking. The regulator can also be tested with a hand-operated vacuum pump. Apply 5 in. Hg to the regulator. A good regulator diaphragm will hold that vacuum.

Checking Voltage Signals

Special Tools

Noid light, test light

When an injector is suspected as the cause of a lean problem, the first step is to determine if the injector is receiving a signal from the PCM to fire. Fortunately, determining if the injector is receiving a voltage signal is easy and requires simple test equipment. Unfortunately, the location of the injector's electrical connector can make this simple voltage check somewhat difficult. For example, on some Chevrolet 2.8-liter V6 engines, the cover must be removed from the cast aluminum plenum chamber that is mounted over the top of the engine before the injector can be accessed.

A hot injector will have a higher resistance than a cold injector.

CAUTION: When performing this test, make sure to keep off of the accelerator pedal. On some models, fully depressing the accelerator pedal activates the clear flood mode in which the voltage signal to the injectors is automatically cut off. Technicians unaware of this waste time tracing a phantom problem.

Once the injector's electrical connector (Figure 11-16) has been removed, check for voltage at the injector using an ordinary test light or a convenient noid light that plugs into the connector. After making the test connections, crank the engine. The noid light flashes if the computer is cycling the injector on and off. If the light is not flashing, the computer or connecting wires are defective. If sufficient voltage is present after checking each injector, check the electrical integrity of the injectors themselves.

An ohmmeter can be used to test the electrical soundness of an injector. Connect the ohmmeter across the injector terminals (Figure 11-17) after the wires to the injector have been disconnected. If the meter reading is infinity, the injector winding is open. If the meter shows more resistance than the specifications call for, there is high resistance in the winding. A reading that is lower than the specifications indicates that the winding is shorted. If the injector is even a little bit out of specifications, it must be replaced. Refer to the service manual for injector specifications.

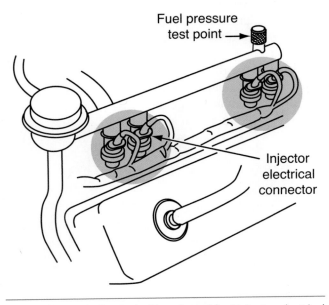

Figure 11-16 Typical location of fuel injector electrical connections.

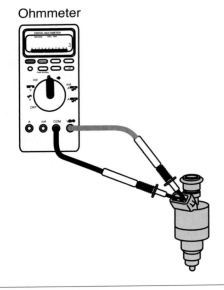

Figure 11-17 An ohmmeter can be connected across the injector's terminals to check injector resistance.

Figure 11-18 Electrical setup for performing an injector balance test.

Injector Balance Test

If the injectors are electrically sound, perform an injector pressure balance test. This test will help isolate a clogged or dirty injector. Photo Sequence 21 shows a typical procedure for testing injector balance. An electronic injector pulse tester is used for this test. Each injector is energized while observing a fuel pressure gauge to monitor the drop in fuel pressure. The tester is designed to safely pulse each injector for a controlled length of time. The tester is connected to one injector at a time (Figure 11-18). The ignition is turned on until a maximum reading is on the pressure gauge. That reading is recorded and the ignition turned off. With the tester, activate the injector and record the pressure reading after the needle has stopped pulsing. This same relative test is performed on each injector.

An injector balance test may be performed to diagnose restricted injectors on MFI and SFI systems. To conduct an injector balance test, follow these steps:

1. Connect the fuel pressure gauge to the Schrader valve on the fuel rail.
2. Connect the injector tester leads to the battery terminals with the correct polarity. Remove one of the injector wiring connectors and install the tester lead to the injector terminals (Figure 11-19).

Special Tools

Injector pulser

It may be wise to start the engine before testing an injector more than one time to prevent flooding.

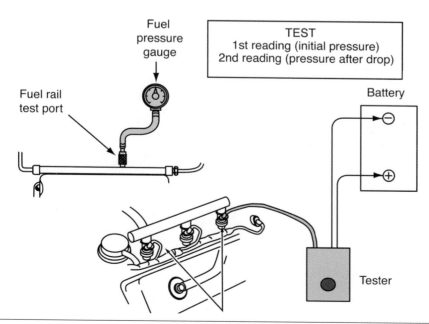

Figure 11-19 Setup for an injector pressure balance test.

Photo Sequence 21
Typical Procedure for Testing Injector Balance

P21-1 Connect the fuel pressure gauge to the Schrader valve on the fuel rail, and then relieve the pressure in the system.

P21-2 Disconnect the number 1 injector and connect the injector pulse tester to the injector's terminals.

P21-3 Connect the injector pulse tester's power supply leads to the battery.

P21-4 Cycle the ignition switch several times until the system pressure is at the specified level.

P21-5 Push the injector pulse tester switch and record the pressure on the pressure gauge. Subtract this reading from the measured system pressure. The answer is the pressure drop across that injector.

P21-6 Move the injector tester to the number 2 injector and cycle the ignition switch several times to restore system fuel pressure.

P21-7 Depress the injector pulse tester's switch and observe the fuel pressure. Again, the difference between the system pressure and the pressure when an injector is activated is the pressure drop across the injector.

P21-8 Move the injector tester's leads to the number 3 injector and cycle the ignition switch to restore system pressure.

P21-9 Depress the switch on the tester to activate that injector and record the pressure drop. Continue the procedure for all injectors, then compare the results of each to specifications and to each other.

470

3. Cycle the ignition switch on and off until the specified fuel pressure appears on the fuel gauge. Many fuel pressure gauges have an air bleed button that must be pressed to bleed air from the gauge. Cycle the ignition switch or start the engine to obtain the specified pressure on the fuel gauge and then leave the ignition switch off.

4. Push the timer button on the tester and record the gauge reading. When the timer energizes the injector, fuel is discharged from the injector into the intake port and the fuel pressure drops in the fuel rail.

5. Repeat steps 2, 3, and 4 on each injector and record the fuel pressure after each injector is energized by the timer.

6. Compare the gauge readings on each injector.

The difference between the maximum and minimum reading is the pressure drop. Ideally, each injector should drop the same amount when opened. A variation of 1.5 to 2 psi (10 kPa) or more is cause for concern (Figure 11-20). If there is no pressure drop or a low pressure drop, suspect a restricted injector orifice or tip. A higher-than-average pressure drop indicates a rich condition. When an injector plunger is sticking in the open position, the fuel pressure drop is excessive. If there are inconsistent readings, the nonconforming injectors either have to be cleaned or replaced.

Injector Power Balance Test

The injector pressure balance test is a good method of checking injectors on vehicles with no-start problems. When the vehicle runs, several other tests are possible. The injector power balance test is an easy method of determining if an injector is causing a miss.

To perform this test, first hook up a tachometer and fuel pressure gauge to the engine. Once all gauges are in place, start the engine and allow it to reach operating temperature. As soon as the idle stabilizes, unplug each injector one at a time. Note the rpm drop and pressure gauge reading. To ensure accurate test results on many electronic systems, it may be necessary to disconnect some type of idle air control device to prevent the computer from trying to compensate for the unplugged injector.

After recording the rpm and fuel pressure drop, reconnect the injector, wait until the idle stabilizes, and move on to the next one. Note the rpm and pressure drop each time. If an injector does not have much effect on the way the engine runs, chances are it is either clogged or electrically defective. It is a good idea to back up the power balance test with an injector resistance check, pressure balance test, or noid light check.

On many vehicles, an injector balance test can be performed with a scan tool. This is helpful when injectors are difficult to access.

CYLINDER	1	2	3	4	5	6
HIGH READING	225	225	225	225	225	225
LOW READING	100	100	100	90	100	115
AMOUNT OF DROP	125	125	125	135	125	110
RESULTS	OK	OK	OK	Faulty, rich (too much fuel drop)	OK	Faulty, lean (too little fuel drop)

Figure 11-20 Pressure readings from an injector pressure balance test.

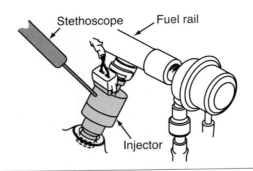

Figure 11-21 Using a stethoscope to check injector action.

Injector Sound Test

Special Tools

Stethoscope

If the injector's electrical leads are difficult to access, an injector power balance test is hard to perform. As an alternative, start the engine and use a technician's stethoscope to listen for correct injector operation (Figure 11-21). A good injector makes a rhythmic clicking sound as the solenoid is energized and de-energized several times each second. If a clunk-clunk instead of a steady click-click is heard, chances are the problem injector has been found. Cleaning or replacement is in order. If an injector does not produce any clicking noise, the injector, connecting wires, or PCM may be defective. When the injector clicking noise is erratic, the injector plunger may be sticking. If there is no injector clicking noise, proceed with the injector ohms test and noid light test to locate the cause of the problem. If a stethoscope is not handy, use a thin steel rod, wooden dowel, or your fingers to feel for a steady on-off pulsing of the injector solenoid.

Injector Flow Testing

Special Tools

12-volt power source, jumper wires, calibrated container

Some vehicle manufacturers recommend an injector flow test rather than the balance test. Follow these steps to perform an injector flow test:

1. Connect a 12-volt power supply to the cigarette lighter socket and disconnect the negative battery cable. If the vehicle is equipped with an air bag, wait 1 minute.
2. Remove the injectors and fuel rail and place the tip of the injector to be tested in a calibrated container. Leave the injectors in the fuel rail.
3. Connect a jumper wire across the specified terminals in the DLC for fuel pump testing.
4. Turn the ignition switch on.
5. Connect a jumper wire from the terminals of the injector being tested to the battery terminals (Figure 11-22).

Use extreme caution when performing a flow test. Fuel is highly flammable, take care to prevent sparks or flame.

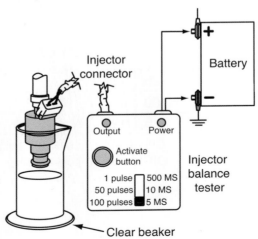

Figure 11-22 An injector used to test an injector's flow rate off the vehicle.

Figure 11-23 Checking for injector leaks with the fuel rail pressurized.

6. Disconnect the jumper wire from the negative battery cable after 15 seconds.
7. Record the amount of fuel in the calibrated container.
8. Repeat the procedure on each injector. If the volume of fuel discharged from any injector varies more than 0.3 cu. in. (5 cc) from the specifications, the injector should be replaced.
9. Reconnect the negative battery cable and disconnect the 12-volt power supply.

Connect the fuel pressure gauge to the fuel system. While the fuel system is pressurized with the injectors removed from the fuel rail after the flow test, observe each injector for leakage from the injector tip (Figure 11-23). Injector leakage must not exceed the manufacturer's specifications. If the injectors leak into the intake ports on a hot engine, the air-fuel ratio may be too rich when a restart is attempted a short time after the engine is shut off. When the injectors leak, they drain all the fuel out of the rail after the engine is shut off for several hours. This may result in slow starting after the engine has been shut off for a longer period of time.

While checking leakage at the injector tips, observe the fuel pressure in the pressure gauge. If the fuel pressure drops off and the injectors are not leaking, the fuel may be leaking back through the check valve in the fuel pump. Repeat the test with the fuel line plugged. If the fuel pressure no longer drops, the fuel pump check valve is leaking. If the fuel pressure drops off and the injectors are not leaking, the fuel pressure may be leaking through the pressure regulator and the return fuel line. Repeat the test with the return line plugged. If the fuel pressure no longer drops off, the pressure regulator valve is leaking.

✔ **SERVICE TIP:** Many technicians use a DSO when testing fuel injectors and fuel injector circuits. If the DSO is equipped with a graphing mode, perform a comparative amperage flow test through the fuel injectors. Discrepancies may be easier to observe and may lead to locating or verifying faulty fuel injectors or related circuit problems. The technician can also record and store various problem situations for future reference.

Oscilloscope Checks

An oscilloscope can be used to monitor the injector's pulse width and duty cycle when an injector-related problem is suspected. The pulse width is the time in milliseconds that the injector is energized. The duty cycle is the percentage of on-time to total cycle time.

To check the injector's firing voltage on the scope, a typical hookup involves connecting the scope's positive lead to the injector supply wire and the scope's negative lead to an engine ground.

Special Tools

Lab scope

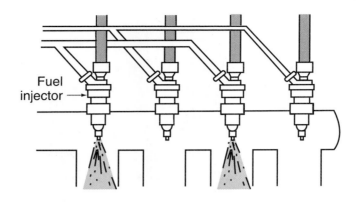

Figure 11-24 Grouped multiport fuel injection.

Fuel injection signals vary in frequency and pulse width. The pulse width is controlled by the PCM which varies it to control the air-fuel ratio. The frequency varies with engine speed. The higher the speed, the more pulses per second there are. Most often, the injector's ground circuit is completed by a driver circuit in the PCM. All of these factors are important to remember when setting a lab scope to look at fuel injector activity. Set the scope to read 12 volts, then set the sweep and trigger to allow you to clearly see the on signal on the left and the off signal on the right. Make sure the entire waveform is clearly seen. Also remember that the setting may need to be changed as engine speed increases or decreases.

Fuel injectors are either fired individually or in groups. When the injectors are fired in groups, a driver circuit controls two or more injectors (Figure 11-24). On some V-type engines, one driver fires the injectors on one side of the engine while another fires the other side. Each fuel injector has its own driver transistor in sequential and throttle body injection. It is extremely important while troubleshooting that you recognize how the injectors are fired. When the injectors are fired in groups, there can be a common or non-common cause of the problem. For example, a defective driver circuit in the PCM would affect all of the injectors in a group, not just one. Conversely, if one injector in the group is not firing, the problem cannot be the driver.

To read the injector waveform on group fuel injection systems, the scope must be connected to one injector harness for each group. Since all of the injectors in the group share the same circuit, a problem in one will affect the entire waveform for the group. The only way to isolate an injector electrical problem is to disconnect the injectors one at a time. If the waveform improves when an injector is disconnected, that injector has a problem. If the waveform never cleans up, the problem is in the driver circuit or the wiring harness.

In sequential fuel injection systems, each injector has its own driver circuit and wiring. To check an individual injector, the scope must be connected to that injector. This is great for locating a faulty injector. If the scope has a memory feature, a good injector waveform can be stored and recalled for comparison to the suspected bad fuel injector pattern. To determine if a problem is the injector itself or the PCM and/or wiring, simply swap the injector wires from an injector that had a good waveform to the suspect injector. If the waveform clears up, the wiring harness or the PCM is the cause of the problem. If the waveform is still not normal, the injector is to blame.

There are three different types of fuel injector circuits. In the conventional circuit, the driver constantly applies voltage to the injector. The circuit is turned on when a ground is provided. To control current flow through the circuit, most injectors have a built-in resistor, while

Injectors that fire individually are SFI injectors.

The side of a V-type engine is commonly referred to as a bank of cylinders.

The conventional driver is also referred to as the saturated driver.

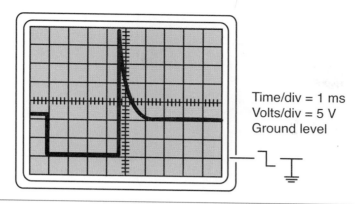

Figure 11-25 The waveform of a conventional fuel injector driver circuit.

others use a resistor in the circuit. The waveform for this type of injector circuit is shown in Figure 11-25. Notice there is a single voltage spike at the point where the injector is turned off. The total on time of the injector is measured from the point where the trace drops on the left to the point where it rises up next to the voltage spike.

Peak and hold injector circuits use two driver circuits to control injector action. Both driver circuits complete the circuit to open the injector. This allows for high current at the injector which forces the injector to open quickly. After the injector is open, one of the circuits turns off. The second circuit remains on to hold the injector open. This is the circuit that controls the pulse width of the injector. This circuit also contains a resistor to limit current flow during on time. When this circuit turns off, the injector closes. When looking at the waveform for this type of circuit (Figure 11-26), there will be two voltage spikes. One is produced when each circuit opens. To measure the on time of this type injector, measure from the drop on the left to the point where the second voltage spike is starting to move upward.

A pulse-modulated injector circuit uses high current to open the injector. This allows for quick injector firing. Once the injector is open, the circuit ground is pulsed on and off to allow for a long on time without allowing high current flow through the circuit. To measure the pulse width of this type of injector (Figure 11-27), measure from the drop on the left to the beginning of the large voltage spike which should be at the end of the pulses.

For all types of injectors, the waveform should have a clean, sudden drop in voltage when it is turned on. This drop should be close to 0 volt. Typically, the maximum allowable voltage during the injector's on time is 600 millivolts. If the drop is not perfectly vertical, either the injector is shorted or the driver circuit in the PCM is bad. If the voltage does not drop to below 600 millivolts, there is resistance in the ground circuit or the injector is shorted. When

Peak and hold injector circuits incorporate unique circuitry that applies higher amperage to initially open the injector, then lower amperage for the remainder of the pulse width.

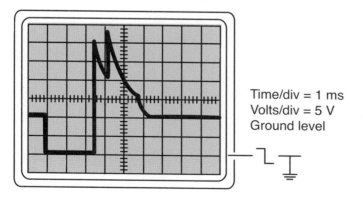

Figure 11-26 The waveform of a peak and hold fuel injector driver circuit.

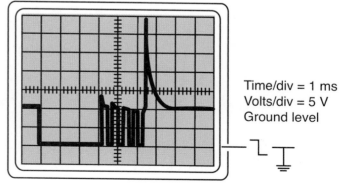

Figure 11-27 The waveform of a pulse modulated fuel injector driver circuit.

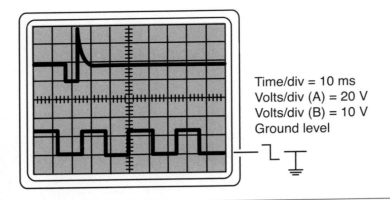

Time/div = 10 ms
Volts/div (A) = 20 V
Volts/div (B) = 10 V
Ground level

Figure 11-28 Compare the ignition reference signal to the injector's ON signal.

comparing one injector's waveform to another, check the height of the voltage spikes. The voltage spike of all injectors in the same engine should have approximately the same height. If there is a variance, the power feed wire to the injector with the variance or the PCM's driver for that injector is faulty.

While checking the injectors with a lab scope, make sure the injectors are firing at the correct time. To do this, use a dual-trace scope and monitor the ignition reference signal and a fuel injector signal at the same time. The two signals should have some sort of rhythm between them. For example, there can be one injector firing for every four ignition reference signals (Figure 11-28). This rhythm is dependent upon several factors—however, it does not matter what the rhythm is, only that it is constant. If the injector's waveform is fine but the rhythm varies, the ignition reference sensor circuit is faulty and not allowing the injector to fire at the correct time. If the ignition signal is lost because of a faulty sensor, the injection system will also shut down. If the injector circuit and the ignition reference circuit shut down at the same time, the cause of the problem is probably the ignition reference sensor. If the injector circuits shuts off before the ignition circuit, the problem is the injector circuit or the PCM.

Another use of the oscilloscope for checking injector operation is viewing the amperage waveform. The amperage waveform requires a low-amps probe. Note the coil charge building over time, the turn-off point, and the pintle opening "gull wing" in Figure 11-29. The gull wing is produced because of the changing inductance of the pintle moving into the body of the injector. This feature is not always as pronounced as it is in this Honda injector.

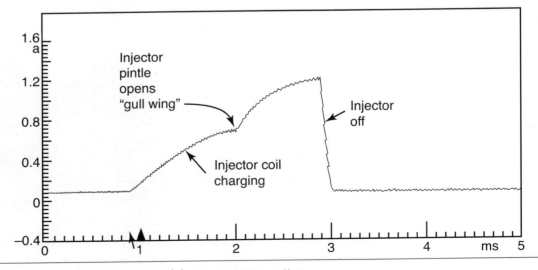

Figure 11-29 Viewing an injector amperage pattern.

Injector Service

Since a single injector can cost up to several hundred dollars, arbitrarily replacing injectors when they are not functioning properly, especially on multiport systems, can be an expensive proposition. If injectors are electrically defective, replacement is the only alternative. However, if the injector balance test indicated that some injectors were restricted or if the vehicle is exhibiting rough idle, stalling, or slow or uneven acceleration, the injectors may just be dirty and require a good cleaning.

Injector Cleaning

Before covering the typical cleaning systems available and discussing how they are used, several cleaning precautions are in order. First, never soak an injector in cleaning solvent. Not only is this an ineffective way to clean injectors, but it most likely destroys the injector in the process. Also, never use a wire brush, pipe cleaner, toothpick, or other cleaning utensil to unblock a plugged injector. The metering holes in injectors are drilled to precise tolerances. Scraping or reaming the opening results in a clean injector that is no longer an accurate fuel-metering device.

The basic premise of all injection cleaning systems is similar in that some type of cleaning chemical is run through the injector in an attempt to dissolve deposits that have formed on the injector's tip. The methods of applying the cleaner can range from single shot, pre-mixed, pressurized spray cans, to self-mix, self-pressurized chemical tanks resembling bug sprayers. The pre-mixed, pressurized spray can systems are fairly simple and straightforward to use since the technician does not need to mix, measure, or otherwise handle the cleaning agent.

Automotive parts stores usually sell pressurized containers of injector cleaner with a hose for Schrader valve attachment. During the cleaning process, the engine is operated on the pressurized container of unleaded fuel and injector cleaner. Fuel pump operation must be stopped to prevent the pump from forcing fuel up to the fuel rail. The fuel return line is to be plugged to prevent the solution in the cleaning container from flowing through the return line into the fuel tank. Follow these steps for the injector cleaning procedure:

1. Disconnect the wires from the in-tank fuel pump or the fuel pump relay to disable the fuel pump. If you disconnect the fuel pump relay on General Motors products, the oil pressure switch in the fuel pump circuit must also be disconnected to prevent current flow through this switch to the fuel pump.
2. Plug the fuel return line from the fuel rail to the tank.
3. Connect a can of injector cleaner to the Schrader valve on the fuel rail and run the engine for about 20 minutes on the injector solution.

Other systems require the technician to assume the role of chemist and mix up a desired batch of cleaning solution for each application. The chemical solution then is placed in a holding container and pressurized by hand pump or shop air to a specified operating pressure (Figure 11-30). The injector cleaning solution is poured into a canister on some injector cleaners and shop air supply is used to pressurize the canister to the specified pressure. The injector cleaning solution contains unleaded fuel mixed with injector cleaner. The container hose is connected to the Schrader valve on the fuel rail (Figure 11-31). The procedure for using pressurized injector cleaner follows:

1. Disable the fuel pump according to the car manufacturer's instructions (for example, pull fuel pump fuse, disconnect lead at pump, etc.). Clamp off the fuel pump return line at the flex connection to prevent the cleaner from seeping into the fuel tank.
2. Before starting the engine, open the cleaner's control valve one-half turn or so to prime the injectors and then start the engine.

Erratic idle speed and engine stalling may also be caused by carbon deposits on the IAC motor pintle and seat or in the IAC motor air passages in the throttle body.

Special Tools

Injector cleaning kit, shop air supply

LTFT can be used to verify clogged injectors. See if the numbers indicate the add on of fuel.

Figure 11-30 Injector cleaner that operates with compressed air.

Figure 11-31 Injector cleaner connected to a fuel rail.

3. If available, set and adjust the cleaner's pressure gauge to approximately 5 psi below the operating pressure of the injection system and let the engine run at 1,000 rpm for 10 to 15 minutes or until the cleaning mix has run out. If the engine stalls during cleaning, simply restart it.

4. Run the engine until the recommended amount of fluid is exhausted and the engine stalls. Shut the ignition off, remove the cleaning setup, and reconnect the fuel pump.

5. After removing the clamping devices from around the fuel lines, start the car. Let it idle for 5 minutes or so to remove any leftover cleaner from the fuel lines.

6. On the more severely clogged cases, the idle improvement should be noticeable almost immediately. With more subtle performance improvements, an injector balance test verifies the cleaning results. Once the injectors are clean, recommend the use of an in-tank cleaning additive or a detergent-laced fuel.

The more advanced units feature electrically operated pumps neatly packaged in roll-around cabinets that are quite similar in design to an A/C charging station (Figure 11-32).

Figure 11-32 A fuel injector cleaner cart.

After the injectors are cleaned or replaced, rough engine idle may still be present. This problem occurs because the adaptive memory in the computer has learned previously about the restricted injectors. If the injectors were supplying a lean air-fuel ratio, the computer increased the pulse width to try to bring the air-fuel ratio back to stoichiometric. With the cleaned or replaced injectors, the adaptive computer memory is still supplying the increased pulse width. This action makes the air-fuel ratio too rich now that the restricted injector problem does not exist. With the engine at normal operating temperature, drive the vehicle for at least 5 minutes to allow the adaptive computer memory to learn about the cleaned or replaced injectors. After this time, the computer should supply the correct injector pulse width and the engine should run smoothly. This same problem may occur when any defective computer system component is replaced.

Fuel Rail, Injector, and Regulator Service

There are service operations that will require removing the fuel injection fuel rail, pressure regulator, and/or injectors. Most of these are not related to fuel system repair. However, when it is necessary to remove and refit them, it is important that it be done carefully and according to the manufacturer's recommended procedures.

Injector Replacement

Photo Sequence 20 in Chapter 10 outlines a typical procedure for removing and installing an injector. Consult the vehicle's service manual for instructions on removing and installing injectors. Before installing the new one, always check to make sure the sealing O-ring is in place (Figure 11-33). Also, prior to installation, lightly lubricate the sealing ring with engine oil or automatic transmission fluid (avoid using silicone grease, which tends to clog the injectors) to prevent seal distortion or damage.

Fuel Rail, Injector, and Pressure Regulator Removal

CAUTION: Cap injector openings in the intake manifold to prevent the entry of dirt and other particles. Also, after the injectors and pressure regulator are removed from the fuel rail, cap all fuel rail openings to keep dirt out of the fuel rail.

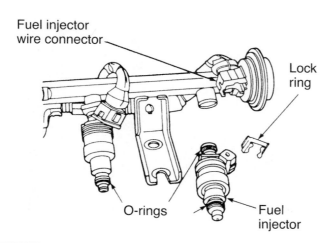

Figure 11-33 When replacing or reinstalling an injector, always replace the O-rings.

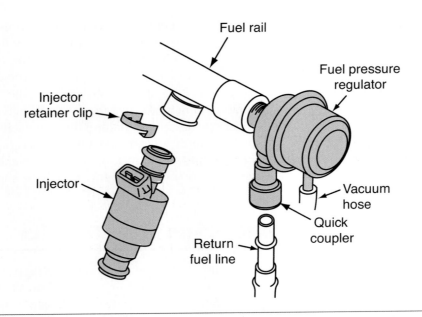

Figure 11-34 Fuel rail components.

The procedure for removing and replacing the fuel rail, injectors, and pressure regulator (Figure 11-34) varies depending on the vehicle. On some applications, certain components must be removed to gain access to these components. The following is a typical removal and replacement procedure for the fuel rail, injectors, and pressure regulator on a General Motors V6 engine:

1. Connect a 12-volt power supply to the cigarette lighter and disconnect the battery negative cable. If the vehicle is equipped with an air bag, wait 1 minute.

2. Bleed the pressure from the fuel system.

3. Wipe excess dirt from the fuel rail with a shop towel.

4. Loosen fuel line clamps on the fuel rail if clamps are present on these lines. If these lines have quick-disconnect fittings, grasp the larger collar on the connector and twist in either direction while pulling on the line to remove the fuel supply and return lines (Figure 11-35 and Figure 11-36).

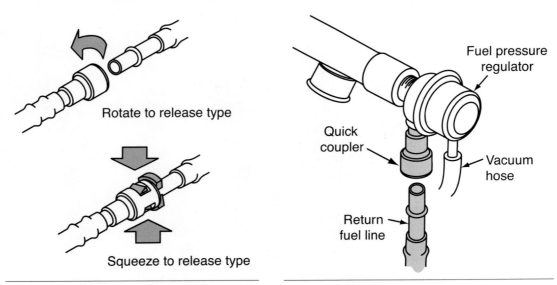

Figure 11-35 Quick-disconnect fuel line fittings.

Figure 11-36 Quick connector fuel line at fuel rail.

5. Remove the vacuum line from the pressure regulator.
6. Disconnect the electrical connectors from the injectors.
7. Remove the fuel rail holddown bolts.
8. Pull with equal force on each side of the fuel rail to remove the rail and injectors.

CAUTION: Do not use compressed air to flush or clean the fuel rail. Compressed air contains water, which may contaminate the fuel rail.

Regulator Cleaning and Inspection

1. Prior to injector and pressure regulator removal, the fuel rail may be cleaned with a spray-type engine cleaner. Approved cleaners are normally listed in the service manual.
2. Pull the injectors from the fuel rail.
3. Use snap ring pliers to remove the snap ring from the pressure regulator cavity. Note the original direction of the vacuum fitting on the pressure regulator and pull the pressure regulator from the fuel rail (Figure 11-37).
4. Clean all components with a clean shop towel. Be careful not to damage fuel rail openings and injector tips.
5. Check all injector and pressure regulator openings in the fuel rail for metal burrs and damage.

CAUTION: Do not immerse the fuel rail, injectors, or pressure regulator in any type of cleaning solvent. This action may damage and contaminate these components.

Installation of Fuel Rail, Injectors, and Pressure Regulator

1. If the same injectors and pressure regulator are reinstalled, replace all O-rings and lightly coat each O-ring with engine oil.
2. Install the pressure regulator in the fuel rail, and position the vacuum fitting on the regulator in the original direction.
3. Install the snap ring above the pressure regulator.

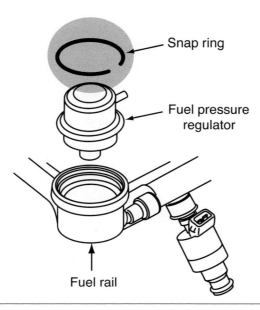

Figure 11-37 Some fuel pressure regulators are retained with a snap ring.

4. Install the injectors in the fuel rail.

5. Install the fuel rail while guiding each injector into the proper intake manifold opening. Be sure the injector terminals are positioned so they are accessible to the electrical connectors.

6. Tighten the fuel rail holddown bolts alternately and torque them to specifications.

7. Reconnect the vacuum hose on the pressure regulator.

8. Install the fuel supply and fuel return lines on the fuel rail.

9. Install the injector electrical connectors.

10. Connect the negative battery terminal and disconnect the 12-volt power supply from the cigarette lighter.

11. Start the engine and check for fuel leaks at the rail and be sure the engine operation is normal.

Idle Speed Checks

In a fuel injection system, idle speed is regulated by controlling the amount of air that is allowed to bypass the airflow sensor or throttle plates. When presented with a car that tends to stall, especially when coming to a stop, or idles too fast, look for obvious problems like binding linkage and vacuum leaks first. If no problems are found, go through the minimum idle checking and setting procedure described on the underhood decal. The instructions listed on the decal spell out the necessary conditions that must be met prior to attempting an idle adjustment. These adjustment procedures can range from a simple twist of a throttle stop screw to more involved procedures requiring circumvention of idle air control devices, removal of casting plugs, or recalibration of the throttle position sensor.

Minimum Idle Speed Adjustment

Special Tools

Jumper wires, tachometer

The minimum idle speed adjustment may be performed on some MFI or SFI systems with a minimum idle speed screw in the throttle body. This screw is factory adjusted, and the head of the screw is covered with a plug. This adjustment should only be required if throttle body parts are replaced. If the minimum idle speed adjustment is not adjusted properly, engine stalling may result. The procedure for performing a minimum idle speed adjustment varies considerably depending on the vehicle. Always follow the vehicle manufacturer's recommended procedure in the service manual. The following is a typical minimum idle speed adjustment procedure for a General Motors vehicle:

1. Make certain there are no vacuum leaks and TP sensor is at spec.

2. Be sure the engine is at normal operating temperature and turn the ignition switch off.

3. Connect terminals A and B in the DLC, and connect a tachometer from the ignition tach terminal to ground.

4. Turn the ignition switch on and wait 30 seconds. Under this condition, the idle air control motor is driven completely inward by the PCM.

5. Disconnect the IAC motor connector.

6. Remove the connection between terminals A and B in the DLC and start the engine.

7. Place the transmission selector in drive with an automatic transmission or neutral with a manual transmission.

8. Adjust the idle stop screw if necessary to obtain 500 to 600 rpm with an automatic transmission or 550 to 650 rpm with a manual transmission. A plug must be removed to access the idle stop screw (Figure 11-38).

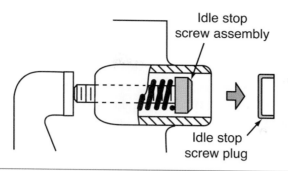

Figure 11-38 Adjusting the idle stop screw for minimum air.

8. Turn the ignition switch off and reconnect the IAC motor connector.

9. Turn the ignition switch on and connect a digital voltmeter from the TP sensor signal wire to ground. If the voltmeter does not indicate the specified voltage of 0.55 volt, loosen the TP sensor mounting screws and rotate the sensor until this voltage reading is obtained. Hold the TP sensor in this position and tighten the mounting screws.

Before an IAC motor is installed in a General Motors throttle body on TBI, MFI, or SFI systems, the distance from the end of the valve to the shoulder on the motor body must not exceed 1.125 in. (28 mm). If the IAC motor is installed with the plunger extended beyond this measurement, the motor may be damaged.

Minimum Idle Speed Adjustment, Throttle Body Injection

A minimum idle speed adjustment is only required if the TBI assembly (Figure 11-39) or TBI assembly components are replaced. If the minimum idle speed adjustment is not adjusted

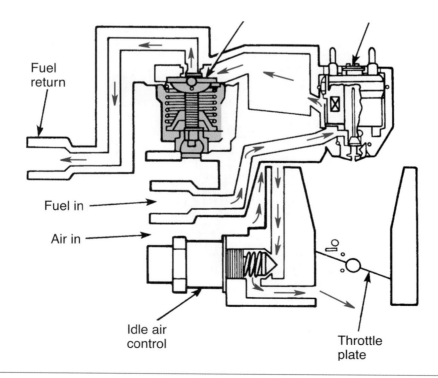

Figure 11-39 A throttle body injection assembly.

properly, engine stalling may result. Proceed as follows for a typical minimum idle speed adjustment on a General Motors TBI system:

1. Be sure that the engine is at normal operating temperature and remove the air cleaner and TBI-to-air cleaner gasket. Plug the air cleaner vacuum hose inlet to the intake manifold.

2. Disconnect the throttle valve (TV) cable to gain access to the minimum air adjustment screw. A tamper-resistant plug in the TBI assembly must be removed to access this screw.

3. Connect a tachometer from the ignition tach terminal to ground and disconnect the IAC motor connector.

4. Start the engine and place the transmission in park with an automatic transmission or in neutral with a manual transmission.

5. Plug the air intake passage to the IAC motor. A special tool is available for this purpose (Figure 11-40).

6. On 2.5L four-cylinder engines, use the appropriate torx bit to rotate the minimum air adjustment screw until the idle speed on the tachometer is 475 to 525 rpm with an automatic transaxle or 750 to 800 rpm with a manual transaxle.

7. Stop the engine and remove the plug from the idle air passage. Cover the minimum air adjustment screw opening with silicone sealant, reconnect the TV cable, and install the TBI gasket and air cleaner.

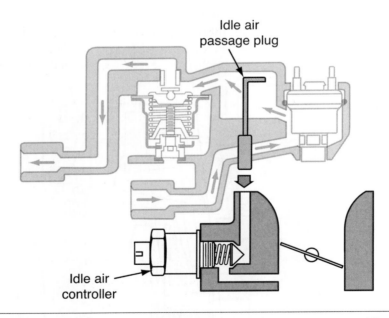

Figure 11-40 Special tool used to plug the air passage to the IAC motor while checking minimum idle speed on a TBI system.

Idle Speed Controls

Some IAC motors, such as those on Chrysler TBIs, have a hex bolt on the end of the motor plunger. However, this hex bolt is not for idle speed adjustment. If this hex bolt is turned, it will not affect idle speed because the PCM will correct the idle speed. If the hex bolt is turned and the length of the IAC motor plunger changed in an attempt to adjust idle speed, the throttle may not be in the proper position for starting, and hard starting at certain temperatures may occur. On some applications, an idle rpm specification is provided with the plunger fully extended. The plunger can be adjusted under this condition. On Chrysler products, the plunger may be fully extended by turning the ignition switch off and then disconnecting the IAC motor connector. The IAC motor hex bolt should only require adjustment if it has been improperly adjusted or if a new motor is installed.

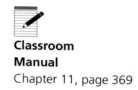

Classroom Manual Chapter 11, page 369

Idle Contact Switch Test

The procedure for diagnosing the idle contact switch varies depending on the vehicle make and model year. The following is a typical idle contact switch test. Refer to the wiring diagram in Figure 11-41 for terminal identification during this procedure. Always refer to a similar diagram for the vehicle being tested.

CAUTION: Never connect a 12-volt source across the idle contact switch terminals on the ISC motor. If these contacts are closed, the contacts will be ruined.

WARNING: Connecting a pair of jumper wires from the terminals of a 12-volt battery to the IAC motor idle contact switch terminals may result in very high current flow and jumper wire heating, which can burn your hands.

1. Backprobe terminal B on the IAC motor and turn the ignition switch on. Connect a digital voltmeter from terminal B to ground and hold the throttle approximately half open. The voltage supplied from the PCM to the idle contact switch should be 4.5 to 8 volts, depending on the system.

2. If the voltage at terminal B on the IAC motor is not within specifications, turn the ignition switch off and disconnect the IAC motor connector and the PCM connector. Connect the ohmmeter leads from terminal B in the IAC motor connector to terminal

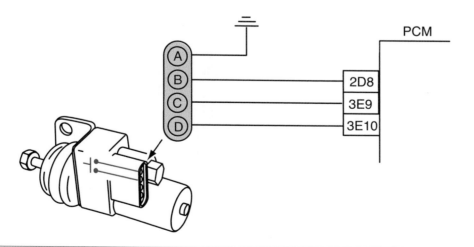

Figure 11-41 Testing the ISC motor switch.

2D8 in the PCM connector. The ohm reading should be less than 0.5 1/2. If the reading is more than this value, repair the resistance problem or open circuit in the wire from the PCM to the idle contact switch.

3. Connect the ohmmeter leads from terminal B on the IAC motor connector to ground. The ohmmeter reading should be infinity. If the reading is not infinity, repair the ground in the wire from the PCM to the idle contact switch.

4. If the voltage at terminal B on the IAC motor is not within specifications and the ohmmeter readings in steps 2 and 3 are satisfactory, reconnect the PCM and IAC motor connectors. Backprobe terminal 2D8 at the PCM, and turn the ignition switch on. Connect a digital voltmeter from terminal 2D8 to ground and observe the voltage. If the voltage is not within specifications, check all the power supply and ground wires to the PCM. When the power supply and ground wires are satisfactory, replace the PCM.

5. If the voltage in step 1 is satisfactory, return the throttle to the idle position and observe the voltmeter. The reading should be less than 1 volt with the throttle in the idle position and the idle contact switch closed. If the voltage reading is within specifications, the idle contact switch is satisfactory.

6. When the voltage in step 5 is not within specifications, connect the voltmeter from terminal A on the IAC motor to ground. If the voltage at this point is above 0.2 volt, repair the resistance problem or open circuit in the ground wire connected to terminal A. When the voltage at terminal A is 0.2 volt or less and the voltage in step 5 is above specifications, replace the IAC motor.

Scan Tester Diagnosis

If the idle speed is not within specifications, the input sensors and switches should be checked carefully with the scan tester. Photo Sequence 22 shows a typical procedure for performing a scan tester diagnosis of an idle air control motor. If the throttle position sensor voltage is lower than specified at idle speed, the PCM interprets this condition as the throttle being closed too much. Under this condition, the PCM opens the IAC or **idle air control bypass air (IAC BPA)** motor to increase idle speed.

If the engine coolant temperature sensor resistance is higher than normal, it sends a higher-than-normal voltage signal to the PCM. The PCM thinks the coolant is colder than it actually is and the PCM operates the IAC or IAC BPA motor to increase idle speed. Many input sensor defects cause other problems in engine operation besides improper idle rpm.

Defective input switches result in improper idle rpm. For example, if the A/C switch is always closed, the PCM thinks the A/C is on continually. This action results in the PCM operating the IAC or IAC BPA motor to provide a higher idle rpm. On many vehicles, the scan tester indicates the status of the input switches as closed or open as well as high or low. Most input switches provide a high-voltage signal to the PCM when they are open and a low-voltage signal if they are closed.

On some vehicles, a fault code is set in the PCM memory if the IAC or IAC BPA motor or connecting wires are defective. On other systems, a fault code is set in the PCM memory if the idle rpm is out of range. On Chrysler products, the actuation test mode (ATM), or actuate outputs mode, may be entered with the ignition switch on. The IAC or IAC BPA motor may be selected in the actuate outputs mode, and the PCM is forced to extend and retract the IAC or IAC BPA motor plunger every 2.8 seconds. When this plunger extends and retracts properly, the motor, connecting wires, and PCM are in normal condition. If the plunger does not extend and retract, further diagnosis is necessary to locate the cause of the problem.

Special Tools

Scan tool, tachometer

Idle air control bypass air (IAC BPA) controls incoming air around the throttle plate to control idle speed.

On some vehicles, the scan tester may be used to force the PCM to operate the IAC or IAC BPA motor while observing the engine rpm.

Classroom Manual
Chapter 11, page 358

On some vehicles, a jumper wire may be connected to specific terminals on the data link connector (DLC) to check IAC BPA valve response.

Photo Sequence 22
Performing a Scan Tester Diagnosis of an Idle Air Control Motor

P22-1 Be sure the ignition switch is off.

P22-2 Connect the scan tester leads to the battery terminals with the correct polarity.

P22-3 Turn the key on but do not start the engine.

P22-4 Program the scan tester as required for the vehicle being tested.

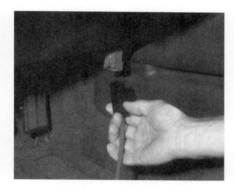

P22-5 Use the proper adapter to connect the scan tester lead to the data link connector (DLC).

P22-6 Select idle air control (IAC) motor test on the scan tester.

P22-7 Press the appropriate scan tester button to increase engine rpm and observe the rpm on the scan tester. The scan tester will automatically perform a non-running actuation test of the IAC hardware and software.

P22-8 At the low range, the IAC motor has a target range of sixteen steps. The scan tester steps the motor through and monitors those steps.

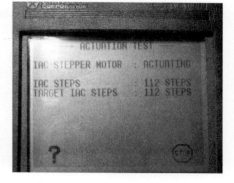

P22-9 At the high range, the scan tester steps the motor through and monitors 112 steps.

WARNING: When performing a set engine rpm mode test on an IAC motor, always be sure the transmission selector is in the park position and the parking brake is applied.

On some IAC or IAC BPA motors, a set engine rpm mode may be entered on the scan tester. In this mode, each time a specified scan tester button is touched, the rpm should increase 100 rpm to a maximum of 2,000 rpm. Another specified scan tester button may be touched to decrease the speed in 100 rpm steps. On some scan testers, the up and down arrows are used to increase and decrease the engine rpm during this test. If the IAC or IAC BPA motor responds properly during this diagnosis, the PCM, motor, and connecting wires are in satisfactory condition and further diagnosis of the inputs is required.

On some systems, the scan tester reads the IAC or IAC BPA motor counts, and the count range is provided in the scan tester instruction manual. Some of the input switches, such as the A/C, may be operated and the scan tester counts should change. If the scan tester counts change when the A/C is turned on and off, the motor, connecting wires, and PCM are operational. When the scan tester counts do not change under this condition, further diagnosis is required.

A scan tool can help detect problems with the IAC valve. If the IAC valve is at or near zero, the idle is too high and the PCM is trying to reduce the idle speed. The IAC valve has totally bottomed out and can travel no farther. If this happens and the idle is still too high, a MIL light and trouble code will result. If the engine is the speed density type (using a MAP sensor instead of a MAF), then the problem could be a manifold vacuum leak, wrong PC valve, or someone has set the minimum air too high. If the IAC counts are higher than they should be, the idle speed is too low. This condition could be caused by a buildup of carbon on the backside of the throttle plate, which cuts down on the amount of air that goes into the engine. If the buildup is bad enough, the engine will die on decel because the IAC valve may not be fast enough to prevent stalling if the throttle is closed too fast.

Another problem that can be detected by the scan tool is intermittent operation of the IAC. Watch the counts at the IAC valve when the engine comes to an idle while warm. Unless the coolant fan or some other load comes on at idle, the IAC counts should be fairly consistent every time you come to a stop, at least within a few counts. If the counts are erratic, then there is a problem with the IAC valve sticking or binding.

Sometimes the complaint is hard starting, but the engine starts easier if the throttle is opened slightly while cranking. Depressing the throttle should not be necessary. If this condition is noted, see if the spring on the IAC valve is overcoming the IAC valve stepper motor when the engine sits for a long period of time. If this happens, the valve will be closed completely, shutting off the bypass air.

IAC Bypass Air Motor and Valve Diagnosis

On some systems, such as Toyota, a jumper wire may be connected to terminals E1 and TE1 in the data link connector (DLC) to diagnose the IAC BPA valve with the engine at normal operating temperature (Figure 11-42). When the engine is started with this jumper wire connection, the engine speed should increase to 1,000 to 1,300 rpm for 5 seconds and then return to idle speed. If the IAC BPA valve does not respond as specified, further diagnosis of the IAC BPA valve, wires, and PCM is required.

On General Motors OBD I vehicles, the IAC BPA motor extends fully when terminals A and B are connected in the DLC and the ignition switch is turned on. With the IAC BPA motor removed from the throttle body, this jumper connection may be completed while observing the IAC BPA motor. If the motor does not extend, further diagnosis of the motor, connecting wires, and PCM is required.

The scan tester can also be used in the snapshot mode on most vehicles to record conditions for playback in the shop. Trouble codes set in OB II will trigger an automatic snapshot (freeze frame) for some code.

The scan tool can also be used to clear any trouble codes set quickly. Remember that the data collected during an OBD II freeze frame will also be cleared.

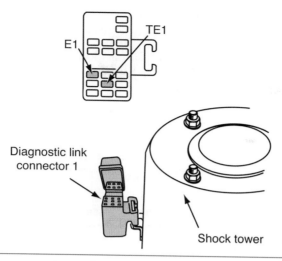

Figure 11-42 The connection at the DLC to check the IAC BPA valve.

IAC Bypass Air Motor Removal and Cleaning

Carbon deposits in the IAC BPA motor air passage in the throttle body or on the IAC BPA motor pintle result in erratic idle operation and engine stalling. Remove the motor from the throttle body and inspect the throttle body air passage for carbon deposits. If heavy carbon deposits are present, remove the complete throttle body for cleaning. Clean the throttle body IAC BPA air passage, motor sealing surface, and pintle seat with throttle body cleaner. Clean the motor pintle with throttle body cleaner.

Some manufacturers do not recommend cleaning of the throttle body. Always check service information.

CAUTION: IAC BPA motor damage may result if throttle body cleaner is allowed to enter the motor.

IAC BPA Motor Diagnosis and Installation

Connect a pair of ohmmeter leads across terminals D and C on the motor connector and observe the ohmmeter reading (Figure 11-43). Repeat this ohmmeter test on terminals A and B on the motor. In each test, the ohmmeter reading should equal the manufacturer's specifications, usually 40 to 80 ohms. If the ohmmeter reading is not within specifications, replace the IAC BPA motor.

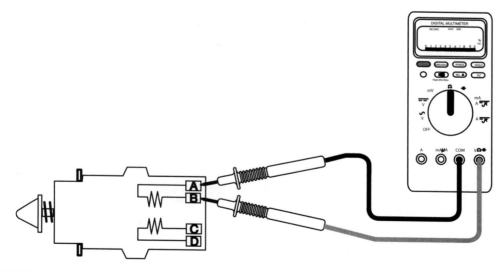

Figure 11-43 The connection at the DLC to check the IAC BPA valve.

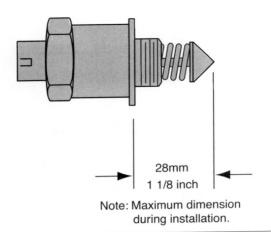

Figure 11-44 Measuring the distance of extension on an IAC BPA motor.

28mm
1 1/8 inch

Note: Maximum dimension
during installation.

CAUTION: IAC BPA motor damage may occur if the motor is installed with the pintle extended more than the specified distance.

If a new IAC BPA motor is installed, be sure the part number, pintle shape, and diameter are the same as those on the original motor. Measure the distance from the end of the pintle to the shoulder of the motor casting (Figure 11-44). If this distance exceeds 1.125 in. (28 mm), use hand pressure to push the pintle inward until this specified distance is obtained.

Install a new gasket or O-ring on the motor. If the motor is sealed with an O-ring, lubricate this ring with transmission fluid and install the motor. If the motor is threaded into the throttle body, tighten the motor to the specified torque. When the motor is bolted to the throttle body, tighten the mounting bolts to the specified torque.

To test the IAC BPA valve, disconnect the IAC BPA valve connector and connect a pair of ohmmeter leads to terminals B+ and ISCC on the valve. Repeat the test between terminals B+ and ISCO on the valve (Figure 11-45). In both these tests on the valve windings, the ohmmeter should read 19.3 to 22.3 ohms. If the resistance of the valve windings is not within specifications, replace the valve.

Disconnect all throttle body linkages and vacuum hoses. Remove the throttle body and the IAC BPA valve mounting bolts. Remove the valve and gasket (Figure 11-46). The cleaning procedure for the IAC BPA motor may be followed on the IAC BPA valve.

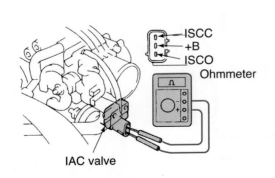

Figure 11-45 Ohmmeter connections to test the IAC BPA valve windings.

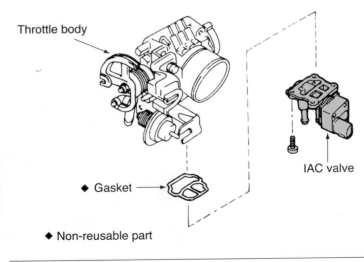

Figure 11-46 IAC BPA valve removed from the throttle body.

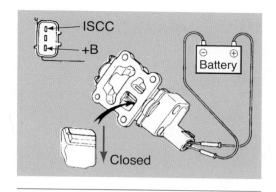

Figure 11-47 Test connections to close the IAC BPA valve.

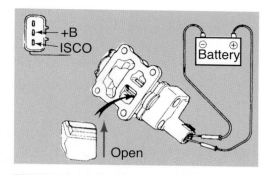

Figure 11-48 Test connections to open the IAC BPA valve.

Connect a jumper wire from the battery positive terminal to the valve B+ terminal. Connect another jumper wire from the battery negative terminal to the valve ISCC terminal. Be careful not to short the jumper wires together. With this connection, the valve must be closed (Figure 11-47). Leave the jumper wire connected from the battery positive terminal to the valve B+ terminal and connect the jumper wire from the battery negative terminal to the ISCO terminal. Under this condition, the valve must be open (Figure 11-48). If the valve does not open and close properly, replace the valve.

Clean the throttle body and valve mounting surfaces and install a new gasket between the valve and the throttle body. Install the valve and tighten the mounting bolts to the specified torque. Be sure the throttle body and intake mounting surfaces are clean and install a new gasket between the throttle body and the intake. Tighten the throttle body mounting bolts to the specified torque. Reconnect the IAC BPA valve wiring connector and all throttle body linkages and hoses.

Diagnosis of Fast Idle Thermo Valve

The **fast idle thermo valve** is factory adjusted and should not be disassembled. Remove the air duct from the throttle body and be sure the engine temperature is below 86°F (30°C). Start the engine and place your finger over the lower port in the throttle body (Figure 11-49). There should be airflow through this lower port and the fast idle thermo valve. If there is no airflow through the lower port, replace the fast idle thermo valve.

As the engine temperature increases, the airflow through the lower throttle body port should decrease. When the engine approaches normal operating temperature, the airflow should stop flowing through the lower port and the fast idle thermo valve. If there is airflow

A **fast idle thermo valve** controls air intake based on temperature.

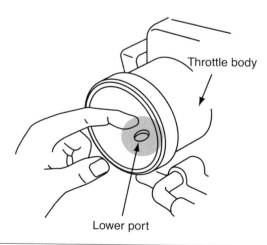

Figure 11-49 Covering the lower port in the throttle body to test the thermo air valve.

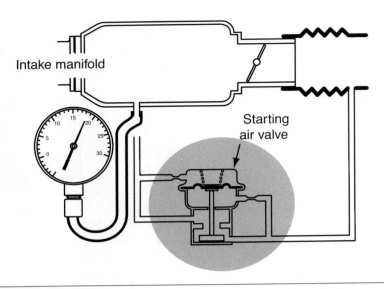

Figure 11-50 IAC BPA valve used with a starting air valve and fast idle thermo valve.

through the lower throttle body port with the engine at normal operating temperature, check the cooling system for proper operation and temperature. The fast idle thermo valve is heated by engine coolant. If the cooling system operation is normal, replace the fast idle thermo valve.

Diagnosis of Starting Air Valve

If the starting air valve is open with the engine running, the idle rpm may be higher than specified. Disconnect the vacuum signal hose from the starting air valve and connect a vacuum gauge to this hose (Figure 11-50). With the engine idling this vacuum should be above 16 in. Hg. If the vacuum is lower than specified, check for leaks or restrictions in the signal hose. When the hose is satisfactory, check for late ignition timing or engine conditions such as low compression that result in low vacuum.

If the signal vacuum is satisfactory, check the hoses from the starting air valve to the intake manifold and the air cleaner for restrictions and leaks. When the hoses are satisfactory, remove the hose from the starting air valve to the air cleaner. With the engine idling, there should be no airflow through this hose. When airflow is present, replace the starting air valve.

With the engine cranking, there should be airflow through the hose from the starting air valve to the air cleaner. If no airflow is present, replace the starting air valve.

CASE STUDY

A customer phoned to say that he was having his SFI Cadillac towed to the shop because the engine had stopped and would not restart. Before the technician started working on the car, he routinely checked the oil and coolant. The engine oil dipstick indicated the crankcase was severely overfilled with oil, and the oil had a strong odor of gasoline. The technician checked the ignition system with a test spark plug and found the system to be firing normally. Of course, the technician thought the no-start problem must be caused by the fuel system and the most likely problems would be the fuel filter or fuel pump.

The technician removed the air cleaner hose from the throttle body and removed the air cleaner element to perform a routine check of the air cleaner element and throttle body. The throttle body showed evidence of gasoline lying at the lower edge of the throttle bore. The technician asked a co-worker to crank the engine while he looked in the throttle body. While cranking the engine, gasoline was flowing into the throttle body below the throttle. The technician thought this situation was impossible. An SFI system cannot inject fuel into the throttle body!

The technician began thinking about how fuel could be getting into the throttle body on this SFI system, and he reasoned the fuel had to be coming through one of the vacuum hoses. Next, he thought about which vacuum hose could be a source of this fuel and remembered the pressure regulator vacuum hose is connected to the intake. The pressure regulator vacuum hose was removed from the throttle body and placed in a container. When the engine was cranked, fuel squirted out of the vacuum hose, indicating the regulator diaphragm had a hole in it.

The technician installed a new pressure regulator and changed the engine oil and filter. After this service, the engine started and ran normally.

Terms to Know

14.7:1 air-fuel ratio

Fast idle thermo valve

Fuel control

Idle air control (IAC)

Idle air control bypass air (IAC BPA)

Peak and hold injector circuits

Stoichiometry

ASE-Style Review Questions

1. While discussing the causes of higher-than-specified idle speeds:
   Technician A says an intake manifold vacuum leak may cause a high idle speed.
   Technician B says if the TP sensor voltage signal is higher than specified, the idle speed may be higher than normal.
   Who is correct?
   A. A only
   B. B only
   C. Both A and B
   D. Neither A nor B

2. *Technician A* says a buildup of carbon on the backside of a throttle body can cause stalling or low idle speeds.
   Technician B says some throttle bodies should not be cleaned.
   Who is correct?
   A. A only
   B. B only
   C. Both A and B
   D. Neither A nor B

3. While discussing IAC BPA motor removal, service, and replacement:
   Technician A says throttle body cleaner may be used to clean the IAC BPA motor internal components.
   Technician B says on some vehicles, IAC BPA motor damage occurs if the pintle is extended more than specified during installation.
   Who is correct?
   A. A only
   B. B only
   C. Both A and B
   D. Neither A nor B

4. While discussing starting air valve diagnosis:
   Technician A says there should be airflow through the valve with the engine running.
   Technician B says if a vacuum leak is present in the vacuum signal hose, the valve will not close properly.
   Who is correct?
   A. A only
   B. B only
   C. Both A and B
   D. Neither A nor B

5. While discussing injector testing:

   Technician A says a defective injector may cause cylinder misfiring at idle speed.

   Technician B says restricted injector tips may result in acceleration stumbles.

   Who is correct?

   A. A only   **C.** Both A and B
   B. B only   **D.** Neither A nor B

6. *Technician A* says that a scan tester is necessary to check trouble codes on OBD I vehicles.

   Technician B says that OBD II vehicles require the use of a scan tool to read diagnostic codes.

   Who is correct?

   A. A only   **C.** Both A and B
   B. B only   **D.** Neither A nor B

7. While discussing scan tester diagnosis of TBI, MFI, and SFI systems:

   Technician A says the scan tester will erase fault codes quickly on many systems.

   Technician B says many scan testers will store sensor readings during a road test and then play back the results in a snapshot test mode.

   Who is correct?

   A. A only   **C.** Both A and B
   B. B only   **D.** Neither A nor B

8. While discussing STFT and LTFT numbers:

   Technician A says that a short-term fuel trim (STFT) is a temporary correction to the fuel mixture by the PCM.

   Technician B says that negative numbers on the long-term fuel trim (LTFT) mean the PCM is adding fuel to the mixture to try and reach 14.7:1.

   Who is correct?

   A. A only   **C.** Both A and B
   B. B only   **D.** Neither A nor B

9. Two technicians are discussing the air-fuel ratio sensor compared to the oxygen sensor:

   Technician A says the air-fuel sensor runs at a lower temperature than the oxygen sensor.

   Technician B says the air-fuel ratio sensor is also known as a wide ratio sensor.

   Who is correct?

   A. A only   **C.** Both A and B
   B. B only   **D.** Neither A nor B

10. While discussing the causes of a rich air-fuel ratio:

    Technician A says a rich air-fuel ratio may be caused by low fuel pump pressure.

    Technician B says a rich air-fuel ratio may be caused by a defective coolant temperature sensor.

    Who is correct?

    A. A only   **C.** Both A and B
    B. B only   **D.** Neither A nor B

ASE Challenge Questions

1. A vehicle emits black smoke in the exhaust. This may be caused by any of the following *except*:
   A. fuel entering the intake via the fuel pressure regulator diaphragm.
   B. a "known good" O_2 sensor generating above-normal voltage signals.
   C. a defective cold enrichment system component(s).
   D. injector(s) registering a high pressure drop during injector balance testing.

2. *Technician A* says the minimum idle speed needs to be adjusted on an MFI engine after the throttle body has been removed and cleaned.
   Technician B says a scan tool is required to properly adjust the minimum idle speed on a TBI system.
   Who is correct?
   A. A only
   B. B only
   C. Both A and B
   D. Neither A nor B

3. An EFI engine with a too high idle speed is being diagnosed. The *least likely* malfunction to cause this condition is:
   A. Retarded ignition timing on system with a starting air valve.
   B. An ECT with higher then specified resistance.
   C. A shorted A/C switch.
   D. Leaking fuel injector(s).

4. *Technician A* says that the current waveform of an injector shows the injector coil charging over time. *Technician B* says that the voltage waveform of an injector shows a sharp peak after the injector is turned off due to inductance in the coil.
   Who is correct?
   A. A only
   B. B only
   C. Both A and B
   D. Neither A nor B

5. Air flow meters and mass airflow sensors are being discussed.
   Technician A says poor idle and surging at cruise may be caused by an air leak in the PCV system.
   Technician B says a MAF failure may be confirmed by unplugging its electrical connection and noting engine performance.
   Who is correct?
   A. A only
   B. B only
   C. Both A and B
   D. Neither A nor B

Job Sheet 27

27

Name _____ Date _____

Visually Inspect an EFI System

Upon completion of this job sheet, you should be able to conduct a preliminary inspection of an electronic fuel injection system.

ASE Correlation

This job sheet is related to the ASE Engine Performance Test's content area: fuel, air induction, and exhaust systems diagnosis and repair; tasks: diagnose hot or cold no-starting, hard starting, poor driveability, incorrect idle speed, poor idle, flooding, hesitation, surging, engine misfire, power loss, stalling, poor mileage, dieseling, and emissions problems on vehicles with injection-type fuel systems; determine needed repairs; inspect and test fuel injectors; clean and replace.

Describe the vehicle being worked on.

Year_____ Make _____ VIN _____

Model_____

Procedures

1. Is the battery in good condition, fully charged, with clean terminals and connections?
   ☐ Yes ☐ No

2. Do the charging and starting systems operate properly? ☐ Yes ☐ No

3. Are all fuses and fusible links intact? ☐ Yes ☐ No

4. Are all wiring harnesses properly routed with connections free of corrosion and are they tightly attached? ☐ Yes ☐ No

5. Are all vacuum lines in sound condition, properly routed, and tightly attached?
   ☐ Yes ☐ No

6. Is the PCV system working properly and maintaining a sealed crankcase?
   ☐ Yes ☐ No

7. Are all emission control systems in place, hooked up, and operating properly?
   ☐ Yes ☐ No

8. Is the level and condition of the coolant/antifreeze good, and is the thermostat opening at the proper temperature? ☐ Yes ☐ No

9. Are the secondary spark delivery components in good shape with no signs of crossfiring, carbon tracking, corrosion, or wear? ☐ Yes ☐ No

10. Are the base timing and idle speed set to specifications? ☐ Yes ☐ No
    Is the engine in good mechanical condition? ☐ Yes ☐ No

11. Is the gasoline in the tank of good quality and not been substantially cut with alcohol or contaminated with water? ☐ Yes ☐ No

12. Does the air intake duct work have cracks or tears? ☐ Yes ☐ No

13. Are all of the induction hose clamps tight and properly sealed? ☐ Yes ☐ No

14. Are there any other possible air leaks in the crankcase? ☐ Yes ☐ No

15. Does the airflow sensor's flap bind, stick, or scrape when rotated through its operating range? ☐ Yes ☐ No

16. Are the electrical connections at the mass air flow meter or manifold pressure sensor good?

☐ Yes ☐ No

17. Is there carbon buildup inside the throttle bore and on the throttle plate? ☐ Yes ☐ No

18. Is there a clicking noise at each injector while the engine is running? ☐ Yes ☐ No

Instructor's Response_____

_____

_____

Job Sheet 28

28

Name _____ Date _____

Check the Operation of the Fuel Injectors on an Engine

Upon completion of this job sheet, you should be able to check the operation of the fuel injectors on an engine using a variety of test instruments and techniques.

ASE Correlation

This job sheet is related to the ASE Engine Performance Test's content area: fuel, air induction, and exhaust systems diagnosis and repair; tasks: diagnose hot or cold no-starting, hard starting, poor driveability, incorrect idle speed, poor idle, flooding, hesitation, surging, engine misfire, power loss, stalling, poor mileage, dieseling, and emissions problems on vehicles with injection-type fuel systems; determine needed repairs; inspect and test fuel injectors; clean and replace.

Tools and Materials

Noid light
DMM
Stethoscope
Lab scope

Describe the vehicle being worked on.

Year_____ Make _____ VIN_____

Model_____

Procedures

Task Completed

1. Remove the electrical connector from each injector. ☐

2. While cranking the engine, check for voltage at each of the connectors. ☐

3. What instrument did you use to check for voltage?

 _____

4. Use an ohmmeter to check each injector winding. ☐

5. Compare resistance readings to the specifications found in the service manual. ☐

6. What are the specifications for resistance? _____ ohms

7. Share your conclusions from the resistance checks:

 _____

 _____

8. Reconnect the electrical connectors to the fuel injectors and run the engine. ☐

9. Using a technician's stethoscope, listen for correct injector operation. ☐

10. Describe what you heard and what is indicated by the noise.

 _____

 _____

Task Completed

☐   **11.** Connect the positive lead of a lab scope to the injector supply wire and the scope's negative lead to a good ground.

☐   **12.** Set the scope on a low voltage scale and adjust the time so that the trace fills the screen.

☐   **13.** Then observe the waveform of each injector, one at a time.

14. Record your findings from each injector.

_____

_____

_____

15. Summarize the results of all of these tests.

_____

_____

_____

_____

Instructor's Response_____

_____

_____

Job Sheet 29

29

Name _____ Date _____

Conduct an Injector Balance Test

Upon completion of this job sheet, you should be able to perform an injector balance test on a port injected engine.

ASE Correlation

This job sheet is related to the ASE Engine Performance Test's content area: fuel, air induction, and exhaust systems diagnosis and repair; tasks: diagnose hot or cold no-starting, hard starting, poor driveability, incorrect idle speed, poor idle, flooding, hesitation, surging, engine misfire, power loss, stalling, poor mileage, dieseling, and emissions problems on vehicles with injection-type fuel systems; determine needed repairs; inspect and test fuel injectors; clean and replace.

Tools and Materials

Fuel pressure gauge
Injector tester

Describe the vehicle being worked on.

Year_____ Make _____ VIN_____

Model_____

Procedures

Task Completed

1. Carefully inspect the fuel rail and injector for fuel leaks. Do not proceed if fuel is leaking. ☐
2. Connect the fuel pressure gauge to the Schrader valve on the fuel rail. ☐
3. Disconnect the electrical connector to the number 1 injector. ☐
4. Connect the injector tester lead to the injector terminals. ☐
5. Connect the injector tester power supply leads to the battery terminals. ☐
6. Cycle the ignition switch several times until the specified pressure appears on the fuel pressure gauge. The specified pressure is _____ psi.
7. Your reading is _____ psi.
8. This indicates:

   _____

   _____

9. Depress the injector tester switch and record the pressure on the pressure gauge. The reading was _____ psi.
10. Move the injector tester to injector 2. ☐
11. Cycle the ignition switch several times until the specified pressure appears on the fuel pressure gauge.
12. Your reading is _____ psi.

13. This indicates:

_____

_____

14. Depress the injector tester switch and record the pressure on the pressure gauge. The reading is _____ psi.

15. Continue the same sequence until all injectors have been tested.

Reading on injector 3 is _____ psi

Reading on injector 4 is _____ psi

Reading on injector 5 is _____ psi

Reading on injector 6 is _____ psi

Reading on injector 7 is _____ psi

Reading on injector 8 is _____ psi

16. Summarize the results:

_____

_____

_____

_____

Instructor's Response_____

_____

_____

Distributor Ignition System Diagnosis and Service

Upon completion and review of this chapter, you should be able to:

❏ Perform a no-start diagnosis and determine the cause of the no-start condition.

❏ Perform a visual inspection of ignition system components, primary wiring, and secondary wiring to locate obvious trouble areas.

❏ Test the components of the primary and secondary ignition circuits.

❏ Inspect and test secondary ignition system wires.

❏ Describe what an oscilloscope is, its scales and operating modes, and how it is used in ignition system troubleshooting.

❏ Identify and describe the major sections of primary circuit and secondary circuit trace patterns, including the firing line, spark line, intermediate area, and dwell zone.

❏ Perform cranking output, spark duration, coil polarity, spark plug firing voltage, rotor, secondary resistance, and spark plug load tests using the oscilloscope.

❏ Test individual ignition components using test equipment such as a voltmeter, ohmmeter, and test light.

❏ Inspect distributor caps and rotors.

❏ Inspect, service, and test ignition coils.

❏ Inspect, service, and test ignition modules.

❏ Remove, service, and replace spark plugs.

❏ Test and set (when possible) ignition timing.

❏ Check ignition spark advance.

Basic Tools

Basic tool set, DMM, fender covers, appropriate service manual

Introduction

This chapter concentrates on testing ignition systems and their individual components. However, it must be stressed that there are many variations in the ignition systems used by auto manufacturers. Due to the vast number of ignition systems used today, it would be difficult to identity every specific system, component, and function of each within this text. Therefore, the broad range of testing procedures examined in this chapter will generally apply to all ignition types. This will aid in developing a thorough overview of complete ignition system testing. In many cases, generic testing is performed to quickly identify problem areas down to identifying a suspect component. From these results, more vehicle-specific details may be required that may not be covered here. However, specific details for each ignition system are found in the appropriate shop manual or information system. This includes manufacturer's specifications, exact test sequences, and service procedures. The fundamentals of ignition system diagnosis will be discussed in this chapter, while the subsequent chapter will focus specifically on the electronic ignition (EI) systems.

Four important precautions should be taken during all ignition system tests:

1. Turn the ignition switch off before disconnecting any system wiring.

2. Do not touch any exposed connections while the engine is cranking or running.

3. Unless thoroughly familiar with both its application to the vehicle and its operation, do not connect or operate any test equipment.

4. Be certain you are familiar with the tests about to be performed as well as the operation of the system or component being tested.

General Ignition System Diagnosis

The ignition system should be tested whenever there is evidence that there is no spark, not enough spark, or when the spark is not being delivered at the correct time to the cylinders. Typically, all ignition problems can be divided into two types: common and uncommon. Common problems are those that affect all cylinders and uncommon problems are those that affect one or more cylinders but not all. Common ignition components include the parts of the primary circuit and the secondary circuit up to the distributor's rotor. Uncommon parts are the individual spark plug terminals inside the distributor cap, spark plug wires, and the spark plugs.

Several indicators can be used to determine uncommon problems. One example may be the symptom or complaint. A rough running engine or cylinder misfire at an idle or under load can be a good indicator of an ignition-related problem affecting one or more cylinders but not *all* cylinders. Another indicator can be the reading from a vacuum gauge. An example would be if a vacuum gauge is connected to the intake manifold of a four-cylinder engine at idle and the needle of the gauge is within the normal vacuum range for three-fourths of the time and drops one-fourth of the time. This indicates that three of the cylinders are working normally while the fourth is not. The cause of the problem is uncommon to the rest of the cylinders. If the cylinder is sealed and all cylinders are receiving the correct amount of air and fuel, the problem must be in the ignition system in the distributor cap, spark plug wires, or spark plugs.

It is usually not necessary to perform all ignition tests on all components on every vehicle all the time. However, it *is* necessary to determine where to begin initial testing. As mentioned, symptoms can help guide the initial steps of a diagnosis as well as performing area tests, system tests, and component or pinpoint tests, in that order. For example, a vehicle is in the shop for a no-start complaint. An *area test* quickly determines an ignition-related problem. During a *system test,* it is discovered that there is no spark at any of the cylinders (problem common to all), but the primary side of the ignition system is functioning normally. There would be no need to continue testing individual primary ignition components. Instead, it is necessary to focus on the secondary system. From this point, individual secondary ignition component tests must be performed to pinpoint the root cause of the failure as well as the part(s) involved.

Determining if the ignition problem is common to all cylinders or is uncommon is the best way to start troubleshooting the ignition system. There are many parts in the system, and many tests can be conducted on them. By dividing the system into common and uncommon components, you can test only those parts that can cause the problem.

These ignition defects may cause a no-start condition or hard starting:

1. Defective coil—primary or secondary (common problem)
2. Defective cap and/or rotor (common problem)
3. Defective primary triggering, e.g., Hall effect or pickup coil (common problem)
4. Open secondary coil wire (common problem)
5. Low or zero primary voltage at the coil or other faulty wiring or grounds (common problem)
6. Fouled spark plugs (common or uncommon problem)
7. Electronic ignition module or control unit (common problem)

If engine misfiring occurs, check these items:

1. Engine compression (uncommon problem)
2. Intake manifold vacuum leaks (common or uncommon problem)

3. High resistance in spark plug wires, coil secondary wire, or cap terminals (uncommon problem)

4. Electrical leakage in the distributor cap, rotor, plug wires, coil secondary wire, or coil tower (uncommon problem)

5. Weak or defective coil (common problem)

6. Defective spark plug(s) (uncommon problem)

7. Low primary voltage and current (common problem)

8. Improperly routed spark plug wires (uncommon problem)

9. Worn distributor bushings (common problem)

10. Erratic primary triggering (common problem)

11. Fuel injector or fuel delivery (common or uncommon problem)

Check these items to diagnose a power loss condition:

1. Engine compression (common problem)

2. Restricted exhaust or air intake (common problem)

3. Late ignition timing (common problem)

4. Insufficient timing advance (common problem)

5. Cylinder misfiring (uncommon problem)

6. Fuel pressure, fuel delivery (common problem)

7. Defective coil (common problem)

If the engine detonates, check for:

1. Higher-than-specified engine compression (common problem)

2. Ignition timing too far advanced (common problem)

3. Spark plug heat range too hot (common problem)

4. Improperly routed spark plug wires (common problem)

5. Incorrect octane fuel being used (common problem)

6. Engine operating temperature too high (common problem)

7. Exhaust gas recirculation (EGR) related (common or uncommon problem)

8. Knock sensor (KS) signal or circuit (common problem)

When the fuel consumption is excessive, check these components:

1. Engine compression (common problem)

2. Late ignition timing (common problem)

3. Lack of spark advance (common problem)

4. Cylinder misfiring (common problem)

5. Rich fuel mixture (common problem)

6. Fuel pressure too high (common problem)

7. Emission control devices (common problem)

8. Erroneous computer input signals or output commands (common or uncommon problem)

Generally, when an engine runs unevenly, the problem is an uncommon problem. Most other times, the problem is a common one. Since the primary circuit of nearly all ignition systems is common to all cylinders, deciding whether to test the primary or secondary circuits comes with deciding if the problem is common or uncommon.

No-Start Diagnosis

Special Tools

12-volt test light, spark tester

Always use a spark tester, do not modify a spark plug. The tester is specially calibrated to stress the ignition system.

When the engine does not run, it is easy to make a decision as to whether the problem is common or uncommon or whether a fault in the primary or secondary ignition circuit is stopping the engine from running. Some quick tests can be conducted to identify the ignition problem that prevents the engine from starting.

Begin with a secondary ignition test. Place a spark tester (Figure 12-1) in the end of a spark plug wire and crank the engine. If there is no spark, try another spark plug wire in the event the first wire or distributor cap is defective. Confirm that there is an adequate spark, which would designate that the ignition is functioning. At times, the ignition coil may be weak but able to deliver adequate spark to occasionally start the engine, depending on the temperature. If there is no spark at the spark plugs, check for coil output at the coil terminal. If spark occurs, the cap and or rotor could be at fault. If there is no spark at the coil tower, connect a test light or dwell-meter to the negative (tach) side of the ignition coil and crank the engine. If the light flashes or there is a dwell reading, the primary is functioning. Therefore, the problem is in the coil. If the test lamp does not flash or there is no dwell reading, check for proper primary feed voltage from the ignition switch to the positive side of the coil. If none is found, check for problems between the coil and ignition switch. If there is power at the positive side of the coil and the test light did not flash or there was no dwell, the problem is narrowed to either the pickup coil or ignition module. On occasion, a defective ignition coil can prevent the module from triggering. First, check the pickup coil or Hall effect switch. The pickup coil, or magnetic pulse generator, can be checked with an ohmmeter, DMM, or scope. The Hall effect switch can be checked with a DMM or scope. If the pickup coil or Hall effect switch is good, inspect the ignition module. A final check can be made on the module using a special module tester. Be sure to check all wiring connections and grounds. As mentioned earlier, an ignition coil can prevent the module from triggering properly, making it appear faulty. Perform a transistor switch test before condemning the module. Simply disconnect the ignition coil, connect the test light clip to the positive battery terminal, and place the probe end of the test light to the coil negative wire harness terminal. Observe the test light while cranking the engine. If the light is flashing, the module is triggering and the coil is at fault. If the light does not flash and all other connections and components are good, then the module is faulty.

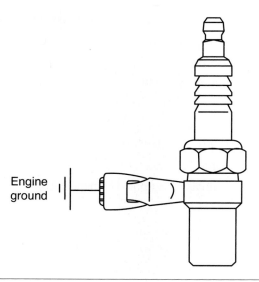

Engine ground

Figure 12-1 A spark tester.

SERVICE TIP: If the engine is fuel injected, check the fuel injector pulse as a quick test. If the injector is pulsing, the primary ignition is probably functioning properly. In most applications, the injector will not pulse if the primary signal is absent.

Visual Inspection

All ignition system diagnosis should begin with a visual inspection. The system should be checked for obvious problems:

❑ Disconnected, loose, or damaged secondary cables

❑ Disconnected, loose, or dirty primary wiring

❑ Loose or damaged distributor cap

❑ Damaged distributor cap or rotor

❑ Worn or damaged primary system switching mechanism or connections

❑ Improperly mounted electronic control unit

Secondary and Primary Wiring Connections

The spark plug wires should be inspected carefully for signs of chafing and cracks that can cause high-voltage leaks and arching. Often there are signs of white or grayish powdery deposits on the secondary cables at the point where they cross or are near metal parts. The deposits are evidence that the high-voltage leak burned dust collected on the cable. These conditions result in an intermittent engine miss, especially under damp conditions. Look at both ends of the plug wires; make sure the boots are neither split nor brittle and that the terminal ends are clean, tight, and fit their connection securely. If the terminal ends are dark or rusty in appearance, replace the plug wires. **Carbon tracking** is carbonized dust that can form on spark plugs, inside wire boots, coil, and distributor cap. The carbon track will form a low-resistance path to ground and cause a misfire. If a spark plug shows signs of carbon tracking, replace the plug and wire together since the carbon track is also on the inside of the plug wire and will transfer to the new plug after a few thousand miles (Figure 12-2). If the vehicle is leaking engine oil on the spark plug wires, this should also be corrected. Oil-soaked plug wire boots will lead to arcing at the cylinder head. Always use the correct plug wire spacers and route plug wires correctly to prevent problems with crossfire and EMI. Also, it makes sense to replace plug wires in sets and that are made specifically for the vehicle. If one plug wire is defective, then the others will probably fail soon. It is usually more cost effective to purchase a set of wires instead of buying two or three.

Spark plug and coil cables are most often referred to as the secondary or spark plug wires.

Classroom Manual
Chapter 12, page 383

Figure 12-2 Carbon tracking on a spark plug.

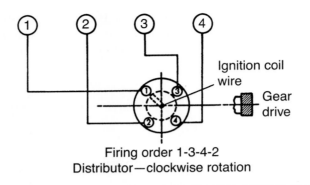

Figure 12-3 Spark plug wire routing.

Secondary cables must be connected according to the firing order. Refer to the manufacturer's service manual to determine the correct firing order and cylinder numbering (Figure 12-3).

An occasional glow around the spark plug cables, known as a **corona effect**, is not harmful but indicates that the cable should be replaced.

Spark plug cables from consecutively firing cylinders should cross rather than run parallel to one another (Figure 12-4). Spark plug cables running parallel to one another can induce firing voltages in one another and cause the spark plugs to fire at the wrong time.

Primary ignition system wiring should be checked for tight connections, especially on vehicles with electronic or computer-controlled ignitions. Electronic circuits operate on very low voltage. Voltage drops caused by corrosion or dirt can cause running problems. Missing tab locks on wire connectors are often the cause of intermittent ignition problems due to vibration or thermal-related failure.

Test the integrity of a suspect connection by tapping, tugging, and wiggling the wires while the engine is running. Be gentle. The objective is to recreate an ignition interruption, not to cause permanent circuit damage. With the engine off, separate the suspect connectors and check them for dirt and corrosion. Clean the connectors according to the manufacturer's recommendations.

The **corona effect** is a glow around the spark plug cables indicating the cable should be replaced.

Many late-model vehicles use dielectric grease on connector terminals to seal out dirt and moisture.

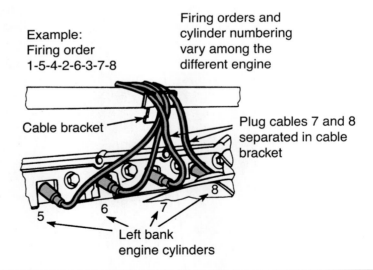

Figure 12-4 Routing spark plug cables to avoid inducing voltages.

Do not overlook the ignition switch as a source of intermittent ignition problems. A loose mounting rivet or poor connection can result in erratic spark output. To check the switch, gently wiggle the ignition key and connecting wires with the engine running. If the ignition cuts out or dies, the problem is located.

Check the battery connection to the starter solenoid. Remember, some vehicles use this connection as a voltage source for the coil. A bad connection can result in ignition interruption.

Moisture can also be a problem in faulty wires, connections, and grounds. Rain or humidity can affect anything electrical that has poor insulation or power or ground connections. If there is electrical leakage, moisture in the air can alter the flow along a conductor. To recreate a moisture problem, spray a light mist of water from a spray bottle on the coil, spark plug wires, and connections. Do not intentionally soak these components. That would not represent a real-world situation. If there is a change in engine speed, particularly idle quality, there is potential for an electrical leak at that point.

Ground Circuits

Keep in mind that to simplify a vehicle's electrical system, automakers use body panels, frame members, and the engine block as current return paths to the battery.

Unfortunately, ground straps are often neglected, or worse, left disconnected after routine service. With the increased use of plastics in today's vehicles, ground straps may mistakenly be reconnected to a non-metallic surface. The result of any of these problems is that the current that was to flow through the disconnected or improperly grounded strap is forced to find an alternate path to ground. Sometimes the current attempts to back up through another circuit. This may cause the circuit to operate erratically or fail altogether. The current may also be forced through other components, such as wheel bearings or shift and clutch cables that are not meant to handle current flow causing them to wear prematurely or become seized in their housing.

Examples of bad ground circuit-induced ignition failures include burnt ignition modules resulting from missing or loose coil ground straps, erratic performance caused by a poor distributor-to-engine block ground, and intermittent ignition operation resulting from a poor ground at the control module.

Electromagnetic Interference

Electromagnetic interference (EMI) can cause problems with the vehicle's onboard computer. EMI is produced when electromagnetic radio waves of sufficient amplitude escape from a wire or conductor. Unfortunately, an automobile's spark plug wires, ignition coil, and alternator coils all possess the ability to generate these radio waves. Under the right conditions, EMI can trigger sensors or actuators. The result may be an intermittent driveability problem that appears to be ignition system related.

To minimize the effects of EMI, check to make sure that sensor wires running to the computer are routed away from potential EMI sources. Rerouting a wire by no more than an inch or two may keep EMI from falsely triggering or interfering with computer operation.

AC generators produce alternating current that is rectified to direct current through a diode set within the AC generator. If a defective diode allows an excessive amount of AC current, the ignition module may be influenced to fire from a falsely produced signal. This can be real yet a rare and intermittent complaint. An AC voltage output check using a DMM or oscilloscope either at the AC generator output terminal or positive terminal of the battery should be no more that 0.12 AC volt. It is a good practice to perform this test whenever the vehicle is in for a charging system check, battery service, or tune up-related services.

Electromagnetic interference (EMI) appears as AC noise on an electrical signal.

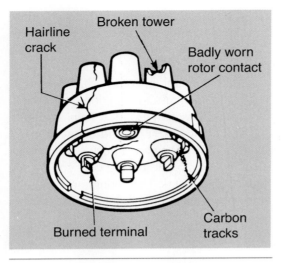

Figure 12-5 Types of distributor cap defects.

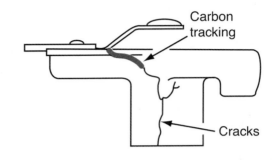

Figure 12-6 Inspect rotor for damage and signs of high-voltage leaks.

Distributor Cap

Classroom Manual
Chapter 12, page 384

The distributor cap should be properly seated on its base. All clips or screws should be tightened securely. The distributor cap and rotor also should be removed for visual inspection. Physical or electrical damage is easily recognizable. Electrical damage from high voltage can include corroded or burned metal terminals and carbon tracking inside distributor caps. Carbon tracking indicates that high-voltage electricity has found a low-resistance conductive path over or through the plastic. The result is a cylinder that fires at the wrong time, misfires, or fires at the same time as another cylinder. Check the outer cap towers and metal terminals for defects. Cracked plastic requires replacement of the unit. Damaged or carbon-tracked distributor caps or rotors should be replaced (Figure 12-5). Also inspect for moisture.

The rotor should be inspected carefully for discoloration and other damage (Figure 12-6). Inspect the top and bottom of the rotor carefully for grayish, whitish, or rainbow-hued spots. Such discoloration indicates that the rotor has lost its insulating qualities. High voltage is being conducted to ground through the plastic.

If the distributor cap or rotor has a mild buildup of dirt or corrosion, it should be cleaned. If it cannot be cleaned, it should be replaced. Small round brushes are available to clean cap terminals. Wipe the cap and rotor with a clean shop towel, but avoid cleaning these components in solvent or blowing them off with compressed air which may contain moisture. Cleaning these components with solvent or compressed air may result in high-voltage leaks.

Check the distributor cap and housing vents. Make sure they are not blocked or clogged. If they are, the internal ignition module will overheat. It is good practice to check these vents whenever a module is replaced.

Control Modules

Electronic and computer-controlled ignitions use transistors as switches. These transistors are contained inside a control module housing that can be mounted to or in the distributor or remotely mounted on the vehicle's fire wall or another engine compartment surface. Control modules should be tightly mounted to clean surfaces. A loose mounting can cause a heat buildup that can damage and destroy transistors and other electronic components contained in the module. Some manufacturers recommend the use of a special heat-conductive silicone

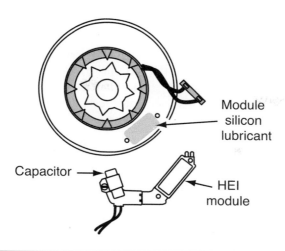

Figure 12-7 When replacing control modules, some manufacturers specify to apply silicon lubricant to the mounting surface.

grease between the control unit and its mounting (Figure 12-7). As noted, this helps conduct heat away from the module, reducing the chance of heat-related failure. During the visual inspection, check all electrical connections to the module. They must be clean and tight.

Sensors

The transistor in the control module is activated by a voltage pulse from a crankshaft position sensor. In most ignition systems, this sensor is either a magnetic pulse generator or Hall effect sensor. These sensors are mounted either on the distributor shaft or the crankshaft.

Magnetic pulse generators are relatively trouble-free. The reluctor or pole piece is replaced if it is broken or cracked. Magnetic pulse generators can also lose magnetism or the ability to produce the proper voltage needed to trigger the module. Magnets in the magnetic pulse generator are subjected to engine vibration and heat. The magnets can crack or break, affecting voltage output production reliability. In some cases, the unit may check within specifications with an ohmmeter yet is not generating enough AC voltage.

Under unusual circumstances, the non-magnetic reluctor can become magnetized and upset the pickup coil's voltage signal to the control module. Use a steel feeler gauge to check for signs of magnetic attraction, and replace the reluctor if the test is positive (Figure 12-8).

According to SAE J19309 standards that were recently mandated to the automotive industry, the term **distributor ignition (DI)** replaces all previous terms for distributor-type ignition systems that are electronically controlled.

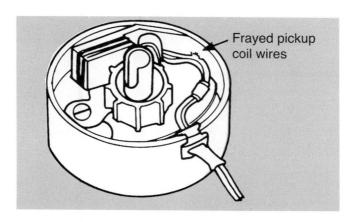

Frayed pickup coil wires

Figure 12-8 Inspect pickup coil wiring for damage.

Hall effect sensor problems are similar to those of magnetic pulse generators. Wires from the sensor that may move and rub as the breaker plate moves should be inspected. Also, check the condition of the distributor shaft drive coupling. Wear could affect the engine timing.

Although not detected by a visual inspection, the Hall effect device can become defective and fail to provide a voltage change to the ignition module. This can be checked with a DMM or oscilloscope. This test, as well as pulse generator testing, will be covered later.

Diagnosing with an Engine Analyzer

If the shop possesses a computerized engine analyzer, the ability to capture large amounts of engine and computer data in a short period has increased significantly. The appealing aspect of using this piece of equipment is that it is versatile and multi-functional with the capability to carry out almost any necessary engine performance-related test. It can be used as a diagnostic analyzer to perform all tests related to mechanical, electrical, ignition, fuel delivery, and computer systems. The technician can use other special test functions within the analyzer to perform individual pinpoint tests. Many analyzers have several automatic test sequences to choose from, or the technician may prefer to create menu-driven customized tests. Other features may include a variety of DMM functions, advanced lab scope capabilities, four- or five-gas exhaust analyzers, advanced on-board computer interface functions, and a service information system (Figure 12-9). Many diagnostic analyzers incorporate waveform libraries or the ability to store patterns and other information for reference. All these tests can be done either independently or as part of a complete test routine.

The objective is to gather as much engine performance-related information as possible in the shortest amount of time. After taking a few minutes to connect to the vehicle for either a complete test sequence or individual tests, the analyzer is ready for data collection. Performing a complete test will quickly gather engine data that otherwise could take a great deal of time. For example, an electronic version of a compression test can be performed without removing a spark plug in less than 15 seconds. Results from this test as well as others can bring a problem to the attention of the technician that may otherwise not have been detected or considered. Of course, there are alternate methods to gather test results without the use of an analyzer. The challenge is to do it quickly to maximize time efficiency and accuracy.

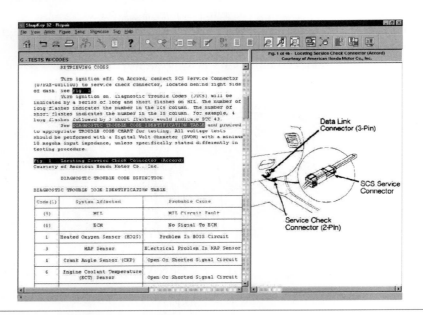

Figure 12-9 Service information system.

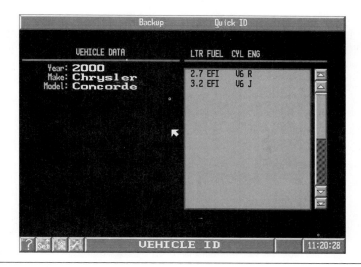

Figure 12-10 Vehicle identification data.

Not all analyzers use the same terminology or have identical test routines or functions. The following is a typical representation of the type of tests performed using a computerized diagnostic engine analyzer, including test examples and explanations of results.

Complete Test

The Complete Test begins with entering information into the analyzer about the vehicle such as year, make, model, and engine (Figure 12-10). Next, the computer loads engine performance limits later compared to actual test results for use in diagnostics. The technician is guided through the tests with screen prompts such as vehicle connections and visual inspections (Figure 12-11).

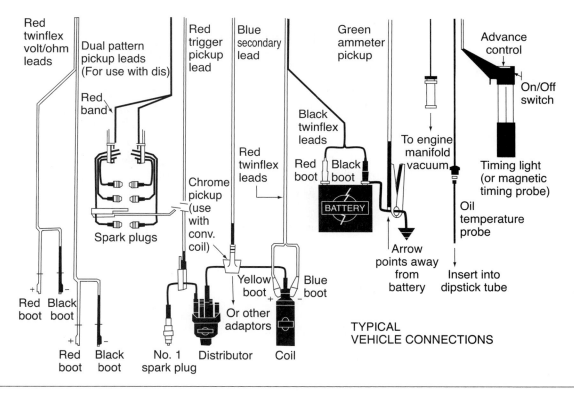

Figure 12-11 Vehicle connection instructions.

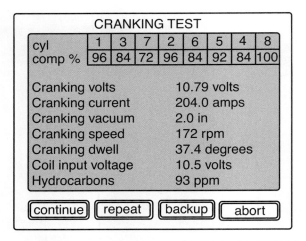

Figure 12-12 Cranking test.

The following engine tests are portions of the complete test:

❏ *Cranking Test*. This 12- to 15-second test is set up to disable the engine from starting. The engine is cranked to capture data for cranking rpm, primary ignition, battery and starter load, hydrocarbons (HC), cranking vacuum, and secondary ignition voltage at the end of the test (Figure 12-12). The electronic compression test measures the amount of amperage the starter requires while cranking the engine. The analyzer knows the location of the number one cylinder in the firing sequence as it begins the test. The amps-per-cylinder is calculated and matched to the firing order. If a particular cylinder requires less amperage at that moment from the starter during the compression stroke, it means there is less resistance or pressure in the cylinder. Low compression in that cylinder when compared to the others was a result of the lower starter draw matched to that cylinder. Of course, a manual compression test or cylinder leakage test will still need to be performed to confirm the problem. However, the problem cylinder was detected early and easily. Other information from the Cranking Test includes battery and starter condition, as well as ignition function. This is a convenient series of tests to perform in the event of a no-start complaint.

❏ *Running Tests*. This test series simulates no-load cruising conditions. It is performed at three speeds: 2,500 rpm, 1,500 rpm, and idle. It simultaneously looks at the timing and emission curves, charging system, and vacuum (Figure 12-13). Using these results, fuel efficiency and performance can be evaluated. This may include emission control devices, fuel delivery, and mechanical condition.

	HIGH CRUISE	LOW CRUISE	IDLE
RPM	2504	1520	661
VOLTS	13.8	14.0	----
AMPS	8	8	8
RIPPLE	0.06	0.04	0.04
HC PPM	43	24	73
CO %	0.90	0.15	0.37
O_2 %	0.00	0.00	0.00
CO_2 %	14.8	15.4	15.2
A/F RATIO	14.3	14.6	14.5
VAC "HG	20.4	20.0	18.6
PEAK AMPS	17		

Figure 12-13 Test data from the running test.

CYL[1]	FIRING KV	SPARK KV	SPARK DURATION
1	9.7	1.12	1.45
8	10.0	1.20	1.37
4	10.6	1.25	1.37
3	10.8	1.25	1.32
6	11.1	1.17	1.40
5	11.5	1.14	1.32
7	10.2	1.09	1.46
2	11.0	1.14	1.40

Figure 12-14 Ignition test data.

❏ *Secondary Ignition or kV Tests.* Depending on the type of engine and ignition system, several tests are performed on both the primary and secondary ignition systems through the evaluation of firing voltages, burn time, and coil performance. These test results can be either digitally or graphically displayed. An oscilloscope function within the unit offers an alternative to numerical data. Digital data is flanked with limits or firing history to help determine problem areas. It displays coil oscillations, firing voltage, spark kV, spark duration, kV change under load, and circuit gap (Figure 12-14). Cylinders with a wide gap or high resistance will show a higher than normal firing kV and a shorter spark duration, while a shorted condition will indicate a low firing voltage and longer spark duration. In addition to the new supporting test adapters, new testing methods have been created to address the newest ignition systems, including the latest distributorless (EI) and coil-over-plug (COP) systems.

❏ *Cylinder Power Balance Tests.* Some engines will have each cylinder cancelled in order to record the rpm change, while other engines will use a different type of power balance. Conventionally, the analyzer interrupts power to the ignition for several seconds as it records the rpm drop for that cylinder. The next cylinder in the firing order is cancelled and so on until all cylinder rpm drops have been recorded. This is a relative test, which means the results of the rpm drop test per cylinder are compared to other cylinders. The analyzer is looking for uniformity. During this test, the analyzer is able to record exhaust gas changes before and after cylinder cancellation. This is valuable in determining fuel delivery-related problems on a per-cylinder basis. Some analyzers will track changes in HC, CO, O_2, and CO_2 during the power balance test, while others will only track HC. The other cylinder power contribution test method uses technology that measures the crankshaft speed between cylinder firings. The crankshaft speed should be uniform throughout all cylinders. In the event that a cylinder has lower power contribution, the crankshaft slows momentarily then speeds up when the next normal cylinder fires. The results of both test methods are displayed numerically and relatively (Figure 12-15).

CYL	MIN	RESULTS	MAX
1	74	100	100
2	74	84	100
3	74	69↓	100
4	74	100	100
5	74	100	100
6	74	100	100

RAPID POWER BALANCE 4:41:57

Figure 12-15 Cylinder power contribution test data indicating low power on cylinder number

Operator Selectable Tests

Other pinpoint tests can be selected either in conjunction with the Complete Test or separately. These tests can be performed as a portion of the tests listed above *or* by selecting any of the following:

- ❑ DMM tests, including all volt, ohm, and low amps
- ❑ Frequency functions and calculations
- ❑ Dual trace lab scope functions for testing an array of input signals and output commands, including pressure and vacuum waveforms (Figure 12-16)
- ❑ Single or dual ignition oscilloscope on any ignition system with or without exhaust analyzer readings (Figure 12-17)
- ❑ Separate battery, starting, and charging system tests with high amps inductive probe and alternator waveform (Figure 12-18)
- ❑ Analyzer-powered vacuum pump with variable vacuum adjustment and gauge
- ❑ Manual cylinder power balance test
- ❑ Emissions testing and air-fuel ratio display
- ❑ Diagnostic reporting

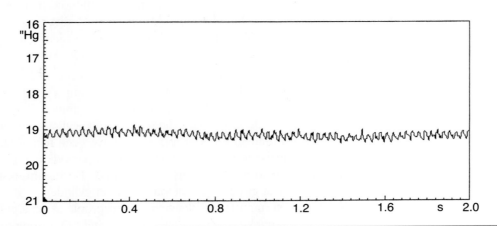

Figure 12-16 Lab scope displaying a typical vacuum waveform pattern.

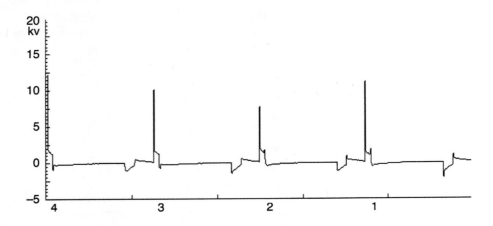

Figure 12-17 Secondary ignition pattern with firing kV.

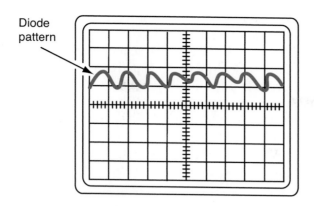

Diode
pattern

Figure 12-18 Alternator waveform.

On-Board Computer Diagnosis

The vehicle's computer system is either incorporated into the test routine or can be viewed independently from any of the engine tests. Test capabilities are typically a function of the vehicle's computer. This determines the type of tests that can be performed via the computer and other functions such as the type of data displayed and refresh rates. The tests typically include:

❏ Standard and bi-directional data communications with vehicle computer system

❏ Diagnostic trouble code (DTC) retrieval and definitions

❏ Diagnostic charts for codes and troubleshooting

❏ Vehicle data stream

❏ Specialty tests

❏ OBD II-enhanced tests

❏ Data capture and storage for playback and graphical analysis (Figure 12-19)

Although considered a valuable diagnostics tool, an engine analyzer is usually financially out of reach of the technician. It is usually the property of the shop. Regardless of whether an

2003 FORD TRUCK

** Scroll for data. OK to drive. **

RPM 2544   O2S1 (MV) 558   O2S2 (MV) 133

TPS (V)	1.77	TP MODE	P/T
TPCT	0.9	ECT (V)	0.48
ECT (dF)	214	IAT ACT (V)	2.19
MASS AIR (V)	2.93	IDLE AIR (%)	98.8
EVR (%)	63.2	DPFE (V)	2.81
INJ PW2 (MS)	11.63	INJ PW1 (MS)	11.60
SFTRIM 2 (%)	0.5	SFTRIM 1 (%)	2.5
LFTRIM 2 (%)	-6.2	LFTRIM 1 (%)	-6.2
VREF (V)	4.97	VPWR BATT (V)	14.0
WAC-WOT A/C	ON	SPARK ADV (D)	29
CANP PURGE	ON	FUEL PUMP	ON
PARK/NEU POS	R-DL	VEH SPEED (MPH)	30
OPEN/CLSDLOOP	CLSD	BRAKE SW	OFF
		ACCS A/C	OFF

Figure 12-19 Input and output data from a 1993 Ford Truck using a Snap-on scan tool.

analyzer of this type is available for use, the basis of the tests described above has been in existence for many years. Technological advances have paved the way for a variety of hand-held scanners, DMMs, graphing power meters, and scopes. Feature-packed test equipment continues to emerge and is affordable enough for the average technician to own. Although these units are considered extremely powerful diagnostic tools, they are somewhat limited as compared to the diagnostic analyzer platform discussed above. Regardless of the product being used, the knowledge of product application and vehicle systems is essential for proper utilization and effective, efficient diagnostics.

Oscilloscope Testing

The oscilloscope can look inside the ignition system by giving the technician a visual representation of voltage changes over time.

No discussion of ignition troubleshooting would be complete without a comprehensive discussion of oscilloscope use. The job of the oscilloscope is to convert the electrical signals of the ignition system into a visual image showing voltage changes over a given period of time. This information is displayed on a CRT screen in the form of a continuous voltage line called a pattern or trace. By studying the pattern, a technician can determine what is happening inside the ignition system.

The information on the design and use of oscilloscopes given in this text is general in nature. Always follow the oscilloscope manufacturer's specific instructions when connecting test leads or conducting test procedures.

Scales

On the face of the CRT screen is a voltage versus time graph (Figure 12-20). The waveform patterns are displayed on the graph. The graph charts the changes in voltage and the time span in which the changes occur. The voltage versus time graph has four scales: two vertical scales (one on the left and one on the right), a horizontal percent of dwell scale, and a horizontal millisecond scale. The technician must select the proper scale for the test being conducted.

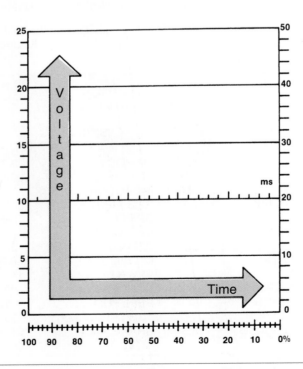

Figure 12-20 An oscilloscope plots a graph of voltage levels versus time.

518

The left and right vertical scales measure voltage. The vertical scale on the left side of the graph is divided into increments of 1 kilovolt (1,000 volts) and ranges from 0 to 25 kilovolts (kV). This scale is useful for testing secondary voltage. It can also be used to measure primary voltage by interpreting the scale in volts rather than kilovolts (0 to 25 volts).

The vertical scale on the right side of the graph is divided into increments of 2 kilovolts and has a range of 0 to 50 kV. This scale is also used for testing secondary voltage. This scale can also be used to measure primary voltage in the 0 to 500-volt range.

The horizontal percent of dwell scale is located at the bottom of the scope screen. This scale is used for checking the dwell angle in both the primary and secondary circuits. The dwell scale is divided into increments of 2 percentage points and ranges from 0 to 100 percent.

The fourth scale—the millisecond scale—is a horizontal line that runs along the center of the voltage versus time graph. Depending on the test mode selected by the technician, the millisecond scale shows a range of 0 to 5 milliseconds (ms) or 0 to 25 milliseconds. The 5 ms scale is often used to measure the duration of the spark (Figure 12-21). In the 25-millisecond mode, the complete firing pattern can normally be displayed. The 5 ms scale selection expands the pattern in order to enhance the detail of the trace. A typical duration should be 1.0 ms to 1.5 ms.

An oscilloscope or scope displays voltage changes from left to right, similar to reading a book. So voltage lines or traces that appear on the left of the CRT screen occurred before those on the right. Oscilloscopes normally have four leads (Figure 12-22): primary pickup that connects to the primary circuit or negative terminal of the ignition coil; a ground lead that connects to a good engine ground; a secondary pickup, which clamps around the coil's high tension wire; and a trigger pickup, which clamps around the spark plug wire of the number one cylinder. The information received from these leads is translated into the scope pattern on the CRT screen.

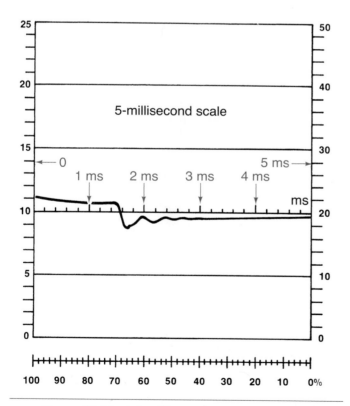

Figure 12-21 A 5-millisecond pattern showing spark duration.

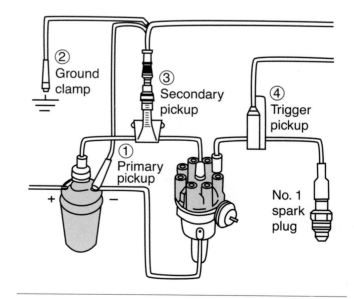

Figure 12-22 Typical oscilloscope pick-up connections to an ignition system.

519

SECONDARY PATTERN

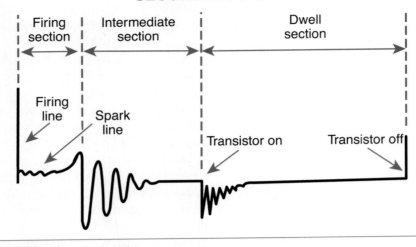

Figure 12-23 Typical secondary pattern.

Pattern Types and Phases

To monitor various phases of ignition system performance, a typical scope pattern is divided into three main sections: firing, intermediate, and dwell. Information received from specific testing in each one of these areas can be used to piece together a complete ignition picture.

Depending on the oscilloscope function selected by the technician, the scope can display the voltage versus time pattern for either the secondary or primary circuit. A typical secondary pattern is shown in Figure 12-23. A secondary pattern displays the following types of information:

- ❏ Firing voltage
- ❏ Spark duration
- ❏ Coil and condenser oscillations
- ❏ Transistor on/off switching or breaker point close and open
- ❏ Dwell and cylinder timing accuracy
- ❏ Secondary circuit accuracy

A typical primary circuit pattern is shown in Figure 12-24. Most problems that could occur in the primary pattern will usually also be seen in the secondary pattern. Conversely, some secondary ignition problems can also be seen in primary patterns. This can be helpful when the secondary connection is limited due to some newer ignition system designs.

PRIMARY PATTERN

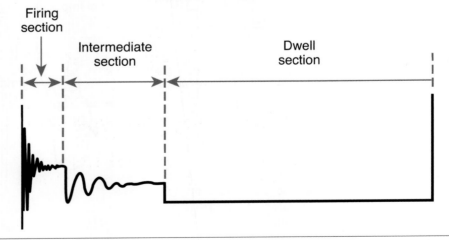

Figure 12-24 Typical primary pattern.

Pattern Display Modes

The oscilloscope has several ways to display the voltage patterns of the primary and secondary circuits. When the **display pattern** is selected, the oscilloscope displays the patterns of all the cylinders in a row from left to right as shown in Figure 12-25. This pattern arrangement is also commonly referred to as a **parade pattern**. It refers to the same information using a different name. Each cylinder's ignition cycle is displayed in the engine's firing order. In the example of a secondary display pattern shown in Figure 12-25, the firing order is 1,8,4,3,6,5,7,2. The pattern begins with the spark line of the number one cylinder and ends with the firing line for the number one cylinder. This display pattern allows the technician to compare the voltage peaks for each cylinder. The raster pattern, sometimes referred to as stacked, is nothing more that the repositioning of the parade pattern. This is done by stretching out each cylinder's pattern from the parade pattern to the width of the screen, then stacking them in the firing order beginning at the bottom (Figure 12-26). Viewing the pattern in this mode helps to enhance detail and vertically compare the cylinders' patterns.

The **display pattern** on an oscilloscope displays cylinder patterns from left to right.

The **parade pattern** starts with the spark line of the number one cylinder and ends with the firing line of the number one cylinder.

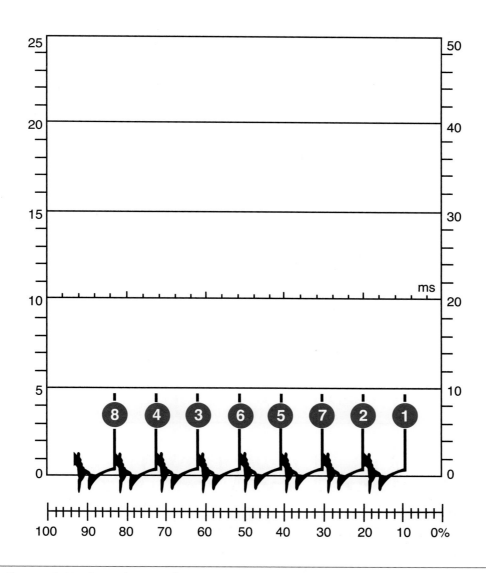

Figure 12-25 Secondary parade pattern.

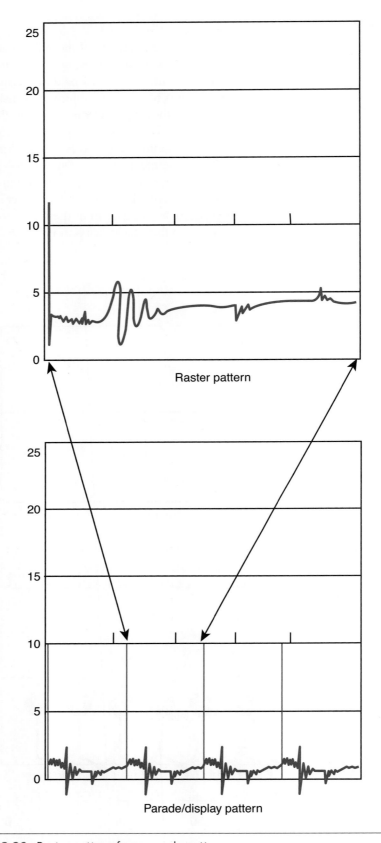

Figure 12-26 Raster pattern from parade pattern.

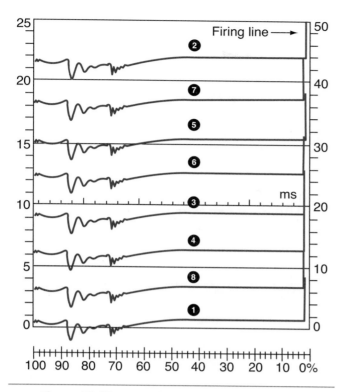

Figure 12-27 Secondary raster pattern.

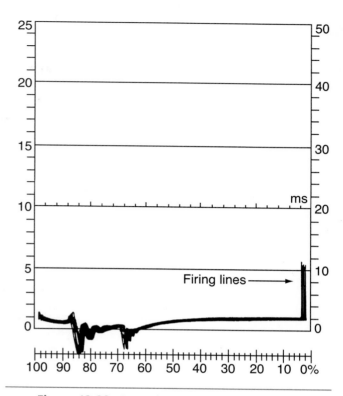

Figure 12-28 Secondary superimposed pattern.

A second choice of patterns is the **raster pattern** (Figure 12-27). A raster pattern stacks the voltage patterns of the cylinders one above the other. The number one pattern is displayed at the bottom of the screen and the rest of the cylinder's firing patterns are arranged above it in the engine's firing order.

In a raster pattern, the patterns for each cylinder are displayed across the width of the graph, beginning with the spark line and ending with the firing line. This allows for a much closer inspection of the voltage and time trends than is possible with the display pattern.

All of the patterns for the cylinders can also be displayed in a **superimposed pattern**. A superimposed pattern displays all the patterns one on top of the other. Like the raster pattern, the superimposed voltage patterns are displayed the full width of the screen, beginning with the spark line and ending with the firing line.

A superimposed pattern allows a technician to detect variations of one cylinder's pattern from the others. A superimposed pattern is shown in Figure 12-28.

Understanding Single Cylinder Patterns

The following section takes a closer look at the secondary circuit pattern (Figure 12-29). As shown, the firing line of the pattern appears at the left side of the CRT screen. An upward line signifies voltage. The firing line indicates the voltage needed to start the spark. The ignition coil's voltage pressure rises up to the point where it overcomes all electrical resistance in the secondary circuit. Typically, it requires around 10,000 volts to overcome this resistance and initiate a spark.

Keep in mind that cylinder conditions have an effect on this resistance. Leaner air-fuel mixtures increase the resistance and increase the required **firing voltage**.

The **raster pattern** stacks voltage patterns one above the other.

The **superimposed pattern** displays all patterns one on top of the other using the full width of the screen.

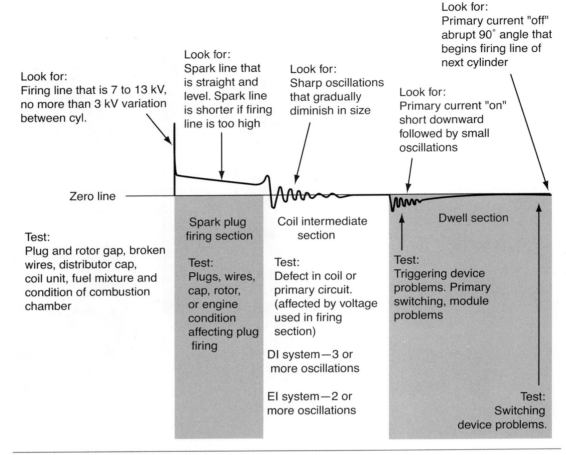

Figure 12-29 Breakdown of a single cylinder secondary circuit pattern.

The firing line and **spark line** of the firing section indicate **firing voltage** and spark duration.

Classroom Manual
Chapter 12, page 385

Once secondary resistance is overcome, the spark jumps the plug gap, establishes current flow, and ignites the air-fuel mixture in the cylinder. The length of time the spark actually lasts is represented by the **spark line** portion of the pattern. The spark line begins at the firing line and continues tracing to the right until the coil's voltage drops below the level needed to keep current flowing across the gap. In other words, the spark is draining the voltage built up in the coil's secondary windings. As long as the coil has sufficient voltage, the spark continues.

The spark line has a slant and length which, combined with the firing line, provide insight into firing performance. The slant of the spark line indicates the amount of voltage required to sustain spark at the spark plug after the coil fires. The steeper the slant, known as spark kV, the more voltage is required to overcome the resistance. The length of the spark line, referred to as spark duration and is measured in milliseconds, measures the amount of time the spark remains between the spark plug electrodes. The physics of the dissipation of the coil's energy proportionally brings the three measurements (firing kV, spark kV, and burn time) together. Evidence of this can be found in common ignition problems such as a wide gap/open secondary or close-gap/shorted spark plug or spark plug wire. In the case of the wide gap condition, the firing line indicates more voltage will be required to initiate the spark to the spark plug, causing a high firing line. Due to the relationship with the spark line, the high resistance added to the circuit will require more voltage to sustain spark across the spark plug. This will result in a steeper slant to the spark line or a higher spark kV. Finally, extra energy from the coil reduces the amount of time the spark can remain between the electrodes of the spark plug, resulting in lower spark duration, as measured as lower millisecond duration. To summarize, if the firing voltage is higher than normal, the burn time will decrease. If the voltage is lower than normal, the burn time will increase.

It is important to remember the elements that affect and determine firing voltage requirements. Internal and external cylinder resistance are two basic areas that influence these requirements. Internal resistance refers to anything inside the cylinder that would cause the firing voltage to increase or decrease. This would include compression, mixture, and temperature. For example, high compression adds to cylinder pressure, which increases resistance. Therefore, higher voltage will be required to fire. A lean mixture has more oxygen molecules between the HC molecules. The additional oxygen does not conduct electricity as easily as HC molecules. Therefore, more voltage will be required to initiate spark and sustain voltage. Even ignition timing, as well as combustion chamber temperature, will affect the firing voltage requirements.

External resistance refers to anything external to the cylinder and usually includes the secondary ignition components. Examples are spark plugs, wires, distributor cap, rotor, coil wire, and ignition coils. Wide air gaps at the spark plugs or faulty spark plug wires will proportionately increase firing voltage requirements. On vehicles equipped with distributor caps and rotors, an incorrect rotor-to-cap air gap will have the same effect. As discussed earlier, it will be necessary to determine if the ignition problem is common to all cylinders or if it affects only one or several cylinders. If all the firing lines indicate high voltage, the problem is common to all cylinders. If only one or a few cylinders have abnormal requirements, then the problem has to do with only those particular cylinders.

Identifying firing-related problems is relatively easy. After determining whether the problem is common to all cylinders or only one, proceed to the next step. For example, if one cylinder has a firing line and burn time different than the other cylinders, the problem is most likely the spark plug, spark plug wire, or distributor cap. It may also be a mixture or mechanical related problem within the cylinder that you have identified. To determine how to locate the problem, refer to Photo Sequence 23 on page 530.

Continuing on, once coil voltage drops below the level needed to sustain the spark, the next major section of the scope trace pattern begins. This section is called the **intermediate section** or coil-condenser zone. It shows the remaining coil voltage as it dissipates or drops to zero. Remember, once the spark has ended, there is quite a bit of voltage stored in the ignition coil. The voltage remaining in the coil then oscillates or alternates back and forth within the primary circuit until it drops to zero. Notice that the lines representing the coil-condenser section steadily drop in height until the coil's voltage is zero.

The next section of the trace pattern begins with the primary circuit current on signal. It appears as a slight downward turn followed by several small oscillations. The slight downward curve occurs just as current begins to flow through the coil's primary winding. The oscillations that follow indicate the beginning of the magnetic field buildup in the coil. This curve marks the beginning of a period known as the dwell section of zone. The end of the dwell zone occurs when the primary current is turned off by the switching device (transistor or breaker points). The trace turns sharply upward at the end of the dwell zone. Turning the primary current off collapses the magnetic field in the coil and generates another high-voltage spark for the next cylinder in the firing order. Remember, the primary current off signal is the same as the firing line for the next cylinder. The length of the dwell section represents the amount of time that current is flowing through the primary.

Most scope patterns look more or less like the one just described. The patterns produced by some systems have fewer oscillations in the intermediate section. Patterns may also vary slightly in the dwell section. The length of this section depends on when the control module turns the transistor on and off. Several variables may affect this timing. These variables can be categorized as follows:

1. variable dwell
2. current limiting

The **intermediate section** shows coil voltage dissipation.

The dwell section shows the activation of primary coil current flow and primary coil current switch off. The primary current of signal is also the firing line for the next cylinder in the firing order.

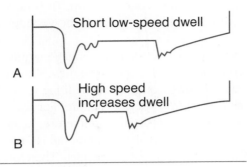

Figure 12-30 Dwell starts earlier at higher rpm.

Figure 12-31 Pattern showing an ignition system limiting current during dwell.

Most electronic ignition systems have a variable dwell function built into their control modules. In these systems, dwell changes significantly with engine speed. At idle and low rpm speeds, a short dwell provides enough time for complete ignition coil saturation. The current on and current off signal appears very close to each other, usually less than 20 degrees.

As engine speed increases, the control module lengthens the dwell time (Figure 12-30). This, of course, increases the available time for coil saturation.

Although not common, it is possible for the variable dwell function of the control module to fail and still allow the vehicle to run. If testing indicates a lack of variable dwell on a system equipped with it, the control module must be replaced.

Many modern electronic ignitions feature **current limiting**. These systems saturate the ignition coil quickly by passing very high current through the primary winding for a fraction of 1 second. Once the coil is saturated, the need for high current is eliminated and a small amount of current is used to keep the coil saturated. This type system extends coil life.

The point at which the control module cuts back from high to low current appears as a small blip or oscillation during the dwell section of the pattern (Figure 12-31). At very high engine speeds, this telltale blip may be missing because the module keeps high current flow going to keep the coil continually saturated for fast firing.

At the opposite extreme, if the primary winding of the coil has developed excessive resistance or the coil is otherwise faulty, the cutback blip may never occur. This is because the primary winding never becomes fully saturated. Further testing of the coil would be needed to pinpoint the cause of the missing blip.

As with variable dwell, it is possible for the vehicle to run with the current limit function of the control module burned out or faulty. Again, the blip would be missing from the trace.

▲ **WARNING:** Never touch the secondary wiring, cap, or coil if the current limiting blip is missing from the pattern of an ignition system that should have it. The secondary voltage available could easily exceed 100,000 volts.

Open Circuit Precautions

Some technicians check coil output by removing a secondary spark plug wire while keeping it from arcing to ground and reading the coil maximum output. Open circuit testing of the secondary is not recommended.

Features such as variable dwell and current limiting circuitry pull out all primary resistance if a spark does not occur. This causes tremendous amounts of current to flow through the primary circuit, producing an extremely high-voltage spark that must search out a path to ground. A frequent path to ground chosen by the spark is through the side of the coil. This results in an insulation breakdown at the coil and a site where arcing to ground is likely to occur in the future.

Current limiting is a feature that saturates the coil with high current for 1 second. This enables a small amount of current to be used to keep the coil saturated.

To prevent these tremendous voltages from occurring during coil output testing, an appropriate spark tester is used when checking the coil voltage output. The spark tester is connected to the coil wiring running to the distributor or to an ignition cable. A special grounding clip on the spark tester is then connected to a good ground source. (Refer to Figure 12-1.)

Some spark testers look like a spark plug, but it has a very large electrode gap. When the engine is cranked to check for voltage output, the ignition coil is forced to produce a higher voltage to overcome the added resistance created by the wide gap. Typically, about 35,000 volts as needed to fire the test plug. This is enough to stress the system without damaging it.

SERVICE TIP: Do not attempt to use an ordinary spark plug to make a test spark plug by breaking the side electrode off. It will not work the same as an actual test plug. The center electrode is too close to the spark plug shell, and the spark will simply jump there with practically no effective stress test. Several types of spark testers are available, including those adjustable to 40kV.

Coil Output and Oscillation Test

To safely perform a coil output test, proceed as follows:

1. Install the test plug in the coil wire or a plug wire if there is not a coil wire.
2. Set the oscilloscope on display and a voltage range of 50 kV.
3. Crank the engine over and note the height of the firing line.

The firing line should exceed 35 kV and be consistent. Lower-than-specified firing line voltages may indicate that the test plug did not fire. This could be the result of lower-than-normal available voltage in the primary circuit. The control module may have developed high internal resistance. The coil or coil cable may also be faulty. Further testing is needed to help pinpoint the problem.

In addition to the coil output test, simply observe the coil oscillations while connected to the scope. The best setting is either raster or 5 ms sweep. Coil oscillations begin at the end of the spark line at the first rise. One cycle is counted when the upswing of that oscillation almost reaches the top of the coil pattern as it begins the second oscillation. Simply count the top of each coil oscillation to determine how many there are. Electronic systems will have three or more oscillations. The more coil oscillations, the more coil capacity in reserve. Some EI and other electronic systems may have only two oscillations because there is only one coil for every one or two cylinders.

Spark Duration Testing

Spark duration is measured in milliseconds using the millisecond sweep of the oscilloscope. Most vehicles have a spark duration of approximately 1.5 milliseconds (Figure 12-32). A spark duration of approximately 0.8 millisecond is too short to provide complete combustion. A short spark also increases pollution and power loss. If the spark duration is too long (over approximately 2.0 milliseconds), the spark plug electrodes might wear prematurely. When the oscilloscope shows a long spark duration, it normally follows a short firing line which may indicate a fouled spark plug, low compression, or a spark plug with a narrow gap.

Spark duration is normally measured two times—during engine cranking, and during engine running.

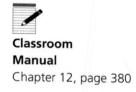

Classroom Manual
Chapter 12, page 380

Spark duration, measured in milliseconds, is useful in determining function of the spark plugs.

Cranking Test. When a vehicle is experiencing a no-start or a hard start condition, perform a spark duration cranking test. This test also helps identify potential future problems by indicating a borderline condition. To perform the crank test, do the following:

1. Disable the ignition system so the engine does not start. Different oscilloscope manufacturers recommend different methods of preventing engine startup during a cranking test. Always consult the scope manufacturer's instructions.

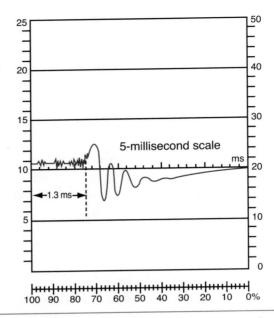

Figure 12-32 A 5-millisecond pattern showing a spark duration of 1.3 milliseconds.

2. Set the pattern selector to the 5-millisecond position, the function selector to secondary, and pattern height control to the 0 to 25 kV scale.

3. Crank the engine and quickly note the spark duration.

Compare the recorded duration times to manufacturer's specifications. If spark duration is too short, look for causes of low coil output voltage. Start with a thorough examination of the primary system after checking for proper operation of the battery, starting, and charging systems. If these items appear to be working properly, search for sources of high primary circuit resistance.

Set the oscilloscope in the primary circuit position and check the height of the initial firing line oscillation. If a lower-than-normal height is observed, the most likely causes are a poor control module ground, or loose, dirty, or corroded connectors in the primary circuit. Remember that a 1-volt loss in the primary circuit can result in a 1,000-volt loss in the secondary circuit.

☑ **SERVICE TIP:** Again, refer to Figure 12-32. Notice the small, hash-like lines along the spark line. Remember that any vertical line on the scope represents voltage. The hash-lines on the spark line are a result of air and fuel movement within the cylinder reacting to a swirl of rapidly changing mixture movement. The molecules are all mixed up and not uniform. This movement, known as combustion turbulence, affects the spark line by literally moving it during the spark duration. This can represent a valuable bit of information in determining an ignition fault. Sometimes a fault can occur in the secondary that will not necessarily greatly affect the firing or spark line. For example, this may be a misfire under load. Look closely at the spark lines while lightly throttling the engine. If all the spark lines except one have combustion turbulence, there is a good possibility that the spark is not jumping across the gap of the spark plug. In this example, the spark may be firing through the spark plug shell to ground, external of the cylinder when the cylinder load increases. Remember, electricity seeks the path of least resistance.

Running Test. The running test measures spark duration while the engine is running at a specific rpm. To perform a running test on the oscilloscope, follow these steps:

1. Set the pattern selector to the 5-millisecond position, the function selector to secondary, and pattern height control to the 0 to 25 kV scale.

	ms	600 rpm	1,200 rpm	2,400 rpm	Spark Duration
8 cylinder	0.5	2%	4%	8%	too short
	1.0	4%	8%	17%	minimum
	1.5	6%	13%	25%	average
	2.0	8%	17%	33%	too long
6 cylinder	0.5	1.5%	3%	6%	too short
	1.0	3%	6%	12%	minimum
	1.5	4.5%	9%	18%	average
	2.0	6%	12%	24%	too long
4 cylinder	0.5	1%	2%	4%	too short
	1.0	2%	4%	8%	minimum
	1.5	3%	6%	13%	average
	2.0	4%	8%	17%	too long

Figure 12-33 Converting percent of dwell to milliseconds.

2. Start the engine and adjust the speed to 1,000 rpm.

3. Note the spark duration.

Some older oscilloscopes do not have a millisecond sweep. Instead, they are equipped with a percent of dwell scale. In that case, the percent of dwell must be converted to milliseconds. A conversion chart is given in Figure 12-33. The table lists percent of dwell at various rpm and the corresponding spark duration in milliseconds. On oscilloscopes without a millisecond sweep, the pattern selector should be set to superimpose or raster when performing the spark duration cranking and running tests.

The raster pattern in Figure 12-34 is a secondary pattern of a 4-cylinder engine at 600 rpm. The spark duration is 3 percent of dwell, or 1.5 milliseconds.

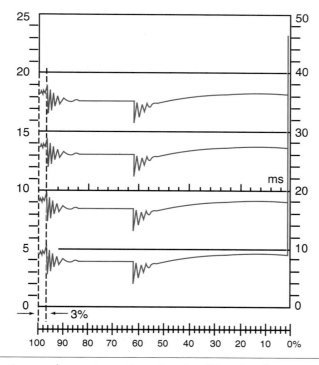

Figure 12-34 A secondary raster pattern showing firing duration as 3 percent of dwell.

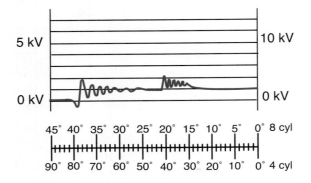

Figure 12-35 Reversed primary coil connections cause reversed secondary polarity and upside-down secondary scope patterns.

Coil Polarity

Less voltage is required to fire the spark plugs with proper coil polarity. If the coil polarity is reversed, 20 to 40 percent more voltage is needed to fire the spark plugs.

If the coil polarity is correct, the firing lines of the display pattern extends upward. If the polarity is reversed, the firing lines extend downward. If reverse polarity is indicated (Figure 12-35), check the coil primary connections. They are probably reversed.

Spark Plug Firing Voltage

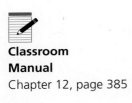

Classroom Manual
Chapter 12, page 385

The coil must generate sufficient voltage in the secondary system to overcome the total resistance in the secondary circuit and to establish a spark across the spark plug electrodes. On the oscilloscope, this spark plug firing voltage is seen as the highest line in the pattern. The firing voltage might be affected by the condition of the spark plugs or the secondary circuit, engine temperature, fuel mixture, and compression pressures. To test the spark plug firing voltage on an oscilloscope, observe the firing line of all cylinders for height and uniformity. The normal height of the firing voltages should be between 7 and 13 kV with no more than a 3 kV variation between cylinders.

If during the test one or more of the firing voltages are uneven, low, or high, consult Figure 12-36 for possible causes and corrections. Refer to Photo Sequence 23 to learn how to identify the cause of an abnormally high firing line. It should be noted that although the following photo sequence demonstrates locating a component causing a high firing voltage, the same elimination process is used to locate a component causing a low firing voltage. To do this, simply follow the same process but use a spark tester instead of a grounding probe. The spark tester will cause a controlled increase in the firing voltage at the point at which the problem part is eliminated.

Condition	Probable Cause	Remedy
Firing voltage lines the same, but abnormally high	1. Retarded ignition timing 2. Fuel mixture too lean 3. High resistance in coil wire 4. Corrosion in coil tower terminal 5. Corrosion in distributor coil terminal	1. Rest ignition timing. 2. Check for vacuum leak. 3. Replace coil wire. 4. Clean or replace coil. 5. Clean or replace distributor cap.
Firing voltage lines the same, but abnormally low	1. Fuel mixture too rich 2. Breaks in coil wire causing arcing 3. Cracked coil tower causing arcing 4. Low coil output 5. Low engine compression	1. Check for plugged air filter. 2. Replace coil wire. 3. Replace coil. 4. Replace coil. 5. Determine cause and repair.
One or more, but not all firing voltage lines higher than the others	1. Idle mixture not balanced 2. EGR valve stuck open 3. High resistance in spark plug wire 4. Cracked or broken spark plug insulator 5. Intake vacuum leak 6. Defective spark plugs 7. Corroded spark plug terminals	1. Readjust idle mixture. 2. Inspect or replace EGR valve. 3. Replace spark plug wires. 4. Replace spark plugs. 5. Repair leak. 6. Replace spark plugs. 7. Replace spark plugs.
One or more, but not all firing voltage lines lower	1. Curb idle mixture not balanced 2. Breaks in plug wires causing arcing 3. Cracked coil tower causing arcing 4. Low compression 5. Defective or fouled spark plugs	1. Readjust idle mixture. 2. Replace spark plug wires. 3. Replace coil. 4. Determine cause and repair. 5. Replace spark plugs.
Cylinders not firing	1. Cracked distributor cap terminals 2. Shorted spark plug wire 3. Mechanical problem in engine 4. Defective spark plugs 5. Spark plugs fouled	1. Replace distributor cap. 2. Determine cause of short and replace wire. 3. Determine problem and correct. 4. Replace spark plugs. 5. Replace spark plugs.

Figure 12-36 Firing line diagnosis.

Photo Sequence 23
Determining the Cause of a High Firing Line

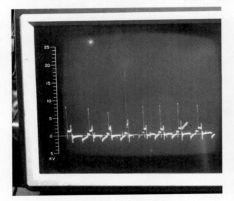

P23-1 Observe the secondary display pattern with the engine at idle.

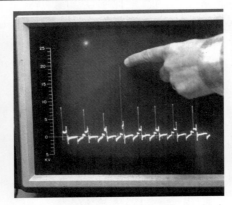

P23-2 Identify the cylinder with the high firing line and record the height of the line on a piece of paper.

P23-3 Turn the engine off and carefully remove the plug wire from the affected cylinder.

P23-4 Connect a jumper wire or grounding probe to the end of the spark plug wire and to a good ground.

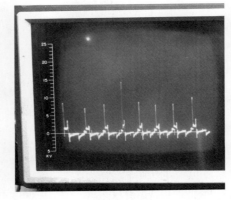

P23-5 Start the engine and observe the height of the firing line now. Do not allow the engine to run very long with the plug bypassed. After you have a reading, turn the engine off.

P23-6 Record the new height and write the difference between the readings off to the side.

P23-7 Move the jumper wire to the correct terminal at the distributor cap.

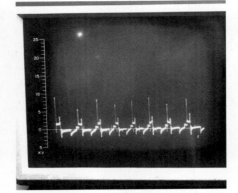

P23-8 Start the engine and observe the height of the firing line. Then turn the engine off.

P23-9 Record the new height and write the difference between the readings off to the side.

Determining the Cause of a High Firing Line (continued)

P23-10 Reconnect the spark plug wire and start the engine. Allow it to idle while analyzing the results. Compare the drops in required voltage during each phase of this test. Notice when C was bypassed, the voltage dropped considerably more than when other parts were bypassed. This indicates high resistance in the spark plug wire.

Rotor Air Gap Voltage Drop

When high firing voltages are present in one or more of the cylinders, perform a rotor air gap voltage drop test. The purpose of this test is to determine the amount of secondary voltage that is required to bridge the rotor gap.

 SERVICE TIP: Excessive rotor air gap can be caused by a bent distributor shaft or worn shaft bearings. Check shaft movement prior to testing the rotor air gap on the scope.

To perform a rotor air gap voltage drop test:

1. Set the pattern selector to the display position, the function selector to secondary, and the pattern height control to the 0 to 25 kV scale.
2. Start the engine and adjust the speed to 1,000 rpm.
3. Observe the height of the firing lines. Record the height and firing order number of any abnormal cylinder.
4. Shut the engine off.
5. Remove the spark plug wire of the abnormal cylinder from the distributor cap. Connect one end of the jumper lead to ground and the other end to the large portion of a grounding probe. Place the other end of the grounding probe in the distributor cap tower terminal.

   CAUTION: Do not remove the spark plug wire from the distributor cap tower terminal while the engine is running. This causes an open circuit and might damage the ignition system components.

6. Start the engine and adjust the speed to 1,000 rpm.
7. Observe the firing line of the abnormal cylinder previously recorded. There should be a slight drop in the firing voltage when using the grounding probe (see Figure 12-37).

Classroom Manual
Chapter 12, page 384

Special Tools

Jumper wire, grounding probe

Notice the difference in spark duration with the plug wire grounded in Figure 12-37.

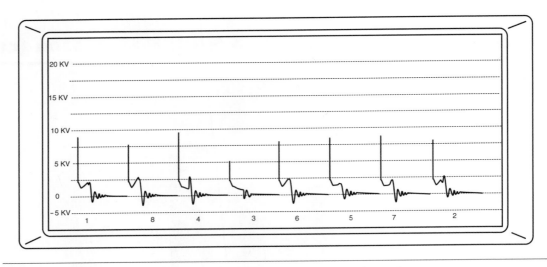

Figure 12-37 Grounded cylinder 3 showing rotor air gap voltage drop.

The observed voltage represents voltage needed to overcome the resistance in the coil wire and of the rotor air gap. The rotor air gap measurements should not exceed manufacturer's specifications. If during the test, the firing line remains high or drops to a level that does not meet manufacturer's specifications, the rotor or distributor cap might be defective. Visually inspect both and replace as necessary. A bent distributor shaft or worn shaft bushings will cause excessive rotor air gaps on about half the cylinders.

Secondary Circuit Resistance

Analysis of the spark line of a secondary pattern reveals the condition of the secondary circuit. The amount of resistance in the secondary circuit is indicated by the slope of the spark line. Excessive resistance in the secondary circuit causes the spark line to have a steep slope with a shorter firing duration. The excessive resistance restricts the current flow necessary to generate a good spark.

Refer to Figure 12-38. The firing kV is very high and the duration is very short. If you look closely at the pattern for number three compared to the rest of the cylinders, the duration

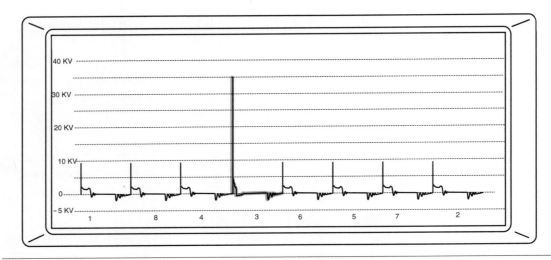

Figure 12-38 High kV showing excessive secondary resistance on cylinder 3. Note firing line height and lack of duration in spark line.

is very short; you might even say non-existent. The reason for this is the fact that the energy from the coil was almost totally spent in jumping the plug gap with very little left for duration. The situation is a plug wire with extremely high resistance and shows the tendency for the spark kV to rise when the spark plug gap is difficult to ionize. The slope on the spark line, as short as it is, is steeply downward.

A good spark line should be relatively even and measure 2 to 4 kV in height. High resistance in the secondary circuit produces a firing line and spark line that are higher in voltage with shorter firing durations. If high resistance is shown during testing, see Figure 12-39.

Symptom	Probable Cause	Remedy
SECONDARY CIRCUIT RESISTANCE TEST		
High resistance in spark lines	1. Rotor tip burned 2. High resistance in the spark plug wire 3. Distributor cap segments burned 4. Faulty spark plug	1. Visually inspect. 2. Perform ohmmeter test. 3. Visually inspect. 4. Perform secondary resistance test using the grounding probe.
After grounding spark plug, abnormal spark lines appear normal	1. Faulty spark plug	1. Substitute spark plug.
After grounding spark plug, abnormal spark lines still show high resistance	1. Corroded distributor cap towers 2. Distributor cap segments burned 3. High resistance in the spark plug wires	1. Visually inspect. 2. Visually inspect. 3. Perform ohmmeter test.
All the spark lines show high resistance	1. Defective coil wire 2. Rotor tip burned	1. Perform ohmmeter test. 2. Visually inspect.
SPARK PLUGS UNDER LOAD TEST		
One or more of cylinders show a voltage rise over 4 kV	1. Spark plug gap too wide 2. Worn spark plug electrodes 3. Open or high resistance spark plug wire 4. Improper fuel mixture 5. Open spark plug resistor 6. Leakage in vacuum system	1. Check gap, then regap plug. 2. Replace the spark plug. 3. Perform ohmmeter test. 4. Adjust mixture. 5. Substitute spark plug. 6. Check for leaky vacuum hoses and diaphragms or gaskets.
One or more cylinders show a voltage rise less than 3 kV or no voltage rise at all	1. Shorted spark plug wire 2. Broken spark plug insulator 3. Fouled spark plug 4. Low compression	1. Perform secondary insulation test. 2. Replace spark plug. 3. Clean or replace spark plug. 4. Perform compression test.
CYLINDER TIMING ACCURACY TEST		
Transistor turn-off signals exceed specifications for engine being tested	1. Bent distributor shaft 2. Worn distributor bushings 3. Worn gear on distributor 4. Worn camshaft gear 5. Worn timing chain	1. Perform distributor test. 2. Perform distributor test. 3. Visually inspect. 4. Visually inspect. 5. Visually inspect.

Figure 12-39 Troubleshooting spark plugs.

To pinpoint the cause of high resistance, use a grounding probe and jumper wire to bypass each component of the secondary circuit on all abnormal cylinders. Connect one end of the jumper lead to ground and the other end to the large portion of a grounding probe. Start the engine and adjust the speed to 1,000 rpm. Touch each secondary connector with the point of the grounding probe and observe the spark lines.

If after grounding the abnormal spark lines appear normal, the part just bypassed is the cause of the problem. For further diagnosis, see Figure 12-39.

While scanning the secondary circuit with the grounding probe, watch the firing lines on the scope. If there is an insulation break in the circuit, the firing lines decrease in height when the high secondary voltage arcs across to the probe.

Any insulated part of the secondary circuit can break down and leak. This includes the coil, coil wire and boots, distributor cap and terminals, and spark plug wires and boots.

Spark Plugs Under Load

The voltage required to fire the spark plugs increases when the engine is under load. The voltage increase is moderate and uniform if the spark plugs are in good condition and properly gapped. However, if any unusual characteristics are displayed on the scope patterns when load is applied to the engine, the spark plugs are probably faulty. This condition is most evident in the firing voltages displayed on the scope. To test spark plug operation under load, note the height of the firing lines at idle speed. Then, quickly open and release the throttle (snap accelerate) and note the rise in the firing lines while checking the voltages for uniformity. A normal rise would be between 3 and 4 kV upon snap acceleration.

If during testing, one or more of the cylinders show a voltage rise of over 4 kV or if one or more cylinders show a voltage rise less than 3 kV or no voltage rise at all, check Figure 12-39.

Coil Condition

The energy remaining in the coil after the spark plugs fire gradually diminishes in a series of oscillations. These oscillations are observable in the intermediate sections of both the primary and secondary patterns. If the scope pattern shows an absence of normal oscillations in the intermediate section, check for a possible short in the coil by testing the resistance of the primary and secondary windings.

Individual Component Testing

Ignition systems have many characteristics unique to the system's manufacturer. It would be impossible in any one textbook to explain all the variations. However, the basic goal of any electrical troubleshooting procedure is always to identify electrical activity under a given set of circumstances. The manufacturer's service manual provides the technician with both the desired activity (specification) and the set of circumstances (procedures).

Figure 12-40 and Figure 12-41 outline a procedure for quickly isolating an ignition related problem when an oscilloscope is not available. The first troubleshooting tree determines whether a spark is generated in the secondary system. If no spark is available, the second troubleshooting tree leads step by step through individual component tests until the problem is located.

The secret to component testing is to use good troubleshooting practices. Work systematically through a circuit, testing each wire, connector, and component. Do not jump around back and forth between components. The component inadvertently overlooked is probably the one causing the trouble. Checks must be made for available voltage, voltage output, resistance of wires and connectors, and available ground. Always compare the readings with specifications given in the manufacturer's service manual.

The following sections briefly outline common test procedures for individual system components. For accurate testing, always refer to service manual wiring diagrams and testing instructions.

Any problem that makes kV higher makes duration shorter and vice versa.

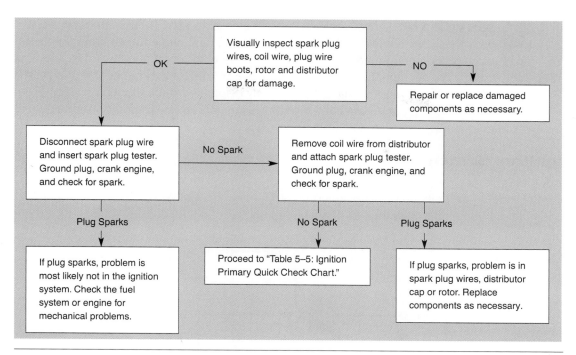

Figure 12-40 Ignition secondary quick-check chart.

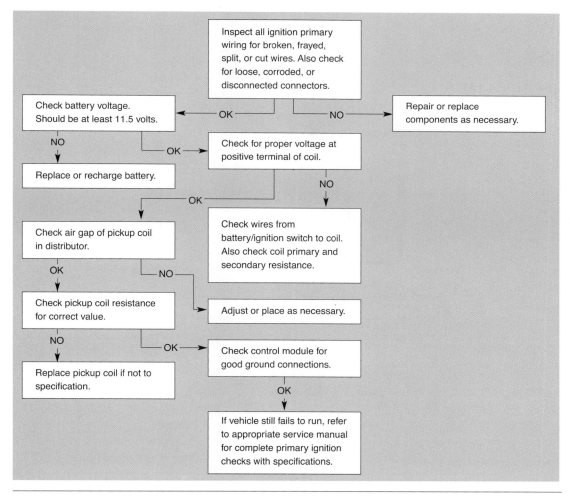

Figure 12-41 Ignition primary quick check chart.

SERVICE TIP: Service manuals are indispensable when troubleshooting ignition problems. They contain such vital information as base and advance timing specifications, color-coded wiring diagrams and terminal connector descriptions, and illustrations showing test connections for voltage, continuity, and resistance checks. The diagnostic charts found in the service manual are particularly helpful in troubleshooting ignition problems. Ignoring the valuable data contained in the ignition service section of these manuals can lead to many frustrating and costly service mistakes.

Ignition Switch

A faulty ignition switch or bad ignition switch wiring does not supply power to the ignition control module or ignition coil. The ignition system shown in Figure 12-42 has two wires con-

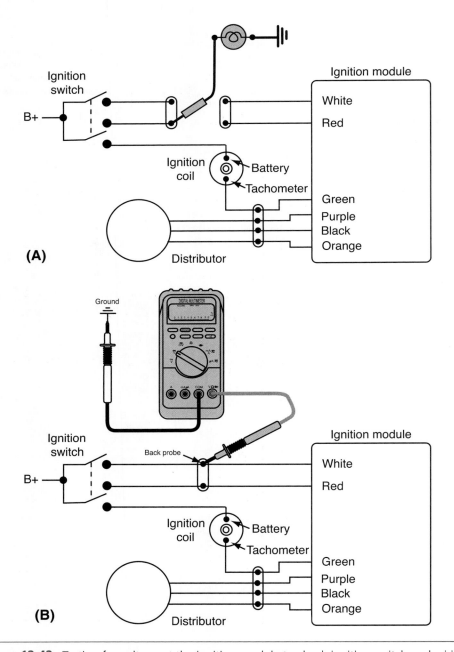

Figure 12-42 Testing for voltage at the ignition module to check ignition switch and wiring (A) test light method, and (B) digital multimeter.

nected to the run terminal of the ignition switch. One is connected to the module. The other is connected to the primary resistor and coil. The start terminal of the switch is also wired to the module.

You can check for voltage using either a 12-volt test light or a digital voltmeter. To use the test light method, turn the ignition key off and disconnect the wire connector at the module. Also, disconnect the S terminal of the starter solenoid to prevent the engine from cranking when the ignition is in the run position. Turn the key to the run position and probe the red wire connection to check for voltage. Also check for voltage at the bat terminal of the ignition coil using the test light. Next, turn the key to the start position and check for voltage at the white wire connector at the module and the bat terminal of the ignition coil.

To make the same test using a digital voltmeter, backprobe the appropriate module connection and ground. The power available to the module should be at lest 90 percent of battery voltage.

SERVICE TIP: Although some manufacturers state in their service procedures to use a straight pin to pierce the wire insulation for voltage tests, the preferred method is to backprobe a connector or use jumper wires instead. Doing anything to connectors or insulation that compromises the circuit's integrity should be avoided.

Scope Testing

One of the best indicators of a primary ignition circuit problem is the waveform on an ignition oscilloscope. These scopes are designed to show either a primary or a secondary pattern of the ignition system. The secondary pattern is mostly comprised of the secondary circuit, but also shows the primary when the secondary is doing nothing. During dwell or the time when there is primary current flow, the secondary circuit is sitting idle with no electrical activity. The secondary pattern shows the electrical activity in the primary during this time.

A primary ignition pattern shows the action of the primary circuit. To be able to spot abnormal sections of a primary waveform, you must know what causes each change of voltage and time in a normal primary waveform. Although the true cycle of the primary circuit begins and ends when the switching transistor is turned on, the displayed pattern begins right after the transistor is turned off. At this moment in time, the magnetic field around the windings collapses and a spark plug is fired.

Looking at the primary pattern shown in Figure 12-43, time moves from left to right. Therefore, the trace at the left represents the collapsing of the primary winding after the transistor turns

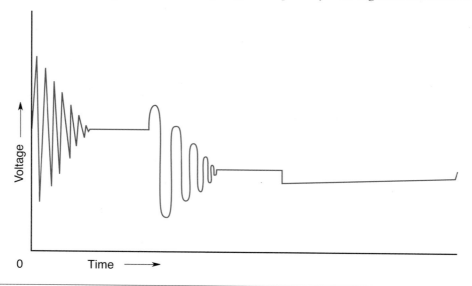

Figure 12-43 A typical primary pattern.

off and primary current flow is interrupted. The height of these oscillations depends on the current that was flowing through the winding right before it was stopped. The amount of current flow depends on the time it was able to flow, the voltage or pressure applied to the winding, and the resistance of the winding. High primary circuit resistance will reduce the maximum amount of current that can flow through the winding. Reduced current flow through the winding will reduce the amount of voltage that can be induced when the field collapses. The most likely sources of unwanted resistance in the primary circuit are shown in Figure 12-44.

During the collapsing of the primary winding, the spark plug is firing. The primary circuit's trace shows sharp oscillations of decreasing voltages. The overall shape of this group of oscillations should be conical and should last until the spark plug stops firing.

After the firing of the plug, some electrical energy remains in the coil. This energy must be released prior to the next dwell cycle. The next set of oscillations shows the dissipation of this voltage. These oscillations should be smooth and become gradually smaller until the zero volt line is reached. At that point there is no voltage left and the coil is ready for the next dwell cycle.

Immediately following this dissipation of coil energy is the transistor on signal. This is when the time current begins to flow through the primary circuit. It is the beginning of dwell. When the transistor turns on, there should be a clean and sharp turn in the trace. A clean change indicates the circuit was instantly turned on. If there is any sloping or noise at this part

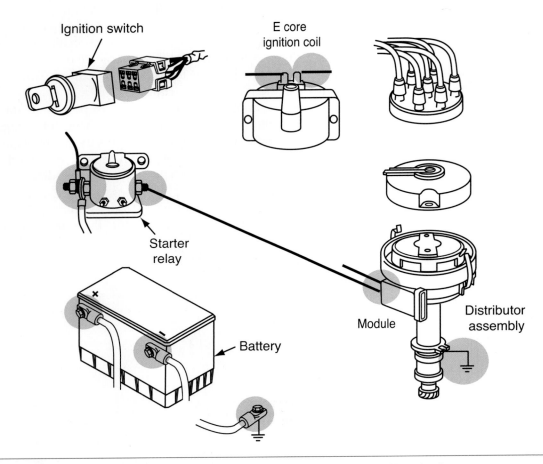

Figure 12-44 Possible sources of high resistance in the primary ignition circuit.

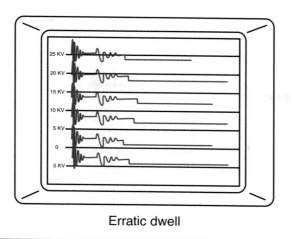

Erratic dwell

Figure 12-45 A primary pattern (raster) showing cylinders with different transistor "ON" times.

of the signal, something is preventing the circuit from being instantly turned on. Also, any variation among cylinders will show up as an erratic dwell in the primary pattern (Figure 12-45).

During dwell, the trace should be relatively flat. However, many ignition systems have features that change current flow during dwell (Figure 12-46). These features are designed to allow complete coil saturation only when that is needed. Other times, the current is limited according to the needs of the engine. By reducing the amount of current, the amount of volt-

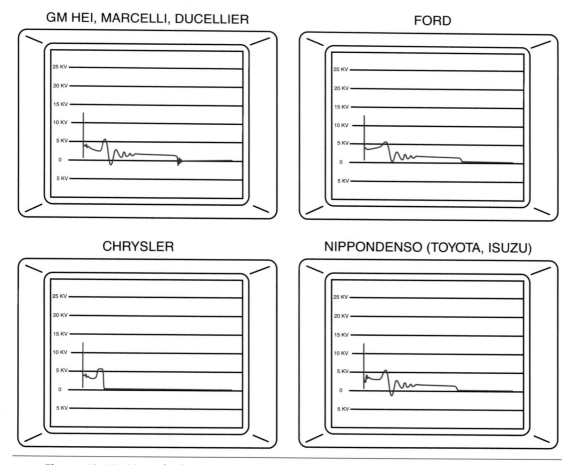

Figure 12-46 Normal primary ignition patterns for various ignition systems.

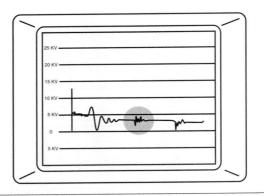

Figure 12-47 Erratic transistor "ON" signal.

age induced in the secondary is also reduced. Since the system uses only a portion of this voltage, the unused voltage must be dissipated by the coil. By controlling the amount of voltage produced by the coil, the amount of voltage the coil must dissipate as heat is also controlled.

If there are erratic voltage spikes at the transistor on signal (Figure 12-47), the ignition module may be faulty, the distributor shaft bushings may be worn, or the armature is not securely fitted to the distributor shaft. The problem is preventing dwell from beginning smoothly and will cause the engine to have a rough idle, intermittent miss, and/or higher-than-normal HC emission levels.

If the transistor on signal looks erratic while viewing the ignition system with a superimposed pattern, change the pattern display to raster. This allows the patterns to be stacked on top of each other. Sometimes the erratic signal is actually transistor on signals that are occurring at different times. Each cylinder has its own and unique time for events. This is not good. It can cause rough engine operation at all speeds, especially idle. It can also cause a general lack of power, poor acceleration, and backfiring. There are a number of factors that can cause this:

1. Worn upper distributor shaft bushing
2. Worn distributor drive gear
3. Worn camshaft sprocket
4. Stretched timing chain or belt
5. Excessive camshaft endplay
6. Loose armature or pickup unit

High resistance or low voltage in the primary circuit can cause high HC levels as well as a miss under load with a normal idle. This problem will be evident in the primary pattern as seen before and will also be apparent in a secondary waveform (Figure 12-48). Weak coil oscil-

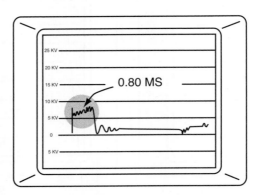

Figure 12-48 Primary resistance affects the secondary waveform.

lations and an erratic spark line that slopes upward show a weak coil. The coil appears to be weak, but the actual problem is low voltage to the coil. Check the AC generator's output and drive belt. Also check for excessive voltage drops throughout the primary circuit.

No-Start Diagnosis

Special Tools

12-volt test light, test spark plug

The cause of a no-start condition can be in the air, fuel, or ignition system. Often manufacturers will include a detailed troubleshooting tree to follow to locate the cause of the no-start condition (Figure 12-49). If the problem is caused by an ignition fault, follow this procedure to determine the exact cause of the problem:

1. Connect a 12-volt test lamp from the coil tachometer (tach) terminal to ground. Turn the ignition switch on. On high energy (HEI) systems, the test light should be on because the module primary circuit is open. If the test light is off, there is an open

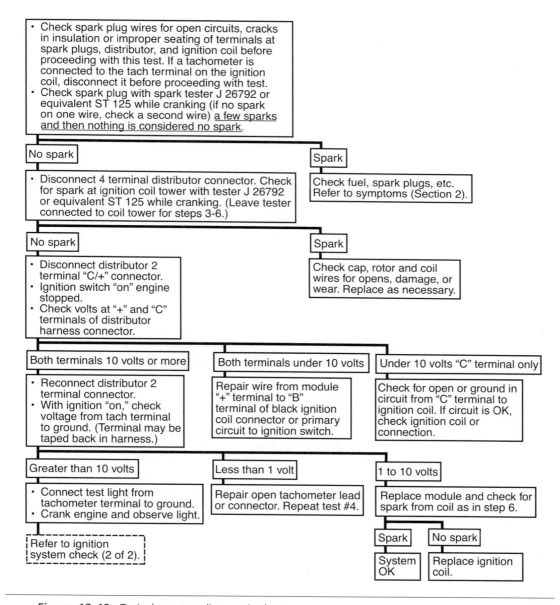

Figure 12-49 Typical no-start diagnostic chart.

circuit in the coil primary winding or in the circuit from the ignition switch to the coil battery terminal. On early Chrysler electronic or Ford Dura-Spark systems, the test light should be off because the module primary circuit is closed. Since there is primary current flow, most of the voltage is dropped across the primary coil winding. This action results in very low voltage at the tach terminal, which does not illuminate the test light. On these systems, if the test light is illuminated there is an open circuit in the module or in the wire between the coil and the module.

2. Crank the engine and observe the test light. If the test light flickers while the engine is cranked, the pickup coil signal and the module are satisfactory. When the test lamp does not flicker, one of these components is defective. The pickup coil may be tested with a DMM. If the pickup coil is satisfactory, the module is defective. Before testing the pickup, check the voltage supply to the positive primary coil terminal with the ignition switch on before the diagnosis is continued.

✔ **SERVICE TIP:** On most Chrysler fuel injected engines, the voltage is supplied through the automatic shutdown (ASD) relay to the coil positive primary terminal and the electric fuel pump. Therefore, a defective ASD relay may cause 0 volts at the positive primary coil terminal. This relay is controlled by the PCM. On some Chrysler products, the relay momentarily closes when the ignition switch is turned on, then closes only while the engine is cranking or running. If the ASD relay closes with the ignition switch on and the engine not cranking or running, it only remains closed for about 1 second. This action shuts the fuel pump off and prevents any spark from the ignition system if the vehicle is involved in a collision with the ignition switch on and the engine stalled. A fault code should be present in the computer memory if the ASD relay is defective.

3. If the test light flickers, connect a test spark plug to the coil secondary wire and ground the spark plug case. The test spark plug must have the correct voltage requirement for the ignition system being tested. For example, test spark plugs for HEI systems have a 25,000-volt requirement compared to a 20,000-volt requirement for many other test spark plugs. A short piece of vacuum hose may be used to connect the test spark plug to the center distributor cap terminal on HEI systems with an integral coil in the distributor cap.

4. Crank the engine and observe the spark plug. If the test spark plug fires, the ignition coil is satisfactory. If the test spark plug does not fire, the coil is likely defective because the primary circuit no-start test proved the primary circuit is triggering on and off.

5. Connect the test spark plug to several spark plug wires and crank the engine while observing the spark plug. If the test spark plug fired in step 4 but does not fire at some of the spark plugs, the secondary voltage and current is leaking through a defective distributor cap, rotor, or spark plug wires, or the plug wire is open. If the test spark plug fires at all the spark plugs, the ignition system is satisfactory.

Ignition Coil Inspection and Tests

Classroom Manual
Chapter 12, page 386

The ignition coil should be inspected for cracks, carbon tracking, and corrosion at the secondary connection, as well as the primary wiring connectors. The ignition coil can also be checked with an ohmmeter. Make sure that the power to the coil is off, and check the coil primary winding as shown in Figure 12-50. Most primary windings range from 0.5 to 2 Ω resistance. The secondary winding can be checked on single secondary terminal coils by checking from one primary terminal to the secondary terminal as shown in Figure 12-51. The expected resistance should be 6,000 to 20,000 Ω. Also, check the wiring for a short to the case as shown in Figure 12-52. Check your service information for exact specifications for these tests. In some HEI integral coils, the secondary winding is connected from the secondary terminal to the coil

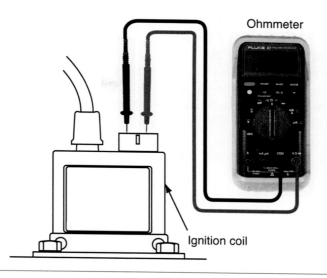

Figure 12-50   Ohmmeter connected to primary coil terminals.

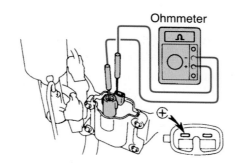

Figure 12-51   Ohmmeter connected from one primary terminal to the coil tower to test secondary winding.

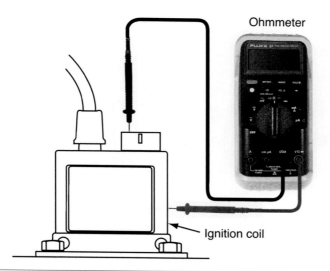

Figure 12-52   Checking for a short to the coil casing.

frame. When the secondary winding is tested in these coils, the ohmmeter must be connected from the secondary coil terminal to the coil frame or to the ground wire terminal extending from the coil frame to the distributor housing.

The ohmmeter tests on the primary and secondary windings indicates the general condition of the coil. However, these tests do not guarantee that the coil will work fine. Some defects will only show up on scope tests, such as defective insulation around the coil windings, which causes high-voltage leaks.

Pickup Coil

Classroom Manual
Chapter 12, page 392

The pickup coil of a magnetic pulse generator or metal detection sensor is also checked for proper resistance using an ohmmeter. Connect the ohmmeter to the pickup coil terminals to test the pickup coil for an open or a shorted condition. While the ohmmeter leads are connected, pull on the pickup leads and watch for an erratic reading indicating an intermittent open in the pickup leads. Most pickup coils have 150 to 900 ohms resistance, but *always* refer to the manufacturer's specifications. If the pickup coil is open, the ohmmeter will display an infinity reading. When the pickup coil is shorted, the ohmmeter will display a reading lower than specifications.

Connect the ohmmeter from one of the pickup leads to ground to test the pickup for a short to ground. If there is no short to ground, the ohmmeter will give an infinity reading. Refer to Figure 12-53. Ohmmeter one is testing for a short to ground, while ohmmeter two is testing the resistance of the pickup coil. Also, while in the number two configuration, the pickup coil's AC voltage can be checked. Set the DMM to AC Volts and crank the engine. Observe the average voltage reading. A rule of thumb for a typical pickup coil is about 500 mV (check the manufacturer's specifications). Before cranking, be sure that there is adequate clearance between the rotating distributor shaft and test leads!

The oscilloscope can be used to check pickup coil operation. The primary scope leads are connected to the pickup coil leads. The scope is set on its lowest scale. When the distributor shaft is spun, an AC sine wave trace should appear on the CRT screen. The trace is not a true sine wave, but it should have both a positive and negative pulse. As each tooth of the reluctor or pole piece approaches, the tooth on the magnet of the pickup coil passes. The AC voltage will proportionately rise, then sharply drop (Figure 12-54).

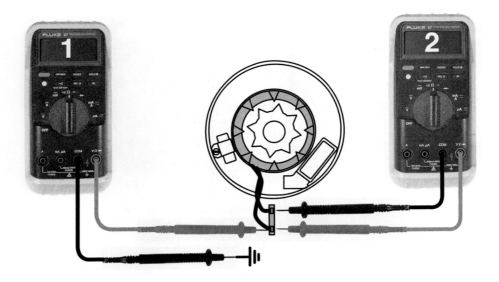

Figure 12-53 Ohmmeter to pickup connections for shorted, open, and high resistance.

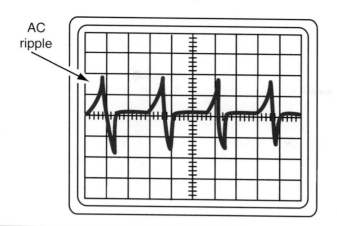

AC ripple

Figure 12-54 The trace of a good PM generator.

Another method of measuring the pickup coil's AC signal is with a simple voltmeter set on its low voltage scale. The meter registers AC voltage during cranking. Measure this voltage as close to the control module as possible to account for any resistance in the connecting wire from the pickup coil to the module.

Hall Effect Sensors

Most Hall effect sensors can be tested by connecting a 12-volt battery across the plus (+) and minus (–) voltage (supply current) terminals of the Hall layer and a voltmeter across the minus (–) and signal voltage terminals.

With the voltmeter hooked up, insert a steel feeler gauge or knife blade between the Hall layer and magnet. If the sensor is good, the voltmeter should read within 0.5 volt of battery voltage when the feeler gauge or knife blade is inserted and touching the magnet. When the feeler gauge or blade is removed, the voltage should read less than 0.5 volt.

It is often possible to observe the voltage levels of the Hall effect sensor using an oscilloscope. Set the scope on its low scale primary pattern position and connect the primary positive lead to the Hall signal lead using a weather pack jumper wire (Figure 12-55). The negative lead should connect to ground or the ground terminal at the sensor's connector.

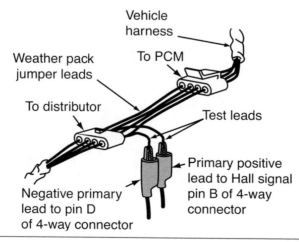

Vehicle harness

Weather pack jumper leads

To PCM

To distributor

Test leads

Negative primary lead to pin D of 4-way connector

Primary positive lead to Hall signal pin B of 4-way connector

Figure 12-55 Jumper leads in place with test leads for checking the Hall effect sensor.

Figure 12-56 Square wave trace produced by properly operating Hall effect sensor.

With the engine running, the pattern should show a square wave pattern ranging from approximately 0 to 12 volts (Figure 12-56). If the range is out of specifications or distorted, replace the sensor.

Control Module

Use an ohmmeter to insure that the control module connection to ground is good. As shown in Figure 12-57, one lead of the meter is connected to the ground terminal at the module. The other is connected to a good engine ground. Zero resistance indicates good continuity in the ground circuit. Any resistance reading during this test is unacceptable.

The most effective method of testing for a defective control module is to use an ignition module tester. This electronic tester evaluates and determines if the module is operating within a given set of design parameters. It does so by simulating normal operating conditions while looking for faults in key components. A typical ignition module tester performs the following tests:

❑ Shorted module test

❑ Cranking current test

❑ Key on, engine off current test

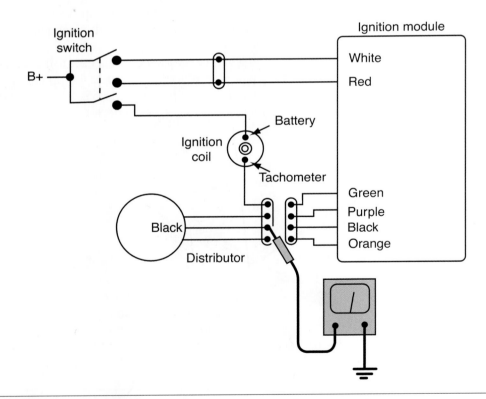

Figure 12-57 Using a voltmeter to check for proper module grounding.

❏ Idle current test

❏ Cruise current test

❏ Cranking primary voltage test

❏ Idle primary voltage test

❏ Cruise primary voltage test

Some ignition module testers are also able to perform an ignition coil-spark test (actually firing the coil) and a distributor pickup test.

Test selection is made by pushing the appropriate button. The module tester usually responds to these tests with a pass or fail indication.

Unfortunately, many module testers are designed to troubleshoot specific makes and models of ignitions. Many shops find it impractical to have testers for every type of system they service.

Keep in mind that control modules are very reliable. They are also one of the most expensive ignition system components. So if a module tester is not available, check out all other system components before condemning the control module.

Testing Secondary Ignition Wires

Inspect all the spark plug wires and the secondary coil wire for cracks and worn insulation which cause high-voltage leaks. Inspect all the boots on the ends of the plug wires and coil secondary wire for cracks and hard, brittle conditions. Replace the wires and boots if they show evidence of these conditions. Some vehicle manufacturers recommend spark plug wire replacement only in complete sets.

The spark plug wires may be left in the distributor cap for test purposes, so the cap terminal connections are tested with the spark plug wires. Connect the ohmmeter leads from the end of a spark plug wire to the distributor cap terminal inside the cap to which the plug wire is connected (Figure 12-58).

If the ohmmeter reading is more than specified by the vehicle manufacturer, remove the wire from the cap and check the wire alone. If the wire has more resistance than specified, replace the wire. When the spark plug wire resistance is satisfactory, check the cap terminal for corrosion. Repeat the ohmmeter tests on each spark plug wire and the coil secondary wire.

Classroom Manual
Chapter 12, page 390

Normally the maximum allowable resistance across a spark plug wire is 10,000 ohms per foot of length.

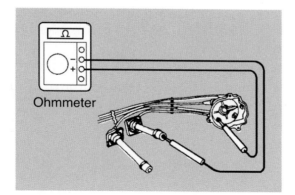

Ohmmeter

Figure 12-58 Ohmmeter connected to the spark plug wire and the distributor cap terminal to test the secondary circuit.

Stress Testing Components

Often, an intermittent ignition problem only occurs under certain conditions such as extremes in heat or cold, or during rainy or humid weather. Careful questioning of the customer should lead to determining if the problem is stress condition related. Does the problem occur on cold mornings? Does it occur when the engine is fully warmed up? Is it a rainy day problem? If the answer to any of these questions is positive, you can reproduce the same conditions in the shop during stress testing.

Cold Testing. With the scope on raster, cool major ignition components such as the control module, pickup coil, and major connections one at a time using a liquid cool-down agent.

> ⚠ **WARNING:** When using cool-down sprays, wear eye protection and avoid spraying your skin or clothing. Use extreme caution.

After cooling a component, watch the pattern for any signs of malfunction, particularly in the dwell zone. If there is no sign of malfunction, cool down the next component after the first has warmed to normal operating temperature. Cooling (or heating) more than one component at a time provides inconclusive results.

Heat Testing. To heat stress components, use a heat gun or hair dryer to direct hot air into the component. Heat guns intended for stripping paint and other household jobs can become extremely hot and melt plastic, wire insulation, and other materials. Use a moderate setting and proceed cautiously. Look for changes in the dwell section of the trace, particularly in the variable dwell or current limiting areas. If connections appear to be the problem, disconnect them, clean the terminals, and coat them with dielectric compound to seal out dirt and moisture.

Moisture Testing. A wet stress test is performed by lightly spraying the components, coil and ignition cables, and connections with water. Do not flood the area; a light mist does the job. A scope set on raster or display helps pinpoint problems, but it is often possible to hear and feel the miss or stutter without the use of a scope. As with heat and cold testing, do not spray down more than one area at a time or results could be misleading. If you suspect a poor connection, clean and seal it, then retest it.

> ● **CUSTOMER CARE:** The term tune-up essentially means replacing spark plugs and other ignition parts and adjusting the air-fuel mixture and setting the timing. Fuel injection systems, and electronic ignition eliminated these once common service procedures over a decade ago. Today, the computer-controlled system normally makes these adjustments. Unfortunately, your customers have probably not kept abreast of these changes. Many shop owners state that over half of the customers coming in for tune-ups still think their vehicles have breaker points.

When a customer approaches you for a standard tune-up, it is the ideal time to explain in simple terms just how the vehicle operates and what is involved in a modern tune-up. These procedures should include these points:

- ❏ An initial test drive with the customer
- ❏ Spark plug change
- ❏ Compression test or dynamic compression test
- ❏ Visual inspection of ignition system
- ❏ Base timing check and idle speed check
- ❏ Inspect and adjust timing belt (if applicable)
- ❏ Inspection of vacuum lines, air intake ducts, and PCV valve operation
- ❏ Check operation of EGR valve and oxygen sensor
- ❏ Pull computer trouble codes—if easily accessible; scan data and verify inputs and outputs
- ❏ Replace air and fuel filters if the service interval is reached

- ❏ Inspect all belts and hoses
- ❏ Check fan operation
- ❏ Inspect water pump seal
- ❏ Service battery terminals and top all fluids
- ❏ Emission control maintenance

Spark Plug Service

Spark plug service is a vital part of a complete engine tune-up. Vehicle manufacturers' recommendations for service intervals can range anywhere between 20,000 and 100,000 miles, depending on a variety of factors:

- ❏ The type of ignition system
- ❏ Engine design
- ❏ Spark plug design
- ❏ Vehicle operating conditions
- ❏ The type of fuel used
- ❏ Types of emission control devices used

Classroom Manual
Chapter 12, page 389

Regardless of what other tools are used, a spark plug socket is essential for plug removal and installation. Spark plug sockets are available in three sizes: 13/16-inch (for 14-millimeter gasketed and 18-millimeter tapered-seat plugs), 5/8-inch (for 14-millimeter tapered-seat plugs), and 9/16-inch. They can be either 3/8- or 1/2-inch drive, and many feature an external hex so that they can be turned using an open end or box wrench.

A spark plug socket has an internal rubber bushing to prevent plug insulator breakage. When the spark plugs are removed, they should be set in order so the technician can identify the spark plug from each cylinder. Spark plug carbon conditions are an excellent indicator of cylinder conditions. If the spark plugs all have light brown or gray carbon deposits, the cylinders have been operating normally with a proper air-fuel ratio.

To properly remove an engine's spark plugs:

1. Remove the cables from each plug, being careful not to pull on the cables. Instead, grasp the boot and twist it off gently.
2. Using a spark plug socket and ratchet, loosen each plug a couple of turns.
3. Use compressed air to blow dirt away from the base of the plugs.
4. Remove the plugs, making sure to remove the gasket as well (if applicable).

SERVICE TIP: To save time and avoid confusion later, use masking tape to mark each of the cables with the number of the plug it attaches to.

Inspecting Spark Plugs

Once the spark plugs have been removed, it is important to read them (Figure 12-59). In other words, inspect them closely, noting in particular any deposits on the plugs and the degree of electrode erosion. A plug in good working condition can still have minimal deposits on it. They

Platinum plugs should not be filed.

Fused spot deposit   Overheating   Carbon fouled   Pre-ignition

Figure 12-59 Normal and abnormal spark plug conditions.

are usually light tan or gray in color. However, there should be no evidence of electrode burning, and the increase of the air gap should be no more than 0.001 inch for every 10,000 miles of engine operation. A plug that exceeds this wear should be replaced and the cause of excessive wear repaired.

It is possible to diagnose a variety of engine conditions by examining the firing end of the spark plugs. If an engine is in good shape, they should all look alike. Whenever plugs from different cylinders look differently, a problem exists somewhere in the engine or its systems. The following are examples of plug problems and how they should be dealt with.

Cold Fouling. This condition is the result of an excessively rich air-fuel mixture. It is characterized by a layer of dry, fluffy black carbon deposits on the tip of the plug. **Cold fouling** is caused by a rich air-fuel mixture or an ignition fault causing the spark plug not to fire. If only one or two of the plugs show evidence of cold fouling, sticking valves are the likely cause. The plug can be used again, provided its electrodes are filed and the air gap is reset. Correct the cause of the problem before reinstalling or replacing the plugs.

Wet Fouling. When the tip of the plug is practically drowned in excess oil, this condition is known as **wet fouling**. In an overhead valve engine, the oil may be entering the combustion chamber past worn valve guides or valve guide seals. If the vehicle has an automatic transmission, a likely cause of wet-fouled plugs is a defective vacuum modulator that is allowing transmission fluid into the chamber. On high-mileage engines, check for worn rings or excessive cylinder wear. The best solution is to correct the problem and replace the plugs with the specified type.

Splash Fouling. Splash fouling occurs immediately following an overdue tune-up. Deposits in the combustion chamber, accumulated over a period of time due to misfiring, suddenly loosen when the temperature in the chamber returns to normal. During high-speed driving, these deposits can stick to the hot insulator and electrode surfaces of the plug. These deposits can actually bridge across the gap, stopping the plug from sparking. Normally, splash-fouled plugs can be cleaned and reused.

Gap Bridging. A plug with a bridged gap is rarely seen in automobile engines. **Gap bridging** occurs when flying carbon deposits within the combustion chamber accumulate over a long period of stop-and-go driving. When the engine is suddenly placed under a hard load, the deposits melt and bridge the gap, causing misfire. This condition is best corrected by replacing the plug.

Glazing. Under high-speed conditions, the combustion chamber deposits can form a shiny, yellow glaze over the insulator. When it becomes hot enough, the glaze acts as an electrical conductor causing the current to follow the deposits and short out the plug. **Glazing** can be prevented by avoiding sudden wide-open throttle acceleration after sustained periods of low-speed or idle operation. Because it is virtually impossible to remove glazed deposits, glazed plugs should be replaced.

Overheating. This condition is characterized by white or light gray blistering of the insulator. There may also be considerable electrode gap wear. **Overheating** can result from using too hot a plug, over-advanced ignition timing, detonation, a malfunction in the cooling system, an overly lean air-fuel mixture, using too-low octane fuel, an improperly installed plug, or a heat-riser valve that is stuck closed. Overheated plugs must be replaced.

Turbulence Burning. When **turbulence burning** occurs, the insulator on one side of the plugs wears away as the result of normal turbulence in the combustion chamber. As long as the plug life is normal, this condition is of little consequence. However, if there is a larger than normal air gap, overheating can be the problem. In this case, check for all of the common causes of overheating discussed previously.

Preignition damage is caused by excessive engine temperatures. Preignition damage is characterized by melting of the electrodes or chipping of the electrode tips. When this problem

If **cold fouling** is present on a vehicle that operates a great deal at idle and low speeds, plug life can be lengthened by using hotter spark plugs.

Wet fouling is a term describing the tip of the spark plug being drowned in excess oil.

Splash fouling refers to loosened deposits found in the combustion chamber due to misfiring.

Gap bridging occurs when carbon deposits accumulate in the combustion chamber due to stop-and-go driving.

Glazing takes place under high speed conditions and can form a glaze over the insulator from combustion chamber deposits.

occurs, look for the general causes of engine overheating, including over-advanced ignition timing, a burned head gasket, and using too-low octane fuel. Other possibilities include loose plugs or using plugs of the improper heat range. Do not attempt to reuse plugs with preignition damage.

A sure sign of reversed coil polarity damage is a slight dishing of the ground electrode. The center electrode does not normally wear badly. Misfiring and rough idling may also be present. In addition to reversal of the coil primary leads, older vehicles can experience reverse coil polarity damage by reversing the battery polarity.

Regapping Spark Plugs

Both new and used spark plugs have their air gaps set to the engine manufacturer's specifications. Use an approved tool not only to measure the gap, but also to bend the side electrode to make the adjustment. The combustion gauge and adjusting tool shown in Figure 12-60 is designed for use on new plugs only. This is because it utilizes flat gauges for the adjustment procedure. The gauges are mounted on the tool like spokes on a wheel. Above the gauges is an anvil, which is used to apply pressure to the electrode. On the opposite end of the tool is a curved seat. This seat performs two functions: it supports the plug shell during the procedure. It also compresses the ground electrode against the gauge, thus setting the air gap.

The tapered regapping tool is simply a piece of tapered steel with leading and trailing edges of different dimensions. Between these two points, the gauge varies in thickness. A scale, located above the gauge, indicates the thickness at any given point. When the gauge is slid between electrodes, it stops when the air gap size reaches the thickness on the gauge. The scale reading is made in thousandths of an inch. Adjusting slots are available to bend the ground electrode as needed to make the air gap adjustment.

In many instances, the electrodes on a used spark plug are no longer flat, rendering a flat gauge useless for regapping purposes. Instead, a round wire gauge is required (Figure 12-61). Combination round and flat feeler gauge sets are available. These multipurpose tools can be used to check air gaps, adjust contact points, bend electrodes, and fire contact points and electrodes.

When regapping a spark plug, always check the air gap of a new spark plug before installing it. Never assume the gap is correct just because the plug is new. Do not try to reduce a plug's air gap by tapping the side electrode on a bench. Never attempt to set a wide gap, electronic ignition type plug to a small gap specification. Likewise, never attempt to set a small gap, breaker point ignition type plug to the wide gap necessary for electronic ignitions. In either case, damage to the electrodes results. Never try to bend the center electrode to adjust the air gap. This cracks the insulation.

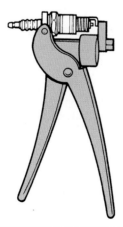

Figure 12-60 Combination spark plug gauge and adjusting tool.

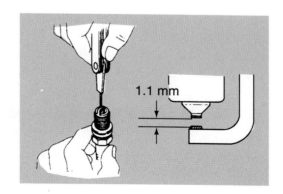

1.1 mm

Figure 12-61 Measuring spark plug gap.

Plug Type	Cast-iron Head	Aluminum Head
14-mm Gasket	25 to 30	15 to 22
14-mm Tapered Seat	7 to 15	7 to 15
18-mm Tapered Seat	15 to 20	15 to 20

Figure 12-62 Plug installation torque values (in foot pounds).

To properly install spark plugs:

1. Wipe dirt and grease from the plug seats with a clean cloth.
2. Be sure the gaskets on gasketed plugs are in good condition and are properly placed on the plugs. If reusing a plug, install a new gasket on it. Be sure that there is only one gasket on each plug.
3. Adjust the air gap as needed.
4. Install the plugs and finger tighten. If the plugs cannot be installed easily by hand, the threads in the cylinder head may require cleaning with a thread-chasing tap. Be especially careful not to cross-thread the plugs when working with aluminum heads.
5. Tighten the plugs with a torque wrench following the vehicle manufacturer's specifications or the values listed in Figure 12-62.
6. Use an anti-seize compound on spark plugs installed in aluminum cylinder heads.

CAUTION: If thread lubricant is used, reduce the torque setting. Keep in mind that many spark plug manufacturers do not recommend the use of thread lubricant.

Spark Plug Wire Installation

When spark plug wires are being installed, make sure they are routed properly as indicated in the vehicle's service manual. When removing the spark plug wires from a spark plug, grasp the spark plug boot tightly and twist while pulling the cable from the end of the plug. When installing a spark plug wire, make sure the boot is firmly seated around the top of the plug, then squeeze the boot to expel any air that is trapped inside.

Two spark plug wires should not be placed side by side for a long span if these wires fire one after the other in the cylinder firing order. When two spark plug wires that fire one after the other are placed side by side for a long span, the magnetic field from the wire that is firing builds up and collapses across the other wire. This magnetic collapse may induce enough voltage to fire the other spark plug and wire when the piston in this cylinder is approaching TDC on the compression stroke. This action may cause detonation and reduced engine power.

Ignition Module Removal and Replacement

Classroom Manual Chapter 12, page 391

The ignition module removal and replacement procedure varies depending on the ignition system. Always follow the procedure in the vehicle manufacturer's service manual. As an example, follow these steps for module removal and replacement on an HEI distributor:

CAUTION: Lack of silicone grease on the module mounting surface may cause module overheating and damage.

1. Remove the battery wire from the coil battery terminal and remove the inner wiring connector on the primary coil terminals. Remove the spark plug wires from the cap.

2. Rotate the distributor latches one-half turn and lift the cap from the distributor.

3. Remove the two rotor retaining bolts and the rotor.

4. Remove the primary leads and the pickup leads from the module.

5. Remove the two module mounting screws and remove the module from the distributor housing.

6. Wipe the module mounting surface clean and place a light coating of silicone heat-dissipating grease on the module mounting surface.

7. Install the module and tighten the module mounting screws to the specified torque.

8. Install the primary leads and pickup leads on the module.

9. Be sure the lug on the distributor shaft fits into the rotor notch while installing the rotor and tighten the rotor, mounting screws to the specified torque.

10. Install the distributor cap and be sure the projection in the cap fits in the housing notch.

11. Push down on the cap latches with a screwdriver and rotate the latches until the lower part of the latch is hooked under the distributor housing.

12. Install the coil primary leads and battery wire on the coil terminals. Be sure the notch on the primary leads fits onto the cap projection. Install the spark plug wires.

Distributor Service

Often it is required to remove the distributor from the engine to perform the necessary services. This is especially true if the distributor shaft bushings are worn or the armature is damaged. Other service procedures that are not related to the ignition system may also require that the distributor be removed. All service procedures for a distributor depend on the manufacturer and design of the distributor. Always follow the recommended procedures given in the service manual.

Before removing the distributor, check the condition of the distributor's bushing. Do this by grasping the distributor shaft and moving it toward the outside of the distributor. If any movement is detected, remove the distributor and check the bushing on the bench. A typical procedure for removing a distributor follows:

1. Disconnect the negative battery cable.

2. Disconnect the distributor wiring connector and the vacuum advance hose, if so equipped.

3. Remove the distributor cap and note the position of the rotor. On some vehicles it may be necessary to remove the spark plug wires from the cap prior to cap removal.

4. Note the position of the vacuum advance, then remove the distributor holddown bolt and clamp (Figure 12-63).

5. Pull the distributor from the engine. Most distributors will need to be twisted as they are pulled out of their bore.

6. Once the distributor is removed, install a shop towel in the distributor opening to keep foreign material out of the engine block.

The armature is also called the trigger wheel, pole piece, or reluctor.

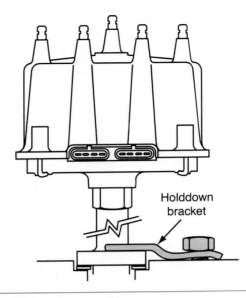

Figure 12-63 Distributor holddown assembly.

Distributor Bushing Check

Special Tools

Soft-jawed vise, pin punch, paint stick or chalk, hand-operated vacuum pump

Before proceeding with the disassembly of the distributor, the shaft's bushings need to be checked. Do this regardless of the type of repair you intended for the distributor. Lightly clamp the distributor housing in a soft-jaw vise. Clamp a dial indicator on the top of the distributor housing. Position the plunger of the indicator so that it rests on the distributor shaft. When the shaft is pushed horizontally, observe the movement of the shaft on the indicator. Compare this movement to the specifications given in the service manual. If the movement exceeds the allowed amount, the distributor bushings and/or shaft are worn. Many manufacturers recommend complete distributor replacement rather than bushing or shaft replacement.

Distributor Disassembly

The procedure for disassembling a distributor will vary slightly according to the design of the distributor. The following steps are a typical distributor disassembly procedure:

1. Mark the distributor drive gear in relation to the distributor shaft so the gear may be reinstalled in its original position (Figure 12-64).

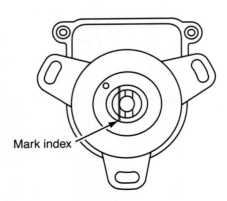

Figure 12-64 Mark the position of the distributor drive gear in relation to the distributor shaft.

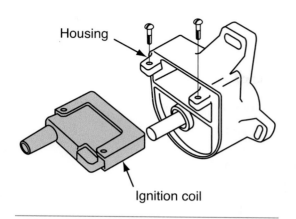

Figure 12-65 On some systems, the ignition coil must be removed from the housing before disassembling the distributor.

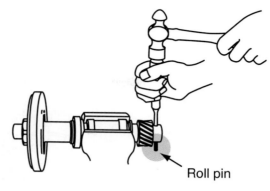

Figure 12-66 Drive out the roll pin to remove the drive gear.

SERVICE TIP: It may be necessary to remove the ignition coil assembly from the distributor prior to removing the distributor shaft (Figure 12-65). Always refer to the manufacturer's procedure for disassembling a distributor.

2. Support the distributor housing in a soft-jaw vise and drive the roll pin from the gear and shaft with a pin punch and hammer (Figure 12-66).

3. Pull the gear from the distributor shaft and remove any spacers between the gear and the housing. Note the position of these spacers so they may be reinstalled in their original position. On some distributors, it may be necessary to use a hydraulic press and fixtures to remove the drive gear (Figure 12-67.

4. Wipe the lower end of the shaft with a shop towel and inspect this area of the shaft for metal burrs. Remove any burrs with fine emery paper.

5. Pull the distributor shaft from the housing.

Sometimes the shutter for the Hall effect unit is part of the rotor assembly.

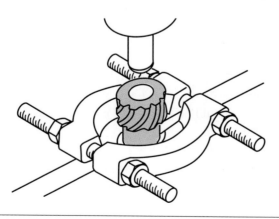

Figure 12-67 On some distributors, the drive gear must be pressed off.

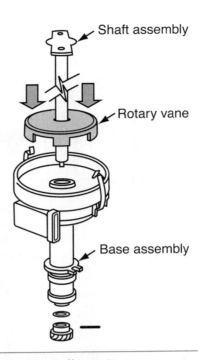

Figure 12-68 Distributor with Hall effect unit.

6. Remove the pickup coil leads from the module and the pickup retaining clip. Lift the pickup coil from the top of the distributor. On some distributors it may be necessary to detach the Hall effect shutter wheel assembly from the distributor shaft (Figure 12-68) before removing the shaft from the distributor housing.

Distributor Inspection

CAUTION: Distributor electrical components may be damaged by washing them in solvent.

Check these items during a typical distributor inspection:

1. Inspect all lead wires for worn insulation and loose terminals. Replace these wires as necessary.
2. Inspect the distributor shaft for wear. Replace the shaft assembly, if necessary.
3. Check the distributor gear for worn or chipped teeth.
4. Inspect the reluctor for damage. If the high points are damaged, the distributor bushing is probably worn, allowing the high points to hit the pickup coil.
5. The entire assembly should be cleaned thoroughly. Make sure all solvent residue is completely removed.

Distributor Assembly

After the complete assembly has been inspected, cleaned, and any service is completed, follow these steps to assemble the distributor:

1. Install the pickup coil and the retaining clip. Connect the pickup leads to the module.
2. Install the module and mounting screws. Many manufacturers require that silicone lubricant be applied to the connectors in the primary circuit and/or to the mounting

base of the ignition module. Some systems, such as the Ford Dura-Spark system, should have the silicone applied to the end of the rotor and on the spark plug terminals inside the distributor. Check the service manual for the recommendations.

3. Place some bushing lubricant on the shaft and install the shaft in the distributor.

4. Install the spacers between the housing and gear in their original position.

5. Install the gear in its original position and be sure the hole in the gear is aligned with the hole in the shaft.

6. Support the housing on top of a vise and drive the roll pin into the gear and shaft.

7. Install a new O-ring or gasket on the distributor housing.

Installing and Timing the Distributor

Photo Sequence 24 shows the correct way to reinstall a distributor on a typical engine. The following procedure may be followed to install the distributor and time it to the engine if all reference marks were made during the removal of the distributor:

1. Lubricate the O-ring on the distributor shaft.

2. Position the rotor so that it is aligned with the mark made to the distributor housing prior to removal.

3. Align the distributor housing to the mark made on the engine block during removal.

4. Lower the distributor into the engine block, making sure the distributor drive is fully seated. Then make sure the distributor housing is fully seated against the engine block. Sometimes it may be necessary to wiggle or rock the distributor to seat it fully into the drive gear. Distributors with **drive lugs** (Figure 12-69) must be mated with the drive grooves in the camshaft. Both are offset to eliminate the possibility of installing the distributor 180 degrees out of time.

Drive lugs are protrusions at the end of the shaft that seat in mating grooves in the end of the camshaft.

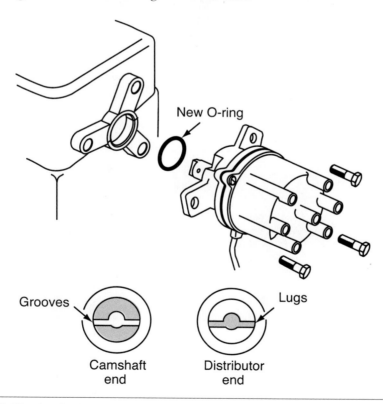

Figure 12-69 The distributor drive must be aligned with the camshaft drive to seat the distributor in the engine block or cylinder head.

Photo Sequence 24
Timing the Distributor to the Engine

P24-1 Remove the number one spark plug.

P24-2 Place your thumb over the number one spark plug opening and crank the engine until compression is felt.

P24-3 Crank the engine a very small amount at a time until the timing marks indicate that the number one piston is at TDC on the compression stroke.

P24-4 Determine the number one spark plug wire position in the distributor cap.

P24-5 Install the distributor with the rotor under the number one spark plug wire terminal in the distributor cap and one of the reluctor high points aligned with the pickup coil.

P24-6 After the distributor is installed in the block, turn the distributor housing slightly so the pickup coil is aligned with the reluctor.

P24-7 Install the distributor clamp bolt, but leave it slightly loose.

P24-8 Connect the pickup leads to the wiring harness.

P24-9 Install the spark plug wires in the cylinder firing order and in the direction of distributor shaft rotation. Set ignition timing.

Many distributors are not equipped with drive lugs but with an oil pump drive shaft which can be installed 180 degrees out of timing. Make certain that the number one cylinder is on TDC compression with this type of distributor drive when you install the distributor.

☑ **SERVICE TIP:** Distributors equipped with a helical drive gear will rotate as the distributor is being installed, causing the distributor to move away from the reference marks. Pay attention to how much the rotor moves, then remove the distributor and move the rotor backwards the same amount. This should allow the shaft to rotate while the distributor is being installed and still be aligned with the reference marks.

5. Rotate the distributor a small amount so the timer core teeth and pickup teeth are aligned.
6. Install the distributor holddown clamp and bolt, and leave the bolt slightly loose.
7. Install the spark plug wires in the direction of distributor shaft rotation and in the cylinder firing order.
8. Connect the distributor wiring connectors.
9. Take the steps necessary to place the vehicle in "base" timing.
10. Set the timing according to manufacturer recommendation.
11. Restore the timing (connectors, jumper wires, etc.) back to normal.
12. Test drive the vehicle to verify the repair.

Ignition Timing

Although the computer assumes control of ignition timing, having the correct base timing is critical for the proper operation of the engine. Because the computer bases its control or change in ignition timing on the base setting, if the base is wrong, all other ignition timing settings will also be wrong. On some systems, the base timing is adjustable. Always refer to the appropriate service manual before checking and attempting to set the ignition timing on an engine with an electronic control system.

Each ignition system has its own set of procedures to check ignition timing. Always refer to the vehicle's emissions underhood label or the appropriate service manual before proceeding. These will give the correct procedure for disabling the computer's control of the timing, the correct timing specifications, and the conditions that must be present when checking or adjusting base ignition timing. Some engines will have two reference marks on the crankshaft pulley. The instructions will tell you which one to use for checking ignition timing. Either an inductive or magnetic probe timing light should be used to observe the timing marks as the engine is run at the specified speed (Figure 12-70).

When the distributor is timed to the engine, the number one piston should be at TDC on the compression stroke with the timing marks aligned and the rotor should be under the number one spark plug wire terminal in the distributor cap with one of the high points aligned with the pick-up coil.

Special Tools

Jumper wires, timing light

The underhood decal is called the Vehicle Emissions Control Information (VECI) label.

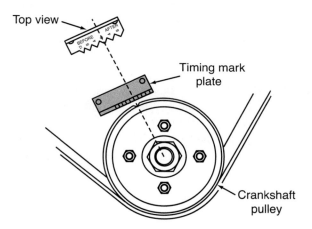

Figure 12-70 When the timing light flashes, the timing mark on the crankshaft pulley must appear at the specified location on the timing indicator above the pulley.

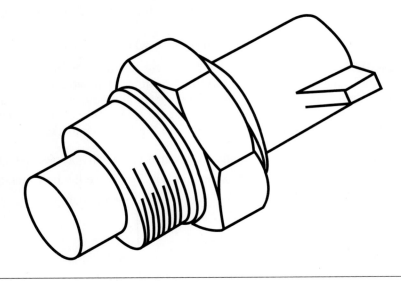

Figure 12-71 A knock (detonation) sensor.

It is important to remember most computerized engine control systems use a knock sensor (Figure 12-71) to retard the ignition timing when detonation occurs. The use of the sensor allows the PCM to set ignition timing with the most advance possible. If the knock sensor is faulty, the engine will lack performance due to lost timing advance, or the engine will have a heavy spark knock because the timing is too far advanced. This spark control system (Figure 12-72) should be checked while checking the ignition system's base timing.

Because so many different types of engine control and ignition systems exist, the exact procedure for checking the timing depends entirely on what systems the engine is equipped with. For example, on Ford EEC-IV systems, there is a base timing or SPOUT connector that needs to be disconnected (Figure 12-73). Doing this prevents the PCM from controlling the timing and keeps the ignition at its base setting. The self-test mode of the EEC-IV system also allows timing advance to be monitored. Again, this procedure varies, and the service manual should be referred to for the correct procedure.

Typically, GM vehicles with fuel injected engines require that an underhood test connector be grounded by connecting two terminals in the DLC or disconnecting a connector at the distributor. Most carbureted engines require the disconnecting of a four-wire terminal at the distributor (Figure 12-74). The latter disconnects the ignition module from the computer but cannot be disconnected on fuel injected engines.

Vehicles with adjustable timing have several methods to place the vehicle into "base" timing. *Always* check for specific information for the vehicle in question. Do not assume that every make, model, and year are the same. For instance, some Toyotas require that two DLC terminals be connected (Figure 12-75). Some Hondas are checked at idle in park or neutral if equipped with a manual transmission. A scan tool is connected and a set timing mode must be entered if checking the timing on some Mitsubishi models. Additionally, if this mode is selected, the technician must exit properly or the vehicle will stay in base timing for a period of time, possibly causing engine damage.

On many vehicles, such as General Motors and Ford vehicles, a timing connector must be disconnected while checking basic timing.

Honda's programmed ignition system is referred to as the PGM-IG system.

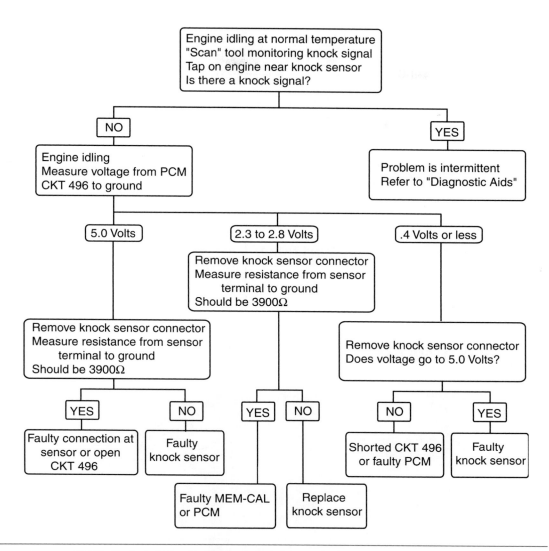

Figure 12-72 Typical electronic spark control diagnostic chart.

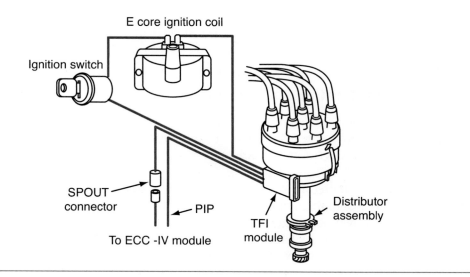

Figure 12-73 TFI system with SPOUT connector.

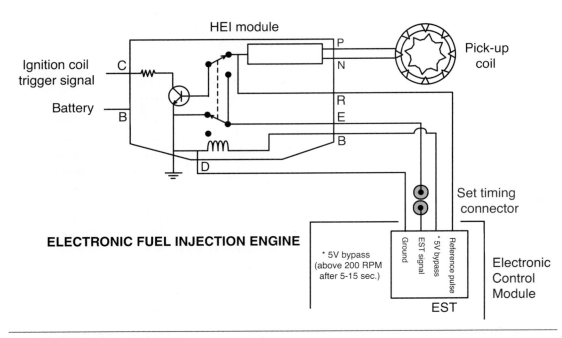

Figure 12-74 GM's set-timing connector from a fuel injected engine.

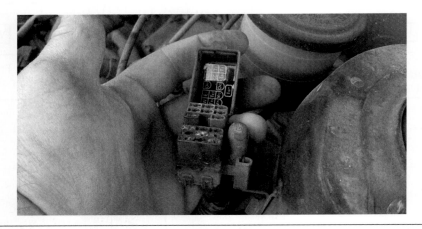

Figure 12-75 Some Toyotas require TI and EI terminals be connected in the DLC to read base timing.

Making Base Timing Adjustments

A **timing meter** shows timing degrees and engine rpm.

Once you have obtained a base timing reading, compare it to manufacturer's specifications. For example, if the specification reads 10 degrees before top dead center (Figure 12-76A) and the reading found is 3 degrees before top dead center (Figure 12-76B), the timing is retarded or off by 7 degrees.

This means the timing must be advanced by 7 degrees to make it correct. To do this, rotate the distributor until the timing marks align at 10 degrees (Figure 12-77), then retighten the distributor holddown bolt (Figure 12-78).

Many late-model vehicles have a magnetic timing probe receptacle near the timing indicator (Figure 12-79). Some equipment manufacturers supply a magnetic **timing meter** with a pickup that fits into the probe hole. The meter pickup must be connected to the number one spark plug wire, and the power supply leads must be connected to the battery terminals. Many timing meters have two scales, timing degrees and engine rpm.

Timing marks aligned at 10

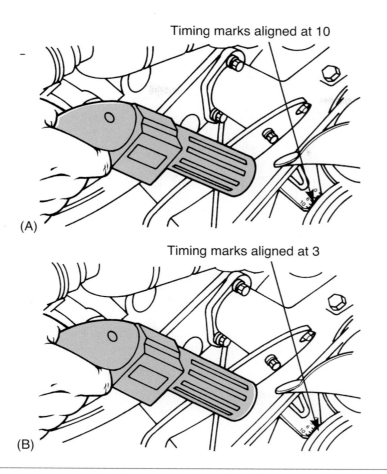

(A)

Timing marks aligned at 3

(B)

Figure 12-76 (A) Timing marks illuminated by a timing light at 10 degrees BTDC, and (B) timing marks at 3 degrees BTDC.

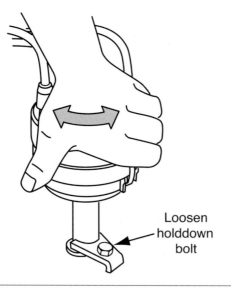

Loosen holddown bolt

Figure 12-77 Turning the distributor to set timing.

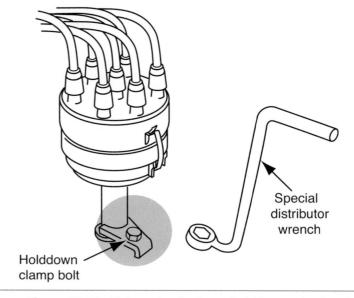

Holddown clamp bolt

Special distributor wrench

Figure 12-78 Tighten the distributor holddown bolt after the timing is adjusted to specifications.

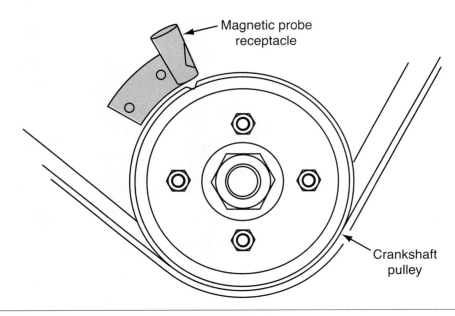

Figure 12-79 Magnetic timing probe receptacle near the timing indicator.

An **offset knob** is adjusted in order for the engine to be started and timing to be indicated on the meter scale.

A magnetic timing **offset knob** on the meter must be adjusted to the manufacturer's specifications to compensate for the position of the probe receptacle. Once the offset is adjusted, the engine may be started and the timing is indicated on the meter scale. The timing adjustment procedure is the same with the timing meter or timing light.

Computer-Controlled Ignition System Service

To check the base timing, the manufacturers give specific instructions for readying the engine. Some require disconnecting certain components, such as many Ford and GM systems which require that a connector located in the engine compartment be disconnected. The emission's decal may provide information about the location of this connector. It is important that these instructions be followed so as to disable the computer's control over ignition timing. After these preparation steps are followed, the engine may be timed in the normal manner.

Distributor pickups can be tested in the way discussed earlier. It is important that all wires and connectors between the distributor and module and module to the engine control computer be visually checked as well as checked for excessive resistance with an ohmmeter.

If the resistance checks are within specifications, the circuit should be checked with a digital voltmeter. Turn the ignition on and connect the voltmeter across the voltage input wire and ground. Compare the reading to specifications. The voltmeter should also be used to check the resistance across the ground circuit of the distributor. Do this by measuring the voltage drop across the circuit; a drop of more than 0.1 volt is excessive.

Connect the voltmeter from the pickup signal wire to ground. If the voltmeter does not fluctuate while cranking the engine, the pickup is malfunctioning. A good pickup will have readings from nearly 0 volts to 9 to 12 volts.

On Chrysler optical distributors, the pickup voltage supply and ground wires may be tested at the four-wire connector near the distributor. With the ignition switch on, connect the voltmeter from the orange voltage supply wire to ground. The reading should be 9.2 to 9.4 volts. Because the expected reading may vary with the model year, always refer to the manufacturer's service manual before conducting the test.

When the voltmeter is connected between the black/light blue ground wire and an engine ground, the reading should be less than 0.2 volt. When the voltmeter is connected from the gray/black reference pickup wire or the tan/yellow SYNC pickup wire to an engine ground, the voltmeter should cycle from nearly 0 volts to 5 volts while the engine is cranking. If the pickup signal is not within specifications, the pickup is defective. A defective SYNC pickup in this distributor should not cause a no-start problem but will affect engine performance.

✔ **SERVICE TIP:** After adjusting the initial timing on computer-controlled systems, be sure to check for, and clear, any trouble codes that may have been set in the computer's memory. Anytime a connector or component is disconnected for initial timing adjustment, a code will likely be stored.

CASE STUDY

A customer complains that his vehicle runs poorly, misfiring and losing power. He is embarrassed because he just performed his first ever preventive maintenance job on the vehicle—replacing spark plugs and ignition cables. He insists he did the job correctly. The problem must be somewhere else in the system.

Since the problem did not occur until after the customer worked on the vehicle, the technician begins troubleshooting by inspecting the plugs and cables. The job appears to be carefully and neatly done. The technician pulls a plug and inspects it. It is properly gapped, cleaned, and has been installed snugly, but not too tightly. The cables are not top quality but have been carefully and neatly installed and routed. Too neatly the technician suspects. At one point, three ignition cables have been wrapped together with several turns of electrical tape. It makes for a neat, tight appearance, but may be causing an inductive crossfire situation. The high secondary voltage in a firing plug's ignition cable may be inducing a voltage in the wire running parallel to it causing it to fire a plug out of turn. The technician removed the tape and routes the wire according to service manual instructions. A test drive confirms this was the problem. Neatly wrapped and routed plug wire may look nice, but engine performance problems could result.

Terms to Know

Carbon tracking	Firing voltage	Spark duration
Cold fouling	Gap bridging	Spark line
Corona effect	Glazing	Splash fouling
Current limiting	Intermediate section	Superimposed pattern
Display pattern	Offset knob	Timing meter
Distributor ignition (DI)	Overheating	Turbulence burning
Drive lugs	Parade pattern	Wet fouling
Electromagnetic interference (EMI)	Raster pattern	

ASE-Style Review Questions

1. Upon inspection, a spark plug reveals a layer of dry, fluffy black carbon deposits on its tip.
   Technician A says the engine may be running too cold.
   Technician B says the mixture may be too rich.
   Who is correct?
   A. A only **C.** Both A and B
   B. B only **D.** Neither A nor B

2. The firing line on an oscilloscope pattern extends below the waveform.
   Technician A says the polarity of the coil is reversed.
   Technician B says the problem is fouled spark plugs.
   Who is correct?
   A. A only **C.** Both A and B
   B. B only **D.** Neither A nor B

3. The firing lines on an oscilloscope pattern are all abnormally low.
   Technician A says the problem could be low coil output.
   Technician B says the problem could be an overly rich air-fuel mixture.
   Who is correct?
   A. A only **C.** Both A and B
   B. B only **D.** Neither A nor B

4. *Technician A* says a Hall effect switch can be used to trigger an ignition module.
   Technician B says a Hall effect switch produces a digital output signal.
   Who is correct?
   A. A only **C.** Both A and B
   B. B only **D.** Neither A nor B

5. *Technician A* says spark plug wires have the ability to generate electromagnetic interference.
   Technician B says the coils in an AC generator have the ability to generate electromagnetic interference.
   Who is correct?
   A. A only **C.** Both A and B
   B. B only **D.** Neither A nor B

6. While discussing lean air-fuel mixtures:
   Technician A says leaner mixtures decrease the electrical resistance inside the cylinder and decrease the required firing voltage.
   Technician B says leaner mixtures shorten the time available for spark duration.
   Who is correct?
   A. A only **C.** Both A and B
   B. B only **D.** Neither A nor B

7. While discussing no-start diagnosis with a test spark plug:
   Technician A says if the test light flickers at the coil tach terminal, but the test spark plug does not fire when connected from the coil secondary wire to ground with the engine cranking, the ignition coil is defective.
   Technician B says if the test spark plug fires when connected from the coil secondary wire to ground with the engine cranking but the test spark plug does not fire when connected from the spark plug wires to ground, the cap or rotor is defective.
   Who is correct?
   A. A only **C.** Both A and B
   B. B only **D.** Neither A nor B

8. While discussing ignition coil ohmmeter tests:
   Technician A says the primary winding should read several hundred ohms.
   Technician B says the secondary resistance cannot be measured because there is only one connection at the output terminal.
   Who is correct?
   A. A only **C.** Both A and B
   B. B only **D.** Neither A nor B

9. While discussing timing of the distributor to the engine with the number one piston at TDC compression and the timing marks aligned:
   Technician A says the distributor must be installed with the rotor under the number one spark plug terminal in the distributor cap and one of the reluctor high points aligned with the pickup coil.
   Technician B says the distributor must be installed with the rotor under the number one spark plug terminal in the distributor cap and the reluctor high points out of alignment with the pickup coil.
   Who is correct?
   A. A only **C.** Both A and B
   B. B only **D.** Neither A nor B

10. While discussing basic ignition timing adjustment on vehicles with computer-controlled distributor ignition (DI):
    Technician A says on some DI systems a timing connector must be disconnected.
    Technician B says the distributor must be rotated until the timing mark appears at the specified location on the timing indicator.
    Who is correct?
    A. A only **C.** Both A and B
    B. B only **D.** Neither A nor B

ASE Challenge Questions

1. Two technicians are discussing the comparison between spark duration and firing kV.
   Technician A says if firing kV is high, duration will be short.
   Technician B says if firing kV is low, duration will be long.
   Who is correct?
   A. A only
   B. B only
   C. Both A and B
   D. Neither A nor B

2. A scope pattern indicates an irregular early firing of number 3 spark plug. The firing order is 1-4-3-2. This may be caused by:
   A. a bad ignition coil.
   B. the number 2 plug wire lying against number 3 wire.
   C. a shorted number 3 spark plug.
   D. the number 3 plug wire lying against number 4 wire.

3. An engine with electronic spark control (ESC) develops a pinging during acceleration.
   Technician A says the timing may be advanced too much.
   Technician B says pinging may be caused by not following proper service procedures.
   Who is correct?
   A. A only
   B. B only
   C. Both A and B
   D. Neither A nor B

4. An ESC-equipped engine quit running. It cranks strong, but will not start. The problem may be:
   A. a ground in the secondary circuit.
   B. corrosion on the battery's positive cable.
   C. no silicone grease on the ignition module.
   D. either A or C.

5. Spark plugs are being discussed.
   Technician A says spark plugs sold by the vehicle's dealership do not need the gap checked before installation.
   Technician B says any type plug can be used in ESC systems if the reach, heat range, and the gap match specifications.
   Who is correct?
   A. A only
   B. B only
   C. Both A and B
   D. Neither A nor B

Job Sheet 30

30

Name _____ Date _____

Scope Testing an Ignition System

Upon completion of this job sheet, you should be able to observe the ignition system activity by observing it on a scope.

ASE Correlation

This job sheet is related to the ASE Engine Performance Test's content areas: general engine diagnosis and computerized engine controls diagnosis and repair; tasks: diagnose mechanical, electrical, electronic, fuel, and ignition problems with an oscilloscope and engine diagnostic equipment; determine needed action.

Tools and Materials

A vehicle
Service manual for the above vehicle
A "tune-up" scope

Describe the vehicle being worked on.

Year _____ Make _____ VIN _____

Model_____

Engine type _____ Size _____ Firing order_____

Procedures

Task Completed

1. Describe the general appearance of the engine.

   _____

   _____

   _____

   _____

2. Connect the scope leads to the vehicle. ☐

3. Set the scope to look at the secondary ignition circuit in the display or parade pattern. ☐

4. Start the engine and observe the height of the firing lines. Are they within 3 kV of each other? _____ If not, which cylinders are the most different from the rest? _____ Are the heights of the firing lines all below 13 kV? _____ Is the height of the firing lines above 7 kV? _____ Based on these findings, what do you conclude?

   _____

   _____

   _____

   _____

5. Switch the scope to show the patterns in raster. Observe the length, height, and shape of the spark line for each cylinder. Describe them.

_____

_____

_____

_____

6. Based on these findings, what do you conclude?

_____

_____

_____

7. Describe the general appearance of the intermediate section of the pattern for each cylinder.

_____

_____

_____

_____

8. Based on these findings, what do you conclude?

_____

_____

_____

9. Describe the general appearance of the dwell section for each cylinder.

_____

_____

_____

_____

10. Based on these findings, what do you conclude?

_____

_____

_____

11. Based on the appearance of all sections of the scope pattern for each cylinder, what are your recommendations and conclusions?

_____

_____

_____

_____

Instructor's Response_____

_____

_____

Job Sheet 31

31

Name _____ Date _____

Testing an Ignition Coil

Upon completion of this job sheet, you should be able to test an ignition coil with an ohmmeter.

ASE Correlation

This job sheet is related to the ASE Engine Performance Test's content area: ignition system diagnosis; task: inspect and test ignition coil(s), and replace as needed.

Tools and Materials

A vehicle or a separate ignition coil   A DMM
Service manual for the above vehicle or ignition coil   A "tune-up" scope

Describe the vehicle being worked on.

Year_____ Make _____ VIN _____

Model_____

Procedures

Task Completed

1. Describe the general appearance of the coil.

   _____

   _____

2. Locate the resistance specifications for the ignition coil in the service manual.

   The primary winding should have _____ ohms of resistance.

   The secondary winding should have _____ ohms of resistance.

3. If the coil is still in the vehicle, disconnect all leads connected to it.   ☐

4. Connect the ohmmeter from the negative (tach) side of the coil to its container or frame. Observe the reading on the meter. The reading was _____ ohms. What does this indicate?

   _____

5. Connect the ohmmeter from the center tower of the coil to its container or frame. Observe the reading on the meter. The reading was _____ ohms. What does this indicate?

   _____

6. Connect the ohmmeter across the primary winding of the coil. The reading was _____ ohms. Compare this to specifications. What is indicated by this reading?

   _____

7. Connect the ohmmeter across the secondary winding of the coil. The reading was _____ ohms. Compare this to specifications. What is indicated by this reading?

   _____

8. Connect the scope and set the screen to raster or 5-ms sweep. Start the engine and observe the coil oscillations. The number of coils oscillations is _____ . Based on your observation, the coil's oscillation performance is _____ .

_____

9. Based on the above tests, what is your conclusion about the coil?

_____

Instructor's Response_____

_____

_____

Job Sheet 32

32

Name _____ Date _____

Individual Component Testing

Upon completion of this job sheet, you should be able to test components of the primary and secondary ignition system.

ASE Correlation

This job sheet is related to the ASE Engine Performance Test's content area: ignition system diagnosis and repair; tasks: inspect and test primary circuit wiring and components, and repair or replace as needed; inspect and test distributor and service as needed; and inspect and test ignition system secondary circuit wiring and components and replace as needed.

Tools and Materials

12-volt test light
DMM
Appropriate service manual
Lab scope

Describe the vehicle being worked on.

Year _____ Make _____ VIN _____

Model_____

Type of ignition system _____

Procedures

Task Completed

1. Check the following components:

   Ignition Switch

   a. Turn the ignition key off and disconnect the wire connector at the module. ☐

   b. Disconnect the S terminal of the starter solenoid to prevent the engine from cranking when the ignition is in the run position. ☐

   c. Turn the key to the run position. ☐

   d. With the test light, probe the red wire connection to check for voltage. Was there voltage?
      ☐ Yes ☐ No

   e. Check for voltage at the bat terminal of the ignition coil. Was there voltage?
      ☐ Yes ☐ No

   f. Turn the key to the start position and check for voltage at the start power wire connector at the module. Was there voltage?   ☐ Yes ☐ No

   g. Check for voltage at the bat terminal of the ignition coil. Was there voltage?
      ☐ Yes ☐ No

   Conclusions:

   _____

   _____

☐   **h.** Turn the ignition switch to the off position.

☐   **i.** Install a small straight pin into the appropriate module's input power wire.

☐   **j.** Connect the digital voltmeter's positive lead to the straight pin and ground the negative lead to the distributor base.

        k. Turn the ignition to the run position. Your voltage reading is _____

        l. Turn the ignition to the start position. Your voltage reading is _____

Conclusions:

_____

_____

Pickup Coil

☐   **a.** Turn the ignition off.

☐   **b.** Remove the distributor cap.

☐   **c.** Connect the ohmmeter to the pickup coil terminals.

        d. Record your readings. _____ ohms

        e. The specified resistance is _____ ohms.

Conclusions:

_____

_____

☐   **f.** Connect the ohmmeter from one of the pickup leads to ground.

        g. Record your readings. _____ ohms

        h. The specified resistance is _____ ohms.

Conclusions:

_____

_____

☐   **i.** Reinstall the distributor cap, then connect the scope leads to the pickup coil leads.

☐   **j.** Set the scope on its lowest scale.

☐   **k.** Spin the distributor shaft by cranking the engine with the ignition disabled.

        l. Describe the trace shown on the scope.

_____

Conclusions:

_____

_____

☐   **m.** Disconnect the scope.

☐   **n.** Connect a voltmeter set on its low voltage scale.

        o. Describe the meter's action.

_____

_____

Conclusions:

_____

_____

Hall Effect Sensors

a. Connect a 12-volt battery across the plus (+) and minus (–) voltage (supply current) terminals of the Hall layer. ☐

b. Connect a voltmeter across the minus (–) and signal voltage terminals. ☐

c. Insert a steel feeler gauge or knife blade between the Hall layer and magnet, then remove the feeler gauge. ☐

d. Describe what happens on the voltmeter.

_____

_____

Conclusions:

_____

_____

e. Remove the 12-volt power source and prepare the engine to run. ☐

f. Set a lab scope on its low scale primary pattern position. ☐

g. Connect the primary positive lead to the Hall signal lead and the negative lead should connect to ground or the ground terminal at the sensor's connector. ☐

h. Start the engine and observe the scope. ☐

i. Record the trace on the scope.

_____

_____

Conclusions:

_____

_____

Control Module

a. Connect one lead of the ohmmeter to the ground terminal at the module and the other lead to a good engine ground. ☐

b. Record your readings. _____ ohms

c. The specified resistance is _____ ohms.

Conclusions:

_____

_____

Secondary Ignition Wires

a. Remove the distributor cap with the spark plug wires attached to the cap but disconnected from the spark plugs. ☐

b. Calibrate an ohmmeter on the X1,000 scale. ☐

c. Connect the ohmmeter leads from the end of a spark plug wire to the distributor cap terminal inside the cap to which the plug wire is connected. ☐

d. Record your readings. _____ ohms

e. The specified resistance is _____ ohms.

Conclusions:

_____

_____

Task Completed

☐

☐

Spark Plugs

a. Remove the engine's spark plugs. Place them on a bench according to the cylinder number.

b. Carefully examine the electrodes and porcelain of each plug.

c. Describe the appearance of each plug.

_____

_____

_____

Conclusions:

_____

_____

_____

d. Measure the gap of each spark plug and record your findings.

_____

_____

_____

e. What is the specified gap? _____ inches

Conclusions:

_____

_____

_____

Instructor's Response _____

_____

_____

Job Sheet 33

33

Name _____ Date _____

Setting Ignition Timing

Upon completion of this job sheet, you should be able to check and set the ignition timing on a distributor-type ignition system.

ASE Correlation

This job sheet is related to the ASE Engine Performance Test's content area: ignition system diagnosis and repair; task: check and adjust ignition system timing and timing advance/retard.

Tools and Materials

Timing light   Appropriate service manual
Tachometer

Describe the vehicle being worked on.

Year _____ Make _____ VIN _____

Model _____

Engine type and size _____ Ignition timing specs _____

Source of specifications: _____

Conditions that must be met before checking the timing:

_____

What should the idle speed be? _____ rpm

What is the idle speed? _____ rpm

Task Completed

If not within specifications, correct the idle before proceeding.   ☐

Procedures

1. Connect the timing light pickup to the number one cylinder's spark plug wire, and the   ☐
    power supply wires on the light should be connected to the battery terminals with the
    proper polarity.

2. Start the engine.   ☐

3. The engine must be idling at the manufacturer's recommended rpm, and all other timing   ☐
    procedures must be followed.

4. Aim the timing light marks at the timing indicator and observe the timing marks. Timing
    found _____ degrees.

5. If the timing mark is not at the specified location, rotate the distributor until the mark is at
    the specified location. Describe any difficulties you had doing this.

    _____

    _____

6. Tighten the distributor holddown bolt to the specified torque. What is the specified torque? _____

☐ 7. Reconnect any hoses, or components that were disconnected for the timing procedure.

Instructor's Response _____

_____

_____

Job Sheet 34

34

Name _____ Date _____

Visually Inspect a Distributor Ignition System

Upon completion of this job sheet, you should be able to visually inspect the components of a distributor-type ignition system.

ASE Correlation

This job sheet is related to the ASE Engine Performance Test's content area: ignition system diagnosis and repair; task: inspect and test distributor; service as needed.

Tools and Materials

Steel feeler gauge set
A vehicle with a distributor

Describe the vehicle being worked on.

Year _____ Make _____ VIN _____

Model _____

Type of ignition _____

Procedures

1. Are the spark plug and coil cables firmly seated in the distributor cap and coil and onto spark plugs? ☐ Yes ☐ No

2. Do the secondary cables have cracks or signs of worn insulation? ☐ Yes ☐ No

3. Are the boots on the ends of the secondary wires cracked or brittle? ☐ Yes ☐ No

4. Are the secondary cables connected according to the firing order? ☐ Yes ☐ No

5. Are there any white or grayish powdery deposits on secondary cables? ☐ Yes ☐ No

6. Are spark plug cables from consecutively firing cylinders running parallel with each other? ☐ Yes ☐ No

7. Is the coil cracked or showing signs of leakage in the coil tower? ☐ Yes ☐ No

8. Is there oil around or on the coil? ☐ Yes ☐ No

9. Is the distributor cap properly seated on its base? ☐ Yes ☐ No

10. Are the mounting clips or screws securely fastened? ☐ Yes ☐ No

11. Is there any electrical damage on the rotor or in the distributor cap? ☐ Yes ☐ No

12. Is there any evidence of carbon tracking in the distributor cap? ☐ Yes ☐ No

13. Are the distributor cap towers damaged? ☐ Yes ☐ No

14. Is the rotor discolored or burnt? ☐ Yes ☐ No

15. Are the distributor cap and housing vents blocked or clogged? ☐ Yes ☐ No

16. Are the hoses to the vacuum advance unit secured to the unit (if so equipped)?
☐ Yes ☐ No

17. With the rotor on the distributor shaft, move the rotor on the distributor shaft clockwise and counterclockwise. Does it move in one direction only? ☐ Yes ☐ No

18. Did it spring back into position after it was released? ☐ Yes ☐ No

19. Are the connections of the primary ignition system wiring tight connections? ☐ Yes ☐ No

20. Is the control module tightly mounted to a clean surface? ☐ Yes ☐ No

21. Are the electrical connections to the module corroded? ☐ Yes ☐ No

22. Are the electrical connections to the module loose or damaged? ☐ Yes ☐ No

23. Is the reluctor of the switching unit broken or cracked? ☐ Yes ☐ No

24. Is the insulation of the pickup coil wire leads worn? ☐ Yes ☐ No

25. Is the non-magnetic reluctor magnetized? ☐ Yes ☐ No

26. Record your summary of the visual inspection. Include what looked good as well as what looked bad.

_____

_____

_____

27. Based on the visual inspection, what are your recommendations?

_____

_____

_____

Instructor's Response_____

_____

_____

Electronic Ignition Systems (EI) Diagnosis and Service

Upon completion and review of this chapter, you should be able to:

❏ Diagnose and determine the cause of a no-start condition on an EI system.

❏ Perform tests on the camshaft and crankshaft sensors as well as EI systems.

❏ Perform coil tests on an EI system.

❏ Install and adjust camshaft and crankshaft sensors.

❏ Perform magnetic sensor tests on EI systems.

Introduction

Electronic ignition (EI) systems should not overwhelm a knowledgeable technician. Basically, the ignition system itself looks very different, but it still accomplishes the same tasks; deliver a spark of the correct intensity and duration at the proper time. Eliminating the distributor replaces wear-prone parts that can cause inaccuracy. A primary system still switches a coil on and off with a module that keeps track of engine position using electronic sensors on the crankshaft and camshaft. The secondary winding of the coils still provides a high voltage surge that is delivered to the spark plugs. One major difference is also apparent—EI systems use more than one coil, even though these coils still operate under the same basic principle. Waste spark systems use one coil for two spark plugs; coil near plug and coil over plug (COP) systems use a coil for each spark plug. The advantage here is simple; there is more time to saturate the coils, which is important with high rpm engines where one coil may not have sufficient time to fully charge, turbocharging and supercharging which effectively raise compression and spark demand and leaner mixtures, which require a hotter spark.

Diagnosis of the primary and secondary ignition systems is very similar to distributor-based systems with some of the variations explained in this chapter. We will not spend as much time explaining ignition patterns because you can reference Chapter 12 as necessary. We will explain some of the differences in diagnosis, and scope connections.

Effects of Improper Ignition Timing

The primary circuit controls the secondary circuit. Therefore, the primary circuit controls ignition timing. Most problems within the primary circuit of computer-controlled ignition systems result in starting problems or poor performance due to incorrect timing.

If the engine does not start, it is usually because the engine is not receiving air, fuel, or spark. To verify if there is spark, observe the spark while cranking the engine. To do this, remove a spark plug wire from a spark plug and insert a spark tester in the spark plug wire terminal. Position the test plug away from throttle body and fuel lines and connect the test plug's grounding clip to a good ground (Figure 13-1). Crank the engine and observe the spark at the test plug. If the spark is weak, the ignition coil should be checked. If there is no spark, the triggering and switching units of the primary circuit should be checked.

If engine performance is poor, the cause of the problem can be many factors. There can be a problem with the engine such as poor compression, incorrect valve timing, overheating, etc. The air-fuel mixture or the ignition timing may be incorrect. When the ignition timing is not correct, many tests will point to the problem. Incorrect ignition timing will cause incomplete combustion at one or all engine speeds. Incomplete combustion will cause excessive oxygen in

Basic Tools

Basic tool set, tachometer, DMM

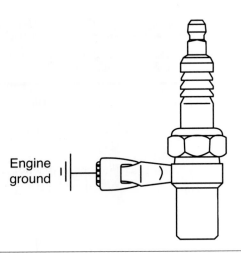

Figure 13-1 Spark plug tester.

the exhaust. This will cause the PCM to always try to correct the apparent lean mixture. Incorrect timing is not a lean condition, but the PCM cannot tell that the timing is wrong. It only knows there is too much oxygen in the exhaust. Under this condition, the waveform from the O_2 sensor will be lean biased (Figure 13-2).

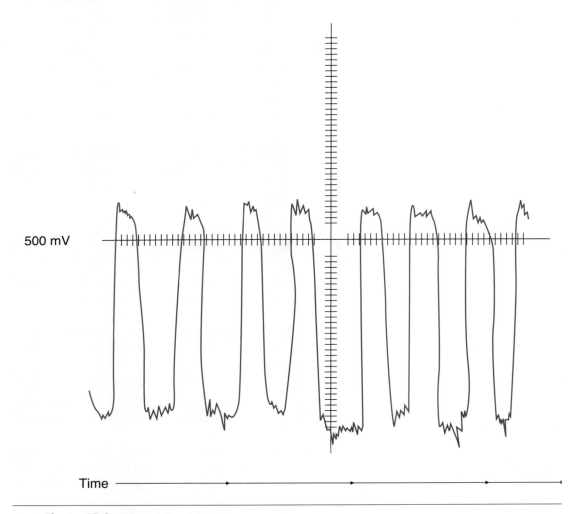

500 mV

Time

Figure 13-2 A lean biased O_2 sensor.

Excessive O_2 in the exhaust will also show up on an exhaust gas analyzer. With incorrect timing you may also see higher than normal amounts of oxygen (O_2) or hydrocarbons (HC). Remember, it takes several seconds for the exhaust to be analyzed. If you slowly accelerate the engine and see the HC and O_2 levels on the exhaust gas analyzer rise, the condition that existed several seconds earlier was the cause of the rise in emissions levels. To make this easier to track, make sure you hold the engine at each test speed for at least 7 seconds. This will enable you to observe the rise or fall of the emission levels at that particular speed.

Incorrect ignition timing will also affect manifold vacuum readings and ignition system waveforms on a scope. When anything indicates a problem with the primary ignition circuit, the suspected parts should be tested. To save time, do not check the components in the secondary circuit until you know all of the primary is working fine.

Visual Inspection

All ignition system diagnosis should begin with a visual inspection. The system should be checked for obvious problems. Although some no-start problems and incorrect ignition timing are caused by the primary circuit, the secondary circuit should also be checked.

1. Spark plug and coil cables should be pushed tightly on the coil and spark plugs.
2. Inspect all secondary cables for cracks and worn insulation, which causes high-voltage leaks.
3. Inspect all of the boots on the ends of the secondary wires for cracks and hard, brittle conditions.
4. The ignition coil should be inspected for cracks or any evidence of voltage leaks in the coil tower.
5. Secondary cables must be connected according to the firing order.
6. Spark plug cables from consecutively firing cylinders should cross rather than run parallel to one another.
7. If dealing with a no-start condition, make sure the "Check Engine" or "Service Engine Soon" (MIL) lamp comes on when the ignition is turned to run. If not, check the PCM power and ground.

Primary Circuit

Primary ignition system wiring should be checked for clean, tight connections, especially on vehicles with electronic- or computer-controlled ignitions. Electronic circuits operate on very low voltage. Voltage drops caused by corrosion or dirt can cause running problems. Missing tab locks on wire connectors are often the cause of intermittent ignition problems due to vibration- or thermal-related failure.

Test the integrity of a suspect connection by tapping, tugging, and wiggling the wires while the engine is running. Be gentle. The object is to recreate an ignition interruption, not to cause permanent circuit damage. With the engine off, separate the suspect connectors and check them for dirt and corrosion. Clean the connectors according to the manufacturer's recommendations.

All wiring connections to and from the distributor and ignition module should be thoroughly inspected for damage.

Do not overlook the ignition switch as a source of intermittent ignition problems. A loose mounting rivet or poor connection can result in erratic spark output. To check the switch, gently wiggle the ignition key and connecting wires with the engine running. If the ignition cuts out or dies, the problem is located.

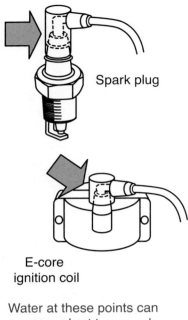

Spark plug

E-core
ignition coil

Water at these points can
cause a short to ground

Figure 13-3 Places to check for moisture in an ignition system.

Check the battery connection to the starter solenoid. Remember, some vehicles use this connection as a voltage source for the coil. A bad connection can result in ignition interruption.

Moisture can cause a short to ground or reduce the amount of voltage available to the spark plugs. This can cause poor performance or a no-start condition. Carefully check the ignition system for signs of moisture. Figure 13-3 shows the common places where moisture may be present in an ignition circuit.

Ground Circuits

Keep in mind that to simplify a vehicle's electrical system, automakers use body panels, frame members, and the engine block as current return paths to the battery.

Unfortunately, ground straps are often neglected, or worse, left disconnected after routine service. With the increased use of plastics in today's vehicles, ground straps may mistakenly be reconnected to a non-metallic surface. The result of any of these problems is that the current that was to flow through the disconnected or improperly grounded strap is forced to find an alternate path to ground. Sometimes the current attempts to back up through another circuit. This may cause the circuit to operate erratically or fail altogether. The current may also be forced through other components, such as wheel bearings or shift and clutch cables that are not meant to handle current flow, causing them to wear prematurely or become seized in their housing.

Examples of bad ground circuit-induced ignition failures include burnt ignition modules resulting from missing or loose coil ground straps, erratic performance caused by a poor ground, and intermittent ignition operation resulting from a poor ground at the control module.

Electromagnetic Interference

Electromagnetic interference (EMI) can cause problems with the vehicle's on-board computer. EMI is produced when electromagnetic radio waves of sufficient amplitude escape from a wire or conductor. Unfortunately, an automobile's spark plug wires, ignition coil, and alternator coils all possess the ability to generate these radio waves. Under the right conditions, EMI can trigger sensors or actuators. The result may be an intermittent driveability problem that appears to be ignition system related.

To minimize the effects of EMI, check to make sure that sensor wires running to the computer are routed away from potential EMI sources. Rerouting a wire by no more than an inch or two may keep EMI from falsely triggering or interfering with computer operation.

Control Modules

Electronic control modules for EI systems can be mounted on the engine block, on top of the valve cover, under the coils or on the firewall. They can also be contained within the PCM itself. Make sure that the mounting of the module is clean and tight, since some modules depend on their mounting surface for a proper ground. During your visual inspection, make sure the wiring connections are also clean and tight.

Sensors

The transistor in the control module is activated by a voltage pulse from a crankshaft position sensor. This sensor is either a magnetic pulse generator or Hall effect sensor. The magnetic reluctor-style crank sensor is shown in Figure 13-4. The crank sensor is mounted approximately 0.050 inch away from the reluctor cast into the crankshaft. The sensor is not adjustable. The waveform of the sensor produces a characteristic waveform that can be checked with an oscilloscope and will look similar to that shown in Figure 13-5. Every notch produces an A/C voltage that rises and falls as the notch passes, telling the module the position of the crankshaft.

Hall effect sensor problems are similar to those of magnetic sensors. This sensor produces a voltage when it is exposed to a magnetic field. The Hall effect assembly is made up of a permanent magnet located a short distance away from the sensor. Attached to the harmonic balancer pulley is a shutter wheel. When the shutter is between the sensor and the magnet, the

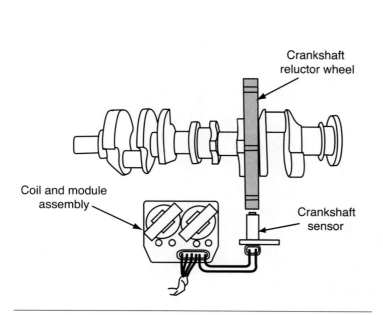

Figure 13-4 The GM system reluctor is machined as part of the crankshaft with notches that pulse the crankshaft sensor.

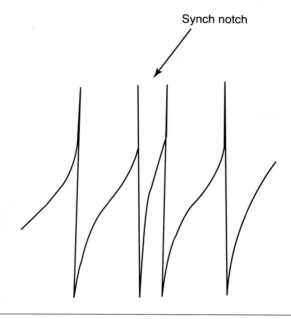

Figure 13-5 A portion of the magnetic sensor's waveform showing the sync notch.

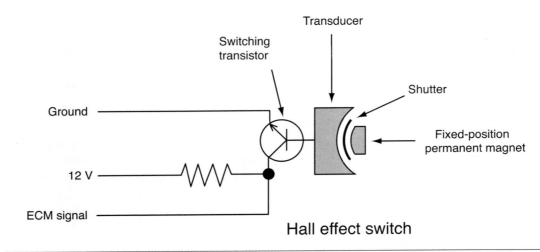

Hall effect switch

Figure 13-6 The position of the shutter blade in or out of the Hall effect sensor will produce either an on or off signal.

One problem with some EI systems is the difficulty of accessing the primary circuit on waste spark systems.

The oscilloscope can look inside the ignition system by giving the technician a visual representation of voltage changes over time.

Special Tools

Tune-up scope

magnetic field is interrupted and voltage immediately drops to zero (Figure 13-6). This drop in voltage is the signal to the ignition module. When the shutter leaves the gap between the magnet and the sensor, the sensor produces voltage again.

The ignition module switches primary current flow on and off in response to the signal from the pickup unit. The module does not act alone in determining how long the dwell should be or when the firing of a spark plug should occur. The PCM sends a signal to the module to control the actual timing of events.

Scope Testing

The EI system secondary ignition can be tested with an oscilloscope, but with either COP or waste spark there are special adaptors that have to be used to read the secondary pattern. The COP adaptors are shown in Figure 13-7, while a typical set-up for a waste spark system and an example pattern is shown in Photo Sequence 25. The interpretation of the scope patterns is very much like that used for a DI ignition as explained in Chapter 12. The primary circuit on

Inductive pick-ups for coil-on-plug systems for engine analyzer

Figure 13-7 Coil-on-plug ignition adapters for diagnostic analyzer connections.

Photo Sequence 25
Coil Current Ramping on a Waste Spark Ignition System

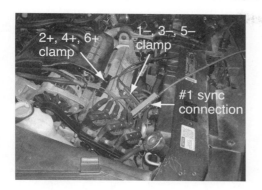

P25-1 For our example, we are connecting a console-style engine analyzer. With this setup, we can check secondary waveforms as well as current ramping the primary. Shown are the necessary levels for the secondary ignition pattern.

P25-3 The low-amp probe used was designed for use with our engine analyzer. Since we have the secondary in place, we can synchronize the current ramp to the correct cylinders.

P25-4 The low-amps probe is connected to the ignition feed at the module. The low-amps probe requires no piercing or back-probing. The amp clamp fits around the wire.

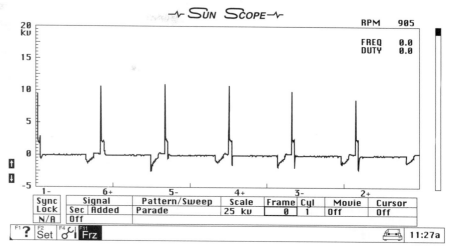

P25-2 There are three leads shown in the previous picture. Half of the spark plugs fire forward (conventionally, shown as 6+, 4+, 2+ on the pattern) and half fire backward. The forward leads are connected to a red cable end and the negative (backward, shown as 1–, 3–, 5–) firing cables are connected to a black cable end. Number one cylinder is also clamped with a "sync" lead. The waste spark is also availasble for viewing on most testers.

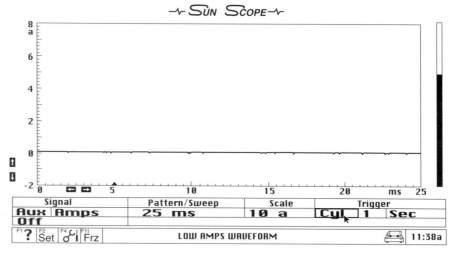

P25-5 Moving to the amp-probe, we have to set the baseline to zero. Note that since the probe is dedicated to our analyzer, it changes the range to amperage automatically. Most probes require the use to convert from mV to mA.

589

Coil Current Ramping on a Waste Spark Ignition System (continued)

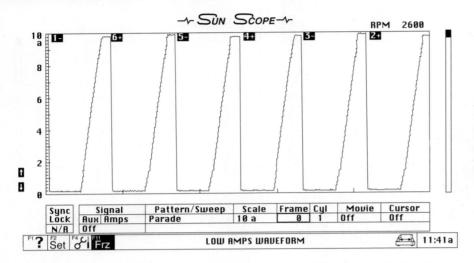

P25-6 The amp-probe synchronized to the correct cylinders. The slope on the front edge is showing the time it takes for the coil to saturate. The flat tops are due to current limiting.

P25-7 There are many low-amp probes available that can be connected either to an oscilloscope or DVOM.

waste spark systems is usually hidden by the fact that the coil pack is on top of the module, making it difficult to obtain a primary pattern using conventional methods. The low-amp probe can be placed on the primary ignition feed wire to the ignition module and a primary current ramp can be watched for irregularities (Figure 13-8). A current ramp of the primary ignition on a typical waste spark system is also shown in Photo Sequence 25. If the vehicle is a coil over plug (COP) ignition, the negative lead on an individual coil can be scoped with a resulting pattern, much like that shown in Figure 13-9. The primary can also be checked by using a test

Figure 13-8   Amperage ramp of a coil primary circuit on an EI system.

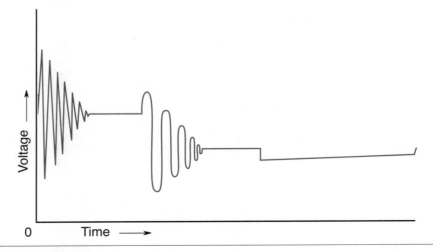

Figure 13-9   A typical primary pattern.

Figure 13-10 Checking coil primary switching with a test lamp.

lamp watching for a flash as the primary is switched on and off as the engine is cranked as in Figure 13-10.

General Diagnostics

Standard test procedures using an oscilloscope, ohmmeter, and timing light can be used to diagnose problems in distributorless ignition systems. Keep in mind, however, that problems involving one cylinder may also occur in its companion cylinder that fires off the same coil. Some oscilloscopes require their pickups placed on each pair of cylinders to view all patterns. Special adapters are available to make these hookups less troublesome. Many newer engine analyzers with an oscilloscope use a single adapter that allows viewing of all cylinder patterns at one time.

Follow the testing procedures outlined in the vehicle's service manual for the engine control system. Specific computer-generated trouble codes are designed to help troubleshoot ignition problems in these systems. The diagnostic procedure for EI systems varies depending on the vehicle make and model year. Always follow the recommended procedure.

When diagnosing these systems, keep in mind there is a separate primary circuit for each coil. If one coil does not work properly, it may be caused by something common or not common to other coils. Regardless of the system's design, there are common components in all electronic ignitions: an ignition module, crankshaft and/or camshaft sensors, ignition coils, secondary circuit, and spark plugs. Most of these components are common to only the spark plug or spark plugs they are connected to. The components that are common to all cylinders (such as the camshaft sensor) will most often be the cause of no-start problems. The other components will cause misfire problems. There are many other factors that can cause a misfire. Check those before moving on to the ignition system. Test for intake manifold vacuum leaks. Also check the engine's compression and the fuel injection system.

Visual Inspection

All ignition system diagnosis should begin with a visual inspection. The system should be checked for obvious problems: disconnected, loose, or damaged secondary cables; disconnected, loose, or dirty primary wiring; worn or damaged primary system switching mechanism; and improperly mounted ignition module.

Classroom Manual
Chapter 13, page 418

According to the SAE J1930 standards, the term electronic ignition (EI) replaces all previous terms for distributorless ignition systems.

15 percent of all driveability complaints are caused by problems in the ignition system.

70 percent of all electrical problems are at the connectors.

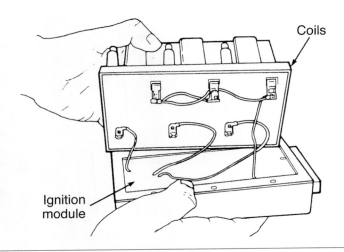

Figure 13-11 Some EI systems have wiring attachments under the coil assembly that need to be inspected.

If a crank (or sometimes cam) sensor fails, the engine will not start. Both of these sensor circuits can be checked with a voltmeter. If the sensors are receiving the correct amount of voltage and have good low-resistance circuits, their output should be a digital signal or a pulsing voltmeter reading while the engine is cranking. If any of these conditions do not exist, the circuit needs to be repaired or the sensor needs to be replaced.

On distributorless or direct ignition systems, visually inspect the secondary wiring connections at the individual coil modules. Make sure all of the spark plug wires are securely fastened to the coil and the spark plug. If a plug wire is loose, inspect the terminal for signs of burning. The coils should be inspected for cracks or any evidence of leakage in the coil tower. Check for evidence of terminal resistance. Separate the coils and inspect the underside of the coil and the ignition module wires (Figure 13-11). A loose or damaged wire or bad plug can lead to carbon tracking of the coil. If this condition exists, the coil must be replaced.

Inspect all secondary cables for cracks and worn insulation, which cause high-voltage leaks. Inspect all of the boots on the ends of the secondary wires for cracks and hard brittle conditions. Replace the wires and boots if they show evidence of these conditions. Some vehicle manufacturers recommend spark plug wire replacement only in complete sets.

If the visual inspection does not reveal obvious problems, test those individual components and circuits that could cause the problem.

No-Start Diagnosis

When an EI system has a no-start problem, begin your diagnosis with a visual inspection of the ignition system. Check for good and tight primary connections. Perform a spark intensity check with a spark tester. A bright, snapping spark indicates that the secondary voltage output is good. If the spark is weak or if there is no spark, check the primary circuit.

If a crank or cam sensor fails, the engine may not start. Both of these sensor circuits (Figure 13-12) can be checked with a voltmeter. If the sensors are receiving the correct amount of voltage and have good low-resistance ground circuits, their output should be a digital signal or a pulsing voltmeter reading while the engine is cranking. If any of these conditions do not exist, the circuit needs to be repaired or the sensor needs to be replaced. Some manufacturers have a fail-safe mode where it is possible the engine can start and run with a faulty camshaft sensor. Each manufacturer may use a different strategy and will vary from one engine family to another. Depending on the strategy, a prolonged cranking time may be a symptom, or perhaps it may take two to three times for the engine to start.

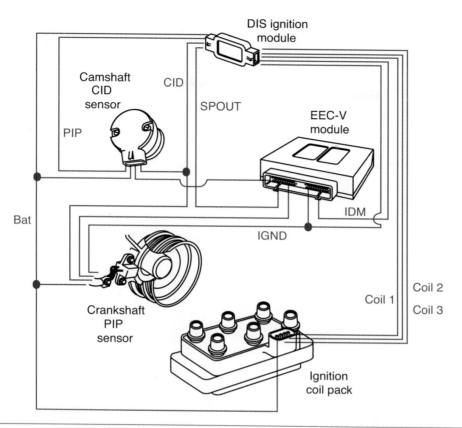

Figure 13-12 An EI system with a camshaft and a crankshaft sensor.

If a crankshaft or camshaft sensor needs to be replaced, always clean the sensor tip and install a new spacer (if so equipped) on the sensor tip. New sensors typically have a new spacer already installed on the sensor. Install the sensor until the spacer lightly touches the sensor ring and tighten the sensor mounting bolt.

The base timing cannot be adjusted on an EI system. However, the air gap between the sensor and the trigger wheel can affect the operation of the ignition system. On many EI systems, this gap is not adjustable. This does not mean that the gap is not important and should not be checked. The gap should be checked whenever possible. If there is no provision for adjusting the gap and the gap is incorrect, the sensor should be replaced. If the gap between the blades and the crankshaft sensor is incorrect, the engine may fail to start, stall, misfire, or hesitate on acceleration.

The primary can also be checked with a logic probe. There are three lights on a logic probe. The red light illuminates when the probe senses more than 10 volts. When the monitored signal has less than 4 volts, the green light flashes. The yellow light will flash whenever the voltage changes. This light is used to monitor a pulsing signal, such as one produced by a digital sensor like a Hall effect switch.

To check the primary circuit with a logic probe, turn the ignition on. Touch the probe to both (positive and negative) primary terminals at the coil. The red light should come on at

If the camshaft sensor is mounted in the previous distributor opening, this sensor must be timed to the engine when it is installed.

Special Tools

Logic probe

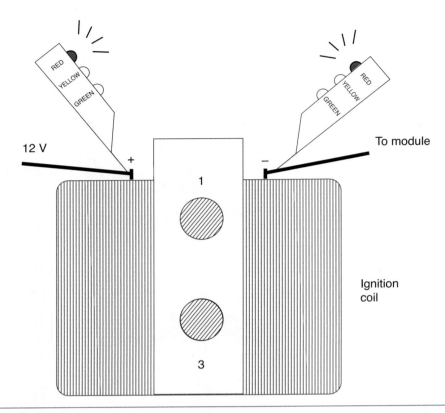

Figure 13-13 The red light on a logic probe should turn on when touched to both sides of the primary winding.

Doublecheck the ground connection for voltage drop when using a logic probe. Since the ground LED comes on with less than 4 volts, the ground could have excessive voltage drop and still light.

both terminals (Figure 13-13). This indicates that at least 10 volts is available to the coil and there is continuity through the coil. If the red light does not come on when the positive side of the coil is probed, check the power feed circuit to the coil. If the light comes on at the positive terminal but not on the negative, the coil has excessive resistance or is open.

Now move the probe to the negative terminal of the coil and crank the engine. The red and green light should alternately flash. This indicates that over 10 volts is available to the coil while cranking and that the circuit is switching to ground. If the lights do not come on, check the ignition power feed circuit from the starter. If the red comes on but the green light does not, check the crankshaft or camshaft sensor. If these are working properly, the ignition module is probably defective.

A Hall effect switch is also easily checked with a logic probe. If the switch has three wires, probe the outer two wires with the ignition on (Figure 13-14). The red light should come

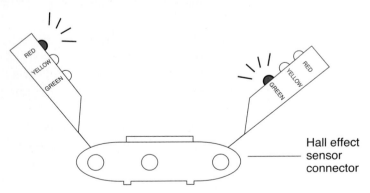

Figure 13-14 One end terminal of a Hall effect sensor connector should cause the red light to come on; the other end terminal should cause the green light to come on.

on when one of the wires is probed, and the green light should come on when the other wire is probed. If the red light does not turn on at either wire, check the power feed circuit to the sensor. If the green light does not come on, check the sensor's ground circuit.

Backprobe the center wire and crank the engine. All three lights should flash as the engine is cranked. The red light will come on when the sensor's output is above 10 volts. As this signal drops below 4 volts, the green light should come on. The yellow light will flash each time the voltage changes from high to low. If the logic probe's lights do not respond in this way, check the wiring at the sensor. If the wiring is okay, replace the sensor.

Keep in mind that the action of a Hall effect switch and an inductive sensor can be monitored with a lab scope. The waveform for a normally operating inductive-type crankshaft sensor is shown in Figure 13-15. Carefully examine the waveform for traces of noise and false pulses.

The pattern of the trace will vary according to the position and number of slots machined into the trigger wheel. The trigger wheel in Figure 13-16 has nine slots. Eight of them are evenly spaced and one slot is placed close to one of the eight. The trace for this type sensor will also have eight evenly spaced pulses. One of the pulses will quickly be followed by another pulse (Figure 13-17). Any waveform that does not match the configuration of the trigger wheel indicates a problem with the sensor or its circuit.

The oddly spaced slot is called the synch slot. This function is sometimes achieved with a camshaft sensor.

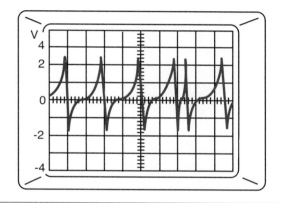

Figure 13-15 Normal waveform for an inductive crankshaft sensor.

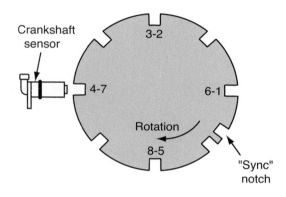

Figure 13-16 An nine-slot trigger wheel for a crankshaft sensor.

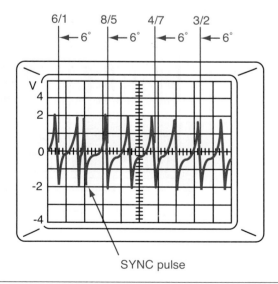

Figure 13-17 Waveform for the sensor shown in Figure 13-16.

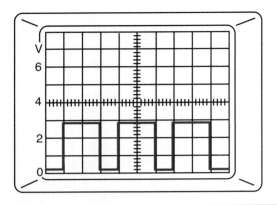

Figure 13-18 Normal waveform for a Hall effect sensor.

If the crankshaft sensor is a Hall effect switch, digital waves should be seen on the scope (Figure 13-18). Each pulse should be identical in spacing, shape, and amplitude. By using a dual trace lab scope, the relationship between the crankshaft sensor and the ignition module can be observed. During starting, the module will provide a fixed amount of timing advance according to its program and the cranking speed of the engine. By observing the crankshaft sensor output and the ignition module, this advance can be observed (Figure 13-19). The engine will not start if the ignition module does not provide for a fixed amount of timing advance.

☑ **SERVICE TIP:** When checking the ignition system with a lab scope, gently tap and wiggle the components while observing the trace. This may indicate the source of an intermittent problem.

☑ **SERVICE TIP:** Another quick test of the primary ignition triggering operation is to check fuel injector pulsing. If a no-start is ignition related, this test will eliminate the primary or secondary as the fault. Connect a fuel injector test light to the fuel injector connector and observe the light while cranking the engine. If the injector test light flashes, the primary triggering circuit is functioning. The injector will not pulse if there is no trigger input to the PCM. This quickly confirms crankshaft sensor operation. If there is no spark and the injector pulses, the fault is probably within the secondary system.

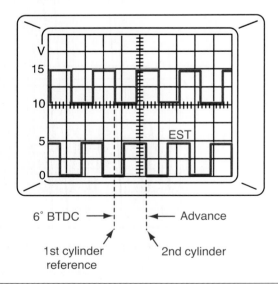

Figure 13-19 EST and crankshaft sensor signals compared on a dual-trace scope.

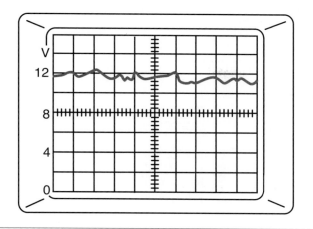

Figure 13-20 A voltage signal caused by a poor ignition module ground.

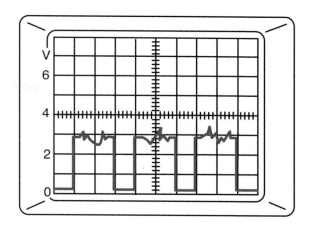

Figure 13-21 A faulty sensor signal caused by a bad ground.

Ground Circuits

Poor grounds can be identified by conducting voltage drop tests and monitoring the circuit with a lab scope. When conducting a voltage drop test, the circuit must be turned on and have current flowing through it. If the circuit is tested without current flow, the circuit will show zero voltage drop, indicating that it is good regardless of the amount of resistance present.

The same is also true when checking a ground with the lab scope. Make sure the circuit is on. If the ground is good, the trace on the scope should be at zero volt and be flat. If the ground is bad, some voltage will be indicated and the trace will not be flat (Figure 13-20).

Often, a bad sensor ground will cause the same symptoms as a faulty sensor. Before condemning a sensor, check its ground with a lab scope. Figure 13-21 shows the output of a good Hall effect switch with a bad ground. Depending on the Hall effect sensor or the circuit design, a potential faulty ground may cause the signal at the top or bottom of the pattern to be abnormal. Check the service manual for specific operation.

Electromagnetic Interference (EMI)

EMI problems can be identified by connecting a lab scope to voltage and ground wires. Common problems, such as poor spark plug wire insulation, will allow EMI. In addition, EMI-induced signals can influence other low-voltage signals, such as inputs to the PCM. These false values can be interpreted by the PCM as an actual signal and cause improper outputs to the engine. Figure 13-22 shows a voltage trace that is contaminated with EMI from the ignition system.

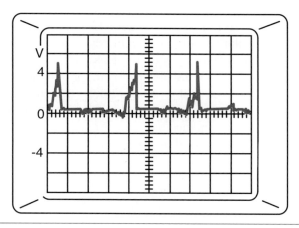

Figure 13-22 A voltage trace with ignition system noise.

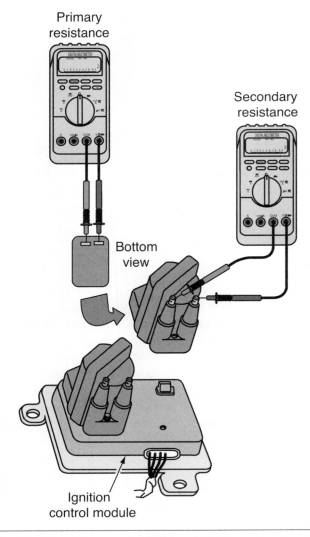

Figure 13-23 Checking primary and secondary resistance of an EI coil.

Ignition Coils

The coils in an EI system can be checked in the same way as conventional coils, however, they all must be checked and inspected (Figure 13-23). When checking the resistance across the windings, pay particular attention to the meter reading. If the reading is out of specifications, even if it is only slightly out, the coil or coil assembly should be replaced.

To check the primary windings, calibrate an ohmmeter on the X1 scale and connect the meter leads to the primary coil terminals to test the winding. An infinity ohmmeter reading indicates an open winding. The winding is shorted if the meter reading is below the specified resistance. Most primary windings have a resistance of 0.5–2 ohms but the exact manufacturer's specifications must be compared to the meter readings.

The ohmmeter tests on the ignition coil windings do not check the coil for insulation leakage.

To check the secondary winding, calibrate the meter on the X1,000 scale and connect it from the coil's secondary terminal to one of the primary terminals. A meter reading below the specified resistance indicates a shorted secondary winding. An infinity meter reading proves that the winding is open.

In some coils, the secondary winding is connected from the secondary terminal to the coil frame. When the secondary winding is tested in these coils, the ohmmeter must be connected from the secondary coil terminal to the coil frame or to the ground wire extending from the coil frame.

Many secondary windings have 8,000 to 20,000 ohms resistance, but the meter readings must be compared to the manufacturer's specifications. The ohmmeter tests on the primary and secondary windings indicate satisfactory, open, or shorted windings. Ohmmeter tests do not indicate such defects as defective insulation around the coil windings, which cause high-voltage leaks. Therefore, an accurate indication of coil condition is the coil maximum voltage output test with a spark tester connected from the coil secondary wire to ground as explained in the no-start diagnosis.

CUSTOMER CARE: When verifying the customer's complaint, be aware of some common terms having more than one meaning that the customer may use. If possible, go for a test drive with the customer in order to witness the symptom together. For instance, the customer may complain of a transmission shifting problem or a shudder at part-throttle. To the customer, this may seem like a transmission-related problem. However, to a trained technician it may be an ignition-related problem. For example, a weak coil that breaks down under load can cause the symptom described. Listen to the customer, but politely decline his or her self-diagnosis unless it is obvious and relevant.

Secondary Circuit

Testing the secondary circuit of an EI system is just like testing the secondary of any other type of ignition system. The spark plug wires and spark plugs should be tested to ensure they have the appropriate amount of resistance. Since the resistance in the secondary dictates the amount of voltage the spark plug will fire with, it is important that secondary resistance be within the desired range.

A quick way to examine the secondary circuit is with a tune-up scope. On the scope, the firing line and the spark plug indicates the resistance of the secondary. While observing the firing line, remember the height of the line increases with an increase in resistance. The length of the spark line decreases as the firing line goes higher. This means that high resistance will cause excessively high firing voltages and reduced spark times.

Excessive resistance is not the only condition that will affect the firing of a spark plug. Spark plug wires and the spark plug itself can allow the high voltage to leak and establish current through another metal object instead of the electrodes of the spark plug. When this happens, the spark plug does not fire and combustion does not take place within the cylinder.

All spark plug wires should be inspected for cracks and worn insulation, which cause high-voltage leaks. Also inspect the boots on the ends of the plug wires for cracks and hard, brittle conditions. Replace the wires and boots if they show evidence of these conditions. Some vehicle manufacturers recommend spark plug wire replacement only in complete sets.

Spark plug wires can be also checked with an ohmmeter. To do this, calibrate the ohmmeter on the X1,000 scale, then connect it across each cable. If the ohmmeter reading is more than specified by the manufacturer, replace the wire. When replacing spark plug wires, be sure they are routed properly.

Classroom Manual
Chapter 13, page 416

Improperly torqued spark plugs can cause a misfire on EI systems because the circuit is completed through ground.

Special Tools

Anti-seize compound, spark plug torque wrench

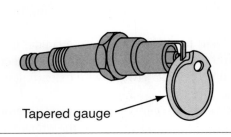

Figure 13-24 Using a taper gauge to check spark plug gap.

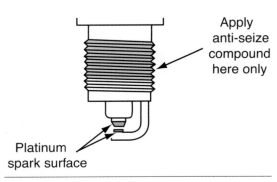

Apply anti-seize compound here only

Platinum spark surface

Figure 13-25 Proper placement of anti-seize compound on the threads of a spark plug.

Never file a platinum-tipped plug.

Most late-model engines are equipped with platinum tip spark plugs. Extra care must be taken when setting the gap on these plugs. A wire gauge is recommended by many manufacturers for checking and adjusting spark plug gap. However, the use of more delicate spark plugs has led many manufacturers to recommend the use of a tapered gauge (Figure 13-24).

Keep in mind while working on waste fire-type systems that the secondary circuit is completed through the metal of the engine. If the spark plugs are not properly torqued into the cylinder heads, the threads of the spark plug may not make good contact and the circuit may offer resistance. Always tighten spark plugs to their specified torque. Many of today's engines have aluminum cylinder heads. Not only do these heads require different torque specifications than iron heads, they also require extra care when installing the plugs. Make sure the threads of the plugs match the threads of the spark plug bores. Take extra care not to cross-thread them. Repairing damaged threads in a cylinder head can be a costly and time-consuming job.

An easy way to help prevent cross-threading of spark plugs is to insert the porcelain end into a vacuum line and then insert the spark plug. It should be difficult to cross-thread a spark plug with the vacuum hose. This can also be helpful if the spark plug is difficult to reach. No method is fail-safe, but this is safer than trying to start a spark plug with the ratchet and socket.

Most manufacturers recommend the use of an anti-seize compound on spark plug threads. This compound must be applied in the correct amounts and at the correct place (Figure 13-25). Too little compound will cause gaps in the contact between the spark plug threads and the spark plug bores. Too much may allow the spark to jump to a buildup rather than the spark plug electrode.

Chrysler EI Systems

Classroom Manual
Chapter 13, page 435

Chrysler's EI systems use the waste spark method for firing the spark plugs. These systems rely on a camshaft and a crankshaft sensor to inform the powertrain control module (PCM) as to the position of piston number one and time at which the other pistons reach top dead center (TDC). Basic ignition timing is not adjustable in these systems.

If an engine misfire problem is caused by the ignition system, the basic operation of the secondary circuits should be checked. This testing should begin with a complete visual inspection of the circuit. With any waste spark system, the secondary circuit should be looked at as groups of two. Each coil fires two cylinders. Each coil and its associated spark plugs should be checked as a group.

For each group of cylinders, the coil, spark plug wires, and spark plugs should be checked on a scope or with an ohmmeter. Resistance checks of coils and spark plug wires are recommended. Make sure the plug wires are fully seated onto the spark plug and the ignition coil.

No-Start Diagnosis

Chrysler EI systems use a crankshaft timing and camshaft reference sensor that are modified Hall effect switches. When the engine fails to start, follow these steps:

1. Check for fault codes which could be caused by a defective camshaft reference signal or crankshaft timing sensor signal or low primary current in coil number one, two, or three.

2. With the engine cranking, check voltage from the orange wire to ground on the crankshaft timing sensor and the camshaft reference sensor (Figure 13-26). Over 7 volts is satisfactory. If the voltage is less than specified, repeat the test with the voltmeter connected from PCM terminal seven to ground. If the voltage is satisfactory at terminal seven but low at the sensor orange wire, repair the open circuit or high resistance in the orange wire. If the voltage is low at terminal seven, replace the PCM. Be sure 12 volts are supplied to PCM terminal three with the ignition switch off or on, and 12 volts must be supplied to PCM terminal nine with the ignition switch on. Check the PCM ground connections on terminals eleven and twelve before replacing the PCM.

3. With the ignition switch on, check the voltage drop across the ground circuit (black/light blue wire) on the crankshaft timing sensor and the camshaft reference sensor. A reading below 0.2 volts is satisfactory.

☑ **SERVICE TIP:** When using a digital voltmeter to check a crankshaft or camshaft sensor signal, crank the engine a very small amount at a time and observe the voltmeter. The voltmeter reading should cycle from almost 0 volts to a higher voltage of 9 to 12 volts. Since digital voltmeters do not react instantly, it is difficult to see the change in voltmeter reading if the engine is cranked continually.

Prior to checking the voltage signal from a crankshaft or camshaft sensor on an EI system, the sensor voltage supply and ground wires should be checked.

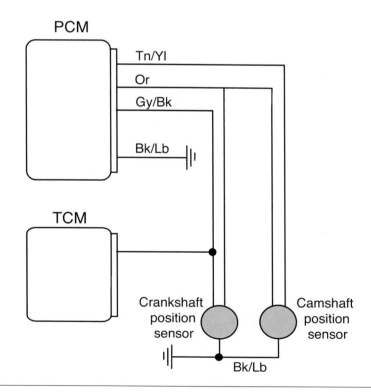

Figure 13-26 Crankshaft timing and camshaft reference sensor terminals.

4. If the reading in steps 3 and 4 are satisfactory, connect a 12-volt test lamp or a digital voltmeter from the gray/black wire on the crankshaft timing sensor and the tan/yellow wire on the camshaft reference sensor to ground. When the engine is cranking, a flashing 12-volt lamp indicates that a sensor signal is present. If the lamp does not flash, sensor replacement is required. Each sensor voltage signal should cycle from low voltage to high voltage as the engine is cranked.

☑ **SERVICE TIP:** A no-start problem on Chrysler 3.3L engines may be a shorted cam or crank sensor. These sensors are three-wire Hall effect sensors that are fed 8 volts by the PCM. If the power feed is shorted to ground, the PCM shuts off causing the engine to stall and not restart. Because the PCM is not working, a scan tool will not work, nor will the system's self-test. To verify this as the cause of the problem, simply disconnect the sensor and try to retrieve data with a scan tool. If data is now available, the sensor or wire to it is shorted. Voltage to the sensor can also be checked with a voltmeter; again, there should be 8 volts available.

If the sensor tests are satisfactory and the engine will not start, proceed with these coil and PCM tests:

Special Tools

12-volt test light

1. Check the spark plug wires with an ohmmeter.
2. With the engine cranking, connect a voltmeter from the dark green/black wire on the coil to ground. If this reading is below 12 volts, check the automatic shutdown (ASD) relay circuit (Figure 13-27).

☑ **SERVICE TIP:** Some Chrysler vehicles have separate fuel pump and ASD relays. Always use the proper wiring diagram for the vehicle being tested.

3. If the reading in step 2 is satisfactory, check the primary and secondary resistance in each coil with the ignition switch off. Primary resistance should be 0.52 to 0.62 ohm and the secondary resistance 11,000 to 15,000 ohms (Figure 13-28). If these ohm readings are not within specifications, replace the coil assembly.
4. With the ignition switch off, connect an ohmmeter across the three wires from the coil connector to PCM terminals seventeen, eighteen, and nineteen (Figure 13-29). These terminals are connected from the coil primary terminals to the PCM. If an infinity ohmmeter reading is obtained on any of the wires, repair the open circuits.
5. Connect a 12-volt test lamp from the dark blue/black wire, dark blue/gray wire, and black/gray wire on the coil assembly to ground while cranking the engine. If the test lamp does not flutter on any of the three wires, replace the PCM. Since the crankshaft and camshaft sensors, wires from the coils to the PCM, and voltage supply to the coils have been tested already, the ignition module must be defective. This module is an integral part of the PCM on Chrysler vehicles; thus the PCM must be replaced.

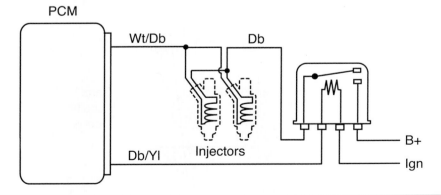

Figure 13-27 ASD relay circuit.

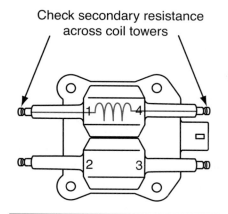

Check secondary resistance across coil towers

1   4

2   3

Figure 13-28 Measure across the companion coil towers to measure secondary winding resistance.

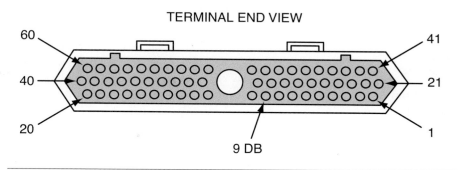

TERMINAL END VIEW

60

40

20

41

21

1

9 DB

Figure 13-29 PCM terminal identification numbers.

■ **CAUTION:** Do not crank or run an EI-equipped engine with a spark plug wire completely removed from a spark plug. This action may cause leakage defects in the coils or spark plug wires.

▲ **WARNING:** Since EI systems have more energy in the secondary circuit, electrical shocks from these systems should be avoided. The electrical shock may not injure the human body, but such a shock may cause you to jump and hit your head on the hood or push your hand into contact with a rotating cooling fan.

6. If the tests in steps in 1 to 4 are satisfactory, connect a spark tester to each spark plug wire and ground and crank the engine. If any of the coils do not fire on the two spark plugs connected to the coil, replace the coil assembly.

Sensor Replacement

If the crankshaft timing sensor or the camshaft reference sensor is removed, follow this procedure to replace the sensor:

1. Thoroughly clean the sensor tip and install a new spacer on the sensor tip. New sensors should be supplied with the spacer installed (Figure 13-30).

2. Install the sensor until the spacer lightly touches the sensor ring and tighten the sensor mounting bolt to 105 inch-pounds.

■ **CAUTION:** Improper sensor installation may cause sensor, rotating drive plate, or timing gear damage.

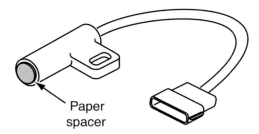

Paper spacer

Figure 13-30 Paper spacer on the tip of a crankshaft sensor.

Ford EI Systems

Classroom Manual
Chapter 13, page 434

Special Tools

Star tester, breakout box

Ford has used several designs of EI through the years. It is important that the design be identified prior to testing the system. Throughout all, certain basic operational features can be applied. The profile ignition pickup (PIP) signal is an indication of crankshaft position and engine speed. The PIP signal is sent to both the ignition module and the PCM. The cylinder identification (CID) signal is used in conjunction with PIP for the identification of cylinder number one for the PCM. Most systems utilize the waste spark method of firing.

Many of Ford's recommended diagnostic procedures for their EI systems involve the use of their special scan tool NGS (New Generation STAR tester) and a breakout box. Both of these concentrate on testing the primary circuit. The primary circuit controls the action of the secondary. Problems in the primary can cause no-start and ignition timing problems. Most misfire problems are caused by problems in the secondary.

Make sure the secondary cables are secured tightly to the spark plugs and the ignition coil. Also check the cables for any damage or signs of arcing. The resistance of the cables can be checked with an ohmmeter. Do this by removing the cable and measuring across the cable. Most Ford EI systems use locking tabs to secure the spark plug cable to the ignition coil. To remove the cable from the coil, squeeze the locking tabs at the spark plug terminal and pull the boot straight up (Figure 13-31). If the cable does not freely release from the coil, squeeze the tabs and twist the boot while pulling the cable up. When reconnecting the cable to the coil, make sure the locking tabs are in place by pressing down on the center of the cable terminal (Figure 13-32).

The voltage signal from crankshaft or camshaft sensors may be checked with a 12-volt test light or a digital voltmeter while cranking the engine.

Make sure the factory-supplied wire separators are used and are in their original position after the spark plug wires are put in place (Figure 13-33). The spark plugs and ignition coils can be tested in the same way as other ignition systems.

Although the ignition timing on Ford's EI system is not adjustable, the air gaps at the crankshaft and camshaft sensors are critical (Figure 13-34). On some engines this gap is adjustable and on others it is an indication that the sensor should be replaced. When checking the gap, make sure there are no signs of damage to the rotating vane assembly.

Ford's **Electronic Distributorless Ignition System (EDIS)** is a high data rate system. The crankshaft position sensor is a variable reluctance-type sensor triggered by a 36-minus-1 tooth trigger wheel pressed onto the rear of the crankshaft dampener. The signal generated by this sensor is called a **variable rate sensor signal (VRS)**. The VRS signal provides engine position and rpm information to the ignition module. Base timing is 10 degrees BTDC and is not adjustable. The ignition module receives information from the crankshaft position sensor and other sensors to calculate the on and off times for the coils in order to achieve the desired dwell and spark advance. The ignition module also synthesizes a PIP signal for use by the PCM's engine control strategy.

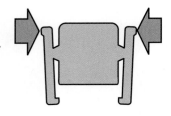

Figure 13-31 Spark plug wire locking tabs.

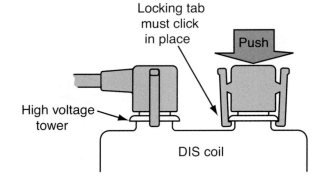

Figure 13-32 When reinstalling the plug wires, firmly push down on the center of the boot to lock it into position.

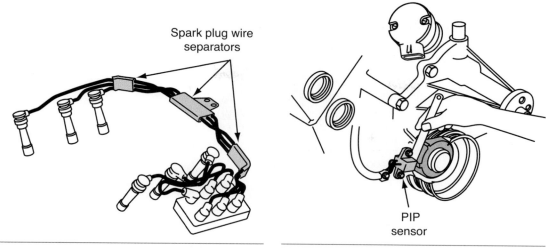

Figure 13-33 Make sure the wire separators are in place when reconnecting the spark plug wires.

Figure 13-34 Crankshaft position (PIP) sensor gap measurement and alignment.

Low Data Rate and High Data Rate EI Service and Diagnosis

The diagnostic procedure for the low data rate and high data rate systems varies according to the system being tested. It is important to follow the correct wiring diagram and procedure for the system being tested. Ford provides separate diagnostic harnesses for the two systems. The diagnostic harness has various leads that connect in series with each component in the system (Figure 13-35). A large connector on the harness is connected to Ford's breakout box

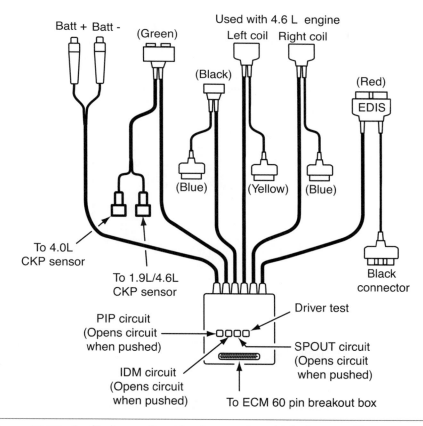

Figure 13-35 Ford's diagnostic wiring harness for a high data rate EI system.

Figure 13-36 Breakout box.

(Figure 13-36). Overlays for the ignition system on each engine are available to fit the breakout box terminals. These overlays identify the ignition terminals connected to the breakout box terminals (Figure 13-37).

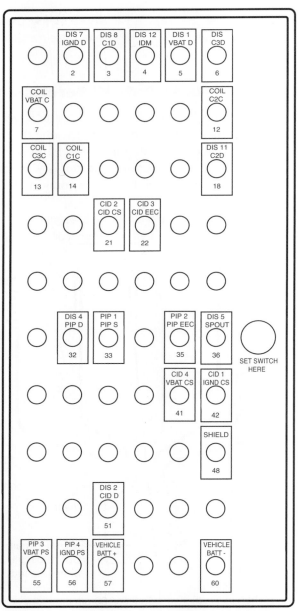

Figure 13-37 A typical breakout box overlay.

The technician must follow the test procedures given in the service manual and measure the resistance or voltage at the specified breakout box terminals connected to the ignition system. If a diagnostic harness, overlay, and breakout box are not available, the voltage and resistance measurements must be taken at the terminals of the individual components. The following is a general no-start diagnostic procedure for a high data rate system:

1. Connect a spark tester to each of the spark plug wires to ground and crank the engine. If the test plug does not fire on a pair of spark plugs connected to the same coil, test the spark plug wires connected to that coil. If these wires are good, the coil is probably bad. When the spark tester does not fire on any plug, continue testing.

2. Connect a voltmeter from terminal two on each coil pack to ground (Figure 13-38). With the ignition switch on, the voltmeter should read 12 volts. If the voltage is less than that, test the wire from the ignition switch to the coils and check the ignition switch.

3. With the ignition off, connect an ohmmeter across the primary terminals of each coil. If the primary winding resistance readings are not within specifications, replace the coil.

4. Connect the ohmmeter across the secondary terminals of each coil pack. Replace the coil pack if the secondary resistance is not within specifications.

5. Connect the ohmmeter between each primary coil terminal to the terminal on the primary terminal of the ignition module. Each wire should have less than 0.5 ohm resistance. If the resistance is greater than that, replace the wire.

6. Connect the voltmeter from terminal six on the ignition module to ground. With the ignition on, the voltmeter should read 12 volts. If the voltage is less than that, test the wire from the power relay to terminal six and the power relay itself. Turn the ignition off.

7. Connect the voltmeter from terminal ten at the ignition module and ground. The voltmeter should read 0.5 volt or less with the ignition on. If the reading is greater than that, repair or place the ground wire.

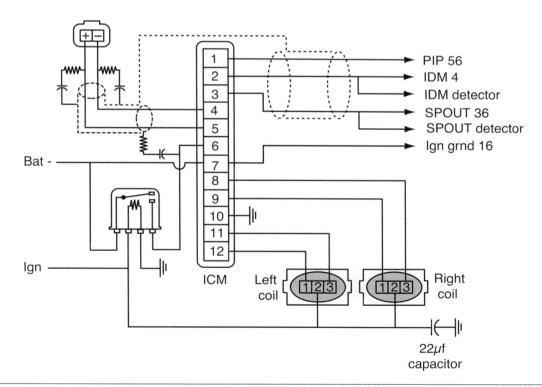

Figure 13-38 Wiring diagram for the EI system on a 4.6L engine.

8. Connect the voltmeter from terminal seven at the ignition module and ground. With the ignition on, the voltmeter should read less than the specified amount. If the voltage is higher, repair the ground wire.

9. Connect the voltmeter from terminals eight, nine, eleven, and twelve at the ignition module to ground and crank the engine. The voltmeter reading should fluctuate on each wire. If the readings do not fluctuate on one wire, replace the ignition module. If the readings do not fluctuate, proceed testing.

10. With the ignition off and the ignition module disconnected, connect an ohmmeter across terminals four and five at the harness. If it reads 2,300 to 2,500 ohms, the crankshaft position sensor and connecting wires are okay. If the readings are outside those specifications, repeat the test at the crankshaft position sensor terminals. If the readings are now within specifications, repair the wires. If the readings are still outside the specified range, replace the sensor.

11. Inspect the trigger wheel behind the crankshaft pulley and the crankshaft position sensor for damage.

12. If the voltmeter did not fluctuate during step 9 but the crankshaft position sensor checks out fine, replace the ignition module.

General Motors EI Systems

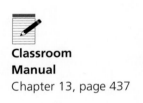

Classroom Manual
Chapter 13, page 437

GM has used several variations of EI systems. Some are waste spark systems, while others are direct or integrated systems. With each basic design there have been different operational designs as well. Each of these variances has its own diagnostic procedure. Make sure you identify the system being worked on prior to doing any exhaustive diagnostics. Each of the diagnostic procedures are based on the premise that the secondary circuit has been inspected and tested before testing the primary. Obviously, if the engine does not start, the secondary should not be suspect. However, if the engine misfires, the secondary circuit should be looked at.

The spark plug wires should be visually inspected and their resistance checked with an ohmmeter. Check the spark plug wires across the entire length of the cable, including the cable ends. Some GM engines are equipped with an integrated ignition coil and module assembly. This assembly must be removed (Figure 13-39) to access the spark plugs. Once the assembly is

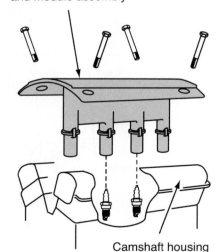

Figure 13-39 Integrated ignition coil and module assembly.

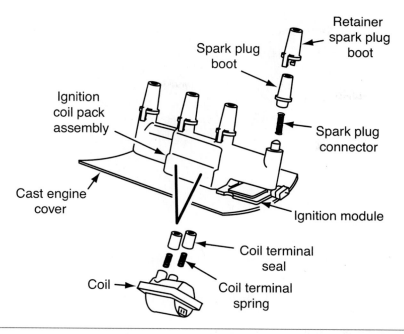

Figure 13-40 All of the ignition coil and module assembly should be inspected.

removed, all of its components should be inspected and tested with an ohmmeter (Figure 13-40). This testing includes the various connectors and wiring harnesses, as well as the ignition coils, mounted to cover for the assembly.

Ignition Coil Tests

With the coil terminals disconnected, an ohmmeter calibrated on the X1 scale should be connected to the primary coil terminals to test the primary winding. The primary winding in any EI coil should have 0.35 to 1.50 ohms resistance. An ohmmeter reading below the specified resistance indicates a shorted primary winding. A higher reading indicates excessive resistance, and an infinity meter reading proves that the primary winding is open.

An ohmmeter calibrated on the X1,000 scale should be connected to each pair of secondary coil terminals to test the secondary windings. The coil secondary winding in an EI type I system should have 10,000 to 14,000 ohms resistance, whereas the secondary windings in EI type 2 systems have 5,000 to 7,000 ohms resistance. If the secondary winding is open, the ohmmeter reading is infinity. A shorted secondary winding provides an ohmmeter reading below the specified resistance.

Crankshaft Sensors

A basic timing adjustment is impossible on any EI system. However, if the gap between the blades and the crankshaft sensor is incorrect, the engine may fail to start, stall, misfire, or hesitate on acceleration. Follow these steps during the crankshaft sensor adjustment procedure:

1. Install the sensor loosely on the pedestal.
2. Position the sensor and pedestal on the special adjusting tool recommended by GM.

Special Tools

GM special sensor adjusting tool

Figure 13-41 Using the special alignment tool to adjust the crankshaft sensor.

3. Position the adjusting tool on the crankshaft surface (Figure 13-41).
4. Tighten the pedestal-to-block mounting bolts to the specified torque.
5. Tighten the pinch bolt to the specified torque.
6. The interrupter rings on the back of the crankshaft pulley should be checked for a bent condition, and the same crankshaft sensor adjusting tool may be used to check these rings. Place the tool on the pulley extension surface and rotate the tool around the pulley (Figure 13-42). If any blade touches the tool, replace the pulley.

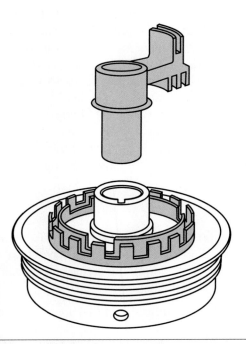

Figure 13-42 Checking the interrupter rings with the crankshaft sensor alignment tool.

If a single-slot crankshaft sensor requires adjustment, follow this procedure:

⚠ WARNING: Always be sure the ignition switch is off before attempting to rotate the crankshaft with a socket and breaker bar. If the ignition switch is on, the engine may start suddenly and rotate the socket and breaker bar with tremendous force. This action may result in personal injury and vehicle damage.

1. Be sure that the ignition switch is off and then rotate the crankshaft with a pull handle and socket installed on the crankshaft pulley nut. Continue rotating the crankshaft until one of the interrupter blades is in the sensor and the edge of the interrupter window is at the edge of the defector on the pedestal.

2. Insert the special adjustment tool between each side of the blade and the sensor. If the tool does not fit between each side of the blade and the sensor, adjustment is required. The gap measurement should be repeated at all three blades.

3. If a sensor adjustment is necessary, loosen the pinch bolt and insert the adjusting tool between each side of the blade and sensor. Move the sensor as required to insert the gauge.

4. Tighten the sensor pinch bolt to the specified torque.

5. Rotate the crankshaft and recheck the gap at each blade.

No-Start Diagnosis

If the engine fails to start, follow these steps for a no-start ignition diagnosis:

1. Connect a spark tester from each spark plug wire to ground and crank the engine while observing the spark tester.

2. If the spark tester does not fire on any spark plug, check the 12-volt supply wires to the coil module; some coil modules have two fused 12-volt supply wires. Consult the manufacturer's wiring diagrams for the car being tested to identify the proper coil module terminals.

3. If the spark tester does not fire on a pair of spark plugs, the coil connected to that pair of spark plugs is probably defective.

4. If the spark tester did not fire on any of the spark plugs and the 12-volt supply circuits to the coil module are satisfactory, disconnect the crankshaft and camshaft sensor connectors and connect short jumper wires between the sensor connector and the wiring harness connector. Be sure the jumper wire terminals fit securely to maintain electrical contact. Each sensor has a voltage supply wire, a ground wire, and a signal wire on 3.8L engines. On the 3.3L and 3300 engines, the dual crankshaft sensor has a voltage supply wire, ground wire, crank signal wire, and SYNC signal wire. Identify each of these wires on the wiring diagram for the system being tested.

5. Connect a digital voltmeter to each of the camshaft and crankshaft sensor black ground wires to an engine connection. With the ignition switch on, the voltmeter reading should be 0.2 volts or less. If the reading is above 0.2 volts, the sensor ground wires have excessive resistance.

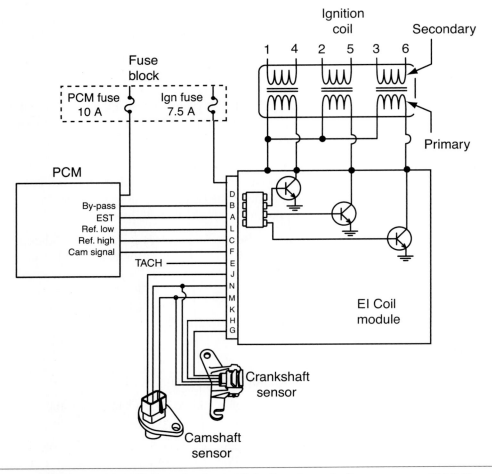

Figure 13-43 Wiring diagram for the crankshaft and camshaft sensors on a GM 3.8L engine.

6. With the ignition switch on, connect a digital voltmeter from the camshaft and crankshaft sensor white/red voltage supply wires to an engine ground (Figure 13-43). The voltmeter readings should be 5 to 11 volts. If the readings are below these values, check the voltage at the coil module terminals that are connected to the camshaft and crankshaft sensor voltage supply wires. When the sensor voltage supply readings are low at the coil module terminals, the coil module should be replaced. If the voltage supply readings are low at either sensor connector but satisfactory at the coil module terminal, the wire from the coil module to the sensor is defective. On the 3.3L and 3300 engines, the crankshaft sensor ground wire and voltage supply wire are checked in the same way as explained in steps 5 and 6.

7. If the camshaft and crankshaft sensor ground and voltage supply wires are satisfactory, connect a digital voltmeter to each sensor signal wire and crank the engine. Each sensor should have a 5-volt and 7-volt fluctuating signal. On the 3.3L and 3300 engines, test this voltage signal on the crank and SYNC signal wires at the crankshaft sensor. If the signal is less than specified, replace the sensor with the low signal.

8. When the camshaft and crankshaft sensor signals on 3.8L engines or crank and SYNC signals on 3.3L and 3300 engines are satisfactory and the spark tester did not fire at any spark plug, the coil module is probably defective.

9. On 3.8L engines where the coil assembly is easily accessible, the coil assembly screws may be removed and the coil lifted up from the module with the primary coil wires still connected. Connect a 12-volt test lamp across each pair of coil primary wires and crank the engine. If the test lamp does not flicker on any of the coils, the coil module is defective, assuming that the crankshaft and camshaft sensor readings are satisfactory.

Some GM EI systems are referred to as fast-start systems. These systems require additional steps during the diagnosis of a no-start condition. When an EI Type 1 Fast-Start System fails to start, complete steps 1,2, and 3 in the previous procedure, then do the following:

1. If the 12-volt supply circuits to the coil module are satisfactory, disconnect the crankshaft sensor connector and connect four short jumper wires between the sensor connector and the wiring harness connector.

2. Connect a digital voltmeter from the sensor ground wire to an engine ground. With the ignition switch on, the voltmeter should read 0.2 volts or less. A reading above this value indicates a defective ground wire.

3. Connect the voltmeter from the sensor voltage supply wire to an engine ground. With the ignition switch on, the voltmeter reading should be 8 to 10 volts. If the reading is lower than specified, check the voltage at coil module terminal N (Figure 13-44).

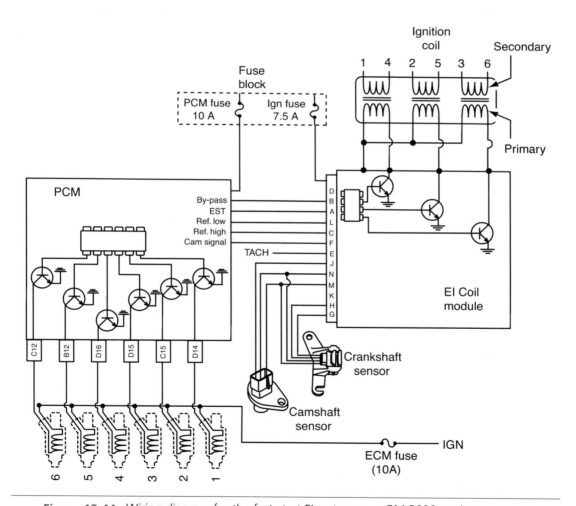

Figure 13-44 Wiring diagram for the fast-start EI system on a GM 3800 engine.

When the voltage at terminal N is satisfactory and the reading at the sensor voltage supply wire is low, the wire from terminal N to the sensor is defective. A low-voltage reading at terminal N indicates a defective coil module.

4. If the readings in steps 3 and 4 are satisfactory, connect a voltmeter from the 3X and 18X signal wires at the sensor connector to an engine ground and crank the engine. The voltmeter reading should fluctuate from 5 to 7 volts. The exact voltage may be difficult to read, especially on the 18X signal, but the reading must fluctuate. If the voltmeter reading is steady on either sensor signal, the sensor is defective.

5. Connect a digital voltmeter from the 18X and 3X signal wires at the coil module to an engine ground and crank the engine. The voltmeter readings should be the same as in step 4. If these voltage signals are satisfactory at the coil module terminals but low at the sensor, repair the wires between the coil module and the sensor.

6. If the 18X and 3X signals are satisfactory at the coil module terminals, remove the coil assembly-to-module screws and lift the coil assembly up from the module. Connect a 12-volt test lamp across each pair of coil primary terminals and crank the engine. If the test lamp does not flash on any pair of terminals, the coil module is defective.

Module Testing

EI module tests are easy to perform. The EI module can be compared to the regular distributor ignition system module in its operation. Previous testing steps in the diagnostic procedure have probably brought you to this point. It was likely that a cylinder or pair of cylinders matched to the same coil was misfiring or had no spark. Next, through an elimination process, it must be determined whether the module or the coil is at fault. Tests can be performed dynamically with the module installed. Identify the affected coil pack and remove it from the module. This will expose the two connector terminals from the module that slide onto the coil unit. If the system you are working on groups the coils into one unit rather than possessing individual coils, remove the coil pack assembly. This will expose the wires and terminal ends from the module to the coil. Slide the wire connectors off the coil assembly. Be sure to note the orientation of the coil and identify the module wires as you remove them from the coil. Once the connectors are open, install a test light in place of the coil pack across the module terminals. The test light acts as a substitute load to the module. Crank the engine and observe the test light. If the light flashes, the module is switching the primary ignition circuit properly, signifying that the problem is within the coil. If the light does not flash, the module is defective. Remember, it is possible for the module to be defective in such a way that one coil of the set may fire while another may not. It is impossible to repair or service a malfunctioning or inoperative ignition module, therefore the entire unit must be replaced.

☑ **SERVICE TIP:** If the module tests discussed above lead to a replacement of the modules, be sure to thoroughly test the disconnected coil before reinstalling it. It is possible the coil has an internal short that caused the module to fail. By reinstalling the affected coil on the new module, a repeat failure is likely to occur. It is recommended that the coil be tested using an approved spark tester that will adequately stress the coil to simulate real driving conditions.

EI Systems with Magnetic Sensors

With the wiring harness connector to the magnetic sensor disconnected and an ohmmeter calibrated on the X10 scale connected across the sensor terminals, the meter should read 900 to 1,200 ohms on 2.0L, 2.2L, 2.8L, 3.1L, and 3.4L engines. The meter should indicate 500 to 900 ohms on a Quad 4 (2.4L) engine, and 800 to 900 ohms on a 2.5L engine (Figure 13-45).

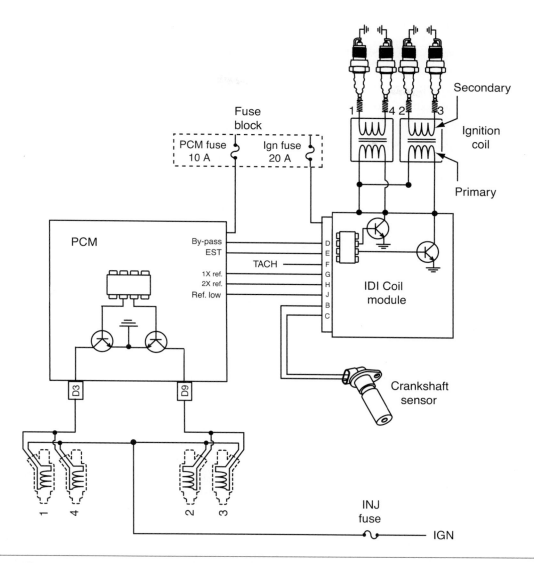

Figure 13-45 Wiring diagram for an integrated EI system.

Meter readings below the specified value indicate a shorted sensor winding, whereas infinity meter readings prove that the sensor winding is open. Since these sensors are mounted in the crankcase, they are continually splashed with engine oil. In some sensor failures, the engine oil enters the sensor and causes a shorted sensor winding. If the magnetic sensor is defective, the engine fails to start.

With the magnetic sensor wiring connector disconnected, an alternating current (AC) voltmeter may be connected across the sensor terminals to check the sensor signal while the engine is cranking. On 2.0L, 2.8L, 3.1L, and 3.4L engines, the sensor signal should be 100 millivolts (mV) AC. The sensor voltage on a Quad 4 engine should be 200 mV AC. When the sensor is removed from the engine block, a flat steel tool placed near the sensor should be attracted to the sensor if the sensor magnet is satisfactory.

No-Start Diagnosis

When an engine with an EI system and a magnetic sensor fails to start, complete steps 1, 2, and 3 of the previously given no-start diagnostic procedure, then follow this procedure:

1. If the spark tester did not fire on any of the spark plugs, check for 12 volts at the coil module voltage input terminals. Consult the wiring diagram for the system being tested for terminal identification.

2. If 12 volts are supplied to the appropriate coil module terminals, test the magnetic sensor.

3. When the magnetic sensor tests are satisfactory and the engine does not start because of an ignition problem, the coil module is probably defective.

Mitsubishi EI Systems

Mitsubishi direct electronic ignition systems are found on some Chrysler products and all Mitsubishi vehicles. This system uses a crank angle sensor to monitor engine speed and the location of cylinder number one. The signals from the sensor are sent to the PCM, which sends ignition signals to a power transistor to control ignition timing and dwell. There is a power transistor for each ignition coil. This system uses the waste spark method of firing cylinders.

Like all EI systems, the ignition coils can be tested with an ohmmeter. The resistance of the primary winding should be 0.77 to 0.95 ohm. To measure the resistance of the secondary winding, connect the meter across the adjacent spark plug terminals at the coil. For example, the towers for cylinders one and four and cylinders two and three. The readings should be 10,300 to 13,900 ohms. Readings outside the specifications indicate that the coil should be replaced.

The power transistor can also be checked with an ohmmeter. However, the transistor must be energized with an external power source. Since there are different system designs used on different Mitsubishi models, always refer to the service manual for proper terminal identification. The following procedure refers to the connector shown in Figure 13-46. Unplug the transistor's electrical connector. Connect the positive lead of the meter to terminal three and the negative lead to terminal one. The reading should be infinity. Now connect the positive side of a 1.5-volt battery to terminal two and the negative side to three. With the ohmmeter still connected to terminals one and three, it should now read low resistance.

Now move the positive lead from the battery to terminal five and the negative lead to terminal three. Connect the positive ohmmeter lead to terminal three and the negative lead to terminal six. There should be good continuity between these two points. If any readings are outside of the specifications, the power transistor assembly should be replaced.

Figure 13-46 Typical Mitsubishi power transistor connector.

Nissan EI Systems

Some Nissan and Infiniti vehicles are equipped with a direct ignition system. The system is similar to other manufacturers, but Nissan does not rely on waste spark firing. Nissan uses coil-over-plug, or COP. Each plug has its own coil. The system uses a crankshaft position sensor to monitor engine speed and the location of cylinder number one. Signals from this sensor and other sensors are sent to the PCM, which controls a power transistor. The power transistor controls the action of the coils. A relay, called the power transistor relay, sends 12 volts to the ignition coils when the relay is energized. The power transistor controls the ground circuit of the primary windings in each coil.

The ignition coils are tested with an ohmmeter in the same way as other coils. Primary winding resistance and secondary winding resistance are measured across the windings on each coil. Make sure the coils are removed from the spark plugs and any source of power before testing them with an ohmmeter. Typically, 0.7 ohm is the desired primary resistance and 7,000 to 8,000 ohms are desired for the secondary. If the resistance readings are outside the specifications stated in the service manual, replace the coil.

If the coils check out with an ohmmeter but still do not spark, check the voltage from the relay. There should be 12 volts. If this voltage is not present, check to see if battery voltage is available to the relay. If voltage is present there, check the relay control wires from the PCM. If the circuit is good, the relay should be replaced. If no voltage is available to the relay, check the relay control circuit from the PCM and the ignition switch. If the wires and the ignition switch are good, continue testing by the PCM.

The power transistor can be checked with an ohmmeter. Refer to the service manual for the proper terminal combinations and meter polarity required for testing the transistor. Figure 13-47 is an example of these testing requirements. If any combination does give the desired readings, replace the transistor assembly.

Infiniti is Nissan's luxury car.

Terminal Combinations				Meter polarity	Continuity?
1	2	3	4	+	Yes
a	b	c	d	−	
1	2	3	4	−	No
a	b	c	d	+	
1	2	3	4	+	Yes
e	e	e	e	−	
1	2	3	4	−	No
e	e	e	e	+	
e	e	e	e	+	Yes
a	b	c	d	−	
e	e	e	e	−	Yes
a	b	c	d	+	

Figure 13-47 Terminal combinations and desired ohmmeter readings at the power transistor connector in Nissan EI systems.

Toyota EI Systems

Some Toyota and Lexus engines are equipped with electronic ignition. One of the systems used by Toyota is very similar to the waste spark systems used by everyone else. It uses multiple camshaft sensors, one crankshaft sensor, an ignition module, and one coil for each pair of cylinders. There is one camshaft sensor for each coil. A major difference with this system is that the ignition coil is mounted directly over one spark plug. This eliminates the second spark plug wire from the ignition coil. The other systems have an individual ignition coil mounted over each spark plug.

Toyota uses a special locking feature to secure the spark plug wires to the spark plugs and the ignition coils. To remove the wires, begin by using a pair of needle nose pliers. Disconnect the cable clamp from the engine wire protector (Figure 13-48).

Disconnect the cables from the spark plugs by firmly holding the boot and, with a twist-pull effort, remove the cable from the plug (Figure 13-49).

Using a screwdriver, lift up the locking tab and separate the spark plug wire from the ignition coils (Figure 13-50).

On some engines it may be necessary to remove a wire protector assembly to remove the spark plug wires. Once the wires are removed, they can be checked with an ohmmeter (Figure 13-51). The maximum allowable resistance across a secondary cable is 26,000 ohms. If the resistance is higher, check the condition of the terminals. If they are fine, replace the cable.

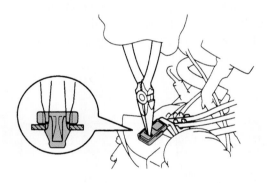

Figure 13-48 Use needle nose pliers to remove the wires from the spark plug cable protector.

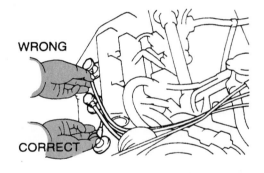

Figure 13-49 Pull the cable off the plug by grasping the boot, not the wire.

Figure 13-50 Use a screwdriver to lift the lock at the cable's connection to the coil.

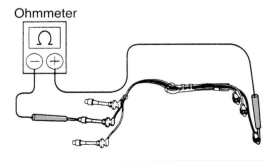

Figure 13-51 Measuring the resistance of the spark plug cable assembly.

To reinstall the spark plug cables, reverse the procedure. Make sure all locking mechanisms are secure and that the routing of the cables is correct (Figure 13-52).

Many Toyota, as well as other import engines, use dual electrode spark plugs. Contrary to common belief, the air gap of this type plug is very important and is adjustable. If the air gap to one of the ground electrodes is larger than the gap to the other electrode, electrons will always jump the smaller gap. The air gaps to both electrodes must be the same. Check and adjust the gap with a wire feeler gauge (Figure 13-53).

Critical to the proper operation of the system is the air gap at the camshaft sensor. Although the gap is not adjustable, it should be checked whenever diagnostics points to a faulty camshaft signal. To measure the gap, use a non-magnetic feeler gauge. The gap should be between 0.0008 and 0.016 inch. If the gap is not within these specifications, the sensor should be replaced.

The camshaft sensor should also be checked with an ohmmeter (Figure 13-54). Connect the negative ohmmeter lead to the ground terminal at the sensor and measure the resistance

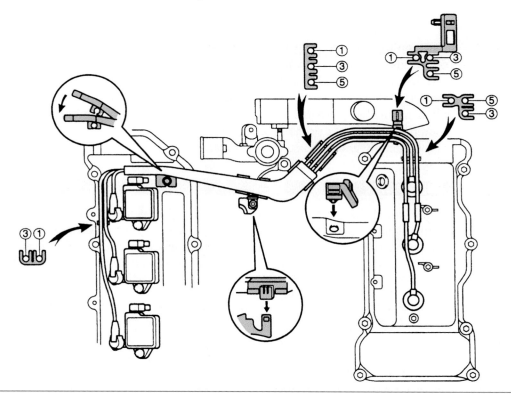

Figure 13-52 Correct routing for the spark plug cables.

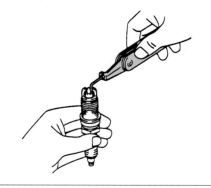

Figure 13-53 Check the air gap at both spark plug electrodes.

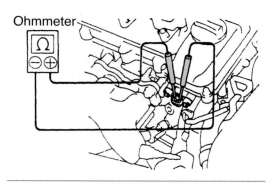

Figure 13-54 Checking a camshaft sensor with an ohmmeter.

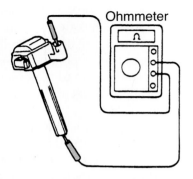

Figure 13-55 Checking the secondary of an ignition coil.

between that terminal and the others, one at a time. The resistance should be within the specified range. If not, replace the sensor.

The secondary ignition coil windings may contain a high-voltage diode, which is why Toyota only recommends and gives specifications for measuring the resistance across the primary windings of some of their coils. When the resistance is measurable, the normal secondary winding resistance is 11,000 to 17,500 (11 to 17.5K) ohms (Figure 13-55). The specified primary winding resistance is normally less than 1 ohm. Also check the coils for shorts to ground by connecting one ohmmeter lead to a primary or secondary terminal and the other to ground. There should be no continuity and the meter should read infinity.

CASE STUDY

A customer complained about a stalling problem on a Pontiac with an EI system. When questioned about this problem, the owner said the engine stalled while driving in the city, but it only happened about once a week. Further questioning of the customer indicated the engine would restart after 5 to 10 minutes, and the owner said the engine seemed to be flooded.

The technician performed voltmeter tests, ohmmeter tests, and an oscilloscope diagnosis on the EI system. There were no defects in the system, and the engine operation was satisfactory. The fuel pump pressure was tested and the filter checked for contamination, but no problems were discovered in these components. The customer was informed that it was difficult to diagnose this problem when the symptoms were not present. The service writer asked the customer to phone the shop immediately the next time the engine stalled without attempting to restart the car. The service writer told the customer that a technician would be sent out to check the problem when she phoned the shop.

Approximately 10 days later, the customer phoned and said her car had stalled. A technician was dispatched immediately, and she connected a spark tester to several of the spark plug wires. The ignition system was not firing any of the spark plugs. The car was towed to the shop without any further attempts to start the engine.

The technician discovered that the voltage supply and the ground wires on the crankshaft sensor were normal, but there was no voltage signal from this sensor while cranking the engine. The crankshaft sensor was replaced and adjusted properly. When the customer returned later for other service, she reported the stalling problem had been eliminated.

Terms to Know

Electronic distributorless ignition system (EDIS)

Variable rate sensor signal (VRS)

ASE-Style Review Questions

1. While discussing the primary and secondary ignition circuits:
   Technician A says the primary controls the secondary.
   Technician B says the primary circuit also controls ignition timing in the secondary circuit.
   Who is correct?
   - **A.** A only
   - **B.** B only
   - **C.** Both A and B
   - **D.** Neither A nor B

2. While discussing EI ignition systems:
   Technician A says base timing is not adjustable on EI systems.
   Technician B says that the air gap is adjustable on some crank sensors but timing is not affected.
   Who is correct?
   - **A.** A only
   - **B.** B only
   - **C.** Both A and B
   - **D.** Neither A nor B

3. *Technician A* says a logic probe can be used to check a Hall effect sensor.
   Technician B says the red LED should light on both end terminals with the key on.
   Who is correct?
   - **A.** A only
   - **B.** B only
   - **C.** Both A and B
   - **D.** Neither A nor B

4. Two technicians are discussing a Nissan EI system:
   Technician A says that the coil used for this system should have a primary resistance of 2 ohms.
   Technician B says that the secondary circuit should have between 7,000 and 8,000 ohms resistance.
   Who is correct?
   - **A.** A only
   - **B.** B only
   - **C.** Both A and B
   - **D.** Neither A nor B

5. While discussing waste spark EI systems:
   Technician A says a high resistance spark plug can affect the firing of its companion spark plug.
   Technician B says improper spark plug torque can cause an engine misfire.
   Who is correct?
   - **A.** A only
   - **B.** B only
   - **C.** Both A and B
   - **D.** Neither A nor B

6. While discussing how to test a crankshaft position sensor:
   Technician A says a logic probe can be used.
   Technician B says a DMM can be used.
   Who is correct?
   - **A.** A only
   - **B.** B only
   - **C.** Both A and B
   - **D.** Neither A nor B

7. While discussing the possible causes for a no-start condition on an EI-equipped engine:
   Technician A says a shorted crankshaft sensor may prevent the engine from starting.
   Technician B says a shorted spark plug may stop the engine from starting.
   Who is correct?
   - **A.** A only
   - **B.** B only
   - **C.** Both A and B
   - **D.** Neither A nor B

8. While discussing the diagnosis of an EI system in which the crankshaft and camshaft sensor tests are satisfactory but a spark tester connected from the spark plug wires to ground does not fire:
   Technician A says the coil assembly may be defective.
   Technician B says the voltage supply wire to the coil assembly may be open.
   Who is correct?
   - **A.** A only
   - **B.** B only
   - **C.** Both A and B
   - **D.** Neither A nor B

9. While discussing EI service and diagnosis:

Technician A says the crankshaft sensor may be rotated to adjust the basic ignition timing.

Technician B says the crankshaft sensor may be moved to adjust the clearance between the sensor and the rotating blades on some EI systems.

Who is correct?

A. A only   C. Both A and B
B. B only   D. Neither A nor B

10. While testing the coils in an EI system:

Technician A says an infinity reading means that the windings have zero resistance and are shorted.

Technician B says the primary windings in each coil should be checked for shorts to ground.

Who is correct?

A. A only   C. Both A and B
B. B only   D. Neither A nor B

ASE Challenge Questions

1. Chrysler EI systems are being discussed.

Technician A says a no-start condition may also shut down the PCM.

Technician B says a scan tool will not retrieve DTCs if the cam sensor feed circuit is shorted.

Who is correct?

A. A only   C. Both A and B
B. B only   D. Neither A nor B

2. GM EI systems are being discussed.

Technician A says a hesitation on acceleration may be caused by a bent interrupter ring.

Technician B says a no-start condition will result if there is an oil leak in the crankshaft's magnetic sensor.

Who is correct?

A. A only   C. Both A and B
B. B only   D. Neither A nor B

3. *Technician A* says some Toyotas' EI waste spark systems use one spark plug wire per coil.

Technician B says loss of engine performance may be traced to improper air gap on one electrode of the dual electrode spark plug.

Who is correct?

A. A only   C. Both A and B
B. B only   D. Neither A nor B

4. *Technician A* says that if the primary ignition system does not receive a trigger, then there will also be no injector pulse.

Technician B says that if there is injector pulse and no ignition, the cause is probably in the secondary ignition system.

Who is correct?

A. A only   C. Both A and B
B. B only   D. Neither A nor B

5. The waveform shown in the figure above is typical of a:

A. Hall effect switch.
B. Hall effect switch with inductive sensor.
C. Hall effect switch sensor being tested with a soldering gun.
D. none of the above.

Job Sheet 35

35

Name _____ Date _____

Visually Inspect and Test an Electronic Ignition (EI) System

Upon completion of this job sheet, you should be able to perform a visual inspection of the EI-type system and perform routine tests.

ASE Correlation

This job sheet is related to the ASE Engine Performance Test's content area: ignition system diagnosis and repair. Tasks: inspect system components, no-start testing, driveability, and emissions problems with electronic ignition systems and determine needed repairs; inspect and test primary circuit wiring and components and repair or replace as needed; and inspect and test secondary ignition system wiring and components and service as needed.

Tools and Materials

Clean rag   A vehicle equipped with an EI system
Appropriate service manual   Scan tool

Describe the vehicle being worked on.

Year _____ Make _____ VIN _____

Model _____

Type of ignition _____

Procedures

1. Are the spark plug cables connected and firmly seated on the spark plugs and coil pack terminals? ☐ Yes ☐ No

2. Do the spark plug cables show signs of cracked, worn, or oil-soaked insulation? ☐ Yes ☐ No

3. Are the spark plug cables routed properly and in their support looms? ☐ Yes ☐ No

4. Are there any white or grayish powdery deposits on the spark plug cables? ☐ Yes ☐ No

5. Is (are) the primary wire connector(s) firmly inserted into the EI module housing? ☐ Yes ☐ No

6. Are the connections to the module corroded? ☐ Yes ☐ No

7. If this system has the coils mounted to the EI module, are the coil(s) tightly bolted and positioned properly? ☐ Yes ☐ No

8. Is the module/coil assembly properly fastened to the engine mounting brackets or as individual components if so designed? ☐ Yes ☐ No

9. Are there any cracks or burn marks in the coil towers? ☐ Yes ☐ No

10. If equipped with a Hall effect crankshaft sensor, is there evidence of scraping, rubbing, or other contact marks on the interrupter blades? ☐ Yes ☐ No

11. Are there oil, antifreeze, or other contaminants on the crankshaft sensor or wire connector? ☐ Yes ☐ No

12. Connect a scan tool and check for any ignition-related or misfire-related fault codes. Are any fault codes present? ☐ Yes ☐ No

13. Record a summary of the visual inspection and tests performed. Include all findings including items that passed as well as items that failed. _____

_____

_____

_____

_____

14. Based on the visual inspection, what are your recommendations? _____

_____

_____

_____

_____

Instructor's Response_____

_____

_____

Emission Control System Diagnosis and Service

Upon completion and review of this chapter, you should be able to:

❏ Describe the different types of state vehicle emissions (I/M) control programs.

❏ Diagnose and service evaporative (EVAP) systems.

❏ Describe the testing of the canister and of the canister purge valve.

❏ Describe the inspection and replacement of positive crankcase ventilation (PCV) system parts.

❏ Diagnose spark control systems.

❏ Diagnose engine performance problems caused by improper exhaust gas recirculation (EGR) operation.

❏ Diagnose and service the various types of EGR valves.

❏ Diagnose EGR vacuum regulator (EVR) solenoids.

❏ Diagnose EGR pressure transducers (EPT).

❏ Check the efficiency of a catalytic converter.

❏ Diagnose and service secondary air injection systems.

Introduction

The three kinds of emissions that are being controlled in gasoline engines today are unburned hydrocarbons, carbon monoxide, and oxides of nitrogen.

Unburned hydrocarbons (HC) are particles, usually vapors, of gasoline that have not been burned during combustion. They are present in the exhaust and in crankcase vapors. Any raw gas that evaporates from the fuel system is classed as HC.

Carbon monoxide (CO) is a poisonous chemical compound of carbon and oxygen. It forms in the engine when there is not enough oxygen to combine with the carbon during combustion. When there is enough oxygen in the mixture, carbon dioxide (CO_2) is formed. CO_2 is not a pollutant and is the gas used by plants to manufacture oxygen. CO is found in the exhaust principally, but can also be in the crankcase.

Oxides of nitrogen (NO_X) are various compounds of nitrogen and oxygen. Both of these gases are present in the air used for combustion. They are formed in the cylinders during combustion and are part of exhaust gas. NO_X is the principle chemical that causes smog.

What comes out of an engine's exhaust depends on two factors. One is the effectiveness of the emission control devices. The other is the effectiveness or efficiency of the engine. A totally efficient engine changes all of the energy in the fuel, in this case gasoline, into heat energy. This heat energy is the power produced by the engine. To run, an engine must receive fuel, air, and heat. Our air is primarily nitrogen and oxygen, but the engine only uses the oxygen.

In order for an engine to run efficiently, it must have fuel mixed with the correct amount of air. This mixture must be shocked by the correct amount of heat (spark) at the correct time. All of this must happen in a sealed container or cylinder. When conditions are met, a great amount of heat energy is produced and the fuel and air combine to form water and carbon dioxide. The nitrogen leaves the engine unchanged.

When the engine does not receive enough air or gets too much fuel, it produces less heat energy, water, and carbon dioxide. It also produces carbon monoxide and releases some fuel and air that was not burned. As the amount of air delivered to the engine decreases, the production of CO increases and more unburnt fuel and air is released in the exhaust.

Basic Tools

Vacuum gauge, exhaust gas analyzer, basic tool set, fender covers, safety glasses

Classroom Manual
Chapter 14, page 449

NO_x is commonly pronounced "knocks."

When the engine receives too much air, the amount of heat energy it produces increases, but complete combustion does not take place and the exhaust has large amounts of fuel and air in it. Because of the increased heat, NO_X is formed. If the cylinder cannot hold the pressures because of a compression leak, the amount of heat it produces decreases and large amounts of fuel and air are released in the exhaust.

Since it is nearly impossible for an engine to receive the correct amounts of everything, a good running engine will emit some amounts of pollutants. It is the job of the emission control devices to clean them up.

Emissions Testing

The first Clean Air Act prompted Californians to create the California Air Resources Board (California ARB). California ARB's purpose was to implement strict air standards. These became the standard for federal mandates. One of the approaches to clean the air by the ARB was to start periodic motor vehicle inspection (PMVI). The purpose for the PMVI, or **inspection and maintenance (I/M)**, is to inspect a vehicle's emission controls once a year. This inspection included a tailpipe emissions test and an underhood inspection. The tailpipe test certifies that the vehicle's exhaust emissions are within the limits set by law. The underhood and vehicle inspection verifies that the pollution control equipment has not been tampered with or disconnected.

The 1970 Clean Air Act made it illegal for professional mechanics to remove, disconnect, or tamper with factory emission controls. The law did not make it illegal for the owner of the vehicle to make the emission control devices inoperable. The Clean Air Act was revised in 1990 and it made it illegal for anyone to remove, disconnect, or tamper with emission controls on any vehicle that would be used on public roads.

Today, states have incorporated an emissions test with their annual vehicle registration procedures. Most states have the IM240 or similar program that was first implemented in California. The IM240 tests the emissions of a vehicle while it is operating under a variety of load conditions. This is an improvement over exhaust testing during idle and high speed with no load.

The IM240 test requires the use of a chassis dynamometer, commonly called a **dyno**. While on the dyno, the vehicle is operated for 240 seconds and under different load conditions. The test drive on the dyno simulates both in-traffic and highway driving and stopping. The emissions tester tracks the exhaust quality through these conditions.

The IM240 program also includes a leakage test (Figure 14-1) and a functional test (Figure 14-2) of the evaporative emission control devices and a visual inspection of the total emission control system. If the vehicle fails the test, it must be repaired and certified before it can be registered.

The exhaust analyzer typically used to certify vehicles is an infrared tester. Specific infrared testers have been approved to certify exhaust levels and must meet specific standards.

Testing a vehicle while it is driven on a chassis dynamometer is called transient testing.

Exhaust Analyzers

Testing the quality of the exhaust is both a procedure for testing emission levels and a diagnostic routine. One of the most valuable diagnostic aids is the exhaust analyzer (Figure 14-3).

A typical infrared exhaust analyzer has a long sample hose with a probe in the end of the hose. The probe is inserted in the vehicle's tailpipe. When the analyzer is turned on, an internal pump moves an exhaust sample from the tailpipe through the sample hose and the analyzer. A water trap and filter in the hose removes moisture and carbon particles.

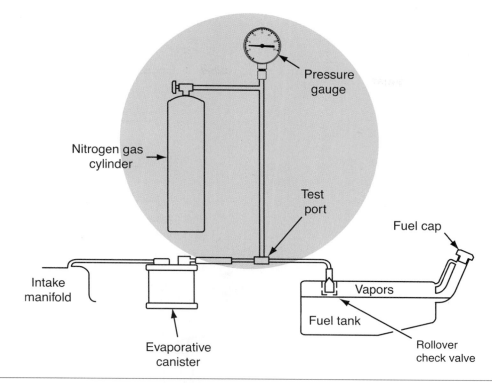

Figure 14-1 Typical fuel and evaporative test setup.

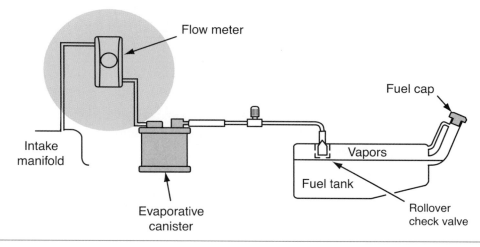

Figure 14-2 Typical evaporative system purge test setup.

Figure 14-3 Typical exhaust gas analyzer screen.

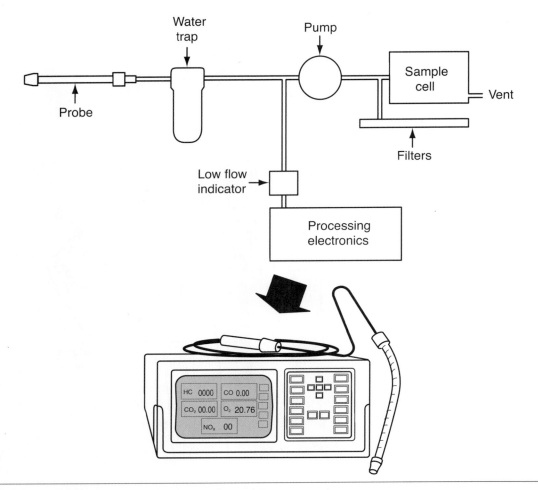

Figure 14-4 The basic operation of an exhaust gas analyzer.

The pump forces the exhaust sample through a sample cell in the analyzer. The exhaust sample is then vented to the atmosphere (Figure 14-4). In the sample cell, a beam of infrared light passes through the exhaust sample. The analyzer then determines the quantities of HC and CO if the analyzer is a two-gas analyzer, or HC, CO, CO_2, and O_2 if it is a four-gas analyzer. Some analyzers can also measure NO_X and are called five-gas analyzers.

When using the exhaust analyzer as a diagnostic tool, it is important to realize that the severity of the problem dictates how much higher than normal the readings will be. When attempting to identify the exact cause of the abnormal readings, disable the secondary air system. Then rerun the tests. Without the secondary air, the readings will give a more accurate look at the air-fuel mixture. This should help to find the cause of the problem.

IM240 Test

The IM240 test is a 240-second test of a vehicle's emissions. During this time, the vehicle is loaded to simulate a short drive on city streets and then a longer drive on a highway. The complete test cycle includes acceleration, deceleration, and cruising. During the test, the vehicle's exhaust is collected by a **constant volume sampling (CVS)** system that makes sure a constant volume of ambient air and exhaust pass through the exhaust analyzer. The CVS exhaust hose covers the entire exhaust pipe. Therefore, it collects all of the exhaust, not just a sample as regular gas analyzers do. The hose contains a mixing tee that draws in outside air to maintain a constant volume to the gas analyzer. This is important for the calculation of mass exhaust emissions. During an IM240 test, the exhaust gases are measured in grams per mile.

I/M 240 TESTING

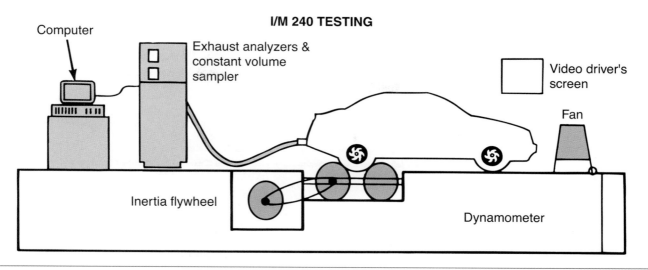

Figure 14-5 Components of a typical IM240 test station.

The test is conducted on a chassis dynamometer (Figure 14-5), which loads the drive wheels to simulate real-world conditions. The rollers of the dyno are loaded by a computer after an inspector inputs basic information about the vehicle to be tested. The computer's program will adjust the resistance on the rollers according to the vehicle's weight, aerodynamic drag, and road friction.

At most I/M testing stations when a vehicle enters an inspection lane, the driver is greeted by a lane inspector. The inspector receives the needed information from the driver, then directs the driver to the customer waiting area. The inspector moves the vehicle to be visually inspected. Upon completion of the inspection, the inspector enters all pertinent information into the computer. After this information has been entered, the inspector moves the car onto the inertia simulation dynamometer.

Once on the dyno, the hood of the vehicle is raised and a large cooling fan is moved in front of the car. The engine is allowed to idle for at least 10 seconds, then testing begins. The transient driving cycle is 240 seconds and includes periods of idle, cruise, and varying accelerations and decelerations. During the drive cycle, the car's exhaust is collected by the CVS system. The inspector driving the car must follow an electronic, visual depiction of the speed, time, acceleration and load relationship of the transient driving cycle. For cars equipped with a manual transmission, the inspector is prompted by the computer to shift gears according to a predetermined schedule.

The IM240 drive cycle is displayed by a trace. The trace is based on road speed versus time. The test is comprised of two phases—Phase 1 and Phase 2—combining the results of these two phases results in the Test Composite (Figure 14-6). Phase 1 is the first 95 seconds of the drive cycle.

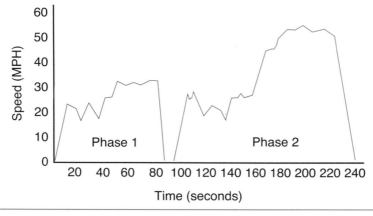

Figure 14-6 An IM240 drive trace.

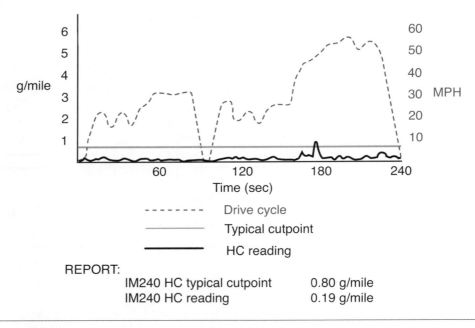

REPORT:

IM240 HC typical cutpoint	0.80 g/mile
IM240 HC reading	0.19 g/mile

Figure 14-7 A hydrocarbon trace from an IM240 test.

The car travels 56 hundredths of a mile. This phase represents driving on a flat highway at a maximum speed of 32.4 miles per hour under light to moderate loads. Phase 2 is from the 96th second of the test to the end, or the 240th second. The vehicle travels 1.397 miles on a flat highway at a maximum speed of 56.7 miles per hour, including a hard acceleration to highway speeds.

The end result of the IM240 test is the measurement of the pollutants emitted by the vehicle during normal, on-the-road driving. After the test, the customer receives an inspection report. If the car failed the emissions test, it must be fixed. This inspection report is a valuable diagnostic tool (Figure 14-7). To correct the problems that caused the vehicle to fail, the inspection report must be studied.

The cutpoint represents the maximum amount or limit of each gas that the vehicle is allowed to emit.

The report shows the amount of gases emitted during the different speeds of the test. It also shows the cutpoint for the various pollutants. Readings above the cutpoint line are above the maximum allowable. Nearly all vehicles will have some speeds and conditions where the emissions levels are above the cutpoint. A vehicle that fails the test will have many abnormal speeds or will have an area of very high pollutant output.

To use the report as a diagnostic tool, pay attention to all of the gases and to the loads and speeds at which the vehicle went over the cutpoint (Figure 14-8). Think about what system or sys-

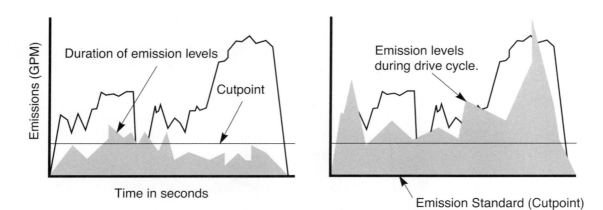

Figure 14-8 Left: The emissions level is acceptable. Right: The emissions level is unacceptable.

tems are responding to the load or speed. Would a malfunction of these systems cause this gas to increase? Pay attention not only to the gases that are above the cutpoint, but also to those that are below. If the HC readings are above the cutpoint at a particular speed and the NO_x readings are slightly below the cutpoint at the same speed, fixing the problem causing the high HC will probably cause the NO_x to increase above the cutpoint. As combustion is improved, the chances of forming NO_x also improves. Consider all of the correlations between gases when diagnosing a problem.

The IM240 measures HC, CO, NO_x, and CO_2. The job of a technician is to determine what loads and engine speeds the vehicle failed. Then test those systems that could affect engine operation and exhaust output at the failed speeds and loads. These tests could include looking at an O_2 waveform, watching a four-gas analyzer, checking data with a scan tool, reading injector pulse width, or doing other tests.

Most of the IM240 drive cycle can be simulated either in the diagnostic bay or on the road. It is hard to duplicate the entire drive cycle on a road test, but you can successfully duplicate part of the test. Hopefully it is the part that failed the vehicle. This may be important for diagnostics and is very important for the verification of the repair. The drive cycle is actually six operating modes. These modes need to be duplicated to identify why the vehicle failed that part of the trace.

- ❏ Mode one—idle, no load at 0 miles per hour
- ❏ Mode two—acceleration from 0 to 35 miles per hour
- ❏ Mode three—acceleration from 35 to 55 miles per hour
- ❏ Mode four—a steady cruise at 35 miles per hour
- ❏ Mode five—a steady high cruise at 55 miles per hour
- ❏ Mode six—decelerations from 35 miles per hour to 0 and from 55 miles per hour to 0

Keep these in mind, along with how many seconds each module lasts, and you will be able to duplicate the actual drive cycle as best as possible with the equipment you have. Once the problem has been diagnosed and repaired, repeat the test you did to find the problem. This will let you know if you fixed the problem.

Recently, many states are implementing other types of testing programs that are in place of the IM240 test.

Other I/M Testing Programs

Many state I/M programs only measure the emissions output from a vehicle while it is idling. The test is conducted with a certified exhaust gas analyzer. The measurements of the exhaust sample are then compared to standards dictated by the state according to the production year of the vehicle.

Some states also include a preconditioning mode at which the engine is run at a high idle (approximately 2,500 rpm) or the vehicle is run at 30 mph on a dynamometer for 20 to 30 seconds, prior to taking the idle tests. This preconditioning mode heats up the catalytic converter, allowing it to work at its best. Some programs include the measurement of the exhaust gases during a low constant load on the dyno or during a constant high idle. These measurements are taken in addition to the idle tests. A visual inspection and functional test of the emission control devices is part of some I/M programs. If the vehicle has tampered, non-functional, or missing emission control devices, the vehicle will fail the inspection.

The California ARB developed a test that incorporates steady-state and transient testing. This test is called the **Acceleration Simulation Mode (ASM)** test. The ASM includes a high-load steady-state phase and a 90-second transient test. This test is an economical alternative to the IM240 test, which requires a dyno with a computer-controlled power absorption unit. The ASM can be conducted with a normal chassis dyno and a five-gas analyzer.

Testing a vehicle at a constant load on a dyno is called steady state testing.

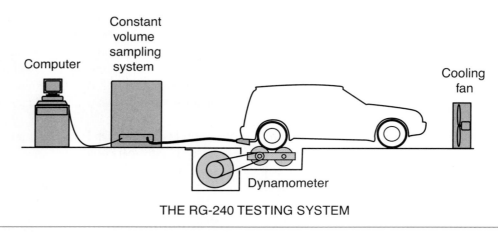

THE RG-240 TESTING SYSTEM

Figure 14-9 Typical components for a RG-240 test station.

**Repair Grade
(RG-240)** is an
economical version
of the standard
I/M 240 test.

Another alternative to the IM240 test is the **Repair Grade (RG-240)** test. This program uses a chassis dynamometer, constant volume sampling, and a five-gas analyzer (Figure 14-9). It is very similar to the IM240 test and is conducted in the same way, however it is more economical. The primary difference between the IM240 and the RG-240 is the chassis dyno. Although they accomplish basically the same task, the RG-240 dyno is less complicated but nearly matches the load simulation of the IM240 dyno.

The **OBD II** emission monitor is performed in conjunction with an exhaust emissions test or as a stand-alone test. More states will likely begin to use the OBD II test in lieu of the IM240 or ASM-type tests as time goes on. The ability of the OBD II system to reveal engine data, performance, and emission control-related history began on all domestic vehicles in 1996, although a few models had some OBD II capability in 1994 and 1995. The diagnostic trouble codes (DTC) and systems monitor data and history are retrieved by connecting a scanner to the under-dash diagnostic connector. When a repeated emissions-related failure occurs within a specific period of time, the PCM will turn the malfunction indicator lamp on to warn the driver of a problem. A DTC is then stored in memory and is used to identify the system at fault.

Evaporative Emission Control System Diagnosis and Service

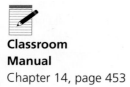

**Classroom
Manual**
Chapter 14, page 453

Evaporative
emissions are
hydrocarbon
emissions.

⚠ **WARNING:** If gasoline odor is present in or around a vehicle, check the EVAP system for cracked or disconnected hoses and check the fuel system for leaks. Gasoline leaks or escaping vapors may result in an explosion causing personal injury and/or property damage. The cause of fuel leaks or fuel vapor leaks should be repaired immediately. Do not overlook a loose fuel tank cap!

EVAP system diagnosis varies depending on the vehicle make and model year. Always follow the service and diagnostic procedure in the vehicle manufacturer's service manual. If the EVAP system (Figure 14-10) is purging vapors from the charcoal canister when the engine is idling or operating at very low speed, rough engine operation will occur, especially at higher atmospheric temperatures. Cracked hoses or a canister saturated with gasoline may allow gasoline vapors to escape to the atmosphere, resulting in gasoline odor in and around the vehicle.

All of the hoses in the EVAP system should be checked for leaks, restrictions, and loose connections. The electrical connections in the EVAP system should be checked for looseness, corroded terminals, and worn insulation. When a defect occurs in the canister purge solenoid and related circuit, a DTC is usually set in the PCM memory. If a DTC related to the EVAP system is set in the PCM memory, always correct the cause of this code before further EVAP system diagnosis.

A scan tester may be used to diagnose the EVAP system. In the appropriate tester mode, the tester indicates whether the purge solenoid is on or off. Connect the scan tester to the DLC and

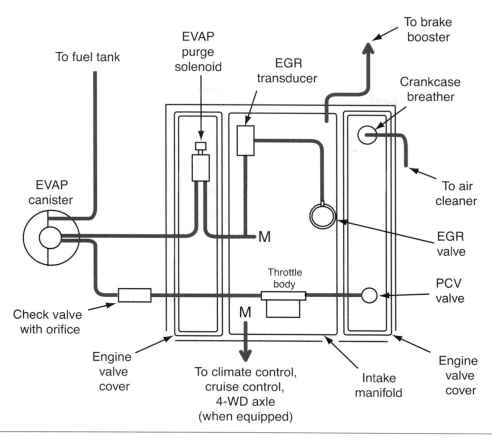

Figure 14-10 A typical EVAP system.

start the engine. With the engine idling, the purge solenoid should be off. Leave the scan tester connected and road test the vehicle. Be sure all the conditions required to energize the purge solenoid are present, and observe this solenoid status on the scan tester. The tester should indicate the purge solenoid is on when all the conditions are present for canister purge operation. If the purge solenoid is not on under the necessary conditions, check the power supply wire to the solenoid, solenoid winding, and the wire from the solenoid to the PCM.

EVAP System Component Diagnosis

Check the canister to make sure that it is not cracked or otherwise damaged (Figure 14-11). Also make certain that the canister filter is not completely saturated. Remember that a saturated char-

Special Tools

Vacuum gauge,
hand-operated
vacuum pump

Figure 14-11 The charcoal canister should be checked for damage.

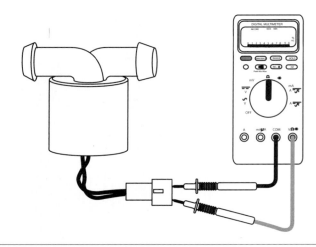

Figure 14-12 Testing the canister purge solenoid.

coal filter can cause symptoms that can be mistaken for fuel system problems. Rough idle, flooding, and other conditions can indicate a canister problem. A canister filled with liquid or water causes backpressure in the fuel tank. It can also cause richness and flooding symptoms during purge or start-up. (Some trucks have intentionally pressurized fuel tank systems. Check the calibration and engine decal before diagnosing.)

To test for saturation, unplug the canister momentarily during a diagnosis procedure and observe the engine's operation. If the canister is saturated, either it or the filter must be replaced depending on its design. Some models have a replaceable filter, others do not.

A vacuum leak in any of the evaporative emission components or hoses can cause starting and performance problems as can any engine vacuum leak. It can also cause complaints of fuel odor. Incorrect connection of the components can cause rich stumble or lack of purging, resulting in fuel odor. To conduct a vacuum-on, valve-open test, consult the vehicle's service manual. Check the service manual for other common evaporative emission control system problems.

Classroom Manual
Chapter 14, page 461

The canister purge solenoid (Figure 14-12) winding may be checked with an ohmmeter. With the tank pressure control valve removed, try to blow air through the valve with your mouth from the tank side of the valve. Some restriction to airflow should be felt until the air pressure opens the valve. Connect a vacuum hand pump to the vacuum fitting on the valve and apply 10 in. Hg to the valve. Now try to blow air through the valve from the tank side. Under this condition, there should be no restriction to airflow. If the tank pressure control valve does not operate properly, replace the valve.

If the fuel tank has a pressure and vacuum valve in the filler cap, check these valves for dirt contamination and damage. The cap may be washed in clean solvent. When the valves are sticking or damaged, replace the cap.

Basically, there are two different EVAP systems that you may encounter on fuel injected vehicles. The first is just a basic EVAP system used during OBD I. The computer controls the purge valve by pulse width modulation of the purge solenoid whenever conditions are appropriate. The basic EVAP system does not use computerized monitoring for leaks in the system.

The second and more recent EVAP system is the **enhanced EVAP** used with OBD II. The vent valve and tank pressure sensor were added for diagnostic purposes, and the fuel level sensor was added to the list of input sensors to the PCM. With enhanced EVAP, the PCM conducts several tests to determine if the system is operational and there are no leaks. Modern (OBD II) evaporative emission systems must be able to detect a leak as small as 0.020 inch in diameter and use tests to confirm that there are no leaks present. (Figure 14-13).

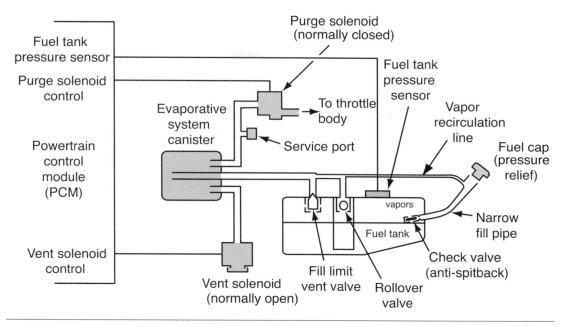

Figure 14-13 The enhanced EVAP system components.

The following information is intended to help you understand the system involved, not to diagnose a particular system, so always use information specific to the vehicle concerned.

The PCM can control the purge and vent valves open and closed. When used in conjunction with the tank pressure sensor, this enables diagnostics on the integrity of the system.

The first test we will describe is the **initial vacuum pulldown**, or the **weak vacuum test**. This will establish if there is a large system leak. Once the enable criteria is set, the PCM will close the vent valve and open the purge valve. This would, of course, cause a vacuum to be indicated on the fuel tank pressure sensor if the system were sealed. If the vehicle should fail to pull a vacuum of up to 7 in. Hg, then the MIL would illuminate after the second consecutive failure under similar conditions and set a code PO455, gross evaporative system leak. Some vehicles will set a "Check Fuel Cap" light and a code PO457 after this test was failed after a refueling event was detected by the PCM using the fuel level sensor. One of the problems associated with the evaporative emissions were customers who failed to install their fuel cap tight enough for a good seal, which alerts the PCM to a problem. Some other problems that may be indicated by this failure are:

❏ A large system leak, such as a vapor line loose or kinked, or a hole in the system
❏ A stuck-open vent solenoid, or loose vent line (basically the PCM cannot close the valve)

If the initial vacuum pull down (weak vacuum test) passes, the system assumes there is no large leak, and begins to check for a small leak. This is done through the **vacuum hold and decay**, or **small leak test**. For this test, the system will look for a leak that is down to 0.020 inch in size. The vent and purge solenoids will close and the PCM will watch for decay in the vacuum being held. If the vacuum drops too quickly, a PO442 is armed and set if the system fails again under similar conditions. With leaks so small the software in the PCM makes allowance for the buildup of pressure inside the tank due to high volatility fuel, and/or high ambient temperatures and excessive fuel slosh. For these reasons, the test will abort if fuel level, engine load, or fuel tank pressure should change abruptly. The pressure from high volatility fuel could reduce the vacuum measured and result in false code triggering.

In order to guard against this false triggering, the vent can be opened, eliminating the vacuum entirely, and then the system vent and purge valves can be closed while the PCM actually measures how much pressure is being added to the tank by the fuel volatility.

Finally, the cold idle test is used to entirely screen out the effects of fuel volatility by running the test only with the fuel cold (engine running less than 10 minutes) and the vehicle at idle or less than 10 mph. Even this test can be cancelled by conditions that may produce excessive fuel slosh conditions.

The sealing of the purge valve is checked by measuring vacuum inside the tank with the vent and purge valve closed. If there is vacuum building under these conditions it is assumed that the purge valve is leaking, setting code P1443 on Ford vehicles.

The most recent use of the vent valve and tank pressure sensor on EVAP systems on some 2003 and 2004 vehicles is the **Engine Off Natural Vacuum (EONV)** test. The system runs a pressure and vacuum test based on the natural volatility of the fuel. After driving, the fuel will be warm, and if the vent to the tank is closed, the fuel pressure will build according to the ambient temperature. After the fuel starts to cool down, again with the vent closed, there will be a natural vacuum build within the tank that the system can measure. The PCM can use this information to determine if there are any small leaks in the system. If the fuel is too volatile, as determined by the rise in pressure after the engine is turned off, the test is cancelled. This test is also a small leak test setting PO442 if failed.

Some systems use a **leak detection pump (LDP)** that pressurizes the system and checks for leaks, along with the fuel tank pressure sensor.

The **charcoal canister** is the heart of the EVAP system. The charcoal canister can be located in the engine compartment or near the fuel tank. The location near the fuel tank is especially true if the vehicle has the ORVR refueling vapor recovery system. The canister has to be large enough on these vehicles to store the vapors from refueling. Charcoal has the ability to store the vapors and then release them when the PCM determines that the engine can handle the extra fuel load, usually after the engine is warm. When the PCM decides to purge the canister, fresh air is allowed into the canister and vacuum is applied from the engine through the purge valve or solenoid.

The **purge valve** has undergone many changes since the inception of EVAP. Earlier purge valves were vacuum operated without computer control, but modern purge valves are computer controlled to allow more precise control of the vapors according to the many sensor inputs at the PCM. The purge solenoid can also be remotely mounted from the purge valve. The canister purge valve is normally closed. It opens the inlet to the purge outlet when vacuum is applied to the valve. Purging is done only when conditions warrant. Some of the conditions typically include such items as:

❏ 150 seconds since the PCM entered closed loop

❏ Coolant temperature above 176°F

❏ PCM not enabling injector shutdown, such as during a traction control

❏ Vehicle speed above 20 mph

❏ Engine speed above 1100 rpm

❏ Temperature sensor not indicating overheating

❏ Low coolant not indicated

The conditions that purge occur will, of course, vary from vehicle to vehicle.

The **vent valve** is opened to allow fresh air through the charcoal canister. Early canisters did not use a vent valve; the vent was left open to the atmosphere. The vent valve is only used for OBD II diagnostics. The vent valve can be closed with the purge valve open while the PCM monitors the fuel tank pressure sensor for vacuum. Vacuum should build with the vent closed and the purge open. The vent is checked for blockage during normal purging by checking for excessive vacuum with the vent valve commanded open. If there is vacuum drawn on the fuel tank with the vent and purge valves open, then the vent must be blocked. Finally, the purge valve can be checked for leaks by closing both the purge and vent valves and determining if a vacuum begins to build. If it does, the purge valve is leaking.

The most recent use of the vent valve and tank pressure sensor on EVAP systems on some 2003 and 2004 vehicles is the Engine Off Natural Vacuum (EONV) test. The system runs pressure and vacuum test based on the natural volatility of the fuel. After driving the fuel will be warm, and if the vent to the tank is closed, the fuel pressure will build in a predictable way according to the ambient temperature. After the fuel starts to cool down, again with the vent closed, there will be a natural vacuum build within the tank that the system can measure. The PCM can use this information to determine if there are any small leaks in the system. If the fuel is too volatile, as determined by the rise in pressure after the engine is turned off, the test is cancelled.

The **fuel tank pressure sensor** is mounted on the fuel tank and can measure vacuum and pressure at the tank. The fuel tank pressure sensor is a PCM input.

The **fuel tank level sensor**, which is actually the same sensor used for the fuel gauge, but is used on OBD II vehicles to determine whether the fuel tank pressure or vacuum test should be run. If the tank is very full or near empty, the test will not be run. The fuel tank sensor can also be checked at startup and during driving to detect refueling which could upset the results of the leak test.

The **fuel tank pressure control valve** is designed to control fuel tank pressure while the vehicle is sitting still. Vapors are stored in the tank with the valve closed. If tank pressure builds too high, the valve opens and lets the vapor into the canister. When the engine is running, the valve has vacuum applied, opening the valve and allowing vapors to be stored in the canister until purging.

Some systems use a leak detection pump (LDP) that pressurizes the system and checks for leaks, along with the fuel tank pressure sensor.

PCV System Diagnosis and Service

No adjustments can be made to the PCV system. Therefore, service to this system involves careful inspection, operation, and replacement of faulty parts. Some engines use a fixed orifice tube in place of a valve. These should be cleaned periodically with a pipe cleaner soaked in carburetor cleaner. Although there is no PCV valve, this type of system is diagnosed in the same way as those systems with a valve. When replacing a PCV valve, match the part number on the valve with the vehicle maker's specifications for the proper valve. If the valve cannot be identified, refer to the part number listed in the manufacturer's service manual. Using an exact replacement PCV valve is critcial to engine performance and emission control.

If the PCV valve is stuck in the open position, excessive airflow through the valve causes a lean air-fuel ratio and possible rough idle operation or engine stalling. When the PCV valve or hose is restricted, excessive crankcase pressure forces blowby gases through the clean air hose and filter into the air cleaner. Worn rings or cylinders cause excessive blowby gases and increased crankcase pressure, which forces blowby gases through the clean air hose and filter into the air cleaner. A restricted PCV valve or hose may result in the accumulation of moisture and sludge in the engine and engine oil.

Leaks at engine gaskets, such as rocker arm cover or crankcase gaskets, will result in oil leaks and the escape of blowby gases to the atmosphere. However, the PCV system also draws

Classroom Manual
Chapter 14, page 466

Figure 14-14 Check the engine for signs of oil leaks.

unfiltered air through these leaks into the engine. This action could result in wear of engine components, especially when the vehicle is operating in dusty conditions. Check all the engine gaskets for signs of oil leaks (Figure 14-14). Be sure the oil filler cap fits and seals properly.

The first step of PCV servicing is a visual inspection. As shown in Figure 14-15, the PCV valve can be located in several places. The most common location is in a rubber grommet in the valve or rocker arm cover. It can be installed in the middle of the hose connections as well as directly in the intake manifold.

Once the PCV valve is located, inspect the system by using the following procedure:

Special Tools

Vacuum gauge

1. Make sure all the PCV system hoses (Figure 14-16) are properly connected and that they have no breaks or cracks.

2. Remove the air cleaner and inspect the carburetor or fuel injector air filter. Crankcase blowby can clog these with oil. Clean or replace such filters. Oil in the air cleaner assembly indicates that the PCV valve or hoses are plugged. Make sure you

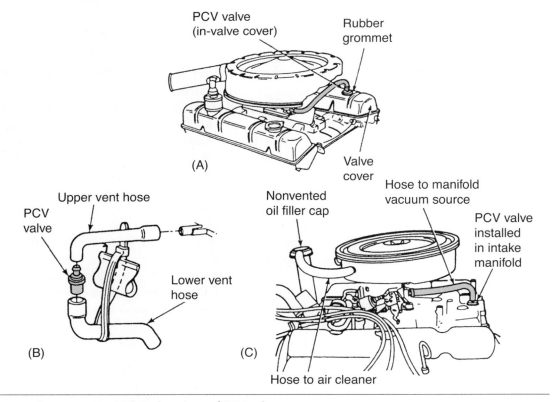

Figure 14-15 Various locations of PCV valves.

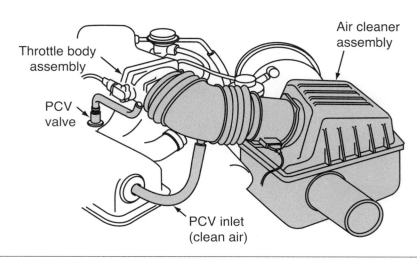

Figure 14-16 Typical PCV system plumbing on a port fuel injection vehicle.

check these and replace the valve and clean the hoses and air cleaner assembly. When the PCV valve and hose are in satisfactory condition and there was oil in the air cleaner assembly, perform a cylinder compression test to check for worn cylinders and piston rings.

3. Check the crankcase inlet air filter, which is usually located in the air cleaner. As the filter does its job, it becomes dirty. If it is oil soaked, it is a good indication that the PCV system is not working the way that it should. When the filter becomes so dirty that it restricts the flow of clean air to the crankcase, it can cause the same problems as a clogged PCV valve. So be sure to check this particular filter when checking the PCV system and replace it as necessary.

4. Inspect for dirt deposits that could clog the passages in the manifold or carburetor base. These deposits can prevent the system from functioning properly, even though the PCV valve, valve filter, and hoses might not be clogged.

CUSTOMER CARE: You may want to discourage the customer from bringing his or her own parts to you for installation. Even well-meaning customers may not understand what may result. For example, if the customer asks the technician to install something as simple as a PCV valve they had purchased elsewhere, other problems may result. True, replacing the PCV valve is simple, but what if it is the wrong one, even though it may look identical? Or what if it is an inferior component or is otherwise defective? Driveability or idle problems may result if it is the wrong calibration or application. Unless you are confident of the source and quality of the component you are installing, it may be better to avoid the practice of installing customer-purchased parts.

Functional Checks of PCV Valve

A rough idling engine can signal a number of PCV valve problems, such as a clogged valve or a plugged hose. But before beginning the functional checks, double check the PCV valve part number to make certain the correct valve is installed. If the correct valve is being used, perform these functional checks:

1. Disconnect the PCV valve from the valve cover, intake manifold, or hose.
2. Start the engine and let it run at idle. If the PCV valve is not clogged, a hissing is heard as air passes through the valve.

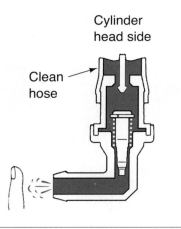

Cylinder
head side

Clean
hose

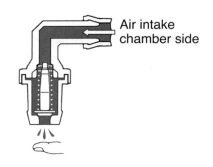

Air intake
chamber side

Figure 14-17 Blowing air through the
PCV valve from the inlet side.

Figure 14-18 Blowing air through the
PCV valve from the outlet side.

3. Place a finger over the end of the valve to check for vacuum. If there is little or no
vacuum at the valve, check for a plugged or restricted hose. Replace any plugged or
deteriorated hoses.

4. Turn the engine off and remove the PCV valve. Shake the valve and listen for the
rattle of the check needle inside the valve. If the valve does not rattle, replace it.

Some vehicle manufacturers recommend removing the PCV valve from the rocker arm
cover and the hose. Connect a length of hose to the inlet side of the PCV valve, and blow air
through the valve with your mouth while holding your finger near the valve outlet (Figure 14-17).
Air should pass freely through the valve. If air does not pass freely through the valve, replace the
valve. Move the hose to the outlet side of the PCV valve and try to blow back through the valve
(Figure 14-18). It should be difficult to blow air through the PCV valve in this direction. When air
passes easily through the valve, replace the valve.

> **WARNING:** Do not attempt to suck through a PCV valve with your mouth. Sludge
> and other deposits inside the valve are harmful to the human body.

Proper operation of the PCV system depends on a sealed engine. Remember that the
crankcase is sealed by the dipstick, valve cover, gaskets, and sealed filler cap. If oil sludging or
dilution is noted and the PCV system is functioning properly, check the engine for oil leaks and
correct them to ensure that the PCV system functions as intended. Also, be aware of the fact that
a very worn engine may have more blowby than the PCV system can handle. If there are symptoms that indicate the PCV system is plugged (oil in air cleaner, saturated crankcase filter, etc.)
but no restrictions are found, check the wear of the engine.

Spark Control Systems

Through the years, many devices and combination of devices have been used to control ignition
timing. Today, most systems rely on the PCM for timing control based on inputs from various
sensors. One of these sensors, the knock sensor, has one purpose—spark control.

**Classroom
Manual**
Chapter 14, page 469

Diagnosis of Knock Sensor and Knock Sensor Module

> **CAUTION:** Operating an engine with a detonation problem for a sufficient num-
> ber of miles may result in piston, ring, and cylinder wall damage.

If the knock sensor system does not provide an engine detonation signal to the PCM, the engine detonates, especially on acceleration. When the knock sensor system provides excessive spark retard, fuel economy and engine performance are reduced.

The first step in diagnosing the knock sensor and knock sensor module is to check all the wires and connections in the system for loose connections, corroded terminals, and damage. With the ignition switch on, be sure 12 volts are supplied through the fuse to the knock sensor module. Repair or replace the wires, terminals, and fuse as required.

Connect a scan tester to the DLC and check for DTCs related to the knock sensor system. If DTCs are present, diagnose the cause of these codes. When no DTCs related to the knock sensor system are present, the system needs to be checked. Photo Sequence 26 shows a typical procedure for diagnosing knock sensors and knock sensor modules. The following is a typical diagnostic procedure. However, always follow the recommended procedure of the manufacturer.

Special Tools

Scanner, 12-volt test light

1. Connect the scan tester to the DLC and be sure the engine is at normal operating temperature.

2. Operate the engine at 1,500 rpm and observe the knock sensor signal on the scan tester. If a knock sensor signal is present, disconnect the wire from the knock sensor and repeat the test at the same engine speed. If the knock sensor signal is no longer present, the engine has an internal knock or the knock sensor is defective. When the knock sensor signal is still present on the scan tester, check the wire from the knock sensor to the knock sensor module for picking up false signals from an adjacent wire. Reroute the knock sensor wire as necessary.

3. If the knock sensor signal is not indicated on the scan tester in step 2, tap on the engine block near the knock sensor with a small hammer. When the knock sensor signal is now present, the knock sensor system is satisfactory.

4. When a knock sensor signal is not present in step 3, turn the ignition switch off and disconnect the knock sensor module wiring connector. Connect a 12-volt test light from 12 volts to terminal D in this wiring connector (Figure 14-19). If the light is off, repair the wire connected from this terminal to ground. When the light is on, proceed to step 5.

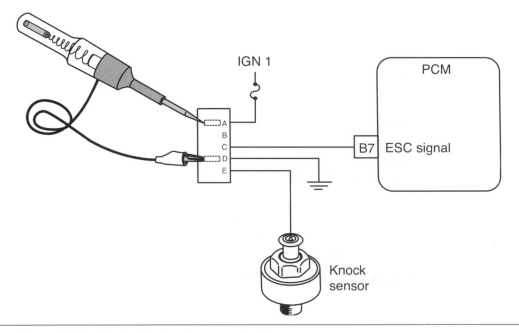

Figure 14-19 Testing the knock sensor circuit.

Photo Sequence 26
Typical Procedure for Diagnosing Knock Sensors and Knock Sensor Modules

P26-1 Be sure the engine is at normal operating temperature and the ignition switch is off.

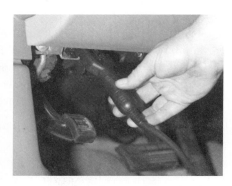

P26-2 Connect the scan tester leads to the battery terminals with the correct polarity.

P26-3 Program the scan tester for the vehicle being tested.

P26-4 Use the proper adapter to connect the scan tester lead to the data link connector (DLC).

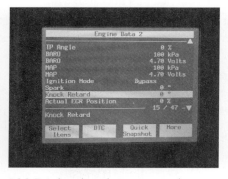

P26-5 Select knock sensor on the scan tester.

P26-6 Observe the knock sensor signal on the scan tester with the engine running at 1,500 rpm. The knock sensor should indicate no signal.

P26-7 Tap on the right exhaust manifold above the sensor with a small hammer and observe the scan tester reading.

P26-8 The knock sensor should now indicate a signal on the scan tester.

P26-9 Shut the engine off and disconnect the scan tester leads.

5. Reconnect the knock sensor module wiring connector and disconnect the knock sensor wire. Operate the engine at idle speed and momentarily connect a 12-volt test light from 12 volts to the knock sensor wire. If a knock sensor signal is now generated on the scan tester, there is a faulty connection at the knock sensor or the knock sensor is defective. When a knock sensor signal is not generated, check for faulty wires from the knock sensor to the module or from the module to the PCM. Check the wiring connections at the module. If the wires and connections are satisfactory, the knock sensor module is likely defective.

☑ **SERVICE TIP:** When installing a knock sensor, make sure it is tightened to the proper amount of torque. If the knock sensor torque is more than specified, the sensor may become too sensitive and provide an excessively high voltage signal, resulting in more spark retard than required. When the knock sensor torque is less than specified, the knock sensor signal is lower than normal, resulting in engine detonation.

EGR System Diagnosis and Service

Manufacturers program the PCM for correct EGR flow. If there is too much or too little, it can cause performance problems by changing the engine breathing characteristics (Figure 14-20). With too little EGR flow, the engine can overheat, detonate, and emit excessive amounts of NO_x. When any of these problems exist and it seems likely that the EGR system is at fault, check the system. Typical problems that show up in EGR systems follow.

Classroom Manual
Chapter 14, page 470

- ❏ Rough idle. Possible causes are an EGR valve stuck open, dirt on the valve seat, or loose mounting bolts. Loose mounting causes vacuum leak and a hissing noise.
- ❏ Surge, stall, or does not start. Probable cause is the valve stuck open.
- ❏ Detonation (spark knock). Any condition that prevents proper EGR gas flow can cause detonation. This includes a valve stuck closed, leaking valve diaphragm, restrictions in flow passages, EGR disconnected, or a problem in the vacuum source.

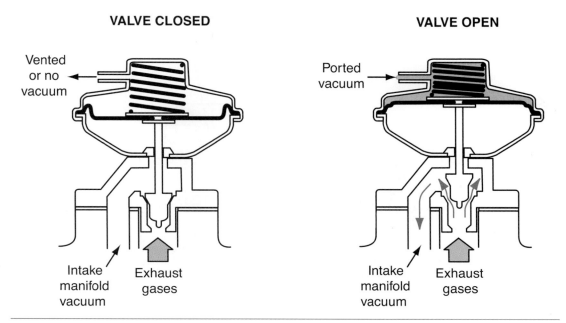

Figure 14-20 The operation of a normal ported EGR valve.

❏ Excessive NO$_x$ emissions. Any condition that prevents the EGR from allowing the correct amount of exhaust gases into the cylinder can cause this problem. High combustion temperatures allow NO$_x$ to form. Therefore, anything that allows combustion temperatures to rise can cause this problem as well.

❏ Poor fuel economy. This is an EGR condition only if it relates to detonation or other symptoms of restricted or zero EGR flow.

EGR System Troubleshooting

Before attempting to troubleshoot or repair a suspected EGR system on a vehicle, the following conditions should be checked and be within specifications.

❏ Engine is mechanically sound.

❏ Injection system is operating properly.

❏ Electronic timing advance is operating properly.

If one or more of these conditions is faulty or operating incorrectly, perform the necessary tests and services to correct the problem before servicing the EGR system.

Most often in the closed-loop electronic control EGR systems, the valve by itself functions the same as a ported EGR valve (Figure 14-21). Apart from the electronic control, the system can have all of the problems of any EGR system. Some EGR valves are fully electrically controlled and do not use a vacuum signal to open and close the valve. This type EGR system can show the same problems as others, with the exception of vacuum leaks and other vacuum-related problems. Sticking valves, obstructions, and loss of vacuum produces the same symptoms as on non-electronic controlled systems. If an electronic control component is not functioning, the condition is usually recognized by the computer. Check the service manual for instructions on how to use computer service codes. The **EGR vent solenoid (EGRV)** and **EGR control solenoid (EGRC)** solenoids (Figure 14-22) or the **EGR vacuum regulator (EVR)** should normally cycle on and off very frequently when EGR flow is being controlled (warm engine and cruise rpm). If they do not, it indicates a problem in the electronic control system or the solenoids. Generally, an electronic control failure results in low or zero EGR flow and might cause symptoms like overheating, detonation, and power loss.

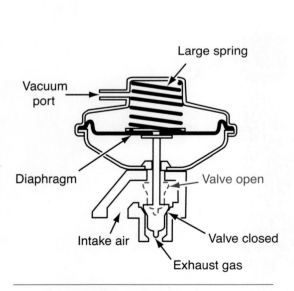

Figure 14-21 Typical vacuum-operated EGR valve.

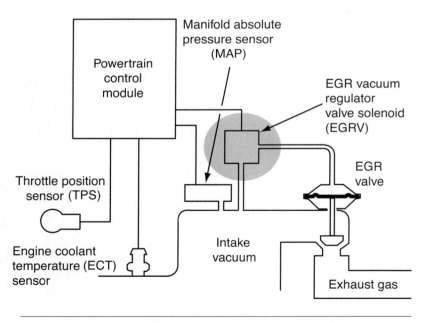

Figure 14-22 An EGR system with an EVR solenoid.

Before attempting any testing of the EGR system, visually inspect the condition of all vacuum hoses for kinks, bends, cracks, and flexibility. Replace defective hoses as required. Check vacuum hose routing. (See the underhood decal or the manufacturer's service manual for correct routing.) Correct any misrouted hoses. If the emissions system is fitted with an EVP sensor, the wires routed to it should also be checked.

If the EGR valve remains open at idle and low engine speed, the idle operation is rough and surging occurs at low speed. When this problem is present, the engine may hesitate on low-speed acceleration or stall after deceleration or after a cold start. If the EGR valve does not open, engine detonation occurs. When a defect occurs in the EGR system, a diagnostic trouble code is usually set in the PCM memory.

In many EGR systems, the PCM uses inputs from the ECT, TPS, and MAP sensors to operate the EGR valve. Improper EGR operation may be caused by a defect in one of these sensors. A scan tester may be connected to the DLC to check for an EGR DTC or a DTC from another sensor which may affect EGR operation. The cause of any DTCs should be corrected before any further EGR diagnosis.

The PCM uses information from the ECT, TPS, and MAP to operate the EGR valve.

EGR Valves and Systems Testing

Test the EGR valve by using a vacuum gauge or hand vacuum pump. Follow these procedures for using either piece of test equipment.

Special Tools

Vacuum gauge, pressure gauge, exhaust gas analyzer, tachometer, scanner

1. Disconnect a vacuum line connected to an intake manifold port.
2. Put a vacuum gauge between the disconnect vacuum line and the intake manifold port.
3. Connect a tachometer.
4. Start the engine and gradually increase speed to 2,000 rpm with the transmission in neutral.
5. The reading from the manifold vacuum gauge (Figure 14-23) should be above 16 inches of vacuum. If not, there could be a vacuum leak or exhaust restriction. Before continuing to test the EGR, correct the problem of low vacuum.

Using a hand vacuum pump to check the operation of the EGR valve:

1. Check all vacuum lines for correct routing. Ensure that they are attached securely. Replace cracked, crimped, or broken lines.
2. Remove the vacuum supply hose from the EGR valve port.

Figure 14-23 Check the engine's vacuum level before testing the EGR.

3. Connect the vacuum pump to the port and supply 18 inches of vacuum. Observe the EGR diaphragm movement. In some applications, a mirror may be held under the EGR valve to see the diaphragm movement. When the vacuum is applied, the diaphragm should have moved. If the valve diaphragm did not move or did not hold the vacuum, replace the valve.

4. With the engine at normal operating temperature, check the vacuum supply hose to make sure there is no vacuum to the EGR valve at idle. Then plug the hose.

5. Install a tachometer.

6. On EFI engines (multi-point injection), disconnect the throttle air bypass valve solenoid.

7. Observe the engine's idle speed. If necessary, adjust idle speed to the emission decal specification.

8. Slowly apply 5 to 10 inches of vacuum to the EGR valve vacuum port using a hand vacuum pump. The idle speed should drop more than 100 rpm (the engine may stall), and then return to normal (± 25 rpm) again when the vacuum is removed.

9. If the idle speed does not respond in this manner, remove the valve and check for carbon in the passages under the valve. Clean the passages as required or replace the EGR valve. Carbon may be cleaned from the lower end of the EGR valve with a wire brush, but do not immerse the valve in solvent and do not sandblast the valve.

10. If the EGR valve is operating properly, unplug and reconnect the EGR valve vacuum supply hose.

11. Reconnect the throttle air bypass valve solenoid if removed.

Differential Pressure Feedback EGR

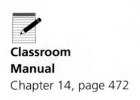

Classroom Manual
Chapter 14, page 472

The differential pressure feedback (DPFE) EGR system is used with Ford OBD II systems to monitor the performance of the EGR systems. This system verifies the integrity of the EGR system controls and flow rate. See Figure 14-24 for the individual components. The following is a brief description of the diagnostic routine for some of the components of the DPFE EGR system. You should always check for specific information for the vehicle you are servicing.

First we will look at the DPFE sensor itself (Figure 14-25). The codes that set for the DPFE sensor are P1400 for low voltage from the DPFE and P1401 for high voltage (Figure 14-26). A low-

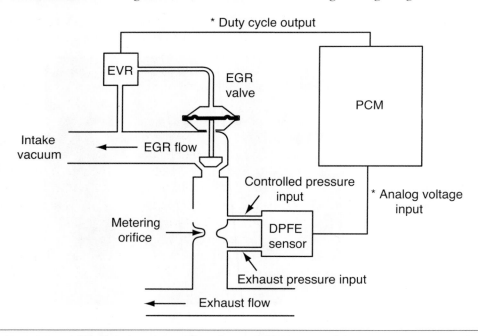

Figure 14-24 This drawing shows the exhaust gas passages of the EGR system, the metering orifice, EVR solenoid, and DPFE sensor.

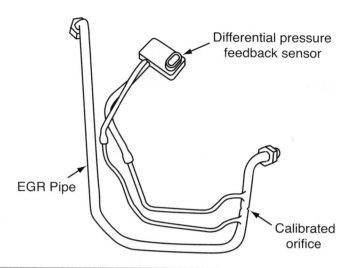

Differential pressure
feedback sensor

EGR Pipe

Calibrated
orifice

Figure 14-25 The DPFE EGR pipe with sensor, calibrated orifice, and upstream and downstream hoses.

Component	Codes
DPFE Sensor	P1400 Low Voltage from sensor P1401 High Voltage from sensor P1405 Upstream DPFE hose unplugged P1406 Downstream DPFE hose unplugged
EGR Self-Test Failure	P1408, P0401
EGR Vacuum Regulator Solenoid Circuit	P1409

Figure 14-26 Ford DPFE code definitions.

voltage condition could occur with a short to ground in the wiring or a defective sensor or PCM. A high-voltage code could be set by a short to power at the wiring, defective sensor, or PCM.

The DPFE sensor's hoses are also checked for disconnection by commanding the EGR valve closed. If EGR flow is still indicated, the test assumes that the upstream DPFE hose is off or plugged or the EGR tube is plugged or damaged, and a code P1405 code is set. If the EGR valve is commanded closed and a negative EGR flow is indicated at the DPFE sensor, then a P1406 is set. The P1406 code is set by a damaged or disconnected downstream hose.

The EGR system runs its monitor for P0401 whenever the engine is at cruise and the rate of EGR is high. The PCM checks feedback from the DPFE sensor and compares it to the rate of EVR control. If the two do not agree, a P0401 is flagged in the PCM. The test for P1408 is only conducted when the KOER test is initiated, but it is similar to the P0401 test. If the KOER test was run after repairs to the EGR system, a technician could verify the repair without waiting for the PCM to conduct the test on a road test. The PCM looks for an expected EGR flow at a fixed rpm. If the amount of EGR expected is not achieved, then the code is set. Finally, the P1409 code is the monitor for the EVR valve, which is set when the expected voltage drop across the valve is out of specification. This helps the PCM realize that there is a problem with the wiring to the valve or the valve itself.

Digital EGR Valve Diagnosis

The digital EGR valve may be diagnosed with a scan tester. With the engine at normal operating temperature and the ignition switch off, connect the scan tester to the DLC. Start the engine and

KOER is the abbreviation for Key On, Engine Running test initiated by the technician with a scan tool.

Never assume that a code indicated means a part must be replaced. Always check the circuit according to the latest diagnostic procedure.

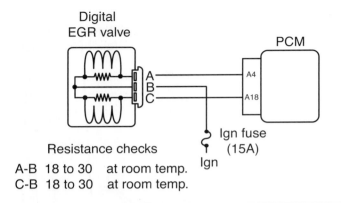

Resistance checks
A-B 18 to 30   at room temp.
C-B 18 to 30   at room temp.

Figure 14-27 Digital EGR valve circuit.

allow the engine to operate at idle speed. Select EGR control on the scan tester and then energize EGR solenoid number one with the scan tester. When this action is taken, the engine rpm should decrease slightly. The engine rpm should drop slightly as each EGR solenoid is energized with the scan tester. When the EGR valve does not operate properly, check these items before replacing the EGR valve:

1. Check for 12 volts at the power supply wire on the EGR valve (Figure 14-27).
2. Check the wires between the EGR valve and the PCM.
3. Remove the EGR valve, and check for plugged passages under the valve.

Linear EGR Valve Diagnosis

The linear EGR valve diagnostic procedure varies depending on the vehicle make and model year. Always follow the recommended procedure in the vehicle manufacturer's service manual. The scan tester may be used to diagnose a linear EGR valve. The engine should be at normal operating temperature prior to EGR valve diagnosis. Since the linear EGR valve has an EVP sensor, the actual pintle position may be checked on the scan tester. The pintle position should not exceed 3 percent at idle speed. The scan tester may be operated to command a specific pintle position, such as 75 percent, and this commanded position should be achieved within two seconds. With the engine idling, select various pintle positions and check the actual pintle position. The pintle position should always be within 10 percent of the commanded position. Refer to Figure 14-28. When the linear EGR valve does not operate properly:

1. Check the fuse in the 12-volt supply wire to the EGR valve.
2. Check for open circuits, grounds, and shorts in the wires connected from the EGR valve winding to the PCM.

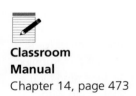

Classroom Manual
Chapter 14, page 473

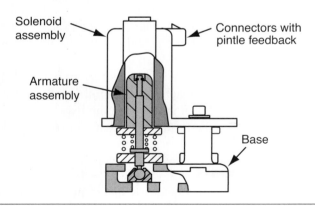

Figure 14-28 A typical linear EGR valve.

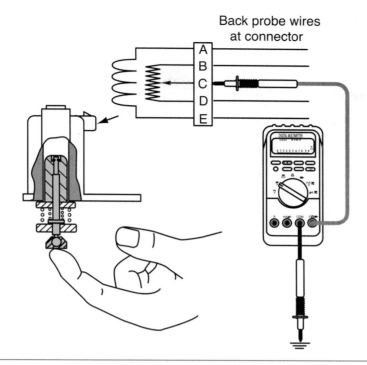

Back probe wires
at connector

Figure 14-29 Checking the pintle position sensor.

3. Use a digital voltmeter to check for 5 volts on the reference wire to the EVP sensor.
4. Check for excessive resistance in the EVP sensor ground wire.
5. Leave the wiring harness connected to the valve, and remove the valve. Connect a digital voltmeter from the pintle position wire at the EGR valve to ground, and manually push the pintle upward (Figure 14-29). The voltmeter reading should change from approximately 1 volt to 4.5 volts.

If the EGR valve did not operate properly on the scan tester and tests 1 through 5 are satisfactory, replace the valve. Remember that not all EGR valves are tested the same way. Check the appropriate service manual for specific tests.

EGR Efficiency Tests

Most testing of EGR valves involves the valve's ability to open and close at the correct time. An EGR system has one job: to control NO_x emissions. The system needs to be tested further to determine if it is doing what it was designed to do. Checking the efficiency of the system may also uncover the cause of other problems. If the exhaust passage in the valve is not the size it was designed to be, can the valve be as effective as it should? EGR efficiency testing is extremely important while diagnosing the cause of high NO_x and detonation.

Many technicians err by thinking that the EGR valve is working properly if the engine stalls or idles very rough when the EGR valve is opened. Actually, the results of this test indicate that the valve was closed and it will open. A good EGR valve opens and closes, but it also allows the correct amount of exhaust gas to enter the cylinder. EGR valves are normally closed at idle and open at approximately 2,000 rpm. This is where the EGR system should be checked. Keep in mind, anything that increases combustion temperatures can cause an increase in NO_x. Common causes of high NO_x are faulty cooling systems, incorrect ignition timing, lean mixtures, and faulty EGRs.

A five-gas exhaust analyzer can be used to check an EGR system. Allow the engine to warm up, then raise the engine speed to around 2,000 rpm. Watch the NO_x readings on the analyzer. The meter measures NO_x in parts per million. In most cases, NO_x should be below 1,000 ppm. It is normal to have some temporary increases over 1,000 ppm. However, the reading should be generally under 1,000 ppm. If the NO_x is above 1,000, the EGR system is not doing its job. The exhaust passage in the valve is probably clogged with carbon.

If only a small amount of exhaust gas is entering the cylinder, NO_x will still be formed. A restricted exhaust passage of only 1/8 inch will still cause the engine to run rough or stall at idle, but it is not enough to control combustion chamber temperatures at higher engine speeds. Keep in mind, never assume the EGR passages are okay just because the engine stalls at idle when the EGR is fully opened.

Electronic EGR Controls

Classroom Manual
Chapter 14, page 474

Special Tools

DMM, scanner

When the EGR valve checks out and everything looks fine visually but a problem with the EGR system is evident, the EGR controls should be tested. Often, a malfunctioning electronic control will trigger a DTC. Service manuals give the specific directions for testing these controls. Always follow them. The tests below are given as examples of those test procedures.

EGR Vacuum Regulator (EVR) Tests

Connect a pair of ohmmeter leads to the EVR terminals to check the winding for open circuits and shorts (Figure 14-30). An infinity ohmmeter reading indicates an open circuit, whereas a lower-than-specified reading means the winding is shorted. Then, connect the ohmmeter leads from one of the EVR solenoid terminals to the solenoid case (Figure 14-31). A low ohmmeter reading indicates that the winding is shorted to ground. If the winding is not shorted, an infinity reading will be observed.

✔ **SERVICE TIP:** The same quad driver in a PCM may operate several outputs. For example, a quad driver may operate the EVR solenoid and the torque converter clutch solenoid. On General Motors computers, the quad drivers sense high current flow. If a solenoid winding is shorted and the quad driver senses high current flow, the quad driver shuts off all the outputs it controls rather than being damaged by the high current flow. When the PCM fails to operate an output or outputs, always check the resistance of the solenoid windings in the outputs before replacing the PCM. A lower-than-specified resistance in a solenoid winding indicates a shorted condition, and this problem may explain why the PCM quad driver stops operating the outputs.

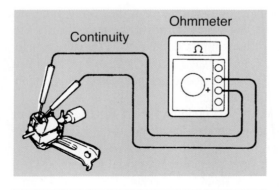

Figure 14-30 Use an ohmmeter to check the windings of an EVR solenoid.

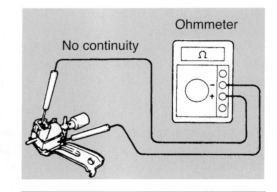

Figure 14-31 Use an ohmmeter to test an EVR solenoid for shorts to ground.

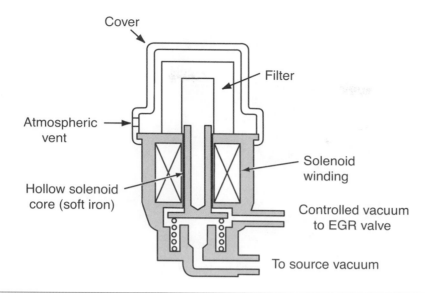

Figure 14-32 Operation of an EGR solenoid.

A scan tester may be used to diagnose the EVR solenoid operation. This procedure is shown in Photo Sequence 27. In the appropriate mode, the scan tester displays the EVR solenoid status as on or off. With the engine idling, the EVR solenoid should remain off. Drive the vehicle until the conditions required to open the EVR solenoid are present. Once these conditions are present, the scan tester should indicate that the EVR solenoid is on.

✔ **SERVICE TIP:** In some EGR systems, the PCM energizes the EVR solenoid (Figure 14-32) at idle and low speeds. Under this condition, the solenoid shuts off vacuum to the EGR valve. When the proper input signals are available, the PCM de-energizes the EVR solenoid and allows vacuum to the EGR valve.

Exhaust Gas Temperature Sensor Diagnosis

Remove the exhaust gas temperature sensor and place it in a container of oil. Place a thermometer in the oil and heat the container (Figure 14-33). Connect the ohmmeter leads to the exhaust gas temperature sensor terminals. The exhaust gas temperature sensor should have the specified resistance at various temperatures.

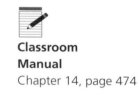

Classroom Manual
Chapter 14, page 474

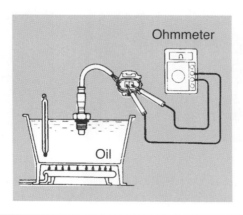

Figure 14-33 Testing an exhaust gas temperature sensor.

Photo Sequence 27
Diagnosing an EGR Vacuum Regulator Solenoid

P27-1 Disconnect the connector to the EGR solenoid and connect the leads of an ohmmeter to the solenoid's terminals.

P27-2 Compare your readings to the specifications for the solenoid.

P27-3 Connect the ohmmeter leads to one of the solenoid's terminals and to ground. An infinite reading means the solenoid is not shorted to ground.

P27-4 Reconnect the wiring to the connector and run the engine to bring it to normal operating temperature. While the engine is running, prepare the scan tool for the vehicle.

P27-5 Turn the engine off and connect the power cable of the scan tool to the vehicle.

P27-6 Enter the necessary information into the scan tool.

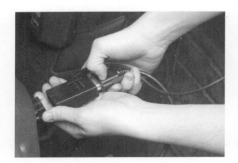

P27-7 Connect the scan tool to the DLC.

P27-8 Start the engine and obtain the EGR data on the scanner. The EGR valve should be off and remain off while the engine is idling.

P27-9 Take the vehicle for a test drive with the scan tool still connected. The EGR solenoid should cycle to ON once the vehicle is at a cruising speed. If it does not, check the solenoid and associated circuits.

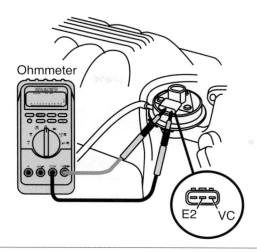

Figure 14-34 Checking the EGR position sensor with an ohmmeter set on K ohms.

Inspecting a Toyota EGR Position Sensor

This procedure is based on a procedure for a 2000 Toyota Camry with a 3.0L engine. Always make sure to use the appropriate service information for the vehicle concerned.

1. Disconnect the EGR position sensor, and with an ohmmeter, measure the connector terminals on the EGR valve VC and E2 (Figure 14-34). The resistance should be between 1.5K and 4.3kΩ. If it is not, replace the EGR valve position sensor. If resistance is within specifications, go to step 2.

2. With a hand-held vacuum pump, apply 5 in. Hg vacuum to the EGR valve (Figure 14-35) while checking the voltage from ECM terminals EGLS and E2. With the vacuum

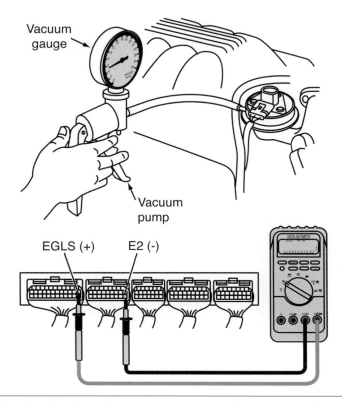

Figure 14-35 Checking for voltage change at PCM with vacuum at the EGR valve.

applied, the voltage should read approximately 3 to 5 volts. With the vacuum released, it should read approximately 0.4 to 1.6 volts. If both of these readings are okay the position sensor is good. If the reading is not within the specified range, replace the sensor and retest.

3. If the voltage readings are incorrect, check for an open from either one of the wires from the EGR valve to the ECM, defective wiring terminals/bad connection to the PCM, PCM connections, and a possible bad PCM.

Catalytic Converter Diagnosis

Classroom Manual
Chapter 14, page 478

Special Tools

Vacuum gauge, pressure gauge, pyrometer, exhaust gas analyzer

Detailed instructions on testing a catalytic converter are given in the chapter on exhaust systems in this book. Although a catalytic converter is definitely an emission control device, it is also an integral part of the exhaust system. What follows is a summary of the tests given in the previous chapter.

A plugged converter or any exhaust restriction can cause loss of power at high speeds, stalling after starting (if totally blocked), a drop in engine vacuum as engine rpm increases, or sometimes popping or backfiring through the intake manifold.

There are many ways to test a catalytic converter (Figure 14-36). One of these is to simply rap the converter with a rubber mallet. If the converter rattles, it needs to be replaced and there is no need to do other testing. A rattle indicates loose catalyst substrate which will soon rattle into small pieces. This is one test and is not used to determine if the catalyst is good.

A vacuum gauge can be used to watch engine vacuum while the engine is accelerated. Another way to check for a restricted exhaust or catalyst is to insert a pressure gauge in the exhaust manifold's bore for the O_2 sensor. With the gauge in place, hold the engine's speed at 2,000 rpm and watch the gauge. The desired pressure reading will be less than 1.25 psi. A substantial restriction will give a reading of over 2.75 psi.

The converter should be checked for its ability to convert CO and HC into CO_2 and water. There are three separate tests for doing this. The first method is the delta temperature test. To conduct this test, use a hand-held digital pyrometer. By touching the pyrometer probe to the exhaust pipe just ahead of and just behind the converter, there should be an increase of at least 100°F (or 8 percent above the inlet temperature reading) as the exhaust gases pass through the converter. If the outlet temperature is the same or lower, nothing is happening inside the converter.

The next test is called the O_2 storage test and is based on the fact that a good converter stores oxygen. Begin by disabling the air injection system. Once the analyzer and converter are warmed up, hold the engine at 2,000 rpm. Watch the readings on the exhaust analyzer. Once the numbers stop dropping, check the oxygen level on the gas analyzer. The O_2 readings should be

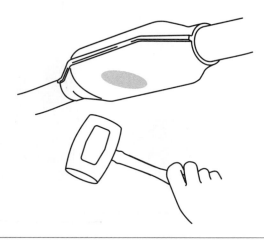

Figure 14-36 "Mallet" test of converter.

about 0.5 to 1 percent. This shows the converter is using most of the available oxygen. It is important to observe the O_2 reading as soon as the CO begins to drop. If the converter fails the tests, chances are that it is working poorly or not at all.

This final converter test uses a principal that checks the converter's efficiency. Before beginning this test, make sure the converter is warmed up. Calibrate a four-gas analyzer and insert its probe into the tailpipe. Disable the ignition. Then crank the engine for 9 seconds while pumping the throttle. Watch the readings on the analyzer. The CO_2 on fuel injected vehicles should be over 11 percent, and carbureted vehicles should have a reading of over 10 percent. As soon as the readings are obtained, reconnect the ignition and start the engine. Do this as quickly as possible to cool off the catalytic converter. If, while the engine is cranking, the HC goes above 1,500 ppm, stop cranking, the converter is not working. Also, stop cranking once the CO_2 readings reach 10 or 11 percent. The converter is good. If the catalytic converter is bad, there will be high HC and low CO_2 at the tailpipe. Do not repeat this test more than one time without running the engine in between. If a catalytic converter is found to be bad, replace it.

The OBD II system checks catalyst efficiency by comparing a pre-catalyst heated oxygen sensor with a post-catalyst heated oxygen sensor (Figure 14-37). The activity on the post-catalyst should show very little activity when compared to the pre-catalyst sensor. The converter, if working properly, should clean the exhaust to the point that the oxygen sensor will have very little response.

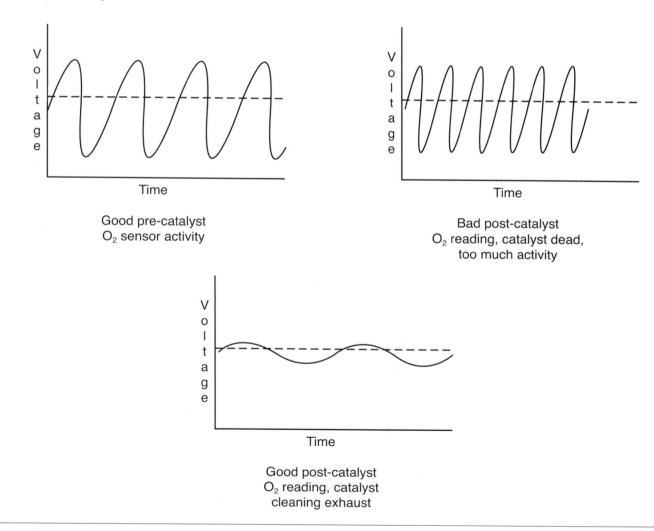

Good pre-catalyst
O_2 sensor activity

Bad post-catalyst
O_2 reading, catalyst dead,
too much activity

Good post-catalyst
O_2 reading, catalyst
cleaning exhaust

Figure 14-37 Comparison of O_2 activity with catalyst efficiency.

Air Injection System Diagnosis and Service

Classroom
Manual
Chapter 14, page 480

Air injection systems are often referred to as **AIR systems** or secondary air systems.

A diverter is often called a gulp valve.

Not all engines are equipped with an air injection system; only those that need them to meet emissions standards have them. Therefore, air injection systems are vital to proper emissions on engines equipped with them. Each system has its own test procedure—always follow the manufacturer's recommendations for testing. There are three basic designs of air injection systems.

There are basically two types of AIR systems you may encounter. The older OBD 1 systems were used before 1996, and OBD II systems have been used since 1996. Since you may still encounter an occasional OBD I system, some information on those systems has been included (Figure 14-38). Typically, as you recall from the classroom manual, the OBD I systems had a belt-driven pump and supplied both the exhaust manifold when cold and the catalytic converter when warm. Most OBD II systems use the AIR injection used with an electrically driven AIR pump that only supplies the exhaust manifold when the engine is cold to help eliminate HC and CO emissions until the converter becomes warm enough to oxidize them.

AIR systems are computer-controlled. When the system is in closed loop, the air from the air injection system must be directed away from the O_2 sensor. Some systems have switching valves that allow a small amount of air to flow past the O_2 sensor. The computer knows how much and adjusts the O_2 input accordingly. Sometimes the amount of air that can move through a closed switching valve is marked on its housing.

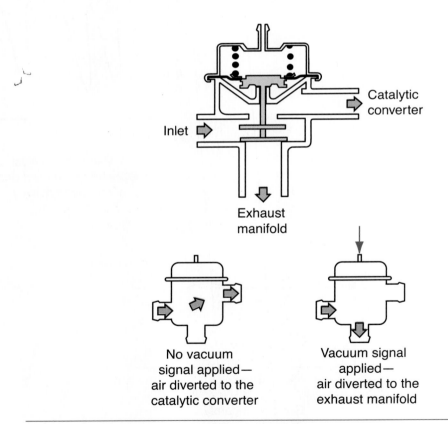

Figure 14-38 A diverter valve for an OBD I system.

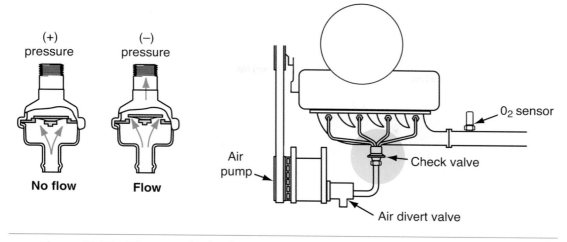

Figure 14-39 AIR system check valves prevent exhaust from venting the system.

Check Valve Testing

All of the types of air injection systems have at least one common element: a one-way check valve (Figure 14-39). The valve opens to let air in but closes to keep exhaust from leaking out. The check valve can be checked with an exhaust gas analyzer. Start the engine and hold the probe of the exhaust gas analyzer near the check valve port. If any amount of CO or CO_2 is read, the valve leaks. If this valve is leaking, hot exhaust is also leaking which could ruin the other components in the air injection system.

✓ **SERVICE TIP:** If the metal container or the clean air hose from this container shows evidence of burning, some of the one-way check valves are allowing exhaust into this container and clean air hose.

Secondary Air Injection System Service and Diagnosis

The first step in diagnosing a secondary air injection system is to check all vacuum hoses and electrical connections in the system. Many AIR system pumps have a centrifugal filter behind the pulley. Air flows through this filter into the pump and the filter keeps dirt out of the pump. The pulley and filter are bolted to the pump shaft, and these two components are serviced separately (Figure 14-40). If the pulley or filter is bent, worn, or damaged, it should be replaced. The pump assembly is usually not serviced.

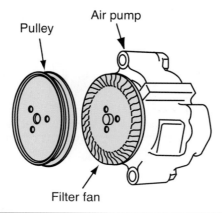

Figure 14-40 AIR pump pulley and filter from an OBD I system.

Special Tools

Belt tension gauge, scanner, vacuum gauge

The secondary air system is a more complex electronic version of the air injection system. In its bypass mode, secondary air is vented to the atmosphere.

Classroom Manual
Chapter 14, page 481

AIR systems were called Thermactor Air Systems by some manufacturers. Therefore, the AIRB was called the TAB and the AIRD was called the TAD.

The air pump belt must have the specified tension. A loose belt or a defective AIR system may result in high emission levels and/or excessive fuel consumption.

In some AIR systems, pressure relief valves are mounted in the AIRB and AIRD valves. Other AIR systems have a pressure relief valve in the pump. If the pressure relief valve is stuck open, airflow from the pump is continually exhausted through this valve, which causes high tailpipe emissions.

If the hoses in the AIR system show evidence of burning, the one-way check valves are leaking, which allows exhaust to enter the system. Leaking air manifolds and pipes result in exhaust leaks and excessive noise.

Some AIR systems will set DTCs in the PCM if there is a fault in the AIRB or AIRD solenoids and related wiring. In some AIR systems, DTCs are set in the PCM memory if the airflow from the pump is continually upstream or downstream. Always use a scan tester to check for any DTCs related to the AIR system and correct the causes of these codes before proceeding with further system diagnosis.

If the AIR system does not pump air into the exhaust ports during engine warmup, HC emissions are high during this mode, and the O_2 sensor takes longer to reach normal operating temperature. Under this condition, the PCM remains in open loop longer. Since the air-fuel ratio is richer in open loop, fuel economy is reduced.

When the AIR system pumps air into the exhaust ports with the engine at normal operating temperature, the additional air in the exhaust stream causes lean signals from the O_2 sensor. The PCM responds to these lean signals by providing a rich air-fuel ratio from the injectors. This action increases fuel consumption. A vehicle can definitely fail an emission test because of air flowing past the O_2 sensor when it should not be. If the O_2 sensor is always sending a lean signal back to the computer, check the air injection system.

Electric AIR Pumps

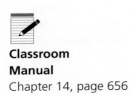

Classroom Manual
Chapter 14, page 656

Some later-model air injection reactor systems use a PCM-controlled electric air pump (Figure 14-41). These electric systems have an AIR solenoid and a solenoid control relay. When the PCM grounds the solenoid relay, battery voltage is applied to both the solenoid and the AIR pump. Usually, DTCs will be set if one of the components fails or the hoses or check valve leak. System checks can be performed with a scan tool.

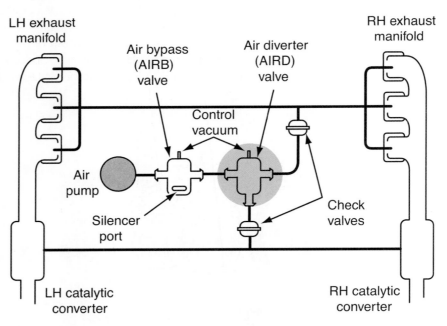

Figure 14-41 Secondary air injection system with AIRB and AIRD valves (OBD I type).

Set the scan tool display to observe the oxygen sensor(s). Start the engine and allow it to reach operating temperature and obtain a normal idle. Enable the AIR system and check the HO₂S voltages. If the voltages are low, the AIR pump, solenoid, and shut-off valve are working properly. If the voltages are not low, perform individual AIR system component tests as directed by the appropriate service manual.

AIRB Solenoid and Valve Diagnosis

When the engine is started, listen for air being exhausted from the AIRB valve (Figure 14-42) for a short period of time. If this air is not exhausted, remove the vacuum hose from the AIRB and start the engine. If air is now exhausted from the AIRB valve, check the AIRB solenoid and connecting wires. When air is still not exhausted from the AIRB valve, check the air supply from the pump to the valve. If the air supply is available, replace the AIRB valve.

During engine warmup, remove the hose from the AIRD valve to the exhaust ports and check for airflow from this hose. If airflow is present, the system is operating normally in this mode. When air is not flowing from this hose, remove the vacuum hose from the AIRD valve and connect a vacuum gauge to this hose. If vacuum is above 12 in. Hg, replace the AIRD valve. When the vacuum is zero, check vacuum hoses, the AIRD solenoid, and connecting wires.

✓ **SERVICE TIP:** With the engine at normal operating temperature, the AIR system sometimes goes back into the upstream mode with the engine idling. It may be necessary to increase the engine speed to maintain the downstream mode.

In an air injection system's upstream mode, the air is directed to the exhaust manifold to aid in the burning of the HC and CO compounds. This also warms the oxygen sensor faster.

Air is directed to the catalytic converter in the downstream mode.

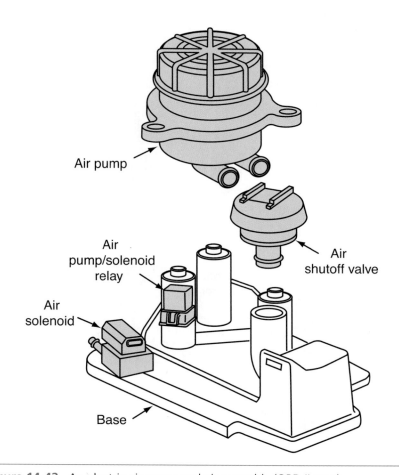

Figure 14-42 An electric air pump and air assembly (OBD II type).

With the engine at normal operating temperature, disconnect the air hose between the AIRD valve and the catalytic converters and check for airflow from this hose. When airflow is present, system operation in the downstream mode is normal. If there is no airflow from this hose, disconnect the vacuum hose from the AIRD valve and connect a vacuum gauge to the hose. When the vacuum gauge indicates zero vacuum, replace the AIRD valve. If some vacuum is indicated on the gauge, check the hose, the AIRD solenoid, and connecting wires.

Air System Efficiency Test

Run the engine at idle with the secondary air system on (enabled). Using an exhaust gas analyzer, measure and record the oxygen (O_2) levels. Next, disable the secondary air system and continue to allow the engine to idle. Again, measure and record the oxygen level in the exhaust gases. The secondary air system should be supplying 2 to 5 percent more oxygen when it is operational (enabled).

OBD II Monitoring of the AIR System

As you recall, the OBD II AIR system does not pump air into the catalytic converter and does not operate at all times like the AIR system used with OBD I systems (Figure 14-43). The OBD II AIR system is meant only to operate long enough to warm the exhaust and help promote catalytic converter light-off by burning leftover HC and CO in the exhaust manifold of a cold engine.

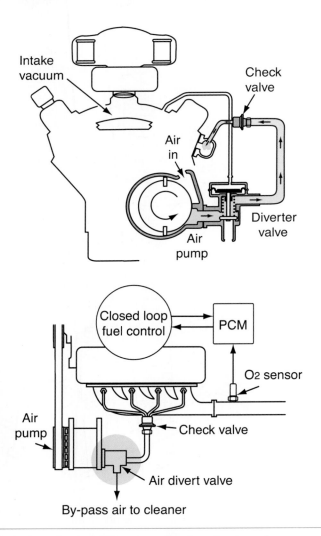

Figure 14-43 Typical mechanical AIR system showing diverter valve, pump, and check valves.

Code	Possible Causes
PO411 Downstream AIR System Flow	Basic mechanical faults such as: Damaged hoses or pipes, stuck open or leaking solenoids, defective AIR pump
PO412 (AIR) Circuit Malfunction	An electrical fault in the solenoids, electric AIR pump, or circuitry involved in the control of the AIR system.

Figure 14-44 Basic OBD II codes for the AIR system.

The OBD II system monitors the AIR system for operation (Figure 14-44). The switching solenoids and relays are monitored for opens and shorts in the wiring. The AIR typically has a passive test that is run by comparing the oxygen sensor voltage shortly after start-up and comparing it to the voltage level at the O_2 sensor just before going into closed loop. The O_2 sensor should show very lean if the AIR pump was working. If the passive test is failed or inconclusive results are obtained, the PCM runs an intrusive test. The AIR pump is commanded on with the system in closed loop. Of course, in this situation the heated oxygen sensors go very lean. It may be important to note that if the AIR system hoses are intended to last for the lifetime of the vehicle, the AIR system does not have to be tested for the amount of air pushed through the system, just that the system is working.

Some codes for the AIR system are included here for the sake of discussion, and these are only some of the generic codes for the OBD II system (Figure 14-44). Make sure and use the specific service information for the vehicle concerned before beginning work.

Terms to Know

Acceleration simulation mode (ASM)	Engine Off Natural Vacuum (EONV)	OBD II
AIR systems	Enhanced EVAP	Purge valve
Charcoal canister	Fuel tank level sensor	Repair grade (RG-240)
Constant volume sampling (CVS)	Fuel tank pressure control valve	Small leak test
Dyno	Fuel tank pressure sensor	Vacuum hold and decay
EGR control solenoid (EGRC)	Initial vacuum pulldown	Vent valve
EGR vacuum regulator (EVR)	Inspection and maintenance (I/M)	Weak vacuum test
EGR vent solenoid (EGRV)	Leak detection pump (LDP)	

ASE-Style Review Questions

1. While discussing catalytic converter operation:
   Technician A says a delta temperature test should be conducted.
   Technician B says a good converter will have an outlet that is 8 percentt cooler than the inlet.
   Who is correct?
   A. A only   **C.** Both A and B
   B. B only   **D.** Neither A nor B

2. While discussing EGR valve diagnosis:
   Technician A says if the EGR valve does not open, the engine may hesitate on acceleration.
   Technician B says if the EGR valve does not open, the engine may detonate on acceleration.
   Who is correct?
   A. A only   **C.** Both A and B
   B. B only   **D.** Neither A nor B

3. While discussing EGR valve diagnosis:
   Technician A says a defective throttle position sensor (TPS) may affect the EGR valve operation.
   Technician B says a defective engine coolant temperature (ECT) sensor may affect the EGR valve operation.
   Who is correct?
   A. A only   **C.** Both A and B
   B. B only   **D.** Neither A nor B

4. While discussing a linear EGR:
   Technician A says that the commanded pintle position and the actual pintle position have to be the same.
   Technician B says that the pintle position at idle cannot exceed 3 percent.
   Who is correct?
   A. A only   **C.** Both A and B
   B. B only   **D.** Neither A nor B

5. While discussing digital EGR valve diagnosis:
   Technician A says a scan tester may be used to command the PCM to open each solenoid in the EGR valve.
   Technician B says the EGR valve should open when 18 in. Hg of vacuum are supplied to the valve at idle speed.
   Who is correct?
   A. A only   **C.** Both A and B
   B. B only   **D.** Neither A nor B

6. While discussing EVR solenoids:
   Technician A says an ohmmeter can be used to check the solenoid for shorted windings.
   Technician B says an ohmmeter can also be used to check for grounded windings.
   Who is correct?
   A. A only   **C.** Both A and B
   B. B only   **D.** Neither A nor B

7. *Technician A* says that a desired backpressure reading is 1.25 psi or less.
   Technician B says that anything less than 2.75 psi is acceptable.
   Who is correct?
   A. A only   **C.** Both A and B
   B. B only   **D.** Neither A nor B

8. While discussing PCV system diagnosis:
   Technician A says an accumulation of oil in the air cleaner indicates the PCV valve is stuck open.
   Technician B says an accumulation of oil in the air cleaner may indicate the piston rings and cylinders are worn.
   Who is correct?
   A. A only   **C.** Both A and B
   B. B only   **D.** Neither A nor B

9. While discussing PCV system diagnosis:
   Technician A says a defective PCV valve may cause rough idle operation.
   Technician B says satisfactory PCV system operation depends on a properly sealed engine.
   Who is correct?
   A. A only   **C.** Both A and B
   B. B only   **D.** Neither A nor B

10. While discussing knock sensor and knock sensor module diagnosis:
    Technician A says with the engine running at 1,500 rpm, if the engine block is tapped near the knock sensor, a knock sensor signal should appear on the scan tester.
    Technician B says with the engine running at 1,500 rpm, if a 12-volt test light is connected from a 12-volt source to the disconnected knock sensor wire, a knock sensor signal should appear on the scan tester.
    Who is correct?
    A. A only   **C.** Both A and B
    B. B only   **D.** Neither A nor B

ASE Challenge Questions

1. IM240 and RG240 tests are basically the same *except*:
   A. They use the same test methods.
   B. They analyze the same gases.
   C. They use the same type chassis dyno.
   D. They use constant volume sampling.

2. *Technician A* says improper canister purge will cause an engine to run rough in hot weather.
   Technician B says the ORVR system is part of the evaporative emissions system.
   Who is correct?
   A. A only
   B. B only
   C. Both A and B
   D. Neither A nor B

3. Knock sensors are being discussed.
   Technician A says false signals may be generated if the sensor's wiring is improperly routed.
   Technician B says a hammer may be used in conjunction with a scan tool to test the sensor.
   Who is correct?
   A. A only
   B. B only
   C. Both A and B
   D. Neither A nor B

4. EGR systems are being discussed.
   Technician A says a too high TP Sensor signal may cause the EGR to open early.
   Technician B says DTCs from other sensors can assist in diagnosing EGR failures.
   Who is correct?
   A. A only
   B. B only
   C. Both A and B
   D. Neither A nor B

5. *Technician A* says an inoperative air injection system will lower catalytic converter temperatures.
   Technician B says a catalytic converter's O_2 storage ability can be tested by removing the air pump drive belt.
   Who is correct?
   A. A only
   B. B only
   C. Both A and B
   D. Neither A nor B

Job Sheet 36

36

Name _____ Date _____

Check the Emission Levels on an Engine

Upon completion of this job sheet, you should be able to measure the emissions levels of an engine and determine the cause of any abnormal readings.

ASE Correlation

This job sheet is related to the ASE Engine Performance Test's content area: general engine diagnosis; task: prepare a 4- or 5-gas analyzer, inspect and prepare vehicle for test and obtain exhaust readings; interpret readings and determine needed actions.

Tools and Materials

Hand tools   Clean engine oil
Exhaust gas analyzer   Spark plug gapper

Describe the vehicle being worked on.

Year_____ Make _____ VIN_____

Model_____

Engine type and size: _____ Vehicle's mileage:_____

Describe the general operating condition of the engine:

_____

_____

Procedures

1. Look up timing specs for this car.

Required timing is _____ATDC or BTDC at _____rpm. What conditions are recommended by the manufacturer for setting timing?

_____

Source of specs_____

2. Check timing. Timing is found to be_____.

3. Connect vacuum gauge. Make sure this is a source for manifold vacuum.

Vacuum reading is:_____ What does this tell you about the engine?

_____

_____

For this engine, what are the HC and CO specs for this car?

_____ HC _____ CO _____ CO_2 _____ O_2 _____ NO_X

What source did you use for these specs?

_____

4. Warm up and calibrate a 4- or 5-gas exhaust analyzer. Disable and block off secondary air to the exhaust (if equipped).

   Prior to running any tests with the engine at idle, what are your readings on the exhaust analyzer?

   _____ HC _____ CO _____ CO_2 _____ O_2 _____ NO_X

   What do these indicate?

   _____

   _____

   With the engine running at about 2,500 rpm, what are your readings on the exhaust analyzer?

   _____ HC _____ CO _____ CO_2 _____ O_2 _____ NO_X

   What do these indicate?

   _____

   _____

5. Now remove the air cleaner assembly.

   With the engine at idle, the vacuum reading now is:_____.

   What does this tell you?

   _____

   _____

   What are your readings on the exhaust analyzer?

   _____ HC _____ CO _____ CO_2 _____ O_2 _____ NO_X

   What do these indicate?

   _____

   _____

   With the engine running at about 2,500 rpm, what are your readings on the exhaust analyzer?

   _____ HC _____ CO _____ CO_2 _____ O_2 _____ NO_X

   What do these indicate?

   _____

   _____

6. Partially restrict the intake airflow.

   With the engine at idle, the vacuum reading now is:_____.

   What does this tell you?

   _____

   _____

   What are your readings on the exhaust analyzer?

   _____ HC _____ CO _____ CO_2 _____ O_2 _____ NO_X

   What do these indicate?

   _____

   _____

   With the engine running at about 2,500 rpm, what are your readings on the exhaust analyzer?

   _____ HC _____ CO _____ CO_2 _____ O_2 _____ NO_X

   What do these indicate?

   _____

   _____

7. Set airflow back to normal.

□

8. Plug off the PCV valve.

With the engine at idle, the vacuum reading now is:_____.

What does this tell you?

_____

_____

What are your readings on the exhaust analyzer?

_____ HC _____ CO _____ CO_2 _____ O_2 _____ NO_X

What do these indicate?

_____

_____

With the engine running at about 2,500 rpm, what are your readings on the exhaust analyzer?

_____ HC _____ CO _____ CO_2 _____ O_2 _____ NO_X

What do these indicate?

_____

_____

_____

9. Return the PCV system to normal.

□

10. Cause a manifold vacuum leak. If fuel injected, the engine may not run with a large vacuum leak. If this happens, cause a small manifold vacuum leak.

Source of leak:_____

With the engine at idle, the vacuum reading now is:_____.

What does this tell you?

_____

_____

What are your readings on the exhaust analyzer?

_____ HC _____ CO _____ CO_2 _____ O_2 _____ NO_X

What do these indicate?

_____

_____

With the engine running at about 2,500 rpm, what are your readings on the exhaust analyzer?

_____ HC _____ CO _____ CO_2 _____ O_2 _____ NO_X

What do these indicate?

_____

_____

11. Correct the vacuum leak.

□

12. Open the EGR valve slightly, if possible. (Use heavy gloves to prevent burns.)

With the engine at idle, the vacuum reading now is:_____

What does this tell you?

_____

_____

What are your readings on the exhaust analyzer?

_____ HC _____ CO _____ CO_2 _____ O_2 _____ NO_X

What do these indicate?

_____

_____

With the engine running at about 2,500 rpm, what are your readings on the exhaust analyzer?

_____ HC _____ CO _____ CO_2 _____ O_2 _____ NO_X

What do these indicate?

_____

_____

13. Allow the EGR valve to close.

14. Turn engine off.

15. Take a tailpipe reading.

What are your readings on the exhaust analyzer?

_____ HC _____ CO _____ CO_2 _____ O_2 _____ NO_X

What do these indicate?

_____

_____

16. Ground a spark plug wire on one of the cylinders.

With the engine at idle, the vacuum reading now is:_____.

What does this tell you?

_____

_____

_____

What are your readings on the exhaust analyzer?

_____ HC _____ CO _____ CO_2 _____ O_2 _____ NO_X

What do these indicate?

_____

_____

With the engine running at about 2.500 rpm, what are your readings on the exhaust analyzer?

_____ HC _____ CO _____ CO_2 _____ O_2 _____ NO_X

What do these indicate?

_____

_____

17. Return the spark plug wire to normal.

18. Bypass or disconnect the electronic timing advance system (if equipped).

With the engine at idle, the vacuum reading now is:_____.

What does this tell you?

_____

_____

What are your readings on the exhaust analyzer?

_____ HC _____ CO _____ CO_2 _____ O_2 _____ NO_X

What do these indicate?

_____

_____

With the engine running at about 2,500 rpm, what are your readings on the exhaust analyzer?

_____ HC _____ CO _____ CO_2 _____ O_2 _____ NO_X

What do these indicate?

_____

_____

19. Reconnect the timing control system. ☐

20. Snap acceleration from idle speed.

    Indicate the vacuum reading after the snap:_____.

    What does this tell you?

_____

_____

    What are your readings on the exhaust analyzer prior to the snap?

    _____ HC _____ CO _____ CO_2 _____ O_2 _____ NO_X

    What do these indicate?

_____

_____

    What are the readings on the exhaust analyzer several seconds after the snap?

    _____ HC _____ CO _____ CO_2 _____ O_2 _____ NO_X

    What do these indicate?

_____

_____

21. Squirt a small amount of water down the throttle plates.

    With the engine at idle, the vacuum reading now is:_____.

    What does this tell you?

_____

_____

    What are your readings on the exhaust analyzer?

    _____ HC _____ CO _____ CO_2 _____ O_2 _____ NO_X

    What do these indicate?

_____

_____

    With the engine running at about 2,500 rpm, what are your readings on the exhaust analyzer?

    _____ HC _____ CO _____ CO_2 _____ O_2 _____ NO_X

    What do these indicate?

_____

_____

22. Insert exhaust gas analyzer probe into the oil filler tube. (Caution—do not immerse the probe into the oil).

    With the engine at idle, what are your readings on the exhaust analyzer?

    _____ HC _____ CO _____ CO_2 _____ O_2 _____ NO_X

What do these indicate?

_____

_____

With the engine running at about 2,500 rpm, what are your readings on the exhaust analyzer?
_____ HC _____ CO _____ CO_2 _____ O_2 _____ NO_X

What do these indicate?

_____

_____

23. Summarize below what affects an engine's vacuum reading and the readings on an exhaust gas analyzer.

_____

_____

_____

_____

_____

_____

_____

_____

_____

_____

_____

_____

_____

_____

Instructor's Response_____

_____

_____

Job Sheet 37

37

Name _____ Date _____

Check the Operation of a PCV System

Upon completion of this job sheet, you should be able to check the PCV valve and hoses to determine if they are working properly.

ASE Correlation

This job sheet is related to the ASE Engine Performance Test's content area: emissions control systems diagnosis and repair. Tasks: diagnose oil leaks, emissions, and driveability problems resulting from failure of the positive crankcase ventilation system, and inspect and test positive crankcase ventilation filter/breather cap, valve, tubes, orifices, and hoses; service or replace as needed.

Tools and Materials

Exhaust gas analyzer   Appropriate service manual
Oil

Describe the vehicle being worked on.

Year_____ Make _____ VIN _____

Model_____

Procedures

Task Completed

1. Describe the location of the PCV valve.

   _____

   _____

2. Run the engine until normal operating temperature is reached.   ☐

3. Record the CO reading from the exhaust analyzer. _____ percent

4. Remove the PCV valve from the valve or camshaft cover.   ☐

5. Record the CO reading. _____ percent

6. Explain why there was a change in CO and what service you would recommend.

   _____

   _____

   _____

7. Place your thumb over the end of the PCV valve.   ☐

8. Record the CO reading. _____ percent

9. Explain why there was a change in CO and what service you would recommend.

   _____

   _____

   _____

☐

10. Remove the valve from its hose and check for vacuum.

11. Record your results.

_____

_____

_____

☐

12. Hold and shake the PCV valve.

13. Record your results.

_____

_____

14. State your conclusions about the PCV system on this engine.

_____

_____

_____

Instructor's Response_____

_____

_____

Job Sheet 38

38

Name _____ Date _____

Check the Operation of an EGR Valve

Upon completion of this job sheet, you should be able to check the operation of an EGR valve and associated circuits.

ASE Correlation

This job sheet is related to the ASE Engine Performance Test's content area: emissions control systems diagnosis and repair. Tasks: diagnose emissions and driveability problems caused by failure of the exhaust recirculation system; inspect and test valve, valve manifold, and exhaust passages of exhaust gas recirculation systems; service or replace as needed; and inspect and test vacuum/pressure controls, filters, and hoses of exhaust gas recirculation systems; service or replace as needed.

Tools and Materials

Hand-operated vacuum pump
Vacuum gauge
Exhaust gas analyzer
Scan Tool for assigned vehicle
A vehicle with an electronic EGR transducer and solenoid

Describe the vehicle being worked on.

Year_____ Make _____ VIN _____

Model_____

Type of EGR system _____

Procedures

_____
Task Completed

1. Inspect the hoses and connections within the EGR system. Record the results.

   _____

   _____

2. Pull the hose to the EGR transducer. ☐

3. Run the engine and check for vacuum at the hose. ☐

4. Turn off the engine and connect the scan tool to the DLC. ☐

5. Insert a tee fitting into the vacuum hose and reconnect the supply line to the transducer. Connect the vacuum gauge to the tee fitting. ☐

6. Start the engine. ☐

7. Record the vacuum reading: _____ in. Hg

8. There should be a minimum of 15 in. Hg. Summarize your readings.

   _____

   _____

9. Actuate the solenoid with the scan tool. ☐

10. Did the solenoid click? ☐ Yes ☐ No
11. Did the vacuum fluctuate with the cycling of the solenoid? ☐ Yes ☐ No
12. Disconnect the vacuum hose and backpressure hose from the transducer.
13. Disconnect the electrical connector from the solenoid.
14. Plug the transducer output port.
15. Apply 1 to 2 psi of air pressure to the backpressure port of the transducer.
16. With the vacuum pump, apply at least 12 in. Hg to the other side of the transducer. How did the transducer react?

_____

_____

What did this test tell you?

_____

_____

17. Run the engine at fast idle.
18. Insert the probe of the gas analyzer in the vehicle's tailpipe.
19. Record these exhaust levels:

_____ HC _____ CO _____ CO$_2$ _____ O$_2$ _____ NO$_X$

What was the injector pulse width at this time? _____ ms

20. Remove the vacuum hose at the EGR valve and attach the vacuum pump to the valve.
21. Apply just enough vacuum to cause the engine to run rough. Keep the vacuum at that point for a few minutes and record the readings on the gas analyzer while the engine is running rough.

_____ HC _____ CO _____ CO$_2$ _____ O$_2$ _____ NO$_X$

What was the injector pulse width at this time? _____ ms

22. Describe what happened to the emissions levels and explain why:

_____

_____

_____

_____

23. Apply the vehicle's parking brake.
24. Disconnect the vacuum pump and plug the vacuum hose to the EGR valve.
25. Firmly depress and hold the brake pedal with your left foot and place the transmission in drive.
26. Raise engine speed to about 1,800 rpm.
27. Record the readings on the gas analyzer.

_____ HC _____ CO _____ CO$_2$ _____ O$_2$ _____ NO$_X$

What was the injector pulse width at this time? _____ ms

28. Return the engine to an idle speed, place the transmission into park, then shut off the engine.
29. Describe what happened to the emissions levels and explain why:

_____

_____

_____

_____

30. Reconnect the EGR valve. ☐

31. Apply the vehicle's parking brake. Start the engine. ☐

32. Firmly depress and hold the brake pedal with your left foot and place the transmission in drive. ☐

33. Raise engine speed to about 1,800 rpm. ☐

34. Record the readings on the gas analyzer.

    _____ HC _____ CO _____ CO_2 _____ O_2 _____ NO_x

    What was the injector pulse width at this time? _____ ms

35. Return the engine to an idle speed, place the transmission into park, then shut off the engine. ☐

36. Describe what happened to the emissions levels and explain why:

    _____

    _____

    _____

    _____

37. Use the scan tool and retrieve any DTCs. Record any that were displayed.

    _____

38. Clear the codes. ☐

Instructor's Response_____

_____

_____

_____

APPENDIX A

ASE PRACTICE EXAMINATION

1. *Technician A* says a hydrometer reading of 1.200 at 80°F means the battery must be recharged before performing a capacity test.
   Technician B says a capacity test can be correctly performed with the battery cables connected.
   Who is correct?
   - **A.** A only
   - **B.** B only
   - **C.** Both A and B
   - **D.** Neither A nor B

2. *Technician A* says engine detonation may be caused by low octane fuel.
   Technician B says that detonation can be caused by engine overheating.
   Who is correct?
   - **A.** A only
   - **B.** B only
   - **C.** Both A and B
   - **D.** Neither A nor B

3. The PCV system is being discussed.
   Technician A says oil in the crankcase breather filter will confirm that the PCV valve is clogged.
   Technician B says the PCV valve must be disconnected to be checked with the engine operating.
   Who is correct?
   - **A.** A only
   - **B.** B only
   - **C.** Both A and B
   - **D.** Neither A nor B

4. An EI-equipped vehicle will not start.
   Technician A says a good first step is to make certain that the MIL is illuminated with the key on.
   Technician B says to check for spark at more than one plug wire during diagnosis.
   Who is correct?
   - **A.** A only
   - **B.** B only
   - **C.** Both A and B
   - **D.** Neither A nor B

5. Electronic fuel injection systems are being discussed.
   Technician A says a hard to start engine may have an open electrical fuel pump relay bypass circuit.
   Technician B says a hesitation when accelerating from idle may be caused by dirty injectors.
   Who is correct?
   - **A.** A only
   - **B.** B only
   - **C.** Both A and B
   - **D.** Neither A nor B

6. A vehicle backfires during almost every deceleration.
   Technician A says a secondary air injection diverter valve may be causing the backfire.
   Technician B says this condition may be caused by a low performing secondary air injection pump.
   Who is correct?
   - **A.** A only
   - **B.** B only
   - **C.** Both A and B
   - **D.** Neither A nor B

7. A vehicle has a consistent slow- or no-crank condition. This may be caused by:
   - **A.** A slipping AC generator drive belt
   - **B.** A defective AC generator
   - **C.** A low regulator voltage
   - **D.** Any of the above

8. A fuel-injected vehicle comes in with a miss on acceleration.
   Technician A says that there could be an ignition problem.
   Technician B says that the intake to MAF duct could be opening up at a tear on acceleration.
   Who is correct?
   - **A.** A only
   - **B.** B only
   - **C.** Both A and B
   - **D.** Neither A nor B

9. *Technician A* says static electricity generated by clothing in contact with vehicle upholstery must be discharged before beginning work on electronic devices.
   Technician B says the best control for static electricity is the wearing of a waist or wrist grounding strap.
   Who is correct?
   - **A.** A only
   - **B.** B only
   - **C.** Both A and B
   - **D.** Neither A nor B

10. *Technician A* says a camshaft installed one tooth retarded can be compensated by adjusting the ignition's base timing.
    Technician B says to use the starter to crank the engine with the timing belt removed to determine if the engine is causing interference or free-wheeling.
    Who is correct?
    - **A.** A only
    - **B.** B only
    - **C.** Both A and B
    - **D.** Neither A nor B

11. A fuel-injected engine runs rough and sluggish on acceleration. When checking the fuel pressure, the technician notices that the fuel pressure is at specifications during idle, but falls off on acceleration.
    Technician A says the fuel pump could be faulty.
    Technician B says the fuel filter could be clogged.
    Who is correct?
    - **A.** A only
    - **B.** B only
    - **C.** Both A and B
    - **D.** Neither A nor B

12. Cylinder leakage testing is being discussed.
Technician A says air escaping through the PCV system indicates worn piston rings.
Technician B says bubbles in the cooling systems during this test indicates a blown head gasket between the tested cylinder and an adjacent cylinder.
Who is correct?
- **A.** A only
- **B.** B only
- **C.** Both A and B
- **D.** Neither A nor B

13. A test light connected between the ignition coil's positive terminal and ground illuminates but does not flash during cranking. This condition is:
- **A.** Caused by a grounded positive contact on the breaker points
- **B.** A normal condition
- **C.** Indicative of a grounded ignition pick-up coil
- **D.** Caused by an open coil primary winding

14. A poorly performing engine with no stored DTCs is being diagnosed.
Technician A says the first diagnostic step is to follow the manufacturer's "diagnose by symptom" routine.
Technician B says gas analyzers and scopes are the best diagnostic method in this instance.
Who is correct?
- **A.** A only
- **B.** B only
- **C.** Both A and B
- **D.** Neither A nor B

15. *Technician A* says the installation of a magnetic crankshaft position sensor may require the use of a special tool to set the air gap.
Technician B says that some crankshaft sensors are not adjustable.
Who is correct?
- **A.** A only
- **B.** B only
- **C.** Both A and B
- **D.** Neither A nor B

16. An electric cooling fan runs with the AC compressor engaged, but not when the engine overheats.
Technician A says a grounded cooling fan temperature switch may be the cause.
Technician B says the PCM may not be grounding the cooling fan relay.
Who is correct?
- **A.** A only
- **B.** B only
- **C.** Both A and B
- **D.** Neither A nor B

17. An automatic ride control system lowers the vehicle after engine shutdown, but before the door is opened and closed.
Technician A says a height sensor may be bad.
Technician B says the ride control module may be at fault.
Who is correct?
- **A.** A only
- **B.** B only
- **C.** Both A and B
- **D.** Neither A nor B

18. A vehicle seems to have a key-off battery drain.
Technician A says a test lamp connected in series with the battery negative terminal can be used to determine if there is a draw.
Technician B says an ammeter connected in the same manner will also determine if the draw is excessive.
Who is correct?
- **A.** A only
- **B.** B only
- **C.** Both A and B
- **D.** Neither A nor B

19. A vehicle has high NO$_x$ emissions.
Technician A says this could be caused by carbon accumulation in the combustion chambers.
Technician B says this can be caused by an engine that is overheating.
Who is correct?
- **A.** A only
- **B.** B only
- **C.** Both A and B
- **D.** Neither A nor B

20. The spray pattern of a fuel injector is being discussed.
Technician A says an open or empty space in the cone indicates a dirty injector.
Technician B says it is normal to have small fuel drips in the center of the cone during injector firing.
Who is correct?
- **A.** A only
- **B.** B only
- **C.** Both A and B
- **D.** Neither A nor B

21. Technician A says a low steady engine vacuum reading may indicate late ignition timing.
Technician B says a leaking intake gasket will cause a low steady vacuum reading.
Who is correct?
- **A.** A only
- **B.** B only
- **C.** Both A and B
- **D.** Neither A nor B

22. A cylinder power balance test is being performed.
Technician A says if there is not a rpm drop on any cylinder, then each cylinder is producing the same amount of power.
Technician B says a cylinder showing a larger rpm drop than the others means less power is being produced by that cylinder.
Who is correct?
- **A.** A only
- **B.** B only
- **C.** Both A and B
- **D.** Neither A nor B

23. While discussing EI ignition systems:
Technician A says some waste spark systems do not need a cam sensor.
Technician B says that some coil-over-plug systems do not need a cam sensor.
Who is correct?
- **A.** A only
- **B.** B only
- **C.** Both A and B
- **D.** Neither A nor B

24. *Technician A* says formed plastic fuel lines may be replaced with flexible fuel line and clamps.
Technician B says a fuel filler cap with a defective vacuum valve may cause a lean fuel mixture under some conditions.
Who is correct?
A. A only
B. B only
C. Both A and B
D. Neither A nor B

25. Turbochargers are being discussed.
Technician A says an exhaust leak between the converter and muffler may cause decreased boost pressure.
Technician B says a binding wastegate linkage may produce higher than specified boost pressure.
Who is correct?
A. A only
B. B only
C. Both A and B
D. Neither A nor B

26. An EGR system is being tested.
Technician A says supplying 5–10 in. Hg of vacuum directly to the valve's port and noting engine operation is a quick method to confirm or eliminate the valve as the malfunctioning component.
Technician B says this method can be used to test a positive backpressure system.
Who is correct?
A. A only
B. B only
C. Both A and B
D. Neither A nor B

27. Secondary air injection is being discussed.
Technician A says air directed to the exhaust manifold during idling at normal operating temperatures may be a normal condition.
Technician B says air directed to the manifold during operation at normal temperatures will cause the O_2 sensor voltage signals to be low.
Who is correct?
A. A only
B. B only
C. Both A and B
D. Neither A nor B

28. The starter motor drags during cranking.
Technician A says corrosion on the battery cables may cause this problem.
Technician B says low voltage in the ignition switch to starter relay (solenoid) circuit can be the cause.
Who is correct?
A. A only
B. B only
C. Both A and B
D. Neither A nor B

29. A vehicle comes into the shop with a code set for a large evaporative emission system leak.
Technician A says the first thing to check for is a loose or missing fuel cap.
Technician B says the next step is to do a visual inspection of the evaporative emission hoses.
Who is correct?
A. A only
B. B only
C. Both A and B
D. Neither A nor B

30. While discussing OBD II systems:
Technician A says that not all emissions-related trouble codes set on the first failure.
Technician B says that these codes are called "B" type codes.
Who is correct?
A. A only
B. B only
C. Both A and B
D. Neither A nor B

31. *Technician A* says the actual catalyst efficiency monitor is the O_2 sensor at the outlet of the catalytic converter.
Technician B says a vacuum gauge can be used to check for a suspected clogged converter.
Who is correct?
A. A only
B. B only
C. Both A and B
D. Neither A nor B

32. *Technician A* says a restricted fuel filter may cause NO_X emissions to increase.
Technician B says the O_2 voltage signals may be high if the fuel pump check valve is stuck closed.
Who is correct?
A. A only
B. B only
C. Both A and B
D. Neither A nor B

33. An engine exhaust has a strong sulphur odor and tends to make the eyes water.
Technician A says an excessively restricted air filter may cause this condition.
Technician B says a bad O_2 sensor may be the cause.
Who is correct?
A. A only
B. B only
C. Both A and B
D. Neither A nor B

34. The vacuum regulator controlling the canister purge valve bleeds (is open) at all engine speeds and temperatures.
Technician A says the engine may tend to flood at idle.
Technician B says the HC emissions may increase above specifications during normal engine operating condition.
Who is correct?
A. A only
B. B only
C. Both A and B
D. Neither A nor B

35. A fully electronic EGR system is not functioning properly. The *least likely* cause is:
A. The PCM is not cycling the solenoids
B. A loss of vacuum at the valve's port
C. Low voltage to the valve
D. No reference voltage to the EVP sensor

36. *Technician A* says a quick test of the secondary air injection system's check valve is to visually inspect the supply hoses for signs of heat damage.
Technician B says on some secondary air systems a badly tuned engine may cause the system to malfunction.
Who is correct?
A. A only **C.** Both A and B
B. B only **D.** Neither A nor B

37. A digital auto-ranging multimeter connected between a power feed conductor and ground is registering up and down the mVDC scale.
Technician A says this could indicate that there is no voltage on that conductor.
Technician B says this could indicate a ground in the conductor prior to the meter connection.
Who is correct?
A. A only **C.** Both A and B
B. B only **D.** Neither A nor B

38. While discussing computer systems grounds:
Technician A says to use an ohmmeter to check the resistance to ground.
Technician B says it is better to check for a voltage drop to ground with the system operating.
Who is correct?
A. A only **C.** Both A and B
B. B only **D.** Neither A nor B

39. No visible leaks are found on an engine that requires periodic topping off of the cooling system.
Technician A says a stuck open vacuum valve in the radiator cap may be the cause.
Technician B says an infrared gas analyzer can be used to check for combustion chamber to cooling system leaks.
Who is correct?
A. A only **C.** Both A and B
B. B only **D.** Neither A nor B

40. While discussing oxygen sensors:
Technician A says the O_2 sensor can be killed by coolant entering the exhaust.
Technician B says that the oxygen sensor can be checked with a digital storage oscilloscope.
Who is correct?
A. A only **C.** Both A and B
B. B only **D.** Neither A nor B

41. The AC generator output measured at the battery is 11 VDC at 5 amps. The output is over 13 VDC when measured at the AC generator output terminal.
Technician A says a bad wire to terminal connection may be the cause.
Technician B says a faulty voltage regulator could be the cause.
Who is correct?
A. A only **C.** Both A and B
B. B only **D.** Neither A nor B

42. Below is the rough data obtained from an exhaust test using a five gas analyzer.

Emission	Idle result	Cruise result
O_2	Low	Low
CO_2	Low	Low
CO	High	Higher
HC	High	Higher
NO_X	Low	Low

Technician A says that a leaking injector may cause these readings.
Technician B says that high fuel pressure could be the cause.
Who is correct?
A. A only **C.** Both A and B
B. B only **D.** Neither A nor B

43. One cylinder produces less power than the other cylinders during a power balance test. This may be caused by any of the following *except*:
A. Excessive resistance in the spark plug wire
B. Excessive resistance in the coil wire
C. Fouled spark plug
D. Cracked spark plug cable boot

44. A vehicle experiences repeated ignition module failures. The *most likely* cause is:
A. Heat buildup from lack of silicone grease
B. High resistance between the module and its mounting
C. Loose primary circuit conductors/connectors
D. Weak magnetic sensor

45. While discussing automotive computer networks:
Technician A says that some computer network languages are used only to talk to a scan tool.
Technician B says that in 2006, the computer language used will be CAN.
Who is correct?
A. A only **C.** Both A and B
B. B only **D.** Neither A nor B

46. While discussing servicing of TBI systems:
    Technician A says removal of the throttle body may involve draining the cooling system.
    Technician B says many parts of the throttle body can be sufficiently cleaned without removing it from the engine.
    Who is correct?
    A. A only
    B. B only
    C. Both A and B
    D. Neither A nor B

47. An engine's performance gets worse as the engine speed increases.
    Technician A says an engine vacuum test may reveal the system causing the problem.
    Technician B says an extended overrich fuel mixture may cause the cruise vacuum to be lower than the idle vacuum.
    Who is correct?
    A. A only
    B. B only
    C. Both A and B
    D. Neither A nor B

48. *Technician A* says a PCV valve can be checked by blowing through each end and noting the results.
    Technician B says air passing freely from the intake side to the outlet side indicates a faulty valve.
    Who is correct?
    A. A only
    B. B only
    C. Both A and B
    D. Neither A nor B

49. The 5 VDC reference signal is not present at the EVP sensor.
    Technician A says the PCM may be registering a cold engine and not supplying the signal.
    Technician B says there may be an open in the PCM to sensor circuit.
    Who is correct?
    A. A only
    B. B only
    C. Both A and B
    D. Neither A nor B

50. An ignition has no spark at the coil wire during testing.
    Technician A says a broken distributor drive gear will cause this condition.
    Technician B says an open in the Hall effect sensor circuit will create this problem.
    Who is correct?
    A. A only
    B. B only
    C. Both A and B
    D. Neither A nor B

APPENDIX B

Metric Conversions

	to convert these	to these,	multiply by:
TEMPERATURE	Centigrade Degrees	Fahrenheit Degrees	1.8 then + 32
	Fahrenheit Degrees	Centigrade Degrees	0.556 after − 32
LENGTH	Millimeters	Inches	0.03937
	Inches	Millimeters	25.4
	Meters	Feet	3.28084
	Feet	Meters	0.3048
	Kilometers	Miles	0.62137
	Miles	Kilometers	1.60935
AREA	Square Centimeters	Square Inches	0.155
	Square Inches	Square Centimeters	6.45159
VOLUME	Cubic Centimeters	Cubic Inches	0.06103
	Cubic Inches	Cubic Centimeters	16.38703
	Cubic Centimeters	Liters	0.001
	Liters	Cubic Centimeters	1000
	Liters	Cubic Inches	61.025
	Cubic Inches	Liters	0.01639
	Liters	Quarts	1.05672
	Quarts	Liters	0.94633
	Liters	Pints	2.11344
	Pints	Liters	0.47317
	Liters	Ounces	33.81497
	Ounces	Liters	0.02957
WEIGHT	Grams	Ounces	0.03527
	Ounces	Grams	28.34953
	Kilograms	Pounds	2.20462
	Pounds	Kilograms	0.45359
WORK	Centimeter-Kilograms	Inch-Pounds	0.8676
	Inch-Pounds	Centimeter-Kilograms	1.15262
	Meter Kilograms	Foot-Pounds	7.23301
	Foot-Pounds	Newton-Meters	1.3558
PRESSURE	Kilograms/Square Centimeter	Pounds/Square Inch	14.22334
	Pounds/Square Inch	Kilograms/Square Centimeter	0.07031
	Bar	Pounds/Square Inch	14.504
	Pounds/Square Inch	Bar	0.06895

APPENDIX C

Engine Performance Special Tool Suppliers

AllData Corp.
9412 Big Horn Blvd., Elk Grove, CA
(800) 829-8727
www.alldata.com

Ferret Instruments Inc.
2128 Yosemite Dr., Lebanon, IN
(800) 627-5655
www.ferretinstruments.com

Fluke Corp.
box 9090 Everett, WA
(800) 44-FLUKE
www.fluke.com

Hickok Inc.
10514 DuPont Ave., Cleveland, OH
(800) 342-5080
www.hickok-inc.com

Kal Equip
15825 Industrial Pkwy., Cleveland, OH
(800) 228-7667
www.kalequip.com

Kleer-Flo Company
15151 Technology Drive
Eden Prairie, MN 55344
(800) 328-7942
Fax: 612-934-3909
www.kleer-flo.com

Mac Tools
4635 Hilton Corp. Dr., Columbus, OH
(800) MAC-TOOLS
www.mactools.com

Matco Tools
4403 Allen Rd., Stow, OH
(800) 433-7098
www.matcotools.com

Micro Processor Systems Inc. MPSI / Nexiq
6405 19 Mile Rd., Sterling Heights, MI
(800) 639-6774
www.nexiq.com

Mitchell Repair Information Co.
14145 Danielson Street, Poway, CA
(858) 391-5000
www.mitchellrepair.com

OTC Tools
655 Eisenhower Dr.
Owatonna, MN
(800) 533-6127
www.otctools.com

SK Hand Tool Corp.
9500 W. 55th St., Ste. B, McCook, IL
(800) 822-5575
www.skhandtool.com

Snap-on Diagnostics
2801 80th Street, Kenosha, WI
(877) 762-7664
http://www.snapondiag.com

Snap-on Tools Co.
2801 80th Street, Kenosha, WI
(262) 656-5200
www.snapon.com

SPX Service Solutions
28635 Mound Road, Warren, MI 48092
(586) 574-2332
Fax: 1-800-578-7375
http://www.servicesolutions.spx.com/

Sun Electric Corporation
One Sun Parkway
Crystal Lake, IL 60014

GLOSSARY

Note: Terms are highlighted in color, followed by Spanish translation in bold.

14.7:1 air-fuel ratio The preferred air to fuel ratio for efficient engine operation 14.7 parts of air to 1 part of fuel.

Relación de aire a combustible 14.7:1 La relación preferida de 14.7 partes de aire a 1 parte de combustible que provee la operación más eficiente del motor.

Abrasive cleaning Any friction method used to clean components.

Limpieza abrasiva Cualquier metodo de fricción que se usa para limpiar los componentes.

Acceleration simulation mode (ASM) A test that incorporates steady-state and transient testing.

Modo de simulación de acceleración (ASM) Una prueba que incorpora las pruebas de los régimenes permanentes y transitorios.

Actuation test mode A scan tester mode used to cycle the relays and actuators in a computer system.

Modo de prueba de activación Instrumento de pruebas de exploración utilizado para ciclar los relés y los accionadores en una computadora.

Actuator Electrical devices that a computer uses to carry out mechanical actions, usually a relay solenoid or lamp.

Actuador Los dispositivos eléctricos que utiliza una computadora para realizar las acciones mecánicas, por lo generral un senioode de relé o una lámpara.

Adjusting pads, mechanical lifters Metal discs that are available in various thicknesses and are positioned in the end of the mechanical lifter to adjust valve clearance.

Cojines de ajuste, elevadores mecánicos Discos metálicos disponibles en diferentes espesore que se colocan en el extremo del elevador mecánico para ajustar el espacio libre de la válvula.

Advance-type timing light A timing light that is capable of checking the degrees of spark advance.

Luz de ensayo de regulación del encendido tipo avance Luz de ensayo de regulación del encendido capaz de verificar la cantidad del avance de la chispa.

After top dead center (ATDC) Any measurement after top dead center.

Después de punto muerto superior (ATDC) Cualquier medida después del punto muerto superior.

AIR bypass (AIRB) solenoid A computer-controlled solenoid that directs air to the atmosphere or to the AIR diverter solenoid.

Solenoide de paso AIR Solenoide controlado por computadora que conduce el aire hacia la atmósfera o hacia el solenoide derivador AIR.

Air charge temperature (ACT) sensor A sensor that sends a signal to the computer in relation to intake air temperature.

Sensor de la temperatura de la carga de aire Sensor que le envía una señal a la computadora referente a la temperatura del aire aspirado.

AIR diverter (AIRD) solenoid A computer-controlled solenoid in the secondary air injection system that directs air upstream or downstream.

Solenoide derivador AIRD Solenoide controlado por computadora en el sistema de inyección secundaria de aire que conduce el aire hacia arriba o hacia abajo.

Air-fuel ratio The proportion of air to fuel that an engine is burning. 14.7:1 is ideal.

Relación de aire-combustible La proporción de aire al combustible que quema en un motor. El 14.7:1 es ideal.

Air-operated vacuum pump A vacuum pump operated by air pressure that may be used to pump liquids such as gasoline.

Bomba de vacío accionada hidráulicamente Bomba de vacío accionada por la presión del aire que puede utilizarse para bombear líquidos, como por ejemplo la gasolina.

AIR system An air injection system.

Sistema de aire Un sistema de inyección de aire.

Alternating current (AC) An electric current that reverses its direction at regularly recurring intervals.

Corriente Alterna Una corriente eléctrica que reserva su dirección a intervalos de repetición regular.

American wire gauge (AWG) A system that designates wire sizes established by the SAE.

Calibre de alambre Americana (AWG) Un sistema que usa la SAE para establecer los tamaños del alambre

Ammeter A device used to measure electrical current.

Amperímetro Un dispositivo para medir la corriente eléctrica.

Amplitude The difference between the highest and lowest voltage in a waveform signal.

Amplitud La diferencia entre el voltaje más alto y el voltaje más bajo en una señal en forma de onda.

Analog meter A meter with a moveable pointer and a meter scale.

Medidor analógico Medidor provisto de un indicador móvil y una escala métrica.

API American Petroleum Institute. An organization that sets standards for petroleum-based products, such as engine oil.

API Instituto Americano del petróleo. Una organización que establece estándares para productos basados en petróleo.

ASE blue seal of excellence A seal displayed by an automotive repair facility that employs ASE-certified technicians.

Sello azul de excelencia de la ASE Logotipo exhibido en talleres de reparación de automóviles donde se emplean mecánicos certificados por la ASE.

ASE technician certification Certification of automotive technicians in various classifications by the National Institute for Automotive Service Excellence (ASE).

Certificación de mecánico de la ASE Certificación de mecánico de automóviles en áreas diferentes de especialización otorgada por el Instituto Nacional para la Excelencia en la Reparación de Automóviles (ASE).

Atomization The process of breaking up a liquid into small particles or droplets.

Atomización El proceso de romper un líquido adentro de partículas pequeñas o gotas.

Automatic shutdown (ASD) relay A computer-operated relay that supplies voltage to the fuel pump, coil primary, and other components on Chrysler fuel injected engines.

Relé de parada automática Relé accionado por computadora que les suministra tensión a la bomba del combustible, al bobinado primario, y a otros componentes en motores de inyección de combustible fabricados por la Chrysler.

Average responding Meters that show the average voltage peak.
Respuesta promedio Los medidores que enseñan la picadura promedio de voltaje.

"Banjo" fitting A round fitting through which a bolt with a passage drilled through it is used to attach the fitting to a housing. Looking at the fitting and line it resembles a banjo.
Conexión banjo Una conexión redondo por la cual un perno perforado por un taladro se usa para conectar la conexión con un cárter. La conexión con la linea parecen un banjo.

Barometric (Baro) pressure sensor A sensor that sends a signal to the computer in relation to barometric pressure.
Sensor de la presión barométrica Sensor que le envía una señal a la computadora referente a la presión barométrica.

Baud rate The rate at which a PCM is able to transfer and receive data. Baud rate is measured in bits per second.
Velocidad Baud La velocidad a la cual el PCM (módulo de control de la potencia del motor) es capaz de transferir y recibir data. La velocidad Baud es medida en mordidas por segundo.

Before top dead center (BTDC) Any measurement, usually in degrees, prior to the top dead center point.
Antes de punto muerto superior (ATDC) Cualquier medida, suele ser en grados, antes del punto muerto superior

Belt tension gauge A gauge designed to measure belt tension.
Calibrador de tensión de la correa de transmisión Calibrador diseñado para medir la tensión de una correa de transmisión.

Bidirectional control Command of a function usually performed by the PCM via a scan tool. (i.e., shifting of an electronic transmission) used for diagnostics.
Control bidireccional Mando de una función que suele controlar el PCM por medio de una herramienta exploradora (por ejemplo, de cambio de marcha de una transmisioón electrónica) que se usa para los diagnósticos.

Blowby Compression and exhaust gases that blow by the piston rings and enter into the engine's crankcase.
Fuga en la cámara de la combustión Compresión y gases de escape que se escapan atraves de los anillos del pistón y entran adentro de la cacerola del aceite.

Blowgun A device attached to the end of an air hose to control and direct airflow while cleaning components.
Soplete Dispositivo fijado en el extremo de una manguera de aire para controlar y conducir el flujo de aire mientras se lleva a cabo la limpieza de los componentes.

Boost pressure The amount of intake manifold pressure created by a turbocharger or supercharger.
Presión de sobrealimentación Cantidad de presión en el colector de aspriación producida por un turbocompresor o un compresor.

Bore The diameter of a cylinder in the engine block.
Calibrado del cilíndro El diámetro de un cilindro en el bloque motor.

Breakout box A terminal box that is designed to be connected in series at Ford PCM terminals to provide access to these terminals for test purposes.
Caja de desenroscadura Caja de borne diseñada para conectarse en serie a los bornes del módulo del control del tren transmisor de potencia de la Ford, con el objetivo de facilitar el acceso a dichos bornes para propósitos de prueba.

Burn time The length of the spark line while the spark plug is firing; measured in milliseconds.

Duración del encendido Espacio de tiempo que la línea de chispas de la bujía permanece encendida, medido en milisegundos.

Calibrator package (CAL-PAK) A removeable chip in some computers that usually contains a fuel backup program.
Paquete del calibrador Pastilla desmontable en algunas computadoras; normalmente contiene un programa de reserva para el combustible.

Cam (camshaft) A straight shaft with lobes machined along it length to act on a valve lifter to open intake and exhaust valves.
Árbol de levas Un árbol recto que tiene lóbulos maquinados por su lengitud que actúan un levantador de válvulas para abrir las válvulas de entrada y salida.

Canister purge solenoid A computer-operated solenoid connected in the evaporative emission control system.
Solenoide de purga de bote Solenoide accionado por computadora conectado en el sistema de control de emisiones de evaporación.

Canister-type pressurized injector cleaning container A container filled with unleaded gasoline and injector cleaner and is pressurized during the manufacturing process or by the shop air supply.
Recipiente de limpieza del inyector presionizado tipo bote Recipiente lleno de gasolina sin plomo y limpiador de inyectores, presionizado durante el proceso de fabricación o mediante el suministro de aire en el taller mecánico.

Carbon dioxide (CO_2) A gas formed as a byproduct of the combustion process.
Bióxido de carbono (CO_2) Gas que es un producto derivado del proceso de combustión.

Carbon monoxide A gas formed as a byproduct of the combustion process in the engine cylinders. This gas is very dangerous or deadly to the human body in high concentrations.
Monóxido de carbono Gas que es un producto derivado del proceso de combustión en los cilindros del motor. Este gas es muy peligroso y en altas concentraciones podría ocasionar la muerte.

Carbon tracking The formation of carbonized dust between distributor cap terminals.
Rastro de carbon La formación de polvo de carbon entre los terminales de la tapa del distribuidor.

Cetane A rating used to classify diesel fuel, refers to the volatility of the fuel.
Cetano Una relación usada para clasificar el combustible diesel, se refiere a la votálidad del combustible.

Charcoal canister A container of activated charcoal where gasoline vapors are stored by the evaporative emissions system until they can be burnt in the engine.
Bote de carbon Un contenedor de carbón activado en donde se almacenan los vapores de gasolina del sistema de emision hasta que se pueden quemar en el motor.

Chassis ground The ground of the circuit.
Tierra del armazón La tierra de un circuito.

Chemical cleaning Removing dirt or buildup from parts using a variety of solvents and other chemicals.
Limpieza por proceso químico Remover la suciedad o la aumentación de las partes usando una variedad de disolventes u otros productos químicos.

Class 2 Communication protocol that toggles a voltage from zero to seven volts. The signal can be sent in long or short pulse widths. Class II is medium speed at 10,400 bits per second.

Clase 2 Un protocolo de comunicación que escoja un voltaje de cero a siete voltíos. El señal se puede mandar en anchuras de pulso largos o cortos. El clase II es de velocidad mediana en 10,400 bits por segundo.

Closed loop A computer operating mode in which the computer uses the oxygen sensor signal to help control the air-fuel ratio.

Bucle cerrado Modo de funcionamiento de una computadora en el que se utiliza la señal del sensor de oxígeno para ayudar a controlar la relación de aire y combustible.

Clutch slippage Occurs when the clutch disc does not firmly grip connect the flywheel and pressure plate.

Deslizamiento del embrague Ocurre cuando el disco del embrague no une el volante y la placa de presión con firmeza.

CNG Compressed Natural Gas. An alternative fuel source for engines.

CNG Gas Natural Comprimido. Una fuente de combustible alternativo para los motores.

Cold fouling The result of an excessively rich air-fuel mixture characterized by a layer of dry, fluffy black carbon deposits on the tip of the plug.

Ensuciamiento en frío El resultado de una mezcla de aire y combustible demasiado rico que se caracteriza por una capa de depósitos vellosos secos y negros en la punta de la bujía.

Combustion chamber The area in the cylinder head above the piston where fuel burning takes place.

Cámara de combustión El área en la cabeza del cilindro arriba del piston en donde se quema el combustible.

Compact discs (CD-ROM) Electronic data systems that can store large amounts of information.

Disco compacto (Cd-ROM) Los sistemas de datos electrónicos que pueden almacenar grandes cantidades de la información.

Comprehensive tests A complete series of battery, starting, charging, ignition, and fuel system tests performed by an engine analyzer.

Pruebas comprensivas Serie completa de pruebas realizadas en los sistemas de la batería, del arranque, de la carga, del encendido, y del combustible con un analizador de motores.

Compression fitting A common type of mating surfaces in fittings.

Montaje de presión Un tipo común de superficies parejos en los montajes.

Compression gauge A gauge used to test engine compression.

Manómetro de compresión Calibrador utilizado para revisar la compresión de un motor.

Compression ratio The amount the air and fuel mixture is compressed in the cylinder during the compression stroke. It is expressed in terms of volume, such as 8:1.

Relación de compresión La cantidad de la mezcla del aire y combustible se comprimen en el cilindro durante la carrera de compresión. Se expresa en términos de volúmen, tal como el 8:1.

Computed timing test A computer system test mode on Ford products that checks spark advance supplied by the computer.

Prueba de regulación del avance calculado Modo de prueba en una computadora de productos fabricados por la Ford que verifica el avance de la chispa suministrado por la computadora.

Computer output circuits Electrical circuits from the PCM that are used to activate the computers actuators.

Circuitos de salida de la computatdora Los circuitos eléctricos del PCM que se usan para activar los actuadores de la computadora.

Constant volume sampling (CVS) Ensures a constant volume of air is sampled by the emission analyzer.

Muestreo de volumen constante (CVS) Asegura que un volumen constante de aire se muestra por el analizador de emisión.

Continuity tester A self-powered test light.

Probador de continuidad Un foco de probador que tiene su propria energía.

Continuous self-test A computer system test mode on Ford products that provides a method of checking defective wiring connections.

Prueba automática continua Modo de prueba en una computadora de productos fabricados por la Ford que proporciona un método de verificar conexiones defectuosas del alambrado.

Coolant hydrometer A tester designed to measure coolant specific gravity and determine the amount of antifreeze in the coolant.

Hidrómetro de refrigerante Instrumento de prueba diseñado para medir la gravedad específica del refrigerante y determinar la cantidad de anticongelante en el refrigerante.

Cooling system pressure tester A tester used to test cooling system leaks and radiator pressure caps.

Instrumento de prueba de la presión del sistema de enfriamiento Instrumento de prueba utilizado para revisar fugas en el sistema de enfriamiento y en las tapas de presión del radiador.

Corona effect A glow around the spark plug cables indicating the cable should be replaced.

Efecto de corona Una incandescencia alrededor de los cables de las bujías indicando que el cable debe ser cambiado.

Crude oil Oil as it is taken from the ground, before it has been refined.

Aceite en crudo El aceite en su estado de salir de la tierra, antes de que se ha refinado.

Current ramp Graphing current on an oscilloscope; usually utilizing current probes.

Rampa de corriente El corriente gráfico en un osciloscópio, normalmente utilizando las sondas de corriente.

Current limiting A feature that saturates the coil with high current for 1 second.

Limitador de corriente Una característica que satura la bobina en una alta corriente por un segundo.

Custom tests A series of tests programmed by the technician and performed by an engine analyzer.

Pruebas de diseño específico Serie de pruebas programadas por el mecánico y realizadas por un analizador de motores.

Customization Refers to the ability of the customer to customize certain accessories such as door looks, power sliding doors, etc. to act according to a set routine (such as all doors lock when the vehicle is put into gear).

Personalización Refiere a la habilidad del cliente a personalizar ciertos acesorios tal como los cierres de puerta, las puertas corredizas de poder, etc. a funcionar según una rutina predeterminada (como todas las puertas se cierran cuando el vehículo se pone en marcha).

Cycle One complete set of changes in a recurring signal.

Ciclo Un juego de cambio completo en una señal recurrente.

Cylinder leakege The amount of air or volume lost from a sealed cylinder measured in percent.

Fuga del cilíndro La cantidad del aire o del volúmen que se pierde de un cilíndro sellado que se mide en porcentaje.

Cylinder leakage tester A tester designed to measure the amount of air leaking from the combustion chamber past the piston rings or valves.

Instrumento de prueba de la fuga del cilindro Instrumento de prueba diseñado para medir la cantidad de aire que se escapa desde la cámara de combustión y que sobrepasa los anillos de pistón o las válvulas.

Data link connector (DLC) A computer system connector to which the computer supplies data for diagnostic purposes.

Conector de enlace de datos Conector de computadora al que ésta suministra datos para propósitos diagnósticos.

Delta pressure feedback EGR sensor (DPFE) A sensor that uses a pressure difference between two passages to verify operation of the EGR valve.

Sensor de retroalimentación EGR de presión Delta (DPFE) Un sensor que utiliza una diferencia de presión entre dos pasajes para verificar la operación de la válvula EGR.

Detonation Abnormal combustion. Refers to the ignition of the air-fuel mixture inside the combustion chamber prior to the firing of a spark plug.

Detonación Combustión anormal. Se refiere a la ignición de la mezcla de aire/combustible adentro de la cámara de combustión antes de que la chispa de la bujía ocurra.

Diagnostic trouble code (DTC) A code retained in a computer memory representing a fault in a specific area of the computer system.

Códigos indicadores de fallas para propósitos diagnósticos Código almacenado en la memoria de una computadora que representa una falla en un área específica de la computadora.

Diesel engine An internal combustion engine that utilizes diesel fuel.

Motor de Diesel Un motor de combustión interna que utiliza el combustible diesel.

Diesel particulates Small carbon particles in diesel exhaust.

Partículas de diesel Pequeñas partículas de carbón presentes en el escape de un motor diesel.

Digital EGR valve An EGR valve that contains a computer-operated solenoid or solenoids.

Válvula EGR digital Una válvula EGR que contiene un solenoide o solenoides accionados por computadora.

Digital meter A meter with a digital display.

Medidor digital Medidor con lectura digital.

Digital multimeter (DMM) The same as a multimeter, but uses a digital readout and can usually be used to test a computerized vehicle.

Multímetro digital (DMM) Lo mismo que un multímetro pero tiene una lectural digital y se puede usar para efectuar una prueba en un vehículo computerizado.

Digital storage oscilloscope (DSO) A device that converts voltage signals to digital information and stores it in memory.

Osciloscopio de almacen digital (DSO) Un dispositivo que convierta las señales de voltaje en la información digital y la deposita en la memoria.

Direct current (DC) An electrical current that flows in one direction.

Corriente Directa Una corriente eléctrica que fluye solamente en una dirección.

Displacement The total volume of the cylinders in an engine, usually expressed in cubic centimeters or liters.

Desplazamiento El volúmen total de los cilíndros en un motor, expresado normalmente en centímetros cúbicos o litros.

Display pattern Cylinder patterns displayed from left to right on an oscilloscope.

Diagrama de presentación Las diagramas de los cilíndros que se presentan de la izquierda a la derecha en un osciloscopio.

Distributor ignition (DI) system SAE J1930 terminology for any ignition system with a distributor.

Sistema de encendido con distribuidor Término utilizado por la SAE J1930 para referirse a cualquier sistema de encendido que tenga un distribuidor.

Double-flare fitting A tubing fitting made with a special tool that has an anvil and a cone. The process is performed in two steps: first, the anvil begins to fold over the end of the tubing, then the cone is used to finish the flare by folding the tubing back on itself doubling the thickness and creating two sealing surfaces.

Montaje de doble abocinado Un montaje de tubo hecho con una herramienta especial que consiste de un yunque y un cono. El proceso se efectúa en dos pasos: primero el yunque pliega de la extremidad del tubo por encima de si mismo, luego se usa el cono para acabar el abocinado plegando el tubo así doblando el espesor para crear dos superficies de estanqueidad.

Downstream air Air injected into the catalytic converter.

Aire conducido hacia abajo Aire inyectado dentro del convertidor catalítico.

Drive cycle A specified set of driving conditions. The drive cycle is important to adaptive strategies of a computer. The proper drive cycle is also required of OBD II systems to complete monitor tests on certain systems.

Ciclo de ejecución Un juego especifico de condiciones de manejar. Estas condiciones son importantes para la estrategia adaptiva de la computadora. El ciclo de manejo apropiado es también requerido para el sistema OBD II (diagnóstico abordo del vehículo II) para competir con las pruebas del monitor en algunos sistemas.

Drive lugs Protrusions at the end of the shaft that seat in mating grooves in the end of the camshaft.

Lenguetas de mando Parte saliente en la extremidad de un árbol que se asientan en las ranuras apareadas en la extremidad del árbol de levas.

Dual overhead camshafts (DOHC) An engine with two camshafts per cylinder head, one each for intake and exhaust.

Árboles de levas elevados duales (DOHC) Un motor con dos árboles de levas por cada cabeza del cilíndro, un para cada entrada y salida.

Dual trace oscilloscope A device that can display two active, independent patterns at the same time.

Osciloscopio de doble muestra Un dispositivo que puede demonstrar dos muestras activas e independientes a la vez.

Duty cycle On-time to off-time ratio, as measured in a percentage of pulse width or degrees of dwell.

Ciclo de duración La relación de apagado y encendido, según es medido en porcentaje de la amplitud del pulso o grados Dwell (tiempo en que los puntos están cerrados medidos en grados).

Dwell The amount of time the current is flowing through a circuit. Most often, this term is applied to ignition systems.

Tiempo en que los puntos están cerrados medidos en grados La cantidad de tiempo que la corriente está fluyendo atraves de un circuito. Más común, este termino es aplicado a sistemas de ignición.

Dwell meter A meter used to indicate the on time of a coil primary in degrees of crankshaft rotation.

Medidor de parada Un medidor que se usa para indicar en grados que la bobina primaria de la rotación del cigueñal esta en tiempo.

Dyno Dynamometer, a device that simulates road-load conditions in the shop.

Dino Dinomómetro Un dispositivo que simula las condiciones de la pista dentro del taller.

Efficiency The measure of the relationship between the amount of energy put into an engine and the amount of energy available from the engine.

Eficiencia La medida de la relación entre la cantidad de energía que se aplica a un motor y la cantidad de energía que es disponible del motor.

EGR control solenoid (EGRC) *See* EGR vent solenoid (EGRV).

Solenoide de control EGR (EGRC) Véase EGR vent solenoid (EGRV).

EGR pressure transducer (EPT) A vacuum switching device operated by exhaust pressure that opens and closes the vacuum passage to the EGR valve.

Transconductor de presión EGR Dispositivo de conmutación de vacío accionado por la presión del escape que abre y cierra el paso del vacío a la válvula EGR.

EGR vacuum regulator (EVR) solenoid A solenoid that is cycled by the computer to provide a specific vacuum to the EGR valve.

Solenoide regulador de vacío EGR Solenoide ciclado por la computadora para proporcionarle un vacío específico a la válvula EGR.

EGR vent solenoid (EGRV) A solenoid that normally cycles on and off frequently when EGR flow is controlled.

Solenoide de respirado (EGRV) Un solenoide que suele ciclarse prendiendose y apagandose frecuentemente cuando se controla el flujo del EGR.

Electromagnetic interference (EMI) Appears as AC noise on an electrical signal.

Interferencia electromagnética (EMI) Aparece como ruido de CA en una señal eléctrica.

Electromotive force (EMF) The force that exists between a positive and a negative point within an electrical circuit.

Fuerza electromotiva (EMF) La fuerza que existe entre un punto positivo y negative dentro de un circuito eléctrico.

Electronic automatic transmission control (EATC) A transmission that is shifted by electronic actuators via computer control.

Control electrónico de la transmisión automática (EATC) Una transmisión que cambia de marcha por medio de los actuadores electrónicos controlado por computadora.

Electronic distributorless ignition system (EDIS) A type of ignition system produced by Ford Motor Company that uses a high data rate signal for crankshaft inputs.

Sistema de ignición electrónica sin distribuidor (EDIS) Un tipo de sistema de ignición producido por la companía Ford Motor que utiliza una señal de información de alta velocidad para las entradas del cigueñal.

Electronic fuel injection (EFI) A generic term applied to various types of fuel injection systems.

Inyección electrónica de combustible Término general aplicado a varios sistemas de inyección de combustible.

Electronic ignition (EI) system SAE J1930 terminology for any ignition system without a distributor.

Sistema de encendido electrónico Término utilizado por la SAE J1930 para referirse a cualquier sistema de encendido que no tenga distribuidor.

Electronic vacuum regulator valve (EVRV) Used to activate the EGR valve by controling the vacuum available via a solenoid.

Válvula electrónica reguladora del vacío (EVRV) Se usa para activar la válvula EGR por CA controlando el vacío disponible por medio de un solenoide.

Enable criteria The specific operating conditions for diagnostic tests.

Criterios de capacidad Las condiciones específicas de operacion para las pruebas diagnósticas.

Energy conserving oils Engine oils designed to reduce friction and therefore save fuel.

Acietes que conservan energía Los aceites de motor diseñados para disminuir la fricción así preservando el combustible.

Engine analyzer A tester designed to test engine systems such as battery, starter, charging, ignition, and fuel plus engine condition.

Analizador de motores Instrumento de prueba diseñado para revisar sistemas de motores, como por ejemplo los de la batería, del arranque, de la carga, del encendido, y del combustible además de la condición del motor.

Engine coolant temperature (ECT) sensor A sensor that sends a voltage signal to the computer in relation to coolant temperature.

Sensor de la temperatura del refrigerante del motor Sensor que le envía una señal de tensión a la computadora referente a la temperatura del refrigerante.

Engine lift A hydraulically operated piece of equipment used to lift the engine from the chassis.

Elevador de motores Equipo accionado hidráulicamente que se utiliza para levantar el motor del chasis.

Engine off natural vacuum (EONV) An EVAP monitoring system that runs pressure and vacuum tests based on the natural volatility of the fuel as the fuel warms up and cools off after driving.

Vacío normal de motor apagado (EEONV) Un sistema monitor del EVAP que efectúa las pruebas de presión y vacío basado en la volatilidad normal del combustible al calentarse y enfriarse el combustible después de estar en marcha.

Enhanced EVAP An EVAP system used on OBD II in which the PCM conducts several tests to determine if the system is operational and there are no leaks.

EVAP intensificado Un sistema EVAP que se usa en OBD II en el cual el PCM ecfectúa varias pruebas para determinar si funciona el sistema y no hay fugas.

Environmental Protection Agency (EPA) The federal agency that enforces laws to prevent activity that may be harmful to the environment.

Agencia de Protección del Medio Ambiente (EPA) Una agencia federal que impone los leyes para prevenir cualquier actividad que podría ser dañosa al medio ambiente.

Evaporative (EVAP) system A system that collects fuel vapors from the fuel tank and directs them into the intake manifold rather than allowing them to escape to the atmosphere.

Sistema de evaporación Sistema que acumula los vapores del combustible que escapan del tanque del combustible y los conduce hacia el colector de aspiración en vez de permitir que los mismos se escapen hacia la atmósfera.

Exhaust gas analyzer A tester that measures carbon monoxide, carbon dioxide, hydrocarbons, and oxygen in the engine exhaust.

Analizador del gas del escape Instrumento de prueba que mide el monóxido de carbono, el bióxido de carbono, los hidrocarburos, y el oxígeno en el escape del motor.

Exhaust gas recirculation (EGR) valve A valve that circulates a specific amount of exhaust gas into the intake manifold to reduce NOx emissions.

Válvula de recirculación del gas del escape Válvula que hace circular una cantidad específica del gas del escape hacia el colector de aspiración para disminuir emisiones de óxidos de nitrógeno.

Exhaust gas recirculation valve position (EVP) sensor A sensor that sends a voltage signal to the computer in relation to EGR valve position.

Sensor de la posición de la válvula de recirculación del gas del escape Sensor que le envía una señal de tensión a la computadora referente a la posición de la válvula EGR.

Exhaust gas temperature sensor A sensor that sends a voltage signal to the computer in relation to exhaust temperature.

Sensor de la temperatura del gas del escape Sensor que le envía una señal de tensión a la computadora referente a la temperatura del escape.

Expansion tank The volume in a fuel tank that cannot be filled with fuel in order to allow for expansion in warm weather.

Tanque de expansión El volumen en un tanque de combustible en el que no se puede llenar con combustible para permitir su expansión en una clima cálida.

Fast idle thermo valve A valve that uses a thermo wax element to control the amount of bypass air the engine receives thus controlling the cold idle speed. Should be closed at normal operating temperature.

Válvula térmico de marcha en vacío rápida Una válvula que usa un elemento de cera térmica para controlar la cantidad de aire de desviación que recibe el motor así controlando la velocidad de marcha en vacío frío. Debe ser cerrado en las temperaturas normales de operación.

Federal Test Procedure A transient-speed mass sampling emissions test conducted on a loaded dynamometer. This is the test used by manufacturers to certify vehicles before they can be sold.

Procedimiento de la Prueba Federal Una prueba de la velocidad transitoria para la prueba de la masa en un dinamómetro cargado. Esta es la prueba usada por los fabricantes para certificar los vehículos antes de que ellos puedan ser vendidos.

Feedback A term used to describe a PCM's ability to control the activity of an actuator in response to input from sensors.

Retroalimentación Un termino usado para describir la habilidad del PCM (módulo de control de la potencia del motor) de controlar la actividad de un actuador en respuesta a la información de los sensores de entra.

Feeler gauge Metal strips with a specific thickness for measuring clearances between components.

Calibrador de espesores Láminas metálicas de un espesor específico para medir espacios libres entre componentes.

Field service mode A computer diagnostic mode that indicates whether the computer is in open or closed loop on General Motors PCMs.

Modo de servicio de campo Modo diagnóstico de una computadora que indica si la misma se encuentra en bucle abierto o cerrado en módulos del control del tren transmisor de potencia de la General Motors.

Firing order The order in which the individual cylinders in an engine are fired.

Orden de encendido El orden en el cual los cilíndros individuales de un motor se encienden.

Firing voltage The voltage needed to start the spark.

Voltaje de encendido El voltaje requerido para prender la bujía.

Flash code diagnosis Reading computer system diagnostic trouble codes (DTCs) from the flashes of the malfunction indicator light (MIL).

Diagnosis con código de destello La lectura de códigos indicativos de fallas para propósitos diagnósticos de una computadora mediante los destellos de la luz indicadora de funcionamiento defectuoso.

Flat-rate manual A book that helps determine the length of time a particular task or job will take.

Manual de precios uniformes Un libro que asista en determinar la cantidad del tiempo que va tomar un trabajo o una tarea específica.

Floor jack A hydraulically operated device mounted on casters used to raise one end or corner of the chassis.

Gato de pie Dispositivo accionado hidráulicamente montado en rolletes y utilizado para levantar un extremo o una esquina del chasis.

Four-gas emissions analyzer An analyzer designed to test carbon monoxide, carbon dioxide, hydrocarbons, and oxygen in the exhaust.

Analizador de cuatro tipos de emisiones Analizador diseñado para revisar el monóxido de carbono, el bióxido de carbono, los hidrocarburos, y el oxígeno en el escape.

Four-wheel alignment A process where the front and rear wheel angles are adjusted to the vehicle's thrust angle.

Alineación de cuatro ruedas Un proceso en el cual los ángulos de las ruedas delanteras y traseras se ajusten al ángulo de empuje del vehículo.

Freeze frame A feature of some scan tools and a requirement of OBD II. This feature takes a snapshot of the operating conditions present when the PCM sets a diagnostic trouble code.

Marco congelado Una característica de algunas herramientas exploratorias y un equipo de OBD II (diagnóstico abordo del vehículo II). Esta característica toma una foto de las condiciones de operaciones presente cuando el PCM (módulo de control de la potencia del motor) establece un código de diagnostico de problema.

Frequency The number of complete cycles that occur in a specific period of time.

Frecuencia El número de ciclos completos que ocurren en un periodo específico de tiempo.

Fuel control Engine running with the PCM in control of fuel mixture based on inputs from the various sensors instead of programmed values (such as during cold start).

Control de combustible El motor en marcha con el PCM en control de la mezcla del combustible basado en las entradas de varios sensores en vez de los valores programados (tal como durante un encendido frío).

Fuel cut rpm The rpm range in which the computer stops operating the injectors during deceleration.

Detención de combustible según las rpm Margen de revoluciones al que la computadora detiene el funcionamiento de los inyectores durante la desaceleración.

Fuel pressure test port A threaded port on the fuel rail to which a pressure gauge may be connected to test fuel pressure.

Lumbrera de prueba de la presión del combustible Lumbrera fileteada que se encuentra en el carril del combustible a la que puede conectársele un calibrador de presión para revisar la presión del combustible.

Fuel pump volume The amount of fuel the pump delivers in a specific time period.

Volumen de la bomba del combustible Cantidad de combustible que la bomba envía dentro de un espacio de tiempo específico.

Fuel tank level sensor The input that the PCM uses to determine the fuel level in the fuel tank important for EVAP leak detection sensing.

Sensor de nivel del combustible del tanque La entrada que usa el PCM para determinar el nivel del combustible en el tanque de combustible que es importante en la detección por sensor de las fugas del EVAP.

Fuel tank pressure control valve The valve designed to control fuel tank pressure while the vehicle is sitting still. If tank pressure builds too high the valve opens and lets the vapor into the canister.

Válvula de control de presión del tanque de combustible La válvula diseñada a controlar la presión del tanque de combustible mientras que el vehículo esta imóbil. Si la presión del tanque sube demasiado abre la válvula y permite que entra el vapor al bote.

Fuel tank pressure sensor Informs the PCM of the pressure inside the fuel tank for EVAP system tests.

Sensor de presión del tanque de combustible Informa al PCM de la presión dentro del tanque de combustible en las pruebas EVAP.

Fuel tank purging Removing fuel vapors and foreign material from the fuel tank.

Purga del tanque del combustible La remoción de vapores de combustible y de material extraño del tanque del combustible.

Gap bridging Occurs when carbon deposits accumulate in the combustion chamber due to stop-and-go driving.

Puesta en cortocircuito de los electrodos Ocurre cuuando los depósitos de carbon se acumulan en la cámara de combustión debido a la marcha con mucha ida-y-parada.

Glazing A glaze over the insulator from combustion chamber deposits.

Satinado Un lustro en un aislador que viene de los depósitos de la cámara de combustión.

Glitch A momentary disruption in an electrical circuit.

Falla Una interrupción momentánea en un circuito eléctrico.

Glow plug An electrically heated point inside the diesel combustion chamber to aid cold starting.

Bujía de incandescencia Un punto calentado electrónicamente dentro de la cámara de combustión diesel para asistir en el arranque frío.

Graphite oil An oil with a graphite base that may be used for special lubricating requirements such as door locks.

Aceite de grafito Aceite con una base de grafito que puede utilizarse para necesidades de lubricación especiales, como por ejemplo en cerraduras de puertas.

Hall effect pickup A pickup containing a Hall element and a permanent magnet with a rotating blade between these components.

Captación de efecto Hall Captación que contiene un elemento Hall y un imán permanente entre los cuales está colocada una aleta giratoria.

Hand-held digital pyrometer A tester for measuring component temperature.

Pirómetro digital de mano Instrumento de prueba para medir la temperatura de un componente.

Hand press A hand-operated device for pressing precision-fit components.

Prensa de mano Dispositivo accionado manualmente para prensar componentes con un fuerte ajuste de precisión.

Hazard communication standard The part of the right-to know-laws that requires employers to train employees about hazardous materials they may encounter on the job.

Norma de comunicación de peligros La parte de los leyes derecho-de-saber que requiere que los patrones entrenan a los empleados con respeto a los materiales peligroso que puedan encontrar en el trabajo.

Heated resistor-type MAF sensor A MAF sensor that uses a heated resistor to sense air intake volume and temperature and sends a voltage signal to the computer in relation to the total volume of intake air.

Sensor MAF tipo resistor térmico Sensor MAF que utiliza un resistor térmico para advertir el volumen y la temperatura del aire aspirado, y que le envía una señal de tensión a la computadora referente al volumen total de aire aspirado.

Hertz A unit of measurement for counting the number of times an electrical cycle repeats every second. One hertz is one pulse each second.

Hertz Una unidad de medida para contar los números de veces que un ciclo eléctrico se repite cada segundo. Un hertz es un pulso cada segundo.

Horsepower The force required to move 550 pounds one foot in one second.

Caballo motor La fuerza requirida para mudar 550 libras el espacio de un pie en un segundo.

Hot wire-type MAF sensor A MAF sensor that uses a heated wire to sense air intake volume and temperature and sends a voltage signal to the computer in relation to the total volume of intake air.

Sensor MAF tipo térmico Sensor MAF que utiliza un alambre térmico para advertir el volumen y la temperatura del aire aspirado, y que le envía una señal de tensión a la computadora referente al volumen total de aire aspirado.

Hybrid Electric Vehicles (HEVs) Vehicles that use both an electric motor and an internal combustion engine.

Vehículo eléctrico híbrido (HEVs) Los vehículos que utilisan un motor eléctrico y tambien un motor de combustión interna.

Hydraulic press A hydraulically operated device for pressing precision-fit components.

Prensa hidráulica Dispositivo accionado hidráulicamente que se utiliza para prensar componentes con un fuerte ajuste de precisión.

Hydraulic valve lifters Round, cylindrical, metal components used to open the valves. These components are operated by oil pressure to maintain zero clearance between the valve stem and rocker arm.

Desmontaválvulas hidráulicas Componentes metálicos, cilíndricos y redondos utilizados para abrir las válvulas. Estos componentes se accionan mediante la presión del aceite para mantener cero espacio libre entre el vástago de válvula y el balancín.

Hydrocarbons (HC) Leftover fuel from the combustion process.

Hidrocarburos El combustible restante después del proceso de combustión.

Hydrometer A tester designed to measure the specific gravity of a liquid.

Hidrómetro Instrumento de prueba diseñado para medir la gravedad específica de un líquido.

Hyperlink An electronic link that provides a short cut to another document or media directly by clicking on a word in the text that is usually set apart in a different color and underlined, Such as engine performance.

Hipervínculo Una conexión electronica que provee una ruta directa a otro documento o pagina directamente al hacer clic en una palabra del texto que suele ser distincto en otro color y subrayado, tal como engine performance.

Idle air control bypass air (IAC BPA) motor An IAC motor that controls idle speed by regulating the amount of air bypassing the throttle.

Motor para el control de la marcha lenta con el paso de aire Un motor IAC que controla la velocidad de la marcha lenta regulando la cantidad de aire que se desvía de la mariposa.

Idle air control bypass air (IAC BPA) valve A valve operated by the IAC BPA motor that regulates the air bypassing the throttle to control idle speed.

Válvula para el control de la marcha lenta con el paso de aire Válvula accionada por el motor IAC BPA que regula el aire que se desvía de la mariposa para controlar la velocidad de la marcha lenta.

Idle air control (IAC) motor A computer-controlled motor that controls idle speed under all conditions.

Motor para el control de la marcha lenta con aire Motor controlado por computadora que controla la velocidad de la marcha lenta bajo cualquier condición del funcionamiento del motor.

Ignition crossfiring Ignition firing between distributor cap terminals or spark plug wires.

Encendido por inducción Encendido entre los bornes de la tapa del distribuidor o los alambres de las bujías.

Ignition module tester An electronic tester designed to test ignition modules.

Instrumento de prueba del módulo del encendido Instrumento de prueba electrónico diseñado para revisar módulos del encendido.

Induction The spontaneous creation of an electrical current in a conductor as the conductor passes through a magnetic field or a magnetic field passes across the conductor.

Inducción La creación espontanea de una corriente eléctrica en un conductor según el conductor atraviesa un campo magnético o un campo magnético atraviesa el conductor.

Initial vacuum pull down A step in the EVAP leak detection cycle where a vacuum is pulled on the EVAP system and then checked for the rate of decay. (*See also* weak vacuum test.)

Rebajada inicial del vacío Un paso en el ciclo de detección de fugas del EVAP en el cual se cree un vacío en el sistema EVAP y luego se revisa para ver la cantidad de degradación. (véase tambien weak vacuum test.)

Injector balance tester A tester designed to test port injectors.

Instrumento de prueba del equilibro del inyector Instrumento de prueba diseñado para revisar inyectores de lumbreras.

Inspection, maintenance (I/M) testing Emission inspection and maintenance programs that are usually administered by various states.

Pruebas de inspección y mantenimiento Programas de inspección y mantenimiento de emisiones que normalmente administran diferentes estados.

Integrator A chip responsible for fuel control in a General Motors PCM.

Integrador Pastilla responsable de controlar el combustible en un módulo del control del tren transmisor de potencia de la General Motors.

Interface module (VIM) An adapter for the Tech-1 scan tool to display fuel trim information.

Módulo de conexión (VIM) Un adaptador para la herramienta explorador Tech-1 que presenta la información del la consumación del combustible.

Intermediate section The section of the scope trace pattern that shows the remaining coil voltage as it dissipates or drops to zero.

Sección intermedio La sección de la diagrama de una muestra de un osciloscopio que enseña lo que queda del voltaje de la bobina en el momento que dispersa o baja hasta el cero.

International Lubricant Standardization and Approval Committee (ILSAC) An organization that rates the quality of a lubricating oil and applies the API Service Symbol to those oils which pass requirements of Japanese and American automobile manufacturers.

Comisión internacional de regularización y aprobación de lubricantes (ILSAC) Un organización que evalúa la calidad de un aceite lubricante y aplica el signo de Servicio API a aquellos aceites que supera los requerimientos de los fabricantes de automóviles japoneses y americanos.

ISO 9141 Specifies the requirements for one of the digital languages used between the PCM and scan tools per the International Standards Organization.

ISO 9141 Especifica los requerimientos de una de las idiomas digitales que se usan entre el PCM y las herramientas exploradoras según la Organización de Regularización Internacional.

International System (SI) A system of weights and measures in which each unit may be divided by 10.

Sistema internacional Sistema de pesos y medidas en el que cada unidad puede dividirse entre 10.

J1850 Specifies the requirements for a communication standard used as a digital language between the PCM, individual modules, and scan tools per SAE. Also known as Class B and SCP.

J1850 Especifica los requerimientos para una comunicación regularizado que se usa como idioma digita entre el PCM, los módulos individuales y las herramientas exploradores según el SAE. Tambien se llaman Clase B y SCP.

Jack stand A metal stand used to support one corner of the chassis.

Soporte de gato Soporte de metal utilizado para apoyar una esquina del chasis.

Key on, engine off (KOEO) test A computer system test mode on Ford products that displays diagnostic trouble codes (DTCs) with the key on and the engine stopped.

Prueba con la llave en la posición de encendido y el motor apagado Modo de prueba en una computadora de productos fabricados por la Ford que muestra códigos indicadores de fallas para propósitos diagnósticos cuando la llave está en la posición de encendido y el motor está apagado.

Key on, engine running (KOER) test A computer system test mode on Ford products that displays diagnostic trouble codes (DTCs) with the engine running.

Prueba con la llave en la posición de encendido y el motor encendido Modo de prueba en una computadora de productos fabricados por la Ford que muestra códigos indicadores de fallas para propósitos diagnósticos cuando el motor está encendido.

Kilopascal (kPa) The metric version of psi.

Kilopascal (kPa) La version métrica de libras por pulgada cuadrada.

Kilovolts (kV) Thousands of volts.

Kilovoltios (kV) Miles de voltios.

Knock sensor (KS) A sensor that sends a voltage signal to the computer in relation to engine detonation.

Sensor de golpeteo Sensor que le envía una señal de tensión a la computadora referente a la detonación del motor.

Knock sensor module An electronic module that changes the analog knock sensor signal to a digital signal and sends it to the PCM.

Módulo del sensor de golpeteo Módulo electrónico que convierte la señal analógica del sensor de golpeteo en una señal digital y se la envía al módulo del control del tren transmisor de potencia.

Leak detection pump (LDP) A pump used on some vehicles to pressurize the EVAP system which is used in conjunction with a pressure sensor to detect leaks.

Bomba detector de fugas (LDP) Una bomba usado en algunos vehículos para presurizar el sistema EVAP que se usa de acuerdo con un sensor de presión para detectar las fugas.

Lift A device used to raise a vehicle.

Elevador Dispositivo utilizado para levantar un vehículo.

Linear EGR valve An EGR valve containing an electric solenoid that is pulsed on and off by the computer to provide a precise EGR flow.

Válvula EGR lineal Válvula EGR con un solenoide eléctrico que la computadora enciende y apaga para proporcionar un flujo exacto de EGR.

Lobes Elliptical projections on the camshaft that contact the valve lifter in order to open and close the valve lifters.

Lóbulos Las proyecciones elípticas en el árbol de levas que hacen contacto con la levanta-válvulas para abrir y cerrar las levanta-válvulas.

Logic probe A test light with three different colored light-emitting diodes (LEDs) indicating voltage levels.

Explorador lógico Un foco de prueba que tiene diodos emisores de luz (LED) con tres luces de colores distinctos indicando los niveles de voltaje.

Long-Term Adaptive Fuel Trim Long-term fuel injector pulse width compensation determined by the PCM according to operating conditions. This fuel trim is set to maintain minimum emissions output and is the base point for short-term fuel trim.

Restricción del combustible por tiempo largo Compensación amplia de los pulsos del inyector de combustible determinado por el PCM (módulo de control de la potencia del motor) de acuerdo con las condiciones de operaciones. Esta restricción de combustible es calibrada para mantener las emisiones a un mínimo y es el punto de base para la restricción del combustible por tiempo corto.

Long-term fuel trim (LTF) The long-term trends in fuel control. Positive numbers mean that fuel is being added, negative numbers mean fuel is being subtracted from the long term fuel calculation.

Eficiencia de combusitible de largo plazo (LTF) La tendencia de control de combustible en largo plazo. Los números positivos quieren decir que se agrega el combustible, los números negativos reflejan que se substrae el combustible de la calculación de largo plazo.

Low-amps probe A probe to measure small amounts of current with an amp clamp.

Explorador de bajo ampério Un explorador para medir las pequeñas cantidades de corriente con una grapa de ampérios.

Magnetic-base thermometer A thermometer that may be retained to metal components with a magnetic base.

Termómetro con base magnética Termómetro que puede sujetarse a componentes metálicos por medio de una base magnética.

Magnetic probe-type digital tachometer A digital tachometer that reads engine rpm and uses a magnetic probe pickup.

Tacómetro digital tipo sonda magnética Tacómetro digital que lee las rpm del motor y utiliza una captación de sonda magnética.

Magnetic probe-type digital timing meter A digital reading that displays crankshaft degrees and uses a magnetic-type pickup probe mounted in the magnetic timing probe receptacle.

Medidor de regulación digital tipo sonda magnética Lectura digital que muestra los grados del cigüeñal y utiliza una sonda de captación tipo magnético montada en el receptáculo de la sonda de regulación magnética.

Magnetic sensor A sensor that produces a voltage signal from a rotating element near a winding and a permanent magnet. This voltage signal is often used for ignition triggering.

Sensor magnético Sensor que produce una señal de tensión desde un elemento giratorio cerca de un devanado y de un imán permanente. Esta señal de tensión se utiliza con frecuencia para arrancar el motor.

Magnetic timing offset An adjustment to compensate for the position of the magnetic receptacle opening in relation to the TDC mark on the crankshaft pulley.

Desviación de regulación magnética Ajuste para compensar la posición de la abertura del receptáculo magnético de acuerdo a la marca TDC en la roldana del cigüeñal.

Magnetic timing probe receptacle An opening in which the magnetic timing probe is installed to check basic timing.

Receptáculo de la sonda de regulación magnética Abertura en la que está montada la sonda de regulación magnética para verificar la regulación básica.

Malfunction indicator light (MIL) A light in the instrument panel that is illuminated by the PCM if certain defects occur in the computer system.

Luz indicadora de funcionamiento defectuoso Luz en el panel de instrumentos que el módulo del control del tren transmisor de potencia ilumina si ocurren ciertas fallas en la computadora.

Manifold absolute pressure (MAP) sensor An input sensor that sends a signal to the computer in relation to intake manifold vacuum.

Sensor de la presión absoluta del colector Sensor de entrada que le envía una señal a la computadora referente al vacío del colector de aspiración.

Mass The measurement of an objects inertia.

Masa La medida de la inercia de un objeto.

Mass air flow (MAF) sensor A sensor that monitors incoming air flow, including speed, temperature, and pressure.

Sensor de flujo de aire en masa (MAF) Un sensor que amonesta el flujo del aire de entrada, incluyendo la velocidad, la temperatura y la presión.

Material safety data sheet (MSDS) An information sheet that specifies the dangers of human contact with substances used in any business that employees may be exposed to on the job. They must be available to all employees.

Hojas de datos de seguridad de materias (MSDA) Un página de información que especifica los peligros del contacto entre los seres humanos con las sustancias que se usan en cualquier empleo que se pueden exponer los empleados en el trabajo. Deben ser disponibles a cada empleado.

Mechanical valve lifters, or solid tappets Round, cylindrical, metal components mounted between the camshaft lobes and the pushrods to open the valves.

Desmontaválvulas mecánicas, o alzaválvulas sólidas Componentes metálicos, cilíndricos y redondos montados entre los lóbulos del árbol de levas y las varillas de empuje para abrir las válvulas.

Memory calibrator (MEM-CAL) A removable chip in some computers that replaces the PROM and CAL-PAK chips.

Calibrador de memoria Pastilla desmontable en algunas computadoras que reemplaza las pastillas PROM y CAL-PAK.

Meter impedance The total internal electrical resistance in a meter.

Impedancia de un medidor La resistencia eléctrica interna total en un medidor.

Metering The process of controlling the flow of something. Metering of gasoline is controlled by the time it is allowed to flow, by the size of the opening it is flowing through, or by the pressure causing it to flow.

Regulación El proceso de controlar el flujo de algo. La regulación de la gasolina es controlada por el tiempo que es permitida que fluya, por el tamaño de la apertura que está fluyendo, o por la presión que la está causando que fluya.

Mild hybrid A hybrid vehicle that has an internal combustion engine and electric motor. The internal combustion engine runs to charge the batteries for the electric motor. The electric motor only runs when the small internal combustion motor needs assistance during acceleration or hill pulling.

Híbrido moderado Un vehículo híbrido que tiene un motor de combustión interno y un motor eléctrico. El motor de combustión interno esta en marcha para cargar las baterías para el motor eléctrico. El motor eléctrico sólo funciona cuando el pequeño motor interno de combustión requiere asistencia durante la aceleración o en jalar una carga en una colina.

Muffler chisel A chisel that is designed for cutting muffler inlet and outlet pipes.

Cincel para silenciadores Cincel diseñado para cortar los tubos de entrada y salida del silenciador.

Multimeter A hand-held tool capable of measuring volts, amps, and ohms.

Multímetro Una herramienta de mano que es capaz de medir los voltíos, los amperios y los ohmios.

Multiport fuel injection (MFI) A fuel injection system in which the injectors are grounded in the computer in pairs or groups of three or four.

Inyección de combustible de paso múltiple Sistema de inyección de combustible en el que los inyectores se ponen a tierra en la computadora en pares o en grupos de tres o cuatro.

National Institute for Automotive Service Excellence (ASE) An organization responsible for certification of automotive technicians in the US.

Instituto Nacional para la Excelencia en la Reparación de Automóviles Organización que tiene a su cargo la certificación de mecánicos de automóviles en los Estados Unidos.

Neutral/drive switch (NDS) A switch that sends a signal to the computer in relation to gear selector position.

Conmutador de mando neutral Conmutador que le envía una señal a la computadora referente a la posición del selector de velocidades.

New Generation Star Tester (NGS) A proprietary scan tool used by Ford dealer technicians.

Probador Estrella de Nueva Generación (NGS) Una herramienta exploradora propietario usado por los técnicos de sucursales Ford.

Noid light Used to determine if a fuel injector is receiving the proper voltage pulse.

Luz noid Usado para determinar si un inyector de combustible recibe el impulso apropiado de voltaje.

Non-sinusoidal A voltage trace that is not uniform in size and shape. Usually a D/C voltage in automotive terms.

No sinusoidal Un razgo de voltaje que no es uniforme en tamaño ni en forma. Suele referirse en términos automativos de voltaje D/C.

Nose switch A switch in the IAC motor stem that informs the computer when the throttle is in the idle position.

Conmutador de contacto de la marcha lenta Conmutador en el vástago del motor IAC que le advierte a la computadora cuándo la mariposa está en la posición de marcha lenta.

OBD II The second generation of on-board diagnostics.

OBD II La segunda generación de los diagnósticos a bordo.

Occupational Safety and Health Act (OSHA) A set of rules and guidelines implemented to protect employees from a dangerous work environment.

Acta de seguridad y salud en el lugar de trabajo (OSHA) Un grupo de reglas y guías implementados para proteger los empleados de los peligros en un medio ambiente del trabajo.

Octane A rating used to classify gasoline, refers to the volatility of the fuel.

Octano Una clasificación usada para clasificar la gasolina, se refiere a que tan volátil es el combustible.

Offset A misfire that results in one and one half times the emission output standard.

Desviado Un fallo del encendido que resulta en un escape que excede lo indicado por la ley de emisiones por uno y medio.

Ohmmeter A tool that measures resistance in an electrical circuit.

Ohmiómetro Una herramienta que mide la resistencia en un circuito eléctrico.

Oil pressure gauge A gauge used to test engine oil pressure.

Manómetro de la presión del aceite Calibrador utilizado para revisar la presión del aceite del motor.

Oil pump pickup The screen in the oil pan that supplies engine oil to the oil pump.

Recobro de l a bomba de aceite La rejilla en el cárter de aceite que suministrs el aceite a la bomba de aceite.

Onboard refueling vapor recovery (ORVR) An EVAP system that is designed to eliminate HC emissions during refueling by storing them in a large canister and burning them during engine operation.

Recobro de vapor de reaprovisionamiento de combustible (ORVR) Un sistema de EVAP que es diseñado a eliminar las emisiones de HC durante el reaprovisionamiento al almacenarlos en un gran bote y quemarlos durante la operación del motor.

Open loop A computer operating mode in which the computer controls the air-fuel ratio and ignores the oxygen sensor signal.

Bucle abierto Modo de funcionamiento de una computadora en el que se controla la relación de aire y combustible y se pasa por alto la señal del sensor de oxígeno.

Optical-type pickup A pickup that contains a photo diode and a light emitting diode with a slotted plate between these components.

Captación tipo óptico Captación que contiene un fotodiodo y un diodo emisor de luz entre los cuales está colocada una placa ranurada.

Oscilloscope A cathode ray tube (CRT) that displays voltage waveforms from the ignition system.

Osciloscopio Tubo de rayos catódicos que muestra formas de onda de tensión provenientes del sistema de encendido.

Output driver A transistor in the output area of a control device that is used to turn various output devices off and on.

Ejecutador de salida Un transistor en el área de salida de un dispositivo de control, que es usado para apagar y prender varios dispositivos de ejecución.

Output state test A computer system test mode on Ford products that turns the relays and actuators on and off.

Prueba del estado de producción Modo de prueba en una computadora de productos fabricados por la Ford que enciende y apaga los relés y los accionadores.

Overhead camshaft (OHC) An engine designed with the camshaft acting directly on the top of the valve spring.

Árbol de levas en cabeza (OHC) Un motor diseñado con el árbol de levas accionando directamente arriba del resorte de la válvula.

Overhead valve (OHV) An engine with the camshaft mounted in the block. Pushrods are utilized to open and close the valves.

Válvula en cabeza (OHV) Un motor que tiene el árbol de levas montado en el bloque. Las varillas empujadoras se usan para abrir y cerrar las válvulas.

Oxygen (O$_2$) A gaseous element that is present in air.

Oxígeno (O$_2$) Elemento gaseoso presente en el aire.

Oxygen (O$_2$) sensor A sensor mounted in the exhaust system that sends a voltage signal to the computer in relation to the amount of oxygen in the exhaust stream.

Sensor de oxígeno (O$_2$) Sensor montado en el sistema de escape que le envía una señal de tensión a la computadora referente a la cantidad de oxígeno en el caudal del escape.

Parade pattern The pattern that starts with the spark line of the number one cylinder and ends with the firing line of the number one cylinder.

Diagrama de desfile La diagrama que comienza con la línea de encendido cilíndro número uno y termina con la línea de encendido del cilíndro número uno.

Parallel hybrid A hybrid vehicle that uses both the electric motor and the internal combustion engine to power the drive wheels. The electric motor in the parallel system is small and uses principally the internal combustion engine except during times of low load, such as in town driving or freeway cruising.

Híbrido paralelo Un vehículo híbrido que usa un motor eléctrico y tambien un motor de combustión interno para propulsar las ruedas motrices. El motor eléctrico en el sistema paralelo es pequeño y usa principalmente el motor de combustión internos salvo cuando hay carga baja, tal como la marcha en la ciudad o en la carretera.

Parameter identification mode (PID) A mode that is accessed through an OBD II compliant scanner that provides manufacturer specific or common vehicle information.

Modo de identificación de parámetro (PID) Un modo al cual hay acceso por medio de un explorador sumiso OBD II que provee la información específica de fabricador o común del vehículo.

Parasitic drain Battery drain on a computer-equipped vehicle.

Descarga parásita La descarga de la batería en un vehículo equipado con una computadora.

Park/neutral switch A switch connected in the starter solenoid circuit that prevents starter operation except in park or neutral.

Conmutador PARK/neutral Conmutador conectado en el circuito del solenoide del arranque que evita el funcionamiento del arranque si el selector de velocidades no se encuentra en las posiciones PARK o NEUTRAL.

Particulates Very small pieces of matter, usually carbon, that can be harmful.

Partículos Pedacitos muy pequeños de la material, normalmente de carbon, que pueden ser peligrosos.

Parts per million (ppm) The volume of a gas such as hydrocarbons in ppm in relation to one million parts of the total volume of exhaust gas.

Partes por millón (ppm) Volumen de un gas, como por ejemplo los hidrocarburos, en partes por millón de acuerdo a un millón de partes del volumen total del gas del escape.

Passive anti-theft system (PATS) An anti theft system that requires no action from the driver to activate.

Sistema pasivo anti-robo (PATS) Un sistema de anti-robo que no requiere acción del conductar para activarse.

Peak and hold injector circuits Circuits that incorporate unique circuitry that applies higher amperage to initially open the injector, then lower amperage for the remainder of the pulse width.

Circuitos de inyector picadura y retención Los circuitos que incorporan la circuitería que aplica un amperaje más alto para abrir el inyector desde un principio, y luego el amperaje más bajo por lo que queda de la duración del impulso.

Pending situation A circumstance that prevents a test from being performed due to an uncorrected fault.

Situación pendiente Una circunstancia que previene que se efectúa una prueba debido a un fallo no corregido.

Photoelectric tachometer A tachometer that contains an internal light source and a photoelectric cell. This meter senses rpm from reflective tape attached to a rotating component.

Tacómetro fotoeléctrico Tacómetro que contiene una fuente interna de luz y una célula fotoeléctrica. Este medidor advierte las rpm mediante una cinta reflectora adherida a un componente giratorio.

Pinging noise A shop term for engine detonation that sounds like a rattling noise in the engine cylinders.

Sonido agudo Término utilizado en el taller mecánico para referirse a la detonación del motor cuyo ruido se asemeja a un estrépito en los cilindros de un motor.

Pipe expander A tool designed to expand exhaust system pipes.

Expansor de tubo Herramienta diseñada para expandir los tubos del sistema de escape.

Pneumatic tools Tools such as impact wrenches that operate with controlled compressed air.

Herramientas neumáticas Las herramientas tal como las llaves neumáticas que operan con el aire comprimido controlado.

Polyurethane air cleaner cover A circular polyurethane ring mounted over the air cleaner element to improve cleaning capabilities.

Cubierta de poliuretano del filtro de aire Anillo circular de poliuretano montado sobre el elemento del filtro de aire para facilitar la limpieza.

Port EGR valve An EGR valve operated by ported vacuum from above the throttle.

Válvula EGR lumbrera Válvula EGR accionada por un vacío con lumbreras desde la parte superior de la mariposa.

Port fuel injection (PFI) A fuel injection system with an injector positioned in each intake port.

Inyección de combustible de lumbrera Sistema de inyección de combustible que tiene un inyector colocado en cada una de las lumbreras de aspiración.

Positive crankcase ventilation (PCV) valve A valve that delivers crankcase vapors into the intake manifold rather than allowing them to escape to the atmosphere.

Válvula de ventilación positiva del cárter Válvula que conduce los vapores del cárter hacia el colector de aspiración en vez de permitir que los mismos se escapen hacia la atmósfera.

Pounds per square inch (psi) A measurement of pressure.

Libras por pulgada cuadrada (psi) Una medida de la presión.

Power balance tester A tester designed to stop each cylinder from firing for a brief time and record the rpm decrease.

Instrumento de prueba del equilibro de la potencia Instrumento de prueba diseñado para detener el encendido de cada cilindro por un breve espacio de tiempo y registrar el descenso de las rpm.

Power tools Tools that are operated by outside power sources.

Herramienta de motor Las herramientas que se operan por medio de un suministro de potencia exterior.

Power train control module (PCM) SAE J1930 terminology for an engine control computer.

Módulo del control del tren transmisor de potencia Término utilizado por la SAE J1930 para referirse a una computadora para el control del motor.

Prelubrication Lubrication of components such as turbocharger bearings prior to starting the engine.

Prelubrificación Lubrificación de componentes, como por ejemplo los cojinetes del turbocompresor, antes del arranque del motor.

Pressure gauge A tool used to measure a gas, a liquid, or air in pounds per square inch.

Manómetro de presión Una herramienta que se usa para medir un gas, un líquido, o el aire en libras por pulgada cuadrada.

Pressurized injector cleaning container A small, pressurized container filled with unleaded gasoline and injector cleaner for cleaning injectors with the engine running.

Recipiente presionizado para la limpieza del inyector Pequeño recipiente presionizado lleno de gasolina sin plomo y limpiador de inyectores para limpiar los inyectores cuando el motor está encendido.

Programmable read only memory (PROM) A computer chip containing some of the computer program. This chip is removable in some computers.

Memoria de solo lectura programable (PROM) Pastilla de memoria que contiene una parte del programa de la computadora. Esta pastilla es desmontable en algunas computadoras.

Protocol The language or method used by a PCM to communicate to other computers.

Protocolo El lenguaje o método usado por una PCM (módulo de control de la potencia del motor) para comunicarse con las otras computadoras.

Pulse A voltage signal that increases from a constant value and then decreases back to its original value.

Pulso Una señal de voltaje que incrementa desde un valor constante y después disminuye de regreso a su valor original.

Pulsed secondary air injection system A system that uses negative pressure pulses in the exhaust to move air into the exhaust system.

Sistema de inyección secundaria de aire por impulsos Sistema que utiliza impulsos de la presión negativa en el escape para conducir el aire hacia el sistema de escape.

Pulse modulated A circuit that maintains average voltage levels by pulsing the voltage on and off.

Pulso modulado Un circuito que mantiene niveles promedios de voltaje pulsando en una frecuencia de voltaje de apagado a encendido.

Pulse rate The number of pulses that take place over a specific period of time.

Velocidad del puso Los números de pulso que toman lugar sobre un periodo de tiempo especifico.

Pulse width The duration from the beginning to the end of a signal's on-time or off-time.

Amplitud del pulso La duración desde el principio al final de una señal prendida y apagada.

Purge valve A valve that is opened electrically by the PCM to allow the EVAP canister to be purged of stored vapors and burned in the engine when conditions dictate.

Válvula de purga Una válvula que es abierto electrónicamente por el PCM para permitir la purga del bote de EVAP de los vapores que se han almacenado y quemado en el motor cuando permiten las condiciones.

Pyrometer An electronic device that measures heat.

Pirómetro Un dispositivo electrónico que mide el calor.

Quad driver A group of transistors in a computer that controls specific outputs.

Excitador cuádruple Grupo de transistores en una computadora que controla salidas específicas.

Quick-disconnect fuel line fittings Fuel line fittings that may be disconnected without using a wrench.

Conexiones de la línea del combustible de desmontaje rápido Conexiones de la línea del combustible que se pueden desmontar sin la utilización de una llave de tuerca.

Radiator shroud A circular component positioned around the cooling fan to concentrate the air flow through the radiator.

Bóveda del radiador Componente circular que rodea el ventilador de enfriamiento para concentrar el flujo de aire a través del radiador.

Radio frequency interference (RFI) Electrical noise that may come from the ignition system.

Interferencia de frecuencia radio (RFI) El ruido eléctrico que puede provenir del sistema del encendido.

Raster pattern A pattern that stacks voltage patterns one above the other.

Diagrama raster Una diagrama que sobrepone las diagramas de voltaje uno arriba del otro.

Recovery tank A reservoir that stores coolant expelled from the radiator by expansion as the coolant warms up. When the radiator cools the coolant is drawn back into the radiator.

Tanque de recobro Un suministro que almacena el líquido refrigerante expulsado del radiador por la expansión que ocurre al calentarse el refrigerante. Cuando se enfría el radiador el líquido refrigerante regresa al radiador.

Reference pickup A pickup assembly that is often used for ignition triggering.

Captación de referencia Conjunto de captación que se utiliza con frecuencia para el arranque del encendido.

Reference voltage A constant voltage supplied from the computer to some of the input sensors.

Tensión de referencia Tensión constante que le suministra la computadora a algunos de los sensores de entrada.

Repair grade (RG-240) An economical version of the standard I/M 240 test.

Grado de reparación (RG-240) Una version económica de la prueba de convención I/M240.

Resource Conservation and Recovery Act (RCRA) A law that basically states hazardous material users are responsible for hazardous materials from the time they are produced until they are properly disposed.

Acta de Conservación y Recobro de los Recursos (RCRA) Un ley que afirma basicamente que los que usan las materias tóxicas son responsables de las materias tóxicas del momento que éstos se producen hasta que se disponen de una manera responsable.

Revolutions per minute (rpm) drop The amount of rpm decrease when a cylinder stops firing for a brief time.

Descenso de las revoluciones por minuto (rpm) Cantidad que descienden las rpm cuando un cilindro detiene el encendido por un breve espacio de tiempo.

Right-to-know-law A law that states employees should be informed about the materials that they are using on the job.

Ley de Derecho en saber Un ley que afirma que los empleados deben ser informado de las propriedades de la materiales que usan en el trabajos.

Ring ridge A ridge near the top of the cylinder created by wear in the ring travel area of the cylinder.

Reborde del anillo Reborde cerca de la parte superior del cilindro ocasionado por un desgaste en el área de carrera del anillo del cilindro.

Room temperature vulcanizing (RTV) sealant A type of sealant that may be used to replace gaskets, or to help to seal gaskets, in some applications.

Compuesto obturador vulcanizador a temperatura ambiente Tipo de compuesto obturador que puede utilizarse para reemplazar guarniciones, o para ayudar a sellarlas, en algunas aplicaciones.

Root mean square (RMS) Meters that convert the AC signal to DC voltage signal.

Raíz de la media del los cuadrados (RMS) Los metros que conviertan los señales AC (corriente alterna a un señal de voltaje DC (corriente continua).

Rotary engine An engine that uses a rotary motion instead of a reciprocating motion. The rotary engine uses a rotor that forms combustion chambers as it turns.

Motor rotatorio Un motor que usa un movimiento rotario en vez de un movimiento reciprocativo. El motor rotario usa un rotor que forma las cámaras de combustión al girar.

Safety glasses Safety Glasses are eyewear that protects eyes from foreign material

Lentes de seguridad Los lentes de seguridad son anteojos que protejan los ojos de materia foránea.

Scan tester A tester designed to test automotive computer systems.

Instrumento de pruebas de exploración Instrumento de prueba diseñado para revisar computadoras de automóviles.

Schrader valve A threaded valve on the fuel rail to which a pressure gauge may be connected to test fuel pressure.

Válvula Schrader Válvula fileteada que se encuentra en el carril del combustible a la que puede conectársele un calibrador de presión para revisar la presión del combustible.

Secondary air injection (AIR) system A system that injects air into the exhaust system from a belt driven pump.

Sistema de inyección secundaria de aire Sistema que inyecta aire dentro del sistema de escape desde una bomba accionada por correa.

Self-powered test light A test light powered by an internal battery.

Luz de prueba propulsada automáticamente Luz de prueba propulsada por una batería interna.

Self-test input wire A diagnostic wire located near the diagnostic link connector (DLC) on Ford vehicles.

Alambre de entrada de prueba automática Alambre diagnóstico ubicado cerca del conector de enlace diagnóstico en vehículos fabricados por la Ford.

Sequential fuel injection (SFI) A fuel injection system in which the injectors are individually grounded into the computer.

Inyección de combustible en ordenamiento Sistema de inyección de combustible en el que los inyectores se ponen individualmente a tierra en la computadora.

Serial data The PCM information sent to the scanner and other control modules.

Información en serie La información PCM enviado al explorador y a otros módulos de control.

Series hybrid Uses a small internal combustion engine to charge batteries when they become discharged. The electric motor provides power to the drive wheels. The series hybrid uses a complex controller to provide seamless acceleration from the electric motor.

Híbrido en serie Usa un pequeño motor de combustión interno para cargar las baterías cuando éstas se discargan. El motor eléctrico provee la fuerza a las ruedas motrices. El híbrido en serie usa un controlador complejo para proveer la aceleración sín interrupción del motor eléctrico.

Shimmy The rapid side-to-side vibration of a wheel/tire assembly.

Abaniqueo La vibración lateral rápida de una asamble de rueda/neumático.

Shop layout The design of an automotive repair shop.

Arreglo del taller de reparación Diseño de un taller de reparación de automóviles.

Short-Term Adaptive Fuel Trim Short-term fuel injector pulse width compensation determined by the PCM according to operating conditions. This fuel trim is set to minimize emissions output and represents minor adjustments to the long-term fuel trim strategy.

Tiempo corto para la adaptación de la restricción del combustible Compensación determinada del pulso corto para la amplitud del inyector de combustible de acuerdo con el PCM (módulo de control de la potencia del motor) para las condiciones de operaciones. Esta restricción del combustible es establecida para minimizar las emisiones y representan pequeños ajustes a la restricción del combustible por tiempo largo.

Silicone grease A heat-dissipating grease placed on components such as ignition modules.

Grasa de silicón Grasa para disipar el calor utilizada en componentes, como por ejemplo módulos del encendido.

Sinusoidal Equal rise and fall from positive and negative.

Sinusoidal La subida y caída del positivo y negativo son iguales.

Slitting tool A special chisel designed for slitting exhaust system pipes.

Herramienta de hender Cincel especial diseñado para hendir los tubos del sistema de escape.

Small leak test An EVAP test run after the vehicle passes the large leak test. The small leak test can detect a hole as small as 0.020".

Prueba de fugas pequeñas Una prueba de EVAP que se efectúa después de que el vehículo haya pasado por la prueba de fugas grandes. La prueba de fugas pequeñas puede detectar un hoyo tan pequeno que de 0.020".

Snap shot testing The process of freezing computer data into the scan tester memory during a road test and reading this data later.

Prueba instantánea Proceso de capturar datos de la computadora en la memoria del instrumento de pruebas de exploración durante una prueba en carretera y leer dichos datos más tarde.

Spark duration A measurement in milliseconds used to determine function of the spark plugs.

Duración de la chispa del encendido Una medida en milisegundos que se usa para determinar la función del las bujías de chispa.

Spark line The display of the voltage level and time of the sparks duration across the sparkplug electrodes.

Línea de la bujía La presentación del nivel del voltaje y el tiempo de duración de la chispa através de los electrodos de la bujía.

Specific gravity The weight of a liquid in relation to the weight of an equal volume of water.

Gravedad específica El peso de un líquido de acuerdo al peso de un volumen igual de agua.

Splash fouling Loosened deposits found in the combustion chamber due to misfiring.

Ensuciamiento por salpicadura Los depósitos desalojados que se encuentran en la cámara de combustión debido a un fallo del encendido.

Standard corporate protocol (SCP) The term used by Ford to describe their J1850 module communication language.

Protocolo establecido de corporación (SCP) El término usado por Ford para describir su idioma de comunicación módulo J1850.

Starting air valve A vacuum-operated valve that supplies more air into the intake manifold when starting the engine.

Válvula de aire para el arranque Válvula accionada por vacío que le suministra mayor cantidad de aire al colector de aspiración durante el arranque del motor.

Stethoscope A tool used to amplify sound and locate abnormal noises.

Estetoscopio Herramienta utilizada para amplificar el sonido y localizar ruidos anormales.

Stoichiometry An air-fuel mixture of 14.7:1 that promotes low emissions.

Stoichiometria Una mezcla de aire-combustible de 14.7:1 que promueba las bajas emisiones.

Strategy A plan. Typically refers to the programs of a PCM that insure low emissions levels.

Estrategia Un plan. Típicamente se refiere a los programas del PCM (módulo de control de la potencia del motor) que aseguran niveles bajos de emisiones.

Sulfuric acid A corrosive acid mixed with water and used in automotive batteries.

Ácido sulfúrico Ácido sumamente corrosivo mezclado con agua y utilizado en las baterías de automóviles.

Superimposed pattern The pattern that displays all patterns one on top of the other using the full width of the screen.

Diagrama superpuesta La diagrama que muestra todas las diagramas una sobre otra usando todo lo ancho de la pantalla.

Switch test A computer system test mode that tests the switch input signals to the computer.

Prueba de conmutación Modo de prueba de una computadora que revisa las señales de entrada de conmutación hechas a la computadora.

Synchronizer (SYNC) pickup A pickup assembly that produces a voltage signal for ignition triggering or injector sequencing.

Captación sincronizadora Conjunto de captación que produce una señal de tensión para el arranque del encendido o para el ordenamiento del inyector.

Tach-dwellmeter A meter that reads engine rpm and ignition dwell.

Tacómetro y medidor de retraso Medidor que lee las rpm del motor y el retraso del encendido.

Tachometer A device used to measure the revolutions per minute of an engine.

Taquímetro Un dispositivo que se usa para medir las revoluciones por minuto de un motor.

Tachometer (TACH) terminal The negative primary coil terminal.

Borne del tacómetro Borne negativo de la bobina primaria.

Temperature switch A mechanical switch operated by coolant or metal temperature.

Conmutador de temperatura Conmutador mecánico accionado por la temperatura del refrigerante o del metal.

Test spark plug A spark plug with the electrodes removed so it requires a much higher firing voltage for testing such components as the ignition coil.

Bujía de prueba Bujía que a consecuencia de habérsele removido los electrodos necesitará mayor tensión de encendido para revisar componentes, como por ejemplo la bobina del encendido.

Thermal cleaning The process of cleaning using high extremely temperatures.

Limpieza termal El proceso de limpieza usando las temperaturas extremadamente altas.

Thermal vacuum switch (TVS) A vacuum switching device operated by heat applied to a thermo-wax element.

Conmutador de vacío térmico Dispositivo de conmutación de vacío accionado por el calor aplicado a un elemento de termocera.

Thermal vacuum valve (TVV) A valve that is opened and closed by a thermo-wax element mounted in the cooling system.

Válvula térmica de vacío Válvula de vacío que un elemento de termocera montado en el sistema de enfriamiento abre y cierra.

Thermostat tester A tester designed to measure thermostat opening temperature.

Instrumento de prueba del termostato Instrumento de prueba diseñado para medir la temperatura inicial del termostato.

Throttle body injection (TBI) A fuel injection system with the injector or injectors mounted above the throttle.

Inyección del cuerpo de la mariposa Sistema de inyección de combustible en el que el inyector, o los inyectores, están montados sobre la mariposa.

Throttle kicker A device used to control throttle movement at an idle, thus controlling idle speed.

Regulador en vacío Un dispositivo que se usa para controlar el movimiento del regulador en marcha lenta, así controlando la velocidad de la marcha lenta.

Throttle position sensor (TPS) A sensor mounted on the throttle shaft that sends a voltage signal to the computer in relation to throttle opening.

Sensor de la posición de la mariposa Sensor montado sobre el árbol de la mariposa que le envía una señal de tensión a la computadora referente a la apertura de la mariposa.

Throttle position switch A switch that informs the computer whether the throttle is in the idle position. This switch is usually part of the TPS.

Conmutador de la posición de la mariposa Conmutador que le advierte a la computadora si la mariposa se encuentra en la posición de marcha lenta. Este conmutador normalmente forma parte del sensor de la posición de la mariposa.

Timing connector A wiring connector that must be disconnected while checking basic ignition timing on fuel injected engines.

Conector de regulación Conector del alambrado que debe desconectarse mientras se verifica la regulación básica del encendido en motores de inyección de combustible.

Timing meter A device that shows timing degrees and engine rpm.

Medidor de tiempo Un dispositivo que muestra los grados del tiempo y las rpm del motor.

Top dead center (TDC) The highest in the cylinder a piston travels.

Punto muerto superior (TDC) Lo más alto que llega un piston en un cilíndro.

Trace See waveform.

Rastro Véase waveform

United States customary (USC) A system of weights and measures.

Sistema usual estadounidense (USC) Sistema de pesos y medidas.

Upstream air Air injected into the exhaust ports.

Aire conducido hacia arriba Aire inyectado dentro de las lumbreras del escape.

Vacuum Any air pressure less than atmospheric.

Vacío Cualquier presión de aire menos de lo atmoférico.

Vacuum delay valve A vacuum valve with a restrictive port that delays a vacuum increase through the valve.

Válvula de retardo de vacío Válvula de vacío con una lumbrera restrictiva que retarda el aumento de vacío a través de la válvula.

Vacuum hold and decay A test used by the EVAP system to determine if there are leaks.

Retención y declina del vacío Una prueba del sistema EVAP para determinar si hay fugas.

Vacuum-operated decel valve A valve that allows more air into the intake manifold during deceleration to improve emission levels.

Válvula de desaceleración accionada por vacío Válvula accionada por vacío que admite una mayor cantidad de aire en el colector de aspiración durante una desaceleración a fin de reducir los niveles de emisiones.

Vacuum pressure gauge A gauge designed to measure vacuum and pressure.

Manómetro de la presión del vacío Calibrador diseñado para medir el vacío y la presión.

Valve overlap The few degrees of crankshaft rotation when both valves are open and the piston is near TDC on the exhaust stroke.

Solape de la válvula Los pocos grados que gira el cigüeñal cuando ambas válvulas están abiertas y el pistón se encuentra cerca del punto muerto superior durante la carrera de escape.

Valve stem installed height The distance between the top of the valve retainer and the valve spring seat on the cylinder head.

Altura instalada del vástago de la válvula Distancia entre la parte superior del retenedor de la válvula y el asiento del resorte de la válvula en la culata del cilindro.

Vane-type airflow sensor A sensor containing a pivoted vane that moves a pointer on a variable resistor. This resistor sends a voltage signal to the computer in relation to the total volume of intake air.

Sensor tipo paleta Sensor con una paleta articulada que mueve un indicador en un resistor variable. Este resistor le envía una señal de tensión a la computadora referente al volumen total de aire aspirado.

Vaporization The process in which a liquid changes to a gas or vapor.

Vaporización El proceso en el cual un líquido cambia de un gas a un vapor.

Variable rate sensor signal (VRS) Provides the ignition module with engine position and rpm information.

Señal del sensor de regimen variable (VRS) Provee la información de la posición del motor y las rpm al módulo del encendido

Vehicle networking The use of several modules to perform tasks on a vehicle that are all interconnected and using a common language.

Intercomunicación del vehículo El uso de varios módulos para efectuar los deberes del vehículo que son interconectados y que usan una idioma común.

Vehicle speed sensor (VSS) A sensor that is usually mounted in the transmission and sends a voltage signal to the computer in relation to engine speed.

Sensor de la velocidad del vehículo Sensor que normalemente se monta en la transmisión y que le envía una señal de tensión a la computadora referente a la velocidad del motor.

Vent valve An EVAP valve opened to allow fresh air through the charcoal canister. It is closed when the PCM needs to measure pressure or vacuum changes.

Válvula de ventilación Una válvula de EVAP que abre para permitir entrar aire fresca entrar al bote de carbón. Se cierra cuando el PCM necesita medir la presión o cuando cambia el vacío.

Viscosity A term used to describe a liquid's ability of a liquid to flow.

Viscocidad Un termino usado para describir la habilidad de un líquido en fluir.

Viscous-drive fan clutch A cooling fan drive clutch that drives the fan at higher speed when the temperature increases.

Embrague de mando viscoso del ventilador Embrague de mando del ventilador de enfriamiento que acciona el ventilador para que gire más rápido a temperaturas más altas.

Volt-amp tester A tester designed to test volts and amps in such circuits as battery, starter, and charging.

Instrumento de prueba de voltios y amperios Instrumento de prueba diseñado para revisar los voltios y los amperios en circuitos como por ejemplo, de la batería, del arranque y de la carga.

Voltmeter A tool used to measure electrical pressure.

Voltímetro Una herramienta que se usa para medir la presión eléctrica.

Wankel engine *See* rotary engine.

Motor Wankel véase motor rotario.

Wastegate stroke The amount of turbocharger wastegate diaphragm and rod movement.

Carrera de la compuerta de desagüe Cantidad de movimiento del diafragma y de la varilla de la compuerta de desagüe del turbocompresor.

Waveform The pattern on an oscilloscope.

Forma de onda La diagrama en un osciloscopio.

Weak vacuum test This will establish if there is a large system leak. Once the enable criteria are set, the PCM will close the vent valve and open the purge valve. This would of course cause a vacuum to be indicated on the fuel tank pressure sensor if the system were sealed. If the vehicle should fail to pull a vacuum of up to 7 in. Hg, then the MIL would illuminate after the second consecutive failure under similar conditions and set a code PO455, gross evaporative system leak.

Prueba de vacío débil Ésto establece si hay una fuga grande en el sistema. Una vez que se establece el criterio autorizado, el PCM cerrará la válvula de ventilación y abrirá la válvula de purga. Ésto por supuesto causaría que se indicara un vacío en el sensor del tanque de combustible si estuviera sellado el sistema. Si el vehículo fallara en sostener un vacío llegando a 7 in. Hg, entonces el MIL iluminaría después del segundo fallo consecutivo bajo condiciones parecidas y pondría un código de PO455, fuga masiva del sistema evaporativo.

Wet compression test A cylinder compression test completed with a small amount of oil in the cylinder.

Prueba húmeda de compresión Prueba de la compresión de un cilindro llevada a cabo con una pequeña cantidad de aceite en el cilindro.

Wet fouling A term describing the tip of the spark plug being drowned in excess oil.

Ensuciamiento en húmedo Un término describiendo que la punta de la bujía de chispas se ha sumerjido en un exceso de aceite.

Wiggle test A test in which the PCM "sees" and indicates a problem with wires or connections while the technician wiggles or moves the wiring.

Prueba de meneo Una prueba en la cual el PCM "vea" e indica un problema con las alambres o las conexiones mientras que el ténico menea o mueva los alambres.

Workplace hazardous materials information system (WHMIS) The Canadian version of MSDS sheets.

Sistema de información de materiales peligrosos del trabajo (WHMIS) La versión canadiense de las hojas MSDS.

Zener diode A special type of diode that allows current to flow in the desired direction only when certain conditions are met.

Diodo Zener Un tipo especial de diodo que permite que la corriente fluya en la dirección deseada solamente cuando ciertas condiciones son cumplidas.

INDEX

A

Abrasive cleaning, 17

Acceleration Simulation Mode (ASM) test, 631

AC generator diagnosis/service, 173–178

    alternator requirement test, 176

    diode pattern test, 177–178

    output test, 174, 176

    regulator voltage test, 176

        photo procedure for, 175

Accelerator pedal position sensor, 299

Actuator test mode (ATM), 264

Actuator

    testing with lab scope, 316–317

    testing with scan tool, 318

Adaptive fuel control (OBD II), 364–365

Advanced engine performance test, 30–31

Aftermarket

    equipment, 335

    suppliers' guides and catalogs, 74

After top dead center (ATDC), 57

AIRB solenoid/valve diagnosis, 659–660

AIRB valve, 659

Air bag safety, 19

Air conditioning system diagnosis, 395–396

AIRD valve, 660

Air filters, 192–194

Airflow sensors, 433–436, 459–460

Air-fuel ratio, 458

Air-fuel (A/F) ratio sensor, 459

Air injection (AIR) system diagnosis and service, 656–661

    AIRB solenoid and valve, 659–660

    check valve testing, 657

    electric pumps, 658–659

    monitoring, 660–661

    secondary AIR system, 657–658

    system efficiency test, 660

AIR systems, 656

Alcohol in fuel test, 412

American wire gauge (AWG), 129

Ammeter, 51–52, 136

Ampere, 134

Analog oscilloscope, 143–144

ASE certification, 30–31, 31 table

ASE practice examination, 677–681

ASE testing, 1

ATDC, 57

Automatic shutdown relay (ASD), 262

Automatic transmission diagnosis, 386–393

    band adjustment, 389

    computer and solenoid valve tests, 393

    controls, 325

    electronically shifted, 391

    lockup torque converters, 388–389

        photo procedure for, 392–393

    torque converters, 388

    transmission fluid, 389

Automotive Service Excellence. *See* ASE

Automotive service information, 73–74

Average responding, 140

B

Band adjustment, 390

Banjo fitting, 428

Base timing, 564

Basic Electrical Tests and Service, 129–190

Battery

    cables, disconnecting, 456

    charging

        drain testing, 169–170

        fast, 166, 168

        maintenance-free, 168–169

        slow, 168

    diagnosis/service, 161–166

        battery capacity test, 165–166

        photo procedure for, 167

        hydrometer testing, 164

        leakage test, 163–164

        open circuit voltage test, 162

    hazards, 6

    safety, 8–9

Before top dead center (BTDC), 57

Belt tension gauge, 49, 173

Brake system diagnosis, 396–398
  inspection and service, 398
Breakout box, 64
  testing, 260–261
Bushing check, 556

C

Camshaft noise, 91
Canister, 443
Carbon tracking, 507
Carrying, 3-4
Catalytic converters, 203–208
  delta temperature test, 206
  diagnosis, 654–656
    photo procedure for, 206
  O_2 storage test, 204, 206
  removal and replacement, 205, 207–208
Caustic liquids, 5
Charcoal canister, 636
CD-ROMs, 75
Charging a battery, 166–170
Chassis ground, 129
Check engine manifold vacuum, 227–229
Check the emission levels on an engine, 665–660
Check the operation of an EGR valve, 673–675
Check the operation of the fuel injectors on an engine, 499–500
Check the operation of a PCV system, 671–672
Check the operation of a TP sensor, 309–310
Check valve test, 657
Chemical cleaning, 16–17
Chrysler Corporation
  diagnostic trouble codes (DTCs), 261–264
  EI system diagnosis and service, 600–603
Circuit breakers, testing, 157–158
Cleaning equipment safety, 16–17
Cleaning solutions, hazardous, 5
Clothing, 2–3, 5
Clutch diagnosis, 384–386
  drag and binding, 385–386
  slippage, 385
  vibration, 385
Clutch switch, 274
Coil condition, 536

Coil output and oscillation test, 527
Coil polarity, 530
Cold fouling, 552
Combustion noise, 91
Compact discs, 75
Component testing, 538–549
  control module, 548–549
  Hall effect sensors, 547–548
  ignition coil, 544–545
  ignition switch, 538–539
  pickup coil, 546–547
  scope testing, 539–543
  secondary ignition wires, 549
  spark plug service, 551
  stress testing, 550–551
Comprehensive component monitor, 366
Compressed air equipment, safe useage, 12–13
Compression gauge, 44, 110
Compression test, 110–114
  interpretation of, 113–114
  photo procedure for, 111–112
Compression tester, 43–44
Computer, 235
Computer control of the fuel injector, 331–332
Computer-controlled ignition system service, 566–567
Computerized engine control problems, 239–240
Computer logic flow, 237–238
Computer networks, diagnosing, 320–325
  automatic transmission controls, 325
  electronic automatic temperature control (EATC), 323
  Ford Passive Anti-Theft System (PATS), 321–322
Computer output circuits, 315
Computer outputs, diagnosing, 315–320
  EGR, 318
  idle air control, 318
  photo procedure for scan tool test, 319
  serial data, 318
  testing actuators with lab scope, 316–317
  testing actuators with scan tool, 318
Computer voltage supply/ground wire diagnosis, 267–273
Connecting rod bearing noise, 89
Connector repair, 372–373
Constant volume sampling (CVS) system, 628
Conduct a cylinder balance test, 125–126
Conduct a cylinder compression and leakage test, 127–128

Conduct a diagnostic check on an engine equipped with OBD II, 379–380

Conduct an injector balance test, 501–502

Continuity tester, 50

Continuous self-test, 260

Control modules, 510–511
    testing, 548–549

Conversion to metric, 42–43

Coolant hydrometer, 48

Coolant leaks, 87–88

Cooling system inspection and diagnosis, 96–103, 123–124
    electric cooling fan, 102–103
    photo procedure for, 97
    visual inspection, 97–102

Cooling system pressure tester, 48, 87

Copper wire
    splicing with crimp and seal splice sleeves, 153
    splicing with splice clips, 153, 155

Corona effect, 508

Corrosive material, 23

Cranking test, 527

Crankshaft sensors, 609–611

Current limiting, 526

Current probes, 137–138

Current ramp, 138

Customer service, 40–41

Customer worksheet, 39

Customization, 324

Cylinder
    leakage, 44
    leakage test, 114–116
    leakage tester, 44–45
    output test, 260

D

Data link connector (DLC), 251

Delta temperature test, 206

Diagnosing Related Systems, 383–410

Diagnosis
    of computer networks, 320–325
    of computer outputs, 315–320
    of distributor ignition, 503–582
    of EFI systems, 453–501
    of EI systems, 583–624

of emission control system, 625–675

of electrical system, 129–190

of engine, 85–128

of engine control system, 237–301

of fuel system, 411–451

of ignition system, 503–582

of intake and exhaust system, 191–233

of related systems, 383–410

Diagnosis, systematic approach to, 38–40
    photo procedure for, 69

Diagnostic equipment safety, 26–27

Diagnostic tools
    electrical, 50–55
    engine, 43–49
    engine analyzer, 68–70
    exhaust analyzer, 67
    fuel system, 59–62
    ignition system, 55–59
    miscellaneous, 70–71
    oscilloscope, 65–66
    scan tools, 62–64

Diagnostic trouble codes (DTCs)
    accessing, 254–246
    actuator test mode, 264
    breakout box testing, 260–261
    for Chrysler Corporation vehicles, 261–264
    complete DLC terminal explanation, 254–255
    continuous self-test, 260
    cylinder output test, 260
    erasing fault codes, 252–254, 260
    for Ford Motor vehicles, 255–257
    for General Motors vehicles, 250–251
    key on, engine off (KOEO) test, 257–259
        photo procedure for, 258
    key on, engine running (KOER) test, 259–260
        photo procedure for, 258
    for Nissan vehicles, 266–267
    OBD II standards, 359–362
    switch test mode, 264
    for Toyota vehicles, 264–266

Digital multimeter (DMM), 53–55
    usage, 139–141

Digital storage oscilloscope (DSO), 143–144

Digital tachometer, 56

Diode testing, 160–161

Display pattern, 521

Distributor cap, 510

Distributor Ignition System Service and Diagnosis, 503–582, 604

  bushing check, 556

  distributor assembly, 556–558

  distributor disassembly, 558–559

  distributor inspection, 558

  distributor installation and timing, 558–561

    photo procedures for, 560

  effects of improper ignition timing, 583–585

  general diagnosis, 504–506

  ignition timing, 561–567

  no-start diagnosis, 506, 543–555

  service, 555–561

  visual inspection, 507–512

Double-flare fitting, 420

Drive cycle, 353

Driveline diagnosis, 394

  differential, 394

  drive shaft and U-joint problems, 384

Drive lugs, 559

Dual trace oscilloscope, 66

Dust, 5

Duty cycle, 140

Dwell meter, 55

Dyno, 626

E

Ear protection, 3

EEPROM, 243

EGR, 318

EGR actuation, 329–330

EGR control solenoid (EGRC), 644

EGR data, 349

EGR system diagnosis and service, 643–650

  catalytic converter, 654–656

  differential pressure feedback, 646–647

  digital EGR valve, 647–648

  EGR efficiency tests, 649–650

  EGR valves and systems testing, 486–487

  linear EGR valve, 648–649

  troubleshooting system, 644–645

EGR vacuum regulator (EVR) tests, 644, 650–651

  photo procedures for, 652

EGR vent solenoid (EGRV), 644

Electrical diagnosis, 128–133

Electrical diagnostic tools, 50–55

Electrical Fuel Injection (EFI) Diagnosis and Service, 453–501

  air system checks, 459–464

  basic checks, 457–459

  component checks, 433

  fuel rail, injector, and regulator service, 479–482

  fuel system checks, 464–467

  idle speed

    checks, 482–484

    controls, 485–493

  injector

    checks, 467–476

    service, 477–479

  preliminary checks, 454–456

    disconnecting battery cables, 456

    service precautions, 455

  pressure relief, 412–413

Electrical problems, 131–133

Electrical repairs, 150–155

Electrical safety, 8

Electrical test

  equipment, 134–139

  meters, 50

Electrical troubleshooting, 149–150

Electrical wiring diagrams, 130–131

Electric cooling fan circuit diagnosis, 102–103

Electric AIR pump, 658–659

Electromagnetic interference (EMI), 509, 586–587, 597

Electromotive force (EMF), 134

Electronic automatic temperature control (EATC0, 323

Electronic control assembly (ECA), 255

Electronic data systems, 75

Electronic distributorless ignition system (EDIS), 604

Electronic EGR controls, 650–654

  EGR vacuum regulator tests, 650–651

  exhaust gas temperature sensor diagnosis, 651

Electronic fuel injection systems. *See* EFI systems

Electronic idle-speed control, 584–493

Electronic Ignition Systems (EI) Diagnostics and Service, 583–624
    Chrysler EI systems, 600–603
    control modules, 587
    electromagnetic interference, 586–587, 597
    Ford EI systems, 604–608
    general diagnostics, 591–600
    General Motors EI systems, 608–616
    ground circuits, 586, 597
    ignition coils, 598–599
    Mitsubishi EI systems, 616
    Nissan EI systems, 617
    no-start diagnosis, 592–596
    primary wiring checks, 585–586
    scope testing, 588, 590
    secondary circuit testing, 599–600
    sensors, 587–588
    Toyota EI systems, 618–620
    visual inspection, 585–591
    waste spark ignition system, photo procedure for, 589–590
Electronically-shifted automatic transmission, 391
Electrostatic discharge (ESD), 236–237, 334–335
Emission Control System Diagnosis and Service, 625–675
    air injection system, 657–661
    catalytic converter diagnosis, 654–656
    EGR system, 643–650
    emissions testing, 626–632
    EVAP system, 632–637
    PCV system, 637–640
    spark control systems, 640–643
Emissions testing, 626–632
    exhaust analyzers, 626–628
    IM240 test, 628–630
    other I/M testing programs, 631–632
Employer/employee obligations, 28–29
Enable criteria, 333, 353
Engine analyzer, 68–70, 512–518
Engine Control System Diagnosis and Service, 235–314
    actuator tests, 316–318
    basic diagnosis, 237–240
    computer voltage supply and ground wires, 267–273
    electronic service precautions, 236–237
    generating sensors, 282–294

    input sensor testing, 273
    mass air flow sensors, 295–300
    scanner use, 246–250
    self-diagnostic systems, 243–246
    service bulletin information, 240–243
    switches, 273–275
    trouble code retrieval, 250–267
    variable resistor-type sensors, 275–281
Engine coolant temperature (ECT) sensor, 275–277
Engine cooling fans, 395–396
Engine diagnosis, 85
    compression test, 110–114
      interpretation of, 113–114
      photo procedure for, 111–112
    coolant leaks, 87–88
    cooling system, 96–103
      photo procedure for, 97
    cylinder leakage test, 114–116
    engine power balance test, 109–110
    exhaust, 92–93
    exhaust gas analyzer, 104–108
    general procedure for, 85
    leaks, 85–88
    noise, 89–91
    oil consumption, 93
    oil pressure tests, 93–95
      photo procedure for, 94
    vacuum tests, 103–104
    valve adjustment, 117–120
    valve timing checks, 117
Engine diagnostic tools, 43–49
Engine idle speed check, 482–494
Engine Off Natural Vacuum (EONV), 636
Engine performance tools, 70–71
    special tools suppliers, 683
Engine tune-up, 76–77
Engine vacuum, 195–201
    diagnosis and troubleshooting, 197–198
    gauge readings, 199
    leak diagnosis, 198, 200–201
    schematic of, 195–196
    test equipment, 198
Enhanced EVAP, 634
EPA. *See* Environmental Protection Agency

EVAP control system, 418, 623–637
  component diagnosis, 633–637
Environmental Protection Agency (EPA), 23
Exhaust diagnosis, 92–93
Exhaust gas analyzer, 67, 88, 104–108, 626–628
Exhaust gas recirculation valve position (EVP) sensor, 280–281
Exhaust manifold, 208–209
Exhaust noise, 92–93
Exhaust pipes and mufflers, 210–211
Exhaust system diagnosis and service, 201–202
  component replacement, 208–211
  system inspection, 202
  system problems, 201–202
Eye protection, 2

F

Failsoft action, 235
Flash DTCs, 343
Fast idle thermo valve diagnosis, 491–492
Fault codes, erasing, 252–254, 260
Feeler gauge, 49, 117
Fire extinguishers, 22, 22 table
Fires, classifications of, 22
Fire safety, 11–12
Firing voltage, 523
First aid kits, 22–23
Flammable liquids/materials, 5
Flat-rate manuals, 74
Floor jack, 15
Flywheel noise, 91
Ford Motor Company
  diagnostic trouble codes (DTCs), 255–257
  EI system diagnosis and service, 604–608
  electronic automatic temperature control (EATC), 323
  ISO 9141, 320
  Passive Anti-Theft System (PATS), 321–322
  photo procedure for testing MAP sensor, 293
  Standard Corporate Protocol (SCP), 320
Four-wheel alignment, 402
Frequency, 140–141
Fuel cap, 445
Fuel control (OBD II), 459

Fuel filters, 422–425
  photo procedure for removing, 424
  servicing, 423, 425
Fuel injection system diagnosis, 432–443
  airflow sensor checks, 433–436
  air system checks, 433
  EFI system component checks, 433
  evaporative emissions, 443–446
  fuel delivery, 437–438
  fuel system checks, 436
  idle adjustment, 443
  injector checks, 438
  injector replacement, 440
    photo procedure for removing and replacing, 441–442
  oscilloscope checks, 440
  preliminary checks, 432
  throttle body, 436
  voltage signal checks, 438–440
Fuel injector cleaner, 61–62
Fuel injectors, 61
Fuel leaks, 86
Fuel lever sensor, 445
Fuel lines, 418–422
Fuel pumps, 425–431
  checking pressure, 428, 430–431
    photo procedure for, 429
  electric, 425–431
  replacement, 431
  troubleshooting, 425
Fuel System Diagnosis and Service, 411–451
  alcohol in fuel test, 412
  fuel delivery, 437–438
  fuel filters, 422–425
  fuel injection systems, 432–443
  fuel lines, 418–422
  fuel pumps, 425–431
  fuel system pressure relief, 412–413
  fuel tanks, 413–418
Fuel system tools, 59–62
Fuel tank level sensor, 637
Fuel tank pressure control valve, 637
Fuel tank pressure sensor, 300, 637
Fuel tanks, 413–418
  draining, 414–415

electric fuel pump removal/replacement, 416–418

    servicing, 415–416

Fuses, testing, 155–156

Fusible links, 156–157

G

Gap bridging, 552

Gasoline, 20

Gasoline safety, 9–10

General Engine Condition Diagnosis, 85–128

General Motors

    diagnostic trouble codes (DTCs), 252–254

    EI system diagnosis and service, 608–616

General repair manuals, 74

Generating sensors, 282–294

    Hall effect switch, 286–288

    knock sensor, 288–290

    magnetic pulse generator, 288–289

    manifold absolute pressure (MAP) sensor, 290–294

    oxygen sensor, 282

    testing with

        DMM, 282–283

        lab scope, 284–285

        scanner, 284

    vehicle speed sensor, 289

General shop safety, 10–11

Glazing, 552

Glitches, 143

Glossary, 685–699

Graphing digital multimeter, 55, 142

Grease, hazardous, 5

Ground circuits, 509, 586

H

Hair, 3, 5

Hall effect sensor, 286–287, 547–548

    testing, 286

Hand tools, safe useage, 26–27

Hard fault, 244

Hazard Communication Standard, 25

Hazardous materials, 23–24

Hazardous waste, 5

    disposal, 23–26

Heated oxygen sensor data, 351

Heating system diagnosis, 395–396

Heat testing, 550

Heavy-duty air cleaner, 194

High data rate, 605–608

High firing line, photo procedure for determining cause, 532–533

High input impedance, 55

High noise levels, 5

High-pressure air, 5

High resistance, 132

HO_2S repair, 369

Hotline services, 75

Hot wire-type MAF sensor, 297–299

Hydraulic press, 7–8

Hydrometer testing, 164

Hyperlinked, 73

I

IAC bypass air motor, 486

    diagnosis and installation, 489–491

    removal and cleaning, 489

    scan tester diagnosis, 489, 491

        photo procedure for, 487

    and valve diagnosis, 318–319, 488

Idle adjustment, 443

Idle air control, 318, 461

Idle contact switch test, 485–486

Idle speed, 482–484

    controls, 485–493

    in fuel injection systems, 482

    minimum adjustment, throttle body injection, 482–484

Ignitable material, 24

Ignition coil, inspection and tests, 544–546, 609

Ignition module, 554–555

    removal and replacement, 554–555

Ignition switch, testing, 538–539

Ignition system service and diagnosis, 508–582. *See also*

    Distributor ignition (DI)

    with engine analyzer, 512–518

    general diagnosis, 504–506

    individual component testing, 536–542

    oscilloscope testing, 518–536

    setting ignition timing, 564–566

    visual inspection, 507–512

Ignition system tools, 55–59
Ignition timing, 561–567
    base timing adjustments, 564–566
    computer-controlled ignition system, 566–567
        improper, effects of, 583–585
IM240 test, 628–630
Infrared exhaust gas analyzer tests, 106–108
Initial vacuum pull down, 635
Injector
    balance test, 59–60, 469, 471
        photo procedure for, 470
    checks, 438, 467–476
    circuit test light, 61
    cleaning, 477–479, 481
    flow test, 472–473
    installing, 481–482
    oscilloscope checks, 473–476
    power balance test, 471
    removing, 479–481
    replacing, 440–442, 479
    sound test, 472
    tester, 60
    voltage signals, 468
Input sensor testing, 273
Inspection and maintenance (I/M), 626
Intake air temperature sensor, 560
Intake Exhaust System Diagnosis and Service, 191–223
Intermediate section, 525
International System. See SI
Isolating computerized engine control problems,
    239–240

J
Jacks, safe useage, 15–16
Jack stand, 15–16
Jewelry, 3, 5
K
Key on, engine off (KOEO) test, 257–259
Key on, engine running (KOER) test, 259–260
Kilopascals, 43
Knock sensor diagnosis, 288–290
kPa, 43

L
L1 test, 30–31
Lab scope
    check of sensors and actuators with, 316–317
    useage, 142–148
Leak detection pump (LDP), 636
Length, 43
Lift, safe useage, 13–14
Lifting, 3–4
Lockup torque converters, 388
Logic probes, 51, 148–149
Long-term fuel trim, 364
Low-amp probes, 137
Low data rate, 605–608

M
Magnetic pulse generator, 288–289
Magnetic timing probe, 58
Main bearing noise, 89
Malfunction indicator lamp (MIL), 355–538, 433
    freeze frame, 359
    history codes, 355–356
Manifold absolute pressure (MAP) sensor, 290–294
    photo procedure for testing, 293
Manual transmission diagnosis, 386
    lockup torque converters, 388
    torque converters, 388
Mass, 43
Mass air flow (MAF) sensors, 295–300
    accelerator pedal position sensor, 299
    fuel tank pressure sensor, 300
    hot wire-type, 297–299
    volume air flow meters, 295–297
Material safety data sheets, 25–26
Maxi-fuses, 157
Measurement, system of, 42–43
Mechanical valve lifters, 117
Metric Conversion Act, 42–43
Metric conversions, 682
Metric system, 42–43
    conversion to AGC wire sizes, 130
Misfire data, 352
Mitsubishi Motors, EI system diagnosis and service, 616

Moisture testing, 550

Monitor the adaptive fuel strategy on an OBD II-equipped engine, 381

MSDS. *See* Material safety data sheets

Multimeter, 53–55

N

National Institute for Automotive Service Excellence (ASE). *See* ASE

New Generation Star (NGS) tester, 321

Nissan Corporation

    diagnostic trouble codes (DTCs), 266–267

    EI system diagnosis and service, 617

Noid light, 61

Noise

    camshaft, 91

    combustion, 91

    connecting rod bearing, 89

    diagnosis, 88–91

    flywheel, 91

    main bearing, 89

    piston pin, 90

    piston ring, 90

    piston slap, 90

    ring ridge, 90

    valve train, 91

    vibration damper, 91

Non-sinusoidal waveform, 144

No-start diagnosis

    distributor ignition system, 506, 543–545

    EI system, 592–596, 601–603, 611–614

O

OBD II diagnosis and service, 333–381

    adaptive fuel control strategy, 364–365

    data link connector, 362

    diagnostics, 337–352

    diagnostic software, 363

    diagnostic trouble codes, 359–362

        clearing, 375

    EEPROMS in, 243

    freeze frame, 353, 359

    intermittent faults, 366–368

    malfunction indicator lamp, 355–358

    monitor test results, 365–366

    photo procedure comparing O_2 signals, 358

    repairing the system, 369–374

        photo procedure for reprogramming, 336–337

    scan tester diagnosis, 340–352

    terminology, 353–355

    verifying vehicle repair, 375

    visual inspection, 337–340

Occupational Safety and Health Act (OSHA), 4–5

Offset knob, 566

Ohmmeter, 51, 138–139

Ohm's law, 150

Oil consumption, 93

Oil, hazardous, 5

Oil leaks, 86

Oil pressure gauge, 48, 93

Oil pressure tests, 93–95

    engine temperature, 95–96

On-board diagnostic II. *See* OBD II

Onboard refueling vapor recovery (ORVR) system, 445–446

Open, 131

Open circuit voltage test, 162

Oscilloscope, 65–66, 143–144

    controls, 146–148

Oscilloscope testing

    coil condition, 536

    coil output and oscillation test, 527

    coil polarity, 530

    open circuit precautions, 526–527

    pattern display modes, 521–523

    pattern types and phases, 520

    rotor air gap voltage drop, 533–534

    scales, 518–519

    secondary circuit resistance, 534–536

    single cylinder patterns, 523–526

    spark duration test, 527–529

    spark plug firing voltage, 530

    spark plugs under load, 536

OSHA. *See* Occupational Safety and Health Act

O_2 storage test, 204, 206

Output state mode (OTM), 363

Overheating, 552
Oxygen (O$_2$) sensor, 70, 282
   diagnosis, 457–459
   testing with a DMM, 282–283
   testing with a lab scope, 284–285
   testing with a scanner, 284

P

Parade pattern, 521–522
Parameter Identification (PID) mode, 363
Parasitic drain, 169
Particulates, 18
Parts replacement, 369
PCM action, 445
PCM quick test, 73
PCV system diagnosis and service, 637–640
   PCV test, 107
   PCV valve functional checks, 639–640
Peak and hold injector circuit, 475
Pending situation, 354
Personal safety, 2–3, 19–20
Pickup coil, testing, 546–547
Pinch-off pliers, 71
Pinging, 91
Piston
   pin noise, 90
   ring noise, 90
   slap, 90
Pneumatic tools, 12–13
Pounds per inch (psi), 43
Power tools, safe useage, 7–12, 27
Powertrain OBD system check, 341
Pressure gauge, 59
Primary wiring connections, 507–509
Program, 235
PROM (programmable read only memory), 240–241
   photo procedure for replacing, 242–243
psi, 44
psia, 102
Pulse width, 140
Purge flow sensor, 444
Purge solenoid, 444
Purge valve, 636
Pyrometer, 204

R

Radiator shroud, 98
Radio frequency interference (RFI), 140
Raster pattern, 523
Reactive material, 23
RCRA. *See* Resource Conservation and Recovery Act
Reference voltage, 268
Refrigerant safety, 10
Relays, testing, 159–160
Relieving pressure in an EFI system, 449–450
Remote starter switch, 117
Repair Grade (RG240) test, 632
Replacing modules, 324
Resource Conservation and Recovery Act (RCRA),
   24–25
Retrieve codes from the computer of an OBD II engine
   control system, 305
Ring ridge noise, 90
Road test a vehicle to check the operation of the auto-
   matic transmission, 407–408
Room temperature vulcanizing (RTV) sealant, 282
Root mean square (RMS), 140
Rotor air gap voltage drop test, 533–534
Running test, 528

S

Safety, 1
   ASE certification, 30–31, 31 table
   cleaning equipment, 16–17
   compressed air equipment, 12–13
   diagnostic equipment, 26–27
   employer/employee obligations, 28–29
   hand tools, 26–27
   hazardous waste disposal, 23–26
   jack and jack stand, 15–16
   lift, 13–14
   lifting and carrying, 3–4
   personal, 2–3
   power tools, 7–12
   vehicle operation, 18–20
   work area, 20–22
Safety glasses, 2
Scan tester, 247
   diagnosis with, 249, 486

features of, 247

initial entries, 248, 250

snapshot testing, 250

Scan tools, 62–64

Secondary AIR system diagnosis and service, 657–658

Secondary circuit resistance, 534–536

Secondary ignition wires, 507–509

testing, 549

Self-diagnostic systems, 243–246

accessing trouble codes, 245–246

visual inspection, 245

Sensors, ignition system, 511–512, 587–588

Serial data, 318, 354

Serpentine belt, 173

Service bulletin information, 240–243

Service information, 71–75

Service manuals, 74

Service port, 444

Servicing Computer Outputs and Networks, 315–332

Shop hazards, 5

Shop safety survey, 33–34

Short, 132

Short-term fuel trim, 364

SI, 42–43

Similar conditions, 354

Sinusoidal waveform, 144

Small leak test, 635

Smart modules, 320

Smoke tester, 47

Solvents, 21

Spanish glossary, 685–699

Spark control systems, 640–643

knock sensor and knock sensor module diagnosis, 640–643

photo procedure for, 642

Spark duration, 527

test, 527–529

Spark line, 524

Spark plug firing voltage, 530

Spark plug service, 551–554

inspecting plugs, 551–553

regapping plugs, 553–554

wire installation, 554

Spark plugs under load test, 536

Spark plug thread taps, 70

Spark testers, 58–59

Special tools suppliers, 683

Specialty repair manuals, 74

Specific gravity, 164

Speed control system diagnosis, 396

Splash fouling, 552

Standard Corporate Protocol (SCP), 320

Starter diagnosis/service, 170–173

current draw test results, 171

Starting air valve diagnosis, 492

Static protection straps, 71

Steering system diagnosis, 399–402

Stepped resistors, testing, 160

Stethoscope, 49, 89

Stoichiometry, 458

Stress testing, 550–551

Substance abuse, 20

Sulfuric acid, 18

Supercharger diagnosis and service, 220–224

diagnosing, 221

history of, 220

removing, 221–224

Superimposed pattern, 523

Suspension system diagnosis, 399–402

Switches, testing, 158–159, 273–275

Switch test mode, 264

System service, 445

T

Tach-dwellmeter, 55

Tachometers, 56

Technician logical diagnosis, 288

Terminal diagnosis/repair, 374

Test a catalytic converter for efficiency, 231–233

Test an ETC sensor, 307–308

Test an O_2 sensor, 311–312

Testing the battery's capacity, 187–188

Testing charging system output, 189–190

Testing, electrical components, 155–161

Testing fuel pressure on an EFI system, 451

Testing a MAP sensor, 313–314

Test light, 50, 1027

Thermal cleaning, 17

Thermal vacuum switch (TVS), 197

Throttle body, 436, 460–461
  inspecting, 461–462
  removal and cleaning, 462–464
Throttle position (TP) sensor, 278–280
Timing light, 56–58
Timing meter, 564
Tires, 402–403
Tools and Safety, 1–36
Torque converters, 388
Toxic material, 24
Toyota
  diagnostic trouble codes (DTCs), 264–266
  EI system diagnosis and service, 618–620
  inspecting EGR position sensor, 653–654
Trace, 65–66
Trip, 354–355
Troubleshooting, electrical problems, 149–150
Turbocharger diagnosis, 211–220
  boost pressure test, 215
  common problems, 212
  component inspection, 217–218
  inspecting, 212
  installing turbocharger, 218
  measuring shaft axial movement, 217
    photo procedure for, 214
  removing turbocharger, 216–217
  start-up and shutdown, 219–220
  troubleshooting guide, 213
  wastegate service, 215–216
Turbulence burning, 552
Two trip monitor, 355
Typical Shop Procedures and Equipment, 37–84

U

United States Customary System (UCS). *See* USC system
Units of measure, 42–43
USC system, 42–43
Use of an ohmmeter, 83–84
Use of a voltmeter, 81–82
Using a DSO on sensors and switches, 183–184

V

Vacuum, 45, 195
Vacuum gauge, 45–46, 103, 191, 199
Vacuum hold and decay, 635

Vacuum leak detector, 47
Vacuum pumps, 46–47
Vacuum tests, 103–104
Valve adjustment, 117–120
Valve timing checks, 117
Valve train noise, 91
Vapor pressure sensor, 444
Variable rate sensor signal (VRS), 604
Variable resistors, testing, 160
Variable resistor-type sensors, 275–281
  engine coolant temperature (ETC) sensor, 275–277
  exhaust gas recirculation valve position sensor, 280–281
  intake air temperature sensor, 277–278
  throttle position sensor, 278–280
Vehicle care, 40–41
Vehicle identification, 71–73
  number, 71, 72
Vehicle networking, 320
Vehicle operation safety, 18–20
Vehicle speed sensor (VSS), 289
Vent solenoid, 444
Vent valve, 636
Vibration damper noise, 91
VIN, location of, 71–72
Visual inspection
  of cooling system, 97–102
  of distributor ignition (DI), 507–512
  of EI system, 585–591
  of ignition system, 507–512
  of OBD II system, 337–340
  of self-diagnostic systems, 245
Visually inspect an EFI system, 497–498
Visually inspect and test an electronic ignition (EI) system, 623–624
Voltage drop test
  circuit control, 173
    photo procedure for, 135
  starting motor circuit insulated side, 172
  starting motor ground side, 172
Voltage signal check, 438–440, 468
Volt/ampere tester (VAT), 52, 165
Voltmeter, 51, 134–135
Volume, 43
Volume air flow meters, 295–297

W

Warm-up cycle, 355
Waveforms, 65, 144–146
Weak vacuum test, 635
Wet fouling, 552
Wheels, 402–403
WHMIS. *See* Workplace Hazardous Materials Information Systems
Wiggle test, 260

Wire repair, 370–372
Wiring diagrams, 130–131
Wiring harness diagnosis/repair, 150–152
Wiring problems, 131–133
Work area safety, 20–22
Working safely around air bags, 35–36
Workplace Hazardous Materials Information Systems (WHMIS), 4, 25